*Mechanisms of
Osmoregulation
in Animals*

Mechanisms of Osmoregulation in Animals

Maintenance of Cell Volume

Edited by

R. GILLES

Laboratoire de Physiologie animale
Institute de Zoologie,
Université de Liège, Belgium

A Wiley–Interscience Publication

JOHN WILEY & SONS
Chichester · New York · Brisbane · Toronto

Copyright © 1979, by John Wiley & Sons Ltd.

All rights reserved.

No part of this book may be reproduced by any means, nor transmitted, nor translated into a machine language without the written permission of the publisher.

Library of Congress Cataloging in Publication Data:

Main entry under title:

Mechanisms of osmoregulation in animals.

   Includes bibliographical references.
   1. Osmoregulation.   I. Gilles, R. [DNLM:
1. Cells—Physiology.   2. Water-electrolyte balance.
QU105 M486]
QP90.6.M4   591.8′76   78–4608

ISBN 0 471 99648 3

Text set in 10/12 pt VIP Times, printed by photolithography and bound in Great Britain at The Pitman Press, Bath.

List of Contributors

M. ABRAMOW — Laboratoire de Biologie médicale, Fondation Reine Elizabeth, Avenue Jean Crocq, 1, 1020 Brussels, Belgium.

J. M. BOUQUEGNEAU — Laboratory of Oceanology, University of Liège, Sart Tilman, Liège, Belgium.

D. A. T. DICK — Department of Anatomy, The University, Dundee DD1 4HN, Scotland.

W. A. DUNSON — Department of Biology, Pennsylvania State University, University Park, Pa 16802, USA.

R. GILLES — Laboratory of Animal Physiology, University of Liège, quai Van Beneden, 22, 4020 Liège, Belgium.

W. N. HOLMES — Department of Biological Sciences, University of California, Santa Barbara, California 93106, USA.

K. JANÀČEK — Laboratory of Cell Membrane Transport, Czechoslovak Academy of Sciences, Budejovicka, 270, Praha 4—KRC, Czechoslovakia.

CH. JEUNIAUX — Laboratoire de Morphologie et systématique animale, Université de Liège, Institut de Zoologie, quai Van Beneden, 22, 4020 Liège, Belgium.

L. B. KIRSCHNER — Department of Zoology, Washington State University, Pullman, Washington 99163, USA.

V. KOEFOED-JOHNSEN — University of Copenhagen, Institute of Biological Chemistry A, 13, Universiteit Parken, DK 2100 Copenhagen, Denmark.

J. J. LEGROS — Centre Hospitalier Universitaire, Institut de Pathologie, Local 4/12, Sart Tilman, Liège, Belgium.

M. PEAKER — The Hannah Research Institute, Ayr, Scotland KA6 5 HL.

R. B. PEARCE — *Department of Biological Sciences, University of California, Santa Barbara, California 93106, USA.*

G. RORIVE — *Clinique Médicale A, Hopìtal de Bavière, 4000 Liège, Belgium.*

S. U. SILVERTHORN — *Department of Physiology and Biophysics, University of Texas Medical Branch, Galveston TX77550, USA.*

F. J. VERNBERG — *Belle W. Baruch Institute for Marine Biology and Coastal Research, University of South Carolina, Columbia, SC 29208, USA.*

Contents

Introductory Comments
R. Gilles ix

PART A: WATER AND SOLUTES IN BIOLOGICAL SYSTEMS

1. Structure and Properties of Water in the Cell
   D. A. T. Dick 3

2. Relations between Solutes and Water: Analysis of Solute Transport
   K. Janáček 47

PART B: CONTROL OF THE INTRACELLULAR FLUID OSMOLARITY

3. Intracellular Inorganic Osmotic Effectors
   G. Rorive and R. Gilles 83

4. Intracellular Organic Osmotic Effectors
   R. Gilles 111

PART C: CONTROL OF THE EXTRACELLULAR FLUID OSMOLARITY

5. Control Mechanisms in Crustaceans and Fishes
   L. B. Kirschner 157

6. Control Mechanisms in Amphibians
   V. Koefoed-Johnsen 223

7. Control Mechanisms in Reptiles
   W. A. Dunson 273

8. Control Mechanisms in Birds
   M. Peaker 323

9. Control Mechanisms in Mammals
   M. Abramow 349

10. Hormones and Osmoregulation in the Vertebrates
    W. N. Holmes and R. B. Pearce 413

PART D: OSMOREGULATION AND ECOLOGY IN THE
        AQUATIC ENVIRONMENT

11. Temperature and Osmoregulation in Aquatic Species
    F. J. Vernberg and S. U. Silverthorn 537

12. Osmoregulation and Pollution of the Aquatic Medium
    J. M. Bouquegneau and R. Gilles 563

13. Osmoregulation and Ecology in Media of Fluctuating Salinity
    R. Gilles and Ch. Jeuniaux 581

PART E: PATHOLOGY OF EXTRACELLULAR FLUID
        REGULATION

14. Pathology of Extracellular Fluid Regulation in Man: Hormonal
    Aspects
    J. J. Legros 611

Addendum Section 637

Taxonomic Index 657

Subject Index 661

Introductory Comments

Life on Earth is strictly dependent on water. Living organisms are highly sophisticated biochemical systems organized around the properties of this molecule. Water indeed provides the principal framework in which most of the molecular interactions indispensable to life take place. It also provides the vehicle carrying the molecules to the different locations where these interactions can proceed.

Besides the many organic compounds found as solutes in living organisms, inorganic ions are also of prime importance: they participate as co-factors in many enzyme reactions, they also provide chemical gradients which can act as stores of potential energy and they influence the permeability of biological membranes to other solutes. The various solutes found in the cells will on the other hand influence the osmotic mobility of water and therefore will play a prominent part in the maintenance of the cell architecture. Besides, many of the enzymatic systems controlling the chemical interactions characteristic of life are located in the cells on highly organized supports. This points to the importance of the maintenance of the cell structure and volume in the reactions which involve these enzymes as catalysts.

Basically, cells can thus be viewed as extremely complex chemical machineries in which the localization and the concentration of the various interacting molecular species must be carefully controlled in order to maintain optimal activity. In such a context, control and maintenance of the cell volume can be considered as an essential requisite of life. Moreover, the problem of cell volume control becomes a crucial one in the conquest of different biotopes and in the establishment of organisms in aquatic environments with fluctuating osmolarities. Accordingly, life originated in some kind of an ocean and the capability of cell volume control is one of the main prerequisites to the invasion of other types of habitat such as the terrestrial or the fresh-water ones. As will be seen throughout this book, organisms living in these media have evolved specific osmotic adaptations enabling the maintenance of their communities. There are several ways in which the problem of cell volume maintenance can be solved. The organism can isolate itself completely from the outside medium, thus avoiding any gain or loss of water. This solution has not been retained widely in the course of evolution. Exchanges with the external medium are indeed necessary to meet the cells needs. Some bacterial spores can survive for very long periods with a very low water content and without exchanges with their environment; in this situation, however, their vital processes are essentially suspended. In most organisms, water appears to cross the cell membrane by diffusion in response to osmolarity gradients. There are two ways of avoiding cell volume changes while still keeping the possibility of exchanges between the intracellular fluid and its environmental medium. The first method consists in controlling the osmolarity of the intracellular fluid in relation to the eventual modifications of the external medium. The second method implies control of the osmolarity of the fluid surrounding the cells whatever the external conditions. The latter solution has been adopted by different eukaryotes and has been termed by Florkin (1962), the 'anisosmotic regulation of the extracellular fluids'. Although the existence of an extracellular fluid different from the external medium appeared quite early in the course of evolution, effective regulation of this medium is an attribute of only a few highly evolved zoological groups. It is found in some worms and molluscs but essentially it occurs in the arthropods and in the vertebrates. Moreover, many of these species are unable to maintain their blood osmotic state when the osmolarity of the environment varies.

The most powerful anisosmotic regulators form the category of the so-called homeosmotic animals; these species can keep their blood osmolarity steady whatever the external conditions. Besides a few crustaceans and fishes, representatives of this group are found among reptiles, birds, and mammals. The inorganic ions Na^+ and Cl^- are the prominent blood osmotic effectors in most of the anisosmotic regulators. Urea is used by some low vertebrates. This organic compound is found essentially in the cyclostomes and the elasmobranchs but also plays a role in various amphibians and reptiles.

Control of the osmolarity of the 'internal medium' in the anisosmotic regulators is achieved by different mechanisms, always involving salt transport, and located in various organs. The 'salt transporting' organs may be morphologically very different ranging, for instance, from the mammalian kidney to the crustacean gills or the anal papillae of some insect larvae. The main physiological features of some of these organs and their complex endocrine controls are described at length in Part C of this volume. It is striking to consider that, with the exception of the nasal salt gland found in birds and reptiles, which appears to be a little different because of its methods of control, the 'salt transporting organs' are built basically on very similar physiological models: they all perform active transport of sodium which is one of the main driving forces for water movements, the counter-ion implicated in the Na^+ transport process is always NH_4^+, H^+ or K^+ and the ultrastructure of these organs is quite similar. Indeed they present important infoldings of the plasma membranes, large intercellular spaces and a high density of mitochondria essentially located close to the folded membranes. It is possible that these mechanisms are derived from the basic mechanism of Na^+ transport which controls the other and more primitive process of maintenance of the cell volume; that is, the control of the osmolarity of the intracellular fluid.

As already stated, there are only a few homeosmotic animals. In all other species, the cells will have to withstand sometimes very important changes in the osmolarity of their environmental medium. Moreover, the powerful mechanisms of blood osmolarity control at work in homeosmotic species may be overwhelmed under certain circumstances or may show a certain time lag before responding to a new situation. This points to the importance of the mechanisms of intracellular fluid osmolarity control in cell volume maintenance.

The processes responsible for the intracellular osmolarity balance can either maintain an osmotic gradient between the intracellular and extracellular fluids or act to keep these two media close to isosmoticity. The first solution is found in plant cells and in fresh-water sponges and protozoans. Plant cells are surrounded by rigid cell walls to avoid swelling due to osmotic intrusion of water. The problem of water inflow in the fresh-water sponges and protozoans is solved by the existence of contractile vacuoles, the primary function of which is to remove any excess of water. In most eukaryotic animal species, the intracellular and extracellular fluids are kept close to isosmoticity. The mechanisms implicated in this process have been termed by Florkin (1962) as mechanisms of 'isosmotic regulation of the intracellular fluid'. They work to maintain the isosmotic equilibrium despite the presence of anionic non-diffusible particles inside the cells; these particles generate an osmotic pressure which would otherwise induce swelling and lysis of the animals' cells having easily distensible membranes. Moreover, these mechanisms are of prime importance in the volume regulation response that cells are able to achieve following changes in the osmolarity of their environmental medium. Part B of this volume deals more specifically with these mechanisms. In all the tissues and cells studied up to now, they implicate active control of the amount of various intracellular osmotic effectors among which the inorganic ions Na^+, K^+, and Cl^- and the free amino acids play a prominent part.

The mechanisms of isosmotic intracellular fluid regulation have been found in tissues or cells of many species from various zoological groups including protozoans, invertebrates, and vertebrates. It is worth noting that many of these organisms either do not have, or have only weakly, the power of anisosmotic regulation of the extracellular fluids. It may thus be concluded that isosmotic regulation at the cell level is a rather primitive process which appeared early and has persisted in the course of evolution. Anisosmotic regulation processes would have been acquired later on, adding to the species possessing them a new range of possibilities.

As will be apparent to the reader of this volume, most of our knowledge on the mechanisms of cell volume maintenance is concentrated on the processes of anisosmotic regulation at the blood level; possibly because these mechanisms are the first an observer can apprehend when looking at the commonly used laboratory mammalian species.

The study of isosmotic regulation at the cell level is probably one of the important pieces of work which can be credited to the so-called comparative physiologists and biochemists. However, this study is only just beginning and the field of research is wide open. It must be hoped that a future book on cell volume control would include, besides new developments on anisosmotic regulation, answers to such fascinating questions as, how a kidney cell manages in its tremendously fluctuating environment or, is the well known particular resistance of the brain neurones to water intoxication related to the power of intracellular osmolarity control of some brain cells' population? These aspects of studies on cell volume control as well as the possible application of these studies, for instance to ecology or pathological situations have only been lightly touched upon or simply not treated at all in this volume because of the lack of pertinent information. More basic research is still needed before these fields of application can be effectively covered.

R. GILLES

Florkin, M. (1962). La régulation isosmotique intracellulaire chez les invertébrés marins euryhalins. *Bull. Acad. R. Belg. Cl. Sci.*, **48**, 687–94.

Part A

Water and Solutes in Biological Systems

Chapter 1

Structure and Properties of Water in the Cell

D. A. T. Dick

I. Introduction	3
II. The Structure of Water	4
A. The water molecule	4
B. Structure of water in mass	6
C. Structure of aqueous solutions	9
III. Association of Water with Macromolecules	10
A. Methods of assessing the amount of water associated with macromolecules	10
B. Experimental values of protein hydration	11
C. Sites of hydration in protein molecules	11
D. Associated water in living cells	12
IV. Water Content of Living Cells	13
A. Methods of measurement	13
B. Water content of cells	14
C. Water content of subcellular organelles	14
V. Water Flow in Living Cells	16
A. Thermodynamic treatment of osmotic equilibrium in cells	16
B. Experimental studies of osmotic equilibrium in cells	21
C. Effect of permeable solutes on apparent equilibrium	23
D. Mechanisms of water flow not directly related to osmotic forces	25
VI. Rate of Water Flow in Cells	26
A. Equations for water flow derived from irreversible thermodynamic and kinetic theory	26
B. Measurements of rate of water flow	30
C. Unstirred layers	32
D. Pore theory of water transfer across cellular membranes	34
E. Water flow in artificial lipid bilayer membranes	36
F. Activation energy of water transfer	37
Glossary of Symbols	38
References	39

I. INTRODUCTION*

Water in liquid form is the most abundant constituent of most living cells and is essential for vital processes. Some organisms such as bacterial spores can survive for long periods with a very low water content; but their vital

* A glossary of the symbols used is presented at the end of this chapter.

processes are essentially suspended, to be restored only when water is supplied. It might be argued that water plays this important part in terrestrial organisms merely because the Earth is the only member of the Solar System known to possess an abundance of liquid water; in the presence of other hydrides such as NH_3 or H_2S, alternative systems of life might be possible. Until planetary exploration has developed further, we cannot exclude such possibilities. What we do know, however, is that water is very different from its closest relations among the hydrides owing to some remarkable consequences of its molecular structure. Vital processes based on other hydrides could not be mere modifications of terrestrial life; lacking the unique features of water, they would have to be so different as to bear little resemblance to life as we know it on Earth. The purpose of this chapter is to examine these unique features of water and to survey its role in living organisms.

II. THE STRUCTURE OF WATER

A. The Water Molecule

The six outer electrons of the oxygen atom lie in four tetrahedrally-oriented hybrid sp^3 orbitals, two of the orbitals being completely filled by a pair of electrons while the other two contain a single electron (Fig. 1(a)). When two

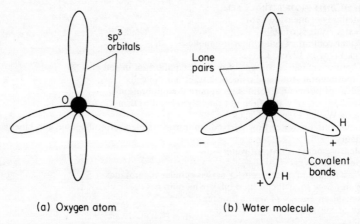

Figure 1 (a) Tetrahedral orientation of sp^3 orbitals around an oxygen atom. (b) Formation of a water molecule. The two covalent bonds to hydrogen atoms and the two lone pairs are tetrahedrally oriented

hydrogen atoms combine with an oxygen atom to form water, the electron of each hydrogen atom enters one of the sp^3 orbitals containing a single electron so as to form a covalent bond, and the new combined electron orbital encloses both H and O nuclei (Fig. 1(b)). The water molecule thus has four electron orbitals, each with a pair of electrons, tetrahedrally-oriented around the O nucleus; two surrounding the H⁺ nuclei and thus positively charged and two each containing only two electrons (the so-called lone pairs) and thus negatively charged. Owing to the near but not exact tetrahedral orientation of the two covalent OH bonds, the angle between them is very close to the tetrahedral angle (109.5°), and is actually 104.5° (Fig. 2). The distance between the O and H atoms is 96 pm. These simple facts, that their shape is not linear and that the distribution of electrical charge within them is not uniform, confer on water molecules remarkable and unique properties both individually and collectively.

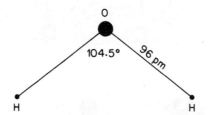

Figure 2 Diagram of OH bonds in a water molecule

1. Dipole Moment

Dipole moment is measured by the product of the charges separated multiplied by the distance of separation. Two charges, one positive and one negative each equal to that of an electron, 1.6×10^{-19} C, and separated by 100 pm, have a dipole moment of 1.6×10^{-29} C m. The dipole moment of a water molecule is 0.6×10^{-29} C m. This is much less than might be expected from the dimensions of the water molecule; the actual amount or the actual separation of the charges must therefore be much less than appears from Fig. 2. However, the dipole moment of the water molecule is nevertheless high compared with similar molecules such as HCl and H_2S.

2. Hydrogen Bond Formation

The positive charge near the hydrogen atom of a water molecule attracts the negative charge in one of the lone pairs of a neighbouring water molecule. The resulting electrostatic attraction is called a hydrogen bond. Since the hydrogen atoms and lone pairs are equal in number and are oriented nearly tetrahedrally with respect to the oxygen atom, four hydrogen bonds can be formed with each water molecule and an extensive three-dimensional lattice structure can be built up (Fig. 3); other hydrides such as HF, H_2S and NH_3 cannot do this, either because of inequality between hydrogen atoms and lone pairs or because of weaker hydrogen bonding. It is because of this lattice

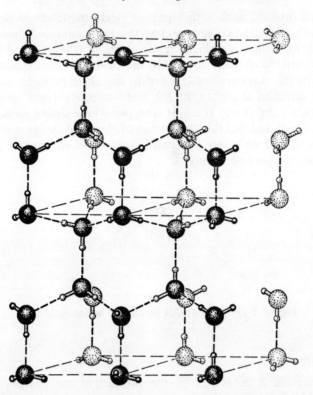

Figure 3 Hexagonal ice lattice. Each oxygen atom is linked to two hydrogen atoms by covalent bonds (solid rods). It is also linked to two hydrogen atoms in neighbouring molecules by hudrogen bonds (dotted lines). From L. Pauling: *The Nature of the Chemical Bond*. 3rd edn. © 1960 by Cornell University. Reproduced by permission of Cornell University Press

structure that water, unlike HF, HCl, H_2S and NH_3, is liquid at normal temperatures. The hydrogen bond is rather weak even in water; for example, the covalent OH bond within the water molecule has an energy of formation of 461 kJ mol^{-1}, while the energy of the hydrogen bond between molecules is only 19 kJ mol^{-1}.

B. Structure of Water in Mass

The high dipole moment and hydrogen bonding give rise to some unusual properties of bulk water which are important in its biological role.

1. Dielectric Constant of Water

The dielectric constant is a measure of the ability of a substance to reduce the intensity of an electric field across it as compared with that across a vacuum. The reduction in field is due in the case of water to an orientation of the molecules in the imposed field so that their small individual fields add together instead of cancelling one another as they do in normal random orientation. The combined field of the molecules opposes and effectively reduces the imposed field; thus the dielectric constant is related to the intensity of the field which the molecules can produce, that is, to the amount and separation of the charges in them or in other words the dipole moment. Since the dipole moment of water is large, its dielectric constant is also large. (This account refers only to orientation polarization which accounts for 95% of the dielectric constant of water.)

The high dielectric constant of water is of great importance in biology for it gives to water the power of dissolving and dissociating electrolytes. The main force holding the ions of an electrolyte together is electrostatic. Since the dielectric constant of water is no less than 78.5 at 25 °C, this attractive force is reduced in water to 1/78.5 of its normal value. It is this effect which allows the ions of electrolytes in aqueous solutions to behave almost as independent molecules. Since in biology it is often the ion rather than the undissociated electrolyte which is the active agent, it is hardly surprising that water is essential to all active biological systems.

2. Structure of Ice

As shown in Fig. 3 ice consists of a hexagonal lattice of water molecules connected by hydrogen bonds. The distance between oxygen atoms is 276 pm. However, it may be noted that if ice consisted of close-packed spherical molecules separated by this distance, its density would be 2.0 g cm^{-3} and not 0.92 g cm^{-3} as is actually the case. The very open tetrahedral structure of ice thus has a great deal of empty space in it. This is the reason for the expansion of water on freezing. This expansion is of course relevant to the effect of freezing on biological systems (though the osmotic effects of freezing are perhaps more important); the fact that ice consequently floats on water is also important to its role as a habitat for aquatic organisms.

3. Structure of Liquid Water

When ice melts a great deal of the lattice structure remains in the liquid water which results. There are several arguments which point to this fact. One of the most obvious is that while the latent heat of vaporization of ice is 51 kJ mol^{-1}, the latent heat of fusion is only 6 kJ mol^{-1}. Since vaporization involves rupture of all the hydrogen bonds, we infer that only 6/51 or approximately 12% of these are broken when ice melts. At the same time, the density rises from 0.92 g cm^{-3} in ice to 1.00 g cm^{-3} in water at 0 °C. There is no widely accepted theory of water structure; however, two current types of theory give a fair view of the possibilities.

(1) Continuum theories of water structure. These assume that liquid water has a homogeneous molecular structure. This is of a lattice-like nature but the lattice is irregular with variations in the number of molecules in a lattice ring (Bernal and Fowler, 1933), with distortion of oxygen–oxygen–oxygen bond angles and oxygen–oxygen distances (Pople, 1951; Kell, 1972). In this model hydrogen bonds are not broken but merely distorted when ice melts.

(2) Mixture models of water. These assume that there are at least two different patterns of molecular structure in water, which differ in their degrees of hydrogen bonding. Early models of this type assumed that most water molecules lay in a tetrahedral lattice with some interstitial molecules occupying spaces in the lattice (Forslind, 1952). Later the 'flickering cluster' model (Frank and Wen, 1957; Nemethy and Scheraga, 1962a, 1962b) assumed that at 20 °C water contains clusters of approximately 50 hydrogen-bonded molecules separated by single non-bonded molecules (Fig. 4). There are difficulties, however, with this model since it is inconsistent with some properties of water, particularly its single dielectric relaxation time (Davis and Jarzynski, 1972). A recent model envisages two kinds of structure in water: (a) open packed, each molecule has four nearest neighbours; (b) close packed, each molecule has more than four near neighbours but is only partially hydrogen bonded. Single water molecules make frequent transitions between the two kinds of environment, changing their number of hydrogen bonds as they do so (Davis and Jarzynski, 1972).

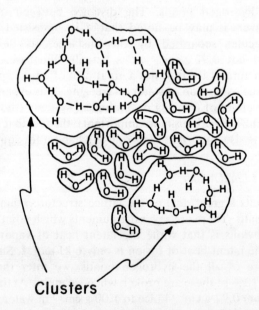

Figure 4 Nemethy–Scheraga 'flickering cluster' model of liquid water. From Nemethy and Scheraga (1962a); reproduced by permission of American Institute of Physics

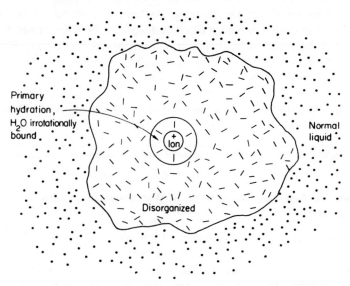

Figure 5 Arrangement of water molecules around an ion. From Klotz (1970); reproduced by permission of Wiley–Interscience

C. Structure of Aqueous Solutions

1. Electrolyte Solutions

According to the classical theory of electrolyte hydration (Frank and Wen, 1957) there are two layers of water surrounding a charged ion (Fig. 5). First, a tightly bound radially oriented monolayer of molecules. Secondly, a much larger zone of water molecules which are paradoxically less organized than normal, since the ionic charge is sufficient to disrupt the normal partial tetrahedral structure of water. However, anions have fewer bound water molecules than cations possibly because they are more free to rotate. Very small ions or multivalent ions such as Li^+, Na^+, Ca^{2+} and Mg^{2+} increase the viscosity of water suggesting increased binding which may be due to ordering of more than one layer of water by the high surface charge. On the other hand with large monovalent ions such as NH_4^+, Rb^+, Br^-, I^-, and NO_3^-, the structure-breaking effect predominates thus reducing the viscosity.

2. Solutions of Non-polar Solutes—Hydrophobic Bonds

Non-polar solutes tend to organize the water around them into box-like structures surrounding the solute, a formation called a clathrate; and the ordering of the water reduces the entropy of the system. A similar action has been proposed for non-polar groups in proteins (Klotz, 1970). When two non-polar molecules approach one another they can diminish the order in the water surrounding them, i.e. increase the entropy, if they draw so close together as to exclude some of the ordered water in between them. Since the entropy must be reduced by a similar amount on reversing the process and thus a certain amount of energy is required to separate the non-polar molecules, effectively a bond between them is formed, called a hydrophobic bond (Nemethy and Scheraga, 1962b). Such bonds are probably important in determining the conformation of protein molecules in aqueous solutions and hence in living cells.

III. ASSOCIATION OF WATER WITH MACROMOLECULES

A. Methods of Assessing the Amount of Water associated with Macromolecules

Water associated with macromolecules has sometimes been called 'bound' water, a term which conjures up a comfortably concrete picture of the macromolecule with an adherent aqueous overcoat; unfortunately the reality is not so simple. In practical terms 'bound' water has to be defined operationally as 'that water in the vicinity of a macromolecule whose properties differ detectably from those of "bulk" water in the same system' (Kuntz and Kauzmann, 1974). Although it may be hoped that different types of measurements are revealing the same 'bound' water, in practice the estimates are often quantitatively very different.

Methods used for estimating associated water are (see Kuntz and Kauzman, 1974):

(a) Preferential hydration. This is the water adjacent to a macromolecule which is more concentrated relative to other solutes than that in the bulk solution—or water from which other solutes are wholly or partially excluded. Difficulties are that any solute binding by the macromolecule will upset the estimate of hydration and that the estimate tends to vary with solute concentration.

(b) Hydrodynamic hydration. This is the water which is carried along with a macromolecule when it moves either by centrifugal force, or by rotary diffusion (measured by fluorescence depolarization, dielectric relaxation, or nuclear magnetic resonance dispersion) or by self-diffusion. Difficulties are that the estimate of hydration is very dependent on the shape, i.e. the axial ratio, of the molecule which is often not known and that hydrodynamic hydration will include any water trapped within crevices in the macromolecule even though this water is not strictly bound according to other criteria.

(c) Structural hydration. Low-angle X-ray diffraction sometimes shows that macromolecules are larger in the presence of water than in the anhydrous state and the difference has been attributed to hydration. Further, large-angle high-resolution X-ray diffraction has revealed the presence of water molecules within or on the surface of macromolecules; this question of the localization of water of hydration will be discussed in Section III-C.

(d) Unfreezable water. When macromolecular solutions are apparently frozen, some unfrozen water can still be detected either by careful calorimetry or by infrared or nuclear magnetic resonance spectroscopy; this unfrozen water is regarded as water of hydration.

(e) Isopiestic hydration. This is merely the water which is absorbed by a macromolecule at a given relative humidity, e.g. 92%, or at high absorption enthalpy or entropy. Estimates are however highly dependent on the level of relative humidity, enthalpy or entropy which is taken to indicate water of hydration.

B. Experimental Values of Protein Hydration

Some values of hydration estimated by various methods are shown in Table 1. The variations in the amount of hydration revealed by different methods

Table 1.

Protein	Hydration (g H_2O/g protein)				
	NMR	Calorimetric	Hydrodynamic	Structural	Isopiestic
Ribonuclease	—	—	0.57	—	0.35
Lysozyme	0.34	0.3	0.46	—	0.25
Myoglobin	0.24	—	0.46	0.11	0.32
β-lactoglobulin	—	0.55	0.45	—	0.30
Chymotrypsinogen	0.34	—	0.38	0.01, 0.07	0.29
Ovalbumin	0.33	—	0.15	—	0.30
BSA	0.40	0.32, 0.49	0.43	—	0.32
Haemoglobin	0.42	0.32	0.50	—	0.37
Subtilisin	—	—	—	0.18	—
Rubredoxin	—	—	—	0.31	—

From Kuntz and Kauzmann (1974); reproduced by permission of Academic Press Inc.

are considerable, e.g. from 0.32–0.49 g H_2O/g protein for bovine serum albumin. The fact of this large variation must always be borne in mind before trying to form pictorial conceptions of water of hydration.

C. Sites of Hydration in Protein Molecules

Combination of data from different techniques leads to the conclusion that there are three main sites for the water associated with protein molecules.

(a) Internal water. This is revealed by high-resolution X-ray diffraction data; from 9 to 13 water molecules are tightly bound often near the metals or ions of active sites where its high dielectric constant probably helps in the dispersal of the local electric charge (Birktoft and Blow, 1972; Liljas and coworkers, 1972).

(b) Surface water. This is revealed by freezing, by hydrodynamic studies, and probably by preferential hydration. Most of the water detected is probably in a monolayer on the surface.

(c) Solvent shell. These are outer layers of water whose orientations are altered by the surface of the macromolecule so as to create an increase in molecular order over that of bulk water (Bernal, 1965). It is very uncertain how much of this water is detected (if any) by the various methods of estimating hydration.

Several theories of protein hydration suggest that one or more molecules of water are bound to polar groups on the side chains of the protein amino acids but not to imino and carbonyl groups of the peptide bonds of the protein (Pauling, 1945; Bull and Breese, 1968). Ling (1972) has suggested, however, that the peptide bonds may bind water in fibrous proteins.

D. Associated Water in Living Cells

Since cells contain large amounts of proteins and other macromolecules, it seems inevitable that some of the water will be so associated with macromolecules as to undergo a modification of its normal properties. There has, however, been much controversy as to the actual extent and importance of water of hydration in living cells. Various lines of evidence have been used.

(a) It has sometimes been suggested that the large apparent non-solvent volumes in cells, i.e. the low values found for Ponder's **R** (see Section V-B), are due to protein hydration. However, as discussed in Section V-B, other explanations besides hydration, such as excluded volume and concentration-dependent ionization might also account for Ponder's **R**.

(b) Some nuclear magnetic resonance (NMR) studies have shown the presence in the cell of water whose NMR spectrum is broadened, presumably by interaction with proteins. However, estimates of the fraction of water so affected vary widely: 75% in erythrocytes (Odeblad, Bhar, and Lindström, 1956); less than 0.2% in muscle (Bratton, Hopkins, and Weinberg, 1965); 27% in muscle and 13% in brain (Cope, 1969); 10% in muscle (Hazlewood, Nichols, and Chamberlain, 1969). Recently Shporer, Haas, and Civan (1976) have claimed to detect a difference in nuclear magnetic resonance between nuclear and cytoplasmic water.

It seems likely that some intracellular water is indeed in a more organized state probably as a result of binding to macromolecules. However, present techniques do not give any reliable estimate as to the extent or significance of this binding.

IV. WATER CONTENT OF LIVING CELLS

A. Methods of Measurement

The simplest way to measure the water content of a cell is to find the loss of weight on drying, usually overnight at 105 °C. However, two problems arise: (a) often we do not know what other cellular constituents are lost besides water; (b) if, as is usual, we are dealing with a tissue and not individual cells, we need to know how much of the lost water was in the extracellular fluid and not in the cells themselves. Estimation of extracellular fluid is usually done by means of a non-penetrating marker, such as inulin, whose concentration is easily estimated. Even when the water content has been measured, there remain problems depending on whether we have measured the weight or volume of the cells and whether we want the water concentration per gram or per cubic centimetre of cells. It is often necessary either to measure the density of the cells (by flotation in fluids of known density) or to assume a specific volume 0.75 $cm^3\ g^{-1}$ for the cell solids, a good approximation for most proteins and probably satisfactory except in cells which contain much lipid.

Indirect methods of estimation of water content involve assumptions about the intracellular solids. Thus cell density may be used to estimate water content if the above value of specific volume of cell solid is assumed. Another indirect method uses the refractive index from which cell solid and hence cell water content can be estimated provided a value for the average specific refraction increment for cell solids is assumed. Like the specific volume, the specific refraction increment tends to be similar for many proteins, in the region of 0.0018. The cellular refractive index may be measured by immersing the cell in an external solution whose refractive index is varied by varying the amount of bovine plasma albumin in the solution until a match is obtained in the phase contrast microscope (Barer and Joseph, 1954, 1955a, 1955b). Alternatively it may be obtained by measuring, with an interference microscope, optical path difference in two solutions of known refractive index (Barer and Dick, 1957).

Estimation of change in water content is often based on changes in cell volume on the assumption that, in the short term at least, the cell membrane is effectively semipermeable and transmits only water and not solutes. Measurement of cell volume can be performed: (1) by direct measurement of the diameter of large spherical of spheroidal cells (remembering that the percentage error in volume is three times the percentage error in diameter); (2) by haematocrit, with allowance for extracellular space; (3) by optical methods, measuring either changes in the optical density of a cell suspension or the intensity of light scattered at right angles to the incident beam—in either case although the theoretical background is extremely complex an empirical relation is found between light intensity and cell volume (Hempling, 1958; Sidel and Solomon, 1957; Farmer and Macey, 1970); (4) by electrical methods, especially measuring the resistance of cell suspensions where the lipid cell membranes create a high-resistance element in the suspension (e.g. the Coulter counter, Coulter Electronics Ltd).

All of these methods may be used in suitable circumstances for measuring changes of cell water content, that is measuring water fluxes in or out of cells, although recently the optical and electrical methods have been most popular especially for cells in which water movements are very rapid, such as erythrocytes.

B. Water Content of Cells

The water content of cells changes, of course, with the external osmotic pressure. However even when cells are maintained at the osmotic pressure of the body fluids in which they normally exist, their water content varies very considerably, from 87% (v/v) in fibroblasts, about 79% in frog skeletal muscle, 75% in human erythrocytes, down to approximately 50% in mature amphibian oocytes (see Table 2). In some cells such as amphibian oocytes, cellular water content diminishes with age as the solid content of the cell increases (Cannon, Dick and Ho-Yen, 1974).

C. Water Content of Subcellular Organelles

1. Nuclei

The water content of the nuclei of most cells is high, usually more than 80% (see Table 2), and responds not merely to general changes in cell water content but to changes in the concentration of substances in the cytoplasm to which the nuclear membrane is selectively permeable, e.g. in the case of amphibian oocyte nuclei, bovine plasma albumin or polyvinylpyrrolidone (Harding and Feldherr, 1959). In other cases such as the salivary gland cell of Drosophilia, the high electrical resistance of the nuclear membrane (Loewenstein and Kanno, 1963) suggests that it may be impermeable and therefore osmotically sensitive to salts also.

2. Mitochondria

Mitochondria have a rather lower water content than nuclei, about 0.67 $\mu l/\mu l$ of rat liver mitochondria suspended in 0.2 M sucrose (272 mosmol) (Bentzel and Solomon, 1967). However a large fraction of this water, about 0.46 μl, lies in a compartment which is accessible to external sucrose and thus appears to be osmotically unresponsive. The remaining 0.21 μl of water lie in a compartment not accessible to sucrose but osmotically responsive. It is not clear which anatomical compartments of the mitochondrion correspond to

Table 2. Water content of cells and organelles.

(a) Cells	Water content (at isotonic osmotic pressure) (cm³/cm³ cell)	Reference
Human erythrocyte	0.76 ± 0.02	Dick and Lowenstein (1958)
	0.74 ± 0.02	Le Fevre (1964)
Chick heart fibroblast	0.87	Dick (1958)
Guinea-pig smooth muscle	0.77[a]	Brading and Setekleiv (1968)
Mouse ascites tumour	0.86[a]	Hempling (1958)
Frog skeletal muscle	0.79	Dydynska and Wilkie (1963)
Toad oocyte (immature)	0.82	Cannon, Dick and Ho-Yen (1974)
Toad oocyte (mature)	0.50	
(b) Organelles		
Nucleus		
Chick heart fibroblast	0.88	Barer and Dick (1957)
Rat hepatocyte	0.88	Schiemer, Gunther and Sina (1967)
Frog oocyte	0.88	Century, Fenichel and Horowitz (1970)
Thymocyte	0.83	Itoh and Schwarz (1957)
Mitochondrion		
Rat hepatocyte	0.67	Bentzel and Solomon (1967)

[a] Calculated assuming specific volume of solid as 0.75 cm³ g⁻¹.

these water fractions. The space between the outer and inner membranes might be the sucrose-accessible compartment but this is normally smaller than the mitochondrial matrix, the opposite of the relation suggested by the functional studies of Bentzel and Solomon. The mitochondrion is also subject to non-osmotic changes of volume which are of two kinds (Lehninger, 1964). Phase-I or low-amplitude changes of 20–40% of mitochondrial volume are produced by changes of respiratory activity, i.e. in the presence of substrate, lack of ADP causes swelling which is reversed by adding ADP. Phase-II or high-amplitude changes involve swelling of up to 200%, produced by Ca^{2+} ions, glutathione, thyroxine, vasopressin, and insulin. It is prevented and partially reversed by ATP. The mechanism of non-osmotic changes is obscure but it appears to be associated with the inner mitochondrial membrane which expands with reduction of cristae (Blondin and Green, 1967; Packer and coworkers, 1968; Stoner and Sirak, 1969).

3. Microsomes

Tedeschi, James, and Anthony (1963) have described osmotic changes in microsomes isolated from rat liver.

V. WATER FLOW IN LIVING CELLS

Water flow appears to be mostly, if not invariably, passive and secondary to other changes, e.g. active transport of ions. There has been argument as to whether in some cases active water flow may occur. Much of this argument turns on what is to be regarded as active water flow. Some cellular mechanisms, such as the contractile vacuole, secrete watery or hypotonic fluid (see Section V-D); however it seems likely that what really occurs is passive leakage of an isotonic saline filtrate, followed by active (and differential) reabsorption of the ionic solutes. It seems generally agreed that this is not to be regarded as active water transport. If one compares the analogous argument regarding active ion transport, perhaps the most persuasive fact is that an enzyme, Na–K–Mg-activated ATPase, is known which appears to link metabolic energy directly to vectorial ion transfer; no such enzyme has so far been demonstrated linking metabolism directly to water transfer. In this chapter, therefore, the water transfer mechanisms to be considered will be passive ones.

A. Thermodynamic Treatment of Osmotic Equilibrium in Cells

The relation between cell volume and external osmotic pressure is best calculated by osmotic equations, but several conditions are required which are not always or not exactly satisfied by individual cells.

(a) Water flow is significantly restricted and therefore governed only by the cell membrane—or, what is the same thing, there are never significant gradients of osmotic pressure either outside or inside the cell membrane. As will be seen later (see Section VI-C), there is evidence that, when water flow is rapid, significant osmotic gradients can exist both outside and inside the membrane and these must be taken into account when calculating the water permeability of cell membranes. In general, however, these gradients do not arise in assessing conditions of osmotic equilibrium.

(b) The cell membrane must be semipermeable, i.e. permeable only to water and not to solutes. This is an extreme situation which is approximately satisfied by cells for limited periods of time. When significant net solute flows occur across the cell membrane (influxes and effluxes which are equal do not matter), then care must be taken not to attribute the resulting deviations from normal osmotic behaviour to supposed abnormalities of intracellular solutes; the flow of water and solutes should then be treated by the equations of irreversible thermodynamics which reduce to the osmotic equations for zero solute flow. (It should be noted that the irreversible thermodynamic equations, since they deal with more complex flows, generally make simplifying assumptions that the intracellular solutions are ideal and dilute; these assumptions are obviously not exactly fulfilled. Although the osmotic treatment is restricted in its application it does deal with non-ideal and non-dilute solutions.)

(c) At equilibrium there is no gradient of osmotic pressure across the cell membrane. In early work it was suggested that the cytoplasm of animals was hypertonic to the environment but this was subsequently shown to be due to the effects of autolysis. When precautions are taken to avoid autolysis by freezing or boiling the cell homogenate to destroy autolytic enzymes then good agreement is found between external and internal osmotic pressure (Buckley, Conway, and Ryan, 1958; Appelboom and coworkers, 1958; Maffly and Leaf, 1959). Certain protozoa are, however, exceptions to this rule; their cytoplasm is hypertonic to the fresh-water environment in which they live and is kept at a constant osmotic pressure by expulsion of water via the contractile vacuole (Kitching, 1938; Schmidt-Nielsen and Schrauger, 1963).

The osmotic pressure is the hydrostatic pressure which must be applied to a solution to bring it into equilibrium with the pure solvent when the two liquids are separated by a semipermeable membrane. Two changes thus occur to the water in the solution: (a) a lowering of chemical potential when solute is added to form the solution; and (b) a raising of chemical potential by applying pressure; since equilibrium with pure solvent is attained the two changes must be equal and opposite. For ideal solutions the change in chemical potential (μ) due to solute is

$$\Delta\mu_1 = RT \ln \left(\frac{n_1}{n_1 + n_2}\right) \quad (1)$$

The subscripts 1 and 2 refer to solvent and solution respectively (see glossary of symbols at the end of this chapter). Since $n_1/(n_1 + n_2)$ is a fraction, the logarithm and hence the chemical potential change is negative.

The pressure change is

$$\Delta\mu_1 = \bar{V}_1 \pi \quad (2)$$

where π is the pressure applied to the solution, the osmotic pressure and $\bar{V}_1$ is the partial molar volume of solvent, in the case of water approximately 18 cm^3.

Then since the two changes are equal and opposite the net change of chemical potential is zero, i.e.

$$\Delta\mu_1 \text{ (solute)} + \Delta\mu_1 \text{ (pressure)} = 0 \quad (3)$$

Thus

$$RT \ln \left(\frac{n_1}{n_1 + n_2}\right) + \pi \bar{V}_1 = 0 \quad (4)$$

or

$$\pi = \frac{RT}{\bar{V}_1} \ln \left(\frac{n_1 + n_2}{n_1}\right) = \frac{RT}{\bar{V}_1} \ln \left(1 + \frac{n_2}{n_1}\right) \approx \frac{RT}{\bar{V}_1} \frac{n_2}{n_1} = \frac{RT}{\bar{V}_1} n_2 \quad (5)$$

Thus

$$\pi V_1 = RTn_2 \tag{6}$$

The usual form in which this equation is applied is that used by Lucke and McCutcheon (1932):

$$\pi (V-b) = \text{constant} = \pi_0 (V_0-b) \tag{7}$$

In this case V refers to total cell volume and the subscript 0 refers to initial conditions, usually of isotonic osmotic pressure and the corresponding cell volume. b is simply $(V - V_1)$, i.e. the non-solvent volume in the cell.

Two reservations must be noted about these equations: (a) equation (1) assumes that the solution is perfect, that is that the mass, size, and intermolecular forces are the same for solute and solvent molecules; (b) the approximation used for the logarithm in equation (5) requires that n_2/n_1 is small, i.e. that the solution is dilute. The importance of these reservations will be seen later.

Equation (7) implies that when V is plotted against π_0/π, a straight line should be obtained. This is usually the case within ordinary experimental

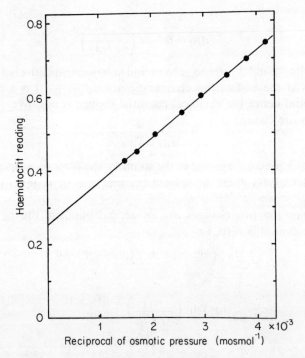

Figure 6 Linear relation between erythrocyte volume (haematocrit) and reciprocal of osmotic pressure. From Le Fevre (1964); reproduced by permission of Rockefeller University Press

error; an example is shown in Fig. 6. A further consequence of equation (7) is that the slope of the plot should be

$$\frac{dV}{d(\pi_0/\pi)} = V_0 - b \tag{8}$$

that is equal to V_1, the water content of the cell at isotonic osmotic pressure. It is here that difficulties arise since the actual water content (W_m), measured by drying, is always found to be larger than (V_0-b). The discrepancy is conventionally measured by the ratio

$$(V_0-b)/W_m = \mathbf{R} \tag{9}$$

This ratio and the symbol **R** given to it were first described by Ponder (see Ponder, 1948) and hence it is called Ponder's **R**. Since $W_m > (V_0-b)$, **R** < 1.0. Values of non-solvent volume, b, and of Ponder's **R** for various cells are shown in Tables 3 and 4. More extensive tables are given by Dick (1966, 1970).

Table 3. Ponder's **R** in cells other than erythrocytes

Cell	Range of relative osmotic pressure	Non-solvent volume (cm³/cm³ cell)	Ponder's R	Reference
Chick heart fibroblast	0.39–1.78	0.18	0.94	Dick (1958)
Guinea-pig smooth muscle	0.75–2.9	0.26	0.96	Brading and Setekleiv (1968)
Mouse ascites tumour	1–2.0	0.33	0.78	Hempling (1958, 1960)
Frog skeletal muscle	1–2.0	0.34	0.83	Dydynska and Wilkie (1963)

How is the discrepancy expressed by Ponder's **R** to be accounted for? In the first place it must be noted that any cells which leak solute, whether for physiological or pathological reasons, will give volume changes less than expected from changes of external osmotic pressure and hence low values of $dV/d(\pi_0/\pi)$, (V_0-b), and **R**. However, in many cases solute leakage can be excluded while Ponder's **R** remains low; then it is necessary to reexamine the equations used.

As stated above, equations (6) and (7) require that the intracellular solutions be perfect and dilute. Since they are not, a correction factor, the osmotic coefficient, is introduced to express the discrepancy:

$$\pi V_1 = \varphi RT n_2 \tag{10}$$

or

$$\pi = \varphi RT m_2 \tag{11}$$

Table 4. Values of Ponder's **R** in human erythrocytes

Range of relative osmotic pressure	Ponder's **R**	Reference
(a) Hypotonic measurements		
1–0.425	0.96–1.05	Guest and Wing (1942)
1–0.5	0.93–0.97	Ponder (1944)
1–0.62	0.99	Ørskov (1946)
1–0.425	0.98	Guest (1948)
1–0.2	0.9	Ponder (1950)
1–0.58	0.97	Hendry (1954)
1–0.62	0.95	Dick and Lowenstein (1958)
1–0.56	0.90	Gaffney (unpublished observations)
1–0.66	0.91^2	Gary-Bobo and Solomon (1968)
1–0.2	0.963^3	From Adair's data (1929)
1–0.69	0.953^3	
(b) Hypertonic measurements		
1–2.39	0.79	Ørskov (1946)
0.5–1.7	0.78	Ponder and Barreto (1957)
1–2.82	0.89^a	Olmstead (1960)
1–1.25	1.02^b	White and Rolf (1962)
0.80–2.33	0.83	Le Fevre (1964)
0.68–1.69	0.80	Savitz, Sidel and Solomon (1964)
0.6–1.8	0.85^b	Cook (1967)
1–1.65	0.67^b	Gary-Bobo and Solomon (1968)
0.7–1.9	$0.80–0.95^d$	Kwant and Seeman (1970)
1–2.4	0.87^c	From McConaghey and Maizels' (1961) data
0.6–1.7	0.92^c	From Adair's (1929) and McConaghey and Maizels' (1961) data

[a] Olmstead's original method of calculation was incorrect and the value shown was recalculated by Le Fevre (1964).
[b] Calculated by present author.
[c] Values computed from Equation (12) for comparison with experimental data.
[d] Kwant and Seeman used for the specific volume of haemoglobin the value 0.85 ml g^{-1} for lyophilized material; the correct value for dissolved haemoglobin is 0.75 ml g^{-1} (Rossi Fanelli, Antonini and Caputo, 1964) and **R** values have been corrected accordingly.

where m_2 is the concentration of solute per litre of solvent. For water where 1 litre weighs approximately 1 kilogram at physiological temperatures, m_2 is virtually the *molal* concentration. Since φ connects an osmotic pressure with a molal concentration it is called the *molal osmotic coefficient*. The value of φ varies with different solutes; for most electrolytes it is 0.93–0.95 at physiological concentrations due to attraction between the ions (the Debye–Hückel effect); for macromolecules such as proteins or nucleic acids it is usually considerably greater than 1.0, and also increases steeply with concentration.

It may be shown (see Dick, 1970) that

$$\text{Ponder's } \mathbf{R} = 1 - \frac{\pi_0}{\varphi_0} \frac{\Delta\varphi}{\Delta\pi} \qquad (12)$$

For the electrolyte part of cell solute, φ does not vary significantly with the range of experimental concentrations; but since φ increases with concentration (i.e. with π) for macromolecules, $\Delta\varphi/\Delta\pi$ is expected to be positive so that $\mathbf{R} < 1.0$.

B. Experimental Studies of Osmotic Equilibrium in Cells

Values of **R** for various cells are shown in Tables 3 and 4. These confirm the general expectation that $\mathbf{R} < 1.0$. When an attempt is made however to account for values of **R** by means of equation (12), difficulties are encountered. The first is that for most cells the bulk of the intracellular proteins are either insufficiently defined or their osmotic properties are insufficiently studied to give adequate estimates of φ or $\Delta\varphi/\Delta\pi$. The only cell whose intracellular protein, haemoglobin, is well defined is the erythrocyte and, even for haemoglobin, data at high concentration and ionic strength are scanty. Values of Ponder's **R** in erythrocytes are shown in Table 4.

There are two encouraging features: (a) in hypotonic experiments reasonable agreement is found between experimental and predicted values of **R** for the erythrocyte; (b) in hypertonic solutions **R** is less than in hypotonic solutions as predicted from equation (12), since for haemoglobin $\Delta\varphi/\Delta\pi$ increases with concentration; although good numerical agreement between experimental and predicted values in hypertonic solution is not obtained. One reason for this might be that the data of McConaghey and Maizels (1961) on concentrated haemoglobin solutions were obtained at low ionic strength unlike the high ionic strength present in an erythrocyte shrunk in hypertonic saline. There are some data which suggest that the osmotic coefficient of haemoglobin rises at high ionic strength (Adair, 1967, unpublished data quoted by Dick, 1967); thus the discrepancy may be due to use of too low a value of $\Delta\varphi/\Delta\pi$ and hence too high a predicted value of **R**. It must, however, always be borne in mind that a small degree of salt leakage will reduce the volume response of cells to osmotic changes and hence lead to low values of **R**; this comment may well apply to some of the data in Table 3 and even to some in Table 4, though efforts have been made to exclude solute leakage in erythrocyte experiments (Dalmark, 1975). Since Kwant and Seeman (1970) have shown that Ponder's $\mathbf{R} = 1.0$ for erythrocyte ghosts, at least it seems clear that the origin of the discrepancy is the erythrocyte contents and not the membrane.

The question remains as to what causes the large positive osmotic coefficients of proteins which give rise to low values of **R**. There have been several explanations:

(a) Preferential hydration. If some water associated with protein is not available to other solutes then the apparent osmotic coefficient will rise with concentration as the free water becomes a progressively smaller fraction of the total water so that solute concentrations are higher than they appear to be. Only preferential hydration is relevant to this, however; other measures of hydration particularly hydrodynamic measurements or water of crystallization used by some authors, e.g. Savitz, Sidel, and Solomon (1964) are not necessarily relevant to the osmotic properties of proteins.

(b) Excluded volume. The volume which determines the osmotic pressure developed by solute molecules in it is the so-called 'free volume'. This is the total volume minus the 'excluded volume'. The latter is the volume into which solute molecules (or strictly the centres of solute molecules) cannot enter. For small solutes the excluded volume is negligible but for macromolecules it includes not merely the volume of the molecules themselves but also a space around each of depth equal to the molecular radius (Fig. 7). Since the

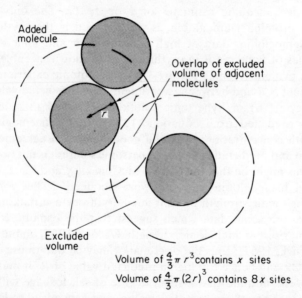

Volume of $\frac{4}{3}\pi r^3$ contains x sites
Volume of $\frac{4}{3}\pi (2r)^3$ contains $8x$ sites

Figure 7 Excluded volume which cannot be entered by the centre of an added molecule; for a sphere this is eight times the molecular volume. From Dick (1966); reproduced by permission of Butterworths & Co. (Publishers) Ltd

excluded volume reduces the free volume it increases the osmotically effective concentration of the molecules so that osmotic pressure rises out of proportion to the overall molecular concentration, i.e. the osmotic coefficient is greater than 1.0 and $\Delta\varphi/\Delta\pi$ is positive.

(c) Concentration-dependent ionization. Gary-Bobo and Solomon (1968) have accounted for pH-dependent changes in **R** in erythrocytes in terms of variations in chloride movements in and out of the cell; they have suggested that these variations are due to the ionization of haemoglobin being concentration dependent so that the amount of gegen-ion (mainly chloride) changes and hence the osmotic effect and the osmotic coefficient of the haemoglobin.

It is not possible at present to assign the observed protein osmotic coefficients quantitatively to each of these effects though it seems likely that all may contribute. Recently Dalmark (1975) has estimated that when the erythrocyte Cl content is varied by changing the external pH, the variation in cellular water content is only 70% of the variation in total intracellular content of molecules (estimated by indirect methods). He has suggested that the ratio 0.7 between the variations in water and molecular contents is analogous to Ponder's **R** and that the variation he has found is not to be accounted for by any of the above possible causes of change in the osmotic coefficient of haemoglobin. However, several problems remain: (a) Dalmark's technique was a very indirect one so that his results depend critically on various assumptions made in the calculation; (b) since his ratio was obtained by varying intracellular solute as well as water, it remains doubtful whether it is really analogous to the Ponder's **R** obtained in osmotic experiments where only cellular water is varied; (c) the osmolarity of the fluid transferred during pH changes in Dalmark's experiments was 37% greater than the osmolarity of the medium and this discrepency remains unexplained.

C. Effect of Permeable Solutes on Apparent Equilibration

If some external or internal solute undergoes net transfer into or out of the cell, then the membrane is not semipermeable and osmotic equilibrium is not strictly possible. However, at a certain external concentration of a permeable solute (usually somewhat higher than if it were impermeable) it is possible to attain a state where for a time no volume change occurs in the cell and a condition of apparent equilibrium is attained. A typical experiment of this type is shown in Fig. 8. This situation is governed by the rules of irreversible thermodynamics. The appropriate equation in this case is (see Section VI-A).

$$J_v = L_p (\Delta p - RT\Delta c_i - \sigma_s RT\Delta c_s) \qquad (13)$$

(see glossary of symbols at the end of this chapter). In the type of experiment illustrated, $J_v = 0$ and $\Delta p = 0$, so that

$$\Delta c_i + \sigma_s \Delta c_s = 0 \qquad (14)$$

or

$$\Delta c_s = -\Delta c_i / \sigma_s \qquad (15)$$

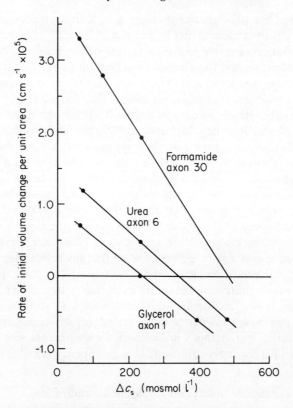

Figure 8 Rate of initial volume change in a squid giant axon plotted against concentration difference of three penetrating solutes. Interpolation gives the concentration difference for zero volume change. (Δc_i = 230 mosmol l^{-1}.) From Villegas and Barnola (1961); reproduced by permission of Rockefeller University Press

that is, for a given concentration difference of impermeable solute (which may for example be wholly within the cell) the concentration of permeable solute required to attain zero volume change is increased by the factor $1/\sigma_s$, where σ_s is the reflection coefficient of the permeable solute. σ ranges from 1.0 for impermeable solutes to 0 for solutes whose permeability is similar to that of water; it can be measured by experiments such as illustrated above or in other ways. Its importance in relation to osmotic equilibrium experiments is a negative one. If permeability of the solute is not considered and its reflection coefficient taken into account, then grossly exaggerated estimates of internal osmotic pressure can be made when permeable external solutes are employed in osmotic experiments. On the other hand if constant cellular volume is to be attained with permeable solutes, e.g. during fixation for microscopy, then concentrations much higher than normal isotonic concentrations must be used.

D. Mechanisms of Water Flow not directly related to Osmotic Forces

1. Pinocytosis

Many animal cells take up droplets of fluid from the exterior by a process of engulfing an invaginated portion of the cell membrane with its contents so as to form an intracellular vacuole which is subsequently absorbed. In initial studies on amoeba, Mast and Doyle (1934) suggested that pinocytosis was a compensatory mechanism following shrinkage of cell volume; however it has been shown that pinocytosis can occur following cell swelling in amoeba (Chapman-Andresen and Dick, 1961). Solute adsorption on the cell membrane precedes pinocytosis (Brandt, 1958; Schumaker, 1958) but the following process of pinocytosis is temperature dependent and prevented by metabolic inhibitors (Chapman-Andresen, 1962, 1967a, 1967b). Stimulants of pinocytosis are basic dyes such as alcian blue, neutral red and acridine orange, proteins such as bovine plasma albumen, lactoglobulin and lysozyme (in basic form only), and concentrated salt solutions. After uptake by pinocytosis in amoeba alcian blue was transferred to secondary lysosomes (Chapman-Andresen, 1967a). Although Ca^{2+} is not itself an active inducer it potentiates the inducing action of other agents and it has been suggested that Ca^{2+} acts as a link in the pinocytotic process (Josefsson, 1968). Pinocytosis is accompanied by increased oxygen uptake (Hansson, Johansson, and Josefsson, 1968) and by a decrease in membrane potential and an increase of membrane conductance (Josefsson, Holmer, and Hansson, 1975). Although fluid is transferred by pinocytosis and this may have some significance in epithelial transport (Karnovsky, 1968), solute or membrane transfer is the more important aspect of pinocytosis. Michl and Spurna (1975) have suggested that pinocytotic uptake of growth-promoting alpha-globulin is important in controlling DNA, RNA and protein synthesis in cultured L cells.

2. Contractile Vacuoles

A contractile vacuole is a fluid-filled vesicle which is regularly emptied at the cell surface and refilled from the cell interior. It occurs in protozoa, algae and sponges. Kitching (1938) demonstrated that in amoeba the contractile vacuole excretes just enough water to balance osmotic inflow from the hypotonic environment and is thus essentially a mechanism for regulating cell osmotic pressure and volume. Schmidt-Nielsen and Schrauger (1963) have shown by freezing point measurements that the vacuole contents have only $\frac{1}{3}$ to $\frac{1}{2}$ of the osmotic pressure of the cytoplasm, and Riddick (1968) has confirmed this and also shown that the vacuole contains a higher Na and a lower K concentration than the cytoplasm. Thus the contractile vacuole produces a relative outflow from the cytoplasm of water and sodium. From the relative concentrations in cytoplasm and vacuole and its potential (about +15 mV relative to cytoplasm) active expulsion of Na into the vacuole and active withdrawal of K from it seem to occur; water flow could, however, be a purely passive process although House (1974) has questioned whether the passive permeability of the vacuolar membrane can be sufficiently high to allow for this.

The function of the contractile vacuole as a regulating mechanism in single cells or simple cell aggregates exposed to hypotonic environments is thus clear; the restricted distribution of vacuoles in nature presumably reflects the takeover of this function by the kidney in higher animals.

3. Electro-osmosis

Since the cell membrane is known to contain fixed electrical charges, it is to be expected that when an electrical potential occurs across the membrane water movement will be created by electro-osmosis. This effect is easily detected in artificial membranes but in cells it is only in those of algae that it has been detected with certainty (Fensom and Dainty, 1963; Barry and Hope, 1969); even in these the contribution of electro-osmotic water flow to turgor pressure has been shown to be negligible (Dainty, 1963). Further Barry and Hope (1969) have pointed out that a large fraction of the water flow accompanying a potential difference is due not to electro-osmosis itself but to unequal solute concentrations in boundary layers of solution due to differences in the transport numbers of ions in the membrane and in solution. Wedner and Diamond (1969) have analysed apparent electro-osmosis in rabbit gall bladder from this point of view and found that much or even all of the water flow may be due to such local concentration differences and not to electro-osmosis. Electro-osmosis is therefore probably not a significant process in animal tissues, if indeed it occurs at all.

VI. RATE OF WATER FLOW IN CELLS

At first sight it is a simple problem to measure the rate of water flow into or out of cells. It will be seen, however, that the subject is in fact extremely confused; measurements may be made by osmotic or tracer flow of water, they may be interpreted by thermodynamic or kinetic theory, unstirred layers must be taken into account.

A. Equations for Water Flow derived from Irreversible Thermodynamic and Kinetic Theory

The first source of equations is the theory of irreversible thermodynamics. It is not possible to give a full account of this but only to give the results which are applicable to water flow. The underlying theory used was developed by

Kedem and Katchalsky (1958, 1961) and simplified accounts of it have been given by Dick (1966) and House (1974).

The basic equations applying to water and solute flow are

$$J_v = L_p(\Delta p - \sum RT\Delta c_i - \sum \sigma_s RT\Delta c_s) \qquad (16)$$

$$J_s = \omega_s RT\Delta c_s + J_v \bar{c_s}(1 - \sigma_s) \qquad (17)$$

Strictly speaking J_v is the rate of change of volume of the cell solution across unit area of cell membrane; however, since in most cases water is the only substance whose volume flux is significant little error is usually involved in regarding J_v as essentially the water flux rate. J_v is controlled by three factors: (1) Δp, the hydrostatic pressure applied across the cell membrane; owing to the weakness of animal cell membranes this is usually virtually zero; (2) $\Sigma \Delta c_i$, the sum of the differences of molecular concentration across the cell membrane of all impermeable solutes; and (3) $\Sigma_s \Delta c_s$ the sum of the differences of concentration across the cell membrane of all permeable solutes, each multiplied by its own reflection coefficient. Several points must be noted, however: (a) the concentrations used are per unit volume of solvent, not of total solution; (b) it is assumed that the solutions are ideal and dilute and no attempt is usually made to correct for deviations by using osmotic coefficients (as seen in Section V-A, this is a sacrifice made to obtain the greater generality and wider application of irreversible thermodynamic equations; it should not, however, be forgotten); (c) the proportionality factor, L_p, called the hydraulic conductivity, is a water permeability coefficient which applies to both hydraulic and osmotic water flow.

In practice equation (16) is used for measuring water permeability in experiments where cells are shrunken or swollen by changing the external osmotic pressure (for technical details see Section III-A) while equation (17) is used where water permeability is measured by measuring passage of tracer water across the cell membrane (tracer may be treated as a solute).

For practical measurements of L_p for water, permeable solutes are avoided ($\Delta c_s = 0$) and either Δp or Δc_i is also made zero, so that

$$J_v = L_p \Delta p \quad (\Delta c_i = 0) \qquad (18)$$

or

$$J_v = L_p RT\Delta c_i \quad (\Delta p = 0) \qquad (19)$$

Equation (18) has been extensively used for measuring L_p in capillary walls which are naturally subjected to hydrostatic pressure and also in large plant cells which are supported by cell walls; it has also been used by Vargas (1968) in an animal cell, the squid axon.

Equation (19) usually has to be expressed in differential form and sometimes integrated before use. Since

$$J_v = \frac{dV}{dt}\frac{1}{A} \text{ (by definition)}$$

then

$$\frac{dV}{dt} = L_p RTA \, (c_{i(c)} - c_{i(m)}) \tag{20}$$

$$= L_p \frac{RT}{\bar{V}_w} A \left(\frac{n_{i(c)}}{n_{w(c)}} - \frac{n_{i(m)}}{n_{w(m)}} \right) \tag{21}$$

since $c_i = n_i/V_w = n_i/V_w = n_i/n_w \bar{V}_w$.

Alternatively since $c_{i(c)} = n_{i(c)}/(V-b)$ (V = cell volume), equation (24) may also be written

$$\frac{dV}{dt} = L_p RTA \left(\frac{n_{i(c)}}{V-b} - c_{i(m)} \right) \tag{22}$$

If equation (22) is to be used directly, then dV/dt must be measured before significant change has occurred in V, or an empirical quadratic relation can be fitted to the initial data so as to estimate dV/dt at $t = 0$ (Dick, Dick, and Bradbury, 1970). At longer times, equation (22) must be integrated for use. If A, the area of cell membrane, remains constant as occurs in flattened cells such as erythrocytes, integration gives

$$L_p RT = \frac{1}{c_{i(m)} At} \frac{n_{i(c)}}{c_{i(m)}} \ln \left(\frac{(V_0 - b) - (n_{i(c)}/c_{i(m)})}{(V_t - b) - (n_{i(e)}/c_{i(m)})} \right) - (V_t - V_0) \tag{23}$$

This may be expressed in an alternative form if it is remembered that at final equilibrium

$$c_{i(m)} = c_{i(c)(e)} = n_{i(c)}/(V_e - b) \tag{24}$$

where $c_{i(c)(e)}$ and $(V_e - b)$ are the cellular solute concentration and solvent volume at equilibrium.

Thus

$$n_{i(c)}/c_{i(m)} = V_e - b \tag{25}$$

Substituting (25) in (23) we obtain

$$L_p RT = \frac{1}{c_{i(m)} At} (V_e - b) \ln \left(\frac{V_e - V_0}{V_e - V_t} \right) - (V_t - V_0) \tag{26}$$

which is a convenient form for experimental use. If the cell surface area does not remain constant, more complex integrals result (see Dick, 1970).

Structure and Properties of Water in the Cell

In the past, water permeability has been measured by a simple equation:

$$\frac{dV}{dt} = P_{osm} A (\pi_{(c)} - \pi_{(m)}) \tag{27}$$

where P_{osm} is the osmotic permeability coefficient. Equation (21) may also be expressed in terms of osmotic pressure by substituting by means of equation (5) giving:

$$\frac{dV}{dt} = L_p A (\pi_{(c)} - \pi_{(m)}) \tag{28}$$

It is thus seen that, provided the osmotic pressure difference is measured in pressure units (atmospheres or kilopascals), P_{osm} and L_p are identical and are expressed in cm^3 cm^{-2} s^{-1} kPa^{-1} or cm s^{-1} kPa^{-1}. Sometimes, however, the water flux is related to a difference of mole fraction of solute as in equation (21). Then the proportionality factor or permeability coefficient (also referred to as P_{osm}) is seen to be $L_p RT/\bar{V}_w$ and is expressed in cm s^{-1}. Since at 20°C $RT/\bar{V}_w$ is equal to $1.354 = 10^5$ kPa or 1336 atm, this unit is considerably smaller than the other.

When tracer water fluxes are measured there is usually no change in cell volume. Equation (17) is thus modified by substituting $J_v = 0$, giving

$$J_s = \frac{dn_s}{dt} \frac{1}{A} = \omega_s RT \Delta c_s \tag{29}$$

A similar equation was formerly derived from kinetic theory by means of Fick's law:

$$\frac{\partial n_s}{\partial t} = DA \frac{\partial c_s}{\partial x} \tag{30}$$

If it is assumed that there is a linear concentration gradient across the cell membrane (30) may be integrated giving

$$\frac{dn_s}{dt} = DA \frac{\Delta c_s}{\Delta x} \tag{31}$$

where Δx is the thickness of the cell membrane. Since Δx is frequently not accurately known it is incorporated with the diffusion coefficient D, giving a permeability coefficient, $P_d = D/\Delta x$ so that equation (31) becomes

$$\frac{dn_s}{dt} = P_d A \Delta c_s \tag{32}$$

On comparing (29) and (32) it is clear that

$$P_d = \omega_s RT \tag{33}$$

In analogy to P_{osm}, if Δc_s is measured in mole fractions then P_d is obtained in cm s^{-1}. The factor RT, connecting P_d and ω_s, is at 20 °C: 2.43×10^6 kPa cm^3 mol^{-1} or 2.40×10^4 atm cm^3 mol^{-1}.

In practice equation (32) is commonly modified and integrated as follows

$$\frac{dc_{s(c)}}{dt} = \frac{P_d A}{V_{1(c)}} (c_{s(m)} - c_{s(c)}) \tag{34}$$

since $c_{s(c)} = n_{s(c)}/V_{1(c)}$. On integration this gives

$$\frac{c_{s(c)}}{c_{s(m)}} = 1 - \exp\left(\frac{P_d A t}{V_{1(c)}}\right) \tag{35}$$

assuming that only $c_{s(c)}$ varies with time and that $c_{s(c)} = 0$ when $t = 0$. This equation may readily be used to calculate P_d from experimental data of uptake of tracer water, either ^2H_2O or ^3H_2O, into cells.

B. Measurements of Rate of Water Flow

1. Osmotic Measurements

Values of P_{osm} or $L_p RT/\bar{V}_w$ (in cm s^{-1}) for a large variety of animal cells are shown in Fig. 9. They have been plotted against the surface/volume ratio of the cells (similar data are tabulated in Dick, 1966 and House, 1974). The range of the permeability coefficients is very large; they also appear to be correlated with surface/volume ratio for all cells except a group of cells

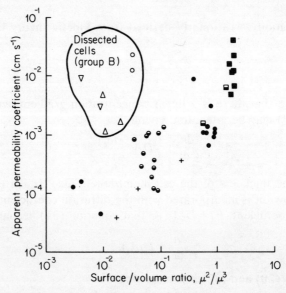

Figure 9 Osmotic permeability coefficient plotted against surface/volume ratio for a variety of cells. From Dick (1966); reproduced by permission of Butterworth & Co. (Publishers) Ltd

isolated by dissection (these will be discussed later). These values might be interpreted by assuming that small cells with large surface/volume ratios either have membranes 1000 times thinner than large cells or the membranes are much more porous. However, electron microscopy shows that cells have a fairly uniform membrane thickness from 7–10 nm. Variations in number of pores are possible but they have not yet been demonstrated in electron micrographs so that they cannot be counted. However, if they exist, there is no reason why pore number should be correlated with cell size; if anything one might have expected large cells to have more pores as a compensation for their low surface/volume ratio. The high permeabilities of dissected cells may be accounted for by solute leakage, the presence of large pores and by surface microvilli (see Dick, 1970).

If the correlation between permeability and surface/volume ratio is neither accidental nor due to membrane changes, then a reasonable explanation is that water permeabilities in large cells are reduced because water is retarded not only by the membrane but by slow diffusion in the large amount of cytoplasm. (This is equivalent to an internal unstirred layer—see Section VI-C.) From some of the data shown, it is possible to estimate (although difficulties of calculation force the use of approximations) that the behaviour of the cells could be accounted for by membrane permeabilities in the range $(3 - 70) \times 10^{-4}$ cm s^{-1} and by diffusion coefficients in the cytoplasm in the range $(0.08-2) \times 10^{-8}$ cm^2 s^{-1}. An attempt was made to explore the relative role of membrane and cytoplasm in controlling water flow in amphibian oocytes which not merely grow in size during development but also increase their surface area by microvilli which grow to a maximum and then diminish during development (Dick, Dick and Bradbury, 1970). Water permeability coefficients did indeed correlate with true total surface area including microvilli. However, even when the permeability coefficients were corrected for microvillar area, the corrected coefficients still correlated with cell size; there was thus evidence of involvement of both membrane and cytoplasm in water flow. In these oocytes the estimated membrane permeability coefficient for water was $(2-30) \times 10^{-4}$ cm s^{-1} and the cytoplasmic diffusion coefficient was $(6-100) \times 10^{-8}$ cm s^{-1}.

Some comments may be made on the above estimates of true membrane permeability and cytoplasmic diffusion.

(a) The estimated membrane permeability coefficients compare well with recent estimates of the water permeability of artificial lipid bilayers with or without addition of antibiotics (see Section VI-D).

(b) The diffusion coefficient in the cytoplasm is a mutual diffusion coefficient and not a self-diffusion coefficient (see Crank, 1956). The mutual diffusion coefficient is largely governed by movement of the slow component, e.g. cytoplasmic proteins, and is much lower than the self-diffusion coefficient for water. For example in a 24% solution of ovalbumin, the mutual diffusion coefficient is 8.7×10^{-8} cm^2 s^{-1} while the self-diffusion coefficient for tracer water in the same solution is 9.78×10^{-6} cm^2 s^{-1} (Wang, Anfinsen and Polestra, 1954). The estimated cytoplasmic diffusion coefficient is thus of the same order of magnitude, 10^{-8}–10^{-7} cm^2 s^{-1}, as mutual diffusion coefficients of protein–water solutions.

In some cell membranes the rate of osmotic water flow is not linear with osmotic gradient; P_{osm} appears to decline with increasing osmotic gradient in erythrocytes (Rich and coworkers, 1967) and in various epithelia, e.g. gall bladder (Diamond, 1966). Diamond suggested that this 'non-linear osmosis' might be due to a direct effect of osmotic pressure on aqueous channels in the cell membrane. Another non-linear phenomenon is rectification of osmotic flow; endosmosis is more rapid than exosmosis in erythrocytes (Blum and Forster, 1970; Farmer and Macey, 1970) and in certain plant cells (Dainty and Ginzburg, 1964), while in certain epithelia, e.g. gall bladder (Diamond, 1966; Wright, Smulders and Tormey, 1972), flow in one direction is greater than in the other for the same osmotic gradient.

2. Tracer Measurements

Values of P_d for tracer water flow are shown in comparison with P_{osm} values in Table 5; in general P_{osm} is considerably larger than P_d. Two main explanations have been offered to account for this difference; the first, unstirred layers in the external solution close to the cell membrane which make P_d appear too low without much effect on P_{osm}; the second, aqueous channels in the cell membrane which increase P_{osm} due to bulk flow.

C. Unstirred Layers

In using equation (36) to measure P_d, it is assumed that Δc_s is the concentration difference across the cell membrane. However, Dainty (1963) and Dainty and House (1966) pointed out that even if the external solution is well stirred (and even more if it is not) there persists an unstirred layer next the cell membrane in which there is a significant concentration gradient of solute. The thickness of the unstirred layer is actually indeterminate but it can be defined as shown in Fig. 10. It varies with the stirring rate and has been estimated for various artificial and biological membranes (see House, 1974); at moderate stirring rates around 60 revolutions per minute the unstirred layer is 50–100 μm thick but without stirring it can increase to 400 μm. In very small cells such as erythrocytes it may be as little as 5 μm (Shaafi and coworkers, 1967).

The unstirred layer affects the apparent value of P_d according to the equation

$$\frac{1}{P_{d(app)}} = \frac{1}{P_{d(true)}} + \frac{\delta}{D} \tag{36}$$

where δ is the thickness of the unstirred layer and D is the diffusion

Table 5. Osmotic and tracer water permeability coefficients in animal cells

Cell	$P_{osm} \times 10^4$ $(L_p RT/\bar{V}_w)$ (cm s^{-1})	$P_d \times 10^4$ (cm s^{-1})	$\dfrac{P_{osm}}{P_d}$	Reference
(a) Free Living Cells				
Frog egg	1.30	0.75	1.7	Prescott and Zeuthen (1953)
Xenopus egg	1.59	0.90	1.8	Prescott and Zeuthen (1953)
Zebra fish egg	0.45	0.36	1.3	Prescott and Zeuthen (1953)
Amoeba (*Chaos chaos*)	0.37	0.23	1.6	Prescott and Zeuthen (1953)
Amoeba proteus	1.2	0.21	5.7	Mast and Fowler (1935); Prescott and Mazia (1954)
Beef erythrocyte	156	51	3.1	Villegas, Barton and Solomon (1958)
Dog erythrocyte	200	44	4.5	Rich and coworkers (1967)
Human erythrocyte	127	53	2.4	Villegas, Barton and Solomon (1958)
Foetal erythrocyte	117	32	3.7	Sjolin (1954); Barton and Brown (1964)
(b) Dissected Cells				
Frog ovarian egg	89	1.28	70	Prescott and Zeuthen (1953)
Zebra fish ovarian egg	29	0.68	43	Prescott and Zeuthen (1953)
Squid axon	11	1.4	7.9	Villegas and Villegas (1960)
Crab muscle	96	1.2	80	Sorenson (1971)
Lobster nerve	—	—	20	Nevis (1958)

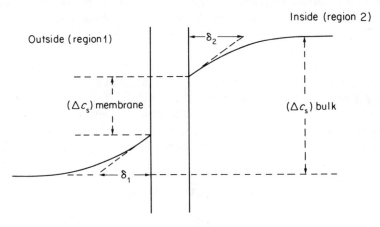

Figure 10 Unstirred layers showing their effect in reducing the true concentration difference across membrane. Notional thicknesses δ_1 and δ_2 are obtained as shown. From Dainty and House (1966); reproduced by permission of the *Journal of Physiology*

coefficient of water in it, usually taken as the same as that in free solution, 2×10^{-5} cm^2 s^{-1}. Thus for an unstirred layer of $100\,\mu$m $D/\delta = 2 \times 10^{-3}$ cm s^{-1} so that for a true value of P_d of this magnitude (similar to that found in artificial lipid bilayer membranes, see Section VI-E), the apparent value of P_d will be halved to 1×10^{-3} cm s^{-1}. The effect of unstirred layers on P_{osm} is much less, probably a reduction of around 2% (see House, 1974).

Unstirred layers are thus clearly important and in some situations they have been found to account for the difference between P_{osm} and P_d, e.g. in some artificial lipid bilayer membranes (see Section VI-E) and in the marine alga, Valonia (Gutknecht, 1967, 1968). In some other experiments where stirring was not performed, it seems possible that the discrepancy between P_{osm} and P_d might also be eliminated by allowing for unstirred layers. However, there remains a residue of experiments (Table 5) where a discrepancy between P_{osm} and P_d remains; it is in these cases that explanation is based on bulk flow through aqueous pores in the cell membrane.

D. Pore Theory of Water Transfer across Cellular Membranes

The pore theory of water transfer explains the difference between P_{osm} and P_d on the basis that P_{osm} is too high, unlike the unstirred layer explanation which suggests that P_d is too low. The reason is that in a water-filled pore in the membrane when there is a net flow of water as in osmotic experiments, water transfer will not be merely by diffusion alone (as in the case of tracer experiments when there is no net flow), but there will be in addition a bulk flow of water limited by viscous forces. The origin of this bulk flow was formerly the object of controversy. It now seems clear that it is the result of a gradient of hydrostatic pressure which exists within a pore even when there is no hydrostatic pressure difference between the bathing solutions (Mauro, 1957; Ray, 1960; Dainty, 1965). The origin of this gradient is explained in Fig. 11 taken from Dainty (1965).

If the difference between P_{osm} and P_d is to be explained by bulk flow in a pore, what is the radius of the pore that would account for the difference? The following simple derivation illustrates the principles involved. If it is assumed that osmotic flow is entirely a bulk viscous flow through a pore (ignoring any relatively small diffusive component) under a hydrostatic pressure equal to the osmotic pressure difference, then the flow is given by Poiseuille's law:

$$\frac{dV}{dt} = \frac{n\pi r^4 \Delta p}{8\eta_w \Delta x} = \frac{A r^2 \Delta p}{8\eta_w \Delta x} \tag{37}$$

where A is the total pore area, equal to $n\pi r^2$ (r = pore radius), or

$$L_p = \frac{dV}{dt} \frac{1}{A\Delta p} = \frac{r^2}{8\eta_w \Delta x} \tag{38}$$

Structure and Properties of Water in the Cell

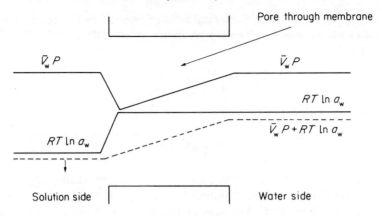

Figure 11 Potential profiles within a pore in a membrane separating a solution from pure water; there is no difference of hydrostatic pressure. Hatched line shows gradual increase of total chemical potential of water ($\bar{V}_w P + RT \ln a_w$) from solution to water. Water activity component of chemical potential ($RT \ln a_w$) rises sharply at entrance to pore owing to exclusion of solute. Since no sudden change of total chemical potential occurs here there must be a sharp fall of the pressure component ($\bar{V}_w P$) at the same point. A gradient of hydrostatic pressure must therefore exist within the pore and cause a bulk flow within it. From Dainty (1965); reproduced by permission of the Company of Biologists Ltd.

Tracer flow through the same pores, is given by Fick's law:

$$\frac{dV}{dt} = D_w A \frac{\Delta c}{\Delta x} \tag{39}$$

or

$$P_d = \frac{dV}{dt} \frac{1}{A \Delta c} = \frac{D_w}{\Delta x} \tag{40}$$

(In both cases it is assumed that the pores are uniform so that there are linear gradients of pressure and concentration respectively.) Combining equations (38) and (40) to eliminate the unknown, Δx, we obtain

$$r^2 = 8\eta_w D_w \frac{L_p}{P_d} \tag{41}$$

It must be noted however that osmotic permeability L_p is here expressed in terms of the pressure difference, i.e. in cm s^{-1} atm^{-1}. Since P_d is expressed in cm s^{-1}, it is more useful for comparison to express L_p also in terms of difference of mole fraction of solute, i.e. as $L_p RT/\bar{V}_w$ in cm s^{-1} (see Section VI-A) so that equation (41) may then be written

$$r^2 = \frac{8\eta_w D_w \bar{V}_w}{RT} \frac{L_p RT}{\bar{V}_w P_d} \tag{42}$$

where $L_p RT/\bar{V}_w P_d$ is now a numerical ratio of osmotic and tracer permeability coefficients as shown in Table 5. A more accurate formulation for r (see House, 1974) is

$$r = \left[\left(\frac{8\eta_w D_w \bar{V}_w}{RT}\right)\left(\frac{L_p RT}{\bar{V}_w P_d} - 1\right)\right]^{\frac{1}{2}} \quad (43)$$

or, evaluating the expression in the first bracket as 0.38 nm (at 25 °C):

$$r = 0.38 \left(\frac{L_p RT}{\bar{V}_w P_d} - 1\right)^{\frac{1}{2}} \quad (44)$$

where r is in nanometres. However, it must be remembered that this derivation rests upon several assumptions and approximations, especially the assumption that Poiseuille's law applies to a pore whose radius appears to be little greater than that of the water molecule itself.

Two further lines of evidence have been used to support the idea that the phospholipid membrane of the living cell does contain water-filled pores. First, the reflection coefficients for non-lipid soluble solutes of varying size have been found to decrease sharply with increasing molecular radius in a way which suggests that their permeability is being limited by pores which appear to vary from 0.4–0.8 nm in radius (see Solomon, 1968), although it is possible that some of the solutes pass at least in part by solution in the membrane (Wright and Diamond, 1969). Secondly, some studies of water permeability in artificial phospholipid bilayers suggest the presence of water-filled pores. This latter topic will be taken up in the next section.

E. Water Flow in Artificial Lipid Bilayer Membranes

Experimental techniques have been devised for producing artificial membranes from phospholipids which appear to be molecular bilayers and thus are fairly good analogues in some respects of cell membranes (Mueller and coworkers, 1962). The water permeability properties of such membranes have therefore been studied for comparison with cellular membranes. The osmotic and tracer permeabilities and their ratio are shown in Table 6. Holz and Finkelstein (1970) and Andreoli and Troutman (1971) have shown that for ordinary phospholipid membranes the ratio of osmotic to tracer permeability after correction for unstirred layers is roughly unity and there is thus no evidence of pores in these membranes. Water appears to move by a solubility–diffusion mechanism (Finkelstein, 1976a). However, such membranes can be treated with macrocyclic antibiotics. These have a hydrophobic exterior capable of interacting with the lipid membrane, and an interior space surrounded by hydrophilic groups and apparently capable of containing one or more water molecules. After antibiotic treatment the ratio of osmotic to tracer water permeability rises to between 3 and 4. The antibiotic thus appears to create bulk water flow; there has been controversy as to whether

Table 6. Osmotic and tracer water permeability coefficients in artifical lipid bilayer membranes

Membrane treatment	$P_{osm} \times 10^4$ $(L_p RT/\bar{V}_w)$ (cm s^{-1})	$P_d \times 10^4$ (cm s^{-1})	$\dfrac{P_{osm}}{P_d}$	Reference
None	11.1	10.6	1.0	Cass and Finkelstein (1967)
None	19.4	21.3	0.9	Everitt and Haydon (1969); Everitt, Redwood and Haydon (1969)
None	2.0	2.0	1.0	Holz and Finkelstein (1970)
None	16.3	13.8	1.2	Andreoli and Troutman (1971)
Nystatin	39.1	12.0	3.3	Holz and Finkelstein (1970)
Amphotericin B	16.6	6.0	2.8	Holz and Finkelstein (1970)
Amphotericin B	388	107.5	3.6	Andreoli and Troutman (1971)

this is achieved by the antibiotic acting simply as a water carrier or by the actual formation of pores; from ion transport studies the latter seems likely in the case of gramicidin and alamethicin (Haydon and Hladky, 1972). In any case the action of the antibiotic offers the possibility that a similar type of mechanism is present in cellular membranes. Finkelstein (1976b) has recently suggested that antidiuretic hormone creates pores 0.2 nm in radius in toad bladder epithelium.

The possible interpretation of osmotic/tracer permeability ratios exceeding 1.0 in terms of pores remains subject to two reservations: (a) there remains a doubt as to the proper corrections for unstirred layers; (b) Thau, Bloch, and Kedem (1966) have obtained ratios up to 2.0 in artificial membranes composed of liquid such as tributyl-phosphate or polyethyl-acrylate on a paper base although such membranes seem unlikely to contain pores. However, ratios above 2.0 probably do indicate the presence of pores.

F. Activation Energy of Water Transfer

By measuring the rate of water transfer through a cellular membrane at different temperatures an estimate of the activation energy can be made by using Arrhenius' equations

$$\ln L_p = \frac{\Delta E}{RT} + \ln K \qquad (45)$$

where ΔE is the activation energy and K is a constant. By plotting $\ln L_p$ against $1/T$, ΔE is obtained from the slope of the plot. Some estimates of activation energy in various cells are shown in Table 7. They vary from figures in the region of the activation energy for diffusion of water in free solution, 19.3 kJ mol^{-1}, to figures which are up to three times this. Low activation

Table 7. Activation energy for water transfer across animal cell membranes

Cell	Activation energy (kJ mol^{-1})	Reference
(a) Osmotic transfer		
Echinoderm egg	54–71	Lucke and McCutcheon (1932)
Erhlich ascites tumour cell	40	Hempling (1960)
Barnacle muscle	31	Bunch and Edwards (1969)
Human erythrocyte	14	Vieira, Shaafi and Solomon (1970)
Dog erythrocyte	16	Vieira, Shaafi and Solomon (1970)
Beef erythrocyte	17	Vieira, Shaafi and Solomon (1970)
(b) Diffusional transfer		
Human erythrocyte	25	Vieira, Shaafi and Solomon (1970)
Dog erythrocyte	21	Vieira, Shaafi and Solomon (1970)
Squid nerve	23	Nevis (1958)
Lobster nerve	10	Nevis (1958)

energies have been interpreted as indicating pores which are wide enough to contain water molecules whose intermolecular forces do not differ significantly from free water. Higher figures have been taken to indicate either that water is being forced to dissolve in a lipid membrane with rupture of some hydrogen bonds (Price and Thompson, 1969) or that it is traversing a pore so narrow that the water in it has a quasi-crystalline character with increased intermolecular forces (Hempling, 1960).

GLOSSARY OF SYMBOLS

		Unit
R	gas constant	J mol^{-1} K^{-1}
T	absolute temperature	K
μ	chemical potential	J mol^{-1}
n	number of moles	mol
$\bar{V}$	partial molar volume	cm^3 mol^{-1}
b	non-solvent volume	cm^3
p	hydrostatic pressure	kPa
π	osmotic pressure (or circumference/diameter)	kPa
R	Ponder's **R**	—
W_m	measured water content	cm^3
φ	molal osmotic coefficient	—

J_v	volume flux (usually equal to water flux)	cm$_3$ cm^{-2} s^{-1}
J_d	solute flux	mol cm^{-2} s^{-1}
c	concentration	mol cm^{-3} or mol solute/mol solution
L_p	hydraulic conductivity	cm s^{-1} kPa^{-1}
ω	solute permeability	mol cm^{-2} kPa^{-1} s^{-1}
σ	reflexion coefficient	—
A	area	cm^2
P_{osm}	osmotic permeability coefficient	cm s^{-1} kPa^{-1} or cm s^{-1}
P_d	tracer permeability coefficient	cm s^{-1}
δ	thickness of unstirred layer	cm
D	diffusion coefficient	cm^2 s^{-1}
η	viscosity	N s cm^{-2}
r	radius	cm
E	activation energy	J mol^{-1}

Subscripts

w	water
v	volume
s	permeable solute
i	impermeable solute
(c)	of cell
(m)	of medium
t	at time t
0	at zero time
e	at equilibrium
1	of solvent
2	of solute

REFERENCES

Adair, G. S. (1929). The thermodynamic analysis of the observed osmotic pressure of protein salts in solutions of finite concentration. *Proc. R. Soc. Lond.* A, **126**, 16–24.

Andreoli, T. E., and S. L. Troutman (1971). An analysis of unstirred layers in series with 'tight' and 'porous' lipid bilayer membranes. *J. Gen. Physiol.* **57**, 464–78.

Appelboom, J. W. T., W. A. Brodsky, W. S. Tuttle, and I. Diamond (1958). The freezing point depression of mammalian tissue after sudden heating in boiling distilled water. *J. Gen. Physiol.*, **41**, 1153–69.

Barer, R., and D. A. T. Dick (1957). Interferometry and refractometry of cells in tissue culture. *Expl Cell Res.*, Suppl. 4, 103–35.

Barer, R., and S. Joseph (1954). Refractometry of living cells. I. basic principles. *Quart. J. Microsc. Sci.*, **95**, 399–423.

Barer, R., and S. Joseph (1955a). Refractometry of living cells. II. The immersion medium. *Quart. J. Microsc. Sci.*, **96**, 1–27.

Barer, R., and S. Joseph (1955b). Refractometry of living cells. III. Technical and optical methods. *Quart. J. Microsc. Sci.*, **96**, 423–47.

Barry, P. H., and A. G. Hope (1969). Electroosmosis in membranes: effects of unstirred layers and transport numbers. II Experimental. *Biophys. J.*, **9**, 729–57.

Barton, T. C., and D. A. J. Brown (1964). Water permeability of the fetal erythrocyte. *J. Gen. Physiol.*, **47**, 839–49.

Bentzel, C. J., and A. K. Solomon (1967). Osmotic properties of mitochondria. *J. Gen. Physiol.*, **50**, 1547–63.

Bernal, J. D. (1965). The structure of water and its biological implications. *Symp. Soc. Expl Biol.*, **19**, 17–32.

Bernal, J. D., and R. H. Fowler (1933). A theory of water and ionic solution with particular reference to hydrogen and hydroxyl ions. *J. Chem. Phys.*, **1**, 515–48.

Birktoft, J. J., and D. M. Blow (1972). Structure of crystalline α–chymotrypsin. V. The atomic structure of tosyl-α-chymotrypsin at 2 Å resolution. *J. Molec. Biol.*, **68**, 187–240.

Blondin, G. A., and D. E. Green (1967). The mechanism of mitochondrial swelling. *Proc. Natn. Acad. Sci. USA*, **58**, 612–19.

Blum, R. M., and R. E. Forster (1970). The water permeability of erythrocytes. *Biochim. Biophys. Acta*, **203**, 410–23.

Brading, A. F., and J. Setekleiv (1968). The effect of hypo- and hypertonic solutions on volume and ion distribution of smooth muscle of guinea-pig taenia coli. *J. Physiol. Lond.* **195**, 107–18.

Brandt, P. W. (1958). A study of the mechanism of pinocytosis. *Expl Cell Res.*, **15**, 300–13.

Bratton, C. B., A. L. Hopkins, and J. W. Weinberg (1965). Nuclear magnetic resonance studies of living muscle. *Science*, **147**, 738–9.

Bull, H. B., and K. Breese (1968). Protein hydration. I. Binding sites. *Arch. Biochem. Biophys.*, **128**, 488–96.

Buckley, K. A., E. J. Conway, and H. C. Ryan (1958). Concerning the determination of total intracellular concentrations by the cryoscopic method. *J. Physiol. Lond.*, **143**, 236–45.

Bunch, W., and C. Edwards (1969). The permeation of non-electrolytes through the single barnacle muscle cell. *J. Physiol. Lond.*, **202**, 683–98.

Chapman-Andresen, C. (1962). Studies on pinocytosis in amoebae. *C. R. Trav. Lab. Carlsberg*, **33**, 73–264.

Chapman-Andresen, C. (1967a). Studies on endocytosis in amoebae. The distribution of pinocytically ingested dyes in relation to food vacuoles in *Chaos chaos*. I. Light microscopic observations. *C. R. Trav. Lab. Carlsberg*, **36**, 161–87.

Chapman-Andresen, C. (1967b). The effect of metabolic inhibitors on pinocytosis in amoebae. *Protoplasma*, **63**, 103–5.

Chapman-Andresen, C., and D. A. T. Dick (1961). Volume changes in the amoeba *Chaos chaos* L. *C. R. Trav. Lab. Carlsberg*, **32**, 265–89.

Cannon, J. D., D. A. T. Dick, and D. O. Ho-Yen (1974). Intracellular sodium and potassium concentrations in toad and frog oocytes during development. *J. Physiol. Lond.*, **241**, 497–508.

Cass, A., and Finkelstein, A. (1967). Water permeability of thin lipid membranes. *J. Gen. Physiol.*, **50**, 1765–84.

Century, T. J., I. R. Fenichel, and S. B. Horowitz (1970). The concentration of water, sodium and potassium in the nucleus and cytoplasm of amphibian oocytes. *J. Cell Sci.*, **7**, 5–13.

Cook, J. S. (1967). Non-solvent water in human erythrocytes. *J. Gen. Physiol.*, **50**, 1311–25.
Cope, F. W. (1969). Nuclear magnetic resonance evidence using D_2O for structured water in muscle and brain. *Biophys. J.*, **9**, 303–19.
Crank, J. (1956). *Mathematics of Diffusion*, Clarendon Press, Oxford.
Dainty, J. (1963). Water relations of plant cells. *Adv. Botanical Res.* **1**, 279–326.
Dainty, J. (1965). Osmotic flow. *Symp. Soc. Expl Biol.*, **19**, 75–85.
Dainty, J., and B. Z. Ginzburg (1964). The measurement of hydraulic conductivity (osmotic permeability to water) of internodal characean cells by transcellular osmosis. *Biochim. Biophys. Acta*, **79**, 102–11.
Dainty, J., and C. R. House (1966). 'Unstirred layers' in frog skin. *J. Physiol. Lond.*, **182**, 66–78.
Dalmark, M. (1975). Chloride and water distribution in human red cells. *J. Physiol. Lond.*, **250**, 65–84.
Davis, C. M., and J. Jarzynski (1972). Mixture models of water. In R. A. Horne (Ed.), *Water and Aqueous Solutions*, Wiley-Interscience, New York. pp. 377–423.
Diamond, J. M. (1966). Non-linear osmosis. *J. Physiol. Lond.*, **183**, 58–82.
Dick, D. A. T. (1958). Osmotic equilbria in fibroblasts in tissue culture measured by immersion refractometry. *Proc. R. Soc. Lond.* B, **149**, 130–43.
Dick, D. A. T. (1966). *Cell Water*, Butterworths, London.
Dick, D. A. T. (1967). In E. B. Reeve and A. C. Guyton (Eds), *Physical Bases of Circulatory Transport*, Saunders, Philadelphia. p. 220.
Dick, D. A. T. (1970). Water movements in cells. In E. E. Bittar (Ed.), *Membranes and Ion Transport*, Vol. 3, Wiley-Interscience, New York. pp. 211–50.
Dick, D. A. T., and L. M. Lowenstein (1958). Osmotic equilibria in human erthrocytes studied by immersion refractometry. *Proc. R. Soc. Lond.* B, **148**, 241–56.
Dick, E. G., D. A. T. Dick., and S. Bradbury (1970). The effect of surface microvilli on the water permeability of single toad oocytes. *J. Cell Sci.*, **6**, 451–76.
Dydynska, M., and D. R. Wilkie (1963). The osmotic properties of striated muscle fibres in hypertonic solutions. *J. Physiol. Lond.*, **169**, 312–29.
Everitt, C. T., and D. A. Haydon (1969). Influence of diffusion layers during osmotic flow across bimolecular lipid membranes. *J. Theor. Biol.*, **22**, 9–19.
Everitt, C. T., W. R. Redwood, and D. A. Haydon (1969). Problem of boundary layers in the exchange diffusion of water across bimolecular lipid membranes. *J. Theor. Biol.*, **22**, 20–32.
Farmer, R. E. L., and R. I. Macey (1970). Perturbation of red cell volume: rectification of osmotic flow. *Biochim. Biophys. Acta*, **196**, 53–65.
Fensom, D. S., and J. Dainty (1963). Electro-osmosis in *Nitella*. *Can. J. Bot.*, **41**, 685–91.
Finkelstein, A. (1976a). Water and nonelectrolyte permeability of lipid bilayer membranes. *J. Gen. Physiol.*, **68**, 127–35.
Finkelstein, A. (1976b). Nature of the water permeability increase induced by antidiuretic hormone (ADH) in toad urinary bladder and related tissues. *J. Gen. Physiol.*, **68**, 137–43.
Forslind, E..(1952). A theory of water. *Acta Polytech.*, **115**, 9–43.
Frank, H. S., and W. Y. Wen (1957). Structural aspects of ion–solvent interaction in aqueous solutions: a suggested picture of water structure. *Disc. Faraday Soc.*, **24**, 133–40.
Gary-Bobo, C. M., and A. K. Solomon (1968). Properties of hemoglobin solutions in red cells. *J. Gen. Physiol.*, **52**, 825–53.
Guest, G. M. (1948). Osmometric behaviour of normal and abnormal human erythrocytes. *Blood*, **3**, 541–55.

Guest, G. M., and M. Wing (1942). Osmometric behaviour of normal human erythrocytes. *J. Clin. Invest.*, **21**, 257–62.

Gutknecht, J. (1967). Membranes of *Valonia ventricosa*: apparent absence of water-filled pores. *Science*, **158**, 787–8.

Gutknecht, J. (1968). Permeability of *Valonia* to water and solutes: apparent absence of aqueous membrane pores. *Biochim. Biophys. Acta*, **163**, 20–29.

Hansson, S. E., G. Johansson, and J. O. Josefsson (1968). Oxygen uptake during pinocytosis in *Amoeba proteus*. *Acta Physiol. Scand.*, **73**, 491–500.

Harding, C. V., and C. Feldherr (1959). Semipermeability of the nuclear membrane in the intact cell. *J. Gen. Physiol.*, **42**, 1155–65.

Haydon, D. A., and S. B. Hladky (1972). Ion transport across thin lipid membranes: a critical discussion of mechanisms in selected systems. *Quart. Rev. Biophys.*, **5**, 187–282.

Hazelwood, C. F., B. L. Nichols, and N. F. Chamberlain (1969). Evidence for the existence of a minimum of two phases of ordered water in skeletal muscle. *Nature, Lond.*, **222**, 747–50.

Hempling, H. G. (1958). Potassium and sodium movements in the Ehrlich mouse ascites tumor cell. *J. Gen. Physiol.*, **41**, 565–83.

Hempling, H. G. (1960). Permeability of the Ehrlich ascites tumor cell to water. *J. Gen. Physiol.*, **44**, 365–79.

Hendry, E. B. (1954). The osmotic properties of the normal human erythrocyte. *Edin. Med. J.*, **61**, 7–24.

Holz, R., and A. Finkelstein (1970). The water and nonelectrolyte permeability induced in thin lipid membranes by the polyene antibiotics nystatin and amphotericin B. *J. Gen. Physiol*, **56**, 125–45.

House, C. R. (1974). *Water Transport in Cells and Tissues*, Arnold, London.

Itoh, S., and I. L. Schwartz (1957). Sodium and potassium distribution in isolated thymus nuclei. *Am. J. Physiol.*, **188**, 490–8.

Josefsson, J. O. (1968). Induction and inhibition of pinocytosis in *Amoeba proteus*. *Acta Physiol. Scand.*, **73**, 481–90.

Josefsson, J. O., N. G. Holmer, and S. E. Hansson (1975). Membrane potential and conductance during pinocytosis induced in *Amoeba proteus* with alkali metal ions. *Acta Physiol. Scand.*, **94**, 278–88.

Karnovsky, M. J. (1968). The ultrastructural basis of transcapillary exchanges. *J. Gen. Physiol.*, **52**, 64–95s.

Kedem, O., and A. Katchalsky (1958). Thermodynamic analysis of the permeability of biological membranes to non-electrolytes. *Biochim. Biophys. Acta*, **27**, 229–46.

Kedem, O., and A. Katchalsky (1961). A physical interpretation of the phenomenological coefficients of membrane permeability. *J. Gen. Physiol.*, **45**, 143–79.

Kell, G. S. (1972). Continuum theories of liquid water. In R. A. Horne (Ed.), *Water and Aqueous Solutions*, Wiley-Interscience, New York, pp. 331–76.

Klotz, I. M. (1970). Water: its fitness as a molecular environment. In E. E. Bittar (Ed.), *Membranes and Ion Transport*, Vol. 1, Wiley-Interscience, New York. pp. 93–122.

Kitching, J. A. (1938). Contractile vacuoles. *Biol. Rev.*, **13**, 403–44.

Kuntz, I. D., and W. Kauzmann (1974). Hydration of proteins and polypeptides. *Adv. Prot. Chem.*, **28**, 239–345.

Kwant, W. O., and P. Seeman (1970). The erythrocyte ghost is a perfect osmometer. *J. Gen. Physiol.*, **55**, 208–19.

Le Fevre, P. G. (1964). The osmotically functional water content of the human erythrocyte. *J. Gen. Physiol.*, **47**, 585–603.

Lehninger, A. L. (1964). *The Mitochondrion: Molecular Basis of Structure and Function,* Benjamin, New York.
Liljas, A., K. K. Kannan, P. C. Bergsten, I. Waara, K. Fridborg, B. Strandberg, U. Carlbom, L. Jarup, S. Lovgren, and M. Petef (1972). Crystal structure of human carbonic anhydrase C. *Nature, Lond. New Biol.,* **235**, 131–7.
Ling, G. N. (1972). Macromolecular hydration. In R. A. Horne (Ed.), *Water and Aqueous Solutions,* Wiley-Interscience, New York. pp. 663–700.
Loewenstein, W. R., and Y. Kanno (1963). The electrical conductance and potential across the membrane of some cell nuclei. *J. Cell Biol.,* **16**, 421–5.
Lucké, B., and M. McCutcheon (1932). The living cell as an osmotic system and its permeability to water. *Physiol. Rev.,* **12**, 68–139.
McConaghey, P. D., and M. Maizels (1961). The osmotic coefficients of haemoglobin in red cells under varying conditions. *J. Physiol. Lond.,* **155**, 28–45.
Maffly, R. H., and A. Leaf (1959). The potential of water in mammalian tissues. *J. Gen. Physiol.,* **42**, 1257–75.
Mast, S. O., and W. L. Doyle (1934). Ingestion of fluid by amoeba. *Protoplasma,* **20**, 555–60.
Mast, S. O., and C. Fowler (1935). Permeability of *Amoeba proteus* to water. *J. Cell. Comp. Physiol,* **6**, 151–67.
Mauro, A. (1957). Nature of solvent transfer in osmosis. *Science,* **126**, 252–3.
Michl, J., and V. Spurna (1975). Pinocytosis as an essential transport mechanism in metazoan cells in culture. *Expl Cell. Res.,* **93**, 39–46.
Mueller, P., D. O. Rudin, H. T. Tien, and W. C. Wescott (1962). Reconstitution of excitable cell membrane structure *in vitro. Circulation,* **26**, 1167–71.
Nemethy, G., and H. A. Scheraga (1962a). Structure of water and hydrophobic bonding in proteins. I. A model for the thermodynamic properties of liquid water. *J. Chem. Phys.,* **36**, 3382–400.
Nemethy, G., and H. A. Scheraga (1962b). The structure of water and hydrophobic bonding in proteins. III. The thermodynamic properties of hydrophobic bonds in proteins. *J. Phys. Chem.,* **66**, 1773–89.
Nevis, A. H. (1958). Water transport in invertebrate peripheral nerve fibres. *J. Gen. Physiol.,* **41**, 927–58.
Odeblad, E., B. N. Bhar, and G. Lindström (1956). Proton magnetic resonance of human red blood cells in heavy water exchange experiments. *Arch. Biochem. Biophys.,* **63**, 221–5.
Olmstead, E. G. (1960). Efflux of red cell water into buffered hypertonic solutions. *J. Gen. Physiol.,* **43**, 707–12.
Ørskov, S. L. (1946). The volume of the erythrocytes at different osmotic pressure. Further experiments on the influence of lead on the permeability of cations. *Acta Physiol. Scand.,* **12**, 202–12.
Packer, L., J. M. Wrigglesworth, P. A. G. Fortes, and B. C. Pressman (1968). Expansion of the inner membrane compartment and its relation to mitochondrial volume and ion transport. *J. Cell Biol.,* **39**, 382–91.
Pauling, L. (1945). The adsorption of water by proteins. *J. Am. Chem. Soc.,* **67**, 555–7.
Ponder, E. (1944). The osmotic behavior of crenated red cells. *J. Gen. Physiol.,* **27**, 273–85.
Ponder, E. (1948). *Hemolysis and Related Phenomena,* Grune and Stratton, New York.
Ponder, E. (1950). Tonicity–volume relationships in partially hemolysed hypotonic systems. *J. Gen. Physiol.,* **33**, 177–93.
Ponder, E., and D. Barreto (1957). The behaviour, as regards shape and volume,

of human red cell ghosts in fresh and in stored blood. *Blood,* **12,** 1016–27.
Pople, J. A. (1951). Molecular association in liquids. II. A theory of the structure of water. *Proc. R. Soc. Lond.* A, **205,** 163–78.
Prescott, D. M., and D. Mazia (1954). The permeability of nucleated and enucleated fragments of *Amoeba proteus* to D_2O. *Expl Cell Res.,* **6,** 117–26.
Prescott, D. M., and E. Zeuthen (1953). Comparison of water diffusion and water filtration across cell surfaces. *Acta Physiol. Scand.,* **28,** 77–94.
Price, H. D., and T. E. Thompson (1969). Properties of liquid bilayer membranes separating two aqueous phases: temperature dependence of water permeability. *J. Molec. Biol.,* **41,** 443–57.
Ray, P. M. (1960). On the theory of osmotic water movement. *Plant Physiol.,* **35,** 783–95.
Rich, G. T., R. I. Shaafi, T. C. Barton, and A. K. Solomon (1967). Permeability studies on red cell membranes of dog, cat and beef. *J. Gen. Physiol.,* **50,** 2391–405.
Rich, G. T., R. I. Shaafi, A. Romualdez, and A. K. Solomon (1968). Effect of osmolality on the hydraulic permeability coefficient of red cells. *J. Gen. Physiol.,* **52,** 941–54.
Riddick, D. H. (1968). Contractile vacuole in the amoeba, *Pelomyxa carolinensis. Am. J. Physiol.,* **215,** 736–40.
Rossi Fanelli, A., E. Antonini, and A. Caputo (1964). Hemoglobin and myoglobin. *Adv. Prot. Chem.,* **19,** 73–222.
Savitz, D., J. W. Sidel, and A. K. Solomon (1964). Osmotic properties of human red cells. *J. Gen. Physiol.,* **48,** 79–94.
Schiemer, H. G., G. Gunther, and D. Sina (1967). Die Wirkung des Cytostaticum Methotrexat aus den Wassergehalt und das Trockengewicht von Kern und Cytoplasma der Leberzellen der Ratte bei der Regeneration nach Teilhepatektomie. *Frank. Z. Path.,* **76,** 427–34.
Schmidt-Nielsen, B., and C. R. Schrauger (1963). *Amoeba proteus*: studying the contractile vacuole by micropuncture. *Science,* **139,** 606–7.
Schumaker, V. N. (1958). Uptake of protein from solution by *Amoeba proteus. Expl Cell Res.,* **15,** 314–31.
Shaafi, R. I., G. T. Rich, V. W. Sidel, W. Bossert, and A. K. Solomon (1967). The effect of the unstirred layer on human red cell water permeability. *J. Gen. Physiol.,* **50,** 1377–99.
Shporer, M., M. Haas, and M. M. Civan (1976). Pulsed nuclear magnetic resonance study of ^{17}O from $H_2^{17}O$ in rat lymphocytes. *Biophys. J.,* **16,** 601–11.
Sidel, V. W., and A. K. Solomon (1957). Entrance of water into human red cells under an osmotic pressure gradient. *J. Gen. Physiol.,* **41,** 243–57.
Sjolin, S. (1954). The resistance of red cells *in vitro*. A study of the osmotic properties, the mechanical resistance and the storage behaviour of red cells of fetuses, children and adults. *Acta Paediat. Stockh.,* **43,** *Suppl.* 98, 1–92.
Solomon, A. K. (1968). Characterisation of biological membranes by equivalent pores. *J. Gen. Physiol.,* **51,** *Suppl.* 335S–64S.
Sorenson, A. L. (1971). Water permeability of isolated muscle fibres of a marine crab. *J. Gen. Physiol.,* **58,** 287–303.
Stoner, C. D., and Sirak, H. D. (1969). Osmotically-induced alterations in volume and ultrastructure of mitochondria isolated from rat liver and bovine heart. *J. Cell Biol.,* **43,** 521–38.
Tedeschi, H., J. M. James, and W. Anthony (1963). Photometric evidence for the osmotic behaviour of rat liver microsomes. *J. Cell Biol.,* **18,** 503–13.
Thau, G., R. Bloch, and O. Kedem (1966). Water transport in porous and non-porous membranes. *Desalination,* **1,** 129–38.

Vargas, F. F. (1968). Water flux and electrokinetic phenomena in the squid axon. *J. Gen. Physiol.*, **51**, *Suppl.* 123S–30S.

Vieira, F. L., R. I. Shaafi, and A. K. Solomon (1970). The state of water in human and dog red cell membranes. *J. Gen. Physiol.*, **55**, 451–66.

Villegas, R., J. C. Barton, and A. K. Solomon (1958). The entrance of water into beef and dog red cells. *J. Gen. Physiol.*, **42**, 355–69.

Villegas, R. and G. M. Villegas (1960). Characterisation of the membranes in the giant nerve fibre of the squid. *J. Gen. Physiol.*, **43**, *No.* 5, *Pt* 2, *Suppl.* 1, 73–103.

Wang, J. H., C. B. Anfinsen, and F. M. Polestra (1954). The self-diffusion coefficients of water and ovalbumin in aqueous ovalbumin solution at 10°. *J. Am. Chem. Soc.*, **76**, 4763–5.

Wedner, H. J., and J. M. Diamond (1969). Contributions of unstirred-layer effects to apparent electrokinetic phenomena in the gall bladder. *J. Membrane Biol.*, **1**, 92–108.

White, H. L., and D. Rolf (1962). Osmometric behaviour of blood cells and of whole body cells. *Am. J. Physiol.*, **202**, 1195–9.

Wright, E. M., and J. M. Diamond (1969). Patterns of non-electrolyte permeability. *Proc. R. Soc. Lond.* B, **172**, 227–71.

Wright, E. M., A. P. Smulders, and J. M. Tormey (1972). The role of the lateral intercellular spaces and solute polarisation effects in the passive flow of water across the rabbit gallbladder. *J. Membrane Biol.*, **7**, 198–219.

Chapter 2

Relations between Solutes and Water: Analysis of Solute Transport

K. JANÁČEK

I. Passive Transport of Non-electrolytes		47
A. Simple diffusion		47
B. Mediated diffusion		51
II. Passive Transport of Ions and Membrane Potentials		54
III. Transport of Water and Coupling between Solute and Water Flows		59
IV. Active Transport		62
V. Transport across Epithelial Cell Layer		65
VI. Experimental Methods		73
References		77

I. PASSIVE TRANSPORT OF NON-ELECTROLYTES

A. Simple Diffusion

Diffusion plays a prominent role in the transport of solutes. We may observe diffusion in its purest form in continuous or even homogeneous media. In chemically homogeneous media the chemical concentration of individual components is everywhere the same and diffusion—in this case often called self-diffusion—can be observed using isotopic labelling. In continuous media where concentration gradients are present, interdiffusion of at least two chemical species, viz. of the solute and of the solvent, takes place. Since the intrinsic diffusion rates of the two species are different, hydrostatic pressure gradients may develop and the resulting mass flow modifies the diffusion rates (Hartley and Crank, 1949). It will be seen that in membranes further complicating factors are operative but let us first briefly consider pure diffusion in order to understand the general nature of diffusional processes.

Diffusion of particles is a result of their permanent translational thermal movement. The movement is random and hence subject to laws of probability; more particles leave a space in which they are concentrated than another in which they are diluted. In this way a net transport of particles from concentrated regions to less concentrated ones is achieved and concentration gradients tend to disappear.

The process of pure diffusion is described by the two laws of Fick. Fick's first law states that in diffusion the flow J of a substance (conveniently expressed as the number of moles of substance diffusing in a unit of time per unit area normal to the direction considered) is proportional to the concentration gradient of the substance. If diffusion proceeds only along the x-axis, the gradient is equal to the derivative of concentration with respect to the x-coordinate and Fick's first law can be written

$$J = -D \frac{\partial c}{\partial x} \qquad (1)$$

Partial derivatives are used to stress the fact that concentration may be a function not only of a spatial coordinate but also of time; Fick's first law applies both to steady and non-steady situations. The minus sign shows that when concentration increases in the direction of positive x-axis, i.e. when the concentration gradient is positive, flow proceeds in the direction of the negative x-axis and is thus by convention negative. It may be seen that if flow is expressed in mol cm^{-2} s^{-1} and concentration in mol cm^{-3}, the diffusion coefficient D will have units cm^2 s^{-1}. Diffusion coefficients encountered in aqueous solutions are of the order of 10^{-5} cm^2 s^{-1}, e.g. 2.5×10^{-5} cm^2 s^{-1} for self-diffusion of water and 0.5×10^{-5} cm^2 s^{-1} for sucrose.

To approach the problem of calculating concentration profiles created by diffusion in regions of non-negligible extension in space, Fick's first law is transformed into Fick's second law or diffusion equation. For the one-dimensional case again, this equation may be written

$$\frac{\partial c}{\partial t} = D \frac{\partial^2 c}{\partial x^2} \qquad (2)$$

Solutions of the diffusion equation are sometimes useful to physiologists, e.g. when diffusion into or out of the intercellular space is considered, or in the case of diffusion of substances which permeate cell membranes so readily that their transport is not rate-limited by them. There is a classical monograph on diffusion processes written primarily for physiologists by Jacobs (1967).

Let us now turn our attention to permeation of cell membranes by simple, i.e. non-mediated, diffusion of molecules. Such a mode of permeation is in general of importance only with substances foreign to the cell, e.g. drugs administered by the research worker or physician; we shall see that for native substrates the membrane is endowed with special mechanisms of permeation and their simple diffusion across the membrane may often be neglected.

In some cases at least the permeation by non-mediated diffusion may be explained rather satisfactorily on the basis of a very simple model, in which the membrane is represented as a thin continuous layer of lipid material with equilibrium distribution of the permeating material at the interfaces between the membrane and the adjacent aqueous solutions. In this model it is not even necessary to solve the diffusion equation and a modified form of Fick's first law can be applied directly. Since the membrane is thin, the derivative in equation (1) may be replaced approximately by a quotient consisting of a finite concentration difference divided by the membrane thickness d:

$$J = D_m \frac{c'_o - c'_i}{d}$$

Here D_m is the diffusion coefficient in the membrane and c'_o and c'_i are concentrations within the membrane, at its outer and inner surfaces (Fig. 1). If equilibrium prevails at the interfaces, the concentrations at the surface of

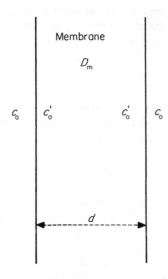

Figure 1 Simple diffusion across a membrane

the membrane (within the membrane) are related to those in the adjacent aqueous medium by a constant partition (distribution) coefficient k:

$$c'_o = kc_o, \quad c'_i = kc_i$$

so that the flow of non-electrolyte across the membrane by simple diffusion is expressed by

$$J = \frac{D_m k}{d}(c_o - c_i) \qquad (3)$$

Experimentally, permeation by simple diffusion can be recognized as non-saturable process, i.e. the flow of solute is related to the concentration difference across the membrane by a constant coefficient P, called the permeability constant:

$$J = P(c_o - c_i) \qquad (4)$$

When comparing equations (4) and (3) it is seen that the above model offers a certain physical interpretation of the permeability constant:

$$P = \frac{D_m k}{d} \tag{5}$$

Oversimplified as the model is, it explains well the positive correlation between the permeability constant of a number of drugs and their oil–water partition coefficients, observed already in the last century by Overton, studied extensively by Collander (see, e.g., Collander, 1949) and discussed thoroughly by Danielli (1952) and Stein (1967). As shown by these authors the correlation is improved if the molecular weight M of the diffusing substance is taken into account and $PM^{1/2}$ (or $PM^{1/3}$ for very large molecules) is plotted against the oil–water partition coefficient, this resulting from the fact that $DM^{1/2}$ (or $DM^{1/3}$) varies much less from molecule to molecule than the diffusion coefficient D itself (see, e.g., Stein, 1967 or Lieb and Stein, 1971 and Jacobs, 1967, for references). With very lipophilic drugs, however, the positive correlation between permeability constant and oil–water partition coefficient breaks down and the permeability constant is actually seen to decrease with extremely high partition coefficients (see Penniston and coworkers, 1969).

A more sophisticated model of penetration of a thin membrane by simple diffusion was developed by Danielli (1952, see also Kotyk and Janáček, 1975). In this model the membrane is represented by a series of energy barriers, one at each interface and n barriers inside the membrane. The interpretation of the permeability constant is then different,

$$P = \frac{ae}{nb + 2e} \tag{6}$$

where a is the rate constant for the transition medium–membrane and b the rate constant for the transition membrane–medium (so that $a/b = k$, the partition coefficient); finally e is the rate constant with which a molecule overcomes each of the n internal symmetrical barriers. When nb is much greater than $2e$, the dependence of the permeability constant on the partition coefficient is seen to be again of a similar character to that in equation (5).

Simple diffusion is encountered also in unstirred layers, regions of laminar flow adjacent to membranes (see, e.g., House, 1974). The effective thickness of unstirred layers (with which the layers would produce the same effects if the concentration gradient in them were linear) ranges from several hundreds of micrometres in solutions with little stirring to several tens of micrometres in solutions agitated vigorously. Each unstirred layer behaves as a membrane in series with the membrane studied and with permeability constant given by the diffusion coefficient of the substance in question divided by the effective thickness of the unstirred layer. A typical permeability constant of an unstirred layer is of the order of 10^{-3} cm s^{-1}. Measured permeability constants as high as that are hence likely to correspond to unstirred layers and one can conclude that the true permeability constant of the membrane is higher still. Measured permeability constants which are not much lower (e.g. of the order of 10^{-4} cm s^{-1}) are to be corrected for the presence of unstirred layers according to the formula

$$\frac{1}{P_{\text{true}}} = \frac{1}{P_{\text{measured}}} - \frac{1}{P_{\text{unstirred layers}}} \tag{7}$$

Diffusion of small hydrophilic molecules is likely to proceed across hydrophilic regions (pores) of the membrane. In this case interactions of diffusing molecules with water flow which can proceed simultaneously are important and equation (4) has to be modified as discussed in Section III of this chapter.

B. Mediated Diffusion

Unlike drugs permeating across cell membranes by simple diffusion, substrates such as sugars and amino acids are transported by mechanisms which display specificity, saturation and, as a result of this, competition between closely related compounds. Their transport is mediated by membrane proteins designated as carriers. It is not yet certain whether the same

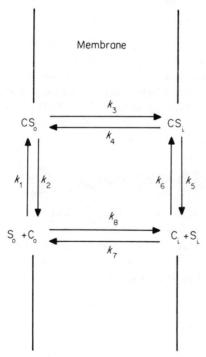

Figure 2 Classical carrier-mediated system

protein is responsible for both the selective binding and the translocation of the substrate across the membrane. Nevertheless, this assumption results in a rather satisfactory classical model of carrier transport (see, e.g. Kotyk, 1973), shown in Fig. 2. There C_o, S_o, CS_o and C_i, S_i, CS_i signify carrier, substrate and carrier–substrate complex at the outer and inner surfaces of the membrane, respectively; in what follows, their concentrations will be denoted by $c, s,$ and cs. The rate constants k are all different so that the model can account for unequal stationary concentrations of substrate, resulting from its active transport. In mediated diffusion which is a passive, equilibrating process it may be assumed that $k_3 = k_4$, $k_7 = k_8$ and $k_1/k_2 = k_6/k_5$. Assuming that the total concentration of the carrier c_t is conserved, the flow of substrate corresponding to the general scheme in Fig. 2 is given by

$$J = c_t \frac{as_o - bs_i}{cs_o + ds_i + es_o s_i + f} \qquad (8)$$

where

$$a = k_1 k_3 k_5 k_7 \qquad b = k_2 k_4 k_6 k_8,$$
$$c = k_1[k_5(k_3 + k_7) + k_7(k_3 + k_4)]$$
$$d = k_6[k_4(k_2 + k_8) + k_8(k_2 + k_3)]$$
$$e = k_1 k_6(k_3 + k_4) \qquad f = (k_7 + k_8)(k_2 k_4 + k_2 k_5 + k_3 k_5)$$

Mediated diffusion results in equilibrium distribution of the substrate: $s_o = s_i$. Hence obviously $a = b$ for mediated diffusion.

The most convenient method of demonstrating mediated diffusion and estimating its characteristics consists of measurement of initial unidirectional fluxes, as described in more detail, e.g., in Kotyk and Janáček (1975). Equation (8) is simplified in this case, since concentration at the *trans*-side (say, s_i) may be neglected during measurement of the initial flow. Equation (8) is then transformed into

$$J_{init} = \frac{(c_t a/c) s_o}{s_o + f/c} = J_{max} \frac{s_o}{s_o + K_{0.5}} \qquad (9)$$

which is of the same form as the Michaelis–Menten formula of enzyme kinetics. Parameters of equation (9), the maximum initial flow J_{max} and the half-saturation constant $K_{0.5}$, can be estimated using linearization procedures, of which the most popular is the Lineweaver–Burk plot. Transformation of equation (9) corresponding to this procedure consists in taking reciprocals of its two sides so that

$$\frac{1}{J_{init}} = \frac{K_{0.5}}{J_{max}} \frac{1}{s_o} + \frac{1}{J_{max}} \qquad (10)$$

Hence $1/J_{init}$ is plotted against $1/s_o$ and the two parameters are evaluated from segments on the axes or from one segment and the slope, as shown in Fig. 3.

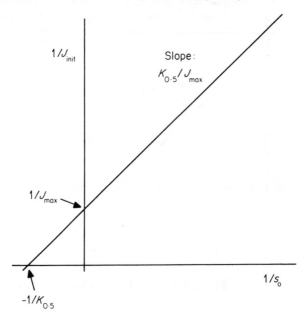

Figure 3 Lineweaver–Burk plot

Another procedure, known as the Woolf–Hofstee plot, corresponds to the transformation

$$J_{init} = J_{max} \frac{s_o}{s_o + K_{0.5}} = J_{max} \left(1 - \frac{K_{0.5}}{s_o + K_{0.5}}\right)$$
$$= J_{max} \left(1 - \frac{K_{0.5}}{s_o} \frac{s_o}{s_o + K_{0.5}}\right) = J_{max} - K_{0.5} \frac{J_{init}}{s_o} \tag{11}$$

Thus, when J_{init} is plotted against J_{init}/s_o, the parameters can again be evaluated as shown in Fig. 4.

The carrier theory of mediated diffusion is strongly supported by the interesting phenomenon of countertransport (Rosenberg and Wilbrandt, 1957) in which passive equilibrium resulting from mediated diffusion is transiently disturbed by the concentration gradient of a compound competing for the same binding site of the carrier. Let us imagine that an isotopically labelled substance is equilibrated across a membrane; its two opposite unidirectional fluxes are then equal. A high concentration of a chemically related (or even chemically identical but unlabelled) substance is now added to one side of the membrane, say, the *cis*-side. Unidirectional flow from the *trans*-side remains unchanged, unidirectional flow from the *cis*-side is decreased by competition. Hence net flow from *trans*- to *cis*-side takes place. This phenomenon appears to require that the binding site in the membrane is

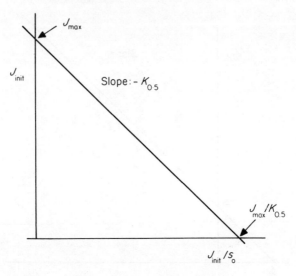

Figure 4 Woolf–Hofstee plot

exposed alternatively to the one or the other side of the membrane and the carrier theory satisfies this requirement.

II. PASSIVE TRANSPORT OF IONS AND MEMBRANE POTENTIALS

The principal difference between the passive transport of non-electrolytes and of ions rests in the fact that flows of ions are strongly influenced by an electrical field (which is the gradient of the electrical potential). As a result of this, equilibrium distribution of an ion across a membrane does not require equality of its concentrations, but rather equality of its electrochemical potentials at the two sides of the membrane. The electrochemical potential of an ion may be expressed by

$$\mu = \mu_0 + RT \ln a + zF\varphi \tag{12}$$

where μ_0 is a constant parameter (called the standard potential) depending on the nature of the solvent but not on the ion concentration or the electrical potential, R is the gas constant (8.314 J mol^{-1} K^{-1}), T the temperature in degrees Kelvin, z the valency of the ion, F the Faraday number (96 490 C mol^{-1}) and φ the electrical potential. Finally, the activity $a = fc$, where c is the molar concentration of the ion and f is the activity coefficient, expressing correction for mutual interactions of ions. Its value is 1 in very dilute solutions and about 0.76 in mammalian physiological salines. From the equality of the electrochemical potential at the inner (i) and the outer (o) sides of the membrane, equilibrium differences of the electrical potentials, $(\varphi_i - \varphi_o)_{eq}$, called Nernst–Donnan equilibrium potentials, can be expressed as

$$(\varphi_i - \varphi_o)_{eq} = \frac{RT}{zF} \ln \frac{c_o}{c_i} = 2.303 \frac{RT}{zF} \log \frac{c_o}{c_i} \tag{13}$$

Thus, e.g.

$$E_K = \frac{RT}{F} \ln \frac{[K^+]_o}{[K^+]_i} = 2.303 \frac{RT}{F} \log \frac{[K^+]_o}{[K^+]_i}$$

$$E_{Cl^-} = -2.303 \frac{RT}{F} \log \frac{[Cl^-]_o}{[Cl^-]_i} = 2.303 \frac{RT}{F} \log \frac{[Cl^-]_i}{[Cl^-]_o}$$

$$E_{Ca^{2+}} = 2.303 \frac{RT}{2F} \log \frac{[Ca^{2+}]_o}{[Ca^{2+}]_i} = 2.303 \frac{RT}{F} \log \frac{[Ca^{2+}]_o^{1/2}}{[Ca^{2+}]_i^{1/2}}$$

etc.

In all the above formulae it has been assumed that the activity coefficients at the two sides of the membrane are practically equal and hence only concentrations appear in the logarithm arguments. The factor $2.303\ RT/F$ to be used with logarithms to the base 10 is equal to 54.2 mV at 0 °C, 58.2 mV at 20 °C and 61.6 mV at 37 °C.

When the equilibrium of an ion is a physical reality, its Nernst–Donnan equilibrium potential is equal to the electrical potential difference across the membrane, the so-called membrane potential. This condition is of assistance in discriminating between the thermodynamic equilibrium of an ion and its stationary distribution resulting from active transport.

To express passive flow of an ionic species use can be made of the general transport equation of Teorell (1953):

flow = mobility × concentration × driving force

with gradient of the electrochemical potential of the ion as the driving force; at equilibrium the gradient disappears and the net flow of the ion also vanishes. In a one-dimensional steady-state case the gradient is equal to the ordinary derivative with respect to x, so that the flow of the ion is expressed by (cf. equation (12))

$$J = cU \left(RT \frac{d \ln a}{dx} + zF \frac{d\varphi}{dx} \right) \tag{14}$$

where c is the local concentration of the ion and U its mobility. Since $d \ln y = dy/y$,

$$d \ln a = \frac{da}{a} = \frac{d(fc)}{fc} = \frac{f\ dc}{f\ c} = \frac{dc}{c}$$

as long as the activity coefficient may be considered to be a constant independent of x. Equation (14) then becomes

$$J = -RTU \frac{dc}{dx} - zFUc\frac{d\varphi}{dx} \tag{15}$$

which is the Nernst–Planck electrodiffusion equation. When $z = 0$, so that we are dealing with a non-electrolyte, the electrodiffusion equation (15) becomes Fick's first law (equation (1)), RTU being Einstein's expression for the diffusion coefficient D. The Nernst–Planck electrodiffusion equation may be seen to be a superposition of Fick's and Ohm's laws; on an infinitesimal scale the movement due to diffusion and the movement due to migration in an electrical field may be added vectorially and, if the direction of the two is the same as in the present case, algebraically.

The electrodiffusion equation (15) can also be considered as an approximation of a very general equation of ionic flow, Schlögl's equation, derived by Schlögl (1964) by methods of irreversible thermodynamics:

$$J = cJ_V - RTU \frac{dc}{dx} - zFUc \frac{d\varphi}{dx} \\ - RTU \frac{d\ln f}{dx} - UMc \left(\frac{\bar{V}}{M} - \frac{\bar{V}_w}{M_w}\right)\frac{dp}{dx} \tag{16}$$

The first term represents the entrainment of the ion by the volume flow J_V, c being again the local concentration of the ion. In the second and third terms we recognize the diffusion and electrical drift known from the Nernst–Planck electrodiffusion equation (15). The fourth term gives the dependence of the ion flow on the gradient of the activity coefficient and the fifth describes pressure diffusion, $\bar{V}$ and $\bar{V}_w$ being partial molal volumes and M and M_w molecular weights of the ion and of the solvent, respectively. In pressure diffusion ions with specific volume greater than that of the solvent molecules are transported from higher to lower hydrostatic pressure, whereas those less voluminous are transported in the opposite direction. Pressure diffusion, however, is rarely of greater quantitative importance. Schlögl's derivation (Schlögl, 1964) explicity neglects possible coupling of the ionic flow J with flows of other solutes and also, obviously, with a chemical reaction in the membrane (active transport).

Comparing equations (15) and (16) it is seen that the Nernst–Planck equation of electrodiffusion neglects further possible entrainment with volume flow, gradient of the activity coefficient and pressure diffusion. As a result of this the electrodiffusion equation can be integrated, at least under some further simplifying assumptions, to obtain practically useful formulae, in which measurable concentration and electrical potential differences replace the derivatives. Approximations of this kind are justified especially in biology by the words of Jacobs (1967): '... in dealing with most biological problems it is not only useless, but actually unscientific, to carry mathematical refinements beyond a certain point, just as it would be both useless and unscientific to employ an analytical balance of the highest precision for obtaining the growth-curve of a rat.'

The electrodiffusion equation (15) can be integrated easily by the procedure of Goldman (1943), where it is assumed that the gradient of the electrical potential is approximately constant so that it may be expressed as the electrical potential difference across the membrane proper, E_m, divided by the membrane thickness d: $d\varphi/dx = E_m/d$. Integration can then be carried out easily, giving ionic flow as a function of E_m and of concentrations c_o and c_i just inside the membrane:

$$J = zFU \frac{E_m}{d} \frac{c_i - c_o \exp(-zFE_m/RT)}{\exp(-zFE_m/RT) - 1} \qquad (17)$$

In the absence of an external circuit the sum of currents carried across the membrane by individual ionic species must be zero. On adding equations (17) for individual ionic species, remarkably simple equations for the electrical potential difference across the membrane proper are obtained, provided that all ions considered are of the same valency. This is actually often the most important case: univalent potassium, sodium and chloride ions are usually the most abundant and most permeable in biological systems and hence it is principally their distribution which generates the cell membrane potential. The E_m is then given by

$$E_m = \frac{RT}{F} \ln \frac{U_{K^+} c_{K^+o} + U_{Na^+} c_{Na^+o} + U_{Cl^-} c_{Cl^-i}}{U_{K^+} c_{K^+i} + U_{Na^+} c_{Na^+i} + U_{Cl^-} c_{Cl^-o}} \qquad (18)$$

It still remains to relate the c's, the concentrations just inside the membrane, to measurable concentrations in the adjacent solutions. It is not correct to relate them by constant partition coefficients, even if equilibria are assumed to prevail at the interfaces. As recognized especially clearly by Johnson, Eyring and Polissar (1954), the steps of the electrical potential at the membrane surfaces (the boundary potentials) E_o and E_i are to be accounted for so that

$$\begin{aligned} c_{K^+o} &= k_{K^+} [K^+]_o \exp(-FE_o/RT) \\ c_{K^+i} &= k_{K^+} [K^+]_i \exp(FE_i/RT) \\ c_{Cl^-o} &= k_{Cl^-} [Cl^-]_o \exp(FE_o/RT) \\ c_{Cl^-i} &= k_{Cl^-} [Cl^-]_i \exp(-FE_i/RT) \end{aligned} \qquad (19)$$

etc.

where k's are partition coefficients that would apply if the boundary potentials were zero, and the symbols in square brackets denote concentrations in aqueous solutions near the membrane. Introducing relations (19) into equation (18), making use of the identity

$$E_o + E_i \equiv \frac{RT}{F} \ln [\exp(FE_o/RT) \exp(FE_i/RT)]$$

and defining permeability coefficient $P = RTUk/d$ we obtain finally for the potential difference across the whole membrane $E = E_o + E_m + E_i$:

$$E = \frac{RT}{F} \ln \frac{P_{K^+}[K^+]_o + P_{Na^+}[Na^+]_o + P_{Cl^-}[Cl^-]_i \exp[F(E_o - E_i)/RT]}{P_{K^+}[K^+]_i + P_{Na^+}[Na^+]_i + P_{Cl^-}[Cl^-]_o \exp[F(E_o - E_i)/RT]} \quad (20)$$

The exponential factors in the chloride terms may be close to 1, if the two boundary potentials are not very different. Relations of the same type as equation (20) are referred to as Goldman equations for membrane potential.

When, e.g., the chloride ion is in thermodynamic equilibrium across the membrane (as seems to be the case in some animal cells), equation (20) will change to

$$E = \frac{RT}{F} \ln \frac{P_{K^+}[K^+]_o + P_{Na^+}[Na^+]_o}{P_{K^+}[K^+]_i + P_{Na^+}[Na^+]_i} = \frac{RT}{F} \ln \frac{[Cl^-]_i}{[Cl^-]_o} \quad (21)$$

since the ion in equilibrium carries no current across the membrane and may be excluded from the summation of expressions (17), described above. Also ions with negligible permeability may be omitted from the Goldman equation.

A different approach to the description of steady-state membrane potentials can be based on the equivalent circuit (or electrical analogue) of the membrane, shown in Fig. 5. The currents in the individual branches are given by

$$I_{K^+} = G_{K^+}(E_{K^+} - E)$$
$$I_{Na^+} = G_{Na^+}(E_{Na^+} - E) \quad (22)$$
$$I_{Cl^-} = G_{Cl^-}(E_{Cl^-} - E)$$

where I's are the current densities of the individual ionic species, G's are the corresponding conductances per unit area, E_{K^+}, etc., the Nernst–Donnan equilibrium potentials, as defined by equation (13) and E the membrane potential.

In a steady state the sum of the currents across the membrane must be zero and hence

$$E = \frac{G_{K^+}E_{K^+} + G_{Na^+}E_{Na^+} + G_{Cl^-}E_{Cl^-}}{G_{K^+} + G_{Na^+} + G_{Cl^-}} \quad (23)$$

which may also be written

$$E = T_{K^+}E_{K^+} + T_{Na^+}E_{Na^+} + T_{Cl^-}E_{Cl^-}$$

where $T_{K^+} = G_{K^+}/(G_{K^+} + G_{Na^+} + G_{Cl^-})$, etc., are transport numbers of individual ionic species. Equation (23) was used explicitly by Hodgkin and

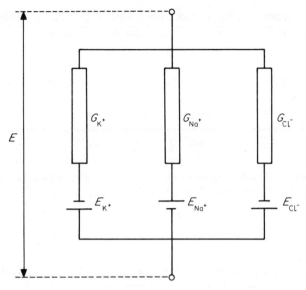

Figure 5 Equivalent circuit of the membrane

Horowicz (1959) and for this reason Jaffe (1974) suggests that equations of this type be called Hodgkin–Horowicz equations.

The choice between the two alternative descriptions of the membrane potential, by the Goldman equation (equation (20)) or by the Hodgkin-Horowicz equation (equation (23)), depends clearly on the condition whether in a given case permeabilities P or conductances G behave as constants. Jaffe (1974) tested a number of voltage–concentration data sets from the literature in order to see whether the dependence of the membrane potential on the logarithm of the external concentration of potassium or chloride ions is curvilinear, as predicted by the Goldman equation, or linear, as predicted by the Hodgkin–Horowicz equation. In all cases studied the data were fitted better by the Hodgkin–Horowicz equation, suggesting that individual ion species are transported across cell membranes by relatively separate pathways of at least formally constant conductance.

III. TRANSPORT OF WATER AND COUPLING BETWEEN SOLUTE AND WATER FLOWS

Since the transport of water is discussed more deeply in Chapter 1, the present section limits itself to a few basic ideas necessary for understanding the description of coupling between solute and water flows. The reflection coefficient is discussed in the following from a slightly different viewpoint to that introduced in Chapter 1: as a phenomenological coefficient describing the extent of the interactions (coupling) between solute and water flows.

Since such coupling implies a common pathway for the two substances, its demonstration adds to the knowledge of the membrane under investigation.

Volume flow from one compartment to another separated by a membrane can be induced: (1) by increasing the hydrostatic pressure in the former (hydraulic flow); or (2) by increasing concentration of solutes, i.e. increasing osmotic pressure, in the latter (osmotic flow).

When the membrane is ideally semipermeable, i.e. only solvent (water) can permeate, whereas solute cannot, the situation is simple: the volume flow J_V is accounted for entirely by the flow of water and may be shown experimentally to be related by the same coefficient L_p (the hydraulic conductivity) to the hydrostatic as well as to the osmotic pressure. The osmotic pressure, which can be calculated by the well known van't Hoff formula:

$$\Pi = RTc \qquad (24)$$

must, of course, be expressed in the same units as the hydrostatic pressure. The new SI unit of pressure is the pascal, Pa (N m^{-2}, i.e. J m^{-3}). Hence if in equation (24) the gas constant $R = 8.314$ J mol^{-1} K^{-1} is used, the pressure will be expressed in pascals if the concentration is introduced in mol m^{-3}, in kilopascals, kPa, if it is in mol l^{-1} (M) and in megapascals, MPa, if it is in mol cm^{-3}. The last choice is probably the best, the MPa (equal to 9.8623 atm) being of rather convenient size. The fact that in the case of a non-permeant solute the coefficient is the same in hydraulic and osmotic flows, finds its expression in the following equation:

$$J_V = L_p(\Delta p - RT\Delta c_{\text{imp}}) \qquad (25)$$

sometimes referred to by physiologists as the Starling hypothesis. The subscript imp emphasizes that for the solute considered in equation (25) the membrane is impermeable; otherwise equation (26) would have to be used. The equality of the coefficients in the two phenomena implies that the mechanisms by which hydraulic and osmotic flows are brought about are in the end the same; thus, e.g., in the case of a porous membrane it follows from the kinetic theory of liquids that a difference in osmotic pressure at the pore openings results in a local step of hydrostatic pressure (see Dainty, 1963).

Ideally semipermeable membranes, however, are an exception, a theoretically interesting limiting case. Membranes which are encountered in biological systems are leaky, i.e. permeable to water as well as to at least some solutes. The osmotic flow induced by a permeating solute is always less than that brought about by a non-permeating one of the same osmolarity (molar concentration taking into account possible dissociation) and hence of the same theoretical, van't Hoff osmotic pressure. There are two reasons for which the volume is diminished: (1) the volume flow is no longer equal to the volume flow of water but rather to the difference between the volume flow of water and the volume flow of solute leaking in the opposite direction. (2) If water flow and solute flow proceed through a common pathway in the membrane the water flow is hindered by frictional interactions with the solute flow in the opposite direction.

The problem was treated theoretically by introducing the so-called Staverman reflection coefficient σ (see Staverman, 1951, for the origin of the concept; Kedem and Katchalsky, 1958, for the complete development in the case of non-electrolytes; and House, 1974, for a complete treatment of water transport in cells and tissues). The theoretical approach is based on thermodynamics of the steady state (thermodynamics of irreversible processes) in which it is assumed that as a result of interactions between individual flows each flow is proportional not only to its conjugate force, but rather to all generalized forces operating in the sytem. Thus in a given case it is assumed that flow of solute is linearly dependent not only on the gradient of the chemical potential of the solute but also on that of water and the same is true about the flow of water. The equations are then integrated and transformed, so that finally they relate easily measurable flows to experimentally available concentration and pressure differences across the membrane. The volume flow in the presence of a single permeating non-electrolyte is then given by a remarkably simple analogy of equation (25):

$$J_V = L_p(\Delta p - \sigma RT \Delta c) \tag{26}$$

where c signifies the concentration of the permeating solute, and the flow of the solute itself is described by

$$J = RT\omega \Delta c + \bar{c}(1 - \sigma)J_V = P\Delta c + \bar{c}(1 - \sigma)J_V \tag{27}$$

$\bar{c}$ being a mean between the concentrations at the two sides of the membrane. The values of the reflection coefficient σ for non-electrolytes are between 0 and 1. The value of 1 corresponds to an ideal semipermeable membrane: it may be seen that in this case equation (26) reduces to equation (25) and similarly equation (27) reduces to equation (4), according to which solute only diffuses across a membrane and is not entrained by simultaneously proceeding volume flow. The value of 0, on the other hand, corresponds to a solute which the membrane cannot distinguish from water molecules—thus concentration difference of labelled water does not induce any osmotic flow. The reflection coefficient for an electrolyte in a charged membrane may even become negative, so that negative osmosis is observed (Katchalsky, 1961).

When both a permeating and a non-permeating solute are present in the system, equations (25) and (26) may be combined to give a more general expression:

$$J_V = L_p(\Delta p - \sigma RT\Delta c - RT\Delta c_{imp}) \tag{28}$$

Some possibilities of measurement of the reflection coefficient are obvious from equation (28). Thus, when $J_V = 0$ and $\Delta c_{imp} = 0$,

$$\sigma = \frac{\Delta p}{RT\Delta c} \tag{29}$$

or, when $J_V = 0$ and $\Delta p = 0$,

$$\sigma = -\frac{\Delta c_{imp}}{\Delta c} \tag{30}$$

Alternatively, hydrostatic pressure difference producing a given volume flow may be compared with osmotic pressure difference for a permeating solute resulting in an equally great flow, etc.

As already mentioned, the reflection coefficient for a permeating solute is always less than one, even if there are no interactions on a common pathway in the membrane. Let us imagine that there are indeed separate pathways for water and solute. The water-permeable pathway behaves as an ideal semipermeable membrane (equation (25)) and the volume flow across it is entirely due to water; hence it is equal to the flow of water J_w multiplied by the partial molal volume of water, $\bar{V}_w$:

$$J_w \bar{V}_w = L_p(\Delta p - RT\Delta c) \tag{31}$$

However, there is also a volume of solute, given by flow of solute J multiplied by the partial molal volume of solute $\bar{V}$. Under the condition of zero total volume flow ($J_V = 0$) equation (27) gives

$$J\bar{V} = \bar{V}RT\omega\Delta c \tag{32}$$

But since the condition of zero total volume flow under which σ is measured means that $J_w \bar{V}_w + J\bar{V} = 0$, the sum of the right-hand sides of equations (31) and (32) is equal to zero. Hence the reflection coefficient in the absence of interactions between solute flow and water flow is equal to (cf. equation (29))

$$\sigma = \frac{\Delta p}{RT\Delta c} = 1 - \frac{\bar{V}\omega}{L_p} \tag{33}$$

In order to demonstrate a common pathway for water and solute (water-filled pores in the membrane) the reflection coefficient for the solute has to be still lower:

$$\sigma < 1 - \frac{\bar{V}\omega}{L_p} \tag{34}$$

IV. ACTIVE TRANSPORT

Rigorous definition of active transport given by Kedem (1961) is based on the concept of coupling: primary active transport is transport coupled to a metabolic reaction and secondary active transport is coupled to flow of substance which is itself subject to a primary active transport. As a result of the coupling the flow of actively transported substance can and mostly does proceed against simple physico-chemical forces, such as concentration gradient of a non-electrolyte or electrochemical potential gradient of an ion. Following the principles of the thermodynamics of irreversible processes Kedem (1961) expresses the flows of various substances J_j ($j = 1, 2, \ldots, n$) and the flow of a metabolic reaction J_r by the following linear relations:

$$J_j = -\frac{1}{R_{jj}} \left(\Delta \bar{\mu}_j + \sum_{\substack{k=1 \\ k \neq j, r}}^{n} R_{jk} J_k + R_{jr} J_r \right) \tag{35}$$

$$J_r = -\frac{1}{R_{rr}} \left(A_r + \sum_{j=1}^{n} R_{jr} J_j \right) \tag{36}$$

in which R's are called resistance coefficients. From the equations it is obvious that the flow of substance j may be linearly related not only to its own conjugate force, the difference of its electrochemical potential $\Delta \bar{\mu}_j$ across the membrane, but also to flows of other substances $J_{k \neq j}$ ($k = 1, 2, \ldots, n$) and to the flow of a metabolic reaction J_r. Likewise, the flow of the metabolic reaction depends linearly not only on its affinity A_r, but also on the flows of substances with which the reaction is coupled. When coupling is a physical reality, the corresponding resistance coefficient is different from zero. Thus when $R_{jr} \neq 0$, we are dealing with a primary active transport of substance j. When $R_{jr} = 0$ but $R_{jk} \neq 0$ and $R_{kr} \neq 0$ we are dealing with a secondary active transport of substance j, dragged in this case across the membrane by substance k, which itself is primarily actively transported. The useful terms primary and secondary active transport were introduced by Stein (1967).

The flow of a substance, J_j, is a vector, whereas the flow of a metabolic reaction, J_r, is a scalar. According to the Curie–Prigogine principle (see, e.g., Katchalsky and Curran, 1965) processes of different tensorial order cannot interact in an isotropic system. Hence the coefficient R_{jr} must be a vector, expressing the anisotropy of the membrane.

The interesting problem of the degree of coupling has been discussed by Essig and Caplan (1968). A linear relation between the flow of actively transported substance and the flow of metabolic reaction does not imply that the coupling is complete (stoichiometric). The degree of coupling q is defined as

$$q = -\frac{R_{jr}}{\sqrt{(R_{jj} R_{rr})}} \tag{37}$$

For complete coupling $q = \pm 1$ while in the absence of coupling $q = 0$. Complete coupling is unlikely because there may be slipping in the active transport mechanism itself, and, moreover, when there are parallel leaks and/or series resistances in the system the coupling is of necessity incomplete (Essig and Caplan, 1968).

A formal kinetic description of active transport is provided by equation (8), discussed in connection with mediated transport; constants a and b are different in this case.

Various criteria of active transport follow more or less directly from its definition and from corollaries of this definition. The following can be listed: (1) stationary distribution of a permeating solute different from thermodynamic equilibrium (cf. the discussion of equations (13)); (2) net movement of a solute which cannot be explained by simple physico-chemical forces (see, e.g., the discussion of the short-circuit current technique in Section VI); (3) flow of solute dependent on metabolism; metabolic inhibitors may be used in this context but care must be taken of their possible primary influence on factors affecting passive transport, such as membrane potential, etc. A similar approach is applicable in the case of secondary active transport where it is even possible to reverse experimentally the gradient of the substance which entrains the solute in question. The formerly popular flux-ratio test for active transport appears to be less satisfactory now, the ratio deviating from the theoretically predicted value for reasons other than coupling with metabolism or with a flow of another species, e.g. mutual drag between particles transported (usually denoted as single-file diffusion) or a competition for a carrier (denoted as exchange diffusion).

Finally we would like to discuss the question of a direct influence of active transport of ions on the membrane potential, i.e. the question of so-called electrogenic pumps. In Section II it has been assumed that the membrane potentials are generated uniquely by a diffusion mechanism, as a result of unequal distribution of one or more ionic species across the membrane displaying selective permeability. Ionic pumps were tacitly assumed to create solely unequal steady-state distributions of ions, rather than to contribute by their driving force to separation of ionic charges across the membrane, i.e. to generation of membrane potentials. A tightly coupled sodium–potassium pump performing forced one-to-one exchange is an example satisfying the above assumption; such a pump is electroneutral. However, at least under certain conditions (e.g., when sodium ions are transported against a relatively small gradient) the pumping ratio is different from one and hence the pump is electrogenic. How can the contribution of an electrogenic pump to the membrane potential be estimated quantitatively? For a steady-state situation in which the membrane potential is governed partly by distribution of potassium and sodium ions, partly by a pump transporting r sodium ions from the cell for each of the potassium ions taken actively in, the problem was solved by Thomas (1972). The formula for the membrane potential derived by Thomas

$$E = \frac{RT}{F} \ln \frac{rP_{K^+}[K^+]_o + P_{Na^+}[Na^+]_o}{rP_{K^+}[K^+]_i + P_{Na^+}[Na^+]_i} \tag{38}$$

is seen to be simply a more general form of equation (21), to which it is reduced for $r = 1$ (one-to-one coupled pump). In another limiting case, that of a purely electrogenic pump, r tends to infinity so that sodium terms may be neglected and equation (38) reduces to the Nernst–Donnan equilibrium potential for potassium ions.

The Thomas equation is instructive: it can be seen that even if there is an electrogenic pump operating, the steady-state membrane potential is still within the limits of Nernst–Donnan potentials for individual ionic species. If it were not so, the steady-state condition (introduced in the derivation as requirement of zero net flows across the membrane) would be violated. In the limiting case of a purely electrogenic sodium pump it is obvious that the steady state of passively distributed potassium ions coincides with their thermodynamic equilibrium.

In non-steady state, however, an electrogenic pump can be demonstrated by the occurrence of membrane potentials higher than the equilibrium potentials of various ionic species present. An electrogenic sodium pump was thus demonstrated in soaked muscle fibres by Kernan (1962) and Keynes and Rybová (1963) and in toad bladder exposed to a high potassium concentration from the serosal side by Frazier and Leaf (1963). Although in the latter case of an epithelial layer separating two reservoirs hardly any concentration changes can be detected due to the vastness of the reservoirs, there is a net flow of sodium across the membrane and hence the whole system is not in a strict steady state postulated above and in such case the electrogenic character of the pump can be recognized on comparing the membrane potential with the equilibrium potentials. Another possibility of demonstrating electrogenic pumps would be to search for rapid membrane potential changes brought about by inhibitors, but since the action of inhibitors itself may be slow, it seems preferable to study the membrane potential changes with rapid changes in the ambient temperature—in this way electrogenic sodium pumping in the ductal epithelium of salivary gland was demonstrated by Augustus (1976).

A mathematical description of non-steady-state membrane potentials is provided by an analogy of the Hodgkin–Horowicz equation (23); it may be sufficient to introduce the electromotive force of the electrogenic pump into the equivalent circuit of the membrane (Fig. 5) and in the case when the membrane potential changes appreciably in time to introduce another branch with the membrane capacity (usually about 1 μF cm^{-2}).

V. TRANSPORT ACROSS EPITHELIAL CELL LAYERS

Transport of ions and of water attained a very high development in epithelial layers which cover the outer surface of multicellular organisms and especially the surfaces of cavities communicating with the external world. In the case of mammalian kidney the structural complexity of the organ makes these processes most efficient but at the same time not easily accessible to experimental studies. For this reason functionally related but morphologically much simpler epithelia such as frog skin and urinary or gall bladder epithelia are studied; often amphibian tissues are used, since they are easily kept under *in vitro* conditions. Flat sheets of these tissues can be used to separate experimentally easily accessible compartments in split chambers. Different as these various epithelial structures are, their function seems to be explained realistically using a simplified structure shown in Fig. 6. Some of these epithelia are indeed formed by a single cell layer, others, such as frog skin, are built as a stratified epithelium in order to sustain heavier wear. However, even in frog skin, there is at least indirect evidence that most of the transepithelial transport is performed by a single cell layer, the first layer of *stratum granulosum*, localized just beneath the outermost cornified cell layer, the *stratum corneum* (cf., e.g., Morel and Leblanc, 1973, 1975).

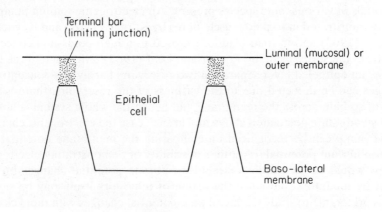

Figure 6 Schematic representation of ion-transporting epithelial layer

The epithelial layer in Fig. 6 represents a planar array of polar cells; the membranes at the opposite poles of the cells are of a different character. The terminology used in the description of epithelial layers is obvious from Fig. 6. The luminal or mucosal membrane, corresponding to the outer membrane of the skin, is sometimes called the apical membrane being localized at the apex of the cells in tubular epithelial structures. The baso-lateral or latero-basal membranes should not be confused with the highly permeable structure situated between epithelial cells and connective tissue and called the basal or basement membrane (*membrane propria*). Also the term serosal membrane is often used for the baso-lateral membrane in the urinary bladder, since it is oriented toward the lining of flat serosal cells at the opposite surface of the connective tissue.

It may be of some interest that the complexity of epithelial layers resulting from the unlike character of the two membranes can be diminished in the so-called non-polar preparation of the epithelial layer, shown schematically in Fig. 7. In this preparation the luminal membranes are closed with liquid paraffin (mineral oil); physiological transport of aqueous salt solutions from the luminal side into the cells favours a close contact between the oil and the membranes. In such preparations the composition of the cell content is

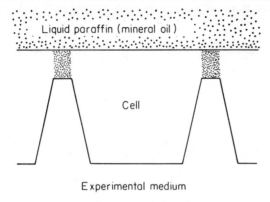

Figure 7 Non-polar preparation of epithelial layer

regulated solely by the baso-lateral membranes and thus the function of these membranes can be conveniently studied (cf., for example, Janáček and Rybová, 1970).

Referring again to Fig. 6, it is seen that the cells are connected together by terminal bars. The term still used for these junctional complexes is tight junctions, but since the terminal bars may be quite permeable, the term limiting junctions should be preferred (cf. DiBona and Civan, 1973). The

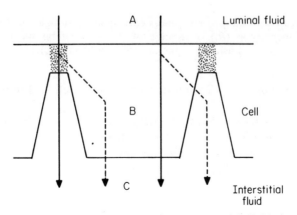

Figure 8 Possible pathways of ions across epithelial layer

relative importance of ionic permeation across the terminal bars and across the cells can be solved by electrical measurements, comparing the electrical resistances of the intercellular and of the transcellular pathways. This was carried out for the epithelium of *Necturus* gall bladder by Frömter (1972) and his approach will be shortly described here.

Various possible pathways of ions across the epithelial layer are shown in Fig. 8. The pathways represented by dashed lines are not experimentally accessible and hence must be omitted from the theoretical treatment, based on a simple electrical circuit with lumped resistances shown in Fig. 9. The

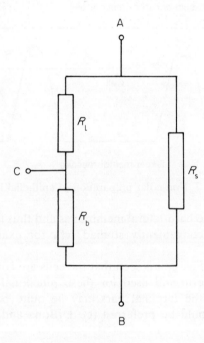

Figure 9 Circuit with lumped resistances corresponding roughly to the resistance of the luminal membrane (R_l), resistance of the baso-lateral membrane (R_b) and resistance of the intercellular shunt (R_s)

individual resistances R_l, R_b and R_s were determined by Frömter (1972) as follows:

(1) Current is passed between compartments A and B and the voltage drop across the epithelium is recorded with a separate pair of electrodes. The ratio of the voltage drop to the current density gives the value R_t, the resistance of the epithelium to transepithelial current, which is related to the individual resistances by

$$\frac{1}{R_t} = \frac{1}{R_s} + \frac{1}{R_l + R_b} \tag{39}$$

(2) Passing again the current between the compartments A and B the individual potential drops across the luminal membrane ΔV_1 and across the baso-lateral membrane ΔV_b are recorded with an intracellular microelectrode. The ratio of the two potential drops is equal to the ratio of the individual resistances of the two membranes:

$$\frac{\Delta V_1}{\Delta V_b} = \frac{R_1}{R_b} \qquad (40)$$

(3) Media A and B are short-circuited and current is injected through a microelectrode into cell C. Resistance between the common point AB and point C is R_z; in an ideal case it would correspond to a parallel combination of resistances R_1 and R_b:

$$\frac{1}{R_z} = \frac{1}{R_1} + \frac{1}{R_b} \qquad (41)$$

In reality the current injected into a single cell flows into media A and B not only via the luminal and baso-lateral membrane of the given cell, but rather enters through intercellular junctions common to all epithelia into the neighbouring cells and passes into the two media also through their membranes. Hence not just a single circuit depicted in Fig. 9 is to be considered, but a whole planar network of such elements. The value of resistance R_z has to be estimated from the spread of electrical potential, which was shown by Frömter (1972) to be described by a modified Bessel zero-order differential equation

$$\frac{d^2V}{dx^2} + \frac{1}{x}\frac{dV}{dx} - \frac{V}{\lambda^2} = 0 \qquad (42)$$

and its solution, satisfying the boundary condition that the potential V spreading from a point source should disappear at large distances, is the so-called zero-order modified Bessel function

$$V = A\ K_0(x/\lambda) \qquad (43)$$

depicted in Fig. 10. The potential fall is more rapid than that corresponding to the common cable equation (exponential) and the required resistance R_z, corresponding to the parallel combination of resistances R_1 and R_b, is related to the applied current i_0 and to the values of A and λ by

$$R_z = \frac{2\pi A \lambda^2}{i_0} \qquad (44)$$

Frömter (1972) examined the potential attenuation with the distance by a number of microelectrode impalements and compared it to a set of Bessel functions $AK_0(x/\lambda)$ plotted for various λ and A. The best fit was found by inspection; from λ and A, R_z was calculated and hence also the values of the individual resistances became available. The figures for the *Necturus* gallbladder were the following:

luminal membrane resistance	4500 Ω cm^2
baso-lateral membrane resistance	2900 Ω cm^2
total transepithelial resistance	310 Ω cm^2

Thus, about 96% of the transcellular current bypasses the cells in the *Necturus* gall bladder, the resistance of the transcellular ionic pathway being

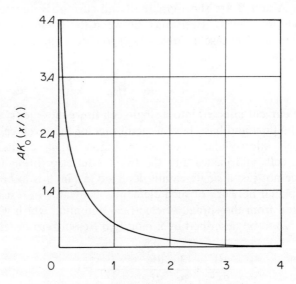

Figure 10 Modified zero-order Besel function

24 times higher than that of the intercellular shunt. The intercellular shunt was identified with the terminal bars by Frömter by scanning the electrical field at the luminal surface of the epithelium with a microelectrode during a transepithelial current flow.

Not all epithelia are as leaky as gall bladders; gall bladders, kidney proximal tubules and intestines belong to the leaky epithelia, whereas distal tubules, urinary bladders and amphibian skins represent tight epithelia with transepithelial resistances of the order of several kΩ cm^2.

Even in tight epithelia like frog skin or toad urinary bladder the terminal bars may become leaky by the 'reversed osmotic gradient' procedure. Strong hypertonicity of urea or other substances (diverse substances such as sodium chloride, creatinine, mannitol, sucrose and even raffinose may be used, but urea is most effective) brings about a profound drop in the transepithelial resistance. Application of large molecules such as inulin or hypertonicity at both sides of the preparation (absence of a gradient) is ineffective. At the same time, the phenomenon of 'anomalous solvent drag', discovered and finally explained by Ussing (1968), is observed. During this phenomenon an asymmetric flow of labelled substances such as sucrose is induced by urea, an uphill movement, proceeding in the direction of the urea flow and against the net osmotic flow of water. Ussing's (1968) explanation is depicted schematically in Fig. 11. It is seen that urea makes the terminal bars leaky, enters the

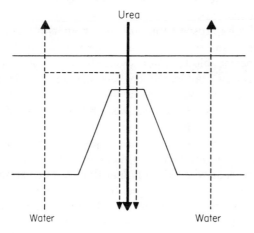

Figure 11 Ussing's (1968) explanation of 'anomalous solvent drag'

intercellular spaces and draws into them water from the neighbouring cells. Hydrostatic pressure is built up in the intercellular spaces and bulk flow results, proceeding against the net water flow taking place mostly through the cells. Bulk flow from the interspaces induces a net flow of all substances, which, like sucrose, permeate the terminal bars more easily than the cell membranes.

The mechanism of opening the terminal bars by a reversed osmotic gradient was examined by DiBona and Civan (1973) by electron microscopy. The drop in the transepithelial resistance induced by the reverse gradient is accompanied by formation of bullous deformations (blisters) in the terminal bars. It appears that solutes diffuse into the terminal bars, attract water osmotically, increase the local osmotic pressure and, thus deforming the structures, they enhance their permeability.

How important is the role of the terminal bars in the water permeability of epithelial layers? According to contemporary evidence the flow of water seems to be mostly transcellular. Thus Van Os and Slegers (1973) compared the permeability of rabbit gall bladder for water calculated from transepithelial diffusion of labelled water molecules with that measured in an osmotic experiment. Applying an adequate correction for the presence of the unstirred layers in the diffusion experiment they found no discrepancy between the two water permeabilities, as would be assumed if large pores represented by the terminal bars were the main pathway for water. Moreover, they called attention to the well documented fact that although various epithelial layers differ enormously in their electrical resistance from various cellular and artificial phospholipid membranes (and we know already that these differences depend on the presence and the character of the terminal bars), there are no conspicuous differences in the osmotic water permeability of all these structures. Hence the authors concluded that ... it is very unlikely that the small junctional area, responsible for the high conductance, contributes significantly to the osmotical water flow, despite the fact that this pathway might be highly permeable to water' (Van Os and Slegers, 1973).

In the toad urinary bladder which becomes permeable to water in the presence of the hormone vasopressin, the situation may be rather similar: Civan and DiBona (1974) showed vasopressin to inhibit the junctional blistering as well as electrical resistance drop brought about by reversed osmotic gradients and discussed above. The phenomenon can be explained by assuming that the hormone increases the water permeability of the luminal cell membranes so that when the luminal fluid is hypertonic the cytoplasm also is hypertonic and there is consequently no water movement from the cells to the terminal bars, that would result in the formation of blisters and in a resistance drop. Thus even here the water permeability of the layer appears to be related to the permeability of cell membranes rather than to that of the terminal bars.

Not only macroscopic but also microscopic osmotic gradients can be the cause of transepithelial water movement. An example of this is the standing-gradient model of fluid transport across gall bladder epithelia developed by Diamond (1971). Solute is actively transported, e.g. into lateral intercellular spaces, water follows osmotically and increases the local hydrostatic pressure. The hydrostatic pressure continually extrudes the fluid from these spaces through their basal openings and thus completes its transport from the lumen toward the blood capillaries at the opposite side of the epithelium. Each individual step of this mechanism could proceed in the opposite direction with the fluid being secreted into the lumen.

At the end of this short account of peculiarities related to the structural complexity of ion-transporting epithelial layers we may mention the dependence of the membrane potential of latero-basal membranes on that of the luminal membranes. Simultaneous changes in the two potentials taking place when the composition of the luminal medium was varied were observed by Reuss and Finn (1975). Since the changes were in the same sense (simultaneous increase or decrease of the two potentials) they could not be explained by the presence of the paracellular shunt and were thus ascribed by the authors to an unknown coupling mechanism. However, as shown by Lindemann (1975), the simultaneous changes may be the result of an interesting artefact: shunting of the luminal membrane at the point of impalement can permit flow

Relations between Solutes and Water: Analysis of Solute Transport 73

of an excess current across the two membranes, the current source being the unimpaled adjacent tissue.

VI. EXPERIMENTAL METHODS

Experimental research in the field of membrane transport may be divided into two categories: the study of transport phenomena and the study of transport mechanisms. The latter approach becomes obviously more and more important as the whole field of membranology develops, but it would be completely outside the scope of the present chapter to deal with intricacies of studies of membrane structure, of the protein chemistry of carriers and related proteins and of the biochemistry of energy transductions. Apart from the valuable information contained in other chapters of the present treatise a concise treatment of the subject by Kotyk and Janáček (1976) may be suggested as introductory reading in this field.

The experimental study of phenomenology of membrane transport consists primarily in the measurement of flows and of driving forces. The most straightforward measurement of flows was already mentioned (together with its mode of evaluation by linearization procedures) in Section I; it consisted of the determination of initial flows, when it was legitimate to assume that the concentration of the substance in question on the opposite side of the membrane was zero (i.e. although already measurable, negligible in comparison with that on the *cis*-side). In such measurements the distinction between the net flow and the unidirectional flow is immaterial, since the backflow is negligible.

In the very simple case of membrane permeation by simple diffusion when, moreover, an uptake into a compartment of a constant volume is considered the value of the permeability constant (determining the flow at each instant) can be deduced from the time course of the changing concentrations. The flow of substance being given by equation (4), the rate of substance amount change in the compartment is given by the flow multiplied by the membrane area A and the rate of its concentration change is obtained by dividing by the compartment volume:

$$\frac{\mathrm{d}c_i}{\mathrm{d}t} = P \frac{A}{V_i} (c_o - c_i) \tag{45}$$

Here the rate of concentration change is written for the 'inner' compartment denoted by subscript i. To describe the time course of the concentration change equation (45) has to be integrated under conditions characterizing the experiment. The procedure is simplified, if the experimentally adjustable concentration in the outer compartment, c_o, is kept constant (a large reservoir is used or the initial medium repeatedly replaced with a fresh one). It can even be kept equal to zero; the inner concentration will then follow an exponential decay

$$c_i = c_{i, t=0} \exp\left(-P\frac{A}{V_i}t\right) \quad (46)$$

A semilogarithmic plot of the concentration against time will then be a straight line with the slope $-P(A/V_i)$.

Similarly, when the experiment starts with $c_i = 0$, the outer concentration being kept constant, the exponential increase of the inner concentration is described by

$$c_i = c_{i, t=\infty}\left[1 - \exp\left(-P\frac{A}{V_i}t\right)\right] \quad (47)$$

where the difference $c_{i,t=\infty} - c_i$ can again be plotted semilogarithmically.

All the above equations neglect the volume flow which can proceed simultaneously, e.g. as a result of the osmotic pressure of the permeating substance. This is always a complicating factor, even in the absence of interactions between solute and solvent flows, since the concentration c_i then changes not only by solute permeation but also by compartment volume changes. The following equation:

$$\frac{dc_i}{dt} = P\frac{A}{V_i}(c_o - c_i) - \frac{c}{V_i}\frac{dV_i}{dt} \quad (48)$$

is then to be used and together with an appropriate equation for the rate of the volume change, solved numerically. When interactions between solute and solvent flows cannot be neglected (i.e. the reflection coefficient of the solute is appreciably less than 1), equations discussed in Section III are to be applied and solved, as done, e.g., by Johnson and Wilson (1967).

In many cases the measurement of flows is simplified by the use of labelled substances (tracers). In the measurement of net flows it is then possible to determine the amounts of transported substances without chemical analyses, but apart from the very simple cases discussed above the mathematical interpretation of concentration time courses is still difficult. This is not the case when unidirectional steady flows are measured: when the total flow of the substance is constant, the flow of tracer is simply equal to this flow multiplied by specific activity in the compartment from which the flow originates. Denoting the concentrations of tracer with an asterisk (so that the specific activity is c^*/c) and the unidirectional flow from the outer to the inner compartment by J_{oi}, that in the opposite direction being J_{io}, we have for the rate of tracer concentration change in the inner compartment the equation

$$\frac{dc_i^*}{dt} = \frac{J_{oi}}{V_i}A\frac{c_o^*}{c_o} - \frac{J_{io}}{V_i}A\frac{c_i^*}{c_i} \quad (49)$$

which can be easily integrated (J's, c's, V_i and A being constants), especially for simple experimental conditions. Thus, when the specific activity in the outer compartment is kept at zero, the decay of the concentration of the labelled substance in the inner compartment is described by

$$c_i^* = c_{i,\,t=0}^* \exp\left(-\frac{A\,J_{io}}{V_i c_i}t\right) \qquad (50)$$

etc. Some more details concerning elementary kinetics of tracer exchange are given, e.g., by Kotyk and Janáček (1975).

Special techniques are required for the measurement of volume flow. Only a short list of basic approaches in this field can be given here (see Kotyk and Janáček, 1975, for references). Volume changes of large and regularly shaped cells can be evaluated from microscopic measurements with a calibrated eyepiece. Thickness of a stretched epithelial layer, proportional to its volume, can be measured by focusing sharply with a water immersion lens on two definite boundaries and reading the difference on the fine screw of the microscope. Volume changes of tissue pieces can be estimated from changes in their weight; prior to weighing the tissue is blotted with a filter paper, preferentially moistened with an appropriate saline. The volume of cells packed by centrifugation can be determined directly in calibrated tubes, a non-permeant substance can be used to determine the extracellular volume in the pellet. Correlations between the volume of cells in suspension and light scattering or absorbance can exist under carefully preserved standard conditions. Transepithelial volume flow is directed into or from an easily accessible compartment, in which the volume changes can be evaluated from changes of concentration of an impermeant substance, by weighing the fluid in the compartment, by measuring the compartment volume using a connected calibrated capillary or by maintaining it constant by a manual or automatic injection of fluid from a microsyringe.

Determination of driving forces, which is another prerequisite of the phenomenological description of transport, is rather straightforward in the case of non-electrolytes. There it usually consists of an estimation of the intracellular content by chemical analysis or from a stationary distribution of a radioactivity labelled substance. The non-electrolyte content can be expressed per dry solids; the water content (evaluated usually from the weight loss by an overnight drying at 95 °C) is expressed in the same way. The quotient of the two figures gives the concentration. The data can be corrected for extracellular solute and water using an extracellular marker, a non-permeant solute, such as inulin. The 'extracellular space' can thus be calculated, in which the concentrations of all solutes are presumably the same as in the bulk of the medium. Care should be taken of possible compartmentation of the substance studied inside the cells; equilibration of labelled substance and its kinetics is often illuminating in this respect.

With ions the situation is complicated by the necessity of determining the membrane potential. To measure a difference of the electrical potential between two fluids, two reference (e.g. calomel) electrodes are required, connected to the two fluids by bridges. The bridges are filled with a concentrated solution of an electrolyte with the same cation and anion mobility (e.g. 3 M potassium chloride)—under this condition the liquid junction potentials at their boundaries are minimized. If one of the two fluids is the intracellular cytoplasm, a microbridge must be used, commonly a glass microelectrode. Glass microelectrodes are prepared from borosilicate capillaries on commerically available pullers. Their tips have to have diameters of rather less than 1 μm. In special cases the tip can be sharpened (and the resistance of the electrode slightly reduced) by a fine microscope-controlled

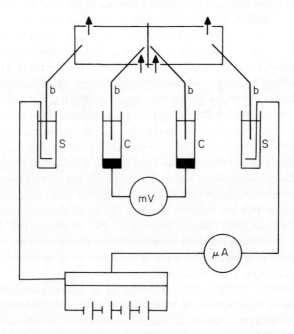

Figure 12 Schematic representation of the short-circuit current technique. C, calomel electrodes; S, silver–silver chloride electrodes; b, bridges

contact with a polished surface of a suitable stone, e.g. agate. The microelectrode is filled, e.g. with methanol (by boiling under reduced pressure), methanol is replaced with water and water with 3 M KCl. To introduce the microelectrode into the cell, a suitable micromanipulator is required.

Intracellular concentrations of many ions can be determined by chemical analysis (flame photometry, potentiometric titrations, etc.), but some uncertainty remains as to their activity coefficients. This problem can be at least partially solved using ion-sensitive microelectrodes (these are real microelectrodes, not microbridges). An excellent up-to-date discussion of this approach can be found in Lev and Armstrong (1975).

This short essay on the measurement of flows and forces would be incomplete without mentioning an ingenious and justly famous technique, with which flow of an ion and a driving force is measured at the same time (the latter is actually kept at zero). In the short-circuit current technique by Ussing and Zerahn (1951), shown schematically in Fig. 12, an epithelial layer separates two halves of a split chamber. The same aerated medium is used at the two sides and the spontaneous electrical potential difference across the preparation is brought to zero by using a current circuit. The difference of the electrochemical potential across the preparation is zero (the electromotive force of the battery serves solely to compensate the potential drops on ohmic resistances) and hence the current in the circuit is equal to the net transepithelial flow of an actively transported ion.

REFERENCES

Augustus, J. (1976). Evidence for electrogenic sodium pumping in the ductal epithelium of rabbit salivary gland and its relationship with ($Na^+ + K^+$)-ATPase. *Biochim. Biophys. Acta*, **419**, 63–75.

Civan, M. M., and D. R. Dibona (1974). Pathways for movement of ions and water across toad urinary bladder. II. Site and mode of action of vasopressin. *J. Membrane Biol.*, **19**, 195–220.

Collander, R. (1949). The permeabiiity of plant protoplasts to small molecules. *Physiol. Plantarum*, **2**, 300–11.

Dainty, J. (1963). Water relations of plant cells. *Adv. Bot. Res.*, **1**, 279–326.

Danielli, J. F. (1952). The theory of penetration of a thin membrane. In H. Davson and J. F. Danielli (Eds) *The Permeability of Natural Membranes*, 2nd ed. Cambridge University Press, Cambridge, New York. pp. 324–35.

Diamond, J. (1971). Standing-gradient model of fluid transport in epithelia. *Fed. Proc.* **30**, 6–13.

Dibona, D. R., and M. M. Civan (1973). Pathways for movement of ions and water across toad urinary bladder. I. Anatomic site of transepithelial shunt pathways. *J. Membrane Biol.* **12**, 101–28.

Essig, A., and S. R. Caplan (1968). Energetics of active transport processes. *Biophys. J.*, **8**, 1434–57.

Frazier, H. A., and A. Leaf (1963). The electrical characteristics of active sodium transport in the toad bladder. *J. Gen. Physiol.*, **46**, 491–503.

Frömter, E. (1972). The route of passive ion movement through the epithelium of *Necturus* gallbladder. *J. Membrane Biol.*, **8**, 259–301.

Goldman, D. E. (1943). Potential, impedance and rectification in membranes. *J. Gen. Physiol.*, **27**, 37–60.

Hartley, G. S., and J. Crank (1949). Some fundamental definitions and concepts in diffusion processes. *Trans. Faraday Soc.*, **45**, 801–18.

Hodgkin, A. L., and P. Horowicz (1959). The influence of potassium and chloride ions on the membrane potential of single muscle fibres. *J. Physiol.*, **148**, 127–60.

House, C. R. (1974). Water transport in cells and tissues. Edward Arnold, London.

Jacobs, M. H. (1967). *Diffusion Processes.* Springer-Verlag, Berlin, Heidelberg, New York.

Jaffe, L. F. (1974). The interpretation of voltage–concentration relations. *J. Theor. Biol.*, **48**, 11–18.

Janáček, K., and R. Rybová (1970). Nonpolarized frog bladder preparation. The effects of oxytocin. *Pflügers Arch.*, **318**, 294–304.
Johnson, F. H., H. Eyring, and M. J. Polissar (1954). The kinetic basis of modern biology. Wiley, New York and Chapman and Hall, London.
Johnson, J. A., and T. A. Wilson (1967). Osmotic volume changes induced by a permeable solute. *J. Theor. Biol.*, **17**, 304–11.
Katchalsky, A. (1961). Membrane permeability and the thermodynamics of irreversible processes. In A. Kleinzeller and A. Kotyk (Eds) *Transport and Metabolism*, Academic Press, New York, London and Publishing House of the Czechoslovak Academy of Sciences, Prague. pp. 69–86.
Katchalsky, A., and P. F. Curran (1965). *Nonequilibrium Thermodynamics in Biophysics*. Harvard University Press, Cambridge, Massachusetts.
Kedem, O. (1961). Criteria of active transport. In A. Kleinzeller and A. Kotyk (Eds) *Membrane Transort and Metabolism*, Academic Press, New York, London and Publishing House of the Czechoslovak Academy of Sciences, Prague. pp. 87–93.
Kedem, O., and A. Katchalsky (1958). Thermodynamic analysis of the permeability of biological membranes to non-electrolytes. *Biochim. Biophys. Acta*, **27**, 229–46.
Kernan, R. P. (1962). Membrane potential of electrolyte-depleted muscle fibres. *Nature*, **204**, 83–4.
Keynes, R. D., and R. Rybová (1963). The coupling between sodium and potassium fluxes in frog sartorius muscle. *J. Physiol.*, **168**, 58P.
Kotyk, A. (1973). Mechanisms of nonelectrolyte transport. *Biochim. Biophys. Acta*, **300**, 183–210.
Kotyk, A., and K. Janáček (1975). *Cell Membrane Transport. Principles and Techniques*, 2nd ed. Plenum Press, New York, London.
Kotyk, A., and K. Janáček (1976). *Membrane Transport. An Interdisciplinary Approach*. Academia, Prague and Plenum Press, New York, London.
Lev, A. A., and W. McD. Armstrong (1975). Ionic activities in cells. In F. Bronner and A. Kleinzeller (Eds) *Current Topics in Membranes and Transport*, Vol. 6, Academic Press, New York, San Francisco, London. pp. 59–123.
Lieb, W. R., and W. D. Stein (1971). The molecular basis of simple diffusion within biological membranes. In F. Bronner and A. Kleinzeller (Eds) *Current Topics in Membranes and Transport*, Vol. 2, Academic Press, New York, London. pp. 1–39.
Lindemann, B. (1975). Impalement artifacts in microelectrode recordings of epithelial membrane potentials. *Biophys. J.*, **15**, 1161–4.
Morel, F., and G. Leblanc (1973). Kinetics of sodium and lithium accumulation in isolated frog skin epithelium. In H. H. Ussing and N. A. Thorn (Eds) *Transport Mechanisms in Epithelia*, Munksgaard, Copenhagen and Academic Press, New York. pp. 73–82.
Morel, F., and G. Leblanc (1975). Transient current changes and Na compartmentalization in frog skin epithelium. *Pflügers Arch.*, **358**, 135–59.
Penniston, J. T., L. Beckett, D. L. Bentley, and C. Hansch (1969). Passive permeation of organic compounds through biological tissue: a non-steady-state theory. *Molec. Pharmacol.*, **5**, 333–41.
Reuss, L., and A. L. Finn (1975). Dependence of serosal membrane potential on mucosal membrane potential in toad urinary bladder. *Biophys. J.*, **15**, 71–5.
Rosenberg, T., and W. Wilbrandt (1957). Uphill transport induced by counterflow. *J. Gen. Physiol.*, **41**, 289–96.
Schlögl, R. (1964). *Stofftransport durch Membranen*. Dr Dietrich Steinkopff Verlag, Darmstadt.
Staverman, A. J. (1951). The theory of measurement of osmotic pressure. *Recl Trav. Chim. Pays-Bas Belg.*, **70**, 344–52.

Stein, W. D. (1967). *The Movement of Molecules across Membranes*, Academic Press, New York, London.
Teorell, T. (1953). Transport processes and electrical phenomena in ionic membranes. *Prog. Biophys. Biophys. Chem.*, **3**, 305–69.
Thomas, R. C. (1972). Electrogenic sodium pump in nerve and muscle cells. *Physiol. Rev.*, **52**, 563–94.
Ussing, H. H. (1968). The effect of urea on permeability and transport of frog skin. *Excerpta Med. Int. Congr. No. 195, Urea and the Kidney.* pp. 138–48.
Ussing, H. H., and K. Zerahn (1951). Active transport of sodium as the source of electric current in the short-circuited isolated frog skin. *Acta Physiol. Scand.* **23**, 110–27.
Van Os, C. H., and J. F. G. Slegers (1973). Path of osmotic water flow through rabbit gall bladder epithelium. *Biochim. Biophys. Acta* **291**, 197–207.

Part B

Control of the Intracellular Fluid Osmolarity

Chapter 3

Intracellular Inorganic Osmotic Effectors

G. RORIVE AND R. GILLES

I. Introduction	83
II. Water Movements and Cell Volume Control	84
A. Basic concepts: volume maintenance and volume readjustment	84
III. Mechanisms of Cell Volume Control	86
A. Water availability, water structuration	86
B. Cation binding and compartmentation	88
C. Ionic movements	89
IV. Conclusions	102
References	104

I. INTRODUCTION

One of the most remarkable properties of living cells is their capacity to maintain a relatively constant volume. However since cells generally contain a large number of charged macromolecules which cannot pass through the plasma membrane, osmotic forces should produce a constant tendency to swell. *A priori,* swelling in such a system could be avoided if the hydrostatic pressure is sufficiently higher inside than outside the cell, or if the cell surface is impermeable to a large fraction of the solutes in the external solution. In some bacterial and plant cells with rigid cell walls, the former mechanism appears to be responsible for volume stability. But animal cells do not have rigid cell walls and they are largely permeable to the major extracellular cations, sodium and potassium. It is thus necessary to look for another mechanism of volume control, i.e. cell water regulation.

Much research has been devoted to this problem during the past decade. Up to now, however, most of the data available come from only a few types of biological material, mainly isolated tissues from euryhaline species, blood cells and kidney cells. Two main experimental approaches were used, either incubation in media of various osmolarities or inhibition of the various mechanisms supposed to be involved in regulation of the intracellular fluid osmolarity.

This review aims essentially at summarizing the nature and characteristics of the main mechanisms actually considered to be involved in the control of the cell's ionic content according to the evidence from prominent studies using the two types of experimental approach.

II. WATER MOVEMENTS AND CELL VOLUME CONTROL

A. Basic Concepts: Volume Maintenance and Volume Readjustment

So far, no conclusive evidence has been obtained in favour of the existence of an active transport of water. Water transport appears to be always subordinated to some other process such as formation of osmotic gradients, action of contractile vacuoles, pinocytosis or electro-osmosis.

Among the mechanisms implicated in water movements, osmotic transfer is by far the most important and the most widely encountered. Indeed, in most animal cells water can move readily across the plasma membrane and will distribute according to concentration gradients achieved by a variety of osmotic effectors. Notable among these are different organic compounds such as amino-acids, large anionic molecules to which the membrane is not permeable and several inorganic ions such as Na^+, K^+ and Cl^-.

At first glance, the permeable and impermeable ions might be distributed on both sides of the plasmatic membrane according to a Gibbs–Donnan equilibrium; from the presence in the intracellular fluid of anionic molecules too large to pass through the membrane, it follows that the total osmotic concentration should be greater in the intracellular fluid than in the extracellular fluid (see Part A of this volume). In such a situation, animal cells, with their easily distensible membranes, should swell until they burst unless specific mechanisms are at work to balance this tendency to colloid osmotic swelling.

On the other hand, there is an important difference between the permeability of animal cells to water and to the major compounds acting as osmotic effectors. If we consider for instance the most important inorganic osmotic effectors Na^+, K^+ and Cl^-, their permeability is generally between five and eight orders of magnitude lower than water permeability (Dick, 1970; House, 1974, Part A of this volume). This means that, at first glance and when considering osmotic relations, the membrane can be considered as roughly semipermeable. In such a view, animal cells should therefore follow the van't Hoff equation (see Part A) and behave as nearly perfect osmometers when submitted to changes in the osmolarity of their environment medium.

In the van't Hoff relation $\Pi_e (V_e - b) = \Pi_i (V_0 - b)$, V is the cell volume, the subscripts e and i defining the experimental and the isotonic conditions; b represents the portion of the volume that does not participate in osmotic phenomena (non-solvent volume); the value of b, also expressed by Ponder's **R** (Ponder, 1948, Chapter 1), can be calculated from the intersect on the ordinate when plotting cell volume against the reciprocal of the relative external osmolarity.

Figure 1 gives results of such an experiment using kidney cortex slices. At 0 °C, when most of the cellular metabolism is suppressed, the tissue apparently behaves as a very good osmometer. On the contrary, under conditions of active metabolism, the kidney slice is a rather poor osmometer and the volume reached for a given change in extracellular osmolarity is smaller than at 0 °C; this induces an apparent increase in the amount of intracellular water behaving as non-solvent volume. Such results are indicative of the existence

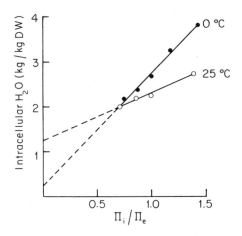

Figure 1 Effect of saline osmolarity on the water content of kidney cortex slices at 0 °C and 25 °C (compiled after data from Kleinzeller, 1960). Π_i/Π_e: osmolarity ratio, subscripts i and e respectively refer to isotonic and experimental salines

of metabolically dependent mechanisms at work to oppose the volume changes occurring following application of osmotic stresses.

Deviations from the osmometer behaviour have been reported many times for a variety of tissues and cells (see for instance the reviews by Dick, 1970; Hoffmann, 1977). In some instances part of the discrepancy from the van't Hoff law observed in hypo-osmotic conditions can be ascribed to forces opposing the swelling and related to the elastic properties of the plasmatic membranes (Mela, 1968). In most cases, however, the tension found at the cell surface is extremely small (see for instance Evans and Hochmuth, 1976a, 1976b) and makes unlikely any theory of volume control implicating development in animal cells of a significant hydrostatic pressure due to elastic properties of membranes. The deviations from the osmometer behaviour however suggest that osmotic effectors distribution do not follow the Gibbs–Donnan equilibrium and lead to the postulation of the existence of cell volume control processes. This is demonstrated by experiments on isolated axons of the crab *Eriocheir sinensis*. In this tissue, the volume reached for a hypo-osmotic shock of given amplitude is dependent on the length of time the tissue is left in the experimental medium; the longer the time is, the smaller is the swelling recorded and therefore the larger the deviation from the osmometer behaviour is (Gilles, 1973, 1974). Further studies of this phenomenon showed that in this tissue, the volume is progressively readjusted to values close to control after an initial phase of osmotic swelling. Volume readjustment following application of hypo-osmotic conditions has been described in many cell types in the past few years (see p. 92). This phenomenon is illustrated schematically in Fig. 2.

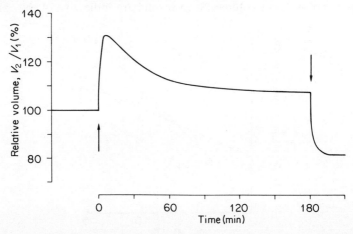

Figure 2 Schematic representation of the volume readjustment process occurring in many tissues and cell types following application of an hypo-osmotic shock (1st arrow). At the 2nd arrow, the tissue is returned to its control saline. Both maximum swelling achieved and length of readjustment period vary with the tissue used and the amplitude of the osmotic shock applied (see in the text for references to original data)

It may thus be considered that, when dealing with volume control, we are in fact dealing with two distinct kinds of process; one related to the maintenance of an intracellular osmolarity suitable for avoiding osmotic swelling in normal conditions despite a Gibbs–Donnan effect and another one related to volume readjustment following application of an osmotic stress. However, up to now there is no clear-cut evidence for the participation of independent systems in these two kinds of processes. In our opinion, it is even reasonable to consider that the same mechanisms are at work in both cell volume maintenance in isotonic conditions and cell volume readjustment under osmotic stress. What could be the mechanisms involved in these cell volume control phenomena?

III. MECHANISMS OF CELL VOLUME CONTROL

A. Water Availability, Water Structuration

The results presented in Fig. 1 might indicate that an important part of the intracellular water behaves as non-solvent in metabolically coupled kidney slices and is therefore not involved in the volume control processes.

Data suggesting that some part of cellular water is non-available for solutes, have been obtained when measuring the volume distribution of various organic molecules such as urea, dimethylsulfoxide, or propylene glycol (Kotyk and Kleinzeller, 1963; De Bruine and Van Steveninck, 1970; Rybová, 1965).

The data obtained are however not very clear. In some tissues not only is the amount of non-solvent water very small but also the volume obtained may be different depending on the technique used. For instance, Savitz, Sidel and Solomon (1964) found that in red blood cells, 20% of the water is apparently not participating in osmotic phenomena and could therefore be considered as non-solvent water. However, Gary-Bobo and Solomon (1968) found that although only 80% of the water is osmotically active, as reported by Savitz, Sidel and Solomon (1964), various non-electrolytes could dissolve in the totality of the red cells' water. In the same biological material, Cook (1967) estimated the non-solvent water to be some 4% of the total water volume considering the water not available for dissolving chloride.

A variety of explanations have been offered for the apparent non-solvent fraction of water; it is generally suggested that cell water may be present in the cell under different structurations (see for instance Dick, 1970; Foster, 1971). This water structuration is directly related to a recent hypothesis put forward to account for the volume adjustment and its associated changes in intracellular composition. In this hypothesis, it is assumed that a so-called 'limited water state' is necessary for maintaining ion selectivity in living cells (Damadian, Goldsmith and Zaner, 1971). In such a system, the changes in ionic composition of the intracellular fluid occurring during application of osmotic shocks (see following sections) are attributed to changes in cell water content arising from the redistribution of water across the cell interface in response to shifts in osmotic equilibrium; concomitantly there is a change in water structuration associated with the change in ionic composition. In the latest, most elaborated models the cell is regarded as an ion exchange system behaving in a way similar to that described by Gregor (1948) for ion exchange resins. In the biological derived models, the physical characteristics of the ion exchange system and the availability of water round about would force selection of more or less hydrated cations (i.e. Na^+ or K^+). The amount of water available would be determined by contractile proteins in the cell membrane. This brings us to the mechanochemical hypothesis proposed by Kleinzeller (see p. 99) to account for an expulsion of isotonic solution during the volume readjustment process.

In this hypothesis, it thus seems that rather than the amount of bound water, it is the state of the intracellular water and its associated cations' selectivity modulation in an ion exchange system which may explain part of the phenomena associated with volume control. This problem is the subject of controversy in the literature. In the present state of knowledge, however, it remains far beyond the scope of this contribution. The interested reader is referred to Damadian, Goldsmith and Zaner, 1971; Minkoff and Damadian, 1976; Cope, 1970, 1976; or Kolata, 1976). Moreover, as we will see later on in this chapter, the presence of contractile elements in membranes, the expelling of water during phases of volume readjustment and the associated modifications in intracellular ionic composition can be explained in a completely different way.

B. Cation Binding and Compartmentation

The usual procedure used to calculate the intracellular concentrations implies that the ions are uniformly distributed between the extracellular and the intracellular water. The extracellular water can be estimated by the determination of the volume distributions of various molecules known to diffuse freely into the extracellular space and not to cross the plasma membrane, nor to be accumulated into the tissues and nor to be metabolized. Numerous contradictions remain concerning the best molecule to label the extracellular water. However, from an extensive study performed with the rat aorta, we got the impression that the determination of the volume of diffusion of inulin still remains the best experimental approach to the measurement of extracellular water (Rorive, 1975). Therefore, intracellular water can be defined as the difference between total water and the volume of water available to inulin:

$$(H_2O)_{ic} = (H_2O)_t - (H_2O)_{in} \qquad (1)$$

If we accept the assumption that the extracellular water has the same ionic composition as the physiological saline, the amount of ions in the intracellular phase can be calculated from the following relation, e.g. for sodium:

$$Na_{ic} = Na_t - Na_{ec} = (H_2O)_{in} \qquad (2)$$

If we assume that the intracellular electrolytes (K^+, Na^+ and Cl^-) are uniformly distributed in the intracellular water, the intracellular concentration can now be calculated from relations (1) and (2).

However, the assumption usually made that the intracellular electrolytes are uniformly distributed in the intracellular water appears as an oversimplification. Compartmentation of Na^+ and K^+ in the cells has been repeatedly established by measurements of steady-state ionic fluxes, e.g. in striated muscle (Keynes and Steinhardt, 1968), in smooth muscle (Hagameijer, Rorive and Schoffeniels, 1965; Daniel and Robinson, 1971) and in kidney cells (Whittam and Davies, 1954; Burg, Grollman and Orloff, 1964). More recently, Siebert and Langendorf (1970) have demonstrated that the ionic composition of nuclei is markedly different from those of the whole cells. In addition, the determination of the activities of intracellular Na^+ and K^+ by cation selective microelectrodes suggests that parts of the intracellular ions are bound by cell constituent proteins (Lev, 1964; Dick and McLaughlin,

1969; Hincke, 1970). Such a binding is also suggested by nuclear magnetic resonance studies (Cope, 1970).

This aspect is of great importance to determine the exact amount of the intracellular ions which can participate in osmotic and electrochemical phenomena. Both physical compartmentation and binding of electrolytes by intracellular components will greatly affect the computed intracellular concentrations of cations thus leading to an overestimate of the intracellular osmotic pressure. Moreover, such compartmentation and/or binding may affect the intracellular concentrations of cations acting as osmotic effectors thus leading to a modulation of the intracellular osmolarity which might play a part in volume control. To our knowledge, this possibility has not yet been investigated deeply. It may however prove to be a rewarding field of research in the future.

C. Ionic Movements

1. *The Pump and Leak Hypothesis*

As already stated (see p. 84) water should enter the cells because of a gradient of osmotic pressure due to the existence of a Gibbs–Donnan equilbrium. This should produce a colloid-osmotic swelling unless the amount of important osmotic effectors can be regulated to achieve isotonicity between the intra- and extracellular fluids. It was soon realized that to assume a stable ionic composition and volume, an active, metabolically dependent, transport of the important inorganic osmotic effectors Na^+ and K^- must be postulated. In a variety of cells and tissues, decrease in metabolic activity and therefore in energy production by cold, anaerobiosis or various pharmacodynamical agents produces a marked swelling whereas restoration of metabolic energy supply allows a net flux of permeable components across the cell membrane. Na^+ and Cl^- will thus diffuse into the cells, whereas K^+ will leak out. As a result of the Gibbs–Donnan system, water and bulk electrolytes flow into the cells as a practically isotonic solution. On the other hand, availability of metabolic energy will reverse the swelling process by promoting active extrusion of Na^+ and active accumulation of K^+, Cl^- following passively. As a consequence of this ionic transport, an osmotic flow is induced and water can be extruded from the cells. In steady-state conditions, the active transport of Na^+ and K^+ thus plays a prominent part in the maintenance of the osmotic equilibrium achieved between the intra- and extracellular fluids. In other words, passive ionic fluxes through leak pathways and cation transport by active pumping are of importance in determining the steady-state volume achieved by the cells.

This concept of 'pump and leak' has been described mainly by Leaf (1959), Tosteson and Hoffman (1960) and Whittam (1969). It states that the existence of an active transport system for Na^+ and K^+ renders the cell membrane functionally impermeable to these ions. The active fluxes coming through the pump are equal and oppositely directed to the leak fluxes. However, it is obvious that in such a system a major volume change can only take place if, other things being equal, there is a net ion flux after inhibition of the pumping activity. This is the case for instance in the diaphragm muscle where, after metabolic inhibition with 2–4 DNP, the net Na^+ flux is practically twice the K^+ leak (Kleinzeller and Knotkova, 1964a, 1964b). On the other hand, in the frog muscle where a 1:1 passive exchange of Na^+ and K^+ should take place, no major changes can be observed at 0°C (Conway, 1957).

Active Na^+ efflux and K^+ influx would thus be implicated in cell volume maintenance. Moreover it appears that at least part of these ionic movements can be ascribed to the activity of an Na^+–K^+ active exchange mechanism.

2. *The Na^+–K^+ Exchange Pump*

The characteristics of this active Na^+ and K^+ transport system and its molecular basis have been studied extensively over the past decade. There have been several excellent reviews dealing with the subject. The reader interested in detailed information on this subject is referred to these publications (Skou, 1965; Dunham and Lunn, 1972; Kleinzeller, 1972; Orringer and Parker, 1973; Whittembury and Grantham, 1976; etc.).

Let us briefly summarize some main characteristics of the Na^+–K^+ exchange pump (see Fig. 3).

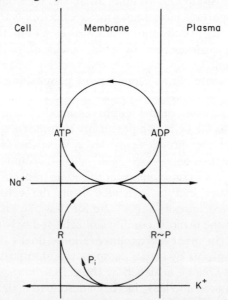

Figure 3 Schematic representation of the Na^+–K^+ exchange pump. R represents a lipoprotein in the membrane. After Orringer and Parker (1973)

(i) Sodium and potassium net movements are in opposite directions. Cell sodium decreases when cell potassium increases.
(ii) Net Na^+ extrusion depends on the K^+ concentration in the extracellular medium (Whittam and Willis, 1963; Whittembury, 1965, 1968; Podevin and Boumendil-Podevin, 1973); and net K^+ entry depends on intracellular Na^+ concentration (Whittam and Willis, 1963; Giebisch and Windhager, 1973; Whittembury and Grantham, 1976).
(iii) Na^+ as well as K^+ movements are both inhibited by ouabain, a cardiac glycoside.

Accordingly, the mechanism of active Na^+–K^+ exchange can be schematized in this way: sodium ions are transported out of the cell in the same time as the terminal phosphate of ATP is transferred to a carboxyl group of a membrane protein (Kahlenberg, Galsworthy and Hoken, 1967). This phosphorylated intermediate is thereafter hydrolysed by a process to which is associated the inward transport of potassium ions (Fig. 3). This last step would be the one inhibited by ouabain.

Recent data have demonstrated that K^+ can compete with Na^+ for outward transport and vice versa, sodium for potassium for inward movement. Under certain conditions, the pump can be made to exchange Na for Na and K for K, a process we know as exchange diffusion (Glynn, Lew and Luthi, 1970; Glynn and Hoffman, 1971).

In the red blood cells, interaction between Na^+–K^+ pump and glucose metabolism has been demonstrated, and Proverbio and Hoffman (1072) have suggested that the ATP used for the pump is synthesized by the phosphoglycerate kinase from the ADP directly derived from the pump.

From studies performed with labelled ouabain, the number of specific binding sites, which may be considered as the number of pumping sites, has been estimated in red blood cells at between 150 and 1200 per cell (Hoffman, 1969; Gardner and Conlon, 1972).

3. *Cation Leak Pathways*

In the previous sections, we have seen that both active movements and passive leaks of Na^+ and K^+ are involved in the maintenance of cell volume under normal conditions. Moreover it appears that these mechanisms and particularly the passive leaks also play an important part in the volume adjustment taking place following changes in the osmolarity of the cells' environmental medium.

a. *Volume control in hypo-osmotic conditions* In this volume adjustment process, a phase during which the cell tends to resume its original volume immediately follows the phase of rapid osmotic swelling. The volume regulatory phase is associated with a decrease in the amount of intracellular osmotic effectors as demonstrated by the fact that the volume shrinks to values lower than control if the tissue is returned to its control saline after volume readjustment in the hypo-osmotic medium (see Fig. 2).

This type of volume regulation response is reminiscent of the isosmotic regulation of the intracellular fluid originally described by Florkin (1962) in euryhaline aquatic eukaryotic species. Apparently, in tissues of invertebrates the control of the intracellular fluid osmolarity associated with the volume adjustment can be essentially related to changes in the intracellular content of low molecular weight organic compounds among which amino acids play a prominent part (see Chapter 4); the changes in inorganic constituents although they do exist, appear to be of less quantitative importance. At the opposite extreme, in vertebrate cells, the volume readjustment phase appears to be essentially related to changes in the inorganic ion content with only minor modifications in organic content (Hendil and Hoffmann, 1974; Chapter 4 of this volume).

First clearly described for vertebrates tissues and cells by Fugelli (1967) on flounder's erythrocytes, this volume regulatory phase following application of hypo-osmotic stresses of reasonable amplitude has now been reported in a variety of tissue and cell types (Kregenow, 1971a, 1971b; Rosenberg, Shank and Gregg, 1972; Roti-Roti and Rothstein, 1973; Shank, Rosenberg and Horowitz, 1973; Dellasega and Grantham, 1972; Hendil and Hoffmann, 1974; Parker and coworkers 1975).

In tissues having normally a high K^+ intracellular content, a K^+ loss appears to be a common mechanism involved in the volume readjustment phase. On the contrary, it seems that in high Na^+ cells such as dog or cat red blood cells, the control of the Na^+ intracellular level is the main mechanism (Parker and coworkers 1975; Parker and Hoffman, 1976). It appears indeed that in dog erythrocytes, volume regulation is essentially achieved by means of a Ca^{2+} dependent sodium extrusion. In this process, Ca^{2+} is accumulated suggesting the possibility of a $Ca^{2+}-Na^+$ exchange.

In normally high K^+ cells, it is now clear that the K^+ loss observed in hypo-osmotic conditions is related at least partly to a modulation of the membrane permeability to that ion (Kregenow 1971a; Sigler and Janáček, 1971; Poznansky and Solomon, 1972; Roti-Roti and Rothstein, 1973; Hendil and Hoffmann, 1974). The basic nature of this K^+ permeability control is however far from being understood. It is probably related to changes in the membrane architecture. Such changes may be, in a mechanicochemical way, induced by the osmotic swelling phase. Modifications in the concentration of ions (particularly Ca^{2+}) prevailing locally at the membrane level might also be implicated. More information is needed to shed some light on this problem.

The problem of knowing whether the changes in K^+ leak are due to a specific mechanism or may be related to a more general increase in permeability also remains an open question. Changes in permeability for amino acids during the volume readjustment phase have been reported (Gerard and Gilles, 1972; Hoffmann and Hendil, 1976; Gilles, 1978). They are moreover of a transient nature since the permeability coefficient returns to normal after the acclimation period at least in the axons of the crab *C. sapidus* (Gilles, 1978). It would thus appear that the osmotic swelling phase following application of an hypo-osmotic shock may first induce a non-specific increase in membrane permeability to a variety of osmotic effectors, organic as well as inorganic. Such permeability changes would be transient and related in some way to the volume evolution during the readjustment phase (Gerard and Gilles, 1972; Gilles, 1978).

It seems however that a specific control of the leak pathway for K^+ must also be taken into consideration (Kregenow, 1974). Such a dissociation between modifications of K^+ and Na^+ permeability during volume readjustment suggests that the channels through which the passive movements of these ions occur can be different. The observations showing that several agents can induce a selective increase in K^+ permeability in red blood cells (Orringer and Parker, 1973) are also in agreement with this hypothesis.

There is thus no doubt about the participation of a membrane permeability modulation during the cell volume readjustment response. As far as the final concentration of intracellular inorganic osmotic effectors is concerned, it would be determined by interactions between the leak pathways and the pump activity. In this respect, it is worth noticing that the volume readjustment phase can be suppressed by different metabolic inhibitors thus indicating the participation of energy consuming mechanisms. With this in mind and considering the pump and leak hypothesis, the Na^+–K^+ exchange pump appears as a clear candidate among the mechanisms implicated in volume readjustment.

Rosenberg, Shank and Gregg (1972) and also Shank, Rosenberg and Horowitz (1973) reported that the readjustment phase in cultured mouse lymphoblasts is inhibited by ouabain. This suggests strongly that part of the Na^+ gained during osmotic swelling is extruded during volume readjustment by means of the ouabain sensitive Na^+–K^+ exchange pump. However, these Na^+ movements arise as isotonic solution, and therefore do not significantly affect the intracellular sodium concentration. However the problem is not that simple and the idea of a direct and sole participation of the Na^+–K^+ exhange pump operating in a pump and leak system suffers several inadequacies when considering either volume readjustment or volume maintenance. This problem will be considered later on (see Section III-4).

b. *Volume evolution in hyperosmotic conditions* Contrary to the situation found in hypo-osmotic media where the cells show a volume readjustment response, many cell types tested *in vitro* in hyperosmotic conditions show no readjustment process. These cells simply respond to increased tonicity by rapid shrinkage and there is no readjustment afterwards. This is true for vertebrate cell types where inorganic ions (especially K^+) are the major osmotic effectors (Kregenow, 1971b; Hendil and Hoffman, 1974) as well as for cells of euryhaline invertebrates in which free amino acids are important osmotic effectors (Lang and Gainer, 1969; Gerard and Gilles, 1972; Gilles, 1973). However, Schmidt-Nielsen (1975) reported experiments by Cala which show volume readjustment after shrinkage in high K+ flounder erythrocytes. In this case, there is both an increase in Na+ and K+ permeability of the shrunken cells; the increase in permeability to Na+ appears however larger than that to K+ and the readjustment can be ascribed to a net gain of osmotic effectors in which an intracellular gain of Na+ is implicated. In high Na+ vertebrate cells, it seems that some regulatory reswelling occurs in the few hours of *in vitro* experiments. In dog red blood cells, the water gain occurring in hyperosmotic media could be related to an increased Na+ permeability of the shrunken cells (Parker and coworkers, 1975).

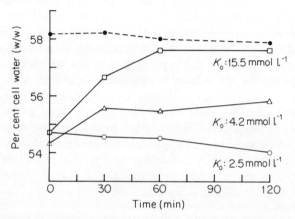

Figure 4 Effect of external K+ on the water content of duck erythrocytes submitted to an hyperosmotic shock ($\Pi_i/\Pi_e = 325/435$); ● - - - ● = water content evolution in control conditions (Π_i: 323 mosmol l⁻¹; K_0: 2.5 mmol l⁻¹. Modified from Kregenow (1974)

Several attempts have been made to obtain volume readjustment after shrinkage in hyperosmotic media. In cells with high amino acid content, addition of such compounds in the external hyperosmotic saline does not induce regulatory reswelling (Gilles, 1973) although an increased uptake of these compounds, related to the increased Na+ content of the external saline, can be observed (Gerard, 1975; Gilles, 1978). Modifications in the Ca²⁺ or Mg²⁺ content of the hyperosmotic saline also seems without significant effect (Gerard, private communication). In duck red cells, Kregenow (1971b) obtained reswelling by increasing the amount of K+ in the incubation medium (fig. 4). In this case, K+ is accumulated against an electro-chemical gradient and there is, as in the flounder's erythrocytes an increase in both Na+ and K+ permeabilities; however, the K+ efflux although elevated remains lower than the K+ influx. This process of K+ accumulation requires external Na+ and since there is no modification of the evolution of the volume readjustment phase upon addition of ouabain, it seems that a Na^+ sensitive but ouabain insensitive active accumulation of K^+ must be involved.

The fact that many cell types show no volume readjustment after shrinkage *in vitro* does not mean that they lack the necessary mechanisms. It would seem rather that these mechanisms are present but are not able to bring about volume readjustment during the short periods of time of the *in vitro* experiments.

Tissues of euryhaline crabs are unable to effect volume readjustment after hyperosmotic shocks *in vitro* (Lang and Gainer, 1969; Gerard and Gilles, 1972; Gilles, 1973), but complete regulation is observed *in vivo* after 15 days of acclimation to a concentrated medium (Gilles, 1977, 1978; Chapter 4).

It would thus appear that in most cases volume readjustment is achieved much more rapidly after swelling than after shrinkage. This may well make sense from a physiological point of view. It may indeed be argued that swelling should be less easily supported by tissues than shrinkage. Distension of membrane structure could produce modifications in the activity of membrane-bound metabolic processes which may not be or may be much less affected by shrinkage. Moreover, swelling may induce leakage and denaturation of intracellular macromolecular constituents if it is not rapidly limited by volume control. This is not the case in the shrinkage process.

4. *Inadequate Aspects of the Na^+–K^+ Exchange Pump System*

If we consider first the volume readjustment process, it has been shown in a variety of tissues and cells that this phenomenon is inhibited by different metabolic inhibitors such as CN^- or 2.4 DNP. However, in most cases, it is not inhibited by ouabain. This is for instance the situation found in cultured mouse lymphoblasts by Roti-Roti and Rothstein (1973), in isolated axons of the crab *C. sapidus* (Gerard, 1975) or in Ehrlich ascites cells (Hendil and Hoffman, 1974). In duck erythrocytes, ouabain has no effect unless these cells are experimentally loaded with sodium. In such condition, however, the volume readjustment observed in the absence of ouabain is no longer a regulatory process since the cells continued to shrink until they reached a minimal volume (Kregenow, 1974). It seems that this shrinkage results from the sodium–potassium exchange pump, responding to the experimentally induced increase in internal sodium. Moreover it appears that in this case, an uncoupling occurs so that more sodium is removed than potassium is taken up.

On the other hand we have seen that the reswelling shown under certain circumstances by duck red cells in hyperosmotic conditions is ouabain independent although it requires active potassium uptake.

In frog skin pretreated with ouabain, there is an inhibition of the volume readjustment response. However this can be explained on the basis of secondary effects of ouabain on permeability rather than by inhibition of the Na^+-K^+ exchange pump (McRobbie and Ussing, 1961).

Up to now there is thus no clear-cut evidence for an essential role of the Na^+-K^+, ouabain sensitive, exchange pump in the volume readjustment process. On the contrary it would appear that this participation is rather limited.

Attention has also been drawn to inadequacies of the Na^+-K^+ pump system in volume maintenance processes as a result of studies on the effect of ouabain on the transport of water and electrolytes in kidney cortex slices and in diaphragm (Kleinzeller and Knotkova, 1964a, 1964b). First of all, the leak and pump hypothesis predicts that the inhibition of the Na^+-K^+ pump by ouabain should lead to a cellular swelling. In kidney slices or isolated tubules such a swelling is only observed when the preparation is incubated in a HCO_3^- buffered saline (Burg, Grollman and Orloff, 1964; Rorive, Nielsen and Kleinzeller, 1972). Secondly, even after inhibition of the Na^+-K^+ pump by ouabain, the cell is able to extrude the excess of water accumulated during incubation at $0\,°C$. As stated before (see p. 90), on immersion in an ice cold isotonic solution of sodium chloride or a balanced physiological saline, the cells swell, taking up water, Na^+ and Cl^- and losing K^+. Upon subsequent incubation at room temperature the cells are capable of extruding H_2O, Na^+ and Cl^- and reaccumulating K^+. These movements of water and ions can be suppressed by a variety of metabolic inhibitors and have been associated with the activity of the Na^+-K^+ pump. However, Kleinzeller and Knotkova (1964a, 1964b) observed that ouabain, even in high concentration, did not inhibit the metabolically dependent extrusion of water as an isotonic solution of NaCl from the cells, but did prevent the K^+ reaccumulation.

These original observations were repeatedly confirmed in kidney tissues (Whittembury, 1968; MacKnight, 1968; Whittembury and Proverbio, 1970; Rorive, Nielsen and Kleinzeller, 1972) as well as in diaphragm (Kleinzeller and Knotkova, 1964b), in smooth muscle (Daniel and Robinson, 1971; Rorive, 1973) and to some extent, in liver slices (Elsehove and von Rossum, 1963; MacKnight, Pilgrim and Robinson, 1974).

Moreover, this metabolically dependent extrusion of water as an isotonic solution appears to be K^+ independent, and not specifically driven by Na^+ movements. Indeed it has been established that a water extrusion can be observed in potassium free media. Willis (1968a) has suggested however that, in such conditions, a small part of the cellular potassium leaking out could be recirculated in the extracellular space, thus providing enough K^+ to activate the Na^+-K^+ pump. The non-specificity of the water extrusion system for Na is demonstrated by the fact that preparations loaded with LiCl, Tris, or choline, are still able to extrude isotonic solutions of these ions, in spite of the fact that they are not actively transported through the Na^+-K^+ ouabain sensitive pump (Kleinzeller, 1972).

5. The ouabain insensitive transport of water

From the key observations very briefly summarized above a functional separation of the Na^+–K^+ exchange pump from the volume controlling processes must be considered. Moreover it appears that ouabain insensitive mechanisms of ionic transport play an important part in these processes. The main physiological characteristics of this water transport system have been obtained initially using preparations of kidney cortex tissues and diaphragm cells and can be summarized as follows:

 (i) The transport of water takes place as a practically isotonic solution of the bulk cellular electrolytes at a constant electrochemical gradient (Kleinzeller and Knotkova, 1964a, 1964b; Whittembury, 1968; Whittembury and Proverbio, 1970).
 (ii) The extrusion of sodium and water from cells previously swollen at 0°C through this ouabain insensitive system is not coupled to reaccumulation of K^+.
 (iii) The ouabain insensitive extrusion of water is dependent on metabolic energy, being completely blocked by uncouplers (e.g. 0.1 mmol l^{-1} dinitrophenol) or anaerobiosis (Kleinzeller and Knotkova, 1964a, 1964b; Whittembury, 1968; Whittembury and Proverbio, 1970; Kleinzeller, 1972).
 (iv) The phenomenon shows a lack of cation specificity (Kleinzeller and Knotkova, 1964a). The tissue is capable of extruding Li^+, Choline or $Tris^+$ to the same extent as Na^+ (Table 1).

Table 1. Effect of 2.4 DNP and of Na^+ substitution on the water content of kidney cortex slices.

Saline	Control	2.4 DNP
Na^+	2.76 ± 0.05	4.52 ± 0.05
Choline	3.06 ± 0.03	3.59 ± 0.05
Tris	2.91 ± 0.05	3.51 ± 0.05

The kidney slices are first leached at 0° C for 2.5 h in one of the salines; they are then aerobically incubated for 1.5 h in the same saline before measurements are made.
The water content is expressed in kg H_2O/kg dry weight (modified from Kleinzeller, 1972).

 (v) The ouabain insensitive extrusion of water and electrolytes is completely blocked by 1.2 mmol l^{-1} ethacrynic acid (Whittembury, 1968; Whittembury and Fishman, 1969; Epstein, 1972a, 1972b; Whittembury and Grantham, 1976). This observation, in conjunction with the inhibition of active Na^+ transport in erythrocytes by this molecule has been interpreted as evidence for a second sodium pump which would be ouabain insensitive and which would operate without K^+ coupling (Whittembury and Proverbio, 1970; Whittembury and Grantham, 1976). This point will be discussed at greater length in another section (see p. 101).

(vi) Moreover, Rorive, Nielsen and Kleinzeller (1972) have demonstrated that the ouabain insensitive control of water and electrolytes in kidney cells is markedly sensitive to variations of external pH and also to external Ca^{2+}. Variations in the saline pH in the range 6.2–8.2 were shown not to affect markedly the steady-state water content of kidney cortex slices. Ouabain and absence of Ca^{2+} produced a marked swelling at pH 8, whereas at lower pH no swelling could be observed in these conditions. At pH 7.2, in the absence of Ca^{2+}, ouabain produced a significant swelling (Fig. 5).

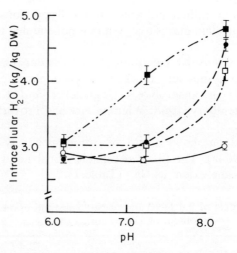

Figure 5 Effect of pH on the water content of kidney cortex slices. ○, control saline; □, Ca^{2+}-free saline; ●, ouabain 0.5 mmol l^{-1}; ■, ouabain 0.5 mmol l^{-1} in Ca^{2+}-free saline. Compiled after data from Rorive and co-workers (1972)

(vii) The ouabain insensitive water transport system may not be present in all cells. In brain cells, considerable swelling was found when the Na^+–K^+ pump was inhibited by ouabain (Pappius, 1964; Bourke and Tower, 1966; Okamoto and Quastel, 1970), indicating that the leak and pump system is an important mechanism implicated in the maintenance of the cell volume in this tissue.

Several hypotheses have been put forward to account for the ouabain insensitive system of ions and water transport.

a. *The cryptic Na^+–K^+ pump hypothesis* For Willis (1968a), the ouabain insensitive fraction of Na^+ extrusion does not necessarily imply some mechanism other than the ordinary cation transport system involving the Na^+–K^+ exchange pump. The effect of K^+ removal or ouabain addition on the Na^+ extrusion observed in kidney cells may be related to the fact that the active surfaces are located upon highly folded surfaces of the cells' membranes. These active sites could thus be in contact with solutions which have a composition somewhat different from that of the bathing medium. The concentration of K^+ or ouabain may therefore be different at the Na^+–K^+ pumping sites located at the outer surface of the membrane and at those sites located deep in the crypts. For instance, Na^+ extrusion in a K^+ free medium can be explained by local recirculation of K^+ in the crypts where the K^+ concentration outside does not really reach zero. Indeed it is not possible to remove all K^+ ions from the extracellular space of kidney cortex slices by a leaching procedure or metabolic inhibition (Willis, 1968b). In the same way, Podevin and Boumendil-Podevin (1973) showed that even with isolated kidney tubules, it is almost impossible to keep the incubation medium completely free from K^+, due to cellular destruction, probably. When recapture of this escaping K^+ is prevented, at least partly by dialysis of tubule suspension against a large amount of K^+ free medium, the extrusion of Na^+ and water becomes significantly lower than in experiments where the tubules are simply incubated in K^+ free saline.

Nevertheless, although the cryptic hypothesis may account for part of the inadequate aspects of the Na^+–K^+ exchange pump system in some tissues, it is far from providing a consistent explanation for many of the observations reported above. Moreover many tissues and cell types on which such inadequacies have been reported do not show important infolding of the plasmatic membranes. Other possibilities have therefore been put forward to account for the ouabain insensitive transport system.

b. *The mechanochemical hypothesis* This hypothesis, originally developed by Kleinzeller and coworkers (see for instance Kleinzeller, 1972), postulates that a mechanochemical system can squeeze out an isotonic solution of intracellular electrolytes. This hypothesis offers the advantage of not requiring any ionic specificity and explains easily the extrusion of an isotonic solution of the intracellular ions. The existence of such a system raises two important correlations:

(1) There must exist some structure in the plasmatic membrane which can induce such a mechanical expulsion of solution.
(2) If this system is the essential one involved in volume maintenance and referring to the Gibbs–Donnan theory, there must be a gradient of hydrostatic pressure across the membrane. This last point does not fit with the now classical observations showing that, in animal cells, there is no significant gradient of hydrostatic pressure (see for instance Conway and McCormack, 1953; Mitchison and Swann, 1955; Rand and Burton, 1964). The measured hydrostatic pressure gradient varies indeed bet-

ween 1 and 5 mm H_2O depending on the cell type used while the osmotic pressure due to the fixed anions could go up to 1 atm in a Gibbs–Donnan equilibrium (Hoffmann, 1977). This of course excludes the possibility for the mechanochemical system to be the main mechanism of volume maintenance. It remains however quite possible that such a system may play a role in the volume readjustment following swelling. It may also be important in limiting the swelling phase during hypo-osmotic shocks in order to avoid a too important distension of the membrane structure. In connection with this, it is interesting to point out that incubation of human red cells in the absence of metabolizable substrate in order to deplete them of ATP, produced a marked increase in the membrane deformability for hydrostatic pressures between 4 mm H_2O and 60 mm H_2O (Weed, La Celle and Merril, 1969). This phenomenon has been attributed by Weed, La Celle and Merril (1969) to a ATP–Ca^{2+} dependent sol–gel change at the interface between the membrane and the cell interior. These results can be related to other observations dealing with interactions of Ca^{2+} and ATP with the cell membrane and volume. In Ehrlich cells, addition of ATP to an incubation medium free of Ca^{2+} and Mg^{2+} produces a marked but reversible swelling (Gasic and Steward, 1968; Steward, Gasic and Hempling, 1969). Addition of ATP also produces a marked swelling of isolated kidney tubules incubated in Ca^{2+} free saline (Rorive and Kleinzeller, 1972). Moreover, a Ca^{2+} activated ATPase can be isolated from the microsomial fraction of this preparation which precipitates at pH 7.2 in the presence of ATP and Ca^{2+} (Rorive and Kleinzeller, 1972). This precipitation phenomenon is reminiscent of the superprecipitation described for actomyosin. A Ca^{2+} activated ATPase which shows superprecipitation has also been isolated from erythrocytes ghosts. This protein can form fibrils in the presence of ATP and Ca^{2+} (Ohmishi, 1962; Rosenthal, Kregenow and Moses, 1970; Palek, Curby and Lionetti, 1971).

Actin-like contractile proteins have now been isolated from membranes of many cell types (Ohmishi, 1962; Marchesi and coworkers, 1969; Tillack and coworkers, 1970; Palek, Curby and Lionetti, 1971; Bettex-Galland, Probst and Behnke, 1972; Aiton and Lamb, 1975; see Marx, 1975 for review). In *E. Coli,* the role of such an actin like contractile protein in ion transport is quite clear. Indeed it shows deficient action and reduced presence in the membranes of K^+ transport deficient mutant strain (Minkoff and Damadian, 1976). In Ehrlich cells, the reversible swelling induced by addition of ATP to a Ca^{2+}, Mg^{2+} free medium is associated with net movements of Na^+ and K^+ (Steward, Gasic and Hempling, 1969). On the other hand, Lew (1971) has shown that the level of ATP in erythrocyte ghosts can affect the permeability to K^+ apparently by controlling the uptake of Ca^{2+}.

From the facts at hand up to now, it may therefore be tentatively concluded that actin-like contractile proteins showing Ca^{2+} dependent ATPase activity are of general occurrence in plasmatic membranes of animal cells and that they are implicated in the volume control. They may be responsible for the expulsion of the isotonic solution of intracellular osmotic effectors postulated in the mechanochemical hypothesis. However, as recently proposed by Hoffmann (1977) these proteins might be connected rather with local, contractile processes in the membrane triggering changes in cation permeability and thus cell volume control. In such a view, these proteins would represent the molecular support of the configuration changes in membrane structure we evoked previously (see p. 93) to account for the ionic permeability modulation occurring during cell volume adjustment.

c. *The ouabain insensitive Na^+ pump* The possibility of the existence of a so-called second Na^+ pump independent of K^+ and ouabain insensitive has also been considered to account for the inadequacies of the Na^+–K^+ exchange pump in volume control in kidney tissue (Whittembury and Fishman, 1969; Whittembury and Proverbio, 1970; Law, 1976). The main support for this hypothesis is that ethacrynic acid inhibits mostly the expulsion of Na^+ and water but has little effect on the K^+ uptake (Whittembury, 1968; Whittembury and Proverbio, 1970). The specificity of the ethacrynic acid effect has however been questioned by various workers who considered that this diuretic does not act exclusively at the membrane level. In their view, the effect of ethacrynic acid could be related to actions of this compound other than direct inhibition of a Na^+ pump such as inhibition of glycolysis and electron transport chain and reduction of the cellular ATP level (Kramar and Kaiser, 1970; Klahr and coworkers, 1971). Studying the binding of ethacrynic acid to rabbit kidney cortex, Epstein (1972a, 1972b) also came to the conclusion that the major part of the binding is not specific to membrane active transport sites.

However, an ouabain insensitive Na^+ and K^+ stimulated ATPase has been described recently in kidney membrane preparation. This enzyme is inhibited by ethacrynic acid and furosemide; ethacrynic acid completely inhibits the enzyme at a concentration (2 mmol l^{-1}) which decreases only by 30% the activity of the Mg^{2+} dependent ATPase and by some 50% the (Na^+–K^+) stimulated ATPase. The activity of this Na^+ ATPase might be related to the ouabain insensitive transport of Na^+ and water (Proverbio, Condrescu-Guidi and Whittenbury, 1975).

The operations of a so-called second Na^+ pump in volume control has also been postulated in tissues other than kidney (Daniel and Robinson, 1971; MacKnight, Pilgrim and Robinson, 1974; MacKnight, Civan and Leaf, 1975). Some observations however suggest that the ouabain insensitive transport of Na^+ could remain under the dependency of the Na^+–K^+ exchange pump which could present different properties under different circumstances (Sachs, 1971), and attempts may be made to relate the ouabain insensitive Na^+ extrusion with the phosphorylated form of the Na^+–K^+ ATPase which has been reported to be present under certain circumstances in red cell membranes and in other tissue preparations (Charnock, Rosenthal and Post, 1963; Albers, 1967; Blostein, 1968). In such a view, the molecular basis of the physiologically different Na^+ active movements would be the same, the change from one form to the other being controlled by some volume dependent allosteric modulation of the structure implicated.

d. *The ouabain insensitive K^+ pump* The possibility of the existence of an ouabain insensitive active accumulation of K^+ has been considered to account for the regulatory reswelling observed in duck erythrocytes under certain circumstances, this active movement of K^+ being sensitive to the external Na^+ level (see p. 94). Schmidt and McManus (1974) described a quite similar movement of K^+ which is inhibited by furosemide.

These observations could be accounted for by the existence of an ouabain insensitive K^+ pump. It is worth noticing that this pump is furosemide sensitive as seems to be the so-called second Na^+ pump also. Could it be then that both the K^+ and Na^+ independent active movements are the expressions of a discoupling due to some volume dependent configuration change of the Na^+–K^+ exchange pump? Obviously, many more results are needed to assess this hypothesis.

IV. CONCLUSIONS

To summarize briefly, it is clear that specific mechanisms implicating active control of the intracellular level of Na^+ and K^+ play an essential role in the cell volume control processes, i.e. cell volume maintenance in steady-state conditions and cell volume readjustment under osmotic stress. The basis of the control of the ionic movements at work in volume control are however far from being understood. Obviously, these movements cannot be ascribed to the activity of a unique mechanism; rather they result from the conjugated effects of processes controlling both the passive and active fluxes of ions. Although many more data would be required to produce a clear picture of the events taking place, the results presented in this review can bring us to several tentative conclusions. They have been summarized in a schematic form in Fig. 6.

Cell volume appears to be controlled by a leak and pump system, the activity of which can be regulated by some volume dependent process.

Cell volume maintenance under steady-state conditions would be essentially achieved by pumping processes, counterbalancing the passive ionic movements and the Gibbs–Donnan forces in order to keep the intracellular fluid and the extracellular fluid close to isosmotic equilibrium, thus avoiding water movements. The activity of an Na^+–K^+ exchange pump appears to be an important mechanism in this process.

Cell volume readjustment following either swelling or shrinkage due to osmotic shocks or to metabolic inhibition would primarily involve modifications in the leak pathways, the specific evolution of the level of both intracellular K^+ and Na^+ being controlled by independent Na^+ and K^+ pumps. The activity of these pumps might come from some volume dependent configuration modulation of the membrane structures responsible for the activity of the Na^+-K^+ exchange pump. As far as the leak pathways are concerned, it seems that both diffusional and specific, volume controlled, channels for Na^+ and K^+ must be considered.

Moreover, it appears that volume dependent conformation changes in membrane contractile elements showing ATPase activity are also implicated

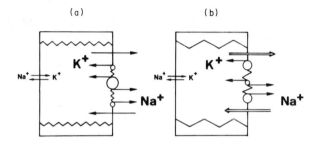

Figure 6 Schematic representation of mechanisms implicated in cell volume control processes. (a) volume maintenance in steady-state conditions. (b) volume readjustment following osmotic stress. ∼, membrane contractile elements related to the leak pathways; →, Na^+ or K^+ leak pathway (more important role in scheme (b) than in scheme (a); → in (a) thus becomes ⇒ in (b); ⇌: diffusional pathway; ⁀Q⁀ Na^+-K^+ exchange pump; O⁻ Na^+ or K^+ specific pump (more important role in scheme (b) than in (a); O⁻ in (a) thus becomes O⁻ in (b). The ionic fluxes are considered to be controlled by directly or indirectly volume dependent, configuration changes in the structures ruling their movements (∼ becomes ⁀⁀).
The exchange pump and the specific pumps may be different configurational expressions of a same molecular structure

in the process. These elements might be involved during volume readjustment following swelling in squeezing the cell thus leading to an expulsion of an isotonic solution of water and electrolytes as proposed in the mechanochemical hypothesis. However, they might be related rather to local, contractile processes triggering changes in cation permeability.

The proposed mechanisms of volume readjustment are by no means mutually exclusive; more results should be needed notably about cation binding processes and water structuration before a more complete picture can be produced. Nevertheless, volume regulatory readjustment appears as a mechanism which plays an important role in the physiology of organisms and probably also in some pathological situations. As stressed in this treatise, many organisms, invertebrates as well as vertebrates, do not possess or have only weak capabilities of blood osmoregulation. These species will thus rely entirely or essentially on their cellular volume regulation processes to ensure survival in adverse osmotic conditions. If we now consider species with very effective blood osmolarity control such as mammals, it is known that the mechanism involved in this process can be overwhelmed under certain circumstances or may show a certain time lag before responding to a new situation. In these cases, the cell volume adjustment processes will be sollicited to avoid bursting of the cells. Cell volume control is also most probably a mechanism by which medullary kidney cells can adapt to the sometimes very high osmolarity of their extracellular space and this mechanism is probably also involved in cell resistance in pathological situations such as sodium depletion or dehydration.

REFERENCES

Aiton, J. F., and J. F. Lamb (1975). The transient effect of ATP on K movement in Hela cells. *J. Physiol.*, **248**, 14–15.

Albers, R. W. (1967). Biochemical aspects of active transport. *A. Rev. Biochem.*, **36**, 727–56.

Bettex-Galland, M., E. Probst and O. Behnke (1972). Complex formation with heavy meromyosin of the isolated actin-like protein from blood platelets. *J. Molec. Biol.*, **68**, 553–5.

Blostein, R. (1968). Relationship between erythrocyte membrane phosphorylation and adensoine triphosphate hydrolysis. *J. Biol. Chem.*, **243**, 1957–65.

Bourke, R. S., and D. B. Tower (1966). Fluid compartmentation and electrolytes of cat cerebral cortex *in vitro*. I. Swelling and solute distribution in mature cerebral cortex. *J. Neurochem.*, **13**, 1071–97.

Burg, M. B., E. F. Grollman, and J. Orloff (1964). Sodium and potassium flux of separated renal tubules. *Am. J. Physiol.*, **206**, 483–91.

Charnock, J. S., A. S. Rosenthal, and R. L. Post (1963). Studies in the mechanism of cation transport. II. A phosphorylated intermediate in the cation stimulated enzymatic hydrolysis of adenosine triphosphate. *Aust. J. Exp. Biol. Med. Sci.*, **41**, 675–86.

Conway, E. J. (1957). Nature and significance of concentration relations of potassium and sodium ions in skeletal muscle. *Physiol. Rev.*, **37**, 84–132.

Conway, E. J. and J. I. McCormack (1953). The total intracellular concentration of mammalian tissues compared with that of the extracellular fluid. *J. Physiol.*, **120**, 1–14.

Cook, J. S. (1967). Non-solvent water in human erythrocytes. *J. Gen. Physiol.* **50**, 1311–25.

Cope, F. W. (1970). The solid state physics of electron and ion transport in biology. *Adv. Biol. Med. Phys.*, **13**, 1–42.

Cope, F. W. (1976). A primer of water structuring and cation association in cells. I. Introduction: the big picture. *Physiol. Chem. Phys.*, **8**, 479–83.

Damadian, R., M. Goldsmith, and K. S. Zaner (1971). Biological ion exchange resins. II. Querp water and ion exchange selectivity. *Biophys. J.*, **11**, 761–72.

Daniel, E., and K. Robinson (1971). Effects of inhibitors of active transport on ^{22}Na

and ^{42}K movements and nucleotide levels in rat uteri at 25 °C. *Can J. Physiol. Pharmacol.,* **49,** 178–204.
De Bruine, A. W., and J. van Steveninck (1970). Apparent nonsolvent water and osmotic behaviour of yeast cells. *Biochim. Biophys. Acta,* **196,** 45–52.
Dellasega, M., and J. J. Grantham (1972). Regulation of renal tubule cell volume in hypotonic media. *Am. J. Physiol.,* **224,** 1288–94.
Dick, D. A. T. (1970). Water movements in cells. In E. E. Bitton (Ed.) *Membranes and Ion Transport,* Vol. 3. Wiley Interscience, London. pp. 211–50.
Dick, D. A. T., and S. G. A. McLaughlin (1969). The activities and concentrations of sodium and potassium in toad oocytes. *J. Physiol. (Lond.),* **205,** 61–78.
Dunham, P. B., and R. B. Lunn (1972). Adenosine triphosphatase and active cation transport in red blood cell membranes. *Arch. Int. Med.,* **129,** 241–7.
Elsehove, A., and L. D. V. von Rossum (1963). Net movements of sodium and potassium and their relation to respiration in slices of rat liver incubated *in vitro. J. Physiol. (Lond.),* **168,** 531–53.
Epstein, R. W. (1972a). The binding of ethacrynic acid to rabbit kidney cortex. *Biochim. Biophys. Acta,* **274,** 119–27.
Epstein, R. W. (1972b). The effects of ethacrynic acid on active transport of sugars and ions and other metabolic processes in rabbit kidney cortex. *Biochim. Biophys. Acta,* **274,** 128–39.
Evans, E. A., and R. M. Hochmuth (1976a). Membrane viscoelasticity. *Biophys. J.,* **16,** 1–11.
Evans, E. A., and R. M. Hochmuth (1976b). Membrane viscoplastic flow. *Biophys. J.,* **16,** 13–26.
Florkin, M. (1962). La régulation isosmotique intracellulaire chez les invertébrés marins euryhalins. *Bull. Acad. R. Belg. Cl. Sci.,* **48,** 687–94.
Foster, R. E. (1971). The transport of water in erythrocytes. In F. Bronner and A. Kleinzeller (Eds) *Current Topics in Membranes and Transport,* Vol. 2. Academic Press, New York. pp. 41–98.
Fugelli, K. (1967). Regulation of cell volume in flounder (*Pleuronectes flesus*) erythrocytes accompanying a decrease in plasma osmolarity. *Comp. Biochem. Physiol.,* **22,** 253–60.
Gardner, J. D., and T. P. Conlon, (1972). The effects of sodium and potassium on ouabain binding by human erythrocytes. *J. Gen. Physiol.,* **60,** 609–29.
Gary-Bobo, C. M., and A. K. Solomon (1968). Properties of hemoglobin solutions in red cells. *J. Gen. Physiol.,* **52,** 825–53.
Gasic, G., and C. Steward (1968). Cell volume regulation in mouse TA$_3$ ascites tumor cells by exogenous ATP as measured by the Coulter counter. *J. Cell. Physiol.,* **71,** 239–42.
Gerard, J. F. (1975). Volume regulation and alanine transport. Response of isolated axons of *Callinectes sapidus* Rathbun to hypo-osmotic conditions. *Comp. Biochem. Physiol.,* **51A,** 225–9.
Gerard, J. F., and R. Gilles (1972). Modifications of the amino-acid efflux during the osmotic adjustment of isolated axons of *Callinectes sapidus. Experientia,* **28,** 863–4.
Giebisch, G., and E. Windhager (1973). Electrolyte transport across renal tubular membranes. In J. Orloff and R. Berliner (Eds) *Renal Physiology,* American Society of Physiologists. pp. 315–76.
Gilles, R. (1973). Osmotic behaviour of isolated axons of a euryhaline and a stenohaline crustacean. *Experientia,* **29,** 1354–5.
Gilles, R. (1974). Métabolisme des acides aminés et contrôle du volume cellulaire. *Arch. Int. Physiol. Biochem.,* **82,** 423–589.
Gilles, R. (1977). Effects of osmotic stresses on the proteins concentration and

pattern of *Eriocheir sinensis* blood. *Comp. Biochem. Physiol.*, **56A**, 109–14.
Gilles, R. (1978). Intracellular free amino acids and cell volume regulation during osmotic stresses. In *Osmotic and Volume Regulation*, A. Benzon Symposium XI. Munksgaard, Copenhagen, pp. 471–491.
Glynn, I. M., and J. F. Hoffman (1971). Nucleotides requirements for sodium–sodium exchange catalysed by the sodium pump in human red cells. *J. Physiol. (Lond.)*, **218**, 239–56.
Glynn, I. M., V. L. Lew, and U. Luthi (1970). Reversal of the potassium entry mechanism in red cells with and without reversal of the entire pump cycle. *J. Physiol. (Lond.)*, **207**, 371–91.
Gregor, H. P. (1948). A general thermodynamic theory of ion exchange processes. *J. Am. Chem. Soc.*, **70**, 1293.
Hagameijer, F., G. Rorive, and E. Schoffeniels (1965). Exchange of ^{24}Na and ^{42}K in rat aortic smooth muscle fibres. *Life Sciences*, **4**, 2141–9.
Hendil, K. B., and E. K. Hoffmann (1974). Cell volume regulation in Ehrlich ascites tumor cells. *J. Cell. Physiol.*, **84**, 115–26.
Hinke, J. A. M. (1970). Solvent water for electrolytes in the muscle fiber of the giant barnacle. *J. Gen. Physiol.*, **56**, 521–41.
Hoffman, J. F. (1969). The interaction between tritiated ouabain and the Na–K pump in red blood cells. *J. Gen. Physiol.*, **54**, 343S–50S.
Hoffmann, E. K., and K. B. Hendil (1976). The role of amino-acids and taurine in isosmotic intracellular regulation in Ehrlich ascites mouse tumour cells. *J. Comp. Physiol.*, **108**, 279–86.
Hoffmann, E. K. (1977). Control of cell volume. In B. L. Gupta, R. B. Moreton, L. Oschman, and B. J. Wall (Eds) *Transport of Ions and Water in Animals*, Academic Press, New York. pp. 285–332.
House, C. R. (1974). Water transport in cells and tissues. Edward Arnold, London.
Kahlenberg, A., P. R. Galsworthy, and L. L. Hoken (1967). Sodium–potassium adenosine triphosphatase: Acylphosphate intermediate shown by L-glutamyl-Y-phosphate. *Science*, **157**, 434–6.
Keynes, R. D., and R. A. Steinhardt (1968). The components of the sodium efflux in frog muscle. *J. Physiol. (Lond.)*, **198**, 581–99.
Klahr, S., J. J. Bourgoignie, Y. Yates and M. S. Bricker (1971). Inhibition of glycolysis by ethacrynic acid and furosemide. *Fed. Proc.*, **30**, 608.
Kleinzeller, A. (1972). Cellular transport of water. In L. E. Hokin (Ed.) *Metabolic Pathways*, Academic Press, New York. pp. 91–131.
Kleinzeller, A., and A. Knotkova (1964a). The effect of ouabain on the electrolyte and water transport in kidney cortex and liver slices. *J. Physiol. (Lond.)*, **175**, 172–92.
Kleinzeller, A., and A. Knotkova (1964b). Electrolyte transport in rat diaphragm. *Physiol. Bohem.*, **13**, 317–26.
Kolata, G. N. (1976). Water structure and ion binding: a role in cell physiology. *Science*, **192**, 1220–2.
Kotyk, A., and A. Kleinzeller (1963). Transport of D-xylose and sugar space in baker's yeast. *Folia Microbiol. (Prague)*, **8**, 156–64.
Kramar, R., and F. Kaiser (1970). Hemmung von SH-enzyman durch ethacrynsaüre. *Experientia*, **26**, 485–6.
Kregenow, F. M. (1971a). The response of duck erythrocytes to hypotonic media. Evidence of a volume control mechanism. *J. Gen. Physiol.*, **58**, 372–95.
Kregenow, F. M. (1971b). The response of duck erythrocytes to hypertonic media. Further evidence for a volume controlling mechanism. *J. Gen. Physiol.*, **58**, 396–412.
Kregenow, F. M. (1974). Functional separation of the Na–K exchange pump from the

volume controlling mechanism in enlarged duck red cells. *J. Gen. Physiol.*, **64**, 393–412.
Lang, M. A., and H. Gainer (1969). Isosmotic intracellular regulation as a mechanism of volume control in crab muscle fibers. *Comp. Biochem. Physiol.*, **30**, 445–6.
Law, R. O. (1976). The effects of ouabain and ethacrynic acid on the intracellular sodium and potassium concentrations in renal medullary slices incubated in cold potassium-free ringer solution and re-incubated at 37 °C in the presence of external potassium. *J. Physiol. (Lond.)*, **254**, 743–58.
Leaf, A. (1959). Maintenance of concentration gradients and regulation of cell volume. *Ann. NY Acad. Sci.*, **72**, 396–404.
Lev, A. A. (1964). Determination of activity and activity coefficients of potassium and sodium ions in frog muscle fibers. *Nature*, **201**, 1132–4.
Lew, V. L. (1971). Effect of ouabain on the Ca^{2+} dependent increase in K^+ permeability in depleted guinea-pig red cells. *Biochem. Biophys. Acta*, **249**, 236–9.
MacKnight, A. D. C. (1968). Water and electrolyte contents of rat renal cortical slices incubated in potassium free media and media containing ouabain. *Biochem. Biophys. Acta*, **150**, 263–70.
MacKnight, A. D. C., M. M. Civan, and A. Leaf (1975). Some effects of ouabain on cellular ions and water in epithelial cells of toad urinary bladder. *J. Membrane Biol.*, **20**, 387–401.
MacKnight, A. D. C., J. P. Pilgrim, and B. Robinson (1974). The regulation of cellular volume in liver slices. *J. Physiol. (Lond.)*, **238**, 279–94.
Marchesi, S. L., G. Steers, V. T. Marchesi, and T. W. Tillack (1969). Physical and chemical properties of a protein isolated from red cell membranes. *Biochemistry*, **9**, 50–7.
Marx, J. L. (1975). Actin and myosin: Role in non muscle cells. *Sciences*, **189**, 34–7.
McRobbie, E. A. C., and H. H. Ussing (1961). Osmotic behaviour of the epithelial cells of frog skin. *Acta Physiol. Scand.*, **53**, 348–65.
Mela, H. J. (1968). Elastic-mathematical theory of cells and mitochondria in swelling process. II. Effect of temperature upon modulus of elasticity of membranous material of egg cells of sea urchin *Strongylocentrotus purpuratus*, and of oyster, *Crassostrea virginica*. *Biophys. J.*, **8**, 83–97.
Minkoff, L., and R. Damadian (1976). Biological ion exchanger resins. X. The cytotonus hypothesis: Biological contractility and the total regulation of cellular physiology through quantitative control of cell water. *Physiol. Chem. Phys.* **8**, 349–87.
Mitchison, J. M., and M. M. Swann (1955). The mechanical properties of the cell surface. III. The sea-urchin egg from fertilization to cleavage. *J. Exp. Biol.*, **32**, 734–50.
Ohmishi, T. (1962). Extraction of actin and myosin like proteins from erythrocyte membrane. *J. Biochem. (Tokyo)*, **52**, 307–8.
Okamoto, K., and J. M. Quastel (1970). Water uptake and energy metabolism in brain slices from the rat. *Biochem. J.*, **120**, 25–36.
Orringer, E. P., and J. C. Parker (1973). Ion and water movements in red blood cells. *Prog. Hematology*, **8**, 1–23.
Palek, J., W. A. Curby, and F. J. Lionetti (1971). Effects of calcium and adenosine triphosphate on volume of human red cell ghosts. *Am. J. Physiol.*, **220**, 19–26.
Pappius, H. M. (1964). Water transport at cell membranes. *Can. J. Biochem.*, **42**, 945–53.
Parker, J. C., and J. F. Hoffman (1976). Influences of cell volume and adrenalectomy on cation flux in dog red blood cells. *Biochim. Biophys. Acta*, **433**, 404–8.
Parker, J. C., H. J. Gitelman, P. S. Glosson, and D. L. Leonard (1975). Role of calcium in volume regulation by dog red blood cells. *J. Gen. Physiol.*, **65**, 81–96.

Podevin, R. A., and E. F. Boumendil–Podevin (1973). Effects of temperature, medium K^+, ouabain and ethacrynic acid on transport of electrolyte and water by separated renal tubules. *Biochim. Biophys. Acta,* **282,** 234–49.

Ponder, E. (1948). Volume changes in hemolytic systems containing resorcinol, taurocholate and saponin. *J. Gen. Physiol.,* **31,** 325–35.

Posnansky, M., and A. K. Solomon (1972). Effect of cell volume on potassium transport in human red cells. *Biochim. Biophys. Acta,* **274,** 111–8.

Proverbio, F., and J. F. Hoffman (1972). Differential behaviour of the Mg ATPase and the Na-MgATPase of human red cell ghosts. *Fed. Proc.,* **31,** 215.

Proverbio, F., M. Condrescu-Guidi, and G. Whittembury (1975). Ouabain-insensitive Na^+ stimulation of an Mg^{2+}-dependent ATPase in kidney tissue. *Biochim. Biophys. Acta,* **394,** 281–292.

Rand, R. P., and A. C. Burton (1964). Mechanical properties of the red cell membrane. Membrane stiffness and intracellular pressure. *Biophys. J.,* **4,** 115–35.

Rorive, G. (1973). Mécanismes impliqués dans le contrôle de la teneur en sodium de l'aorte de rat. *Arch. Int. Physiol. Biochem.,* **81,** 771–5.

Rorive, G. (1975). *Parois artérielles et Hypertension Expérimentale.* Masson, Paris.

Rorive, G., and A. Kleinzeller (1972). The effect of ATP and Ca^{2+} on the cell volume in isolated kidney tubules. *Biochim. Biophys. Acta,* **274,** 226–39.

Rorive, G., R. Nielsen, and A. Kleinzeller (1972). Effect of pH on the water and electrolyte content of renal cells. *Biochim. Biophys. Acta,* **266,** 376–96.

Rosenberg, H. M., B. B. Shank, and E. C. Gregg (1972). Volume changes of mammalian cells subjected to hypotonic solutions *in vitro*: evidence for the requirement of a sodium pump for the shrinking phase. *J. Cell. Physiol.,* **80,** 23–32.

Rosenthal, A. S., F. M. Kregenow, and H. L. Moses (1970). Some characteristics of a Ca^{2+} dependent ATPase. Activity associated with a group of erythrocyte membrane proteins which form fibrils. *Biochim. Biophys. Acta,* **196,** 254–62.

Roti-Roti, L. W., and A. Rothstein (1973). Adaptation of mouse leukemic cells (L.R178 Y) to anisosmotic media. *Expt. Cell Res.,* **79,** 295–310.

Rybova, R. (1965). The free and xylose space in the rat diaphragm and the effect of potassium on it. *Physiol. Bohem.,* **14,** 412–6.

Sachs, J. R. (1971). Ouabain insensitive sodium movements in the human red blood cell. *J. Gen. Physiol.,* **57,** 259–82.

Savitz, D., V. W. Sidel, and A. K. Solomon (1964). Osmotic properties of human red cells. *J. Gen. Physiol.,* **48,** 79–94.

Schmidt-Nielsen, B. (1975). Comparative physiology of cellular ion and volume regulation. *J. Exp. Zool.,* **194,** 107–220.

Schmidt, W. F., and T. J. McManus (1974). A furosemide sensitive cotransport of Na^+ plus K into duck red cells activated by hypertonicity or catecholamines. *Fed. Proc.,* **33,** 1457.

Shank, B. B., H. M. Rosenberg, and C. Horowitz (1973). Ionic basis of volume regulation in mammalian cells following osmotic shock. *J. Cell Physiol.,* **82,** 257–66.

Siebert, L., and H. Langendorf (1970). Ionen hausholt in Zellkerm. *Naturwissenschaften,* **57,** 119–24.

Sigler, K., and K. Janáček (1971). The effect of non electrolyte osmolarity on frog oocytes. I. Volume changes. *Biochim. Biophys. Acta,* **241,** 528–38.

Skou, J. C. (1965). Enzymatic basis for active transport of Na^+ and K^+ across cell membrane. *Physiol. Rev.,* **45,** 596–617.

Steward, C. C., G. Gasic, and H. G. Hempling (1969). Effect of exogenous ATP on the volume of TA3 ascites tumor cells. *J. Cell. Physiol.,* **73,** 125–32.

Tillack, T. W., S. T. Marchesi, V. T. Marchesi, and G. Steers (1970). A comparative study of spectrin: a protein isolated from red blood cell membranes. *Biochim. Biophys. Acta,* **200,** 125–31.

Tosteson, D. C., and J. F. Hoffman (1960). Regulation of cell volume by active cation transport in high and low potassium sheep cells. *J. Gen. Physiol.*, **44**, 169–94.
Weed, R. I., P. L. La Celle, and E. W. Merril (1969). Metabolic dependence of red cell deformability. *J. Clin. Invest.*, **48**, 795–809.
Willis, J. S. (1968a). The interaction of K^+, ouabain, and Na^+ on the cation transport and respiration of renal cortical slices. *Biochim. Biophys. Acta*, **163**, 516–30.
Willis, J. S. (1968b). Water and electrolyte contents of rat renal cortical slices of ground squirrels and hamsters. *Biochim. Biophys. Acta*, **163**, 506–15.
Whittam, R. (1969). Transport and diffusion in red blood cells. In *The Physiological Society Series*, No. 13, Edward Arnold, London.
Whittam, R., and R. E. Davies (1954). Relations between metabolism and the rate of turnover of sodium and potassium in guinea pig kidney cortex slices. *Biochem. J.*, **56**, 445–53.
Whittam, R., and J. S. Willis (1963). Ion movements and oxygen consumption in kidney cortex slices. *J. Physiol. (Lond.)*, **168**, 158–77.
Whittembury, G. (1965). Sodium extrusion and potassium uptake in guinea pig kidney cortex slices. *J. Gen. Physiol.*, **48**, 699–717.
Whittembury, G. (1968). Sodium and water transport in kidney proximal tubular cells. *J. Gen. Physiol.*, **51**, 303S–14S.
Whittembury, G., and J. Fishman (1969). Relation between cell Na extrusion and transtubular absorption in the perfused total kidney: the effect of K, ouabain and ethacrynic acid. *Pflügers Arch. Ges. Physiol.*, **307**, 138–53.
Whittembury, G., and J. J. Grantham (1976). Cellular aspects of renal sodium transport and cell volume regulation. *Kidney Int.*, **9**, 103–20.
Whittembury, G., and F. Proverbio (1970). Two modes of Na extrusion in cells from guinea-pig kidney cortex slices. *Pflügers Arch. Ges. Physiol.*, **316**, 1–25.

Chapter 4

Intracellular Organic Osmotic Effectors

R. GILLES

I. Introduction	111
II. Amino Acids as Intracellular Osmotic Effectors	113
III. Mechanisms of Control of the Amino-acid Pool	119
A. Mechanisms at the cellular level	119
B. Mechanisms in the whole animal	139
IV. Conclusions	146
Acknowledgements	148
References	148

I. INTRODUCTION

An important condition for the liberation of organisms from the constraints of the external medium is the possibility of maintaining intracellular medium suitable for the various molecular interactions, which are the support of life, to take place. In this context, cell volume regulation appears as a fundamental mechanism.

It is clear from our knowledge of solute transport processes at the membrane level, solute concentration on both sides of biological membranes and physicochemical state of water that the relation between the thermodynamical activity of water inside and outside the cell cannot be maintained only by a simple transfer of water across the plasmatic membrane.

Several research works have shown that the intracellular fluid is isosmotic or only slightly hyperosmotic to the extracellular fluid (Conway and McCormak, 1953; Appelboom and coworkers, 1958; Maffly and Leaf, 1959; see also Gilles, 1974a for a review). This is true regardless of the presence of anionic molecules inside the cell too large to pass through the membrane and which should bring about an osmotic pressure difference between the inside and the outside of the plasma membrane (Gibbs–Donnan equilibrium; see Chapters 1, 2 and 3). From the Gibbs–Donnan theory, it follows that animal cells which have easily distensible plasma membranes should necessarily undergo swelling and lysis unless some active mechanisms are present to maintain the activity of water in the intracellular medium close to that of the extracellular fluid; these mechanisms have been termed by Florkin (1962) mechanisms of intracellular fluid isosmotic regulation. They appear to be of general occurrence in animal cells and are not only at work to maintain the cellular volume despite the osmolarity difference induced by the Gibbs–Donnan equilibrium but are also of importance in the volume regulation response that many cells are able to achieve following changes in the osmolarity of their external medium. This cell volume regulation behaviour has been described in a recent review by Hoffmann (1977).

A priori, several mechanisms can be involved in the cell volume control which occurs in tissues withstanding osmotic stress. Some of them are intracellular while others may be dependent on relations between the intra- and extracellular fluids. As far as this last category of mechanisms is concerned, the plasma membrane could be impermeable to some ions of the external medium in the same way as it is impermeable to large anions of the intracellular fluid. Under such conditions, the cell would be in a 'double Donnan equilibrium' which might contribute to the removal of an osmolarity difference between the inside and outside media. Van Slyke and coworkers (1925) and Boyle and Conway (1941) published equations for such a system based on an impermeability for Na^+. Such an impermeability for Na^+ cannot of course be retained any longer. However, it may be that some blood macromolecules are implicated in a 'double Donnan equilibrium'. To our knowledge this possibility has never been investigated in detail although some recent studies show modifications in the protein content of the blood serum of animals subjected to osmotic stresses (see p. 142). Nevertheless, such a mechanism, if it exists, should play only a minor part in the cellular volume control process. Cell volume regulation can indeed be observed in isolated tissues incubated in artificial salines containing only ions to which the plasma membrane is permeable (see below).

Another phenomenon concerned with extracellular–intracellular fluid interactions is related to the evolutive decrease in the ratio of the volume of extracellular fluid to the volume of intracellular fluid (V_e/V_i). As discussed at length in another review (Gilles, 1974a), this evolutive process is of importance in decreasing the magnitude of the changes in volume observed following a given modification in the osmolarity of the extracellular fluid. No doubt, the evolutive decrease in V_e/V_i helps the mechanisms of cell volume regulation; however it cannot be considered as an active process of volume control.

Among the mechanisms of strictly intracellular origin, two main possibilities have been considered up to now:

(i) An hydrostatic pressure could be initiated in the cell in such a way that osmotic flow of water could be prevented. Such a mechanism appears to be at work in bacteria and plant cells which have rigid cell walls, but seems to be absent in animal cells. However Kleinzeller (1965) has suggested that development of an hydrostatic pressure due to contractile elements of the plasmatic membranes contributes to the maintenance of cell volume in some animal cells. Up to now, this theory has never been confirmed but has never really been proven incorrect either (for more detailed discussion, refer to Chapter 3 and to the review by Hoffmann, 1977).

(ii) The amount of various intracellular osmotic effectors could be regulated in such a way that the intracellular fluid always remains close to isosmotic equilibrium with the extracellular fluid. The compounds which first appeared as potent osmotic effectors are the inorganic ions Na^+ and K^+. Leaf (1959) and also Tosteson and Hoffmann (1960) were the first to consider the mechanisms of ion transport in the light of a volume control process. This concept has been very successful and the main lines of thought in this area are discussed in Chapter 3. It is only recently that some organic molecules have been studied in relation to their possible role as intracellular osmotic effectors. Among these, the free amino acids appear of particular importance in animal cells. This chapter is thus essentially devoted to the study of the role that amino acids play as osmotic effectors and to the study of the mechanisms implicated in the regulation of their intracellular concentration.

II. AMINO ACIDS AS INTRACELLULAR OSMOTIC EFFECTORS

Excepting some insects and fishes, the osmolarity of the body fluids in animals is essentially made up by inorganic ions. However, as early as 1901

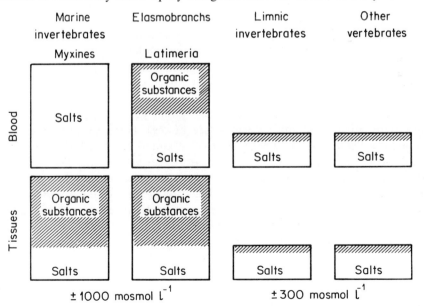

Figure 1 Osmolarity and osmotic effectors in blood and tissues of various animals

Fredericq noticed that the total tissue concentration in organic ions is lower than that in the surrounding extracellular fluid. Figure 1, inspired and extended from an original drawing by Fredericq shows that the cell's osmotic deficit is compensated by organic substances. It is worth noting that, in the zoological groups studied, the amount of intracellular inorganic ions (salts) remains at a very similar, relatively low level. In most vertebrates, where the blood osmolarity is fairly well regulated at a low level, the amount of intracellular organic substances is also low. These substances generally account for about 10 to 20% of the osmolarity of the intracellular fluid. On the other hand, organic substances may account for as much as 60 to 70% of the intracellular fluid osmolarity in species having a high blood osmolarity. This is essentially the case in most marine invertebrates, myxines and elasmobranch fishes in which blood osmolarity reaches values as high as 950 to 1400 mosmol l^{-1}, depending on the osmolarity of the environmental sea-water. Two groups of animals can thus be distinguished on this basis: those having high amounts of intracellular organic osmotic effectors and those having low levels of these compounds. As we will see, organic substances participate in isosmotic intracellular regulation in both groups. This participation may however be expected to be quantitatively larger in animals with high intracellular concentrations. Most studies on the role of organic osmotic effectors in cell volume regulation up to now have thus been achieved on animals or isolated tissues of animals belonging to this group.

The question now arises of knowing what the prominent compounds acting as organic osmotic effectors are.

Osmotic balance sheets, obtained from the muscle tissue of two molluscs are listed in Table 1. In the marine *Sepia officinalis,* organic compounds account for more than half the muscle tissue osmolarity. Organic phosphates and amino acids, together with some other amino compounds appear as important organic osmotic effectors. Moreover, amino compounds account alone for more than 60% of the osmolarity due to the organic substances.

In all the cases studied up to now, amino acids appear as the most important organic osmotic effectors (see below). Hitherto, most investigations have been concerned with the role of amino acids in the intracellular fluid isosmotic regulation. This role is suggested by the fact that the amino-acid content is higher in tissues of marine invertebrates than in tissues of limnic ones. This is illustrated in Table 1 for the case of the marine species *Sepia officinalis* and of the fresh-water species *Anodonta cygnea*. To assess the role of amino acids in isosmotic regulation, therefore, it appears of interest to compare the intracellular amount of free amino acids in euryhaline species having poor capabilities for extracellular fluid anisosmotic regulation. Indeed, in such species the blood osmolarity can be decreased by half upon transfer from sea-water to a diluted medium (see Chapter 5 for details). This may lead to the development of an enormous swelling pressure in the tissues if there is no active mechanism involved in the control of the osmolarity of the

Table 1. Organic and inorganic osmotic effectors in the muscle tissue of two molluscs

Osmotic effector	*Sepia officinalis*[a] mantle muscle	*Anodonta cygnea*[b] fast adductor muscle
Organic compounds		
1. Adenosine triphosphate	8.2	1.7
2. α amino-nitrogen	483	10.5[c]
3. Arginine phosphate	4.3	8.4
4. Betaïne	108.2	—
5. Glycerol	4.2	—
6. Phosphate (acido-soluble other than 1 and 3)	12.2	2.8
7. Taurine	—	0
8. Trimethylamine oxide	86.1	—
Inorganic compounds		
1. Ammonium	2.2	—
2. Calcium	1.9	11.5
3. Chloride	45.0	3.9
4. Magnesium	19.0	3.9
5. Phosphate	92.0	6.9
6. Potassium	189	18.4
7. Sodium	30.8	6.6
8. Sulphate	2.0	—
Osmolarity due to inorganic compounds (mosmol kg^{-1} water)	588.76	80.40
Osmolarity due to organic compounds (mosmol kg^{-1} water)[d]	787.88	26.00
Total cellular osmolarity (mosmol kg^{-1} water)	1376.64	106.40

[a] Robertson (1965).
[b] Potts (1958).
[c] Only carboxylic acids have been determined according to the method of Van Slyke, McFadyen and Hamilton (1941).
[d] Osmolarity of organic compounds has been calculated assuming that the Δ of a molar solution of organic compounds, such as amino acid, is -1.86 °C (Prosser and coworkers, 1950).

intracellular fluid. For instance, when the euryhaline crab *Eriocheir sinensis* is acclimated from sea-water to fresh-water, its blood osmolarity decreases in a few hours from about 1100 mosmol l^{-1} to some 550 mosmol l^{-1}. The van't Hoff–Arrhenius equation tells us that this decrease should induce the development of a swelling pressure of some 12 atm if one considers that the osmolarity of the intracellular fluid remains at its initial value and is not regulated during the acclimation process (Gilles, 1974a).

As shown in Fig. 2, acclimation of *E. sinensis* from sea-water to fresh-water leads to a transient increase in the muscle water content. However, tissue hydration resumes its control value within the first day. Concomitantly, there is a sharp decrease in the muscular content of the free amino compounds. On the contrary, during acclimation from fresh-water to sea-water, the muscle tissue shrinks and resumes its control value more slowly. This volume regulation is paralleled by a slow increase in the tissue amount of free amino compounds. It thus appears that the volume regulation process which occurs in the muscle tissue of the chinese crab is associated with a modification in the content of free amino compounds. This raises the question of knowing what the amino compounds implicated in this process are.

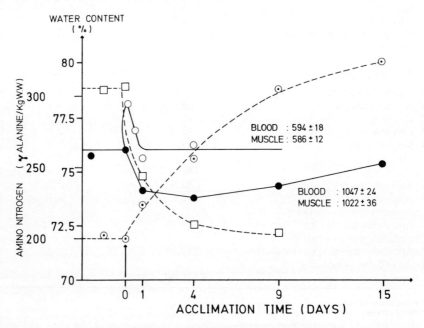

Figure 2 Changes in water content (———), osmolarity and amino nitrogen concentration (- - -) in the muscle of *E. sinensis* during acclimation from fresh-water to sea-water (●, ⊙) or from sea-water to fresh-water (○, □)

Florkin and coworkers were the first to analyse the free amino compounds present in the tissues of an animal subjected to an osmotic shock (Bricteux-Grégoire and coworkers, 1962). They showed that in the muscle of the chinese crab *E. sinensis*, the free amino acids plus taurine, betaine and trimethylamine oxide account for some 40% of the total osmolarity of the tissue. Moreover, after acclimation of the animals from fresh-water to sea-water, the amount of free amino acids is approximately doubled. This increase is particularly important at the level of glycine, arginine, proline, alanine and glutamine acids. An increase can also be recorded at the level of taurine and trimethylamine oxide but not at the level of betaïne. Many studies of this type have been done during the last two decades on euryhaline invertebrates. As an example, Table 2 gives an osmotic balance done on the muscle tissue of the blue crab *Callinectes sapidus* acclimated to sea-water or to twice diluted sea-water.

In the tissue of the sea-water acclimated animals, the amino-acid concentration, when expressed in mosmol kg^{-1} of intracellular water, accounts for about 70% of the total osmolarity of the tissue. The remaining 30% can be accounted for almost completely by the inorganic ions Na$^+$, K$^+$ and Cl$^-$. After acclimation of the crab to the diluted medium, the inorganic ion

Table 2. Contribution of the intracellular free amino acids to the osmolarity of the muscle tissue of *Callinectes sapidus* acclimated to sea-water (SW) and to 50% sea-water (SW/2).

Compound	SW	SW/2
Amino acids		
Alanine	37.60	22.57
Arginine	136.44	95.32
Aspartate	7.80	3.03
Cysteine	2.12	1.61
Glutamate	10.16	3.38
Glycine	361.98	282.01
Histidine	2.47	0.82
Isoleucine	1.77	1.61
Leucine	3.42	3.50
Lysine	3.43	0.08
Methionine	5.89	5.60
Phenylalanine	4.25	0.69
Proline	74.01	48.76
Serine	52.37	5.12
Taurine	69.41	37.30
Tyrosine	2.00	1.15
Valine	5.30	5.60
Total	780.42	513.71
Inorganic ions		
Na	39.5	28.4
K	186.0	162.2
Cl	45.6	25.9
Total	271.1	216.5
Blood osmolarity (osmol l^{-1})	1100	850
Water content (%H$_2$O)	77.1	77.8

Modified from Gérard and Gilles (1972a).
The intracellular amounts of compounds are given in mosmol kg^{-1} intracellular water.

concentration decreases only slightly except for chloride which undergoes a larger variation than sodium and potassium. In contrast, there is an important modification in the amount of amino acids. This decrease accounts for most of the change observed in the total osmolarity of the intracellular fluid. All the amino acids appear to participate in this process. However, it must be noticed that, except for arginine, the so-called essential amino acids play only a minor role in the adjustment of the intracellular fluid osmolarity. On the other hand, it is also apparent that the variation in the amino-acid concentration can only depend on an active process and is not simply due to a dilution of the intracellular fluid caused by tissue swelling. The percentage variation indeed varies from one amino acid to the other and moreover, the tissue water content does not change significantly after acclimation from the concentrated to the diluted medium.

Many studies in recent years have dealt with the free amino-acid content of animals, isolated tissues and cells exposed to osmotic stresses. The results obtained fully corroborate the conclusions drawn above; that is to say the tissue level of free amino acids and taurine is actively regulated according to the variations of the osmolarity of the extracellular fluid. Moreover, except for arginine it is always the non-essential amino acids proline, glycine, alanine, glutamic acid, aspartic acid and serine which show the largest variations. This phenomenon has been recorded in all the invertebrate phyla studied so far (for a review see Schoffeniels and Gilles, 1970, 1972; Gilles, 1975). It has also been documented in fishes (Cholette, Cagnon and Germain, 1970; Huggins and Colley, 1971; Lasserre and Gilles, 1971; Venkatachari, 1974; Vislie and Fugelli, 1975) in batracians (Gordon, 1965; Baxter and Ortiz, 1966), in reptiles (Gilles-Baillien, 1973) as well as in several protozoans such as the ciliates *Miamensis avidus* (Kaneshiro, Holz and Dunham (1969), and *Tetrahymena pyriformis* (Stoner and Dunham, 1970: Dunham and Kropp, 1973). Recently, the role of amino acids in cell volume regulation has also been demonstrated in non-halophilic bacteria (Measures, 1975) and in Ehrlich ascites tumour cells (Hendil and Hoffmann, 1974; Hoffmann and Hendil, 1976). Moreover, several studies indicate that amino acids could play a part in an osmolarity control of the intracellular fluid in some plants (Bernstein, 1963; Webb and Burley, 1965) and small-celled algae (Gilles and Péqueux, 1977). These findings suggest the possibility that amino-acid participation in the intracellular fluid isosmotic regulation is a phenomenon of general occurrence in living organisms. However, more data, particularly about mammalian tissues and cells, would be needed to assess this hypothesis. It is also worth noticing in these studies that the eurhalinity of several species depends only on the efficiency of their mechanisms of intracellular fluid isosmotic regulation. For instance, the sea-star *Asterias rubens* does not show measurable extracellular fluid osmotic regulation, neither do *Arenicola marina, Mytilus edulis, Libinia emarginata* nor the hagfish *Myxine glutinosa*. In these species, intracellular fluid isosmotic regulation is found and involves control of the free amino-acid intracellular content. Thus it may be that regulation of the intracellular fluid osmolarity is a basic, primitive mechanism to which, in various species, osmoregulation of the extracellular fluid has added, in the course of evolution, a new range of possibilities (see also Chapters 5 and 13).

The question now arises of knowing what are the mechanisms involved in the control of the intracellular content of amino acids in relation to the cell volume regulation process.

III. MECHANISMS OF CONTROL OF THE AMINO-ACID POOL

A priori, such mechanisms may depend on intracellular processes or may be concerned with interactions between the cells and their extracellular fluid (hormonal effects for instance). Moreover, these two types of possible processes may coexist. They will be discussed separately hereafter.

A. Mechanisms at the Cellular Level

To be of physiological interest to us such mechanisms, whatever they are, should be related to a volume control process. Therefore, we may first ask the question can isolated tissues subjected to osmotic stresses achieve volume regulation? As a matter of fact, such behaviour has now been described in a number of cases. Figure 3 shows as an example, the volume evolution following

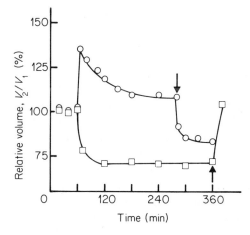

Figure 3 Effects of an hyperosmotic stress ($\Pi_1/\Pi_2 = 0.5$, □) and of an hypo-osmotic stress ($\Pi_1/\Pi_2 = 2$, ○) on the volume of *E. sinensis* isolated axons. Modified from Gilles (1973a)

application of an osmotic shock to isolated axons of the chinese crab *Erisocheir sinensis*. It can be seen that, in hypo-osmotic conditions, the tissue volume resumes almost completely its control values in the time course of the experiment. On the contrary, after application of an hyperosmotic stress, the tissue shrinks and no regulation can be observed. Similar behaviour has been described for a variety of tissues and cell types. This is the case for the red blood cells of the fishes *Pleuronectes flesus* (Fugelli, 1967) and *Pseudopleuronectes americanus* (Cala, 1974; quoted in Schmidt-Nielsen, 1975), the red blood cells of duck (Kregenow, 1971) and of man (Posnansky and Solomon, 1972), the skin epithelia cells (McRobbie and Ussing, 1961) and the oocytes (Sigler and Janáček, 1971) of the frog *Rana temporaria*, the isolated muscle fibres of the crab *Callinectes sapidus* (Lang and Gainer, 1969), the isolated axons of the crabs *C. sapidus* and *E. sinensis* (Gérard and Gilles, 1972b; Gilles, 1973a), the chick blood lymphocytes (Doljanski and coworkers, 1974), the mouse lymphoblasts (Buckhold, Adams amd Gregg, 1965), the L5178Y lymphocytic leukaemia cells of mouse (Roti-Roti and Rothstein, 1973) and Ehrlich ascites cells (Rosenberg, Shank and Gregg, 1972; Hendil and Hoffmann, 1974).

In these various tissues and cell types, a volume regulation following a swelling transitory phase can be demonstrated in hypo-osmotic conditions while no volume control is observed in hyperosmotic media. To our knowledge, volume regulation in an hyperosmotic situation has only been observed in duck red blood cells incubated in the presence of high, non-physiological concentrations of potassium (Kregenow, 1971). On the other hand, it must be noticed that the absence of volume control in hypo-osmotic conditions has been reported in a few cases although the experiments have been performed on a time scale long enough for a regulation to be observed. This is the case for the isolated muscle fibres of frog *sartorius* (Blinks, 1965) and the isolated muscle fibres and axons of the lobster *Homarus gammarus* (Gainer and Grundfest, 1968; Gilles, 1973a). However, such results may not be indicative of the absence of volume regulation in these tissues. They may be due to the fact that the osmotic shock applied has been of a too large amplitude for the cell types considered. The amplitude of the hypo-osmotic stress a tissue is able to withstand would indeed appear to be variable depending on the tissue considered and, for the same tissue, on the species examined. In this respect, euryhaline crustaceans are probably among the best regulators. Tissues isolated from such species can achieve volume regulation following application of hypo-osmotic stresses of an amplitude $\Pi_1/\Pi_2 = 2$. Many tissues, if not all, would thus appear to be capable of volume regulation in hypo-osmotic conditions provided that the amplitude of the applied stress remains within tolerable limits for the species considered. On the contrary, isolated tissues incubated in artificial 'ionic' salines are unable to achieve volume control on the time scale of an *in vitro* experiment following application of an hyperosmotic shock. In relation to this phenomenon, it must be recalled that volume regulation following hyperosmotic stress can be demonstrated on tissues of intact animals but that this process is much slower than the volume control observed in hypo-osmotic conditions (see Fig. 2). These considerations are indicative of the presence of different mechanisms in the two types of process. That these mechanisms are concerned with the control of the amount of intracellular osmotic effectors is clearly shown in Fig. 3. Replacement of the hypo-osmotic saline by the control after the phase of swelling regulation indeed brings about a tissue shrinkage to values lower than control. This indicates that not only water but also intracellular solutes have been lost from the cells during the regulation process. Therefore it becomes of interest to follow the fate of the intracellular amino acids in isolated tissues subjected to osmotic stresses.

As early as 1960, Schoffeniels showed that the total concentration of ninhydrine positive substances in isolated axons of the crab *Eriocheir sinensis* was increased following application of a hyperosmotic shock. Although part of this result may be explained on the basis of a shrinkage of the tissue, it supports the idea that intracellular mechanisms of control of the amino-acid

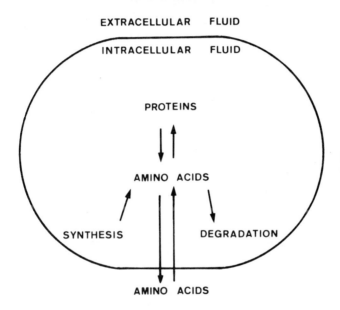

Figure 4 Possible mechanisms of control of the intracellular amino-acid pool

level are involved in the isosmotic regulation process shown during volume readjustment. Further experiments on Ehrlich ascites tumour cells of mouse (Hoffman and Hendil, 1976) and also on isolated axons of the crustaceans *Callinectes sapidus* and *Eriocheir sinensis* (Gilles, 1974a; Gilles and Gérard, 1974) demonstrated that the changes in the amino-acid content induced by osmotic shocks cannot be ascribed to tissue volume modifications and must therefore be due to some intracellular active mechanism.

What could these mechanisms be? *A priori,* three different possibilities can be considered: the amino-acid pool may be controlled either by a variation in the steady state between synthesis and degradation of these compounds, by a modification in the amino acid–protein equilibrium or by a change in the rate of transport of the amino acids through the cell membranes (Fig. 4).

1. *Amino-acid Fluxes*

The concentration of amino acids in isolated axons of *E. sinensis* decreases considerably after 3 hours of incubation in hypo-osmotic conditions (Table 3). Concomitantly, increases in the amounts of several amino acids can be

Table 3. Effect of osmotic shocks on the amino-acid concentration in isolated axons of *Eriocheir sinensis*

Amino acids	Controls	Hypo-osmotic shock ($\Pi_1/\Pi_2 = 2$)	Controls	Hyperosmotic shock ($\Pi_1/\Pi_2 = 0.5$)
Alanine	12.16	8.07	5.61	8.81
Arginine	3.03	1.45	3.78	3.64
Aspartate	23.65	10.25	16.22	18.23
Cysteine	Tr	Tr	Tr	0.50
Glutamate	3.03	2.62	3.72	4.52
Glycine	3.84	3.24	2.27	2.22
Histidine	0.51	0.49	0.15	0.15
Isoleucine	0.50	0.30	0.27	0.33
Leucine	0.45	0.32	0.42	0.42
Lysine	0.39	0.42	0.46	0.42
Methionine	0.10	Tr	0.23	0.44
Phenylalanine	0.30	0.24	0.15	0.17
Proline	17.78	6.94	2.50	4.73
Serine	3.34	3.10	1.56	1.33
Taurine	7.03	5.38	6.25	6.79
Tyrosine	0.30	0.21	0.15	0.18
Valine	0.58	0.52	0.40	0.43

Values are expressed in μmol/100 mg of tissue dry weight.
Tr stands for traces.
Modified and recalculated from original data by Gilles and Schoffeniels (1969a).

recorded in the incubation saline (Table 4). However, this is far from being a general phenomenon. During hyperosmotic stress on the contrary, there is only a slight increase in the tissular amount of amino acids which cannot be completely accounted for by the decrease recorded in the amount of amino acids found in the incubation medium. The changes in the intracellular concentration cannot therefore be explained solely on the basis of modifications in amino-acid fluxes in or out of the cells.

In another connection, it is worth noticing that the differences just mentioned between the changes in amino-acid tissular level induced by hypo- or hyperosmotic stresses are reminiscent of the differences observed in the volume regulation behaviour of the tissues depending on whether they are submitted to an hypo- or to an hyperosmotic shock. Nevertheless, these results show that changes in fluxes, if they cannot completely explain the modifications in the amino-acid intracellular level are undoubtedly implicated in the regulation of their concentration.

The mechanisms controlling the fluxes of amino acids in response to osmotic stresses are far from clear. Ninhydrine positive substances appear to be released from various isolated tissues during volume adjustment following hypo-osmotic stress. This is the case in *Pleuronectes flesus* red blood cells

Table 4. Effect of osmotic shocks on the amount of amino acids released in the incubation saline by isolated axons of *Eriocheir sinensis*

Amino acids	Controls	Hypo-osmotic shock ($\Pi_1/\Pi_2 = 2$)	Controls	Hyperosmotic shock ($\Pi_1/\Pi_2 = 0.5$)
Alanine	6.06	5.99	2.63	2.14
Aspartate	11.11	11.17	9.04	8.40
Cysteine	Tr	Tr	Tr	Tr
Glutamate	2.43	1.43	2.50	2.14
Glycine	1.09	1.25	1.51	1.30
Isoleucine	0.15	0.30	0.43	0.29
Leucine	0.20	0.36	0.54	0.38
Methionine	Tr	Tr	Tr	Tr
Phenylalanine	0.19	0.27	0.25	0.15
Proline	7.93	19.34	0.79	1.80
Serine	0.88	0.83	0.47	0.46
Taurine	0.70	0.83	0.48	0.48
Threonine	0.22	0.18	Tr	Tr
Valine	0.29	0.38	0.26	0.09

Values are expressed in μmol/100 mg of tissue dry weight.
Tr stands for traces.
Modified and recalculated from original data by Gilles and Schoffeniels (1969a).

(Fugelli, 1967), *Callinectes sapidus* muscle fibres (Lang and Gainer, 1969), isolated heart muscle of *Modiolus modiolus* (Pierce and Greenberg, 1970) and in Ehrlich ascites tumour cells (Hendil and Hoffmann, 1974).

The fluxes of labelled alanine in relation to different osmotic conditions has been studied on isolated axons of *C. sapidus* by Gilles and coworkers (Gérard and Gilles, 1972b; Gérard, Dandrifosse and Gilles, 1974; Gérard, 1975). During hypo-osmotic stress, there is an increase in the efflux of radioactive material while under hyper-osmotic conditions, no significant changes in the efflux can be recorded. These results have been confirmed and extended recently by studying the unidirectional fluxes of ^{14}C-alamine. Table 5 summarizes some of the results obtained. Hypo-osmotic stress appears to induce an increase in the alanine efflux and a decrease in the influx. This implies an inhibition of the inward active transport of alanine. During hyperosmotic shock on the other hand, there is an increase in the active uptake of alanine while no significant changes can be recorded in the passive efflux. Comparable effects of hypo- and hyperosmotic salines on the efflux of other non-electrolytes (urea and erythritol) have been recorded on isolated rat portal veins (Jonsson, 1971). As will be discussed later, these differences in changes of transport activity following application of hypo- and hyperosmotic shocks might be related to the

Table 5. Effect of osmotic stresses on the unidirectional fluxes and the permeability coefficient to alanine in isolated axons of *Callinectes sapidus*

	Efflux	Influx	Active component of influx	Permeability coefficient
	(pmol cm^{-2} s^{-1})			($\times 10^{-8}$ cm s^{-1})
Steady-state conditions (SW saline)	0.81		0.79	2.13
Hypo-osmotic shock (75% SW saline)	2.17	0.57	0.52	5.72
Steady-state conditions (75% SW saline)	0.71		0.66	4.87
Hyperosmotic shock (SW saline)	0.70	0.95	0.91	4.38

After Gérard (1975) and Gérard and Gilles (unpublished).
For experimental conditions and applied equations refer to Gérard (1975).
The alanine concentration in the incubation medium is 1 mmol l^{-1}.

differences recorded in volume regulation behaviour under these experimental conditions.

The modifications induced in the active transport of alanine by the osmotic stresses are probably related to changes in the Na$^+$ concentration of the incubation saline. Substitution of sodium by choline in the external medium does indeed induce an inhibition of the alanine influx which results in a higher K_m and a slight but significant lowering of V_{max} (Gérard, 1975). We are thus dealing with a carrier-mediated uphill transport which is sodium dependent. It is thus reasonable to assume that the decrease in sodium external content caused by application of an hypo-osmotic shock induces a decrease in the alanine influx. On the contrary, the increased sodium content of the hyperosmotic saline should bring about an increased active transport activity. Up to now, there is no indication as to the primary causes of the important permeability change observed after application of a hypo-osmotic stress. It might be related to configuration changes in the cell plasmatic membrane induced by the swelling of the tissue (Jonsson, 1971; Gérard and Gilles, 1972b).

2. Equilibrium between Proteins and Amino Acids

Modifications in the equilibrium between proteins and amino acids might also play a part in the control of the amino-acid pool. Some studies give tentative support to this hypothesis (Bedford, 1971; Venkatachari, 1974). However, the results obtained are more indicative than demonstrative. On the other hand, Siebers and coworkers (1972) found no significant changes in protein concentration in whole *Oreonectes limosus* during acclimation from fresh-water to a salinity of 650 mosmol l^{-1}. In the same way, there is no significant variation of the protein concentration in isolated axons of *Eriocheir sinensis* submitted to osmotic stresses (Gilles and Schoffeniels, 1969a). It would thus appear that the equilibrium between proteins and amino acids plays only a minor role, if any, in the adjustment of the amino-acid pool occurring during the cell volume regulation process. This view is also supported by results of Florkin and coworkers (1964) who measured the variation in proline and alanine concentration before and after protein hydrolysis in muscles of *E. sinensis* submitted to hyperosmotic conditions. These authors showed that the increase in the amount of these amino acids in the free pool is paralleled by a similar increase in their total amount. This indicates that the amount of proline and alanine obtained from protein hydrolysis does not vary significantly during acclimation to hyperosmotic conditions. In a similar way, Bedford (1971) showed that the increase in amino nitrogen occurring in isolated foot muscle of the mollusc *Melanopsis trifasciata* submitted to hyperosmotic stress, is paralleled by an increase in total nitrogen. This indicates that the increase in amino acids is not related to a decrease in other intracellular nitrogenous compounds.

The information published up to now thus appears contradictory and more results are needed to assess the role that changes in the equilibrium between proteins and amino acids could play in the regulation of the amino-acid pool.

3. Equilibrium between Amino-acid Synthesis and Degradation

The last mechanism of intracellular origin which might be involved in the active adjustment of the free amino-acid amount is concerned with control of the synthesis and/or degradation of amino acids.

In isolated axons of the blue crab *Callinectes sapidus* subjected to an hypo-osmotic shock, there is a decrease in the concentration of all the measured free amino acids. At the same time, an increase in specific activity of the amino acids labelled from radioactive glucose can be recorded (Table 6). Such a result is unlikely to be due to a change in the availability of the labelled marker for metabolic purposes. Indeed, hypo-osmotic conditions generally lead to a leakage of intracellular solutes during the swelling phase (see Chapter 3 and p. 122 in this chapter). In such a situation and if there is no change in the metabolic utilization of the ^{14}C of glucose, a loss of radioactive

Table 6. Effect of osmotic stresses on the concentration and specific activity of amino acids in isolated axons of *Callinectes sapidus* with glucose u-^{14}C.

Amino acid	Controls		Hypo-osmotic shock		Controls		Hyperosmotic shock	
	concentration	specific activity	concentration	specific activity	concentration	specific activity	concentration	specific activity
Alanine	14.79	11,722	6.37	23,192	4.47	31,280	9.77	28,202
Aspartate	148.05	—*	61.25	—*	86.23	—*	93.06	—*
Glutamate	17.83	2,448	9.55	5,945	10.39	4,408	11.64	4,906
Glycine	6.45	—	3.36	—	3.85	—	5.14	—
Proline	16.09	—	4.04	—	Tr	—	2.04	—
Serine	1.82	18,185	0.94	24,783	1.35	10,604	2.27	12,871
Taurine	10.16	—	6.37	—	9.35	—	11.64	—

Concentrations are given in µmol/100 mg tissue dry weight and specific activities in ct/min/µmol of amino acid.
Dashes indicate no detectable radioactivity.
* Radioactivity is present, specific activity cannot be calculated because of the presence of an undetermined ninhydrine positive compound together with asparate.
Modified from Gilles and Gérard (1974).

glucose from the intracellular medium should result in a decreased disposibility of ^{14}C and therefore a decrease in specific activity should be recorded. On the other hand, there is no noticeable modification in the uptake of glucose from the incubation saline under the hypo-osmotic conditions to which the tissue is subjected (Gilles, 1974a). The increase in specific activity observed may thus be due to an increased synthesis of amino acids from glucose. It may also be the result of a rapid removal of amino acids from the metabolic pool at the beginning of the experiment. In this situation, synthesis from glucose would occur in a pool of smaller size, thus leading to increased specific activity. Such a mechanism would also account for the decrease in the concentration of all the amino acids determined and not only of those which can be labelled from glucose.

During hyperosmotic shock, the concentration of the amino acids increases while there is no significant change in the specific activity of those which are labelled from glucose. These results can be interpreted in terms of a decreased release in the presence of continuous addition of amino acids to the metabolic pool. Increased synthesis of amino acids from glucose is unlikely to account for the increase in amino-acid concentration even if one considers only the case of the non-essential amino acids. Increased synthesis of these amino acids from labelled glucose would indeed lead to an increase in their specific activity. Moreover, no labelling can be found at the level of proline and glycine although they are non-essential amino acids and also their concentration is increased.

Control of the intracellular amount of amino acids thus appears to be achieved mainly by regulation of the removal of these compounds from the metabolic pool.

Release of amino acids from the metabolic pool can be achieved either by degradation, efflux from the cell or protein synthesis (Fig. 4). Modifications in protein synthesis from amino acids is unlikely to be a significant process in short-term experiments on isolated tissues submitted to osmotic stresses (see also p. 125). Control of the outward movement of amino acids occurs during hypo-osmotic stress (Table 4 and p. 122). This process can however only account for an increased removal of amino acids from the metabolic pool and not for a decreased removal as can be expected in hyperosmotic conditions. Moreover, it appears that efflux control cannot account entirely for the decrease in amino-acid level observed in hypo-osmotic conditions. Regulation of amino-acid catabolism may thus play a significant role in the adjustment of their intracellular concentration that occurs during osmotic shocks. In order to test this hypothesis, the effects of osmotic stresses on the $^{14}CO_2$ production by isolated axons of *C. sapidus* preloaded with various ^{14}C amino acids have been studied (Table 7). Hypo-osmotic conditions induce an increased $^{14}CO_2$ production from the different labelled amino acids used. Concomitantly there is an increase in oxygen consumption and in total CO_2 production. Inversely, there is a decrease in $^{14}CO_2$ production, O_2 consumption and CO_2 total production

Table 7. Effect of osmotic stresses on the O_2 consumption, CO_2 production and $^{14}CO_2$ production of *Callinectes sapidus* isolated axons incubated with different (u-^{14}C) amino acids.

Amino acid	QO_2 (l h^{-1}(mg DW)$^{-1}$)			QCO_2 (l h^{-1}(mg DW)$^{-1}$)			$^{14}CO_2$ ((ct/min) h^{-1}(mg DW)$^{-1}$)		
	C	HO	C	HR	C	HO	C	HR	C	HO	C	HR
Alanine	4.88	10.38	2.89	2.09	3.86	7.96	2.17	1.53	371.11	595.96	192.94	140,02
Aspartate	4.08	5.62	4.06	3.34	3.00	4.30	2.92	2.58	165.40	219.35	223.74	179.81
Arginine	2.16	2.42	2.35	1.60	1.62	1.91	1.84	1.20	36.31	44.81	32.67	23.87
Glutamate	5.56	6.38	2.65	1.92	4.16	4.86	2.05	1.43	85.91	113.08	114.03	75.'5
Leucine	2.77	3.46	2.53	1.71	2.09	2.62	2.04	1.30	17.76	21.13	17.50	14.34
Serine	3.60	5.35	3.34	1.98	2.68	4.10	2.59	1.49	31.76	51.18	127.61	84.31

C: controls
HO: hypo-osmotic shock
HR: hyperosmotic shock
DW: dry weight
Modified from Gilles (1972).

during application of a hyperosmotic shock. These findings indicate a decreased amino-acid catabolism during hyperosmotic stress and an increased degradation in the reverse experimental situation. It is also worth noticing that, in both experimental conditions, no significant changes in CO_2 specific activity can be recorded. This indicates that the modification in $^{14}CO_2$ production is not due to a mechanism controlling the catabolic activity of some specific amino acids but rather to a general modification of the oxidative metabolism. The appearance of radioactive CO_2 from the exogenous substrates obviously indicates the entrance of the utilized amino acids into oxidative pathways. However, their contribution to the total CO_2 production remains small and other substrates, the natures of which remain to be determined, must be considered to account for the largest part of the tissue oxidative activity. In this respect, it is interesting to notice the increase in $^{14}CO_2$ production from labelled glucose and pyruvate in isolated axons of *C. sapidus* or *E. sinensis* submitted to hypo-osmotic conditions (Gilles, 1974a). Nevertheless, the results discussed above are indicative of modifications in the tissues oxidative metabolism during osmotic stresses. The participation of amino acids in this process can account for part of the changes recorded in their intracellular concentration under such conditions. The increased utilization of labelled glutamic acid in the intermediary metabolism as well as the increased $^{14}CO_2$ production from this substrate observed in the muscle of *Carcinus maenas* after acclimation from sea-water to 40% sea-water (Chaplin, Huggins and Munday, 1970) is also indicative of an increased amino-acid catabolism in hypo-osmotic conditions.

Modifications in catabolic activity during osmotic stresses can also be traced down to the level of redox changes occuring in respiratory chain components. Gilles and Jobsis (1972) reported on such changes in isolated muscle fibres of *C. sapidus*. Reversible, specific modifications in the pyridine nucleotide oxido-reduction level occur in that tissue when submitted to osmotic stresses. If the tissue is submitted to a hypo-osmotic shock, the pyridine nucleotides are reduced whereas they are oxidized when the tissue is placed back in the control medium. On the other hand, hyperosmotic conditions induce an oxidation of the pyridine nucleotides. These results can be integrated in the framework of variations in oxidative metabolism. Increase in oxidative activity during hypo-osmotic shock, should indeed lead to the formation, via the Krebs cycle, of reducing equivalents which can account for the reduction of the pyridine nucleotides. On the other hand the decrease in oxidative activity observed in hyperosmotic conditions may be related, through a decrease in the availability of reducing equivalents, to the fact that the pyridine nucleotides become more oxidized.

It also seems that, at least in crustaceans, the modifications in oxidative activity can be associated with changes in the general pattern of intermediary metabolism. Carbon distribution patterns between the oxidative and non-oxidative pathways can be estimated by measuring the ratio C_1/C_6 of the $^{14}CO_2$ produced from glucose-1-^{14}C to the production of $^{14}CO_2$ from glucose-6-^{14}C. First, it is worth noticing that this ratio is strikingly high in isolated tissues of crustaceans when compared to the values reported for mammalian tissues (Hoskin, 1962; Gilles and Schoffeniels, 1969b; Hochachka and coworkers, 1970; Thabrew, Poat and Munday, 1971; Gilles, 1974b). Such high values have often been considered as indicating a high hexose monophosphate shunt activity (Hochachka and coworkers, 1970; Schatzlein and coworkers, 1973). Theoretical considerations however show that such high C_1/C_6 ratios mainly reflect a high activity of the anabolic pathways starting from 3-carbon units of glycolysis (Fig. 5a; theoretical treatment can be found in Katz and Wood, 1960; it has been extended by Gilles, 1974a). Among these pathways are those

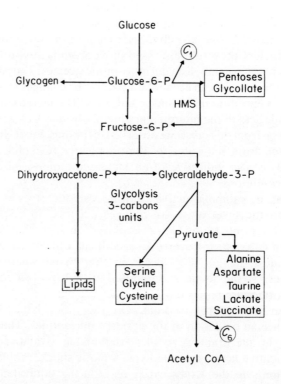

Figure 5 Fate of C_1 and C_6 of glucose in glycolysis and related pathways

involved in the synthesis of the ketoprecursors of alanine, glycine, serine or aspartic acid. This is particularly interesting to consider in view of the large amount of these amino acids which are normally found in crustacean tissues. Further experiments would be needed however in order to see if the high C_1/C_6 ratios can be ascribed specifically to high metabolism of these amino acids or if some other metabolic pathway also takes part in the phenomenon.

The effect of osmotic stresses on the C_1/C_6 ratio has been studied by Gilles (1974b) on isolated axons of the chinese crab *E. sinensis*.

Hypo-osmotic stress increases, while an hyperosmotic condition decreases, the $^{14}CO_2$ production from glucose-6-^{14}C. In both experimental conditions there is no significant change in the $^{14}CO_2$ production from glucose-1-^{14}C (Table 8)

Table 8. Effect of osmotic stresses on $^{14}CO_2$ production form glucose-1-^{14}C and glucose-6-^{14}C in isolated axons of *Eriocheir sinensis*

	glucose-1-^{14}C	glucose-6-^{14}C	C_1/C_6
Control	1271 ± 445	406 ± 32(1)	3.15 ± 1.08(5)
Hypo-osmotic shock	951 ± 393	507 ± 53(2)	1.91 ± 0.86(6)
Control	1289 ± 307	388 ± 87(3)	3.48 ± 1.14(7)
Hyperosmotic shock	1192 ± 508	237 ± 79(4)	4.99 ± 1.12(8)

Results are given in (ct/min) h^{-1} (mg DW)$^{-1}$ ± standard deviation.
Student's *t* test—two-tailed probability; $n = 5$.
1–2 $0.01 < P < 0.005$.
3–4 $0.025 < P < 0.02$.
5–6
7–8 } $0.10 < P < 0.05$.
Modified from Gilles (1974b).

It follows that the C_1/C_6 ratio is decreased during hypo-osmotic shocks and increased in hyperosmotic situations. These changes in $^{14}CO_2$ production together with the variation in the C_1/C_6 ratio can be interpreted in terms of differential utilization of the glucose carbons in the oxidative Krebs cycle pathway and in the 3-carbon units anabolic pathways. These results indicate that during hypo-osmotic stress, there is an increase in oxidative metabolism which leaves less 3-carbon units of glycolysis available for anabolic purposes. On the contrary more 3-carbon units appear to enter the anabolic pathways during hyperosmotic shock; this can be related to the decrease in oxidative metabolism which occurs under these experimental conditions. In this respect it is interesting to consider that succinate metabolism is mainly achieved through the propionyl CoA pathway in the crab *Carcinus maenas* when in sea-water while it is essentially oxidized in the Krebs cycle when the animal is acclimated to diluted media (Thabrew, Poat and Munday, 1973). Such changes in metabolic activity patterns are in agreement with the modifications observed when studying the C_1/C_6 ratio in isolated axons.

The results described above show that the changes in the amino-acid level occurring in isolated tissues concomitantly to the application of an osmotic stress can be ascribed at least partly to a control of a metabolic removal of these compounds from the intracellular pool. This control is essentially achieved by a regulation of the tissue oxidative activity. During hypo-osmotic shock, the amino-acid concentration decreases; this can be associated with an increased catabolism which can also account for the increased CO_2 production and oxygen consumption as well as for the observed increase in the level of reduction of the pyridine nucleotides. In hyperosmotic conditions, the intermediary metabolism is geared towards more anaerobic conditions. This may leave more 3-carbon units of glycolysis available for synthesis of amino-acid ketoprecursors; the decreased oxidative metabolism which is implicated in the increase in amino-acid concentration can also account for the decrease in CO_2 production and in oxygen consumption as well as for the fact that the pyridine nucleotides are becoming more oxidized.

These considerations are summarized in Fig. 6 which relates the intermediary metabolism with the transamination sequence governing the equilibrium between amino acids and their ketoprecursors.

The question arises now of knowing what are the primary causes of the changes in metabolic activity that occur during osmotic stresses.

It is obvious from the experiments on isolated tissues and cells reported above that an hormonal control mechanism is not primarily implicated in the regulation of the intracellular fluid osmolarity. Such a process might however be at work in the whole animal (see p. 142).

On the other hand, increase in saline osmolarity by addition of sucrose

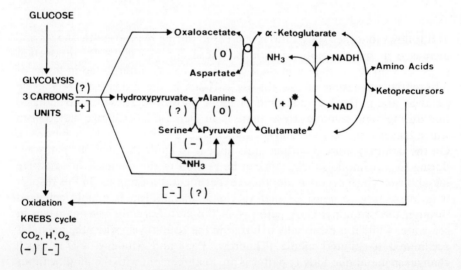

Figure 6 Relations between glycolysis, oxidative metabolism and the transamination sequence which rules the equilibrium between amino acids and their ketoprecursors.
Effect of NaCl addition to the incubation medium of enzymes implicated in these pathways: (+), activation; (−), inhibition; (O), no effect; (?), unknown; *, depends on the amount of NADH in the medium. Signs between [ ] refer to effects obtained by studies other than enzymatic; essentially metabolic studies on isolated tissues. Refer to text for further comments (above and p. 134)

causes a decrease in the amino-nitrogen level of isolated axons of *E. sinensis* instead of the increase recorded when the hyperosmotic shock is achieved by increasing the ionic concentration (Schoffeniels, 1960). Similarly, incubation of isolated foot muscle of the mollusc *Melanopsis trifasciata* in media of increasing salt concentration but of constant osmolarity induces an increase in the intracellular amino-nitrogen content (Bedford, 1971).

Measures (1975) also showed that in non-halophilic bacteria, an increase in the amino-acid content can be induced either by addition of sucrose or of NaCl to the culture medium. The increase is more important however when NaCl is used to achieve the hyperosmotic shock. Furthermore, it may be that the increase recorded when sucrose is added results from changes in intracellular ionic concentration due to the modifications of the external medium osmolarity.

Unfortunately, intracellular ionic concentrations have not been recorded in the different experimental conditions tested. Nevertheless, these findings indicate that it is not the osmolarity of the incubation medium which is responsible, *per se*, for the increase in the amino-acid intracellular content which occurs during hyperosmotic stress. Whether or not the situation is the same following application of a hypo-osmotic stress remains to be demonstrated. Hoffmann and Hendil (1976) showed that decrease in osmolarity achieved by omission of sucrose in the incubation medium while keeping steady its ionic concentration induces a decrease in the intracellular amino-acid content of Ehrlich ascites tumour cells. It may be however that the observed decrease is related only to changes in amino-acid efflux which play an important part in the regulation following hypo-osmotic stress (see p. 121) and not at all to variation in metabolic activity.

It appears reasonable therefore to assume that the changes in metabolic activity which are associated with both hypo- and hyperosmotic conditions are triggered by modification in the intracellular ionic concentration. This raises two important questions:
(1) What are the mechanisms implicated in the control of the metabolic activity by the inorganic ions?
(2) What are the ions essentially involved in these mechanisms?

The ionic composition may control the level and/or the activity of key enzymes involved in the metabolic pathways of deamination and oxidation of amino acids. A first possible control process would be related to genetically controlled modifications in the level or in the kinetic characteristics of key enzymes. From another point of view, the ionic concentration could directly modulate the activity of some key enzymes.

As to the first possibility, Chaplin, Huggins and Munday, (1965) have not been able to demonstrate a significant difference in glutamate dehydrogenase activity between tissues of *Carcinus maenas* acclimated to 50% sea-water or to 100% sea-water. In the same way, there are no changes in the specific activity and in the K_m for a-ketoglutarate of glutamate dehydrogenase or aspartate aminotransferase from tissues of *Eriocheir sinensis* acclimatized to fresh-water or to sea-water (Table 9). These findings indicate that the changes in amino-acid concentration are not related primarily to modifications in the genetically controlled level or kinetic characteristics of key enzymes. Moreover, such a possibility appears unlikely in view of the rate at which the changes in metabolic activity can be affected in experiments performed on isolated tissues.

Wickes and Morgan (1976), however, showed an important increase in specific activity of glutamate dehydrogenase and aspartate aminotransferase in the oyster *Crassostrea virginica* upon acclimation to concentrated media. Moreover, Reddy and his coworkers found changes in the level of alanine and aspartate aminotransferases in aquatic gastropods adjusting to different salinities (private communication). Thus it may be that changes in enzyme level participate in the metabolic modification of the amino-acid concentration in some zoological groups and not in others. More information would be needed however before definite conclusions could be drawn as to the importance of such a mechanism in different species and tissues.

As to the possibility of a direct control of enzyme activity by the inorganic composition of the incubation medium, several studies report on the effect of NaCl or KCl on enzymes implicated in:
 (i) amino-acid metabolism; such as glutamate dehydrogenase, serine hydrolyase, alanine and aspartate aminotransferases;
 (ii) the Krebs cycle and its anaplerotic pathways; such as malate hydrolase, malate dehydrogenases (E.C.1.1.1.37 and 1.1.1.40), isocitrate dehydrogenase, oxalocetate decarboxylase, succinate dehydrogenase;
 (iii) the fate of reducing equivalents such as lactic dehydrogenase, 3-glycerophosphate dehydrogenase or glyoxylate reductase.

The results of these experiments have been presented in a number of papers (see below) and have been discussed in the light of metabolic osmotic adjustments in recent review articles (Schoffeniels and Gilles, 1970; Gilles 1974a, 1975; Schoffeniels, 1976). We shall thus only briefly summarize the main conclusions drawn in incorporating some recent findings about the modification of glutamate dehydrogenase activity by NaCl and NADH. A summarizing scheme is presented in Fig. 6, p. 132.

As far as the oxidative metabolism is concerned, increasing amounts of NaCl in the incubating medium cause a decrease in the activity of all the enzymes studied up to now whatever their source is (Massey, 1953; McLeod and coworkers, 1958; Katunuma, Okada and Nishii, 1966; Weinberg, 1967; Gilles, 1969; Gilles, Hogne and Kearney, 1971). This indicates a decreased activity of the Krebs cycle and of its anaplerotic pathways (at least in the direction Krebs cycle intermediates—3-carbon units of glycolysis) with increasing NaCl concentration in the intracellular medium. Such interpretation is in agreement with the inhibition of the cycle observed when incubating different tissues in the presence of high NaCl concentrations or under

Table 9. Kinetics characteristics of glutamate dehydrogenase (GDH) and aspartate aminotransferase (GTO) from tissues of *Ericheir sinensis* acclimated to fresh-water (FW) or to sea-water (SW)

Tissue	GOT				GDH			
	FW		SW		FW		SW	
	V_{max}	$K_m (\times 10^{-4} \text{ mol l}^{-1})$	V_{max}	$K_m (\times 10^{-4} \text{ mol l}^{-1})$	V_{max}	$K_m (\times 10^{-5} \text{ mol l}^{-1})$	V_{max}	$K_m (\times 10^{-5} \text{ mol l}^{-1})$
Muscle	12.14	9.42 ± 1.53	11.94	7.44 ± 1.25	3.89	4.25 ± 2.67	3.54	3.63 ± 1.14
Gill	3.27	5.28 ± 1.18	3.70	4.79 ± 1.72	2.16	4.46 ± 1.89	2.82	4.12 ± 0.98
Hepatopancreas	1.13	4.41 ± 2.48	1.18	7.81 ± 3.02	—	—	—	—

V_{max} is given in optical density units per minute per mg of tissue DNA.
K_m values are those for α-ketoglutarate.
Modified from Gilles (1974a).

conditions inducing a modification of the intracellular ionic content (Tustanoff and Stewart, 1965; Gilles and Schoffeniels, 1968; Huggins and Munday, 1968; Sernka and Jackson, 1976). It is also in agreement with the decreased oxygen consumption and CO_2 production observed in isolated axons of *Callinectes sapidus* subjected to hyperosmotic stress (see p. 127).

As far as the sequence of amination–deamination of the amino acids is concerned, the activity of serine hydrolyase is decreased by increasing concentrations of NaCl (Gilles, 1969). On the other hand, the activity of alanine and aspartate aminotransferase is not affected (Turano, Fasella and Giartosio, 1962; Chaplin, Huggins and Munday, 1967; Huggins and Munday, 1968; Gilles, 1969). Increasing amounts of NaCl or KCl also induce an increase in the activity of reductive amination of glutamate dehydrogenase (Schoffeniels and Gilles, 1963; Chaplin, Huggins and Munday, 1965; Schoffeniels, 1966; Corman and Kaplan, 1967; Measures, 1975). It is worth noticing that under the same experimental conditions oxidative deamination of glutamate by the enzyme is inhibited (Snoke, 1956; Chaplin, Huggins and Munday, 1965) or unaffected (Measures, 1975). Thus it may be concluded tentatively that under conditions of increased ionic intracellular concentration, glutamate formation from a-ketoglutarate is enhanced. Under the same conditions, the activity of serine hydrolase is decreased thus leading to reduced deaminating activity and since the aminotransferase activity does not change while the oxidative activity of the Krebs cycle is decreased, it may be assumed that the equilibrium between amino acids and their ketoprecursors will be displaced, leading to a lower rate of reaction in the direction amino acids–ketoprecursors. Such an effect, when considered together with an unmodified or increased entrance of the amino acids in the metabolic pool, should result in an intracellular accumulation of these compounds. This concept is in agreement with the increase in amino acids and the decrease in ammonia production observed in different isolated tissues withstanding hyperosmotic stress. It should be born in mind however that the proposed scheme is only tentative, more information as to the intracellular localization of the enzymes, the ionic composition prevailing locally in the cell and the nature of the effect of salts on the activity of other enzymes (particularly on those implicated in ketoprecursor formation from 3-carbon units of glycolysis) are obviously needed before a more detailed picture can be produced.

Of importance also in this respect is the study of the possible interplay between ions and other different effectors in the modulation of the activity of key, allosteric enzymes. An example of such interactions can be found in the effects of NaCl on the reductive amination of a-ketoglutarate by glutamate dehydrogenase at different concentrations of a-ketoglutarate and NADH (Gilles, 1974c). Some of the results obtained are presented in Fig. 7. It can be seen that at low a-ketoglutate concentrations, NaCl is inhibitory as well as at higher substrate concentrations when the amount of NADH is low. On the contrary, NaCl becomes an activator at the high concentrations of NADH which induce a depression of the enzyme activity. The effect of NaCl on the formation of glutamate by glutamate dehydrogenase can thus go from inhibitory to activatory depending on the amount of substrate and coenzyme. On the other hand, serine is a positive modulator of the enzyme activity and the activations induced by both NaCl and serine are cumulative at high inhibitory concentrations of NADH (Fig. 7).

It is of interest to recall here that the concentration of NADH at which these effects are observed are in the range of the concentration found in tissues of crustaceans (Thabrew, Poat and Munday, 1971). These effects may

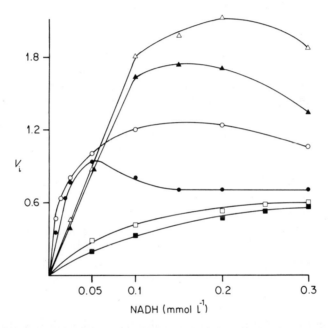

Figure 7 Modulation by NaCl and serine of the effect of NADH on the reductive amination activity of *E. sinensis* muscle glutamate dehydrogenase: □, control; a-ketoglutarate 5×10^{-4} mol l^{-1}; ■, a-ketoglutarate 5×10^{-4} mol l^{-1}; NaCl 300 mmol l^{-1}; ●, control; a-ketoglutarate 5×10^{-3} mol l^{-1}; ▲, a-ketoglutarate 5×10^{-3} mol l^{-1}; NaCl 300 mmol l^{-1}; ○, a-ketoglutarate 5×10^{-3} mol l^{-1}; serine 20 mmol l^{-1}; △, a-ketoglutarate 5×10^{-3} mol l^{-1}; serine 20 mmol l^{-1}; NaCl 300 mmol l^{-1}. Modified from Gilles (1974c)

thus be of physiological significance and provide the cell with feedback control mechanisms of the modulation of NaCl on glutamate formation depending on the cellular level of different amino acids or on the cellular availability of reducing equivalents and therefore on the energetic status of the cell. Besides, such interactions may be of interest in order to understand species differences observed in the evolution of oxygen consumption or of deamination activity following application of osmotic stresses (see also p. 141). Some of these differences seem at first sight to be best fitted in the framework of the fate of reducing equivalents between the amino-acid metabolism and the respiratory chain (Schoffeniels and Gilles, 1970; Schoffeniels, 1976). It is not the purpose of this review to enter into deep discussion about these hypotheses. Too many questions, notably about enzyme–modulator interactions, remain unsolved to produce a clear picture of the molecular events responsible for these phenomena.

A last question now arises: how can NaCl affect the activity of enzymes implicated in amino-acid metabolism? Since the first demonstrations of an NaCl effect at physiological concentrations on the activity of glutamate dehydrogenase (Snoke, 1956; Schoffeniels and Gilles, 1963), different studies have been made of this problem. The effect of NaCl appears to be independent of a variation of the ionic strength of the incubating medium. It must therefore be related to a specific effect of the cationic and/or of the anionic species. In fact, it appears that it is essentially the anionic species which is implicated in the modulation of enzyme activity (Chaplin, Huggins and Munday, 1965; Gilles, 1969; Gilles, Hogue and Kearney, 1971). However, a slight difference in effect between Na^+ and K^+ has been reported in the case of glutamate dehydrogenase (Snoke, 1956; Schoffeniels, 1966). On the other hand, it is worth noting that the series obtained when placing the various anions tested in order of increasing effectiveness is the same for all the enzymatic system studied up to now. This series is $Ac- < Cl- < B_2^- < NO_3$ and it has been obtained in the course of studies on some 23 enzymatic systems from plant or animal sources and including dehydrogenases, hydrolases and lyases (Massey, 1953; Walaas and Walaas, 1956; Rutter, 1957; Richards and Rutter, 1961; Fridovich, 1963; Chaplin, Huggins and Munday, 1965; Susz and coworkers, 1966; Warren and Cheatum, 1966; Warren, Stowring and Morales, 1966; Weinberg, 1967; Chimoskey and Gergely, 1968; Gilles, 1919; Gilles, Hogue and Kearney, 1971).

This order is similar to the one in which these anions disrupt the structure of various macromolecules as determined by physical methods. This, as well as the changes in enzyme thermosensitivity brought about by addition of NaCl to the incubation medium, are arguments to consider the anion effect as resulting from changes in configuration of the enzyme protein rather than from some indirect effect on association ion-substrate, substrate pK or electrostatic shielding (Von Hippel and Wong, 1964; see Gilles, 1974a, 1975 for a review).

4. *Conclusions*

To sum up the information reviewed it may be considered that the following mechanisms are at work in the control of the amino-acid intracellular concentration which occurs in isolated tissues subjected to osmotic stresses.

Increases in the plasmatic membrane's permeability and in the oxidation of amino acids play an essential part in the decrease in the amino-acid intracellular concentration which is observed in hypo-osmotic conditions. The permeability changes would be induced by configuration modifications in the membrane associated with tissue volume changes.

During hyperosmotic shock there is an increase in the intracellular amino-acid content. Up to now, this increase may be primarily associated with a decreased catabolism of the amino acids in the presence of an unchanged or eventually increased entrance of these compounds into the metabolic pool. An increased active influx of amino acids may help this process in situations in which amino acids are present in the incubation saline.

The changes in catabolic activity which are recorded under both hypo- and hyperosmotic conditions are most probably triggered by variations in the intracellular content of inorganic ions, particularly Cl^-.

The fact that two mechanisms are involved in the amino-acid adjustment during hypo-osmotic stress and that only one of these appears to be active in hyperosmotic shock might be related to the fact that the volume regulation process observed is much faster in the first experimental condition than in the second.

B. Mechanisms in the Whole Animal

Two important questions remain to be discussed in the context of this review. Are the mechanisms described at the level of isolated tissues also at work in the whole animal? On the other hand, could some additional mechanisms also play a part in the osmotic regulation processes of intact organism?

Metabolic adjustments and variations in transport activity are the basis of the control of the amino-acid pool in isolated tissues subjected to osmotic stresses.

Increased efflux of amino acids from tissues to blood in euryhaline species undergoing hypo-osmotic stresses is indicated by several studies showing that an increase in the blood amino-acid content takes place concomitantly to the decrease observed at the tissue level (Jeffries, 1966; Clark, 1968; Binns, 1969; Vincent-Marique and Gilles, 1970a, 1970b; Gérard and Gilles, 1972a; Gilles, 1977). The changes observed at the blood level are only transitory and, at least in the case of the chinese crab *Ericheir sinensis* (Gilles, 1977), the evolution of the blood amino-acid content can be related to the evolution of the tissue hydration. The concentration changes observed at the level of several amino acids are however far too large to be accounted for by a simple variation of the extracellular space volume due to modifications in cellular volume. It is thus reasonable to assume that the modification of the blood amino-acid level are due to an increased cell permeability as reported in various studies on isolated tissues.

Transitory decrease in blood amino-acid content during acclimation to concentrated media has been demonstrated in the case of two crustaceans: *Carcinus maenas* (Siebers and coworkers, 1972) and *Eriocheir sinensis* (Gilles, 1977). Such modifications might be related to the increased uptake of amino acids shown on isolated tissues subjected to hyperosmotic stress. In the case of *E. sinensis* however, the variations observed are small and follow an evolution comparable to that of tissue hydration. Thus it could be that they are related to the tissue volume changes occurring during acclimation to the concentrated medium. The transitory shrinkage of the tissues observed in these conditions should indeed induce a concomitant increase in the extracellular space which might account for a transitory dilution of the various blood constituents. Rao and coworkers (1969, quoted in Venkatachari and Keshavan, 1973) and Venkatachari and Keshaven (1973) reported a slight increase in blood amino nitrogen in two crustacean species *(Sesarma plicatum* and *Barytelpusa guerini)* after acclimation to concentrated media. Unfortunately no time scale experiments have been performed and tissue hydration or extracellular space volume have not been measured. It is therefore difficult to discuss these results in the light of the eventual importance of an amino-acid uptake mechanism during acclimation to an hyperosmotic environment in these species. Nevertheless, in the case of *C. maenas* the decrease in the blood amino-acid content seems too large to be accounted for by a simple dilution process.

More experiments would be needed however, particularly on the rate of incorporation of labelled amino acids from blood into the intracellular pool *in vivo*, to assess the importance of the active uptake mechanism in the adjustment of the tissular amino-acid content. Indeed this mechanism appears as the most likely process to account for the origin of the amino acids participating in the intracellular increase observed during acclimation to concentrated media. It is indeed unlikely that a *de novo* synthesis is involved for such a process cannot account for the increase in the level of the essential amino acids. Moreover, the existence of this mechanism should have been demonstrated during *in vitro* experiments on isolated tissues. On the other hand, as we have already discussed (see p. 125), a modification in the equilibrium between intracellular proteins and amino acids would not appear to be an important mechanism, neither would a modification between other cell macromolecular components and the ketoprecursors of amino acids. As a matter of fact, there is no significant variation in the concentration of lipids and polysaccharides during acclimation of the crayfish *Orconectes limosus* to a concentrated medium (Siebers, 1972).

As far as modifications in metabolic activity are concerned, several studies report on changes in NH_3 excretion of euryhaline species subjected to osmotic stresses (Needham, 1957; Jeuniaux and Florkin, 1961; Emerson,

1967; 1969; Mangum and coworkers, 1976). These results may be indicative of modifications in tissue deamination activity provided that they are concomitant to parallel changes in ammonia blood and tissue level. That it is indeed the case is indicated by the increase in NH_3 level of *Callinectes sapidus* blood (Gérard and Gilles, 1972a; Mangum and coworkers, 1976) or *Eriocheir sinensis* muscle tissue (Vincent-Marique and Gilles, 1970a) recorded during acclimation to diluted media. It thus seems that changes in the intracellular fate of NH_3 occur in whole animals subjected to osmotic stresses. In the light of the experiments performed on isolated tissues, these variations can be related to modifications in the activity of the amination–deamination sequence of the amino acids.

Changes in metabolic activity can also be deduced from experiments on whole animals showing modifications in CO_2 production and in oxygen consumption.

In the crayfish *Orconectes limosus* acclimated to a concentrated medium, CO_2 production decreases by 20% (Siebers, 1972). On the other hand, $^{14}CO_2$ production from injected labelled glutamate increases by 50% in *Carcinus maenas* acclimating to a diluted medium (Chaplin, Huggins and Munday, 1970). Such results are in agreement with the modifications in oxidative activity demonstrated on isolated tissues.

The modifications observed in the oxygen consumption cannot be interpreted so easily. Moreover, the variability of these results remains the subject of much controversy. This problem has been discussed at length by Gilles (1973b; 1974a). Summarizing the available data it can be stated that most euryhaline species show a higher oxygen consumption in diluted media than in concentrated ones. In some species, the oxygen consumption remains unaffected or is lower in a diluted or concentrated medium than in the normal one. As stated by Lee and Pyung (1970) the few crustaceans belonging to this last category, such as *Paratelplusa hydrodromus* or *Macrophtalmus japonicus*, always display a relatively higher oxygen consumption in diluted media than in concentrated ones. To our knowledge, three cases have been reported in which there is no change in oxygen consumption; these are the crustaceans *Eriocheir sinensis* (Schwabe, 1933); *Artemia salina* (Gilchrist, 1956) and *Palaemonetes varians* (McFarland and Pickens, 1965). Lee and Pyung (1970), however, showed that in *Eriocheir sinensis* the oxygen consumption is higher in diluted media. On the other hand, Kuenen (1939) measured higher oxygen consumption in *A. salina* acclimated to concentrated media while Eliassen (1953) reported that the oxygen consumption is decreased in this species in concentrated waters. It thus seems that, with an exception for these last few cases, oxygen consumption is always higher in diluted media than in concentrated ones. To our knowledge, this is also true for the few species which have been studied, not after acclimation, but during the acclimation period (Siebers and coworkers 1972). These modifications therefore become indicative of a higher oxidative metabolism in low salinities.

It would thus appear that the mechanisms, whose existence has been demonstrated on isolated tissues, can be traced down through the more complex physiological interactions occurring in whole organisms. However, could other mechanisms be implicated in tissue volume control and in the related control of the intracellular free amino-acid pool in whole animals? Data concerning this problem are almost totally lacking in the literature.

Early experiments by Duchâteau-Bosson and Florkin (1962) showed that removal of the eyestalk gland does not affect the modifications in intracellular amino-acid content in *Eriocheir sinensis* exposed to salinity stress. This might be an argument for considering that an hormonal control is not primarily involved in the amino-acid level adjustment. Schoffeniels (1976), however, recently showed that the concentration of the cyclic nucleotide 3'-5'-AMP undergoes variations in various tissues of *E. sinensis* under osmotic stress. Whether or not these modifications are implicated in the regulation of the amino-acid content or in the cell volume regulation process remain to be demonstrated however.

Up to the present there is thus no clear-cut indication of a possible role for hormonal interaction in the isosmotic regulation of the intracellular fluids.

On the other hand, the fate of amino acids in whole animals subjected to osmotic stress is most likely dependent on interactions between the different tissues.

An example of such interactions can be found in the reversible modifications of blood protein concentration that are observed following application of osmotic stresses to whole animals. Such changes have been reported so far in *Carcinus maenas* (Siebers and coworkers, 1972; Gilles, 1977), in *Astacus fluviatilis* and in *Eriocheir sinensis* (Gilles, 1977). In the three species studied, acclimation to a diluted medium causes a large increase in the blood serum protein concentration (Table 10). This phenomenon has been more particularly studied in the case of *E. sinensis* (Gilles, 1977). The variation can be traced down to the level of all the protein bands as revealed by acrylamide gel electrophoresis and particularly at the level of the proteins containing Cu^{2+}. This last result may be of interest since the variation in the content of these proteins might be part of a physiological adaptation related to the changes in oxygen demand that euryhaline species show upon acclimation to media of different salinities. As discussed previously (see p. 141), most euryhaline species show a higher oxygen consumption when acclimated to diluted media; this increase can be related, at least partly, to the activity of the Na^+ transport mechanism at work in the maintenance of the blood hyperosmotic state (see also Chapter 5). In such conditions of acclimation to diluted media, the increase in the blood level of proteins containing Cu^{2+}, essentially respiratory pigments, may help the animals in meeting the increased tissular oxygen demand. On the other hand, osmotic stresses may induce modifications in the efficiency of the oxygen carrier in many euryhaline species (Mangum, 1976). Thus it may be that the observed changes in Cu^{2+} blood proteins are also part

Table 10. Effect of acclimation to different media on the protein content of the blood serum and on the hydration of the muscle tissue of three euryhaline crustaceans

	Blood serum proteins (g l^{-1})		Muscle water content (% wet wt)	
	SW	SW/2	SW	SW/2
Carcinus maenas	28.10 ± 12.60(56)	48.78 ± 20.20(28) HS	76.77 ± 1.63(5)	76.26 ± 0.92(5) NS
	SW/2	FW	SW/2	FW
Astacus fluviatilis	33.20 ± 5.80(10)	50.50 ± 12.60(6) S	80.14 ± 0.60(10)	82.68 ± 1.30(10) NS
	SW	FW	SW	FW
Eriocheir sinensis	34.64 ± 8.12(9)	65.29 ± 14.20(17) HS	76.1 ± 1.7(10)	76.6 ± 1.8(10) NS

The animals have been acclimated for 1 month to sea-water (SW), fresh-water (FW) or to 50% sea-water (SW/2). Figures represent mean values ± standard deviation. HS: $P > 0.01$; S: $0.005 < P < 0.01$; N.S.: non-significant. After Gilles (1977); reproduced by permission of Pergamon Press.

of an adaptation to modifications in oxygen carrier efficiency. That acclimation to media of different salinities causes such modifications in the three species studied remains to be demonstrated however.

The observed changes in blood protein content cannot be related to variations in tissue hydration and therefore to modifications in the extracellular space volume. The temporal evolution of both phenomena is moreover completely different (see Table 10 and also Gilles, 1977). Therefore, the changes in the blood serum content of peptidic material are likely to be due to changes in their degradation and/or synthesis activity. In such a view, they could also be related to the variations in the amino-acid transport that occurs at the level of the plasma membranes following application of osmotic stresses. During hypo-osmotic shocks, part of the amino acids leaking out of the cells could be stored in the form of blood serum peptides and proteins. On the contrary, the hyperosmotic stress would induce an increased degradation and/or decreased synthesis of blood serum peptidic material. The free amino acids left available could then be transported actively into the cells. On the other hand it is interesting to consider that the modifications in blood protein content may help volume regulation at the cell level. The cells are indeed in a double Donnan equilibrium involving non-diffusible compounds on both sides of the plasmatic membrane. The increase in the number of negative, non-diffusible, charges in the extracellular fluid observed upon acclimation to a diluted medium may therefore reduce the swelling pressure the tissues have to withstand.

The amino acids leaking out of the cells during hypo-osmotic stress as well as the metabolic products of the increased catabolic activity which occurs in such condition (essentially NH_3 and CO_2) may also participate in other physiological sequences related to the overall osmoregulation processes in whole animals. A first interaction of this kind can be found in the modifications of the O_2 carrier efficiency that may induce the blood pH changes caused by the important variations in blood ammonia level occuring during osmotic stresses (readers interested in this problem are referred to the recent review by Mangum, 1976). When considering more particularly euryhaline fishes and crustaceans, another possible interaction is related to the intervention of the NH_3 and CO_2 coming from tissue catabolic activity, in inorganic ion transport at the gill level. Gills are known to be the essential tissue involved in the regulation of blood Na^+ and Cl^- concentrations in these animals (see Gilles, 1975; Chapter 5 of this volume). Accordingly, the active uptake of Na^+ that occurs at the gill level of animals acclimated to diluted media would imply an exchange with NH_4^+ ions and also possibly with H^+ ions. On the other hand, the active influx of Cl^- appears to be related to a mechanism of Cl^-/HCO_3^- exchange (see Chapter 5). During acclimation to diluted media, there is an increase in blood ammonia level and an increased tissular oxidative activity which should result in a larger CO_2 production. Both metabolic products can then be used at the gill level as counter-ion in the mechanisms of active uptake of Na^+ and Cl^-. The fact that the gill tissue of the euryhaline crab *Eriocheir sinensis* shows an important (Na^+-NH_4^+) ATPase activity may be considered as an argument favouring this hypothesis (Péqueux and Gilles, 1977). In the same way, part of the amino acids released from the tissues during the adjustment to hypo-osmotic conditions could be used at the gill level to provide this tissue with a supplement of ammonia and CO_2 required for Na^+ and Cl^- active transport. Recent experiments performed in our laboratory however show that a coupling Na^+/NH_4^+ can only account for a small part of the Na^+ uptake occurring in isolated perfused preparations of the three posterior pairs of gills of the chinese crab *Eriocheir sinensis*. This is illustrated in Fig. 8 where it can be seen that the influx of Na^+

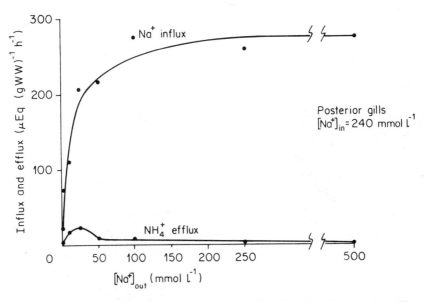

Figure 8 Influx of Na^+ and efflux of NH_4 in isolated perfused posterior gills of *Eriocheir sinensis* acclimated to fresh-water. The internal (perfusion) saline contains 240 mEq Na^+ l^{-1}. After Péqueux, Warnier and Gilles, unpublished

is much larger than the efflux of NH_4^+ whatever the salinity of the external medium. On the other hand, the total influx of Na^+ is not dependent on the addition of NH_4^+ to the perfusion saline. It thus appears that an important part of the Na^+ active uptake occurring in the gills of this species must be accounted for by another type of transport mechanism. An electrogenic Na^+ transport or an Na^+/H^+ coupling are the most likely processes which may be evoked. In this last hypothesis, oxidation of amino acids at the gill level could provide this tissue not only with the ammonia but also with at least part of the H^+ ions required for Na^+ uptake.

Oxidation of amino acids in the gills may also contribute in providing the Na^+ and Cl^- transport mechanisms with part of the energy they need. In the crab *E. sinensis*, acclimation to fresh-water results in a transitory increase in blood amino-acid level. This increase is particularly important at the level of proline (Vincent-Marique and Gilles, 1970b). Moreover, the gill tissue of this crustacean shows an important proline oxidative activity, this activity is particularly high in the posterior gill pairs which are responsible for the active movements of Na^+ and Cl^- in this species (Schoffeniels and Gilles, 1970). This indicates that amino-acid oxidation in gills may be related to the ionic active transport processes.

Further experiments should be needed however to assess critically the role of amino acids in ion transport mechanisms.

IV. CONCLUSIONS

To sum up, control of the amino-acid level, which participates in the adjustment of the intracellular fluid osmolarity during osmotic stresses and therefore in the volume regulation process that occurs under such experimental conditions, can be associated with two different basic mechanisms. The first involves control of the transport of amino acids in and out of the cells. The other is involved in the regulation of the catabolism of the amino acids. These mechanisms have been studied essentially at the level of isolated tissues but various results indicate that they are also at work at the level of whole animals. Under hypo-osmotic conditions, there is an increase in cell membrane permeability to the amino acids. This process is probably related to changes in membrane configuration due to tissue volume modifications. Concomitantly, there occurs an increase in the tissue oxidative metabolism and in the related amino-acid catabolism. During hyperosmotic stress, the active uphill transport of amino acids is enhanced and the amino-acid catabolism decreased.

The fact that in isolated tissues as well as on whole animals, volume regulation is much faster following application of an hypo-osmotic stress than in the reverse situation may be tentatively related to a difference in the rate of transport of the amino acids in and out of the cells in the two different experimental conditions.

The modifications in catabolic activity can be explained, at least partly, on the basis of modulation, by the intracellular ionic concentration, of the activity of key enzymes implicated in the amino-acid oxidative metabolism. However, this scheme has to be considered as tentative. More information on the intracellular localization of enzymes, on the ionic composition prevailing in their direct environment, on the nature of the salt effect on enzyme activity and also on the modulation of this effect by other effectors is obviously needed before a more detailed picture can be produced.

On the other hand, the proposed scheme leaves unanswered the question of

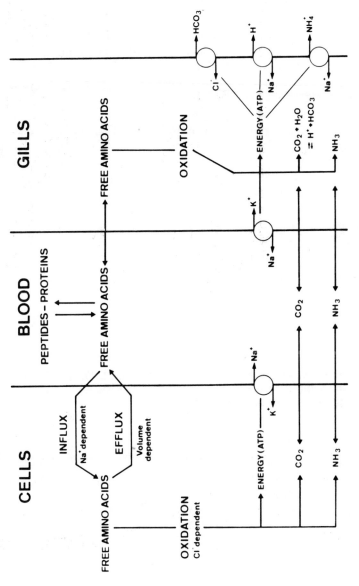

Figure 9 Possible fate of free amino acids during osmotic adjustment in euryhaline crustaceans and fishes

the origin of the amino acids participating in the increase of their intracellular concentration recorded during hyperosmotic shock. Blood peptidic material acting as a reservoir of amino compounds may be important in this respect at the level of the whole animal. When considering the whole organism, it also appears that the cellular mechanisms implicated in the control of the amino-acid concentration can be related with other physiological processes involved in the overall osmoregulation sequence. We have for instance pointed out the possible role of amino acids and of their catabolic products in the active transport of inorganic ions.

These different mechanisms are schematized in Fig. 9. The relationships suggested, if they explain satisfactorily most of the experimental observations, by no means exclude other regulatory processes. In this context, we have discussed the possibility of an endocrine control of the isosmotic regulation of the intracellular fluid. Studies in this field are totally lacking and information about such a mechanism should be of interest.

Acknowledgments

Part of the work described in this review and coming from the author's laboratory has been aided by grants 'Crédit aux Chercheurs' from the Fonds National de la Recherches Scientifique and be a grant No. 204511.76 from the Fonds de la Recherche Fondamentale Collective.

REFERENCES

Appelboom, J. W. T., W. A. Brodsky, W. S. Tutle, and I. Diamond (1958). The freezing point depression of mammalian tissues after sudden heating in boiling distilled water. *J. Gen. Physiol.,* **41,** 1153–69.

Baxter, C. F., and C. L. Ortiz (1966). Amino acids and the maintenance of osmotic equilibrium in brain tissue. *Life Sci.,* **5,** 2321–9.

Bedford, J. J. (1971). Osmoregulation in *Melanopsis trifasciata*. IV. The possible control of intracellular isosmotic regulation. *Comp. Biochem. Physiol.,* **40A,** 1015–27.

Bernstein, L. (1963). Osmotic adjustment of plants to saline media. II. Dynamic phase. *Am. J. Bot.,* **50,** 360–70.

Binns, R. (1969). The physiology of the antennal gland of *Carcinus maenas* (L.). IV. The resorption of amino acids. *J. Exp. Biol.,* **51,** 29–39.

Blinks, J. R. (1965). Influence of osmotic strength on cross-section and volume of isolated single muscle fibers. *J. Physiol.,* **177,** 42–57.

Boyle, P. J., and E. J. Conway (1941). Potassium accumulation in muscle and associated changes. *J. Physiol.,* **100,** 1–63.

Bricteux-Gregoire, S., Gh. Duchateau-Bosson, Ch. Jeuniaux, and M. Florkin (1962). Constituants osmotiquement actifs des muscles de crabe chinois *Eriocheir sinensis*, adapté à l'eau douce ou à l'eau de mer. *Arch. Int. Physiol. Biochim.,* **70,** 273–86.

Buckhold, B., R. B. Adams, and E. C. Gregg (1965). Osmotic adaptation of mouse lymphoblasts. *Biochim. Biophys. Acta,* **102,** 600–8.

Chaplin, A. E., A. K. Huggins, and K. A. Munday (1965). Ionic effects on glutamate dehydrogenase activity from beef liver, lobster muscle and crab muscle. *Comp. Biochem. Physiol.,* **16,** 49–62.

Chaplin, A. E., A. K. Huggins, and K. A. Munday (1967). The distribution of L-α aminotransferases in *Carcinus maenas. Comp. Biochem. Physiol.,* **20,** 195–8.

Chaplin, A. E., A. K. Huggins, and K. A. Munday (1970). The effect of salinity on the metabolism of nitrogen-containing compounds by *Carcinus maenas. Int. J. Biochem.,* **1,** 385–400.

Chimoskey, J. E. and J. Gergely (1968). Effect of ions on sarcoplasmic reticulum fragments. *Arch. Biochem. Biophys.,* **128,** 601–5.

Cholette, C., A. Cagnon, and P. Germain (1970). Isosmotic adaptation in *Myxine glutinosa.* I. Variations of some parameters and role of the amino acid pool of the muscle cells. *Comp. Biochem. Physiol.,* **33,** 333–46.

Clark, M. E. (1968). A survey of the effect of osmotic dilution on free amino acids of various polychaetes. *Biol. Bull.,* **134,** 252–60.

Conway, E. J., and J. I. McCormack (1953). The total intracellular concentration of mammalian tissues compared with that of the extracellular fluid. *J. Physiol.,* **120,** 1–14.

Corman, L., and N. O. Kaplan (1967). Kinetics studies of dogfish liver glutamate dehydrogenase with diphosphopyridine nucleotides and the effect of added salts. *J. Biol. Chem.,* **242,** 2840–6.

Doljanski, F., S. Ben-Sasson, M. Reich, and N. B. Grover (1974). Dynamic osmotic behaviour of chick blood lymphocytes. *J. Cell. Physiol.,* **84,** 215–24.

Duchâteau-Bosson, Gh., and M. Florkin (1962). Régulation isosmotique intracellulaire chez *Eriocheir sinensis* après ablation des pédoncules oculaires. *Arch. Int. Physiol. Biochim.,* **70,** 393–6.

Dunham, P. B., and D. L. Kropp (1973). Regulation of solute and water in *Tetrahymena.* In A. M. Elliot (Ed.) *Biology of Tetrahymena,* Dowben, Hutchinson and Ross Inc. pp. 165–97.

Eliassen, E. (1953). The energy metabolism of *Artemia salina* in relation to body size, seasonal rhythms and different salinities. *Natur. Vitensk. bekke,* **11,** 3–17.

Emerson, D. N. (1967). Some aspects of free amino acid metabolism in developing encysted embryos of *Artemia salina,* the brine shrimp. *Comp. Biochem. Physiol.,* **20,** 245–61.

Emerson, D. N. (1969). Influence of salinity on ammonia excretion rates and tissue constituents of euryhaline invertebrates. *Comp. Biochem. Physiol.,* **29,** 1115–33.

Florkin, M. (1962). La régulation isosmotique intracellulaire chez les invertébratés marins euryhalins. *Bull. Acad. R. Belg. Cl. Sci.,* **48,** 687–94.

Florkin, M., Gh. Duchateau-Bosson, Ch. Jeuniaux, and E. Schoffeniels (1964). Sur le mécanisme de la régulation de la concentration intracellulaire en acides aminés libres, chez *Eriocheir sinensis,* au cours de l'adaptation osmotique. *Arch. Int. Physiol. Biochim.,* **72,** 892–906.

Fredericq, L. (1901). Sur la concentration moléculaire du sang et des tissus chez les animaux aquatiques. *Bull. Acad. Belg. Cl. Sci.,* 428.

Fridovich, I. (1963). Inhibition of acetoacetic decarboxylase by anions. *J. Biol. Chem.,* **238,** 592–8.

Fugelli, K. (1967). Regulation of cell volume in flounder *(Pleuronectes flesus)* erythrocytes accompanying a decrease in plasma osmolarity. *Comp. Biochem. Physiol.,* **22,** 253–60.

Gainer, H., and H. Grundfest (1968). Permeability of alkali metal cations in lobster muscle. A comparison of electrophysiological and osmometric analyses. *J. Gen. Physiol.,* **51,** 399–425.

Gérard, J. F. (1975). Volume regulation and alanine transport. Response of isolated axons of *Callinectes sapidus* Rathbun to hypo-osmotic conditions. *Comp. Biochem. Physiol.*, **51A**, 225–9.

Gérard, J. F., and R. Gilles (1972a). The free amino-acid pool in *Callinectes sapidus* (Rathbun) tissues and its role in the osmotic intracellular regulation. *J. Exp. Mar. Biol. Ecol.*, **10**, 125–36.

Gérard, J. F., and R. Gilles (1972b). Modification of the amino-acid efflux during the osmotic adjustment of isolated axons of *Callinectes sapidus*. *Experientia*, **28**, 863–4.

Gérard, J. F., G. Dandrifosse, and R. Gilles (1974). Changes in the extracellular space of *Callinectes sapidus* axons during osmotic shocks. *Comp. Biochem. Physiol.*, **47A**, 315–21.

Gilchrist, B. M. (1956). The oxygen consumption of *Artemia salina* (L.) in different salinities. *Hydrobiologia*, **8**, 54–63.

Gilles, R. (1969). Effect of various salts on the activity of enzymes implicated in amino-acid metabolism. *Arch. Int. Physiol. Biochim.*, **77**, 441–64.

Gilles, R. (1972). Amino-acid metabolism and isosmotic intracellular regulation in isolated surviving axons of *Callinectes sapidus*. *Life Sci.*, **11**, pt II, 565–72.

Gilles, R. (1973a). Osmotic behaviour of isolated axons of a euryhaline and a stenohaline crustacean. *Experientia*, **29**, 1354–5.

Gilles, R. (1973b). Oxygen consumption as related to the amino-acid metabolism during osmoregulation in the blue crab *Callinectes sapidus*. *Net. J. Sea Res.*, **7**, 280–9.

Gilles, R. (1974a). Metabolisme des acides aminés et contrôle du volume cellulaire. *Arch. Int. Physiol. Biochim.*, **82**, 423–89.

Gilles, R. (1974b). Glucose metabolism during osmotic stress in isolated axons of *Eriocheir sinensis*. *Life Sci.*, **15**, 1363–9.

Gilles, R. (1974c). Studies on the effect of NaCl on the activity of *Eriocheir sinensis* glutamate dehydrogenase. *Int. J. Biochem.*, **5**, 623–8.

Gilles, R. (1975). Mechanisms of iono and osmoregulation. In O. Kinne (Ed.). Vol. 2, pt 1, chap. 5, Wiley-Interscience, London, New York. pp. 259–347.

Gilles, R. (1977). Effects of osmotic stresses on the proteins concentration and pattern of *Eriocheir sinensis* blood. *Comp. Biochem. Physiol.*, **56A**, 109–14.

Gilles, R., and J. F. Gérard (1974). Amino-acid metabolism during osmotic stress in isolated axons of *Callinectes sapidus*. *Life Sci.*, **14**, 1221–9.

Gilles, R., and F. F. Jobsis (1972). Isosmotic intracellular regulation and redox changes in the respiratory chain components of *Callinectes sapidus* isolated muscle fibers. *Life Sci.*, **11**, Pt. II, 877–86.

Gilles, R., and A. Péqueux (1977). Effect of salinity on the free amino acids pool of the red alga *Porphyridium purpureum* (= *P. cruentum*). *Comp. Biochem. Physiol.*, **57A**, 183–5.

Gilles, R., and E. Schoffeniels (1968). Influence du NH_4Cl et du sulfate de vératrine sur la synthèse et le pool des acides aminés au niveau de la chaîne nerveuse ventrale de deux crustacés (*Homarus vulgaris*, M.Edw. et *Astacus fluviatilis*, F.). *Arch. Int. Physiol. Biochim.*, **76**, 452–64.

Gilles, R., and E. Schoffeniels (1969a). Isosmotic regulation in isolated surviving nerves of *Eriocheir sinensis* Milne Edwards. *Comp. Biochem. Physiol.*, **31**, 927–39.

Gilles, R., and E. Schoffeniels (1969b). Metabolic fate of glucose and pyruvate in the nerve cord of *Homarus vulgaris* (M.Edw.) and *Astacus fluviatilis* (F.). *Comp. Biochem. Physiol.*, **28**, 417–23.

Gilles, R., P. Hogue, and E. B. Kearney (1971). Effect of various ions on the succinic dehydrogenase activity of *Mytilus californianus*. *Life Sci.*, **10**, pt II, 1421–7.

Gilles-Baillien, M. (1973). Isosmotic regulation in various tissues of the diamondback terrapin *Malaclemys centrata centrata* (Latreille). *J. Exp. Biol.*, **59**, 39–43.

Gordon, M. S. (1965). Intracellular osmoregulation in skeletal muscle during salinity adaptation in two species of toads. *Biol. Bull.,* **128,** 218–29.

Hendil, K. A., and E. K. Hoffmann (1974). Cell volume regulation in Ehrlich ascites tumor cells. *J. Cell. Physiol.* **84,** 115–26.

Hochachka, P. W., G. N. Somero, D. E. Schneider, and J. M. Freed (1970). The organization and control of metabolism in the crustacean gill. *Comp. Biochem. Physiol.,* **33,** 529–48.

Hoffmann, E. K. (1977). Control of cell volume. In Gupta, B. L., Moreton, R. B., Oschman, L. and Wall, B. J. (Eds) *Transport of Ions and Water in Animals,* Academic Press, London, New York. pp. 285–332.

Hoffmann, E. K., and K. B. Hendil (1976). The role of amino acids and taurine in isosmotic intracellular regulation in Ehrlich ascites tumour cells. *J. Comp. Physiol.,* **108,** 279–86.

Hoskin, F. C. G. (1962). Specificity of the stimulation by quinones of direct oxidation of glucose by brain slices. *Biochim. Biophys. Acta,* **62,** 11–16.

Huggins, A. K., and L. Colley (1971). The changes in non-protein nitrogenous constituents of muscle during the adaptation of the eel *Anguilla anguilla* L. from fresh water to sea water. *Comp. Biochem. Physiol.,* **38B,** 537–41.

Huggins, A. K., and K. A. Munday (1968). Crustacean metabolism. *Adv. Comp. Biochem. Physiol.,* **3,** 271–378.

Jeffries, H. P. (1966). International conditions of a diminishing blue crab population (*Callinectes sapidus*). *Chesapeake Sci.,* **7,** 164–70.

Jeuniaux, Ch., and M. Florkin (1961). Modification de l'excrétion azotée du crabe chinois au cours de l'adaptation osmotique. *Arch. Int. Physiol. Biochim.,* **69,** 385–6.

Jonsson, O. (1971). Effects of variations in the extracellular osmolarity on the permeability to nonelectrolytes of vascular smooth muscle. *Acta Physiol. Scand.,* **81,** 528–39.

Kaneshiro, E. S., G. G. Holz, and P. B. Dunham (1969). Osmoregulation in a marine ciliate, *Miamiensis avidus.* II. Regulation of the intracellular free amino acids. *Biol. Bull.,* **137,** 161–9.

Katunuma, N., M. Okada, and Y. Nishii (1966). Regulation of the urea and TCA cycle by ammonia. *Adv. Enz. Reg.,* **4,** 317–35.

Katz, J., and H. G. Wood (1960). The use of glucose-C^{14} for the evaluation of the pathways of glucose metabolism. *J. Biol. Chem.,* **235,** 2165–77.

Kleinzeller, A. (1965). The volume regulation in some animal cells. *Arch. Biol.,* **76,** 217–32.

Kregenow, F. M. (1971). The response of duck erythrocytes to non-hemolytic hypotonic media. Evidence for a volume controlling mechanism. *J. Gen. Physiol.,* **58,** 372–95.

Kuenen, D. J. (1939). Systematical and physiological notes on the brine shrimp, *Artemia. Arch. Nerland. Zool.,* **3,** 365–449.

Lang, M. A., and H. Gainer (1969). Isosmotic intracellular regulation as a mechanism of volume control in crab muscle fibers. *Comp. Biochem. Physiol.,* **30,** 445–56.

Lasserre, P., and R. Gilles (1971). Modification of the amino acid pool in the parietal muscle of two euryhaline teleosts during osmotic adjustment. *Experientia,* **27,** 1434–5.

Leaf, A. (1959). Maintenance of concentration gradients and regulation of cell volume. *Ann. NY Acad. Sci.,* **72,** 396–404.

Lee, B. D., and C. Pyung (1970). On the oxygen consumption of gill tissue of certain crab in relation to differences in habitat. *Pub. Mar. Lab. Pusan Fish. Coll.,* **3,** 53–6.

Maffly, R. H., and A. Leaf (1959). The potential of water in mammalian tissues. *J. Gen. Physiol.,* **42,** 1257–75.

Mangum, C. P. (1976). The function of respiratory pigments in estuarine animals. In *Estuarine Processes* Vol. 1: *Uses, Stresses and Adaptation to the Estuary*. Academic Press, New York, London. pp. 356–80.

Mangum, C. P., S. U. Silverthorn, J. L. Harris, D. W. Towle, and A. R. Krall (1976). The relationship between blood pH, ammonia, excretion and adaptation to low salinity in the blue crab *Callinectes sapidus*. *J. Exp. Zool.*, **195**, 129–36.

Massey, V. (1953). Studies on fumarase. 2. The effect of inorganic anions on fumarase activity. *Biochem. J.*, **53**, 67–71.

McFarland, W. N., and P. E. Pickens (1965). The effects of season, temperature and salinity on standard and active oxygen consumption of the grass shrimp *Palaemonetes vulgaris*. *Can. J. Zool.*, **43**, 571–81.

McLeod, R. A., C. A. Claridge, A. Hori, and J. F. Murray (1958). Observations on the function of sodium in the metabolism of a marine bacterium. *J. Biol. Chem.*, **232**, 829–34.

McRobbie, E. A. C., and H. H. Ussing (1961). Osmotic behaviour of the epithelial cells of frog skin. *Acta Physiol. Scand.*, **53**, 348–65.

Measures, J. C. (1975). Role of amino acids in osmoregulation of non-halophilic bacteria. *Nature*, **257**, 398–400.

Needham, A. E. (1957). Factors affecting nitrogen excretion in *Carcinides maenas* (Pennant). *Physiologica Comp. Oecol.*, **4**, 209–39.

Péqueux, A., and R. Gilles (1977). Osmoregulation of the chinese crab *Eriocheir sinensis* as related to the activity of the (Na^+–K^+) ATPase. *Arch. Int. Physiol. Biochim.*, **85**, 41–2.

Pierce, S. K., and M. J. Greenberg (1970). Free amino acid efflux from muscle hearts: a demonstration of volume regulation. *Am. Zool.*, **10**, 518.

Posnansky, M., and A. K. Solomon (1972). Regulation of human red cell volume by linked cation fluxes. *Membrane Biol.*, **10**, 259–66.

Potts, W. T. W. (1958). The inorganic and amino acid composition of some lamellibranch muscles. *J. Exp. Biol.*, **35**, 749–64.

Prosser, C. L., D. W. Bishop, F. A. Brown, T. L. Jahn, and V. J. Wulff (1950). *Comparative Animal Physiology*. Saunders, Philadelphia.

Richards, O. C., and W. J. Rutter (1961). Preparation and properties of yeast aldolase. *J. Biol. Chem.*, **236**, 3177–84.

Robertson, J. D. (1965). Studies on the chemical composition of muscle tissue. III. The mantle muscle of cephalopod molluscs. *J. Exp. Biol.*, **42**, 153–75.

Rosenberg, H. M., B. B. Shank, and E. C. Gregg (1972). Volume changes of mammalian cells subjected to hypotonic solutions *in vitro*: evidence for the requirement of a sodium pump for the shrinking phase. *J. Cell. Physiol.*, **80**, 23–32.

Roti-Roti, L. W., and A. Rothstein (1973). Adaptation of mouse leukemic cells (L5178Y) to aniostonic media. I. Cell volume regulation. *Expt. Cell Res.*, **79**, 295–310.

Rutter, R. W. J. (1957). The effect of ions on the catalytic activity of enzymes yeast glucose-6-phosphate dehydrogenase. *Acta Chem. Scand.*, **11**, 1576–86.

Schmidt-Nielsen, B. (1975). Comparative physiology of cellular ion and volume regulation. *J. Exp. Zool.*, **194**, 207–20.

Schatzlein, F. C., H. M. Carpenter, M. R. Rogers, and J. L. Sutko (1973). Carbohydrate metabolism in the striped shore crab, *Pachygrapsus crassipes*. I. The glycolytic enzymes of gill, hepatopancreas, heart and leg muscles. *Comp. Biochem. Physiol.*, **45B**, 393–405.

Schoffeniels, E. (1960). Rôle des acides aminés dans la régulation de la pression osmotique du milieu intérieur des insectes aquatiques. *Arch. Int. Physiol.*, **68**, 507–8.

Schoffeniels, E. (1966). The activity of the L-glutamic acid dehydrogenase. *Arch. Int. Physiol. Biochim.*, **74**, 665–76.
Schoffeniels, E. (1976). Biochemical approaches to osmoregulatory process in crustacea. In S. Davies (Ed.) *Perspectives in Experimental Biology*, Vol. 1, Pergamon Press, Oxford, New York. pp. 107–23.
Schoffeniels, E., and R. Gilles (1963). Effect of cations on the activity of L-glutamic acid dehydrogenase. *Life Sci.*, **2**, 834–9.
Schoffeniels, E. and R. Gilles (1970). Osmoregulation in aquatic arthropods. In M. Florkin and B. Scheer (Eds) *Chemical Zoology*, Vol. V, Academic Press, New York, London. pp. 255–86.
Schoffeniels, E. and R. Gilles (1972). Ionoregulation and osmoregulation in Mollusca. In M. Florkin and B. Scheer (Eds) *Chemical Zoology*, Vol. VII, Academic Press, New York, London. pp. 393–420.
Schwabe, E. (1933). Über die osmoregulation verschiedener krebse (Malacostracen). *Z. Vergl. Physiol.*, **19**, 183–236.
Sernka, T. J. and A. F. Jackson (1976). Hyperosmotic oxidation of glucose and decarboxylation of histidine in the gastric mucosa. *Physiol. Chem. Phys.*, **8**, 309–17.
Siebers, D. (1972). Mechanismen der intrazellulären isosmotischen regulation der aminosäurekonzentration bei dem flusskrebs *Orconectes limosus*. *Z. Vergl. Physiol.*, **76**, 97–114.
Siebers, D., C. Lucu, K. R. Sperling, and K. Eberlein (1972). Kinetics of osmoregulation in the crab *Carcinus maenas*. *Mar. Biol.*, **17**, 291–303.
Sigler, K., and K. Janáček (1971). The effect of non-electrolyte osmolarity on frog oocytes. I. Volume changes. *Biochim. Biophys. Acta*, **241**, 528–38.
Snoke, J. E. (1956). Chicken liver glutamic dehydrogenase. *J. Biol. Chem.*, **223**, 271–6.
Stoner, L. C., and P. B. Dunham (1970). Regulation of cellular osmolarity and volume in *Tetrahymena*. *J. Exp. Biol.*, **53**, 391–9.
Susz, J. P., B. Haber, and E. Roberts (1966). Purification and some properties of mouse brain L-glutamic decarboxylase. *Biochem.*, **5**, 2870–6.
Thabrew, M. I., P. C. Poat, and K. A. Munday (1971). Carbohydrate metabolism in *Carcinus maenas* gill tissue. *Comp. Biochem. Physiol.*, **40B**, 531–41.
Tosteson, D. C., and J. F. Hoffman (1960). Regulation of cell volume by active cation transport in high and low potassium sheep red cells. *J. Gen. Physiol.*, **44**, 169–94.
Turano, C., P. Fasella, and A. Giartosio (1962). On the effect of small ions on the activity of glutamic-aspartic transaminase. *Biochim. Biophys. Acta*, **58**, 255–261.
Tustanoff, E. R., and H. B. Stewart (1965). The effect of neutral salts on particle preparations from rat liver. II. Effects of neutral salts on tricarboxylic acid cycle reactions. *Can. J. Biochem.*, **43**, 359–72.
Van Slyke, D. D., A. B. Hastings, C. D. Murray, and J. Sendroy (1925). Studies of gas and electrolyte equilibria in blood. 8. The distribution of hydrogen, chloride and bicarbonate ions in oxygenated and reduced blood. *J. Biol. Chem.*, **65**, 701–28.
Van Slyke, D. D., D. H. McFadyen, and P. Hamilton (1941). Determination of free amino acids by titration of the carbon dioxide formed in the reaction with ninhydrin. *J. Biol. Chem.*, **141**, 671–80.
Venkatachari, S. A. T. (1974). Effect of salinity adaptation on nitrogen metabolism in the freshwater fish *Tilapia mossambica*. I. Tissue protein and amino acid levels. *Marine Biol.*, **24**, 57–63.
Venkatachari, S. A. T., and R. Keshavan (1973). Blood amino acid levels in the freshwater crab, *Barytelphusa guerini* H. Milne Edwards, as a function of salinity adaptation. *Ind. J. Mar. Sci.*, **2**, 65–8.
Vincent-Marique, C., and R. Gilles (1970a). Modification of the amino acid pool in

blood and muscle of *Eriocheir sinensis* during osmotic stress. *Comp. Biochem. Physiol.*, **35**, 479–85.

Vincent-Marique, C., and R. Gilles (1970b). Changes in the amino-acid concentration in blood and muscle of *Eriocheir sinensis* during hypoosmotic stress. *Life Sci.*, **9**, Pt II, 509–12.

Vislie, T., and K. Fugelli (1975). Cell volume regulation in flounder (*Platichthys flesus*) heart muscle accompanying an alteration in plasma osmolarity. *Comp. Biochem. Physiol.*, **52A**, 415–8.

Von Hippel, P. H., and K. Y. Wong (1964). Neutral salts: the generality of their effects on the stability of macromolecular conformations. *Science*, **145**, 577–80.

Walaas, E., and O. Walaas (1956). Kinetics and equilibria in flavoprotein systems. V. The effects of pH, anions and partial structural analogues of the coenzyme on the activity of D-amino acid oxidase. *Acta Chem. Scand.*, **10**, 122–33.

Warren, J. C., and S. G. Cheatum (1966). Effect of neutral salts on enzyme activity and structure. *Biochemistry*, **5**, 1702–7.

Warren, J. C., L. Stowring, and M. Morales (1966). The effect of structure-disrupting ions on the activity of myosin and other enzymes. *J. Biol. Chem.*, **241**, 309–16.

Webb, K. L., and J. W. A. Burley (1965). Dark fixation of C^{14}–O_2 by obligate and facultative salt marsh halophytes. *Can. J. Bot.*, **43**, 281–5.

Weinberg, R. (1967). Effect of sodium chloride on the activity of a soluble malate dehydrogenase from pea seeds. *J. Biol. Chem.*, **242**, 3000–6.

Wickes, M. A., and R. P. Morgan II (1976). Effect of salinity on three enzymes involved in amino acid metabolism from the american oyster, *Crassostrea virginica*. *Comp. Biochem. Physiol.*, **53B**, 339–43.

Part C

Control of the Extracellular Fluid Osmolarity

Chapter 5

Control Mechanisms in Crustaceans and Fishes

L. B. KIRSCHNER

I. Introduction	157
II. Passive Fluxes and Forces	158
III. The Invasion of Fresh-water: Hyperosmotic Regulation	162
A. Marine osmoconformers	162
B. Hyperosmotic regulation	165
C. Characteristics of active ion transport	170
D. Role of the kidney	181
IV. Hypo-osmotic/ Hypo-ionic Regulation	185
A. Problems faced by hyporegulators	185
B. Transport mechanisms in hyporegulators	191
C. Role of the kidney	199
D. Water intake	203
V. Isosmotic, Hypo-ionic Regulation	206
VI. Appendix	210
Acknowledgments	212
References	212

I. INTRODUCTION

Optimal cell function in active animals requires that the cellular environment have a well defined, relatively constant composition. In aquatic crustaceans and fishes this environment comprises the extracellular fluids (ECF) which are separated from the external medium by epithelia that must be permeable for purposes of nutrient, waste and gas exchange. Constant composition of the ECF can be maintained without energy expenditure if the potential energy of water and of each solute is the same as in the external environment. Differences in potential energy constitute forces that will cause dissipative or diffusive flows which will, in turn, change the composition of the ECF. In this case, a steady state can be maintained only if the organism generates a counterflow exactly equal to the diffusive leak, and such a counterflow requires an energy input. Regulation of the osmotic pressure, which is a measure of the chemical potential of water, and of the ionic composition of the body fluids involves just such a balance between diffusive leaks and active processes. The following discussion will deal with the adaptive strategies that allow animals to succeed in one environment and how these may limit or prohibit their success in another. The current state of our understanding of the regulatory mechanisms will be described and some outstanding questions explored briefly. However, we will see that the adaptive modifications are limited in number and that the same ones appear to be used by both crustaceans and fishes in coping with a particular combination of evolutionary history and current environmental composition. Since the emphasis will be on problems and their solutions, no attempt will be made to discuss the two groups of animals separately.

II. PASSIVE FLUXES AND FORCES

Although any modality of energy is capable of causing net solute or water movement the most common forces in living organisms are gradients of concentration and electrical potential across permeable interfaces; membranes and epithelia. For solutes, their effects are described by the well-known diffusion equation (refer to Chapters 1 and 2)

$$J_{12} = -AD \left(\frac{dc}{dx} + zFc_1 \frac{dE}{dx} \right) \tag{1}$$

where J_{12} is the net flux of the solute, A is the area of the permeable surface, D is the diffusion coefficient, z is the valence of the compound, F is the Faraday constant, c_1 is the concentration of the solute in compartment 1, and the derivatives are gradients of concentration and electrical potential. Another mechanism of dissipative solute movement occurs when the latter is carried from one compartment to another (as from animal to environment) as part of the bulk movement of fluid. Filtration of fluid out of compartment 1, for example, will cause a loss of solute equal to cJ_{12}^v where c is its concentration in the fluid, and J_{12}^v is the filtration rate from compartment 1 to compartment 2. Thus, the passive transfer of solute between the two compartments is described by the sum of these two events:

$$J_{12}^{net} = -AD \left(\frac{dc}{dx} + zFc_1 \frac{dE}{dx} \right) + cJ_{12}^v \tag{2}$$

The solution of this equation, i.e. the ability to predict the dissipative flow in a given circumstance, is sufficiently complex to tax the ingenuity of biophysicists working with such relatively simple preparations as isolated frog skins. It is a much more difficult task with intact organisms. However, the events described by equation (2) *are* treated in holistic animal studies, and they are worth making explicit.

1. Transfer of solute can be caused by gradients in electrochemical potential or by bulk fluid transfer as in loss through the urine. Either might provide the more important route in a particular circumstance, and neither should be neglected without convincing evidence that it is negligible.
2. In diffusive transfer the effective force is the sum of a concentration and an electrical gradient. For non-electrolytes $z = 0$ and the latter can be neglected, but for ions a potential difference across the exchange surface has the same qualitative effect as a concentration gradient. At room temperature a potential of 0.058 V (58 mV) causes the same net flux as a ten-fold difference in concentrations.
3. For a given gradient of electrochemical potential, the magnitude of diffusive transfer is conditioned both by the characteristics of the exchange surface (described by the diffusion coefficient, D) and the area (A) available for diffusion. The product $A \times D$ is effectively a permeability (though not the 'permeability coefficient' usually used). It is not always appreciated that an apparent permeability change can be caused by modifying the pattern of blood flow (not merely its rate), so that more or less of an exchange surface is available for material transfer.
4. Net movement of solute may also be caused if bulk movement of fluid occurs. Such a flux is independent of the electrochemical gradient between an animal and environment. Its rate depends solely on the rate of fluid movement and its concentration in the excreted fluid. In most animals solute loss in the urine is described by this term, as is salt influx by absorption of fluid from the intestine.

The net diffusive transfer of water is described by a different expression, although the underlying physical relationship is the same: net movement is proportional to a difference in chemical potential. The chemical potential of water in a solution can be expressed by the solution osmotic pressure (Π), which is related to the solute concentration ($\Pi = RTc$, where R is the universal gas constant, T is the absolute temperature of the solution, and c is its concentration). For two solutions separated by a membrane impermeable to the solute, osmotic water flow is given by

$$J^v_{12} = AL_p(\Delta \Pi) = AL_p RT(c_2 - c_1) \qquad (3)$$

Again, we have an expression in which flow is proportional to force. Here the force, a difference in solution osmotic pressures, is expressed by the difference in solute concentrations, and the proportionality constant $(AL_p RT)$ is a permeability to water. If the exchange surface is permeable to the solute as well as to water the situation is more complex. Such a complication occurs in most animals; it will be alluded to later, but need not be treated here (see also Chapters 1 and 2).

Another general point needs emphasis. Since the main osmotically active solute in both animal body fluids and most natural bodies of water is NaCl the existence of an osmotic gradient between ECF and environment means that there are equal but oppositely oriented gradients for salt and water diffusion. As a result, the problem of osmotic regulation is intimately coupled with that of regulation of the major ionic constituents in the body fluids of most animals. The two are also coupled in another fashion suggested by the last member of equation (1). For an animal to remain in a volume steady state in the face of osmotic water transfer, the passive flow must be balanced by some active mechanism capable of moving water at exactly the same rate in the opposite direction. As we will see, the necessary 'active' flux (excretion of urine or intestinal fluid absorption) is invariably coupled to solute movement ($= cJ_v$) in the same direction as diffusive solute leak across the body surface. As a consequence the mechanisms that have developed to cope with osmotic water movement actually aggravate the problem of maintaining an ionic steady state by creating a second path for passive leakage of NaCl.

Finally, equations (2) and (3) provide a compact description of the range of strategies available for maintaining hydromineral homeostasis. These fall into two categories, the first which we will call *evasive*, and which includes all changes tending to minimize diffusive leaks. This can be accomplished by minimizing the diffusion gradients between ECF and the external medium. Indeed, diffusive flows of both water and NaCl will, to a first approximation, be abolished if the ionic concentration of ECF is the same as that in the environment. This is exactly the situation in most marine invertebrates which have little or no osmoregulatory problem. It remains true for some fairly euryhaline invertebrates as they invade dilute environments; ECF salt concentration simply drops to match that of the more dilute environment. But the range of such osmotic conforming animals is restricted, and they are certainly excluded from fresh-water: no animals are known with bloods more dilute than about 5–10% SW. Alternatively, for a given gradient passive leaks across the body surface will be lower if the diffusion coefficient (D for solute, L_p for water) is reduced and/or if the area open to exchange is small. As shown in the equations both factors enter into what is called 'the permeability', and the question to what extent animals alter one or the other is both interesting and poorly explored. In any event, a complete description of osmotic and ionic regulation requires that the permeability to water and to NaCl be characterized. This can be done by measuring their fluxes and the existing gradients; the flux/gradient ratio is a permeability coefficient, usually denoted P. A number of techniques have been used to estimate P. Unfortunately, the values obtained appear to be at least partly a function of the measurement technique. This problem will receive further treatment in the Appendix to this chapter. In media which are very hypo- or hyperosmotic to the ocean the ECF is always regulated out of equilibrium with the environment and the body surface, which must remain open for gas and nutrient exchange, is a path for solute and water leak. Evasive mechanisms can minimize these leaks, but offer only a partial solution to the problem which must therefore be solved by a second group of mechanisms that we will call *compensatory*. These are required to create flows of ions and water exactly equal in magnitude to the diffusive flows, but in the opposite direction. Since the compensatory movements must occur against gradients of potential, they are energy requiring, and the underlying mechanisms have been studied intensively for the past 40 years.

The above description of the osmoregulatory problem and the components of its resolution is well illustrated by studies on aquatic crustaceans. Indeed, the general principles emerge from such studies. The group is predominantly aquatic and only a few species can be described as moderately successful in terrestrial habitats. While most are marine animals crustaceans can be found in virtually any body of natural water, from soft fresh to crystalizing brine. Consequently, the group includes members with remarkably different degrees of tolerance to osmotic demands. Some are confined to the ocean (stenohaline, marine) others to fresh-water (stenohaline, fresh-water). A number of species can cope with a moderately broad range of salinities, and hence can range from the ocean into an estuarine environment and may even penetrate into rivers. And a few, like the brine shrimp *Artemia salina* can cope with environments as extreme as nearly fresh-water on the one hand and saturated NaCl on the other. In each case the ability to cope with a particular environmental situation is governed by the physical principles described above. A disproportionate fraction of the analytical work has been done on members of the order Decapoda, and that is why I have placed what may appear to be an undue emphasis on that group. Important, if less extensive research has been done on one group in the Amphipoda (the Gammaridea) and on one group of Anostraca and will be described below. Other members of this large class have been studied only sporadically or not at all.

The aquatic vertebrates are also successful in environments ranging from the most dilute to bodies as concentrated as the Dead Sea. In some instances their osmoregulatory behaviours are remarkably similar to those of crustaceans in the same habitat. Thus, the mechanisms permitting fresh-water (FW) teleosts and crayfish to cope with dilute habitats are almost indistinguishable. The same thing can be said of the adaptive behaviours that permit the brine shrimp *Artemia salina* and a teleost like *Aphanius dispar* to cope with a range of salinities extending from nearly fresh-water to bodies like the Dead Sea or the Great Salt Lake which are much more concentrated than the ocean. In fact, the sole feature that is unique to some members of one group is the ability of the elasmobranch and crossopterigian fishes to maintain themselves in osmotic equilibrium with sea-water by elevating the blood concentration of a solute other than NaCl. No other fish and none of the crustaceans appear to have developed such a capability.

III. THE INVASION OF FRESH-WATER: HYPEROSMOTIC REGULATION

A. Marine Osmoconformers

Most marine crustaceans are osmoconformers. In these species, hemolymph concentrations of Na⁺ and Cl⁻, are essentially the same as in SW as is shown in Table 1. Although other, less abundant, solutes may be maintained out of equilibrium (Robertson, 1957) total osmotic pressure of the blood is about the same as in the environment. A few measurements have been made of potential differences across the body surface of such animals, and all have been reported to be within 1–2 mV of zero. Thus, there is virtually no electrochemical gradient for NaCl or for H_2O across the body surface, and hence no diffusive flows. As long as these animals are in SW there exists no osmoregulatory problem regardless of whether or not the body surface is permeable to salt and water. The chief constraints on the range of salinity tolerance in osmoconformers are their inablity to maintain even a relatively fixed ECF composition in the face of changes in the medium, coupled with an inability of their cells to function when the ECF changes. If the crab *Maja* sp. is abruptly transferred from SW to 80% SW gradients are created favouring osmotic water entry and diffusive salt loss. The results of such an experiment are shown in Fig. 1. The animal gained weight rapidly indicating that the body surface is water permeable. At the same time the osmotic pressure of the blood fell, and dilution was faster than could be accounted for by the water entering, hence salt was lost from the body (Schwabe, 1933; this type of experiment has been performed on many crustaceans. The literature is summarized in Potts and Parry, 1964b and Prosser, 1973). It was not always appreciated that in such an experiment urine flow is markedly increased and represents a route for part of the salt loss observed ($= c_{urine} J_v^{urine}$). However, even when this is taken into account there remains a substantial loss of salt across the body surface. The fact that they come to equilibrium rapidly with more or less concentrated environments means that they lack mechanisms for maintaining the ECF at any concentration other than that of the environment. As a result, these animals are stenohaline and in experiments like that shown in Fig. 1, 5 our of 8 crabs were dead within 16 hours. Permeability coefficients for ions (P_{Na} or P_{Cl}) and water (P_{osm}) are shown for several marine crabs in Table 1; the values for these very permeable animals provides a basis for comparison with others.

It is probable that the excretory organ, the antennal gland, could handle an increased water load when these animals are in a dilute medium. It has been shown to increase its output with increasing dilution (Cornell, 1976), and hence has the capacity to compensate for osmotic water entry. The limiting factor is lack of an active ion transport system capable of absorbing salt as fast as it is lost across the body surface and in the urine when a gradient exists.

Only one vertebrate order, the Myxinoidea (hagfishes), comprising a few

Table 1. Permeability of conformers and hypertonic regulators

Animal	Medium	$\Delta\Pi^b$ (mosmol l^{-1})	P_{osm}^a (l mosmol^{-1} h^{-1})	ΔNa^b (mmol l^{-1})	P_{Na}^a (l kg^{-1} h^{-1})	Reference
Crustaceans						
Maja verrucosa	SW	~0	0.100 (1)	0	0.30 (1)	Schwabe (1933)
Libinia emarginata	SW	~0	0.185 (1)	0	—	Cornell (1976)
Pugettia producta	SW	~0	0.330 (1)	0	0.30 (1)	Cornell (1976)
Porcellana platycheles	SW	~0	0.160 (1)	0	0.028 (1)	Davenport (1972)
Eupagurus bernhardus	SW	~0	0.430 (1)	0	0.068 (1)	Davenport (1972)
Carcinus maenas	SW	~0	—	0	0.054 (2)	Shaw (1961a)
Carcinus maenas	40% SW	210	0.042 (2)	120	0.047 (2)	Shaw (1961a)
Cancer magister	50% SW	250	0.020 (2)	120	0.041 (2)	Hunter (1973)
Eriocheir sinensis	FW	600	0.003 (2)	290	0.009 (2)	Shaw (1961b)
Pseudothelphusa jouyi	FW	470	0.0004 (2)	260	0.005 (2)	Thompson (1970)
Potamon niloticus	FW	540	0.0004 (2)	260	0.003 (2)	Shaw (1959a)
Astacus fluviatilis	FW	450	0.009 (2)	200	0.001 (2)	Bryan (1960)
Vertebrates						
Eptatretus stoutii	SW	~0	0.034 (1)	—	—	McFarland and Munz (1965)
Carassius auratus	FW	350	0.039 (2)	160	0.003 (3)	Maetz (1963)
Salmo gairdneri	FW	310	0.015 (2)	150	0.001 (3)	Fromm (1963); Kerstetter, Kirschner and Rafuse (1970); Sharratt, Chester-Jones and Bellamy, (1964)
Anguilla anguilla	FW	260	0.016 (2)	140	0.002 (3)	Motais (1967)
Platichthys flesus	FW	270	0.009 (2)	150	0.002 (3)	Motais (1967)

[a] The numbers in parentheses refer to one of the methods described in the Appendix.
[b] Concentration differences between ECF and external medium.

species of marine animals, has blood with nearly the same composition as SW (Table 1). Blood Na^+ is slightly higher and divalent ions significantly lower, but the latter contribute little to the total ECF osmotic pressure which is virtually the same as ambient. When the external medium is diluted or concentrated over the range 600–1500 mosmol l^{-1}, ECF concentration changes and comes into osmotic equilibrium with the environment (Cholette,

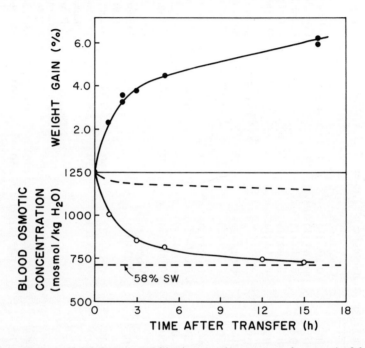

Figure 1 Water and solute permeability in a marine osmoconformer. At 0 hours the crab (*Maja*) was transferred from SW to 58% SW. Weight gain reflects water entry. Decrease in blood osmotic concentration is due primarily to loss of solute, but dilution by the water entering contributes. The initial slope of the lower curve, corrected for water entry, gives the net efflux of solute (primarily NaCl). Drawn from data in Schwabe (1933)

Cagnon and Germain, 1970). Thus, these fishes are osmoconformers. The rate of water entry is considerably slower than in *Maja verrucosa*, and P_{osm}, calculated from net water influx, is about an order of magnitude lower (McFarland and Munz, 1965). Salt efflux is much smaller than in the marine crabs. The reason for this difference is not known, but it can be viewed as preadaptive for the invasion of dilute media. That these animals are confined in nature to SW is apparently due to lack of a mechanism, active inward NaCl transport, needed to compensate for salt loss. In this regard they resemble the osmoconforming stenohaline crabs.

B. Hyperosmotic Regulation

While crustaceans and vertebrate osmoconformers can tolerate only modest dilution in the laboratory, and may be confined to SW in nature, other species have developed mechanisms enabling them to maintain hemolymph concentrations above that of a dilute environment and at levels compatible with normal tissue functions. Under these conditions animals face diffusive

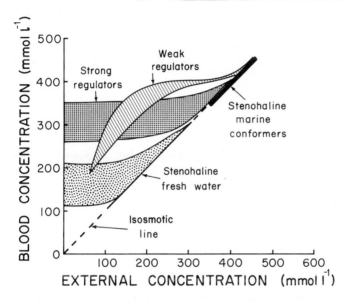

Figure 2 Patterns of hyperosmotic regulation of the extracellular fluids. The range for each group is suggested by the termini of the corresponding pattern. Note that only a few species may extend into the extreme of the range. For example, most 'strong regulators' cannot live in very dilute FW, but the group includes potamonid crabs which can live in media as dilute as stenohaline FW animals

loss of solute and osmotic water loading for reasons outlined above. To maintain the ECF in a steady state requires the expenditure of energy. In turn, the amount of energy required will depend on the degree to which an animal uses the evasive mechanisms; lowering body fluid concentrations to minimize diffusion gradients and reducing the ion and water permeability of the body surface. We will see that their range in dilute media is conditioned in large measure by their ability to manipulate these parameters, and that among them the water permeability of the body surface is probably the most important. Crustaceans capable of living in water more dilute than about 80% SW are able to regulate ECF concentrations above those of the environment. Figure 2 is a synthesis of many studies on ECF concentrations in animals adapted to a range of media from SW to FW. It suggests that there are four patterns. Osmoconformers, as noted, simply remain in osmotic equilibrium with their environment; among crustaceans and vertebrates they are virtually confined to SW. A second group can invade more dilute media and maintain ECF osmotic and NaCl concentrations above ambient but below SW. Below about 40% SW regulation breaks down and body fluid concentrations decrease in parallel with those of the environment. A third group, called strong regulators, has representatives found over the complete range of environments from SW, in which they are isosmotic and isoionic, to FW. The range for individuals varies. Some extend from SW into potable water (*Callinectes sapidus*), others from SW into reasonably soft FW (*Eriocheir sinensis*). A few are even confined to FW, having lost the ability to live in SW. But all of them regulate their ECF NaCl between 250–350 mmol l^{-1} in dilute media. The fourth group has ECF concentrations between 100–200 mmol l^{-1}, substantially lower than any of the others, and most are capable of excreting a urine more dilute than hemolymph, which is unique to these animals. Most of these animals inhabit soft FW (NaCl < 1 mmol l^{-1}) and regulate blood solute concentration within very narrow limits when exposed to waters of increasing salinity until the latter are about the same as blood. At this point regulation breaks down and ECF osmolarity rises with that of the medium. But they are unable to survive even in 50% SW, hence the term stenohaline FW. While the division may be less sharp in nature than in Fig. 2 the latter has both ecological and physiological correlates that make it useful for discussion.

The weak regulators include many estaurine and shore crabs including *Carcinus maenas* (Schwabe, 1933; Shaw, 1961a) and *Cancer magister* (Hunter and Rudy, 1975). Body fluid concentrations remain relatively stable near 400 mmol l^{-1} at ambient concentrations between SW and 40–60% SW. In more dilute media steady-state ECF concentration drops more or less in parallel with the isosmotic line. In one sense this represents a failure of the regulatory mechanism, but the outcome is a reduction in diffusion gradients. This is one of the evasive mechanisms for reducing diffusive leaks, and hence is adaptive for the organism. The efficacy of this change will be limited only by the degree of ECF dilution compatible with tissue function. Table 1 shows that for two representatives in this group, *Carcinus maenas* and *Cancer magister*, there is also a substantial reduction in osmotic permeability of the body surface (P_{osm}) compared to that of osmoconformers. In contrast, salt permeability (P_{Na}) appears to be within the range of stenohaline conformers and is unchanged when the crabs are transferred from SW to dilute SW. Some of the consequences of the changes can be illustrated by considering salt and water flows in *C. maenas* in 40% SW. Diffusive salt loss under these conditions amounts to about 5.6 mmol kg^{-1} h^{-1}. In fact, the total loss of salt is greater than this. Osmotic water entry occurs at the rate of 8–10 ml kg^{-1} h^{-1}, and this is excreted through the antennal gland to maintain the animal's volume constant (Nagel, 1934; Binns, 1969). The urine of *C. maenas* has been shown to have essentially the same NaCl concentration as blood (about 300 mmol l^{-1} in 40% SW), and salt loss due to volume regulation, about 2–3 mmol kg^{-1} h^{-1}, may account for as much as $\frac{1}{4}$ of the total (21% in measurements by Shaw, 1961a). The implications are interesting and emphasize the extreme importance of reducing P_{osm} in hyper-regulators. On the one hand, *C. maenas* would still be a 'leaky' animal even if ion permeability were reduced to zero, and the demands on the active transport mechanism, although reduced, would be of the same order of magnitude. However, if the osmotic permeability remained in the range found for osmoconformers urine volume would be nearly an order of magnitude larger, and salt loss through the urine, about 15 mmol kg^{-1} h^{-1}, would exceed diffusive loss to such an extent that a reduction in salt permeability would be meaningless.

If the parameters for diffusive leaks are important those governing active inward transport are equally so, since the two fluxes must be equal in the steady state. The rate of active ion transport across epithelia is known to increase in a non-linear fashion with external Na$^+$ concentration and to approach an upper limiting value, the maximum transport rate (J_{max}). This behaviour of transport systems will be discussed in more detail later. It suffices here to note that Shaw's (1961a) estimate of J_{max} in *C. maenas* was about 6.8 mmol Na$^+$ kg^{-1} h^{-1}, or about the same as the combined diffusive and urinary salt loss in 40% SW, hence the transport system must be operating near its maximum. Any increase in permeability will result in uncompensated salt loss and hence a fall in blood concentration. Or, if the animal moves into a more dilute environment the diffusive and urinary fluxes will increase because the gradients are larger, but the active influx will actually fall to the value determined by the new (lower) external Na$^+$ concentration. The more dilute the medium the greater will be net salt loss and the attendant fall in blood concentration. Since these animals are rarely found below about 25% SW it can be presumed that their cells cease to function at ECF concentrations around 200 mmol l^{-1}.

Another, larger group of crustaceans displays a somewhat different behaviour. This group, called strong regulators in Fig. 2, includes many decapods and some amphipods. The animals are isosmotic and iso-ionic in sea-water; indeed, many come into equilibrium with the environment at concentrations well below SW and remain at equilibrium as the medium becomes more concentrated. But they have the capacity to maintain internal concentrations more nearly constant over a range of external salinities extending, for some of them, into fresh-water. The blue crab (*Callinectes sapidus*) can be found in river mouths, and *Eriocheir sinensis* spends most of its adult life in rivers with a very low solute concentration, but returns to the sea to breed. A similar range is exhibited by the amphipod, *Gammarus duebeni*. The crabs have some characteristics which are basically marine, notably a relatively high NaCl concentration in the blood and the excretion of urine that is isosmotic and nearly isoionic with blood. Nevertheless, some are well adapted to life in fresh-water and at least one, *Potamon niloticus*, is unable to live in media more concentrated than about 50% SW. The fresh-water crabs *Pseudothelphusa jouyi* and *Metopaulias depressus* survived in 80% SW but not in SW (Thompson, 1970). Blood concentrations for these strongly regulating FW crabs, although reduced, generate formidable gradients across the body surface. Nevertheless, passive loss of Na^+ across the body surface in *E. sinensis* is only about 1.6 mmol kg^{-1} h^{-1} when the crab is in FW; this is less than half that for *C. maenas* in 40% SW in spite of the fact that the gradient is much steeper in the FW crab. The value of P_{Na} must be much reduced, and this is shown in Table 1. Table 1 also shows that P_{osm} for *Eriocheir* is less than 10% that for *C. maenas* and nearly 2 orders of magnitude lower than in SW animals. Water uptake is correspondingly restricted, even though the osmotic gradient is large. Again, the real significance of the reduced P_{osm} is in limiting the loss of solute as the following example shows. When *E. sinensis* is in FW urine volume is about 1.7 ml kg^{-1} h^{-1} (Scholles, 1933), and its NaCl concentration is about the same as in blood; only about 0.4 mmol NaCl kg^{-1} h^{-1} is lost via the urine; this is only about 15% of the total loss rate (Shaw, 1961b). If P_{osm} in *E. sinensis* were as large as in *C. maenas* urine would have to flow about 10 times faster, and salt excretion would be about 5 mmol kg^{-1} h^{-1}. The total loss rate would be 3 times faster (about 8 mmol kg^{-1} h^{-1}), nearly 2/3 via the urine, and would far exceed the maximum rate at which the ions could be recovered by active transport, since J_{max} for Na^+ is only about 2.4 mmol kg^{-1} h^{-1} (Table 2). The same evasive strategies are used by FW crabs like *Potamon niloticus* and *Pseudothelphusa jouyi*. Salt gradients are slightly smaller than in *E. sinensis*, but in the same range. P_{Na} is substantially lower (about 50%) in *P. jouyi* and lower still in *P. niloticus* (Table 1). And water permeability is lower by an order of magnitude in the latter. Thus, salt loss across the body surface and in the urine are both less than in *E. sinensis*. The total loss rate in *P. niloticus* is only 0.8 mmol kg^{-1} h^{-1}, and only 20% as much energy need be expended by active transport mechanisms in maintaining a steady state. As a consequence, *P. niloticus* and *P. jouyi* can invade softer water than can *E. sinensis*. The minimum environmental concentration at which these animals can remain in salt balance is about 0.05 mol l^{-1} for *P. niloticus* (Shaw, 1961a) and 0.1 mmol l^{-1} for *P. jouyi* (Thompson, 1970), but about 0.5 mmol l^{-1} for *E. sinensis* (Koch, Evans and Schicks, 1957).

Another extensively studied pattern of osmoregulatory behaviour is shown by decapod crustaceans like the crayfish and some amphipods (*Gammarus pulex, G. locusta*), and also most FW fish, which are confined to dilute media and cannot live in SW. In this characteristic they differ from weak and most strongly regulating animals. However, as noted, a few crabs are confined to FW and are equally successful there. For example, the minimum ambient NaCl concentration in which FW fish or crayfish can maintain an ionic steady state is in the range 10^{-5}–10^{-4} mol l^{-1}. This is also about the minimum concentration range for *P. niloticus* and *P. jouyi*; if there are differences, they must be small and probably insignificant in natural waters. But the fourth group is distinct in two regards. First, as shown in Fig. 2, their blood concentrations are lower than the others, usually between 100–200 mmol l^{-1}. Thus, gradients for salt loss and water entry are reduced. In addition, the urine produced by their kidneys is very dilute; the osmotic concentration is usually less than 10% that of plasma. Since the presumptive urine begins as a plasma ultrafiltrate most of the filtered solute must be reabsorbed from a relatively water-impermeable region, and this appears to be a distal tubular segment found only in these animals. The two characteristics are, in fact, diagnostic of complete adaptation to FW.

It is curious that although these characteristics have long been known to be associated with life in FW their adaptive significance is by no means clear. Reduction of the ECF concentrations significantly minimizes the gradients for water and NaCl diffusion. However, as shown in Table 1, P_{osm} is higher than in FW crabs, and osmotic inflow and concomitant urine volume is actually larger. Thus, the low ECF osmotic concentration is not accompanied by a reduction in the water load. However, P_{Na} is smaller in this group than in any other, and together with the low internal concentration, limits diffusive ion loss. In the crayfish it is only about 25% that in *P. niloticus* (Bryan, 1960), and the active transport systems are spared a corresponding amount of work.

The contribution of a dilute urine to life in FW is also less obvious than is sometimes assumed. It clearly contributes less to the total rate of salt loss than would an isoionic urine. In the crayfish *Astacus fluviatilus*, urine NaCl concentration is only about 3% that of hemolymph, and urinary loss is about 0.02 mmol kg^{-1} h^{-1}, only about 10% of the total. In fishes a greater proportion of the total is urinary. For example, Na$^+$ loss across the body surface in rainbow trout is about 0.24 mmol kg^{-1} h^{-1} (Kerstetter, Kirschner and Rafuse, 1970), while that in the urine is about 0.4 mmol kg^{-1} h^{-1} (Fromm, 1963). The difference is due to the amount of urine produced by the more *water*-permeable vertebrate. In both cases salt loss would be much more rapid if the urine were iso-ionic. But what is often overlooked is the fact that reabsorption of Na$^+$ and Cl$^-$ from an ultrafiltrate in the kidney requires active transport, and there is no compelling evidence that transport in the excretory organ is energetically less expensive than across the body surface. This question will be discussed further below. Another aspect relates to the curious, but general observation that P_{osm} is larger in both crustaceans and fish in this group, in some cases approaching that of estaurine animals (cf. Table 1). If the energetic advantage of a dilute urine is obscure its correlation with increased water, but not ion, permeability is clear. A larger water permeability is obviously tolerable but is not, in itself, adaptive. It may reflect a change in an important surface characteristic, such as gas exchange, not related to osmotic regulation.

C. Characteristics of Active Ion Transport

Early analysis of the mechanisms of ion transport developed to a large extent in the laboratories of August Krogh in Copenhagen, and the first stages are described in his monograph (Krogh, 1939). The analysis was continued by one of his students, Hans Ussing, who used the isolated, but viable, frog skin as a powerful tool for studying active Na^+ transport. It is not too much to say that Ussing and his colleagues developed the framework within which much of modern transport research takes place. Their contributions include such critical conceptual tools as a practical method for delimiting active transport (Ussing, 1949), coupled solute–solute (Ussing, 1947) and solute–solvent fluxes (Ussing, 1952), and the real (as contrasted to thermodynamic or minimum) energetic cost of Na transport (Zerahn, 1956). The first complete model of Na movement across the frog skin (Ussing, 1960) is still useful. Much of the research directed at transport mechanisms in hyper-regulating fish and crustacea has drawn on the behaviour of the isolated frog skin, but some of the specific characteristics described below for intact animals differ from those in the isolated epithelium. This is not due to a species difference; transport in intact frogs is remarkably similar to that in fresh-water fishes and crustaceans but differs in several details from transport in the isolated frog skin.

1. Criteria of Active Ion Transport

Before embarking on a study of the active transport of an ion it is obviously necessary to be sure that the movement cannot be accounted for by diffusion gradients. The criterion used most frequently is the Ussing flux-ratio equation which relates the ratio of unidirectional fluxes of an ion to its concentrations in the blood and medium and to any electrical potential across the body surface. The approach was described in an earlier review (Kirschner, 1970). It suffices here to note that both Na^+ and Cl^- are actively transported from medium to blood in all FW animals that have been studied to date (except perhaps Cl^- in the eel *Anguilla anguilla*; Krogh, 1937, Motais, 1967).

2. Independence of the Ion Transport Systems

Krogh (1938) showed that both crayfish and goldfish can absorb either Na^+ or Cl^- from solutions containing impermeant counter-ions. For example, Cl^- is absorbed from KCl leaving K^+ in the medium, and Na^+ is absorbed from Na_2SO_4 leaving SO_4^{2-}. In order to maintain electrical neutrality the unbalanced absorption (J_{in}) of sodium must be accompanied by the equivalent excretion (J_{out}) of an ion of the same sign. Krogh proposed that either NH_4^+ or H^+, both metabolic end products, could serve. In a number of animals, including the crayfish (Shaw, 1960), goldfish (Maetz and Garcia Romeu,

1964) and trout (Kerstetter, Kirschner and Rafuse, 1970), $J^{NH_4}_{out}$ is approximately equal to J^{Na}_{in} in dilute media. Thus, it could be the necessary counter-ion, and the possibility is strengthened by the observation that injection of NH_4^+ stimulates Na^+ influx in the goldfish and trout. In addition, high concentrations of NH_4^+ (10 mmol l^{-1}) in the external medium inhibit Na^+ transport in crayfish (Shaw, 1960) and fish (Maetz and Garcia Romeu, 1964), possibly by competition for an 'exchange carrier, in the apical (outer) membrane. Other data are at variance with such a model. For example, frogs excrete practically no NH_4 across the skin when in water. Instead the uptake of Na^+ is balanced by J^H_{out} (Garcia Romeu, Salibian and Pezzani-Hernandez, 1969). Secondly, $J^{NH_4}_{out}$, in some fish, is unaffected by procedures that cause large changes in J^{Na}_{in} (Kerstetter, Kirschner and Rafuse, 1970; de Vooys, 1968; Maetz, 1973). Some of these observations were extended to the crayfish as well (Kirschner, Greenwald and Kerstetter, 1973), and it was proposed that nitrogen was excreted as NH_3, not as NH_4^+, and that Na transport was coupled with excretion of H^+ (Kerstetter, Kirschner and Rafuse, 1970).

Maetz (1973) tested this proposition by measuring ammonia excretion into media more alkaline than the blood, and hence containing higher NH_3 than blood. Since efflux was little affected he concluded that it must be excreted as NH_4^+, not as NH_3. Studies on a recently developed, perfused trout gill show that if NH_4^+ is absent from the ventral aortic perfusate J^{Na}_{in} is markedly reduced. In addition, if J^{Na}_{in} is inhibited in the perfused gill $J^{NH_4}_{out}$ decreases (Payan, Matty and Maetz, 1975). These results are hard to reconcile with those in intact trout, but they support the idea that NH_4^+ may be exchanged for Na^+ in FW gills. On balance the present evidence suggests that coupling may be flexible and either metabolic end product may be used as a partner for active Na^+ uptake.

The exchange ion for J^{Cl}_{in} was proposed by Krogh (1938) to be HCO_3^-. Evidence for this was found in the frog (Garcia Romeu, Salibian and Pezzani-Hernandez, 1969) as well as in goldfish (Maetz, 1956; Maetz and Garcia Romeu, 1964). Again, it is difficult to be sure whether the coupling is actually with HCO_3^- or with OH^-, because the latter, together with respiratory CO_2, would appear in the medium as bicarbonate.

One of the outstanding differences between isolated and *in vivo* preparations, such as the frog skin, is the loss of these independent ion transport mechanisms based on exchange with a metabolic counter-ion. Sodium transport has been studied in the isolated frog skin for many years, and it is clear that it is not linked to the counter transport of either NH_4^+ or H^+, but rather is accompanied by passive Cl^- influx. Transport of Cl^- in the isolated skin has been described recently (Kristensen, 1972, 1973), and it has some of the characteristics of the *in vivo* system, but there is, as yet, no suggestion of HCO_3^- excretion.

3. Ion Transport and Concentration in the Medium

In the original work on the concentration dependence of active Na^+ transport it was found that J_{in}^{Na} in isolated frog skin could be described by

$$J_{in}^{Na} = \frac{J_{max}[Na^+]_o}{K_m + [Na^+]_o}$$

where $[Na]_o$ is the concentration in the external medium and the parameters J_{max} and K_m describe the system's behaviour (Kirschner, 1955).* Since this relationship is formally identical with that for the substrate dependence of reaction rate in an enzyme system, I suggested that the parameters might have the same significance: the maximum rate is a function of the amount of the transport 'carrier', while K_m, the concentration at which the rate is exactly half maximum, is a measure of the affinity of the transport system for its substrate, Na^+.

This model was extended to transport in intact animals in a series of important papers on Na^+ absorption in crustacean hyper-regulators,* and attention was directed to the fact that there was a striking correlation between the value of the parameter K_m and habitat (Shaw, 1964, Greenwald, 1971). It was shown that animals adapted to very dilute media had uptake systems with high affinity for Na^+ (low K_m) while those in brackish water had lower affinities (higher K_m). The studies in decapods were then extended to gammarids, and the same correlation was found (Shaw and Sutcliffe, 1961; Sutcliffe, 1968). Since then the parameter has been evaluated for many animals without disturbing the correlation which is illustrated in Table 2. A more extensive list, which includes animals from other phyla, has been published (Greenwald, 1971). While much attention has been devoted to the lowering of K_m in the invasion of more dilute media, it is perhaps the least important of the flow parameters and will condition an animal's success in dilute medium only when constraints are placed on the parameter J_{max} and those governing dissipative leaks. To illustrate this we consider three animals whose active transport (AT) systems have the kinetics shown in Fig. 3. Animal 1 has a maximum transport rate of 10 per hour (arbitrary units), but a relatively high affinity for Na^+ ($K_m = 1$ mmol l^{-1}). Animals 2 and 3 have much higher maximum transport rates ($J_{max} = 100$ per hour for both) but lower affinities. Animal 2 has a very low affinity; $K_m = 25$ mmol l^{-1}. Animal 3 is intermediate; $K_m = 5$ mmol l^{-1}. To maintain a steady state with body fluids more concentrated than the environment the transport system must be capable of generating an influx exactly equal to the rate of ion loss (LR) from the animal, and the latter will be governed by the leak parameters P_{osm} and P_{Na}. By varying the diffusive parameters a series of hypothetical loss rates

---

* Actually, a term should be added to account for diffusion as well. This may be negligible in many FW animals, but in one as permeable as *Carcinus* it becomes appreciable at high external concentrations.

Table 2. Sodium transport parameters and loss rate in hyper-regulators

Animal	K_m (mmol l^{-1})	J_{max} (mmol kg^{-1} h^{-1})	LR[a]	Habitat or adaption medium	Reference
		Crustaceans			
Carcinus maenas	20	~6–8	17.8	Estuarine	Shaw (1961a)
Marinogammarus finmarchius	6–10	~20	~20	Estuarine	Sutcliffe (1968)
Gammarus duebeni	2.0	~20	4.0	BW	Sutcliffe (1967a)
G. duebeni	0.4	15.0	3.5	FW	Shaw and Sutcliffe (1961)
Eriocheir sinensis	~0.5	~2.4	~2.0	FW	Shaw (1961b)
Pseudothelphusa jouyi	0.15	2.9	1.3	FW	Thompson (1970)
Gammarus pulex	0.10	3.1	1.7	FW	Sutcliffe (1967b)
Astacus pallipes	0.15	1.0	0.15	FW	Shaw (1959a)
		Vertebrates			
Fundulus heteroclitus	2.0	1.7	0.6	Estuarine	Potts and Evans (1967)
Poecilia latipinna	8	12	~0.7	Estuarine	Evans (1973)
Salmo gairdneri	0.5	0.4	0.2	FW	Kerstetter, Kirschner and Rafuse, (1970)
Carassius auratus	0.3	0.7	0.5	FW	Maetz (1972)

[a] LR: net sodium loss to fresh-water (FW). For FW and brackish-water animals this can be measured with the animal in FW. For estuarine forms it can be estimated from the isotope efflux into a more concentrated medium.

(LR_1–LR_5) can be generated. If the animals were very leaky ($LR_1 = 110$ units per hour) none of them could maintain a steady state in any medium except sea-water, because the loss rate exceeds the maximum possible influx in all three. If the rate of loss were reduced to LR_2 (= 75 units per hour), by reducing the permeabilities and/or gradients, animals 2 and 3 could regulate, but animal 1 could not, because the loss rate still exceeds the maximum influx that it can generate. Further reductions of loss (LR_3 and LR_4) still find animal 1 unable to regulate, and it is only when the loss is reduced to LR_5 that it would be able to maintain a steady state at *any* external concentration, even though its transport system has the highest affinity of the three. Thus, as surmised by Beadle and Cragg (1940) the first step in the invasion of dilute

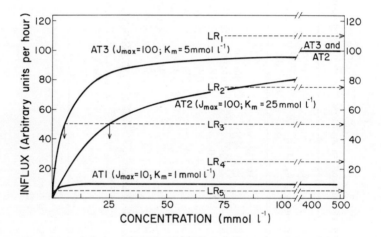

Figure 3 Sodium influx as a function of external concentration. Three patterns are shown, all of them exhibiting 'saturation kinetics'. They differ in values of the parameters J_{max} and K_m. The relationship shown by AT1 is characteristic of animals found in FW, although most have a K_m lower than 1 mmol l^{-1}. AT2 and AT3 show the effect of K_m in animals with the same J_{max}. The vertical arrows show minimum external concentrations for Na^+ balance if both animals are losing Na^+ at the rate LR_3

media required developing mechanisms for maintaining high blood NaCl. Shaw (1961b) extended this by pointing out that solute loss had to be reduced sufficiently to fall below the maximum capacity of the active transport system, and it is this transport parameter (J_{max}), rather than the K_m, which is critical for determining whether an animal is able to regulate at all. We have seen that permeabilities, hence loss rates, decrease as animals become more successful in dilute media (Table 1). If animals 1 and 2 had loss rates of 5 units per hour (LR_5) both could remain in a steady state down to an environmental concentration of 1.3 mmol l^{-1}. Below this concentration the influx in animal

2 would fall below 5 units per hour, but animal 1 could do little better, maintaining a steady state only to 1.0 mmol l^{-1}. Thus, if P_{Na} and P_{osm} were as small in *C. maenas* as in *E. sinensis* it could live in water nearly as dilute (the environmental concentrations would be 0.4 mmol l^{-1} for *E. sinensis* and 5 mmol l^{-1} for *C. maenas*, both fairly fresh-water) and *C. maenas* could invade an even more dilute medium simply by synthesizing more of its low affinity transport system. In fact, it is the leak parameter that precludes this. As shown in Table 2 the LR for *C. maenas* in FW far exceeds J_{max} for absorption.

What is interesting is that no FW animal appears to have a J_{max} as high as in estuarine crabs. Table 2 shows that J_{max}, as well as K_m, is reduced as dilute media are occupied. The significance of this is not apparent, but the correlation is clear. And it is when comparing animals with similar J_{max} values and permeabilities that differences in K_m become important. If we compare animals 2 and 3, both with $J_{max} = 100$ per hour, and assume that both have a loss rate of 50 per hour (i.e. LR$_3$), it is apparent that animal 2 can generate the influx necessary to balance loss at environmental concentrations down to 25 mmol l^{-1}. On moving into water more dilute than this the influx would fall along curve AT2, it will lose salt, and ECF concentration will fall. On the other hand animal 3 can generate the necessary active influx at concentrations down to 5 mmol l^{-1}, and if the K_m were smaller it would be able to maintain a steady state in still more dilute media. The adaptive significance of this for invading new habitats is apparent. Thus, all four parameters are important (and interrelated) in conditioning the penetration of dilute media.

Fewer studies have been undertaken on the Cl$^-$ transport systems. They show the same type of saturation kinetics described for Na$^+$ transport, and when parameters are compared for the two ions in the same animal they appear to be similar. For example, in the rainbow trout J_{max} for Na$^+$ and Cl$^-$ are 0.33 and 0.31 mmol kg^{-1} h^{-1} and K_m values are 0.5 mmol l^{-1} and 0.2 mmol l^{-1} (Kerstetter, Kirschner and Rafuse, 1970; Kerstetter and Kirschner, 1972).

One curious point that has emerged from the studies above is that the affinity of the Na$^+$ transport system lies in the range 0.1–1.0 mmol l^{-1} for nearly all good hyper-regulators but values from epithelia studied *in vitro* are much higher. Neither species nor organ differences are involved as is illustrated by the fact that the K_m for Na$^+$ transport across the skin in intact frogs is 0.2 mmol l^{-1} (Greenwald, 1971), but in the isolated frog skin it is about 10 mmol l^{-1} (Kirschner, 1955; Biber and Curran, 1970). Functional, isolated transport epithelia are almost non-existent among crustaceans and fish, but the gill of *E. sinensis* is capable of vigorous Na$^+$ transport for hours after isolation (Koch, Evans and Schicks, 1954), and it would be very interesting to compare the K_m of this preparation with that obtained by Shaw (1961b) on the intact animal.

4. The Outer (Apical) Membrane of the Transporting Cells

It has come to be realized during the past two decades that transport across these epithelia is a two stage process. The first step involves transfer of an ion from medium into cell across the apical membrane, after which it is extruded from cell to body fluids across the basal or baso-lateral membrane. Since the apical membrane is accessible to manipulations of the external medium, studies on intact animals are usually directed at it rather than at the second, baso-lateral stage.

For many years entry of Na^+ across the apical membrane was considered to be a diffusive process of little chemical note, except that the membrane was selectively permeable to Na^+, excluding all other ions except Li^+ (Ussing,

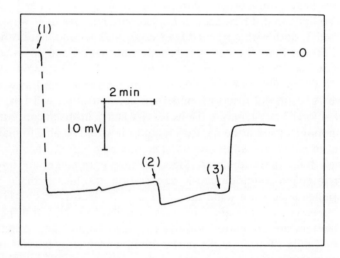

Figure 4 Potential difference across the apical membrane of an interlamellar cell in the FW trout gill. A rainbow trout was anaesthetized and the operculum removed from one side. Both microelectrode and reference were initially in the external medium (1 mmol l^{-1} NaCl). At (1) the microelectrode was advanced with a micromanipulator into the interlamellar region. Penetration gave a PD of 29 mV which was relatively stable. At (2) 1 mmol l^{-1} KCl was added, and a small hyperpolarization occurred. Addition of 0.5 mmol l^{-1} $CaCl_2$ depolarized the preparation. After Kerstetter unpublished

1960). The experiments on which this conclusion was based were done on isolated frog skins and toad urinary bladders usually bathed on the apical (outside) surface with Ringer's solution (NaCl = 110 mmol l^{-1}). Since the cell interior is typically high K^+, low Na^+ a concentration gradient favouring net Na^+ influx exists. And although the potential difference across the apical membrane opposes entry (cell interior positive; Engbaek and Hoshiko, 1957), the electrical behaviour of the skin (Koefoed-Johnsen and Ussing,

1958) was consistent with passive Na⁺ movement into the cell (see Chapter 6 for details). One difficulty in extending this conclusion to the intact animal is that the experimental conditions differ. Except for estuarine forms animals in dilute media seldom encounter such a concentrated environment; ordinary FW has an NaCl ~ 1 mmol l⁻¹, and net uptake in goldfish and crayfish can take place from water much more dilute than this (i.e. 10^{-2} mmol l⁻¹). It is difficult to believe that an inward concentration gradient exists under these conditions. However, we have made some measurements of the apical potential difference (PD) under these conditions. Figure 4 shows that the interlamellar cells in trout gills are negative to the dilute medium by about 30 mV. This could provide a net inward electrochemical gradient even if cellular [Na⁺] were 3 times higher than medium. However, more measurements both of PD and cell concentration are needed before we can decide whether the entry step is passive or active. If it is active, an energy input would be required. This might take the form of an ATP requiring system, but it might also be supplied by the H⁺ (or NH_4^+) gradient generated between cell and environment. Such energetic coupling, with the diffusive flux of one solute driving an 'uphill' movement of another, has been described for several solute pairs in cells and epithelia. Recent work suggests that, whether passive or active, the entry step is rate limiting for the entire transepithelial transfer, and that it is this step which is saturable. Some of the crucial work was done on isolated frog skins (Biber, Chez and Curran, 1965; Biber, 1971), but the results are consistent with the saturation kinetics (Fig. 3) observed in intact crustaceans and fish. The present interpretation is that Na⁺ from the external medium combines with a protein ('carrier'?) in the outer membrane, and the complex formed is transferred across the membrane where it dissociates releasing Na⁺ into the cytoplasm. Since Na⁺ uptake is coupled to the extrusion of NH_4^+ or H⁺ the entire process of apical transfer may be represented by the first step in Fig. 5. The use of two inhibitors of Na⁺ transport permits additional detail to be added to this model. One of these is acetazolamide (diamox) which is an inhibitor of the enzyme carbonic anhydrase. The enzyme catalyzes the hydration of CO_2 in the cell (CO_2 + $H_2O \leftrightarrows H_2CO_3 \leftrightarrows H^+ + HCO_3^-$) thus providing a source of H⁺ and HCO_3^-. The H⁺ may provide the exchange partner for Na⁺ movement across the apical membrane, or it may be used to convert NH_3, entering the cell from blood, to NH_4^+ which can serve as the exchange partner. In either case, inhibition of the enzyme will reduce the availability of an internal cation and should slow Na⁺ entry. This is exactly what is observed in trout (Kerstetter, Kirschner and Rafuse, 1970) and goldfish (Maetz and Garcia Romeu, 1964).

A second compound used for probing the characteristics of the apical membrane is amiloride (N-amidino-3,5-diamino-6-chloropyrazinecarboxamide) which was first developed as a diuretic and naturetic compound in the mammalian kidney, and which was subsequently shown to block Na⁺ transport across toad bladder (Bentley, 1968) and frog skin (Eigler, Kelter and Renner, 1967; Biber, 1971). It also inhibits the active transport of Na^+ across the trout gills, crayfish gills and the body surface of intact frogs (Kirschner, Greenwald and Kerstetter, 1973). It appears, in fish gills, to compete with Na^+ for the saturable component on the apical membrane (Greenwald and Kirschner, 1976). Its mode of action was also one of the reasons for questioning the role of NH_4^+ as an exchange partner in Na^+ uptake, since it had only a small effect on NH_4^+ excretion in fish and crayfish gills at concentrations which completely blocked Na^+ uptake (Kirschner,

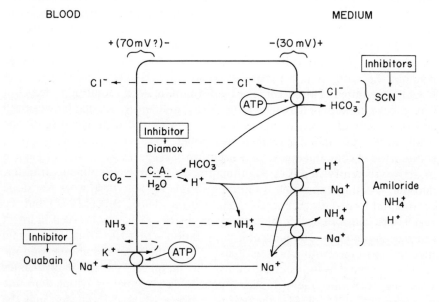

Figure 5 A model of active Na^+ and Cl^- transport across the gill of hyper-regulators. Na^+/H^+ and Na^+/NH_4^+ exchanges are shown as independent and in parallel. They may actually be the same system. Experimental details are described in the text

Greenwald and Kerstetter, 1973). Finally, both H^+ and NH_4^+ in the external medium inhibit Na^+ transport in crayfish (Shaw, 1960) and goldfish (Maetz and Garcia Romeu, 1964). The mechanisms of inhibition have not been analysed but may involve competition with Na^+ for the combining site or 'carrier' in the apical membrane. If so, the inhibitory concentrations suggest that the affinity of these sites is greatest for H^+, about an order of magnitude lower for Na^+, and an order of magnitude lower still for NH_4^+.

Much less attention has been paid to the transepithelial transfer of Cl^- than of Na^+, and the question of movement across the individual membranes is virtually untouched. The following picture can therefore be considered only provisional. Intracellular Cl^- has never been measured in gills but is probably low as in other animal cells. However, since net transfer across the apical membrane can occur from solutions with less than 10^{-4} mol l^{-1} it is unlikely that there is much, if any, chemical gradient favouring entry. The intracellular potential (-30 mV) opposes diffusive entry, and hence it is likely that transport of Cl^- is active and energy-requiring. Influx of Cl^- is completely inhibited by low concentrations of SCN^- in the external medium (Kerstetter and Kirschner, 1972; De Renzis, 1975), and this too suggests that a chemical mechanism exists in the outer membrane. Since Cl^- transfer is coupled obligatorily with the extrusion of HCO_3^- (possibly OH^-) the process could be driven by an HCO_3^- gradient between cell and medium. The availability of HCO_3^- thus may be crucial, and a role for carbonic anhydrase is suggested by the observation that diamox inhibits Cl^- uptake by goldfish (Maetz and Garcia Romeu, 1964), although it had no effect on Cl^- influx in trout at concentrations that blocked Na^+ transport (Kerstetter, Kirschner and Rafuse, 1970). Recently, several laboratories have described an ATPase in fish gills which is stimulated by HCO_3^-. It did not appear to require Cl but was inhibited by SCN^- at about the same concentration as is required for inhibition of transport (Kerstetter and Kirschner, 1974). Much, perhaps most, of this enzyme appears to be associated with mitochondria but some is microsomal and is probably membrane associated in the intact cell. This enzyme might represent the test tube equivalent of the energy source for a Cl^-–HCO_3^- exchange mechanism on the apical membrane.

5. The Inner (Baso-lateral) Membrane of the Transporting Cells

The Na^+ and Cl^- that have entered the cell must be extruded across the inner membrane into the ECF. This membrane has not been accessible in intact animals, and as a result, few experiments bearing directly on its function have been performed; our ideas are based largely on information drawn from isolated amphibian epithelia. The intracellular concentrations of both ions are almost certainly below those in the body fluids hence extrusion from the cell is against a chemical gradient. In amphibian skins the cell is negative to blood (Engbaek and Hishiko, 1957; Biber, Chez and Curran, 1965), and this is probably also true in gills. This means that baso-lateral Na^+ transfer is against gradients of both chemical and electrical potential and is therefore active. It is believed to be mediated by an exhaustively studied 'pump' which transports K^+ from ECF into cell coupled with the extrusion of Na^+ from cell into ECF. The characteristics of this system, which exists in nearly all animal cell membranes, are well known. The coupling of K^+ and Na^+ movements is obligatory; if either one is inhibited the other stops as well. The system is ATP-dependent and is inhibited by a group of compounds called cardiac glycosides which act on the ECF side of the membrane. In the test tube this system is studied as a membrane-bound ATPase that requires both Na^+ and K^+ for activity and which is inhibited by cardiac glycosides (e.g.

ouabain; for reviews of this system see Skou, 1965; Schwartz, Lindenmayer and Allen, 1972; Glynn and Karlish, 1975). Participation of this transport mechanism is based largely on three observations. First, it is present in most transporting epithelia, including many fish gills (Kamiya and Utida, 1969; Jampol and Epstein, 1970), in gills of the crab *C. sapidus* (Towle, Palmer and Harris, 1976) and *Cardisoma guanhumi* (Quinn and Lane, 1966), both strong hyper-regulators in dilute media, and the crayfish (unpublished experiments). Second, when ouabain is added to a solution bathing a frog skin or toad bladder Na^+ transport stops. Transport inhibition by ouabain has recently been shown in perfused trout gill (Payan, Matty and Maetz, 1975). And finally if K^+ is deleted from the inner bathing solution Na^+ transport is inhibited, presumably because the coupling partner for the Na/K mechanism is unavailable.

The mechanism of chloride extrusion is unknown. Although intracellular Cl^- is probably lower than the ECF concentration, the PD across the baso-lateral membrane favours extrusion of an anion, and the electrochemical gradient may be adequate to account for net movement from cell to ECF. More work is needed to verify this surmise.

A model for the transfer of NaCl from medium to blood is shown in Fig. 5. It encompasses most of the evidence summarized above. In dilute solutions the apical membrane is almost impermeable to the main ions bathing both surfaces (K^+ and organic anions within the cell, Na^+ and Cl^- in the medium). Thus, diffusive leaks must be small even when relatively large gradients are present. However, it contains the independent ion exchange mechanisms described above, and these will permit transfer of external Na^+ and Cl^- so long as an exchangeable ion is available from inside the cell. The source of the latter is blood CO_2 (or metabolic CO_2 produced within the epithelial cell). Hydration of CO_2 and dissociation of the H_2CO_3 produces the H^+ and HCO_3^- required. Either the H^+ or NH_4^+ (formed from NH_3 and H^+) provides the counter-ion for Na^+ influx. Although there are some exceptions, carbonic anhydrase is required to insure that hydration of CO_2 occurs at a rate commensurate with normal ion transfer. Inhibition of this enzyme reduces inward NaCl transport by reducing the supply of exchangeable ions. The transfer systems, which are based on membrane proteins, are responsible for the saturation kinetics observed. They are also vulnerable to certain inhibitors added to the external medium; i.e. amiloride for the Na^+ system and SCN^- for the Cl^- system.

Sodium movement from cell to blood occurs via the Na/K transport system, and the K^+ that enters the cell in this exchange simply diffuses back into the ECF. In the absence of evidence to the contrary, we assume that Cl^- diffuses from cell to ECF because the baso-lateral membrane is permeable and the electrochemical gradient suitable. Sodium, but not chloride, transport should also be blocked if inhibitors of the Na/K system are introduced into the blood. It can be seen that an energy input is required for at least two steps.

Extrusion of Na^+ from cell to ECF is one, and since the Na/K system is ATP-dependent, active inward transport of Na^+ must be coupled to oxidative metabolism. The other energy dependent step is Cl^- uptake from medium, and if the HCO_3-dependent ATPase plays a role, this step is also coupled to oxidative metabolism. Whether the entry of Na^+ is active or passive still remains to be determined.

D. Role of the Kidney

The structure and function of the fish kidney was reviewed in great detail recently (Hickman and Trump, 1969), and both its morphology and behaviour suggest analogies with the crustacean antennal gland (Kirschner, 1967; Reigel, 1972). The primary role of these organs in FW animals is to maintain ECF volume constant in the face of an osmotic water influx. Its efficacy in matching output with inflow has been demonstrated in many studies on both fish and crustaceans. The output of the antennal gland in *Carcinus maenas* increases nearly linearly with dilution of the medium (hence with water entry) from 1.8 ml kg^{-1} h^{-1} in SW to 8.8 ml kg^{-1} h^{-1} in 40% SW. There is no sign, in such data, that any upper limit exists for the rate of fluid excretion, and this may well be characteristic of an organ that initiates fluid formation by a physical process like filtration. It also emphasizes a point suggested earlier. The reduction in P_{osm}, seen in FW animals, is more significant in limiting solute loss through the kidney than in limiting the volume of water that must be excreted.

A great deal of evidence suggests that the initial process of urine formation in these animals involves filtration across the most proximal structure of the tubular system, the glomerulus-Bowman's capsule in the vertebrate nephron and the coelomosac in the crustacean antennal gland (Kirschner, 1967; Riegel, 1972, reviews the evidence, and suggests a novel mechanism for filtration). That the proximal end is specialized for ultrafiltration is suggested by the fact that its blood supply derives from a large branch of a main systemic artery, the dorsal aorta in vertebrates and the antennary artery in decapod crustaceans. Thus blood is conveyed to this site with a minimum loss of pressure which is an important force in filtration. The ultrastructure of the glomerulus-Bowman's capsule complex in vertebrates resembles that of the coelomosac in crustaceans (Schmidt-Nielsen, Gertz and Davis, 1968; Tyson, 1968), and the composition of tubular fluid in this region has the characteristics of an ultrafiltrate of blood (reviewed in Riegel, 1972). The rest of the organ functions to process the ultrafiltrate. The region immediately distal to the filtration site in fish contains recognizably different segments probably mediating different transfer functions (Hickman and Trump, 1969), but no corresponding histological analysis exists for the antennal gland (the labyrinth), and we shall consider this region as a single entity in both groups. Many of the nutrient solutes filtered into the tubule are reabsorbed in the proximal segment (e.g. glucose, amino acids). But little net transfer of the main solutes, NaCl, or of water occurs, and fluid exits from this region essentially isosmotic with blood. It may be somewhat reduced in volume as indicated by the fact that inulin (a small polymer that is filtered but not reabsorbed from the tubule) concentration was about 25% higher in tubular fluid from the crayfish labyrinth than in the hemolymph (Riegel, 1965). Some proximal fluid reabsorption is also indicated by the fact that inulin is more concentrated in the final urine than in the blood of FW crabs (e.g. *Pseudothelphusa jouyi*; Thompson, 1970) and several marine crustaceans (summarized in Riegel, 1972); these animals have only a proximal tubular segment. Although not part of the system regulating the osmotic composition of the body fluids, there exists a powerful organic acid excreting mechanism in the proximal segment of both decapod and fish kidneys (e.g. Burger, 1957; Forster and Hong, 1958). The physiological role of this system is not known; it is mentioned here to emphasize the similarity in function between crustacean and vertebrate excretory systems. The antennal gland of all of the hyper-regulating crabs comprises only the coelomosac and labyrinth, and the isosmotic fluid exiting from the latter enters the urinary bladder from which it is excreted. A number of marine teleosts also lack a distal tubular segment. Several of them can, nevertheless, live in FW. In a recent study the toadfish *Opsanus tau* (which is also aglomerular) was shown to be a good hyper-regulator in very dilute media, although it could not adapt to soft FW (Lahlou and Sawyer, 1969). Toadfish urine, like that of crabs with only a proximal tubule, is isosmotic with blood and has a high NaCl concentration. The osmoregulatory function of such kidneys is accomplished if the volume of urine excreted exactly matches osmotic water entry. However, as emphasized earlier, high urine NaCl concentrations are maladaptive for ionic regulation, and ionic losses would be intolerable unless P_{osm}, and in consequence urine volumes, were small. Some measurements of urine concentrations and excretion rates for these animals are shown in Table 3.

Nearly all fresh-water and euryhaline fishes and many FW crustaceans, including the crayfish and all FW and brackish-water gammarids, are capable of producing urine that is substantially more dilute than the ECF. In the vertebrate kidney and decapod antennal gland the ability to produce a dilute urine from an ultrafiltrate is correlated with the appearance of a morphologically distinct distal tubular segment. In the Gammarids a greater capacity to dilute the urine is correlated with a longer tubular system, although no histological distinctions were drawn between different regions (Hynes, 1954). The correlation has, in fact, been noted in FW representatives from several phyla (Schmidt-Nielsen and Laws, 1963; Kirschner, 1967), and a relationship between the ability to dilute the urine and the presence of a distal tubular segment can hardly be doubted. This region is known, from the analysis of tubular fluid, to reabsorb NaCl in mammalian and amphibian (e.g. Walker and coworkers, 1937) nephrons, and in the absence of antidiuretic hormone,

Table 3. Kidney function in osmoconformers and hyper-regulators

	Medium	Flows[a]		Sodium[b]			Osmotic	Reference
		FR	J_v^u	U/B	J_{Na}^u		U/B	
				Osmoconformers				
Homarus sp.	SW	4.4	3.4	0.99	1.5		—	Bryan and Ward (1962)
Pugettia producta	SW	3.2	2.7	0.98	1.2		1.0	Cornell (1976)
Myxine glutinosa	SW	—	0.2	0.89	0.02		0.97	Morris (1965)
				Weak regulators				
Carcinus maenas	SW	3.6	1.8	0.92	0.69		0.98	Shaw (1961b)
	40% SW	14.8	8.8	—	3.76		—	Binns (1969)
								Riegel and Lockwood (1961)
Cancer magister	SW	—	0.6	0.83	0.27		0.96	Hunter (1973)
	50% SW	—	4.9	0.85	1.54		0.97	
				Strong regulators				
Potamon niloticus	FW	—	0.05	0.93	0.03		0.86	Shaw (1959a)
Eriocheir sinensis	FW	—	1.67	1.07	0.54		0.99	de Leersnyder (1967)
Pseudothelphusa jouyi	FW	0.13	0.13	1.03	0.03		0.99	Thompson (1970)
				Completely adapted				
Gammarus pulex	FW	—	15	—	—		0.18	Lockwood (1961)
Astacus fluviatilis	FW	—	3.4	0.03	0.02		—	Bryan (1960)
Carassius auratus	FW	2.0	1.4	0.08	0.02		0.12	Maetz (1963)
Salmo gairdneri	FW	7.6	4.7	0.07	0.05		0.15	Beyenbach (1974)
Anguilla anguilla	FW	1.0	0.6	0.04	0.004		—	Sharrat, Chester-Jones and Bellamy (1964)
Platichthys flesus	FW	4.2	1.8	0.03	0.03		0.37	Lahlou (1967)

[a] The filtration rate (FR) and urine flow (J^u) are in ml kg^{-1} h^{-1}.
[b] U/B: the ratio of urine to blood sodium concentration. J_{Na}^u: renal sodium efflux in mmol kg^{-1} h^{-1}.

to limit water reabsorption, so that the tubular fluid becomes more dilute than the original filtrate. Micropuncture analyses have not been published for fish, but Peters (1935) indicated that both Cl^- concentration and osmotic pressure were reduced in the distal tubule of the crayfish antennal gland. This was confirmed for 'summer animals' (Riegel, 1972) and Na^+ absorption was also shown to occur in this region. It is interesting that the specific activity of the Na/K-ATPase is highest in the distal tubular region of the crayfish antennal gland (Peterson and Loizzi, 1974), which, in the light of its proposed role in Na^+ transport, is consistent with solute reabsorption here. There is evidence that the urinary bladder may be even more important in crayfish than the tubular segment in reabsorbing NaCl and diluting the urine. Kamemoto, Keister and Spalding (1962) showed that Na^+ transport from bladder lumen to blood took place. Riegel's data indicated that the solute concentration fell in the distal tubule only to 60% of the original, and that final dilution (to about 10%) occurred in bladder. In winter and spring animals even less dilution occurred in the tubule, although the final urine was about as dilute as in summer (Riegel, 1972). There is no question that the urinary bladder in some species is capable of vigorous NaCl absorption. However, it will require more data to determine the relative importance of the distal tubule and the urinary bladder in reabsorbing the filtered solute to produce a dilute urine.

An interesting question arises in connection with the adaptive significance of a dilute urine. Potts (1955) showed that the *thermodynamic* cost of ionic regulation is less, per mole of solute transferred by active transport, in the kidney than it is across the body surface in FW. This is because the thermodynamic requirement is based entirely on the magnitude of the electrochemical gradient against which the transfer occurs. In FW the gradient across the body surface is large and constant. In the kidney the magnitude of the gradient varies. It is initially small (the filtered fluid has about the same original composition as ECF), becomes larger as solute is withdrawn and is steepest in the terminal part of the process when the tubular (or bladder) fluid is most dilute. However, the actual energetic cost of transport, based on the amount of ATP used or oxygen consumed by the transport process is probably much higher than the thermodynamic estimate (Kirschner, 1961). More important, there has been no suggestion that the actual energy input in ion transport bears a relationship to the magnitude of the gradient. For example, where the transfer of Na^+ is based on the Na/K transport system it is accepted that an ATP is split for every 3 Na^+ transported (Zerahn, 1956). If this cost is fixed then it should make little difference whether the Na^+ is reabsorbed by active transport in the kidney to produce a dilute urine, or whether it is excreted in iso-ionic urine and then transported across the body surface. The same is true for Cl^- transport. The success of crabs like *Eriocheir sinensis* and *Potamon niloticus* which excrete an iso-ionic urine shows that dilution is not absolutely essential for life in FW. But most fully adapted, stenohaline FW animals have a distal tubular segment and can produce a dilute urine. The exceptions are so few as to emphasize its probable importance (Schmidt-Nielsen and Laws, 1963). This in turn suggests that the thermodynamic calculation may be a model for the energy consumption of ion pumps, which may somehow transfer more solute per ATP when the electrochemical gradient is small than when it is large. Such behaviour has not been described in transport systems, but would be analogous to the 'Fenn principle' in skeletal muscle wherein the amount of energy evolved during contraction varies with the mechanical work done (Fenn, 1924). The proposition is worth exploring.

IV. HYPO-OSMOTIC, HYPO-IONIC REGULATION

A. Problems Faced by Hyporegulators

Although most animals found in SW are stenohaline conformers, some are not; a surprising number regulate ECF solute concentration well below that in the environment. Some of these are confined to the ocean because they do not have the capacity to hyper-regulate in more dilute media (many marine fish). However, a number of estuarine and shore crabs, some prawns and some isopods, and many species of fish are more euryhaline and regulate the ECF concentration over an extended range of ambient concentrations from potable water to media more concentrated than SW. A few species are extremely euryhaline. The brine shrimp *Artemia salina* is a good hyper-regulator in dilute media (NaCl = 40 mmol l^{-1}) with a hemolymph Na$^+$ concentration of 80 mmol l^{-1}. But it can also regulate in 25–30% NaCl (NaCl concentration about 4–5 mol l^{-1}), and the ECF concentration rises only to about 300 mmol l^{-1} (Croghan, 1958a; for *Parartemia zeitziana*, Geddes, 1975a, 1975b). Thus, this group comprises animals in a variety of ecological situations. However, when one examines the regulation of hemolymph concentration in media of different salinities (Fig. 6) they appear to fall into two reasonably distinct categories, and it may be useful to discuss them from this point of view. One group, all crustaceans, includes animals that regulate at a relatively high ECF concentration (300–350 mmol l^{-1}). Most are fairly good hyper-regulators in dilute media. A few of these are isosmotic with SW and regulate only slightly below ambient at higher external concentrations; the prawn *Crangon crangon* (Flügel, 1960) and the shore crab *Pachygrapsus crassipes* (Rudy, 1966) are examples. Others appear to be isosmotic in 65–75% SW and are better regulators in the sense that the body fluid concentration changes less as the medium becomes more concentrated. The shore crab *Uca sp.* is an example (Baldwin and Kirschner, 1976). It maintains the ECF NaCl at 399 mmol l^{-1} in 10% SW (~50 mmol l^{-1}), and this rises only to 447 mmol l^{-1} in 175% SW. This group will be referred to as weak to moderate hyporegulators. The other pattern is found in marine fishes and the branchiopods mentioned above. In SW blood concentrations are 200 mmol l^{-1}

or less. In *P. zeitziana* hemolymph NaCl was 130 mmol l^{-1} in 16% SW (74 mmol l^{-1}) and increased only to 157 mM in SW and 330 mmol l^{-1} in very concentrated media (3900 mmol l^{-1}). We will call these animals strong hyporegulators.

Hyperosmotic regulators present a reasonably orderly series of stages of adaptation to FW, and it may be supposed that the invasion of FW involved such a serial development of the mechanisms described (Beadle and Craig,

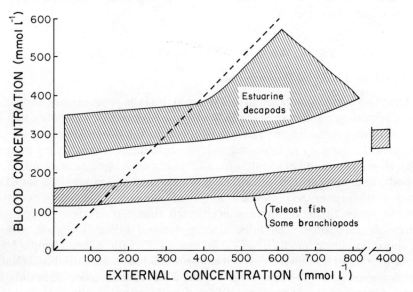

Figure 6 Patterns of hypo-osmotic regulation. Both groups include euryhaline species capable of regulating as shown. However, some members are incapable of hyper-regulation (many marine fish, for example)

1940; Shaw, 1961a, 1961b; Potts and Parry, 1946b). Such a series of stages does not emerge from studies on hyporegulators. Indeed, the adaptive significance of the phenomenon is not entirely clear, and it has probably developed for very different reasons in the two groups. The strongly regulating animals may be descendents of animals completely adapted to FW. Their blood concentrations are equivalent to those of the strongest of the FW regulators, and this character probably became genetically fixed during a long sojourn in this milieu. Such a history is also suggested for fish, by comparative studies on the vertebrate kidney which show that a distal tubule, capable of diluting the urine, is a basic feature, although it may be lost in animals returning to the sea (Smith, 1932, 1951). The maxillary gland of *Artemia salina* also shows two histologically distinct tubular segments (Tyson, 1969). Although nothing is known about the osmotic concentration of its urine, it may be dilute when the animal is in FW. If these are originally FW animals, then secondary radiation into more concentrated habitats leaves them one of two alternatives for coping with increasing external salinity. They can allow ECF concentration to rise and become isosmotic and iso-ionic with the medium. This would be the least costly modification. But, as we have seen, very strong hyper-regulators are unable to survive when the blood becomes more concentrated than about 50% SW. Instead they regulate blood composition within narrow limits until the external concentration is about equal to internal. If the former continues to rise regulation fails, ECF rises with the medium (as shown in Fig. 2), and the animal will not survive. Thus, if this group is to return to SW it must develop mechanisms for maintaining the ECF essentially at FW levels.

The same argument might be applied to weak or moderate hyporegulators; that ECF composition is fixed genetically at a level compatible with optimal tissue function, and this is about 60–70% SW. Most of these crustaceans are estuarine or shore animals, and hence are associated with a dilute milieu. But the argument otherwise lacks force. They are basically marine animals with none of the diagnostic features of the stenohaline FW forms. They cannot even compete with *Eriocheir sinensis* in penetrating FW, yet the latter can return to SW by becoming iso-ionic, which costs nothing energetically, while hyporegulators have the costly problem of maintaining the ECF out of equilibrium. A different possibility is based on recognition that the estuarine or littoral habitat is not merely more dilute than SW, but that its composition may be extremely variable, differing from place to place with the proximity to sources of FW, and from time to time with the tides. Active animals inhabiting such a changeable environment for long periods may have developed the mechanisms necessary for maintaining the body fluid concentrations at some fixed value (65–70% appears to be common) at which the average osmotic work may be minimized (Baldwin and Kirschner, 1976).

Strong Hyporegulators

These animals, which include all actinopterigean fishes found in the ocean (practically all of them teleosts), and a few branchipod crustaceans, face osmotic dessication and salt loading by diffusion across the body surface. The adaptive behaviour in hypotonic regulators is more complex than in hyperosmotic and is only now being worked out. That the animals lose water by osmosis across the body surface has been shown both in fishes (Smith, 1930) and in the brine shrimp (Croghan, 1958a, Geddes, 1975a), where preventing the animal from drinking causes it to lose weight. The same experiment indicates how volume is regulated in the face of osmotic water loss; the animals drink the hyperosmotic medium, and fluid is absorbed from the gut at a rate equal to loss across the body surface and through the kidney. To maintain a steady state, water loss (urine plus osmotic) must equal water intake (drinking). Urine production is usually low (see below) and, to a first approximation drinking rates (Table 4) measure osmotic flow across the body surface (J_v^{osm}), and the ratio $J_v^{osm}/\Delta\Pi$ is a measure of P_{osm}. By comparing P_{osm} in Table 4 with those in Table 1 it is obvious that the osmotic permeability of these animals is lower than that of marine osmoconformers or any of the hyper-regulators except for fresh-water crabs. It is also notable that in fishes capable of regulating both hyper- and hypo-osmotically P_{osm} is considerably less in SW than in FW. The adaptive significance is apparent, since the

Table 4. Drinking and P_{osm} in hypotonic regulators

	J_v^g (ml kg^{-1} h^{-1})	$\Delta\Pi$ (osmol)	P_{osm}	References
	Strong hyporegulators			
Salmo gairdneri	4.5	0.63	7.1	Shehadeh and Gordon (1969)
Anguilla anguilla	3.3	0.82	4.0	Motais and coworkers (1969)
Platichthys flesus	1.5	0.83	1.8	Motais and coworkers (1969)
Pholis gunnellus	0.51	0.6	0.9	Evans (1969)
Serranus sp.	2.1	0.82	2.6	Motais and coworkers (1969)
Paralichthys lethostigma	4.6	0.65	7.9	Hickman (1968b)
Artemia salina	21.6	0.85	36.7	Smith (1969b)
Parartemia zietziana	13.9	0.65	21.4	Geddes (1975b)
	Weak hyporegulators			
Uca sp. (175% SW)	5.0	0.41	12.2	Baldwin and Kirschner (1976)
Penaeus duorarum	17.3	0.16	108	Hannan and Evans (1973)
Metapenaeus bennetlae	6	0.35	17.1	Dall (1967)

osmotic gradient is much steeper in SW (about 300 mosmol l^{-1} in FW, but 600–700 mosmol l^{-1} in SW). However, the mechanism by which permeability is reduced in SW is obscure.

There is a steep salt gradient, favouring inward diffusion across the body surface, in these animals. Any net inward movement must be balanced by active extrusion. In addition, absorption of water from a hyperosmotic solution in the gut is linked to the absorption of nearly all of the NaCl it contains. Details of the ion-coupled water movement will be discussed later. Here it is necessary only to note that it involves substantial salt-loading for the animal, and that the NaCl absorbed must be extruded back into SW. It has been shown that the gills are the site of active extrusion in fish (Keys, 1931) and brine shrimp (Croghan, 1958b; Geddes, 1975b). The unidirectional fluxes of Na$^+$ and Cl$^-$ are very rapid in these animals. Table 5 gives values for some euryhaline animals capable of regulating both in FW and SW. Compari-

Table 5. J_{in}^{Na}, J_{in}^{Cl} and P_{Na} in some hyporegulators

Animal	Fluxes (mmol kg^{-1} h^{-1})		P_{Na}	References
	Na$^+$	Cl$^-$		
Pholis gunnellus	5.6	1.8	20.0	Evans (1969)
Tilapia mossambica	16.5	15.4	69.8	Dharmamba, Bornancin and Maetz (1975)
Gillichthys mirabilis	10.2	7.6	33.4	Thompson (1972)
Serranus scriba	26.5	19.6	88.0	Maetz and Bornancin (1975)
Anguilla anguilla	13.1	12.1	40.7	Motais (1967)
Platichthys flesus	22.1	9.4	91.3	MacFarlane and Maetz (1975)
Mugil capito	78.0	50.8	241.9	Maetz and Pic (1975)
Salmo gairdneri	3.7	3.8	9.3	Kirschner, Greenwald and Sanders (1974); Gordon (1963)
Artemia salina	135	140	440.1	Smith (1969b)
Uca sp. (175% SW)	65.3	81.8	—	Baldwin and Kirschner (1976)
Palaemonetes varians	150	50	—	Potts and Parry (1964)
Pachygrapsus crassipes	19.8	—	—	Rudy (1966)

son with Table 1 shows that P_{Na} is several orders of magnitude larger in the latter. In fact, isotope fluxes in hyporegulators are about as fast as those in weak hyper-regulators like *Carcinus maenas*. Unfortunately, it is nearly impossible to analyse the extant data in a manner that permits one to estimate passive ion permeabilities with confidence. The reason is as follows: total efflux of either Na^+ or Cl^-, measured with tracers, comprises unknown proportions due to active transport, passive diffusion and 'exchange diffusion'. The latter, first described by Ussing (1947), is a membrane-mediated exchange of an ion from one solution (e.g. ECF) for an identical ion from the other (e.g. SW). It is insensitive to the electrochemical gradient, is rate-limited by the ion on the low concentration side, and can be abolished by omitting the ion from either solution; i.e. bathing a gill with Cl^--free SW should abolish Cl^- exchange diffusion, but not ordinary (leak) diffusion or active transport from the blood. The process is incapable of generating a net flux (as in Fig. 1, for example), but it can produce very high rates of isotope movement, high enough to make accurate estimation of the diffusive leak or the transport component difficult. Ion influx comprises only passive diffusion and exchange, but its measurement in intact animals is technically more difficult, and most experiments describe efflux data. In principle, the first step in estimating P_{Na} or P_{Cl} should be to eliminate any exchange by replacing Cl^- in SW with a non-penetrating anion or Na^+ with a non-penetrating cation. When this was done in *Artemia salina* it appeared that 70% or more of the unidirectional Cl^- flux was due to exchange, but the rapid movement of Na^+ was largely or entirely diffusive (Smith, 1969b), and similar conclusions were reached for the flounder (Potts and Eddy, 1973) and trout (Kirschner, Greenwald and Sanders, 1974). One direct measurement of the electrical resistance of a gill (in *A. salina*; Smith, 1969a) gave a value of 40 Ω cm^2. This value is compatible with an ion permeable epithelium; the resistance of frog skin, for example, is nearly one hundred times larger. Values for P_{Na} in Table 5 are calculated on the assumption that the total influx across the gill is by diffusive leak. In contrast, the real diffusive permeability to anions appears to be low. Unfortunately, flux measurements in fish cannot be used to estimate P_{Cl} because the exchange component has never been assessed. However, electrical measurements indicate that the ratio $P_{Cl}:P_{Na}$ is only about 0.03 in the flounder *Platichthys flesus* (Potts and Eddy, 1973), 0.1 in *A. salina* (Smith, 1969a), and 0.3 in the trout (Kirschner, Greenwald and Sanders, 1974). Thus, the gill appears to be a cation-selective epithelium with a very low anion permeability. If this is so, the tendency for Na^+ to diffuse from SW to blood unaccompanied by a counter-ion, should generate a voltage across the gill with the blood becoming positive to the external medium. Such a voltage will reduce the rate of inward movement of cations and accelerate their outward movement. In a gill completely impermeable to Cl^- the voltage will be just sufficient in magnitude to cause inward and outward Na^+ movements to be equal, and hence to prevent net inward movement of the ion. The potential difference across the gill (gill TEP) will be discussed in a different context later. The point to emphasize here is that, although the gill is very permeable to cations, the electrochemical gradient for Na^+ is close to zero, and hence there is little tendency for net movement to occur. In effect, the combination of very low anion permeability and very small gradient for cations means that little diffusive leak occurs from the more concentrated SW into the blood (cf. Potts, Fletcher and Eddy (1973) for an excellent discussion of the energetics of this situation).

Since hypo-osmotic regulators must cope with water loss across the body surface any urine excreted obviously aggravates the problem of volume regulation. These animals also face salt loading by absorption of SW from the gut and by diffusion across the gills. If they could excrete a urine containing a higher concentration of NaCl than the blood the kidney might be an important organ in steady-state osmotic regulation. However, it has been repeatedly shown that their urine is isosmotic with blood (cf. Potts and Parry, 1964b; Prosser, 1973), and in many the NaCl concentration is lower, not higher. Thus, from the point of view of water and NaCl regulation the kidney is a liability, and as a consequence urine volume is usually low in these animals. Its role is probably the same as in other marine animals, including osmoconformers. This includes the excretion of divalent ions which is indicated by their high urine concentration (Table 7).

B. Transport Mechanisms in Hyporegulators

1. *Strong Regulators*

a. *Criteria for Active Transport* As noted earlier, most attempts to delimit active transport involved comparing observed fluxes with those predicted by some solution of the diffusion equation (cf. Kirschner, 1970). One solution that is often used is the following. Total (unidirectional) fluxes are very rapid in all hyporegulators, and drinking and urine contribute very little (the SW salmonids are an exception); practically all influx and efflux are across the gill, and since the steady state requires that uptake and loss be the same, $J_{in} = J_{out}$ across the gill, to a first approximation. This can only be true for a diffusing ion when

$$E = \frac{RT}{ZF} \ln \frac{C_{sw}}{C_b} \qquad (5)$$

where E is the TEP across the gill, C_{sw} and C_b are concentrations in the medium and blood and the other quantities were defined earlier. That is, equation (5) is a solution of the general diffusion equation for an ion with the stated constraints: it must be in a steady state, so that $J_{in} = J_{out}$ and C_{sw} and C_b are constant. The voltage E, called the ion's equilibrium potential, is that necessary to keep the concentrations from changing without the need for active transport. The test is simply to measure the actual TEP. If it is equal to E, calculated from C_{sw} and C_{bl}, the ion is distributed by diffusion; if not then active transport must play a role in maintaining the steady state. For most unanaesthetized fish and for the brine shrimp *Artemia salina* TEP values fall in the range 18–35 mV, blood positive. In most cases they are close to the equilibrium potential for sodium (E_{Na}) as can be seen in Table 6. Although most of the measured values are a few millivolts smaller than predicted, the correspondence is striking. For a few fish the value of the TEP is much smaller than E_{Na}, and if these animals were shown to be in a steady state during the electrical measurements there is no question that active extrusion of Na$^+$ would be required. Unfortunately, this condition was not assessed; we do not know whether the fish were in a steady state, and the significance of the observations must await further work. In contrast, there is no question that Cl$^-$ is extruded actively against a fairly steep electrochemical potential difference. It would require that the blood be negative to SW by 25–40 mV for Cl$^-$ to be extruded passively; Table 6 shows that it is usually positive by this amount. Thus, there exists in these animals, a transport mechanism for extruding Cl$^-$ from the body fluids into the medium.

b. *The 'Chloride cell'* Keys and Wilmer (1932) described a type of cell present in greater number in SW-adapted eel gills than in eels adapted to FW, and they suggested that it was responsible for salt extrusion from the fish. The idea received a mixed reception, but is now generally accepted. The role of these salt-excreting cells in gills and salt glands was reviewed recently (Kirschner, 1977). Ultrastructural studies have shown that cells in salt excreting organs share one character, extensive infolding of the basolateral membrane. In fish (Philpott and Copeland, 1963; Bierther, 1970; Shirai and Utida, 1972) the infoldings, mainly in the basal membrane, form a series of intracellular channels open to the ECF (Philpott, 1966) and ending blindly near the apical pole of the cell. A study of *Artemia salina* gill (Copeland, 1967) shows cells with a highly vesiculated interior suggesting that the same basal membrane infoldings occur in this animal. The extent of this internal tubular system is truly remarkable and since it is really an extension of the ECF, it presumably encloses a fluid, high in Na$^+$ but low in K$^+$, which interdigitates with cytoplasm of the opposite cation composition. Two other observations are of some interest. The interior of the tubular system appears to be filled with a polyanionic material (Philpott and Copeland, 1963), that would confer interesting properties on it. For one thing, the intracellular tubule would behave like a cation exchange column and could concentrate cations like Na$^+$. As a corollary the tubule should have a very low Cl$^-$ concentration. If the tubular system plays *any* role in salt excretion this inverse relationship suggests different pathways for Na$^+$ and Cl$^-$ extrusion. The second observation is that this system of intracellular membranes contains most of the alkali metal-activated ATPase in fish gills (Karnaky and coworkers, 1976; Shirai and Utida, 1972), and since this enzyme is implicated

Table 6. Gill TEP and E_{Na} in hyporegulators

Animal	Na+ concentration mmol l^{-1}		Potential difference (mV)		References
	Plasma	SW	E_{Na}	TEP	
			Strong regulators		
Mugil capito	175	520	27.1	22.4	Maetz and Pic (1975)
Serranus sp.	188	520	25.3	25.2	Maetz and Bornancin (1975)
Anguilla anguilla	160	520	29.2	22.5	House and Maetz (1974)
Gillichthys mirabilis	164	480	26.6	20.6	Thompson (1972)
Tilapia mossambica	159	520	31.5	35.2	Dharmamba, Bornancin and Maetz (1975)
Blennius pholis	170	470	24.8	23	House (1963)
Pholis gunnellus	176	410	20.6	18	Evans (1969)
Platichthys flesus[a]	202	470	20.5	19	Potts, Fletcher and Eddy (1973)
Platichthys flesus[a]	170	520	27.8	33.9	MacFarlane and Maetz (1975)
Artemia salina	172	468	24.3	23	Smith (1969b)
Salmo gairdneri	180	450	~50	10.5	Kirschner, Greenwald and Sanders (1974)
Achirus lineatus	—	—	—	−6.8	Evans and Cooper (1976)
Hippocampus erectus	—	—	—	−3.5	Evans and Cooper (1976)
Opsanus beta	—	—	—	−8.1	Evans and Cooper (1976)
			Weak regulators		
Palaeomonetes varians	328	460	8.4	−25	Potts and Parry (1964a)
Uca sp.	448	800	14.9	−1.3	Baldwin and Kirschner (1976)
Pachygrapsus crassipes	680	750	2.8	−1.5	Rudy (1966)

[a] Values for the flounder are taken from two different studies to show the range that can be encountered in the same animal. It is significant that the TEP is smaller than E_{Na} in one study, but substantially larger in the other. The first would require active extrusion to maintain a steady state, the second would not. For the first 9 animals it is safe to conclude only that the electrical measurement provides no evidence for active extrusion. E_{Na} for the rainbow trout (about 50 mV) is not an equilibrium potential. The flux ratio across the gill of this fish is about 2.

in Na⁺ transport in other epithelia its presence in chloride cells (Kamiya, 1972) is provocative.

c. *The Characteristics of Na⁺ Extrusion Across the Gill* If the use of the equilibrium potential has been inconclusive in resolving the question of active or passive Na⁺ transport, the same can be said of most other experimental approaches. The subject has been reviewed repeatedly during the last five years (Maetz, 1971, 1973, 1974; Motais and Garcia Romeu, 1972; Maetz and Bornancin, 1975; Potts, 1976; Evans, 1977). The description that follows will therefore be compressed, and some emphasis will be placed on different conclusions that might be drawn from the data in critical experiments. The analysis will end with no resolution to the question 'active or passive' for Na⁺ transfer. For more extended treatments and a different conclusion the reviews cited above should be consulted.*

Shortly after the pioneering work of Motais on ion fluxes in marine teleosts (Motais, Garcia Romeu and Maetz, 1966; Motais 1967) there appeared a report that Na/K-ATPase levels were much higher in gills of *Fundulus heteroclitus* adapted to SW than to FW, and it was proposed that the enzyme played a role in the active extrusion of Na from the animal (Epstein, Katz and Pickford, 1967). The observation has been extended to many marine teleosts (Kamiya and Utida, 1969; Jampol and Epstein, 1970) and is clearly part of the adaptive response of these animals to live in a hyperosmotic environment. One model (Maetz, 1969, 1971) placed the enzyme on the apical membrane of the chloride cell where it could extrude Na⁺ from the cell into SW in exchange for K⁺ from the environment. The model received strong support when Maetz showed that Na⁺ extrusion from fish was dependent on the K⁺ concentration in the medium. The key experiment is worth describing because it still forms, with variations, a principle mode of attack on the problem. When the medium bathing a marine teleost (or *Artemia salina*) is changed from SW to FW sodium efflux (J^{Na}_{out}) drops to about 25% of the SW value as shown in Fig. 7. If K⁺ is now added to the external medium (at the arrow) so that its concentration is equivalent to that in SW, J^{Na}_{out} is markedly stimulated. In fact, J^{Na}_{out}, in such an experiment, can be shown to be a non-linear function of external K⁺ concentration approaching what appears to be an upper limiting value at high K⁺ concentrations (Maetz, 1969). However, no account was taken of gill permeability or the electrical gradient in the original experiments. When the latter was measured it was found that each change, from SW to FW, and from FW to FW + K⁺, was accompanied

---

* The controversy may be resolved by a recently developed preparation. The skin covering the inner surface of the operculum often contains chloride cells, and in *Fundulus heteroclitus* adapted to 200% SW, these comprise a very large fraction of the cell population (Karnaky and coworkers, 1976). The skin can be isolated, clamped in a chamber and studied like other isolated epithelia (Karnaky and coworkers, 1976). With Ringer's solution bathing both sides it develops a TEP of about 30 mV, in the same range as the gill; measurements of active ion movement show that this consists solely of chloride extrusion. Since this epithelium has neither the complex structure nor respiratory epithelium found in the gill, it should provide data that are much simpler to interpret.

by marked changes in the gill TEP, and the electrical changes can account for much of the observed variation in sodium efflux (Potts and Eddy, 1973; Kirschner, Greenwald and Kerstetter, 1974). In more recent work the electrical, as well as chemical gradient, was measured but two limitations make it difficult to draw unequivocal conclusions even from these data. In many experiments (e.g. Kirschner, Greenwald and Kerstetter, 1974; Evans, Carrier and Bogan, 1974; Evans, Cooper and Bogan, 1976) the electrical measurements are done on one group of animals and flux measurements on a different group, and correlation of data requires the assumption that implantation of an electrical bridge into the body cavity has no effect on ion movement; this is possible, but untested. In addition, in no experiment has the passive permeability been estimated; in fact, no method for evaluating permeability in these animals existed until recently. Instead, the parameter is assumed to remain constant in experiments like the one shown in Fig. 7, and a

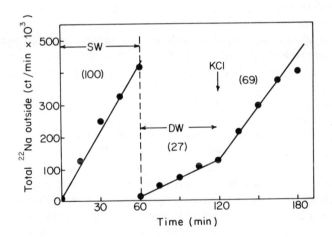

Figure 7 Na^+ efflux from SW-adapted teleost. The gill was initially exposed to SW and $^{22}Na^+$ injected into the fish. The first panel shows the appearance of isotope in the medium. When the medium was changed to FW efflux was suppressed to 27% of the SW value. Addition of KCl (10 mmol l^{-1}) at the arrow stimulated Na^+ efflux. From Kirschner and co-workers (1974), reproduced by permission of the *Journal of General Physiology*

flux change is determined to be active or passive according to whether its magnitude can be calculated from the change in driving force. But passive solute movement is the product of permeability and gradient, and unless both are evaluated the cause of a change in flux remains uncertain.

Histochemical localization of the Na/K-ATPase provides fairly compelling evidence against the original, apical membrane model. Most of the enzyme is associated with the extensive intracellular tubular system described earlier. This has been shown in the fish gill (Shirai and Utida, 1972; Karnaky and coworkers, 1976) although the results are even clearer in the avian (Ernst, 1972a, 1972b) and reptilian (Ellis and Goertemiller, 1974) salt glands, and the rectal gland of the dogfish (Ellis, unpublished observations). In no case is there convincing evidence for any enzyme on the apical membrane. Given its intracellular location, the enzyme would not have direct access to K^+ in the external medium, and hence manipulations of the latter (Maetz, 1969) are unlikely to act on it. Similarly the enzyme must extrude Na^+ into the intracellular channels of the chloride cell, and the latter opens to the ECF, not to the external medium. Alternative roles for the enzyme in Na^+ extrusion (Shirai, 1972; Maetz and Bornancin, 1975) and control of epithelial water permeability (Kirschner, 1977) have been suggested but will not be detailed here because they are entirely speculative.

   d. *The Characteristics of Cl^- Extrusion Across the Gill*   Some aspects of Cl^- movement are clearer than for Na^+. Active extrusion is surely involved, and the minimum *net* efflux must equal Cl^- absorbed in the gut. For many fish this is in the range 1–5 mmol kg^{-1} h^{-1}; it is even higher in *Artemia salina* which is a much smaller animal. To this must be added a quantity equal to any net influx across the body surface by diffusion. The latter may be small for the reasons described above: because the electrochemical gradient is modest and P_{Cl} appears to be very small (Smith, 1969a; Potts and Eddy, 1973; Kirschner, Greenwald and Sanders, 1974). The chloride pump is probably electrogenic since a perfused flounder gill, with Ringer's solution on both sides of the epithelium (hence no concentration gradient), generated a small TEP (Shuttleworth, Potts and Harris, 1974). The TEP was abolished by anoxia, and by ouabain and SCN^- in the vascular perfusate. Ouabain in the external medium, where it would have access to an apical Na/K pump, was without effect. Such electrogenic behaviour of the transport system may indicate that Cl^- extrusion is not accompanied by coupled uptake of another anion (like HCO_3^-) or that the coupling is such that more Cl^- is extruded than counter ion absorbed. Total (tracer) efflux of Cl^- was reduced by SCN^- injected into the blood of eels. The extent of unidirectional flux inhibition was much greater than net Cl^- extrusion in this fish, hence the inhibitor may have reduced an exchange component, but the transport system was clearly blocked, because blood Cl^- concentrations rose (Epstein, Maetz and De Renzis, 1973). It is worth noting that the anion-stimulated ATPase described in FW adapted fish has also been found in gills of SW animals (Kerstetter and Kirschner, 1974; Solomon and coworkers, 1975).

   Attempts to elucidate transport mechanisms from flux studies are as equivocal for Cl^- as for Na^+, and for the same reason. In most studies the total isotope efflux is measured, and in fish it has never been partitioned into fractions comprising transport, diffusive leak (which requires assessing both the gradient and permeability), and exchange diffusion. Smith (1969b) concluded that as much as 70% of J_{out}^{Cl} in *Artemia salina* may be due to Cl–Cl exchange. Such an exchange system probably also exists in fish. For example, when a SW-adapted fish is transferred to FW J^{Cl}_{out} decreases by 75%. That is, it behaves much like sodium in the experiment shown in the two panels of Fig. 7. Much of the decrease in cation efflux can be attributed to the TEP which changes from +20 to about −40 mV as a result of the transfer. But this obviously cannot account for a decrease in anion efflux which should actually be accelerated by the electrical change. The magnitude of the depression is far greater than can be accounted for by complete suppression of transport and almost certainly shows that a large fraction of the unidirectional efflux is due to exchange (which would be abolished because FW is nearly Cl-free). Efflux of Cl⁻ is also sensitive to changes in K⁺ concentration in the external medium. This is shown by experiments like the one in Fig. 7 (but in which J^{Cl}_{out} is measured instead of J^{Na}_{out}). The results are similar. With efflux suppressed in FW (panel 2) KCl is added to the external medium, and J^{Cl}_{out} is stimulated, although less than J^{Na}_{out}. The TEP in such a manoeuvre changes, but in a direction that should suppress diffusive efflux of Cl⁻, hence the stimulation does not result from modifying the electrochemical gradient. However, an unknown fraction may be attributed to Cl–Cl exchange since the K⁺ is usually added as a chloride salt, and the parameter P_{Cl} may also be modified by the experiment. Since these factors were not estimated it cannot be concluded that stimulation of tracer efflux represents an effect on the transport system, although it is one possibility.

A model for salt extrusion, based on the data that appear to be simplest to interpret is presented in Fig. 8. A Cl⁻ transport system is shown in parallel with a Cl–Cl exchange, but they may be different expressions for the same

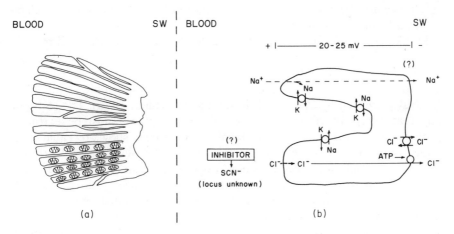

Figure 8 Ion transport across gills of hyporegulators. (a) A highly schematized representation of a 'chloride cell'. The Na⁺/K⁺-ATPase is located on the infolded basal membranes. (b) A model of NaCl extrusion across the chloride cell. Chloride transport is active. Sodium extrusion is shown as passive (dashed line), but the question mark indicates that this aspect is not yet settled

mechanism. There is no evidence for exchange against an external anion, hence electrical neutrality can be maintained only by parallel flow of a cation. Since the gill is very permeable to cations electrical coupling might account for the passive net extrusion. Since Na^+ is the most abundant cation in the ECF its participation has received the most attention. But the gill is relatively non-selective among cations; P_K is even higher than P_{Na} in some fish (Potts and Eddy, 1973; Kirschner, Greenwald and Sanders, 1974), and some net extrusion of K^+ might be expected. It is interesting that just such a phenomenon was reported for the eel (Maetz, 1969). More complex models, based largely on flux experiments such as those described above, have been proposed. For example, the dependence of both J_{out}^{Cl} and J_{out}^{Na} on external K^+ concentration suggests that the transport mechanism involves linked extrusion of Na^+ and Cl^- and can be stimulated, in an unknown manner, by K^+ (Maetz and Bornancin, 1975). As noted, these are plausible interpretations of the data, but until the latter are analysed rigorously the simpler model is as reasonable.

2. Weak and Moderate Regulators

The mechanisms of ion and water regulation in this group have only just begun to receive attention, and the number of studies is small. Values of P_{osm}, calculated from drinking rates in a crab of the genus *Uca* and two shrimps, are shown in Table 4 (urine volumes, where measured, are negligible). These are higher than in most fish, but the crustaceans weigh only a few grams while the fish shown weigh in the order of 100 g. Much of the difference may be due to the allometric relationship between flows, which are surface-dependent, and weight.

Drinking data also permit us to estimate the rate at which NaCl must be excreted. Assuming that nearly all the NaCl ingested is absorbed in the gut (which has been shown for some fish but not for custaceans), an equivalent *net* efflux must occur. Excretion in the urine accounts for only about 2% of the NaCl ingested by *Uca sp.* (Baldwin and Kirschner, 1976). The rest, plus any that enters by diffusion across the body surface, must be extruded, probably across the gills. Unidirectional (tracer) fluxes, shown in Table 5, are larger than the net movement by two orders of magnitude; net efflux is therefore a small difference between rapid unidirectional influx and efflux. The components of the electrochemical gradient (concentrations and TEP) are also shown in Table 6. If influx of Na^+ and Cl^- were due solely to passive diffusion, values for P_{Na} and P_{Cl} could be calculated from these data. Unfortunately, two studies on *Uca sp.* indicate that a substantial part of the flux may be due to exchange diffusion (Evans, Cooper and Bogan, 1976, Baldwin and Kirschner, 1976). When *Uca* sp. was in 175% SW the exchange component may have exceeded 90% of the total, but its exact value could not be established (Baldwin and Kirschner, 1976). The fact that the TEP in *Uca* sp. was near zero suggests that P_{Na} and P_{Cl} are similar, i.e. these gills are not cation selective as are those in fish and *Artemia salina*. A TEP between 13–40 mV, hemolymph negative, was reported for *Palaemonetes varians* in SW (Potts and Parry, 1964a); the shrimp gill may be differentially ion permeable (perhaps $P_{Cl} > P_{Na}$). But quantitative estimates cannot be made until the diffusive fluxes can be measured.

The ambiguity in flux measurements also precludes a complete analysis of active transport in *Palaemonetes varians*. Since SW is positive to blood both electrical and concentration gradients oppose net Na^+ extrusion which must therefore be active. However, the TEP favours Cl^- extrusion and, at the concentrations shown, any value in excess of about 10 mV could generate net outward movement passively. Since the range of TEP's was 13–40 mV there appears to be no need for active Cl^- transport. But good estimates of the passive components of Cl^- influx and efflux would make this conclusion more convincing. In *Uca* sp. the TEP is negligible and both ions must be actively extruded against the electrochemical gradient.

C. Role of the Kidney

Marine Osmoconformers and Hyporegulators

Until 1960 there existed only a handful of studies on excretory physiology in invertebrates. The field was pre-empted by vertebrate physiologists whose studies led to the conclusions that: (1) a filtration kidney is admirably adapted to excreting water loads; (2) since osmotic loading occurs in FW, such a kidney is adaptive there; but (3) that it would be a liability in SW. In this view, such a kidney could only have developed during the evolution of a FW group; the vertebrates, on this evidence, must have arisen in FW. Implicit to the argument was a second premise, namely, that invertebrates, especially marine invertebrates, do not filter a presumptive urine but form an excretory fluid by secretion. However, within a few years it was shown that polymers like inulin, which enter the vertebrate kidney only by filtration and cannot be absorbed from the tubular fluid, are excreted by the kidneys of invertebrates from several phyla, including some, like the cephalopod molluscs, that have no known history nor any modern representatives in FW. In contrast, it is excluded, or nearly so, from the urines of aglomerular fishes and insects which are known to be formed by secretion. Evidence that inulin appears in the urine of secretory kidneys like the insect malpighian tubule (Ramsay and Riegel, 1961) and aglomerular fish nephron (Lahlou and Sawyer, 1969) shows that the tubule is not totally impermeable as was supposed. But urine concentrations are very low in these animals and do not affect the thrust of the argument. The literature through 1966 was summarized in a review (Kirschner, 1967), and more recent studies were summarized by Riegel (1972). Many crustaceans have been shown to excrete inulin (and other filtration markers) with urine concentrations at least as high as those in hemolymph, and there are no reports describing the contrary. We can conclude that the crustacean antennal gland acts as a filtration organ in the same sense as does the kidney in fishes. The mechanism is indeed admirably suited to handling large and variable water loads (cf. increasing filtration rates and urine volumes in *Carcinus maenas* in a series of media of increasing hypotonicity; Binns, 1969), but this is a by-product, and not an 'evolutionary cause', of its development.

Published values of filtration rates show that even marine osmoconformers form a presumptive urine at respectable rates (Tables 3 and 7). This has raised the question of what driving force causes water to enter the animal at a rate sufficient to operate the excretory organ. One possibility is a small osmotic pressure difference generated by the presence of protein in the hemolymph and its absence from the environment, while another suggestion, supported by some data, is that the necessary water movement is coupled to active inward transport of salt (Lockwood and Inman, 1973).

Filtration rates in hyporegulating animals are variable but appear to be smaller than in FW animals. Low filtration rates would be adaptive, since osmotic water loss is a severe problem, and the reduction in marine telosts has become part of the orthodox lore of comparative vertebrate physiology. However, recent studies of the behaviour of filtration markers (e.g. inulin, polyethylene glycol) in SW teleosts indicate that they may be reabsorbed in the distal tubule and urinary bladder after being filtered (Beyenbach and Kirschner, 1976). This would result in underestimation of the GFR, and there is presently no assurance that filtration rates are substantially lower in SW fish than in FW. In contrast, urine volume is undoubtedly markedly reduced (Table 7), and the general thesis that renal water loss is low in marine fish is correct. To what extent the reduction is due to reduced GFR, and to what extent to water reabsorption from the filtrate is presently uncertain.

A number of studies suggest several analogies between the kidney of marine fishes and crustaceans, as well as other marine invertebrates. For example, both groups have the tubular transport system responsible for excreting organic acids (Kirschner, 1967; Riegel, 1972), and the processes of nutrient reabsorption are also similar. The organs also share another characteristic: none play more than a minor role in NaCl regulation. Urine concentrations of Na^+ and Cl^- are never higher than hemolymph, and often are lower (Table 7), and rates of excretion are much smaller than absorption from the gut. Clearly, the branchial extrusion system is much more important in ionic homeostasis. However, comparison of blood and urine ion concentrations (Table 7) show that the concentrations of Mg^{2+}, Ca^{2+}, and SO_4^{2-} are higher in the urine than the blood. Since many marine invertebrates reabsorb little water from the filtrate, the high urine concentration must be due to secretion. This was pointed out some years ago both for crustaceans (Prosser, Green and Chow, 1955) and fishes (Hickman and Trump, 1969, Smith,

Table 7. Kidney function in osmoconformers and hyporegulators

			Urine/blood ratios							
	Medium	J_v^u (ml kg^{-1} h^{-1})	Na	K	Cl	Mg	Ca	osmo-larity	Reference
			Isosmotic animals							
Homarus	SW	3.4	0.99	0.91	1.0	1.8	0.61	—	Bryan and Ward (1962)
Pugettia producta	SW	2.7	0.98	1.1	1.0	1.5	1.1	1.0	Cornell (1976)
Carcinus maenas	SW	3.6	0.92	0.98	1.0	4.2	1.2	0.98	Riegel and Lockwood (1961)
Eptatretus stoutii	SW	0.31	0.97	1.57	1.0	1.2	0.80	—	Munz and McFarland (1964)
Squalus acanthias	SW	3.5	0.96	0.5	1	33	1	0.80	Burger (1967)
			Weak hyporegulators							
Uca pugnax	175% SW	0.22	0.58	1.3	1.2	4.6	1.4	1.2	Green and coworkers (1959)
Pachygrapsus crassipes	170% SW	1.6	0.31	0.77	—	9.8	1.5	0.96	Prosser, Green and Chow (1955)
Palaemonetes varians	120% SW	4.2	0.91	0.83	1.3	4.5	0.98	1.0	Parry (1955)
			Strong hyporegulators							
Salmo gairdneri	SW	<1.5	0.06	—	—	88	3.3	—	Beyenbach (1974)
Anguilla anguilla	SW	0.3	0.35	0.5	0.72	7.2	3	—	Chester-Jones, Chan and Rankin (1969)
Platichthys flesus	SW	<0.6	0.25	0.82	0.61	—	2.8	0.95	Lahlou (1967)
Paralichthys lethostigma	SW	0.3	0.10	0.49	0.83	125	7.1	0.96	Hickman (1968a)

1930). Convincing evidence for tubular secretion is provided by the observation that aglomerular fish excrete urine with a high divalent ion concentration, and injection of Mg^{2+} resulted in a marked diuresis and increased Mg^{2+} clearance. The transport mechanism is located in cells of the proximal tubular segment, since many marine fish and all marine crustaceans lack a distal tubular segment. In the flounder *Paralichthys lethostigma* about 16% of the Mg^{2+} and about 11% of the SO_4^{2-} swallowed are absorbed from the intestine. The ions absorbed are completely eliminated in the urine, and blood concentrations of these ions are maintained constant by the kidney (Hickman, 1968b). Urine excreted by the crab *Uca sp.* in SW has a high Mg^{2+} concentration (Table 7), and blocking the nephropores in 175% SW resulted in a two-to-four-fold rise in hemolymph Mg^{2+} concentration (Baldwin and Kirschner, 1976). Infusion of Mg into fish was tolerated as long as plasma concentration did not exceed about 6 mmol l^{-1}. Such infusions in both flounder (Hickman, 1968b) and trout (Beyenbach and Kirschner, 1975) produced a transient elevation in plasma concentration, but this was accompanied by an almost immediate increase in the rate of secretion, and the load was excreted within a few hours. An inverse correlation is often noted between Mg^{2+} and Na^+ concentrations in the urine of crustaceans and fish, and it was proposed that this might be the result of a coupled ion exchange system that transported Mg^{2+} into the tubule and reabsorbed Na^+ that had entered by filtration (Prosser, Green and Chow, 1955; Natochin and Gusev, 1970). The proposition was examined in the rainbow trout and shown to be incorrect (Beyenbach and Kirschner, 1975). Magnesium is secreted in the tubule, but Na^+ is absorbed primarily from the urinary bladder (see below). Similar studies are needed on the mechanism of divalent ion transport in crustacean antennal glands, and further work on the fish kidney could be undertaken profitably using modern micropuncture and tubular perfusion techniques.

The excretory systems in these animals also have developed mechanisms for conserving water, even though they cannot produce a urine more concentrated than blood. The adaptive significance is apparent, since water loss is a major problem in hypo-osmotic organisms. There is some indication that water may be reabsorbed in the tubular system at least in some fish, but research has focused on the urinary bladder. Observations indicating that the urinary bladder in marine fish appears to play an important role in water conservation are recent. In SW-adapted flounder and trout NaCl concentration in bladder urine was much lower than in ureteral urine, but the total osmotic concentrations of the two fluids were about the same (Lahlou, 1967; Beyenbach and Kirschner, 1976). In the trout Mg^{2+} concentration in bladder urine was higher than in ureteral. Such observations suggest that NaCl and water are reabsorbed from the bladder leaving Mg^{2+} more concentrated in the residual fluid. This was confirmed, and the mechanism analysed in an isolated, perfused urinary bladder from the flounder *Pseudopleuronectes americanus*. The data showed that both Na$^+$ and Cl$^-$ were actively transported from bladder lumen to a solution bathing the basal side of the epithelium (Renfro, 1975). Transport of the two ions was tightly coupled, and deleting either one (e.g. substituting Na$_2$SO$_4$ for NaCl) inhibited transport of the other. Absorption was unaffected by an imposed electrical gradient. Water movement accompanied the transport of solutes, and divalent ions were concentrated in the luminal fluid during the process. Fluid absorption was abolished by compounds known to inhibit Na$^+$ (ouabain, ethacrynic acid) or Cl$^-$ (furosemide) transport systems. Thus, water reabsorption is linked to the active transport of ions. The adaptive significance of this phenomenon can be shown by the following considerations. Water reabsorbed here need not be replaced by drinking sea water and absorbing it in the intestine. Renfro's data showed that about 400 mosmol of solute is transferred per millilitre of water moved from bladder to ECF. In the gut the absorption of one millilitre of water requires transport of nearly five times as much solute (House and Green, 1965; Skadhauge, 1969). Since energy is expended in active transport of the ions with water following passively (cf. Schultz and Curran, 1968; or Chapters 1 and 2 of this book for a discussion of solute-linked water transfer), the process of recovering water from the urine is clearly more economical than replacing it in the gut. It is interesting that the rate of NaCl transport in the SW-adapted flounder bladder was very high, but the activity of the Na/K-ATPase was lower in SW than in FW bladders (Utida and coworkers, 1974). But since ouabain inhibited salt transport in the SW bladder the Na/K system appears to play a role here, too.

D. Water Intake

Smith (1932) used a non-absorbable marker (phenol red) to demonstrate that the SW-adapted eel swallows the medium and absorbs much of the NaCl and water from the intestine. The fish lost weight (dehydrated) when the esophagus was blocked, and he proposed that volume regulation required ingestion and absorption of the medium at a rate commensurate with osmotic and urinary water loss. Similar observations have been extended to the amphipods *Artemia salina* (Croghan, 1958b) and *Parartemia zeitziana* (Geddes, 1975b). In the brine shrimp it was necessary to block both ends of the alimentary canal, indicating that anal as well as oral drinking occurred. Smith's observations were confirmed and amplified on the rainbow trout (Shehadeh and Gordon, 1969) and flounder (Hickman, 1968b). Both animals absorbed nearly all the NaCl and about 75% of the ingested water. Hickman's studies showed that smaller quantities of the divalent ions were also absorbed, but about 85–90% remained in the intestinal lumen in the form of a concentrated solution to be excreted. In *P. zeitziana*, adapted to SW, Mg^{2+} did not become concentrated, although most of the fluid and NaCl were absorbed. This is puzzling, for if Mg^{2+} were absorbed at the indicated rate the transport system responsible for excreting it (presumably in the maxillary gland, the excretory organ) must have an exceptional capacity.

Drinking rates have now been estimated in many hyporegulators, and some values are shown in Table 4. The rates are variable from one animal to another, but where comparisons are possible are always much higher than urine flow (cf. Table 7), and hence they provide a reasonable estimate of osmotic loss across the body surface. It is interesting that the intestine of the European eel absorbs salt faster when the fish is adapted to concentrated sea-water than to normal SW (Skadhauge, 1974), and drinking rates are considerably higher in the more concentrated medium (Skadhauge, 1969) reflecting a more rapid osmotic water loss. However, in the cyprinodont *Aphanius dispar*, which can live in very concentrated salines, drinking rates were the same in SW, $2 \times$ SW and $3.5 \times$ SW which indicates that an alternative adaptive strategy involves reducing the osmotic permeability of the body surface in more concentrated media (Skadhauge and Lotan, 1974). Sea-water entering the alimentary canal is diluted in the anterior intestine by osmotic water inflow from the more dilute ECF. Net fluid absorption then occurs in the more posterior part of the gut; the diluted sea-water is still hyperosmotic to ECF in the eel (Skadhauge, 1969, 1974), and the brine shrimp *Artemia salina* (Croghan, 1958b) but the intestinal fluid was reported to be isosmotic in *Parartemia zeitziana*. The observation that NaCl is absorbed suggests that net water movement against an osmotic gradient may be coupled to solute transport, a phenomenon which had already been demonstrated in the mammalian intestine by Curran (reviewed in Schultz and Curran, 1968). Using isolated gut sacs bathed on both sides with Ringer's solution, House and Green (1965) were able to show that both Na^+ and Cl^- were transported from lumen to ECF side, and that when ion transport was reduced water movement decreased as well. An intestinal perfusion technique has allowed the coupled solute–water transport to be studied quantitatively *in vivo* (Skadhauge, 1969, 1974).

In addition to confirming that both salt and water were absorbed against their gradients these studies established the following. As long as the luminal fluid contained NaCl, water was reabsorbed and the rate of water absorption (J_v) was related linearly to the rate of salt transport. The ratio J_{NaCl}/J_v gives the concentration of the reabsorbed fluid, which was about 1100 mosmol l^{-1} for the eel, close to the value suggested by House and Green for *Cottus scorpius*. If the luminal salt solution was replaced by a non-penetrating non-electrolyte water movement simply followed the impressed osmotic gradient; uphill transfer did not occur. This is clear evidence for the solute linkage proposed in earlier studies. Although solute-linked water flow can occur in FW-adapted fish, the gradient against which flow can be induced is much smaller than in animals adapted to more concentrated media, and the key variable was the rate of NaCl transport which increased in SW-adapted animals. Increased salt transport in SW-adapted eels had also been reported by Oide and Utida (1967). At least two interesting points emerge from these studies, in addition to those concerning the molecular mechanism of fluid transfer. First, the ion concentration of the intestinal absorbate is nearly three times that of the fluid reabsorbed from the urinary bladder. If the transport mechanisms are similar in the two epithelia this means that energy expenditure per millilitre of water reabsorbed is considerably higher in the intestine. Even were this not the case water reabsorbed from the bladder need not be drunk and absorbed from the gut. This in turn reduces the salt load that must be excreted through the gills with a corresponding reduction in the total cost of osmotic regulation. The other point concerns a commonly accepted proposition that fish adapted to fresh-water do not drink. A number of recent studies suggest that appreciable drinking may occur in FW fish (Evans, 1967, 1969; Potts and Evans, 1967; Potts and coworkers, 1967; Motais and coworkers, 1969; Skadhauge, 1969; Lotan, 1969; Kirsch and Mayer-Gostan, 1973; Thompson, 1972), and the crab *Uca sp.* in 10% SW (Baldwin and Kirschner, 1976). While the process appears to be maladaptive, increasing the water load generated by osmotic influx across the body surface, the measurements can hardly be doubted. There are no balance studies comparable to those of Hickman on the SW-adapted flounder, and it is possible that only a fraction of the water ingested is absorbed. Solute-linked water movement should be negligible because of the virtual absence of salt in FW, and osmotic water absorption could be reduced if transit time through the gut is small. More work is clearly needed before we can draw any general conclusion about whether the intestine contributes appreciably to the water load in hyper-regulating animals.

In a few studies the molecular mechanisms underlying fluid transport have been investigated in isolated fish gut. The pioneer work of House and Green (1965) indicated that the transport of Na^+ and Cl^- were only partly independent in isolated gut sacs from *Cottus scorpius*. Thus, replacing Na^+ with choline (a non-penetrating, organic cation) reduced the transfer of Cl^-, and if Cl^- was replaced by the non-penetrating SO_4^{2-} the absorption of Na^+ was reduced. In both cases H_2O transfer was inhibited. Similar observations were made on the perfused eel intestine (Skadhauge, 1974). The effects of deletion of one ion on the flux of its partner may be partly due to changes in the electrical gradient. For example, when the flounder intestine was bathed on both sides by NaCl the TEP across the epithelium was only -3 mV (lumen positive), and both ions were transported at relatively high and equal rates. But when Cl^- was replaced by SO_4^{2-} the movement of Na^+ was markedly reduced. At the same time the TEP across the gut changed to $+1$ mV, ECF side positive, and this would oppose net movement of Na^+ toward the ECF (Huang and Chen, 1971; Ando, Utida and Nagahama, 1975). When sodium was replaced by choline, Cl^- absorption was reduced, but at least some of the change was accounted for by an increase in TEP from -3 to -8 mV. However, even when the TEP was abolished (using a technique known as 'short-circuiting') net movements of both Na^+ and Cl^- were observed from solutions containing non-penetrating counter-ions, so the gut appears to have transport systems for both Na^+ and Cl^-. Under short-circuit conditions the Cl^- flux from lumen to ECF was much larger than the Na^+ flux, hence, the anion pump is the more powerful. At least part of each ion flux may occur by exchange mechanisms similar to those described in the gills of hyperosmotic regulators, i.e. Na^+/H^+ and Cl^-/HCO_3^-. The presence of exchange systems is supported by the observation that carbonic anhydrase is present in the gut, and that the carbonic anhydrase inhibitor, Diamox, suppresses both Na^+ and Cl^- transport in isolated intestine (Huang and Chen, 1971). The Na/K-ATPase is also found in fish intestine, and in eels its concentration increases (as does ion transport) after transfer from FW to SW (Oide, 1967; Jampol and Epstein, 1970). When ouabain is present in the lumen water absorption is reduced (Oide, 1973). In addition, a Na^+-stimulated alkaline phosphatase also increases in eels adapted to SW (Oide, 1970). Water absorption is enhanced when the gut lumen is alkaline (the pH is normally 8.7), and inhibitors of this enzyme (cysteine, borate) block fluid absorption (Oide, 1973).

V. ISOSMOTIC, HYPO-IONIC REGULATION

While the problems of regulating water and NaCl concentrations in the body fluids are conceptually separable, they are, for most animals, intimately related. An adaptive strategy for minimizing or solving one problem usually aggravates the other. However, members of at least two groups of fishes, the elasmobranchs and the sole surviving crossopterygean, *Latimeria chalumnae*, and one amphibian, the frog *Rana cancrivora*, have succeeded in separating the two problems by eliminating one. Most of the research has been directed toward elasmobranchs, and this group will provide the general model. The adaptive strategies employed by elasmobranchs living in sea-water were recognized many years ago (Smith, 1936), and only a modest amount of detail has been added since then.

Table 8 shows concentrations of the major constituents of the ECF, as well as those in urine and rectal gland (see below) fluid. Blood NaCl concentration is well below ambient. It is a little higher than in most marine teleosts, and represents the upper extreme of the range for all vertebrates. However, unlike the teleosts, total ECF solute concentration in elasmobranchs is higher than in sea-water. As shown in Table 8, the osmotic pressure of the body fluids is elevated as a consequence of the presence of high concentrations of urea and trimethylamine oxide (TMAO). This method of avoiding the problem of hyposmotic regulation in sea-water in unique to the vertebrates among aquatic animals, but not to the elasmobranchs. The coelacanth fish *Latimeria chalumnae* shows the same pattern of reduced Na^+ (181 mmol l^{-1}) and Cl^- (199 mmol l^{-1}) and elevated blood urea (355 mmol l^{-1}) and must be

Table 8. The composition of plasma and excretory fluids in *Squalus acanthias*

Fluid	Concentrations (mmol l^{-1})							J_v
	Na	K	Cl	Mg	Ca	urea	TMAO	osmolarity	(ml kg^{-1} h^{-1})
Sea-water	440	9	490	50	10	0	0	930	—
Plasma	250	4	240	1	3	350	70	1000	—
Urine	240	2	240	40	3	100	10	800	0.5[a]
Rectal gland fluid	500	7	500	0	0	18	—	1000	0.5[b]

[a] Published J_v^u values vary greatly, from 0.025 (Chan, Philips and Chester-Jones, 1967) to 1.5 ml kg^{-1} h^{-1} (Forster and coworkers, 1972). In addition to normal species variation low values may reflect the tendency of these animals to become antidiuretic when disturbed (Payan and Maetz, 1973).
[b] Burger notes that this is a long-term average. Flow was actually intermittent.

at or close to osmotic equilibrium with sea-water (Pickford and Grant, 1967), and this is also characteristic of the ECF in the frog *Rana cancrivora* which can live in 80% SW (Gordon, Schmidt-Nielsen and Kelly, 1961).

Since the key to isosmotic regulation is maintenance of elevated urea levels the mechanisms involved have attracted modest attention for many years. Maintenance of steady-state ECF concentrations represents a balance between excretion across the gills and in the urine (Smith, 1931; recent work is reviewed by Schmidt-Nielsen, 1972; Hickman and Trump, 1969) and synthesis by the ornithine-urea pathway in the liver (Goldstein, 1972). As shown in Table 8 the concentrations of urea and TMAO are lower in the urine than in plasma, hence they must be actively reabsorbed from the filtrate. The site of reabsorption is the distal tubule (Schmidt-Nielsen, 1972) but the mechanism is not yet understood. Branchial excretion is considerably larger than urinary, but the rate of loss is restricted by the very low permeability of the gill epithelium (Payan and Maetz, 1973; Payan, Goldstein and Forster, 1973).

The immediate consequence of this adaptation is that the osmotic gradient, though small, is oriented to drive water into the animal. This is reflected by higher rates of urine flow than in marine teleosts, to which is added appreciable water excreted by the rectal gland. P_{osm}, estimated from the sum of urine and rectal gland flow and the osmotic gradient, is probably about 20 l mol^{-1} h^{-1}, which is higher than in marine teleosts. Eliminating the need to osmoregulate also has an important by-product in that the problem of maintaining a low ECF NaCl concentration is not magnified by the large salt transfers accompanying volume regulation in teleosts.

There is still much to be learned about the mechanisms involved in hypo-ionic regulation of NaCl in the blood. The inward-oriented concentration gradient is nearly as large in elasmobranchs as in teleosts. There is no information about whether a TEP exists across the body surface, and hence the magnitude of the diffusion gradient is unknown. But it is clear that a gradient exists for at least one, and perhaps both, of the ions, so diffusive leak inward must occur. Unidirectional fluxes have been measured for a few species (Burger and Tosteson, 1966; Maetz and Lahlou, 1966; Horowicz and Burger, 1968; Chan, Phillips and Chester-Jones, 1967; Payan and Maetz, 1973; Carrier and Evans, 1972). Most of the values have been in the range of 0.3–1 mmol kg^{-1} h^{-1}. This is only about 1% of the rate found in most marine teleosts, and is comparable to unidirectional fluxes in fresh-water fish. Since we do not have a value for the TEP and also lack information about the components of the tracer fluxes it is impossible to calculate values for the Na$^+$ and Cl$^-$ permeabilities. However, a limiting value can be estimated by assuming that there is no voltage across the gill and no exchange component in the isotope flux. With these constraints P_{Na} is about 1×10^{-3} l kg^{-1} h^{-1}, which is similar to the trout (Table 1). Net diffusive uptake, under these conditions, might be about 0.25–0.50 mmol kg^{-1} h^{-1}. If there is an appreciable TEP or if a substantial fraction of the unidirectional fluxes is due to exchange diffusion the net inward leak might be lower.

Part of the NaCl entering is excreted in the urine. Table 8 shows that urine and blood NaCl concentrations are similar, and renal salt flux is probably in the range 0.1–0.3 mmol kg^{-1} h^{-1} (cf. Hickman and Trump, 1969). Thus, a substantial fraction of the diffusive influx is excreted through the kidney.

Burger and Hess (1960) showed that the rectal gland of the spiny dogfish, *Squalus acanthias* is a route for extra renal and extra branchial excretion of NaCl. The gland, found generally among elasmobranchs, is an appendage of the posterior intestine emptying through the cloaca. It secretes a fluid the primary solutes of which are NaCl (~500 mmol l^{-1}) and K$^+$ (~7 mmol l^{-1}). Since the volume of fluid excreted was comparable to that of urine it appeared to play a significant role in maintaining a steady-state NaCl concentration in the ECF (Burger and Hess, 1960; Burger, 1962). However, subsequent studies showed that plasma ion concentrations were unchanged for weeks after the gland was removed (Burger, 1965; Chan, Phillips and Chester-Jones, 1967). This appears to show that the rectal gland is unimportant in maintaining normal body fluid homeostasis. Glandless dogfish excreted a salt load much more slowly than controls (Burger, 1965; Chan, Phillips and Chester-Jones, 1967), suggesting that the main function of the rectal gland may be to excrete such salt loads. However, Burger (1965) made another observation that points to a different conclusion. Extirpation of the rectal gland was followed by a marked diuresis, with urine volumes elevated two-to-four-fold. Since urinary NaCl concentrations were unchanged the salt clearance must have increased correspondingly. One might conclude from this that the rectal gland excretes an appreciable fraction of the resting salt influx, and that the failure of plasma NaCl to rise when it is removed is due to a compensatory increase in renal excretion. The problem clearly needs more attention.

Maetz and Lahlou (1966) found that only 15% of the ^{22}Na$^+$ efflux from the shark *Scyliorhinus canicula* could be accounted for in the urine and rectal gland fluid, and they proposed that most of the Na$^+$ entering was excreted across the body surface, perhaps by the gills. Similar results were found in the lip shark *Hemiscyllium plagiosum* in which the fraction excreted by the kidney and rectal gland was only about 1/3 of the total (Chan, Phillips and Chester-Jones, 1967). Here again, a conclusion is rendered uncertain by incomplete analysis of the unidirectional fluxes across the body surface. If they consist primarily of diffusion plus an active transport component then the conclusion is valid. On the other hand, if a substantial fraction involves Na$^+$–Na$^+$ exchange the conclusion, while formally correct, is physiologically meaningless. Such exchange diffusion, as pointed out earlier, contributes nothing to the net flux, and its contribution (if any) must be subtracted before we can decide whether net influx is balanced by excretion through the kidney and rectal gland. This is an important question that deserves more attention.

VI. APPENDIX

The 'Apparent Permeability' to Ions and Water

Under most conditions the passive flow of solute or water between an organism and its environment is proportional to driving force, where the proportionality constant, called the permeability (P), has the properties of a conductance. In a simple biological system, consisting of an isolated epithelium bathed by two well-stirred solutions, P can be calculated from measurements of three quantities; a flux, the magnitude of the gradient (grad), and the area (A) available for exchange.

$$P = \frac{J}{A} \cdot \frac{1}{\text{grad}}$$

In measuring a flux one might use the total, unidirectional movement (tracer) or the net movement if the system is not in a steady state. For each type of flux an appropriate expression must be chosen for the gradient, but the calculation is otherwise straightforward. However, problems arise when P must be assessed in intact animals. Some of these are outlined below where the basis for calculating P_{Na} and P_{osm} in this paper is described.

The Exchange Area (A)

The total external surface of gill has been estimated for some fish (e.g. Hughes, 1966) and crustaceans (Gray, 1957) but not for most. As a result, flux data in most studies are expressed in terms of body weight. Even when the external area is known, or can be determined, the effective exchange surface probably varies with the pattern of microcirculatory flow in the gill. This distribution is not necessarily related to the total volume flow (i.e. the cardiac minute volume), and it surely must differ in an unknown manner, not merely among animals, but from time to time within an individual. In view of these uncertainties I have preferred to use body weight as the basis for normalizing fluxes, and this choice is reflected in the dimensions and magnitudes of P_{Na} and P_{osm}. While this allows us to compare animals for which no estimates of A are available, it has a marked limitation. Flows are surface-, not volume-limited, and the relationship between gill surface and body weight is undoubtedly allometric. This was clearly shown for both fish and crustaceans (Hughes, 1966; Gray, 1957). That fluxes behave similarly is apparent from the relationship between urine flow (reflecting osmotic water uptake) and body weight in *Gammarus duebeni* (Lockwood and Inman, 1973) and D_2O exchange in *C. maenas* (Smith, 1970). As a result, when the flux per unit weight is used P will apparently be larger in smaller animals, and caution must be exercised in comparing the permeabilities of animals differing in size by orders of magnitude.

Sodium Permeability (P_{Na})

In most of the older literature, and some recent studies, net movement was estimated after sudden transfer of an animal to a more dilute medium. Some data exist in the form of changes in blood concentration (e.g. for *Maja sp.*). I have used these and an approximate value for total ECF Na$^+$ to calculate J_{net}^{Na} and P_{Na}. A correction was applied for dilution of the blood by water entry from the dilute medium. In other studies (e.g. *Porcellana sp.*, *Eupagurus sp.*) J_{net}^{Na} was obtained from changes in Na$^+$ concentration in the medium. Only a few measurements of the TEP have been reported for crabs, but in all cases the value has been in the range 0–5 mV (usually blood negative). This would have only a small effect on ion movement and has been disregarded. Therefore, using method 1

$$P_{Na} = \frac{J_{net}^{Na}}{kg} \cdot \frac{1}{\Delta c_{Na}}$$

where Δc_{Na} is the sodium concentration difference between blood and the dilute medium (mEq l^{-1}).

In most recent experiments isotopes (^{22}Na, ^{24}Na) are used to measure the unidirectional influx or efflux. When the TEP can be neglected the driving force is simply the concentration in the compartment from which the flux originates. For the efflux this is the blood concentration (c_{bl}) and

$$P_{Na} = \frac{J_{out}^{Na}}{kg} \cdot \frac{1}{c_{bl}}$$

which I refer to as method 2. Sometimes the TEP is not negligible; it may be 20–40 mV in marine teleosts, for example. For these cases I have used a variant of the constant field equation (Goldman, 1943) suggested by House (1963) to express the gradient. If influx is measured using method 3

$$P_{Na} = \frac{J_{in}^{Na}[1 - \exp(zFE/RT)]}{zFEc_{sw}/RT}$$

where E is the TEP and c_{sw} is the Na$^+$ concentration in the medium.

Water Permeability (P_{osm})

The same variables must be known to calculate water permeability, and the use of body weight instead of surface area to normalize fluxes creates the same problem as for ions. Again, either net water movement or a tracer (^{2}H$_2$O or ^{3}H$_2$O) flux may be measured, and an appropriate expression for the gradient must be used in the calculation. Unfortunately, net water movement, measured directly, often does not agree with net movement calculated from the difference between tracer influx and efflux. In many, perhaps most, cases the former is larger, sometimes by an order of magnitude. Whether this is due to some fundamental difference between diffusive (tracer) and osmotic (net) movements, or is an artefact caused by the presence of unstirred layers adjacent to the exchange surfaces is much discussed (e.g. Dainty, 1963; Schafter, Patlak and Andreoli, 1974). Whatever the reason for the difference, it introduces uncertainty in comparing flows (and permeabilities) estimated by the two methods. For this reason I have chosen to draw, for illustrative data, only on measurements of net water movement, even though there is a large body of information on tracer fluxes. Some problems engendered by tracer measurement of water movement in crustaceans and annelids have been reviewed by Smith (1976).

In the steady state net water inflow must equal outflow and where the body surface, gut and kidney are the sole pathways

$$J_v^{osm} + J_v^g + J_v^u = 0$$

In FW animals it has been accepted that drinking is negligible, hence urine flow (J_v^u) is a measure of osmotic uptake (J_v^{osm}). Recently, drinking has been observed in both fish and crustacean hyper-regulators. Whether this occurs in nature is unknown, and we do not know what fraction of the ingested water is absorbed. Hence, drinking has been neglected and J_v^u used as equivalent to J_v^{osm}, but this clearly may overestimate flow across the gill. The P_{osm} is

$$P_{osm} = \frac{J_v^u}{kg} \cdot \frac{1}{\Delta c_{osm}}$$

Where Δc_{osm} is the total osmotic concentration difference between blood and medium.

In hyporegulators J_v^{osm} and J_v^u are routes of water loss, but the latter is small and usually disregarded. Drinking is the only known route of intake, and

$$P_{osm} = \frac{J_v^g}{kg} \cdot \frac{1}{\Delta c_{osm}}$$

ACKNOWLEDGMENTS

Most of the ideas expressed germinated during or following discussions with many colleagues, among them R. H. Alvarado, T. H. Kerstetter, L. Greenwald, G. Baldwin, and D. H. Evans. I am especially indebted to J. Maetz for providing the facilities and collaboration that eased my way into the thorny study of hyporegulators.

REFERENCES

Ando, M., S. Utida, and H. Nagahama (1975). Active transport of chloride in eel intestine with special reference to sea water adaptation. *Comp. Biochem. Physiol.*, **51A**, 27–32.

Baldwin, G., and L. B. Kirschner (1976). Sodium and chloride regulation in *Uca* adapted to 175% sea water. *Physiol. Zool.*, **49**, 158–71.
Beadle, L. C., and J. B. Cragg (1940). Studies on adaptation to salinity in *Gammarus* spp. I. Regulation of blood and tissues and the problem of adaptation to fresh water. *J. Exp. Biol.*, **17**, 153–63.
Bentley, P. J. (1968). Amiloride: A potent inhibitor of sodium transport across the toad bladder. *J. Physiol. (Lond.)*, **195**, 317–30.
Beyenbach, K. W. (1974). *Ph.D. Thesis*. Magnesium excretion by the rainbw trout, *Salmo gairdneri*. Washington State University (University Microfilms, Ann Arbor, Michigan).
Beyenbach, K. W., and L. B. Kirschner (1975). Kidney and urinary bladder functions of the rainbow trout in Mg and Na excretion. *Am. J. Physiol.*, **229**, 389–93.
Beyenbach, K. W., and L. B. Kirschner (1976). The unreliability of mammalian glomerular markers in teleostean renal studies. *J. Exp. Biol.*, **64**, 369–78.
Biber, T. U. L. (1971). Effect of changes in transepithelial transport on the uptake of sodium across the outer surface of the frog skin. *J. Gen. Physiol.*, **58**, 131–44.
Biber, T. U. L., and P. F. Curran (1970). Direct measurement of uptake of sodium at the outer surface of the frog skin. *J. Gen. Physiol.*, **56**, 83–99.
Biber, T. U. L., Chez, R. A., and Curran, P. F. (1965). Na transport across frog skin at low external Na concentrations. *J. Gen. Physiol.*, **49**, 1161–76.
Bierther, M. (1970). Die chloridzellen des stichlings. *Z. Zellforsch.*, **107**, 421–46.
Binns, R. (1969). The physiology of the antennal gland of *Carcinus maenas* (L.) II. Urine production rates. *J. Exp. Biol.*, **51**, 11–16.
Bryan, G. W. (1960). Sodium regulation in the crayfish *Astacus fluviatilis*. II. Experiments with sodium-depleted animals. *J. Exp. Biol.*, **37**, 83–99.
Bryan, G. W., and Ward, E. (1962). Potassium metabolism and accumulation of ^{137}caesium by decapod *Crustacea*. *J. Mar. Biol. Ass. (UK)*, **42**, 199–241.
Burger, J. W. (1957). The general form of excretion in the lobster, *Homarus*. *Biol. Bull.*, **113**, 207–23.
Burger, J. W. (1962). Further studies on the function of the rectal gland in the spiny dogfish. *Physiol. Zool.*, **35**, 205–17.
Burger, J. W. (1965). Roles of the rectal gland and the kidneys in salt and water excretion in the spiny dogfish. *Physiol. Zool.*, **38**, 191–6.
Burger, J. W. (1967). In P. W. Gilber, R. G. Mathewson, and D. P. Rall (Eds) *Sharks, Skates and Rays*, Johns Hopkins Press, Baltimore.
Burger, J. W., and W. N. Hess (1960). Function of the rectal gland in the spiny dogfish. *Science*, **131**, 670–1.
Burger, J. W., and D. C. Tosteson (1966). Sodium influx and efflux in the spiny dogfish *Squalus acanthias*. *Comp. Biochem. Physiol.*, **19**, 649–53.
Carrier, J. C., and D. H. Evans (1972). Ion, water and urea turnover rates in the nurse shark *Ginglymostoma cirratum*. *Comp. Biochem. Physiol.*, **41A**, 761–4.
Chan, D. K. O., J. G. Phillips, and I. Chester-Jones (1967). Studies on electrolyte changes in the lip-shark, *Hemiscyllium plagiosum* (Bennet), with special reference to hormonal influence on the rectal gland. *Comp. Biochem. Physiol.*, **23**, 185–98.
Chester-Jones, I., D. K. O. Chan, and J. C. Rankin (1969). Renal function in the European eel (*Anguilla anguilla*, L.): Changes in blood pressure and renal function of the freshwater eel transferred to sea water. *J. Endocrinol.*, **43**, 9–19.
Cholette, C., A. Cagnon, and P. Germain (1970). Isosmotic adaptation in *Myxine glutinosa*. Variations of some parameters and role of the amino acid pool of the muscle cells. *Comp. Biochem. Physiol.*, **33**, 333–46.
Copeland, D. E. (1967). A study of salt secreting cells in the brine shrimp (*Artemia salina*). *Protoplasma*, **63**, 363–84.

Cornell, J. (1976). *Ph.D. Thesis*. Aspects of salt and water balance in two osmoconforming crabs, *Libinia emarginata* and *Pugettia producta* (*Brachyura: Majidae*). University of California, Berkeley (University Microfilms, Ann Arbor, Michigan).
Croghan, P. C. (1958a). The mechanism of osmotic regulation in *Artemia salina* (L.): The physiology of the branchiae. *J. Exp. Biol.*, **35**, 234–42.
Croghan, P. C. (1958b). The mechanism of osmotic regulation in *Artemia salina* (L.): The physiology of the gut. *J. Exp. Biol.*, **35**, 243–9.
Dainty, J. (1963). Water relations of plant cells. *Adv. Bot. Res.*, **1**, 279–327.
Dall, W. (1967). Hypo-osmoregulation in crustacea. *Comp. Biochem. Physiol.*, **21**, 653–78.
Davenport, J. (1972). Volume changes shown by some littoral anomuran crustacea. *J. Mar. Biol. Ass.* (*UK*), **52**, 863–77.
Dharmamba, M., M. Bornancin, and J. Maetz (1975). Environmental salinity and sodium and chloride exchanges across the gill of *Tilapia mossambica*. *J. Physiol.* (*Paris*), **70**, 627–36.
De Renzis, G. (1975). The branchial chloride pump in the goldfish *Carassius auratus*: relationship between Cl^-/HCO_3^- and Cl^-/Cl^- exchanges and the effect of thiocyanate. *J. Exp. Biol.*, **63**, 587–602.
Eigler, J., J. Kelter, and E. Renner (1967). Activity of a new acylguanidine, Amiloride HCl (MK 870) on isolated amphibian skin. *Klin. Wochsch.*, **45**, 737–8.
Ellis, R. A., and C. C. Goertemiller (1974). Cytological effects of salt-stress and localization of transport adenosine triphosphatase in the lateral nasal glands of the desert iguana, *Dipsosaurus dorsalis*. *Anat. Rec.*, **180**, 285–98.
Engbaek, L., and T. Hoshiko (1957). Electrical potential gradients through frog skin. *Acta Physiol. Scand.*, **39**, 348–55.
Epstein, F. H., A. I. Katz, and G. E. Pickford (1967). Sodium and potassium-activated adenosine triphosphatase of gills: role in adaptation of teleosts to salt water. *Science*, **156**, 1245–7.
Epstein, F. H., J. Maetz, and G. De Renzis (1973). Active transport of chloride by the teleost gill: inhibition by thiocyanate. *Am. J. Physiol.*, **224**, 1295–9.
Ernst, S. A. (1972a). Transport adenosine triphosphatase cytochemistry. I. Biochemical characterization of a cytochemical medium for the ultrastructural localization of ouabain-sensitive, potassium-dependent phosphatase activity in the avian salt gland. *J. Histochem. Cytochem.*, **20**, 13–22.
Ernst, S. A. (1972b). Transport adenosine triphosphatase cytochemistry. II. Cytochemical localization of ouabain-sensitive, potassium-dependent phosphatase activity in the secretory epithelium of the avian salt gland. *J. Histochem. Cytochem.*, **20**, 23–38.
Evans, D. H. (1967). Sodium, chloride and water balance of the intertidal teleost. *Xiphister atropurpureus*. II. The role of the kidney and the gut. *J. Exp. Biol.*, **47**, 519–23.
Evans, D. H. (1969). Sodium, chloride and water balance of the intertidal teleost, *Pholis gunnellus*. *J. Exp. Biol.*, **50**, 179–90.
Evans, D. H. (1973). Sodium uptake by the sailfin molly, *Poecilia latipinna*: kinetic analysis of a carrier system present in both fresh-water-acclimated and sea-water-acclimated individuals. *Comp. Biochem. Physiol.*, **45A**, 843–50.
Evans, D. H. (1977). In G. M. O. Maloiy (Ed.) *Comparative Physiology of Osmoregulation in Animals*, Academic Press, New York.
Evans, D. H., and K. Cooper (1976). The presence of Na–Na and Na–K exchange in sodium extrusion by three species of fish. *Nature*, **259**, 241–2.
Evans, D. H., J. C. Carrier, and M. B. Bogan (1974). The effect of external potassium ions on the electrical potential measured across the gills of the teleost, *Dormitator maculatus*. *J. Exp. Biol.*, **61**, 277–83.

Evans, D. H., K. Cooper, and M. B. Bogan (1976). Sodium extrusion by the sea-water-acclimated fiddler crab *Uca pugilator*: Comparison with other marine crustacea and marine teleost fish *J. Exp. Biol.*, **64**, 203–19.

Fenn, W. O. (1924). The relation between the work performed and the energy liberated in muscular contraction. *J. Physiol.*, **58**, 373–95.

Flügel, H. (1960). Uber den einflus der temperatur auf die osmotische resistenz und die osmoregulation der decapoden garnele *Crangon crangon* L. *Kieler Meeresforsch.*, **16**, 186–200.

Forster, R. P., L. Goldstein, and J. K. Rosen (1972). Intrarenal control of urea reabsorption by renal tubules of the marine elasmobranch *Squalus acanthias*. *Comp. Biochem. Physiol.*, **42A**, 3–12.

Forster, R., and S. K. Hong (1958). *In vitro* transport of dyes by isolated renal tubules of the flounder as disclosed by direct visualization. Intracellular accumulation and transcellular movement. *J. Cell Comp. Physiol.*, **51**, 259–72.

Fromm, P. O. (1963). Studies on renal and extrarenal excretion in fresh water teleost, *Salmo gairdneri*. *Comp. Biochem. Physiol.*, **10**, 121–8.

Garcia Romeu, F., A. Salibian, and S. Pezzani-Hernandez (1969). The nature of the *in vivo* sodium and chloride uptake mechanisms through the epithelium of the Chilean frog *Calyptocephalella gayi*. *J. Gen. Physiol.*, **53**, 816–35.

Geddes, M. C. (1975a). Studies on an Australian brine shrimp *Parartemia zeitziana* Sayce. II. Osmotic and ionic regulation. *Comp. Biochem. Physiol.*, **51A**, 561–71.

Geddes, M. C. (1975b). Studies on an Australian brine shrimp *Parartemia zeitziana* Sayce. III. The mechanisms of osmotic and ionic regulation. *Comp. Biochem. Physiol.*, **51A**, 573–8.

Glynn, I. M., and S. J. D. Karlish (1975). The sodium pump. *A. Rev. Physiol.*, **37**, 13–56.

Goldman, D. E. (1943). Potential, impedance and rectification in membranes. *J. Gen. Physiol.*, **27**, 37–60.

Goldstein, L. (1972). In J. W. Campbell and L. Goldstein (Eds) *Nitrogen Metabolism and the Environment*, Academic Press, New York.

Gordon, M. S. (1963). Chloride exchanges in rainbow trout (*Salmo gairdneri*) adapted to different salinities. *Biol. Bull.*, **124**, 45–54.

Gordon, M. S., K. Schmidt-Nielsen, and H. M. Kelly (1961). Osmotic regulation in the crab-eating frog (*Rana cancrivora*). *J. Exp. Biol.*, **38**, 659–78.

Gray, I. E. (1957). A comparative study of the gill area of crabs. *Biol. Bull.*, **112**, 34–42.

Green, J. W., M. Harsch, L. Barr, and C. L. Prosser (1959). The regulation of water and salt by the fiddler crabs, *Uca pugnax* and *Uca pugilator*. *Biol. Bull.*, **116**, 76–87.

Greenwald, L. (1971). Sodium balance in the leopard frog (*Rana pipiens*). *Physiol. Zool.*, **44**, 149–61.

Greenwald, L., and L. B. Kirschner (1976). The effect of poly-L-lysine, amiloride and methyl-L-lysine on gill ion transport and permeability in the rainbow trout, *J. Memb. Biol.* **26**, 371–83.

Hannan, J. V., and D. H. Evans (1973). Water permeability in some euryhaline decapods and *Limulus polyphemus*. *Comp. Biochem. Physiol.*, **44A**, 1199–1213.

Hickman, C. P. (1968a). Urine composition and kidney function in southern flounder *Paralichthys lethostigma* in sea water. *Can J. Zool.*, **46**, 439–55.

Hickman, C. P. (1968b). Ingestion, intestinal absorption and elimination of sea water and salts in the southern flounder, *Paralichthys lethostigma*. *Can. J. Zool.*, **46**, 457–66.

Hickman, C. P., and B. F. Trump (1969). In W. S. Hoar and D. J. Randall (Eds) *Fish Physiology*, Vol. I, Academic Press, New York.

Horowicz, P., and J. W. Burger (1968). Undirectional fluxes of sodium ions in the spiny dogfish, *Squalus acanthias*. *Am. J. Physiol.*, **214**, 635–42.

House, C. R. (1963). Osmotic regulation in the brackish water teleost, *Blennius pholis*. *J. Exp. Biol.*, **40**, 87–104.

House, C. R., and K. Green (1965). Ion and water transport intestine of the marine teleost, *Cottus scorpius*. *J. Exp. Biol.*, **42**, 177–89.

House, C. R., and J. Maetz (1974). On the electrical gradient across the gill of the sea water-adapted eel. *Comp. Biochem. Physiol.*, **47A**, 917–24.

Huang, K. C., and T. S. T. Chen (1971). Ion transport across intestinal mucosa of winter flounder, *Pseudopleuronectes americanus*. *Am. J. Physiol.*, **220**, 1734–8.

Hughes, G. M. (1966). The dimensions of fish gills in relation to their function. *J. Exp. Biol.*, **45**, 177–95.

Hunter, K. C. (1973). Ph.D. Thesis. Salt and water balance in the Dungeness crab. University of Oregon, Eugene (University Microfilms, Ann Arbor, Michigan).

Hunter, K. C., and P. P. Rudy (1975). Osmotic and ionic regulation in the Dungeness crab. *Comp. Biochem. Physiol.*, **51A**, 439–47.

Hynes, H. B. N. (1954). The ecology of *Gammarus duebeni* Lilljeborg and its occurrence in fresh water in Western Britain. *J. Animal Ecol.*, **23**, 38–84.

Jampol, L. M., and F. H. Epstein (1970). Sodium-potassium-activated adenosine triphosphatase and osmotic regulation by fishes. *Am. J. Physiol.*, **218**, 607–11.

Kamemoto, F. I., S. M. Keister, and A. E. Spalding (1962). Cholinesterase activities and sodium movement in the crayfish kidney. *Comp. Biochem. Physiol.*, **7**, 81–7.

Kamiya, M. (1972). Sodium-potassium-activated adenosinetriphosphatase in isolated chloride cells from eel gills. *Comp. Biochem Physiol.*, **43B**, 611–7.

Kamiya, M., and S. Utida (1969). Sodium-potassium-activated adenosinetriphosphatase activity in gills of fresh-water, marine and euryhaline teleosts. *Comp. Biochem. Physiol.*, **31**, 671–4.

Karnaky, K. J., L. B. Kinter, W. B. Kinter, and C. E. Stirling (1976). Teleost chloride cell. II. Autoradiographic localization of gill Na, K-ATPase in killifish *Fundulus heteroclitus* adapted to low and high salinity environments. *J. Cell Biol.*, **70**, 157–77.

Kerstetter, T. H., and L. B. Kirschner (1972). Active chloride transport by the gills of rainbow trout (*Salmo gairdneri*). *J. Exp. Biol.*, **56**, 263–72.

Kerstetter, T. H., and L. B. Kirschner (1974). HCO_3^--dependent ATPase activity in the gills of rainbow trout (*Salmo gairdneri*) Comp. Biochem. Physiol., **48B**, 581–9.

Kerstetter, T. H., L. B. Kirschner, and D. Rafuse (1970). On the mechanisms of sodium ion transport by the irrigated gills of rainbow trout (*Salmo gairdneri*). *J. Gen. Physiol.*, **56**, 342–59.

Keys, A. B. (1931). Chloride and water secretion and absorption by the gills of the eel. *Z. Vergl. Physiol.*, **15**, 364–88.

Keys, A. B., and E. N. Wilmer (1932). Chloride secreting cells in the gills of fishes with special reference to the common eel. *J. Physiol.*, **76**, 368–78.

Kirsch, R., and N. Mayer-Gostan (1973). Kinetics of water and chloride exchanges during adaptation of the European eel to sea water. *J. Exp. Biol.*, **58**, 105–21.

Kirschner, L. B. (1955). On the mechanism of active sodium transport across the frog skin. *J. Cell Comp. Physiol.*, **45**, 61–87.

Kirschner, L. B. (1961). Thermodynamics and osmoregulation. *Nature*, **191**, 815–6.

Kirschner, L. B. (1967). Comparative physiology: invertebrate excretory organs. *A. Rev. Physiol.*, **29**, 169–196.

Kirschner, L. B. (1970). The study of NaCl transport in aquatic animals. *Am. Zool.*, **10**, 365–76.

Kirschner, L. B. (1977). In B. L. Gupta, R. B. Moreton, L. Oschman and B. J. Wall (Eds) *Transport of Ions and Water in Animals*, Academic Press, London.

Kirschner, L. B., L. Greenwald, and T. H. Kerstetter (1973). Effect of amiloride on

sodium transport across body surfaces of fresh water animals. *Am. J. Physiol.*, **224**, 832–7.
Kirschner, L. B., L. Greenwald, and M. Sanders (1974). On the mechanism of sodium extrusion across the irrigated gill of the sea water adapted rainbow trout (*Salmo gairdneri*). *J. Gen. Physiol.*, **64**, 148–65.
Koch, J. J., J. Evans, and E. Schicks (1954). The active absorption of ions by the isolated gills of the crab *Eriocheir sinensis* (M. EDW.). *Konink. Vlaamse Akad. Wetenschap.*, **16**, 3–16.
Koefoed-Johnsen, V. and H. H. Ussing (1958). The nature of the frog skin potential. *Acta Physiol. Scand.*, **42**, 298–308.
Kristensen, P. (1972). Chloride transport across isolated frog skin. *Acta Physiol. Scand.*, **84**, 338–50.
Kristensen, P. (1973). In H. H. Ussing and N. Thorne (Eds) *Transport Mechanisms in Epithelia*, E. Munksgaard, Copenhagen.
Krogh, A. (1937). Osmotic regulation in fresh water fishes by active absorption of chloride ions. *Z. Vergl. Physiol.*, **24**, 656–66.
Krogh, A. (1938). The active absorption of ions in some fresh water animals. *Z. Vergl. Physiol.*, **25**, 335–50.
Krogh, A. (1939). *Osmotic Regulation in Aquatic Animals*, Cambridge University Press, London.
Lahlou, B. (1967). Excrétion rénale chez un poisson euryhalin, le flet (*Platichthys flesus* L.): caracteristiques de l'urine normale en eau douce et en eau de mer et effets des changements de milieu. *Comp. Biochem. Physiol.*, **29**, 925–38.
Lahlou, B., and W. H. Sawyer (1969). Sodium exchanges in the toadfish, *Opsanus tau*, a euryhaline aglomerular teleost. *Am. J. Physiol.*, **216**, 1273–8.
de Leersnyder, M. (1967). Le milieu interieur d'*Eriocheir sinensis* H. Milne-Edwards et ses variations. II. Études experimentale. *Cah. Biol. Mar.*, **8**, 295–321.
Lockwood, A. P. M. (1961). The urine of *Gammarus duebeni* and *G. pulex*. *J. Exp. Biol.*, **38**, 647–58.
Lockwood, A. P. M., and C. B. E. Inman (1973). Water uptake and loss in relation to the salinity of the medium in the amphipod crustacean *Gammarus duebeni*. *J. Exp. Biol.*, **58**, 149–63.
Lotan, R. (1969). Osmotic adjustment in the euryhaline teleost *Aphanius dispar*. *Z. Vergl. Physiol.*, **65**, 455–62.
McFarland, W. N., and F. W. Munz (1965). Regulation of body weight and serum composition by hagfish in various media. *Comp. Biochem. Physiol.*, **14**, 383–98.
MacFarlane, N. A. A., and J. Maetz (1975). Acute response to a salt load of the NaCl excretion mechanisms of the gill of *Platichthys flesus* in sea water. *J. Comp. Physiol.*, **102**, 101–13.
Maetz, J. (1956). Les échanges de sodium chez le poisson *Carassius auratus* L. Action d'un inhibiteur de l'anhydrase carbonique. *J. Physiol.* (*Paris*), **48**, 1085–99.
Maetz, J. (1963). Physiological aspects of neurohypophysial function in fishes with some reference to the amphibia. *Symp. Zool. Soc. Lond.*, **9**, 107–40.
Maetz, J. (1969). Sea water teleosts: evidence for a sodium potassium exchange in the branchial sodium-excreting pump. *Science*, **166**, 613–5.
Maetz, J. (1971). Fish gills: mechanism of salt transfer in fresh water and sea water. *Phil. Trans. R. Soc. B*, **262**, 209–49.
Maetz, J. (1972). Branchial sodium exchange and ammonia excretion in the goldfish *Carassius auratus*. Effects of ammonia-loading and temperature changes. *J. Exp. Biol.*, **56**, 601–20.
Maetz, J. (1973). In H. H. Ussing and N. A. Thorn (Eds) *Transport Mechanisms in Epithelia*, E. Munksgaard, Copenhagen.

Maetz, J. (1974). In D. C. Malins and J. R. Sargent (Eds) *Biochemical and Biophysical Perspectives in Marine Biology*, Vol. 1, Academic Press, London.

Maetz, J., and M. Bornancin (1975). Biochemical and biophysical aspects of salt excretion by chloride cells in teleosts. *Fortschr. Zool.*, **23**, 322–62.

Maetz, J., and F. Garcia Romeu (1964). The mechanism of sodium and chloride uptake by the gills of a fresh-water fish, *Carassius auratus*. II. Evidence for NH_4^+/Na^+ and HCO_3^-/Cl^- exchanges. *J. Gen. Physiol.*, **47**, 1209–27.

Maetz, J., and B. Lahlou (1966). Les échanges de sodium et de chlor chez un Élasmobranch, *Scyliorhinus*, mesurés à l'aide des isotopes ^{24}Na et ^{36}Cl. *J. Physiol. (Paris)*, **58**, 249.

Maetz, J., and P. Pic (1975). New evidence for a Na/K and Na/Na exchange carrier linked with the Cl^- pump in the gill of *Mugil capito* in sea water. *J. Comp. Physiol.*, **102**, 85–100.

Morris, R. (1965). Studies on salt and water balance in *Myxine glutinosa* (L.). *J. Exp. Biol.*, **42**, 359–71.

Motais, R. (1967). Les mécanismes d'échanges ioniques branchiaux chez les téléostéens. Leur rôle dans l'osmorégulation. *Ann. Inst. Ocean. (Monaco)*, **45**, 1–84.

Motais, R., and F. Garcia Romeu (1972). Transport mechanism in the teleostean gill and amphibian skin. *A. Rev. Physiol.*, **34**, 141–76.

Motais, R., F. Garcia Romeu, and J. Maetz (1966). Exchange diffusion effect and euryhalinity in teleosts. *J. Gen. Physiol.*, **50**, 391–422.

Motais, R., J. Isaia, J. C. Rankin, and J. Maetz (1969). Adaptive changes of the water permeability of the teleostean gill epithelium in relation to external salinity. *J. Exp. Biol.*, **51**, 529–46.

Munz, F. W., and W. N. McFarland (1964). Regulatory function of a primitive vertebrate kidney. *Comp. Biochem. Physiol.*, **13**, 381–400.

Nagel, H. (1934). Die aufgaben der exkretionsorgane und der kiemen bei der osmoregulation von *Carcinus maenas*. *Z. Vergl. Physiol.*, **21**, 468–91.

Natochin, Y. V., and G. P. Gusev (1970). The coupling of magnesium secretion and sodium reabsorption in the kidney of teleost. *Comp. Biochem. Physiol.*, **37**, 107–11.

Oide, M. (1967). Effects of inhibitors on transport of water and ion in isolated intestine and Na^+-K^+ ATPase in intestinal mucosa of the eel. *Annot. Zool. Japan*, **40**, 130–5.

Oide, M. (1970). Purification and some properties of alkaline phosphatase from intestinal mucosa of the eel adapted to fresh water or sea water. *Comp. Biochem. Physiol.*, **36**, 241–52.

Oide, M. (1973). Role of alkaline phosphatase in intestinal water absorption by eels adapted to sea water. *Comp. Biochem. Physiol.*, **46A**, 639–45.

Oide, M., and S. Utida (1967). Changes in water and ion transport in isolated intestine of the eel during salt adaptation and migration. *Mar. Biol.*, **1**, 102–6.

Parry, G. (1955). Urine production by the antennal glands of *Palaemonetes varians*. *J. Exp. Biol.*, **32**, 408–22.

Payan, P., and J. Maetz (1973). Branchial sodium transport mechanisms in *Scyliorhinus canicula*: evidence for Na^+/NH_4^+ and Na^+H^+ exchanges and for a role of carbonic anhydrase. *J. Exp. Biol.*, **58**, 487–502.

Payan, P., L. Goldstein, and R. P. Forster (1973). Gills and kidneys in ureosmotic regulation in euryhaline skates. *Am. J. Physiol.*, **224**, 367–72.

Payan, P., A. J. Matty, and J. Maetz (1975). A study of the sodium pump in the perfused head preparation of the trout *Salmo gairdneri* in fresh water. *J. Comp. Physiol.*, **104**, 33–48.

Peters, H. (1935). Über den einfluss des salzgehaltes im aussenmedium auf den bau und die funktion der exkretionsorgane dekapodor crustacean (Nach unter-

suchungen an Potamobius fluviatilis und Homarus vulgaris). *Z. Morph. Ökol. Tiere.*, **30**, 355–81.
Peterson, D. R., and R. F. Loizzi (1974). Biochemical and cytochemical investigations of (Na^+–K^+)-ATPase in the crayfish kidney. *Comp. Biochem. Physiol.*, **49A**, 763–73.
Philpott, C. W. (1966). The use of horseradish peroxidase to demonstrate functional continuity between the plasmalemma and the unique tubular system of the chloride cell. *J. Cell Biol.*, **31**, 86A.
Philpott, C. W., and D. E. Copeland (1963). Fine structure of chloride cells from three species of *Fundulus*. *J. Cell Biol.*, **18**, 389–404.
Pickford, G. E., and F. B. Grant (1967). Serum osmolality in the coelacanth, *Latimeria chalumnae*: urea retention and ion regulation. *Science*, **155**, 568–70.
Potts, W. T. W. (1955). The energetics of osmotic regulation in brackish and fresh-water animals. *J. Exp. Biol.*, **31**, 618–630.
Potts, W. T. W. (1976). In S. Davies (Ed.) *Perspectives in Experimental Biology*, Pergamon Press, London.
Potts, W. T. W., and G. Parry (1964a). Sodium and chloride balance in the prawn, *Palaemonetes varians*. *J. Exp. Biol.*, **41**, 591–601.
Potts, W. T. W., and G. Parry (1964b). *Osmotic and Ionic Regulation in Animals*, Pergamon Press, London.
Potts, W. T. W., and B. Eddy (1973). Gill potentials and sodium fluxes in the flounder *Platichthys flesus*. *J. Comp. Physiol.*, **87**, 29–48.
Potts, W. T. W., and D. H. Evans (1967). Sodium and chloride balance in the killifish *Fundulus heteroclitus*. *Biol. Bull.*, **133**, 411–25.
Potts, W. T. W., C. R. Fletcher, and B. Eddy (1973). An analysis of the sodium and chloride fluxes in the flounder *Platichthys flesus*. *J. Comp. Physiol.*, **87**, 21–8.
Potts, W. T. W., M. A. Foster, P. P. Rudy, and G. Parry-Howells (1967). Sodium and water balance in the cichlid teleost, *Tilapia mossambica*. *J. Exp. Biol.*, **47**, 461–70.
Prosser, C. L. (1973). *Comparative Animal Physiology*, 3rd edn, Saunders, Philadelphia.
Prosser, C. L., J. W. Green, and T. S. Chow (1955). Ionic and osmotic concentrations in blood and urine of *Pachygrapsus crasspeis* acclimated to different salinities. *Biol. Bull.*, **109**, 99–107.
Quinn, D. J., and C. E. Lane (1966). Ionic regulation and Na^+–K^+ stimulated ATPase activity in the land crab *Cardiosoma guanhumi*. *Comp. Biochem. Physiol.*, **19**, 533–43.
Ramsay, J. H., and J. A. Riegel (1961). Excretion of inulin by malpighian tubules. *Nature*, **191**, 1115.
Renfro, J. L. (1975). Water and ion transport by the urinary bladder of the teleost *Pseudopleuronectes americanus*. *Am. J. Physiol.*, **228**, 52–61.
Riegel, J. A. (1965). Micropuncture studies of the concentrations of sodium, potassium and inulin in the crayfish antennal gland. *J. Exp. Biol.*, **42**, 379–84.
Riegel, J. A. (1972). *Comparative Physiology of Renal Excretion*, Oliver and Boyd, Edinburgh.
Riegel, J. A., and A. P. M. Lockwood (1961). The role of the antennal gland in the osmotic and ionic regulation of *Carcinus maenas*. *J. Exp. Biol.*, **38**, 491–9.
Robertson, J. D. (1957). In B. T. Scheer (Ed.) *Physiology of Invertebrate Animals*, University of Oregon Press, Eugene.
Rudy, P. P. (1966). Sodium balance in *Pachygrapsus crassipes*. *Comp. Biochem. Physiol.*, **18**, 881–907.
Schafer, J. A., C. S. Patlak, and T. E. Andreoli (1974). Osmosis in cortical collecting tubules. ADH-independent osmotic flow rectification. *J. Gen. Physiol.*, **64**, 228–40.

Schmidt-Nielsen, B. (1972). In J. W. Campbell and L. Goldstein (Eds) *Nitrogen Metabolism and the Environment*, Academic Press, New York.
Schmidt-Nielsen, B., and D. Laws (1963). Invertebrate mechanisms for diluting and concentrating the urine. *A. Rev. Physiol.*, **25**, 631–58.
Schmidt-Nielsen, B., K. H. Gertz, and L. E. Davis (1968). Excretion and ultrastructure of the antennal gland of the fiddler crab *Uca mordax*. *J. Morph.*, **125**, 473–95.
Scholles, W. W. (1933). Über die mineral regulation wasserlebender evertebraten. *Z. Vergl. Physiol.*, **19**, 522–54.
Schultz, S. G., and P. F. Curran (1968). In C. F. Code and W. Heidel (Eds) *Handbook of Physiology*, Section 6, Vol. III, American Physiological Society, Washington, DC.
Schwabe, E. (1933). Über die osmoregulation verschiedener krebse (malacostracen). *Z. Vergl. Physiol.*, **19**, 183–236.
Schwartz, A., G. E. Lindenmayer, and J. C. Allen (1972). In F. Bronner and A. Kleinzeller (Eds) *Current Topics in Membranes and Transport*, Vol. III, Academic Press, New York.
Sharrat, B. M., I. Chester-Jones, and D. Bellamy (1964). Water and electrolyte composition of the body and renal function of the eel (*Anguilla anguilla* L.). *Comp. Biochem. Physiol.*, **11**, 9–18.
Shaw, J. (1959a). The absorption of sodium ions by the crayfish. I. The effect of external and internal sodium concentrations. *J. Exp. Biol.*, **36**, 126–44.
Shaw, J. (1959b). Salt and water balance in the east African fresh water crab, *Potamon niloticus* (M. EDW.). *J. Exp. Biol.*, **36**, 157–76.
Shaw, J. (1960). The absorption of sodium ions by the crayfish. II. The effect of external anion. *J. Exp. Biol.*, **37**, 534–47.
Shaw, J. (1961a). Studies on ionic regulation in *Carcinus maenas* (L.). *J. Exp. Biol.*, **38**, 135–52.
Shaw, J. (1961b). Sodium balance in *Eriocheir sinensis* (M. EDW.). Adaptation of the crustacea to fresh water. *J. Exp. Biol.*, **38**, 153–62.
Shaw, J. (1964). The control of salt balance in the crustacea. *Soc. Exp. Biol. Symp.*, **18**, 237–54.
Shaw, J., and D. W. Sutcliffe (1961). Studies on sodium balance in *Gammarus duebeni* Lilljeborg and *G. pulex pulex* (L.). *J. Exp. Biol.*, **38**, 1–15.
Shehadeh, Z. H., and M. S. Gordon (1969). The role of the intestine in salinity adaptation of the rainbow trout, *Salmo gairdneri*. *Comp. Biochem. Physiol.*, **30**, 397–418.
Shirai, N. (1972). Electron-microscope localization of sodium ions and adenosinetriphosphatase in chloride cells of the Japanese eel, *Anguilla japonica*. *J. Fac. Sci. Tokyo University (Section IV)*, **12**, 385–403.
Shirai, N., and S. Utida (1972). Development and degeneration of the chloride cell during sea water and fresh water adaptation of the Japanese eel, *Anguilla japonica*. *Z. Zellforsch.*, **103**, 247–64.
Shuttleworth, T. J., W. T. W. Potts, and J. N. Harris (1974). Bioelectric potentials in the gills of the flounder *Platichthys flesus*. *J. Comp. Physiol.*, **94**, 321–9.
Skadhauge, E. (1969). The mechanism of salt and water absorption in the intestine of the eel (*Anguilla anguilla*) adapted to waters of various salinities. *J. Physiol.*, **204**, 135–58.
Skadhauge, E. (1974). Coupling of transmural flows of NaCl and water in the intestine of the eel (*Anguilla anguilla*). *J. Exp. Biol.*, **60**, 535–46.
Skadhauge, E., and R. Lotan (1974). Drinking rate and oxygen consumption in the euryhaline teleost *Aphanius dispar* in waters of high salinity. *J. Exp. Biol.*, **60**, 547–56.

Skou, J. C. (1965). Enzymatic basis for active transport of Na$^+$ and K$^+$ across cell membrane. *Physiol. Rev.*, **45**, 596–617.
Smith, H. W. (1930). The absorption and excretion of water and salts by marine teleosts. *Am. J. Physiol.*, **93**, 480–505.
Smith, H. W. (1931). The absorption and excretion of water and salts by the elasmobranch fishes. II. Marine elasmobranchs. *Am. J. Physiol.*, **98**, 296–310.
Smith, H. W. (1932). Water regulation and its evolution in fishes in the Elasmobranchii. *Q. Rev. Biol.*, **7**, 1–26.
Smith, H. W. (1936). The retention and physiological role of urea. *Biol. Rev.*, **11**, 49–82.
Smith, H. W. (1951). *The Kidney*, Oxford University Press, London.
Smith, P. G. (1969a). The ionic relations of *Artemia salina* (L.). I. Measurements of electrical potential difference and resistance. *J. Exp. Biol.*, **51**, 727–38.
Smith, P. G. (1969b). The ionic relations of *Artemia salina* (L.). II. Fluxes of sodium, chloride and water. *J. Exp. Biol.*, **51**, 739–57.
Smith, R. I. (1970). The apparent water-permeability of *Carcinus maenas* as a function of salinity. *Biol. Bull.*, **139**, 351–62.
Smith, R. I. (1976). In S. Davies (Ed.) *Perspectives in Experimental Biology*, Vol. I, Pergamon Press, Oxford.
Solomon, R. J., P. Silva, J. R. Bend, and F. H. Epstein (1975). Thiocyanate inhibition of ATPase and its relationship to anion transport. *Am. J. Physiol.*, **229**, 801–6.
Sutcliffe, D. W. (1967a). Sodium regulation in the amphipod *Gammarus duebeni* from brackish-water and fresh-water localities in Britain. *J. Exp. Biol.*, **46**, 529–550.
Sutcliffe, D. W. (1967b). Sodium regulation in the fresh-water amphipod, *Gammarus pulex* (L.). *J. Exp. Biol.*, **46**, 499–518.
Sutcliffe, D. W. (1968). Sodium regulation and adaptation to fresh water in gammarid crustaceans. *J. Exp. Biol.*, **48**, 359–380.
Thompson, L. (1970). Ph.D. Thesis. Osmoregulation of the fresh water crabs *Metopaulias depressus* (*Grapsidae*) and *Pseudothelphusa jouyi* (*Pseudothelphusidae*). University of California, Berkeley (University Microfilms, Ann Arbor, Michigan).
Thompson, R. (1972). Ph.D. Thesis. Mechanisms of osmoregulation in a euryhaline goby, *Gillichthys mirabilis*: the role of active and passive transport of sodium and chloride ions across the gills. University of Califonia, San Diego (University Microfilms, Ann Arbor, Michigan).
Towle, D. W., G. E. Palmer, and J. L. Harris (1976). Role of gill Na$^+$ + K$^+$ dependent ATPase in acclimation of blue crabs (*Callinectes sapidus*) to low salinity. *J. Exp. Zool.*, **196**, 315–22.
Tyson, G. E. (1968). The fine structure of the maxillary gland of the brine shrimp, *Artemia salina*: the end sac. *Z. Zellforsch.*, **86**, 129–38.
Tyson, G. E. (1969). The fine structure of the maxillary gland of the brine shrimp, *Artemia salina*: the efferent duct. *Z. Zellforsch.*, **93**, 151–63.
Ussing, H. H. (1947). Interpretation of the exchange of radiosodium in isolated muscle. *Nature*, **160**, 262.
Ussing, H. H. (1949). The distinction by means of tracers between active transport and diffusion. *Acta Physiol. Scand.*, **19**, 43–56.
Ussing, H. H. (1952). Some aspects of the application of tracers in permeability studies. *Adv. Enzymol.*, **13**, 21–65.
Ussing, H. H. (1960). The frog skin potential. *J. Gen. Physiol.*, **43**, 135–47.
Utida, S., M. Kamiya, D. W. Johnson, and H. A. Bern (1974). Effects of fresh water adaptation and of prolactin on sodium-potassium activated adenosine triphosphatase activity in the urinary bladder of two flounder species. *J. Endocr.*, **62**, 11–14.

de Vooys, C. G. N. (1968). Formation and excretion of ammonia in teleostei. I. Excretion of ammonia through the gills. *Arch. Int. Physiol. Biochem.*, **76,** 268–73.

Walker, A. M., C. L. Hudson, T. Findley, and A. N. Richards (1937). The total molecular concentration and the chloride concentration of fluid from different segments of the renal tubule of amphibia. *Am. J. Physiol.*, **118,** 121–9.

Zerahn, K. (1956). Oxygen consumption and active sodium transport in the isolated and short-circuited frog skin. *Acta Physiol. Scand.*, **36,** 300–18.

Chapter 6

Control Mechanisms in Amphibians

V. KOEFOED-JOHNSEN

I. Introduction	223
II. Salt and Water Movements *in vivo*	224
A. Ionic movements across the skin	224
B. The kidney	231
C. The urinary bladder	232
D. Regulation of salt and water balance	233
III. Salt and Water Movements *in vitro*	235
A. The isolated amphibian skin	235
B. The isolated urinary bladder	252
C. Interdependence of active transport and metabolism	255
D. Control mechanisms at the cellular level	256
References	258

I. INTRODUCTION

Among the lower vertebrates the amphibians are probably the group in which mechanisms of osmotic and ionic regulation have been most thoroughly studied.

Isolated preparations of frog skin and more recently toad urinary bladder have served as extremely useful tissues for studies of the properties of epithelia, and have played a key role in the major generalizations about epithelial transport.

Amphibians comprise both terrestrial and aquatic species, and are generally accepted as a freshwater group, but it has also been pointed out that a number of amphibian species are found in brackish or marine environments (Schmidt, 1957; Neill, 1958).

The blood of the frog is hyperosmotic to the pond water, and the urine is hypotonic to the blood. Thus the amphibians have problems in common with the freshwater fish: maintenance of the tonicity of the blood and coping with water following the osmotic gradient.

It has been known for several years that salt can be taken up from very dilute solutions by salt-depleted frogs (Krogh, 1937, 1938), through the isolated skin (Huf, 1935), and that the skin of the frog is very permeable to water (Overton, 1904; Adolf, 1933).

As in other freshwater animals the kidneys manufacture a hypotonic urine, and thus are particularly effective in removing excess water. The hypertonicity of the body fluid is maintained by active reabsorption of ions from the renal tubule.

Besides the skin and the kidneys the amphibian urinary bladder plays an important part in salt and water regulation. Dehydrated toads (Ewer, 1952) and frogs (Sawyer and Schisgall, 1956) may reduce their urine volume by reabsorption from the bladder, and in the hydrated state sodium reabsorption from the bladder is stimulated (Middler, Kleeman and Edwards, 1968).

Thus the main sites where exchanges of water and solutes between body fluids and external environment take place are the skin, kidneys, and bladder.

In this article no attempt has been made to review the literature comprehensively. Most emphasis has been put on the aspects of ion transport and water transfer across the amphibian membranes as exemplified by the isolated frog skin.

The allotted space has permitted quotation of only a fraction of the existing relevant literature; to the many authors of the excellent but unfortunately unmentioned papers, I apologize.

II. SALT AND WATER MOVEMENTS *IN VIVO*

A. Ionic Movements Across the Skin

The ability for active uptake of salt through the skin of living frogs has originally been demonstrated on the frog *R. esculenta* by Krogh (1937). Frogs which have been previously salt depleted for weeks in distilled water are able to take up salt from extremely dilute solutions, i.e. $1/100$ mmol l^{-1} NaCl solution which is 50–100 times more dilute than ordinary pond water. The uptake of NaCl takes place by two separate mechanisms, one for Na^+ and one for Cl^-, as chloride is absorbed with sodium from NaCl solutions, but from KCl, NH_4Cl, and $CaCl_2$ without being accompanied by cations, while sodium can be absorbed from Na_2SO_4 solutions without being accompanied by an anion (Krogh, 1937, 1938, 1939, 1946).

Not only frogs depleted of salt, but starving frogs adapted to tap water will keep their salt content at a constant level despite the inevitable loss of salt through the kidneys and skin. As frogs normally do not drink (Overton, 1904) an equivalent amount of salt must be taken up through the skin.

From the net uptake of NaCl which is found in frogs after prolonged exposure to distilled water it is impossible though to tell whether the influx of salt is increased or salt loss is decreased.

The introduction of readily available radioisotopes of almost every inorganic ion has now made it possible to measure the movement of an ion, e.g.

sodium, into an animal that is losing sodium at the same rate as it is replaced.

Such measurements of Na influx into *R. temporaria* and *Bufo bufo* (Jørgensen, 1950) have shown that Na uptake is increased in Na-depleted animals, the average influx from a 3 mmol l^{-1} NaCl solution being 9.9 μEq h^{-1} (100 cm^2)$^{-1}$ for *R. temporaria* while the controls have an influx of 5.6 μEq Na h^{-1} (100 cm^2)$^{-1}$. A reduced salt loss is also partially responsible for the net uptake of NaCl in salt-depleted anurans.

Studies on the sodium balance in the Leopard frog, *Rana pipiens* (Greenwald, 1971), show that the frog adapted to pond water, i.e. 1.4 mmol l^{-1} NaCl, take up sodium at a rate of 4.5 μEq g^{-1} day^{-1} and loses sodium at a rate of 5.1 μEq g^{-1} day^{-1} in the steady state. The efflux across the skin accounts for 10% of the total sodium loss while the rest is lost through the kidneys.

A determination of the effects of the external Na concentration in the range 0.11–2.4 mmol l^{-1} Na on the Na influx has also made it possible to make an estimation of a K_m for Na uptake. K_m turns out to be 0.2 mmol l^{-1}, reflecting a very high affinity for Na. The K_m for Na remains constant in salt-depleted frogs in which Na uptake rates are enhanced and Na loss rates are decreased.

This constant affinity for Na has also been found in larval *Ambystoma gracile*, which, too, is capable of adjusting Na uptake and loss according to its needs (Alvarado and Dietz, 1970b).

Chloride ions are also actively transported across the skin of living amphibia, as originally shown by Krogh who suggested an exchange of chloride against bicarbonate. The transport of Cl$^-$ takes place against an electrochemical gradient, and a net chloride uptake from 3 mmol l^{-1} KCl solutions is always observed (Jørgensen, Levi and Zerahn, 1954).

As expected, independent mechanisms for sodium and chloride uptake are also demonstrated in this study, as chloride-depleted frogs take up Cl$^-$ much faster than Na$^+$, whereas the reverse is true in frogs selectively depleted of Na$^+$.

Most of our knowledge about osmotic and ionic regulation in amphibia comes from extensive studies on anurans, but the relatively few investigations which have been made on urodeles show that on the whole the general features of salt and water balance in urodeles appear to be very similar to those found in anurans (Krogh, 1939; Jørgensen, Levi and Ussing, 1946; Alvarado and Kirschner, 1963; Dietz, Kirschner and Porter, 1967; Alvarado and Dietz, 1970a; Kirschner and coworkers, 1971, 1973; Greenwald, 1972).

1. *Separate Na$^+$ Uptake*

It has been pointed out originally by Krogh that uptake of an ion, the counter-ion of which is impermeant, can only take place if the active influx is balanced by an efflux of an ion with the same charge. He suggested that Na$^+$ uptake might take place by exchange with NH$_4^+$ or H$^+$, and Cl$^-$ uptake against HCO_3^- output, this, of course being quite independent of the mechanism.

As most freshwater species excrete their nitrogenous wastes as ammonia, the ammonium ion would seem a very likely partner for the exchange process, and evidence has been presented where the rate of NH_4^+ excretion is similar to the Na^+ uptake in larval urodeles sitting in pond water (Dietz, Kirschner and Porter, 1967). In these animals the major route of ammonia excretion is through the skin (Fanelli and Goldstein, 1964).

It is improbable, though, that Na^+ and NH_4^+ movements are coupled in anurans, since frogs excrete practically no ammonia across the body surface. Thus simultaneous measurements of transepithelial net fluxes of Na^+ and NH_4^+ in two South American frogs, *Leptodactylus* sp. and *Callyptocephallea* sp., show an extremely low NH_4^+ excretion compared to Na^+ uptake (Garcia-Romeu, Salibian and Pezzani-Hernandez, 1969). So in this species Na^+/NH_4^+ exchange is ruled out.

The other likely possibility, Na^+/H^+ exchange, has been shown to exist in *Calyptocephallea gayi* (Garcia-Romeu, Salibian and Pezzani-Hernandez, 1969), in *Rana esculenta* (Garcia-Romeu and Ehrenfeld, 1972), and in *Rana pipiens* (Kirschner, Greenwald and Kerstetter, 1973). When kept in dilute Na_2SO_4 solution the Na^+ influx is exactly equivalent to the total titratable acidity of the bathing solution, suggesting a one to one relationship. Furthermore, the universal Na transport inhibitor the diuretic compound amiloride, when added to the external solution (dilute sodium sulphate) at a concentration of 10^{-4} mol l^{-1} reduces Na influx and H^+ excretion by about 50% in *Rana pipiens* (Kirschner, Greenwald and Kerstetter, 1973), while a dose of 5×10^{-5} mol l^{-1} inhibits Na^+ absorption and H^+ excretion 90% in *Rana esculenta* (Garcia-Romeu and Ehrenfeld, 1975a).

These data suggest that the initial step in active sodium transport across the amphibian body from dilute sodium salt solutions with a non-permeant anion requires a coupled Na^+/H^+ exchange to maintain electrical neutrality, and that possibly NH_4^+ excretion can balance the Na influx in some animals.

When placed in NaCl solutions salt-depleted frogs take up Na^+ and Cl^- simultaneously (Krogh, 1939; Salibian, Pezzani-Hernandez and Garcia-Romeu, 1968). Under these circumstances it is difficult to decide whether the uptake is a coupled movement of Na^+ and Cl^-, or whether the separate exchange mechanisms claimed for Na^+ and Cl^- uptake are functioning, as the proposed partners in the exchange process, H^+ and HCO_3^- (or OH^-) will react in the outer solution.

2. *Separate Cl^- Uptake*

When selectively depleted of Cl^- by pretreatment in dilute Na_2SO_4 it has been shown that the uptake of Cl^- in *Leptodactylus* sp. is three times faster than that of Na^+, and that the chloride uptake from choline chloride solutions is accompanied by a corresponding increase in alkalinity in the outer solution (Garcia-Romeu, 1971). The same observation has been made in *Rana esculenta* (Garcia-Romeu and Ehrenfeld, 1975a) and *Rana pipiens* (Alvarado, Poole and Mullen, 1975a).

Whether HCO_3^- or OH^- is the ion exchanged for Cl^- is impossible to decide on the basis of titration of the outer bath, as OH^- will react with CO_2 and result in formation of HCO_3^-. It has been argued against the Cl^-/HCO_3^- process that the passive permeability of the frog skin to HCO_3^- is very low (Smith, Hughes and Huf, 1971). However, it seems widely accepted that Cl^- exchanges with endogeneous HCO_3^-, as injections with the carboanhydrase inhibitor acetazolamide inhibits Cl^- uptake. Until recently frog skin has been thought to be lacking carboanhydrase (Maren, 1967), but more sensitive techniques seem to have revealed the presence of the enzyme (Rosen and Friedley, 1973). No compelling evidence, though, relates enzyme activity to intensity of chloride transport (Alvarado, Poole and Mullen, 1975a).

3. *Transepithelial Potentials*

Measurements of transepithelial potential differences on amphibia sitting in pond water have shown that the PD (positive inside) is generated by active inward transport of Na^+, and that a chloride pump does not contribute to the formation of the PD in larval salamanders (Dietz, Kirschner and Porter, 1967) and in frogs (Alvarado, Dietz and Mullen, 1975b). In these dilute solutions the body wall has a very low permeability to anions, the chloride permeability being comparable to that of sulphate ions.

Over the range 0.1–10 mmol l^{-1} NaCl or Na_2SO_4 the potential varies approximately logarithmically with the sodium concentration, and is largest in salt-depleted animals, and smallest in the salt-loaded ones (Brown, 1962; Dietz, Kirschner and Porter, 1967; Alvarado and Stiffler, 1970; Alvarado, Poole and Mullen, 1975a).

4. *Effects of High Salinity*

Amphibians are generally considered as a freshwater group, and most species have a poor tolerance to salt solutions. When leopard frogs (*Rana pipiens*), are placed in sea-water they die within an hour. The main reason for this is not the cutaneous osmotic water loss, but a massive accumulation of sodium taken up partly through the skin, and partly from drinking (Bentley and Schmidt-Nielsen, 1971).

However, some anurans that normally live in brackish water, e.g. *Rana cancrivora* (Gordon, Schmidt-Nielsen and Kelly, 1961; Schmidt-Nielsen and Lee, 1962), and *Bufo viridis* (Gordon, 1962; Katz, 1975a) can tolerate salinities as high as 29‰.

Bufo viridis and *Rana cancrivora* adjust to the saline environment by increasing the osmolality of their body fluids to levels that are slightly hypertonic to the bathing solution. Increases in plasma concentration above freshwater levels are due partly to increased NaCl concentration, and partly to increased urea concentration associated with an accumulation of urea (Schoffeniels and Tercafs, 1966; Gordon and Tucker, 1968; Dicker and Elliot, 1970). By using the highly diffusible substance urea for raising the plasmal osmotic pressure these animals avoid the complications, e.g. a decrease of intracellular fluid phase, that otherwise would be the consequence of hypertonic extracellular fluid.

The tadpoles of *Rana cancrivora* like all other amphibian larvae are not able to synthesize urea, and use a different osmoregulatory mechanism when living in water of high salinity. They keep their body fluids hypo-osmotic to sea-water by excreting excess salt by an extrarenal pathway, probably through their gills (Gordon and Tucker, 1965).

Even freshwater amphibian species can survive for shorter or longer periods in high salinities. Frogs have thus been shown to tolerate hypertonic saline for several days (Adolph, 1933), and isotonic saline for many weeks (Jørgensen, 1950; Maetz, Jard and Morel, 1958; Crabbé, 1961; Ackrill, Hornby and Thomas, 1969).

The physiological response to the osmotic stress is an increase in plasma osmolality, due to an increase partly in sodium concentration, partly in urea concentration, while the urinary sodium concentration increases towards, but not beyond plasma concentration.

Thus it seems that the retention of urea as an osmotic adaptation found in brackish water amphibia may also occur in various terrestrial and freshwater species: *Rana esculenta* (Ackrill, Hornby and Thomas, 1969), *Xenopus laevis* (Balinsky, Cragg and Baldwin, 1961), *Rana pipiens* (Scheer and Markel, 1962), *Rana ridibunda* (Katz, 1975a); and it seems likely that it represents a physiological mechanism for osmotic adaptation potentially available to all anurans, contributing to the maintenance of body fluid hypertonicity (Tercafs and Schoffeniels, 1962) and allowing retention of water (Maffly and coworkers, 1960).

The freshwater urodele *Amphiuma tigrinum*, too, is able to survive in Ringer solution for months (Kirschner and coworkers, 1971) and maintain the concentration of sodium and chloride in plasma at an only slightly elevated level.

The adaptive changes in this species are a depression of active sodium transport across the skin, and a depression of renal reabsorption of sodium.

5. Nitrogen Metabolism

The close relationship between the nature of the end products of nitrogen metabolism, and the availability of water is very clearly seen in the amphibia. The present day amphibia are adapted to a wide range of habitats from completely aquatic to terrestrial in the adult, though the larvae are almost always aquatic. This range is reflected in the relative amounts of ammonia and of urea which they excrete. Fully aquatic species like *Xenopus laevis* are ammoniotelic, while terrestrial species like *Rana temporaria* and *Bufo bufo* are ureotelic (Cragg, Balinsky and Baldwin, 1961; Schmid, 1968). As ammonia is very toxic it needs the presence of large amounts of water for its excretion. The aquatic urodele *Necturus maculosus* thus excretes 90% of its total waste nitrogen as ammonia through the skin (Fanelli and Goldstein, 1964).

It is interesting, though, that *X. laevis* is not fully aquatic, but is able to aestivate by burrowing in the mud during dry seasons. When returning to water the animals excrete large amounts of urea. They are thus able to adapt to limited water supply by switching from excreting ammonia to synthesis of urea which then accumulates in the tissues (Balinsky, Cragg and Baldwin, 1961; Balinsky and coworkers, 1967b). The same effect can be produced in *X. laevis* by placing the animal in hypertonic NaCl solutions (Balinsky and coworkers, 1961, 1972b; McBean and Goldstein, 1967, 1970b), and in *Rana cancrivora* (Colley and coworkers, 1972; Balinsky, Dicker and Elliott, 1972a). In *X. laevis* the urea accumulation is supported by a reduction in the glomerular filtration rate (McBean and Goldstein, 1970a).

So it seems that amphibia are able to control their waste nitrogen with respect to the type of end product. All amphibia presumably possess the genetic information for the synthesis of ornithine–urea cycle enzymes, but are able to change and control the levels of the enzymes in response to the requirements of a changing environment.

Urea is predominantly excreted by the kidneys, the skin presumably being rather impermeable to this substance. The renal elimination of urea is in most amphibians a passive process, but in certain aquatic anurans of the genus *Rana, R. pipiens, R. clamitans* and *R. catesbiana* urea is actively secreted by the renal tubules (Forster, 1954, 1970). In the developing tadpoles the urinary tubules begin to secrete urea actively exactly at the same developmental stage at which the rise occurs in the activity of the liver enzymes involved in urea synthesis (Forster, Schmidt-Nielsen and Goldstein, 1963). In *Rana cancrivora* where the ability to concentrate urea in the body fluids is vital for its osmoregulation about 99% of the filtered urea is reabsorbed across the renal tubules (Schmidt-Nielsen and Lee, 1962).

The transition from ammoniotelism to ureotelism during metamorphosis is accompanied by a marked increase in the hepatic levels of the ornithine–urea cycle enzymes, and is dependent on a *de novo* synthesis of the enzyme carbamylphosphate synthetase (Cohen and Brown, 1960; Cohen, 1966, 1970; Frieden, 1968; Balinsky, Coetzer and Matheyse, 1972b; Wixom, Kumudavelli Reddy and Cohen, 1972). There is a good correlation, too, between the hepatic levels of carbamylphosphate synthetase I and the degree of adaptation for terrestrial life (Cohen, 1966; Balinsky and coworkers,

1967a). Thus it seems that carbamylphosphate synthetase plays a key role for the transition from an aquatic to a terrestrial existence.

6. Water Exchange

As amphibians do not drink unless they are placed in hyperosmotic salt solutions, water uptake takes place across the skin only. The movement of water across the body wall is a passive process, and is directly proportional to the osmotic gradient across the skin of frogs and toads (Sawyer, 1951). Thus frogs and toads show no net uptake of water from isotonic glucose or sucrose solutions (Kalman and Ussing, 1955; Dicker and Elliott, 1970).

Water transfer increases with increasing temperature (Hevesey, Hofer and Krogh, 1935). The Malayan toad, *Bufo melanostictus*, when kept in water thus increases its rate of water transfer 5–6 times in the temperature range from 12–37 °C (Dicker and Elliott, 1967).

The permeability to water differs in amphibia from different habitats being bigger in terrestrial species than in more aquatic ones (Schmid, 1965). The skin from anurans with a low permeability to water has been shown to have a higher lipid content than skins with a high permeability (Schmid and Barden, 1965).

Tolerance to changes in osmotic concentration of body fluids varies in different species with their ability to withstand dehydration. Amphibians normally living in aquatic habitats withstand loss of their body fluids poorly compared to those living in drier areas (Thorson and Svihla, 1943; Schmid, 1965; Main and Bentley, 1964).

When dehydrated, amphibians generally take up water much faster than when they are normally hydrated. The difference is small in aquatic species, being about 4 μl cm^{-1} h^{-1} in *X. laevis*, and large in terrestrial species, as much as 420 μl cm^{-1} h^{-1} in the ventral pelvic skin of *Bufo punctatus* (McClanahan and Baldwin, 1969). The rate of rehydration has been measured in several frogs and toads in relation to the aridity of the habitats. There does not always, though, seem to be a correlation (Thorson, 1955; Bentley, Lee and Main, 1958; Warburg, 1967; Claussen, 1969).

When exposed to air amphibians lose considerable amounts of water by evaporation through their skin at a rate comparable to that from a free water surface (Overton, 1904; Rey, 1937; Heatwole and coworkers, 1969; Machin, 1969). The only means they possess for diminishing the water loss is to hide away under rocks or the bark of trees. Many anurans have been observed to press their bodies against moist substrates for rehydration after a rainfall. This rehydration posture might indicate that a special part of the ventral skin takes up water more rapidly than other parts. Investigating the desert toad, *Bufo punctatus*, McClanahan and Baldwin (1969) have shown that the ventral pectoral integument takes up insignificant quantities of water, whereas the ventral pelvic integument can take up water at a mean rate of

420 μl cm^{-2} h^{-1} which accounts for nearly 70% of the water uptake in toads totally submerged. Such specialized areas of skin can be found, too, in a number of other anuran species. The high degree of vascularization in the pelvic integument is probably adding to the rapid water uptake found in this region (Christensen, 1974).

Several species have the ability for burrowing into the soil where the humidity is greater, and thereby avoiding excessive water loss from evaporation (Mayhew, 1965). Amphibians may aestivate for months underground. The North American spadefoot toad, *Scaphiopus couchi*, survives in deserts 10 months out of a year by burrowing into the ground. During such dry periods the osmolarity of their body fluids is doubled, from 300–600 mosmol l^{-1}, due largely to accumulation of urea. During hibernation they metabolize stored fat, and utilize water from the large volumes of stored dilute bladder urine. After a rainfall the toads emerge, rehydrate rapidly, excrete urea, and restore the concentrations of the body fluids to normal levels (McClanahan, 1967).

B. The Kidney

Amphibians excrete surplus water in considerable amounts as a dilute urine. A semiaquatic frog like *Rana esculenta* forms daily an amount of urine corresponding to about 60% of its body weight (Jard and Morel, 1963). The amount of urine excreted per day by amphibians immersed in freshwater is generally proportional to the ratio between surface and body weight. Thus the mudpuppy, *Necturus maculosus*, excretes a daily volume of urine corresponding to 25% of its body weight, while the small alpine newt, *Triturus alpestris*, produces 160% of its body weight (Bentley and Heller, 1964).

Both glomerular filtration rate and tubular reabsorption of water play an important role in the regulation of the final urine volume. In *Rana esculenta* the urine volume is nearly directly proportional to the glomerular filtration rate which is doubled in frogs forming urine at twice the average rate (Jard, 1966), while in *Rana catesbiana* urine flow is directly proportional to filtration rate at low and intermediate urine flows only (Forster, 1938). During dehydration or after injection of neurohypophyseal hormone the water permeability of the anuran nephron increases (Sawyer, 1957; Jard and Morel, 1963). The kidney of *Rana cancrivora*, a frog in danger of desiccation, decreases its urine flow by increased tubular water reabsorption; more than 90% of the filtered water is reabsorbed (Dicker and Elliott, 1970). In amphibians antidiuresis follows both dehydration (Adolph, 1933) and transfer to hyperosmotic solutions (Schmidt-Nielsen and Lee, 1962). In *Xenopus laevis* urea accumulation is supported by a reduction in glomerular filtration rate when the animal is transferred to a saline solution (McBean and Goldstein, 1970a, 1970b). As amphibians do not form a hypertonic urine the glomerular filtration rate has to be reduced during dehydration in order to avoid severe renal water losses. When the osmotic balance is improved because of accumulation of urea, urine flow increases, though far below that in freshwater. Injections of hyperosmotic NaCl solutions into *Bufo marinus* causes antidiuresis which is similar to that caused by dehydration (Shoemaker, 1964; Shoemaker and Waring, 1968). For the extreme degrees of antidiuresis in dehydrated animals the decreased glomerular filtration rate is probably more important than tubular reabsorption (Schmidt-Nielsen and Forster, 1954; Sawyer, 1957). The mechanism that links elevated extracellular sodium concentration to antidiuresis is not known.

Although the urine of amphibians cannot reach isotonicity it can approach that under certain circumstances. Thus the urine of *R. pipiens* and *R. temporaria* tends to become isosmotic with both plasma and external medium when the osmotic concentration of the latter is raised above 200 mosmol l^{-1}. In contrast to these frogs the urine of *R. cancrivora* is always less concentrated than the plasma and the bathing fluid (Gordon, Schmidt-Nielsen and Kelly, 1961; Dicker and Elliott, 1970). The elevated plasmal concentration serves to maintain the uptake of water through the skin, and in this way the frogs avoid the dehydration which would otherwise result from swimming in water of high salinity.

C. The Urinary Bladder

In order to adapt to terrestrial life the amphibians have developed mechanisms necessary for conserving the body fluids, that is the ability to regulate water and sodium uptake through the skin, and control water loss from the kidney; but apart from these possibilities many amphibians are able to store and reabsorb water and ions from their urinary bladder.

When living in freshwater amphibians are mostly dependent on extrarenal mechanisms for osmoregulation, and the need for a storage organ for water to be used during osmotic stress is not great. When living on land many amphibians are able to stand periods of dehydration, as their distensable bladder can be used as a storage organ for large amounts of water. Thus the bladder capacity of desert anurans (like *Bufo cognatus*) may equal 50% of their body weight (Ruibal, 1962), in tropical *Bufo marinus* 25% (Shoemaker, 1964), and in aquatic *Xenopus laevis* 1% only (Bentley, 1966).

It is well established that anurans store urine in their bladder from which it may be reabsorbed subsequently during periods of water deprivation, or after injection of hypertonic solutions, or neurohypophyseal hormones (Ewer, 1952; Levinsky and Sawyer, 1953; Ruibal, 1962; MacClanahan, 1967; Middler, Kleeman and Edwards, 1967). The ureteral urine of *Bufo marinus* is dilute, even in dehydrated toads with very low rates of urine flow. The bladder urine approaches isotonicity as water is reabsorbed much faster than solutes. So it seems that the toad bladder is the site of formation of the maximally concentrated urine (Middler, Kleeman and Edwards, 1968).

In addition to its role in water metabolism the urinary bladder is important in regulating sodium excretion and sodium conservation. During severe salt restriction the bladder of *Bufo marinus* reabsorbs a significant proportion of the sodium excreted by the kidneys. The final sodium concentration can be reduced to about one tenth of that in ureteral urine without a change in urine flow (Middler, Kleeman and Edwards, 1968).

It is questionable whether the urinary bladder of urodeles has a similar regulatory role in water and salt conservation (Alvarado and Kirschner, 1963; Bentley and Heller, 1964; Alvarado and Johnson, 1965). However, in the metamorphosed terrestrial *Ambystoma mexicanum* the bladder urine is about eight times more concentrated than the bladder urine of the aquatic larva, and about three times more concentrated than the ureteral urine. But if metamorphosed animals are kept in an aquatic environment the sodium concentration of the bladder urine is about the same as that in the ureteral urine. These results seem to indicate that in the axolotl as in anurans the relative rates of water and sodium reabsorption from the urinary bladder are regulated in response to variations in water available to the animal (Aceves, Erlij and Whittembury, 1970).

D. Regulation of Salt and Water Balance

The water and electrolyte balances of Amphibia are regulated with precision, and the literature contains much information on endocrine control of hydromineral balance (see e.g. the following reviews: Bentley, 1963, 1971; Maetz, 1963, 1968; Deyrup, 1964; Barrington and Jørgensen, 1968).

Much research has shown that the neurohypophyseal hormone in Amphibia, arginine vasotocin, increases water and sodium permeability of the different sites for osmoregulation: skin, kidney, and urinary bladder.

The effect on the water permeability of the skin is more pronounced in terrestrial anurans, and insignificant or non-existent in fully aquatic species like *Xenopus laevis* (Ewer, 1952) and in urodeles (Heller, 1965).

In all amphibia arginine vasotocin induces antidiuresis by decreasing glomerular filtration rate, and by increasing water reabsorption from the renal tubules and urinary bladder (Sawyer, 1967). In urodeles the water retention after injection of AVT seems to be a purely renal effect. The only known exception is *Salamandra maculosa*, in which the effect is due to an increased absorption of urine from the bladder (Bentley and Heller, 1965).

Neurohypophyseal hormone stimulates active transport of sodium across the skin of frogs (Jørgensen, 1950) and toads (Kalman and Ussing, 1955), bladder (Ewer, 1952; Sawyer and Schisgall, 1956), and renal tubules of anurans (Jard and Morel, 1963; Jard, 1966).

Again urodeles behave differently. Although sodium transport occurs in the bladder of urodeles, it does not seem to be stimulated by AVT (Bentley, 1973). However, enhanced cutaneous uptake of sodium after injection of neurohypophyseal hormone has been demonstrated in the axolotl, *Ambystoma mexinanum* (Jørgensen, Levi and Ussing, 1946), and in *Ambystoma tigrinum* (Alvarado and Johnson, 1965). The response might perhaps be partly due to an interference from hormones from the adenohypophysis, as ACTH also elicits sodium uptake in the axolotl (Koefoed-Johnsen and Ussing, 1949), and hypophysectomized Ambystoma are able to maintain their hydromineral balance in pond water (Kerstetter and Kirschner, 1971). Furthermore, aldosterone has been shown to stimulate sodium uptake in *Ambystoma tigrinum* (Alvarado and Kirschner, 1964).

On the other hand, hypophysectomy of *R. temporaria* and *R. esculenta* (Jørgensen, 1947; Jørgensen and Rosenkilde, 1957), of *Bufo bufo*, and of *Bufo marinus* (Jørgensen and Larsen, 1963; Middler and coworkers, 1969) results in a progressive loss of sodium and chloride through the skin. The survival time of the animals can be prolonged by injection of corticotropin (Jørgensen and Larsen, 1967; Middler and coworkers, 1969). The salt balance of the hypophysectomized *Bufo marinus* can be partially maintained when injected with cortisol or corticosterone while aldosterone is without any effect (Middler and coworkers, 1969). So it seems that corticosterone—which is generally considered to be a glucocorticoid—in these animals acts as a mineralocorticoid, too.

Despite much information in the literature on the effect of injected neurohypophyseal hormone on the water balance in the intact Amphibia it is not fully clarified to what extent and how the neurohypophysis participates in the water balance, especially in aquatic species.

For the water balance in terrestrial Anurans it seems that the neurohypophysis plays the important part. When plasma from dehydrated toads is injected into normal hydrated animals it gives rise to a weight gain in these toads by the combined effects of increased cutaneous water uptake and antidiuresis elicited by the released vasotocin (Shoemaker, 1965; Shoemaker and Waring, 1968). It appears that an important fraction of the water uptake in the toad *Bufo bufo* kept in isotonic saline is secondary to the hormonal effect on the active Na transport. When kept in isotonic sucrose solution there is no water response after injection of neurohypophyseal hormone (Kalman and Ussing, 1955).

To what extent the neurohypophysis is involved in sodium regulation of the live Amphibia under ordinary and stress conditions is difficult to decide. Frogs and toads immersed in hypertonic saline release considerable amounts of vasotocin into their circulation (Maetz, Morel and Race, 1959; Bentley, 1969). This does not quite make homeostatic sense with respect to sodium regulation. For mineral balance corticoids and regulating endocrine systems like the adenohypophysis (Middler and coworkers, 1969) and the renin–angiotensin system (Coviello, 1970; Johnston and coworkers, 1967) may be of equal or major importance. Thus optimal rates of sodium transport *in vivo* may be the result of interaction of several hormones.

The fact that the hydro-osmotic and antidiuretic response of vasotocin is absent in aquatic Amphibia, but a natriferic effect can be observed has given rise to speculations whether the two effects are dissociated, and dependent on two different hormonal receptors. It has been suggested that the 'natriferic receptor' had already evolved in the earliest aquatic Amphibia, but that the 'hydro-osmotic receptor' only appeared later in relation to increased exploitation of terrestrial life (see Maetz, 1968).

III. SALT AND WATER MOVEMENTS *IN VITRO*

When salt and water metabolism is studied in intact animals it can be rather difficult to make 'clean' experiments and proofs of action on the different sites for osmoregulation. When designing the experiment one has to take into account that, for example, the mere handling of the experimental animals can cause disturbances in their water and salt balance, and that homeostatic adjustments from feedback mechanisms can easily lead to misinterpretation of observed results. The study of the 'water balance effect' exemplifies the difficulties of *in vivo* experiments. The effect is the weight gain shown by frogs immersed in water after injection of neurohypophyseal hormone (Brunn, 1921), and it is probably due to an action on three osmoregulatory sites: skin, bladder, and kidney. However, for the understanding of the 'Brunn effect' it has been very advantageous that it is possible to study the action of the hormones on the separate target organs *in vitro*.

In order to get thermodynamic information about the underlying mechanisms for salt and water transport, e.g. active transport and passive movements, it is necessary to use isolated preparations of organs or skin, correllate the results obtained to the functional state of the intact animal, and thereby get knowledge about the ability of the animal to maintain homeostasis. Furthermore, when studying inhibiting or activating effects of hormone and drugs on the net transfer of an ion it is important to realize that an agent can affect either the active or passive part of the net transcellular ion movement.

A. The Isolated Amphibian Skin

The overall transport functions of metazoans are carried out by epithelial tissues. As permeability barriers they form a protection against the outside environment, and regulate the composition of body fluids by special absorptive and secretory processes.

Among epithelial membranes the isolated abdominal skin of frogs and toads has undoubtedly been studied most intensively. The frog skin has been the classical object for electrophysiological and permeability studies, and for very good reasons. It is a sturdy preparation, easy to keep alive, can be obtained in large sheets, and offers several advantages for studies of transport phenomena. The parameters that are necessary for distinguishing between active and passive movements of a substance are more readily obtained in such preparations than for instance in cases concerned with transport into and out of cells. The concentrations of the bathing solutions can be determined, and the potential difference across the skin is easily measured, so that forces acting on transfer can be controlled.

Another attractive feature which the frog skin has in common with many epithelia is the fact that when bathed on both sides with Ringer solution the skin develops a potential difference which can be short-circuited, and that in this situation there is equivalence between sodium transport and short-circuit current. This property has proven the short-circuited skin a very useful object not only for examining the physicochemical laws governing ion transfer, but also for studying the effects of hormones and drugs on active sodium transport.

1. Transport of Salt and the Electrical Potential Difference across the Isolated Frog Skin

The history of the isolated frog skin dates back to the middle of the last century when it is mentioned by Du Bois Reymond in his study on 'animal electricity' that the isolated frog skin is able to maintain an electrical potential difference between its inside and outside. The next important observations are made fifty years later when Galeotti (1904) shows that the isolated frog skin can only maintain a potential difference when Na or Li are present in the outer solution, and that the skin is more permeable to Na and Li from the outside to the inside than vice versa. The relation between ionic concentrations and the generated potentials is the object of later studies (e.g. Hashida, 1922; Motokawa, 1935a, 1935b). When the potential difference across the skin is (partially) short-circuited it is able to generate electrical energy for hours (Francis, 1933; Lund and Stapp, 1947). Another milestone is the demonstration of net uptake of salt by the *in vitro* frog skin which, when bathed in Ringer solution, transports chloride ions from the outside to the inside, a transport that can be stopped by metabolic inhibitors (Huf, 1935). Huf assumes that the process is a transport of NaCl towards the inside. With the introduction of radioactive isotopes a new era starts. It is now possible to demonstrate by means of radioactive sodium that the isolated frog skin bathed with identical Ringer solution on both sides transports Na ions faster from the outside to the inside solution than in the opposite direction (Katzin, 1939).

The availability of radioactive tracers for nearly any kind of substance means a turning point, and has a considerable stimulating effect on the study of active transport. Basic concepts about ion transport are formulated, starting with the definition of active transport as a process where the substance in question is moved against its electrochemical gradient (Rosenberg, 1948; Ussing, 1948), and the distinction between active and passive transport by means of flux ratio analysis (Ussing, 1949b, 1972; Teorell, 1949). The short-circuit technique is introduced as a handy tool for measuring active ion transport by means of the short-circuit current (Ussing and Zerahn, 1951). The short-circuiting of the skin is achieved by reducing the spontaneous potential to zero by an opposing EMF, and the short-circuit current is now generated by the skin when it is shorted through an outer circuit of negligible effective resistance. The short-circuited condition is thus a situation where electrochemical potential gradients are eliminated, and the only contribution to the measured current will be from net active ion transport.

2. Transepithelial Movements of Ions and the Active Transport of Sodium as the Source of the Electrical Potential

When bathed with Ringer solution on both sides the isolated frog skin transports Na$^+$ actively from the outside to the inside generating the potential difference across the skin (Ussing, 1949a). Both influx and efflux increase with increasing outer sodium concentration approaching asymptotically a limiting value (Ussing, 1949a; Linderholm, 1952) (Fig. 1). Measurements of

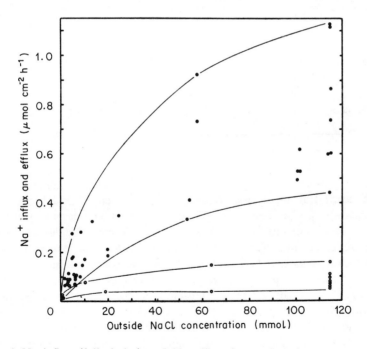

Figure 1 Na influx (full circles) and Na efflux (open circles) through the open-circuited isolated frog skin (*Rana temporaria*) as functions of outside NaCl concentration. From Ussing (1949a); reproduced by permission of *Acta Physiologica Scandinavica*

influx and efflux of sodium together with the short-circuit current show that the net flux of sodium can account for all the current drawn from the skin, in other words sodium is the only actively transported ion (Ussing and Zerahn, 1951). When the net flux of sodium is determined by means of double labelling with the two isotopes ^{24}Na and ^{22}Na (Table 1) along with measurements of the short-circuit current this conclusion has been confirmed unequivocally (Koefoed-Johnsen, Ussing and Zerahn, 1952b).

Table 1. Na fluxes determined by double labelling with ^{22}Na and ^{24}Na on short-circuited isolated skins (*Rana temporaria*) bathed on both sides with Ringer solution (one hour periods). Active Na transport accounts for the total current drawn from the skin. (After data from Koefoed-Johnsen, Ussing and Zerahn, 1952b; reproduced by permission of *Acta Physiologica Scandinavica*)

μA cm^{-2}			
Na influx	Na efflux	Δ Na	Current
19.3	0.9	18.4	20.2
11.8	3.3	8.5	9.5
21.1	1.4	19.7	23.2
29.7	0.9	28.8	29.3
23.9	1.4	22.5	21.2
39.0	0.9	38.1	33.3
29.8	2.5	27.3	24.7
43.6	1.3	42.3	38.8
26.4	2.1	24.3	25.6

Chloride ions are passively transferred along their electrochemical gradient when measured at an outside concentration between 10 and 115 mmol l^{-1} (Koefoed-Johnsen, Levi and Ussing, 1952a) (Fig. 2), and do not contribute to the short-circuit current (Table 2). A small active inward chloride transport can be revealed in several anuran species when the external chloride concentration is so low (1–2 mmol l^{-1}) that the active component is not masked by the large passive fluxes of chloride (Martin and Curran, 1966; Erlij, 1971; Kristensen, 1972; Garcia-Romeu and Ehrenfeld, 1975b; Alvarado, Dietz and Mullen, 1975b). A considerable active inward transport of chloride is found in the skin of the South American frog *Leptodactylus ocellatus* (Zadunaisky, Candia and Chianrandini, 1963). An active outward chloride transport can be demonstrated when the inside of the skin is treated with adrenaline an effect that is probably due to a stimulation of the skin glands (Koefoed-Johnsen, Ussing and Zerahn, 1952b; Lindley, 1969).

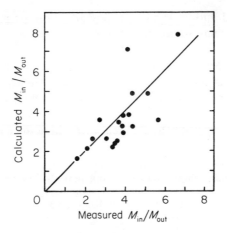

Figure 2 Calculated versus found flux ratios for chloride in the isolated frog skin (*Rana temporaria*). Outside medium 1/10 Ringer, inside Ringer. Chloride ions are moving passively under these circumstances. From Koefoed-Johnsen and co-workers (1952a); reproduced by permission of *Acta Physiologica Sandinavica*

Table 2. Cl fluxes determined by double labelling with ^{36}Cl and ^{38}Cl on short-circuited isolated skins (*Rana temporaria*) bathed on both sides with Ringer solution (one hour periods). Chloride ions do not contribute to the short-circuit current. (After data from Koefoed-Johnsen, Ussing and Zerahn, 1952b; reproduced by permission of *Acta Physiologica Scandinavica*)

μA cm^{-2}			
Cl influx	Cl efflux	ΔCl	Current
16.9	14.9	−2.0	45
12.0	11.5	−0.5	41
6.7	6.1	−0.6	57
4.4	5.3	0.9	59
31.3	32.3	1.0	43

3. The Origin of the Frog Skin Potential

Active sodium transport is solely responsible for the electrical asymmetry of the frog skin, lithium being the only cation so far studied which can substitute to some extent (Zerahn, 1955).

Potassium ions are not transported by the frog skin even if the potassium concentration is raised from normally 2 mmol l^{-1} to 35 mmol l^{-1} (Ussing, 1954). On the other hand both potential (Fukuda, 1942) and sodium transport (Koefoed-Johnsen, 1954) are dependent on the presence of potassium in the inner bathing solution, and are highly reduced in the absence of K+ (Table 3).

The spontaneous potential across the isolated frog skin bathed on both sides with Ringer solution is thus the result of active transport of sodium ions and the shunting effect of passively diffusing ions like chloride ions. The 'true' electromotive force of the sodium battery, the maximal obtainable potential, will consequently develop only in the ideal case where the shunt is reduced to

Table 3. Na fluxes determined by double labelling with ^{22}Na and ^{24}Na on isolated short-circuited skins (*Rana temporaria*); one hour periods. The results show: (1) lack of potassium inhibits the active transport of sodium; (2) potassium ions do not contribute to the short-circuit current. (After Koefoed-Johnsen, 1954, unpublished, quoted by and reproduced from Ussing, 1954. Reproduced by permission of Academic Press, Inc.)

		μA cm^{-2}			
		Na influx	Na efflux	ΔNa	Current
Control		47.0	2.4	44.6	43.5
K-free	Ringer	11.9	4.7	7.2	6.5
K-free	Ringer	6.8	3.7	3.1	2.3
33%	K-Ringer	12.0	2.1	9.9	10.2
33%	K-Ringer	12.8	1.8	11.0	10.5

zero. By substituting the very slowly penetrating SO_4^{2-} ion for Cl^- or by reducing chloride permeability by treating the outside of the skin with $CuSO_4$ (10^{-5} mol l^{-1}) very high and stable potentials can be obtained.

With such skins it is possible to show that the outside of the skin behaves like a sodium electrode towards changes in the outer sodium concentration, and the inside like a potassium electrode in response to changes in potassium concentration of the inner solution. From these observations and on the basis of knowledge about ion transport in epithelial systems and in single cells the two-membrane hypothesis has been put forward to explain the origin of the frog skin potential (Koefoed-Johnsen and Ussing, 1958).

The hypothesis is based on the following assumptions: (1) A continuous cell layer forms the barrier between the inside and outside solutions. (2) The inward and outward facing membranes have different selectivities. (3) The outward facing membrane is selectively but passively permeable to sodium and virtually impermeable to all other cations (except lithium), and permeable to chloride and other small anions. (4) The inward facing membrane is highly permeable to potassium and small anions while the passive permeability to sodium is very low. (5) The active transport mechanism is located at the inner membrane, and is visualized as a Na/K exchange pump which keeps the cellular sodium low, and potassium high.

Consequently the electrical potential which the sodium pump develops across the epithelium in the absence of penetrating anions would be the sum of two Nernst potentials: one between the outside medium and the cell interior determined by the respective sodium concentrations, and one between the cell interior and the inside solution determined by the respective potassium concentrations.

Various phenomena can be explained satisfactorily in terms of the model. The osmotic behaviour of the frog skin thus follows the predicted volume changes when the K/Na ratio of the inner solution is changed in the presence of permeant and non-permeant anions (McRobbie and Ussing, 1961). Likewise in agreement with the model the resistance of the frog skin increases considerably when potassium is substituted for sodium in the outer medium (Ussing and Windhager, 1964). The effect of many inhibitors and stimulators of Na transport can also be explained in terms of the model. Cardiac glycoside ouabain which is generally accepted as an inhibitor of Na/K-ATPase pumps inhibits Na transport violently when added to the solution bathing the inside of the frog skin at a concentration of 10^{-5} mol l^{-1}, but has no effect when added to the outside bathing solution (Koefoed-Johnsen, 1957). The diuretic, amiloride, added to the outside bathing solution of a frog or toad skin is a very strong inhibitor of Na transport, and has no such effect when added at the inside (Eigler, Keiter and Renner, 1967; Baba and coworkers, 1968; Crabbé and Ehrlich, 1968). The action of the drug is attributed to a restriction of the entry path for sodium (Ehrlich and Crabbé, 1968; Salako and Smith, 1970a, 1970b; Nielsen and Tomlinson, 1970; Nagel and Dörge, 1970; Biber, 1971; Cuthbert and Shum, 1974).

The stimulating effect of neurohypophyseal hormones on sodium transport has also been explained according to the model (McRobbie and Ussing, 1961; Ussing, 1973). When added to the solution bathing the inside of the frog skin the hormones increase the permeability of the outside facing border (probably via their intracellular mediator, cyclic-AMP) leading to an increase in cellular sodium which stimulates the pump to an increased activity.

The polyene antibiotic amphotericin B, is known to be able to increase the permeability of many biological membranes (e.g. toad bladder, see Lichstenstein and Leaf, 1965) in a rather non-specific way. The effect is probably due to an interaction with the sterols in the biological membranes as indicated by experiments with artificial lipid membranes with and without cholesterol (e.g. Andreoli and Monahan, 1968; Andreoli, 1973). When the drug is added to the solution bathing the outside of the skin it leads to an increase in the permeability of the outside border which loses its sodium permselectivity and becomes permeable to potassium, too. In such a skin it can be shown that potassium is transported actively from the inside to the outside solution. These observations are also in agreement with the two-membrane hypothesis on the assumption that the pump is a Na/K exchange pump, and the outer membrane is normally not permeable to potassium (Nielsen, 1971). The coupling ratio Na/K is not obtainable from such experiments as the biological response to amphotericin is very variable (cf., e.g., Singer and coworkers, 1969). But the findings do suggest that the cell possesses a potassium pump which maintains the cellular potassium above equilibrium level.

4. *The Anatomical Localization of Ion-selective Membranes in the Frog Skin*

Although many observations have been shown to be in satisfactory agreement with the hypothesis other findings do not fit the pattern so well.

Originally the outer and inner borders of the basal cell layer, str. germinativum, have been thought to be the ion-selective membranes, as the str. germinativum contains the largest cells. This was also in agreement with measurements of intracellular potentials by means of Ling–Gérard type electrodes (Engbaek and Hoshiko, 1957). In most cases only one stable plateau within the epithelium was found. However, in later micropuncture studies several stable plateaux have been found indicating that the epithelial cell layers are coupled through cell junctions of rather high resistance (Ussing and Windhager, 1964).

Potassium exchange studies furthermore seem to indicate that the Na-selective K-impermeable membrane must be situated nearer the outside of the skin, as all epithelial cell layers exchange their potassium at the same rate (Koefoed-Johnsen, 1964; Koefoed-Johnsen and Ussing, 1971).

In the revised model (Ussing and Windhager, 1964) the sodium-selective membrane has been moved to the outside facing membrane of the first living cell layer, while the potassium-selective membrane with the sodium pump will be the total surface of all epithelial cells facing the intercellular space where all ATPase can also be found (Farquhar and Palade, 1964). The interspace is more or less closed to the outside by zonulae occludentes, the 'tight' seals between the cells of the first living cell layer. This model has much in common with that proposed on the basis of electron-microscopy studies (Farquhar and Palade, 1964) where junctions of the conducting type (Loewenstein and Kanno, 1964) have also been observed.

Another finding that supports the idea of coupling between the cells is the fact that lithium coming from the outside solution is found in all epithelial cell layers (Hansen and Zerahn, 1964; Leblanc, 1972).

Moreover, studies of the frog skin capacitance as a function of the frequency of an applied alternating current can be explained satisfactorily if it is assumed that there are two membranes in series, an outer one with high resistance and a capacitance of about $2\ \mu F\ cm^{-2}$, and an inner one with low resistance and a capacitance of about $50\ \mu F\ cm^{-2}$. This latter value is so large that it can hardly be accounted for by only one single cell layer, but would be more conceivable if several cell layers were coupled (Smith, 1971).

However, it seems doubtful that all cell layers participate to the same extent in the transport of sodium. There are many experiments correlating morphological features with functional state that provide evidence against such a coupling being of major importance for active transport.

When the rate of Na transport is increased by short-circuiting the skin the cells in the outer layer of str. granulosum (the first living cell layer right underneath the cornified layer) swell, while the cells in the lower part of the epithelium show no such reaction. The swelling is reversible, and is even more pronounced if the current is increased beyond the short-circuited state by passing an ingoing electric current (Voûte and Ussing, 1968).

If the skin potential is kept fixed at a suitable high value the active transport of sodium stops, and an outgoing current runs through the skin. This reversed current is carried exclusively by passive ion movements. Under such conditions there is a shrinkage of the first living cell layer (Voûte and Ussing, 1970a).

These volume changes due to in- and outgoing currents are predictable on the basis of the original two-membrane model (Koefoed-Johnsen and Ussing, 1958). Only one cell layer (connected with tight seals) seems to dominate the active transport of sodium. Thus as the cells of the outer layer of str. granulosum seem to possess the ionic selectivities proposed for that model, they should constitute the active cell layer rather than str. germinativum as originally proposed.

5. The Intercellular Space

Whether only one cell layer (the outermost living cells) or several layers contribute to the transport, the transported sodium has to leave the epithelium via the interspace system. Consequently the size of the extracellular volume might influence the transport rate, or conversely the transport rate might influence the volume of the interspace.

When small hydrostatic pressures (of 5–30 cm H_2O) are applied to the inside of a frog skin the intercellular volume is expanded. The expansion is reversible, and there is a linear relationship between the pressure gradient and the volume of the interspace. The transport rate of sodium measured as short-circuit current is unaffected by the treatment while the potential falls in some cases indicating that stretching of the 'tight' seals between the cells of the outermost layer leads to a leak of the zonulae occludentes.

On the other hand when the size of the interspace is measured with and without transport (i.e. when choline is substituted for sodium in the outer solution) it is clearly demonstrated that it is the active transport that brings about the expansion of the interspaces. In the absence of sodium in the outer solution the intercellular space is collapsed, whereas a measurable volume persists in the presence of sodium. Actually there is a strong positive correlation between the short-circuit current and percentage of interspace area indicating that sodium is transported into the interspace system (Voûte and Ussing, 1970b).

6. The Na-selective Membrane

There is good evidence that the Na-selective membrane can be identified as the apical membrane of the cells of the outer layer of str. granulosum. Thus lanthanum ions (that do not penetrate normal cells) when present in the solution bathing the outside of the skin pass through the cornified layer and can be found in the subcorneal space indicating that str. granulosum cells form the outer barrier (Martinez-Palomo, Erlij and Bracho, 1971). The same result has been obtained by another quite different technique. By comparing the time course of the potential change following a very rapid change in sodium concentration at the outside of a skin with that of an ion exchange membrane it can be shown that the skin reacts much slower than the ion exchange membrane. The difference in time course can be accounted for if it is assumed that a 5 μm thick additional diffusion layer were interposed between the Na-selective membrane and the bulk solution. This layer is most probably the cornified layer, as freshly moulted skins show nearly the same time course of potential response as the ion exchange membrane (Fuchs, Gebhardt and Lindemann, 1972).

However, it has been a matter of dispute whether sodium enters the cells by an entirely passive process.

Measurements of Na influx with increasing outer Na concentration show a saturation curve (Ussing, 1949; Kirschner, 1955), and short-circuit current saturates like the flux suggesting a decreasing permeability to sodium with increasing outer Na concentration (Cerejido and coworkers, 1964; Salako and Smith, 1970). Tracer kinetic studies of short-time Na uptake by the epithelial cells have also shown that influx saturates with increasing outer sodium, and that the curve can be described by a Michaelis–Menten type equation, which allows a determination of K_m. The value of K_m turns out to be within the range of 4–40 mmol l^{-1} (Rotunno and coworkers, 1970; Biber, Cruz and Curran, 1972b; Erlij and Smith, 1973). As uptake of sodium at the outer surface is influenced by the transepithelial potential, it has been suggested that the saturating flux should procede via a charged carrier system (Biber and Sanders, 1973). However, saturation kinetics need not necessarily involve carrier transport, since transport through narrow pores can saturate also (Heckman, Lindemann and Schnackenberg, 1972).

In order to find out whether the outer membrane behaves as an ideal Na electrode under all circumstances (very low and high outer Na concentrations) a knowledge of the intracellular Na concentration is required together with measurements of the potential across the outer membrane. However, microelectrode studies are very difficult to make due to the rigid structure of the frog skin. Impalement of the microelectrode can easily produce diffusional leakages and changes in cellular concentration leading to an attenuation of the membrane potential (Lindemann, 1975). Furthermore, chemical determinations of intracellular sodium by measuring the sodium concentration in the epithelium are dependent on a reliable marker for the intercellular volume, and may not be representative for the concentration in one single cell layer (Biber and Curran, 1970; Biber, Cruz and Curran, 1972b; Erlij and Smith, 1973).

Recently another method has been developed based on the idea of using the Na-selective membrane itself as an electrode for measuring the intracellular Na activity. Thus the use of microelectrodes can be avoided, and the transepithelial potential can be identified with the voltage across the outer membrane if the series membranes are completely depolarized. This can be done satisfactorily by substituting potassium for sodium in the inner solution (Rawlins and coworkers, 1970; Morel and Leblanc, 1973), and keeping the cellular volume constant by using an impermeable anion like gluconate. When the current–voltage curves of the voltage-clamped Na-selective membrane are recorded before and after blocking the Na channels with amiloride, the I–V curve of the Na channels can be obtained as the difference between the total current and the unspecific leak current. Use of a special fast flow chamber (Gebhardt, Fuchs and Lindemann, 1972) ensures that the outside of the skin can be exposed to solutions with varying Na concentrations for such a short time (20 ms) that the cellular sodium concentration will stay nearly unchanged.

By means of this technique it has been found that the current–voltage curves at a given outer Na concentration can be fitted with the constant field equation (Goldman, 1943; Hodgkin and Katz, 1949) for the physiological voltage range, in other words, the Na transport through the Na-selective channels may be simple electrodiffusion, and this fit also allows an estimation of the permeability coefficient, for sodium, P_{Na}, and the cellular sodium concentration, $[Na]_c$ at the given outer Na concentration. There is a linear relationship between the steady-state $1/P_{Na}$ and $[Na]_o$, the Na concentration in the outer solution. This saturation phenomenon might have been the result of channel saturation (sodium piled up inside the channels) in the case where the transfer process at the inside of the membrane is the rate limiting step. But following the time course of current in response to a steep increase in $[Na]_o$ it can be demonstrated that the currents are transiently much larger than their steady-state plateau. This would not be the case if the sodium channels were saturated, but could rather be explained by assuming that sodium acts as a negative effector on a modifying site, provided that the modifying process—the conformational change closing the Na channel—is relatively slow. Amiloride and sodium may thus compete for the same modifying site, for which, though, amiloride has a much higher affinity (Fuchs, Larsen and Lindemann, 1977).

It is interesting that certain guanidinium compounds (e.g. benzamidazolyl-guanidin, BIG) when added to the outside solution reversibly increase the sodium permeability, and saturation of the steady-state current disappears when $[Na]_o$ is increased. The action of these compounds can be explained in terms of the allosteric model by assuming that BIG when bound to the modifying site prevents sodium from reaching it, and thereby the closing of the channels is avoided (Zeiske and Lindemann, 1974).

The result of these kinetic studies, viz. that sodium current varies with voltage in agreement with the constant field equation, would be in accordance with the view that the movement of sodium ions through the sodium-selective membrane is purely passive.

Many authors, though, are of the opinion that active forces are necessary for the transfer. As the isolated frog skin is capable of transporting sodium ions from very dilute solutions, 1 mmol l^{-1} (Biber, Chez and Curran, 1966) it requires a cellular Na activity of the same order or lower in the case of passive transfer. But it has been found in several cases that the transport pool of sodium (i.e. the cellular sodium with access to the pump) is much smaller than the total cellular amount of sodium (e.g. Cerejido and Rotunno, 1967; Zerahn, 1969) and of lithium (Morel and Leblanc, 1973), a finding that has led to the proposal of a new model, according to which sodium transport takes place along the membranes of the epithelial cells rather than through the cytoplasma (Cerejido and Rotunno, 1968).

However, determinations of $[Na]_c$ and $[Li]_c$ using the technique described by Fuchs, Larsen and Lindemann (1977) have shown that neither $[Li]_c$ nor $[Na]_c$ exceeds $[Li]_o$ or $[Na]_o$ respectively as zero membrane voltage, and thus there is no need for an active entry step (Cirne and Lindemann, 1975).

The discrepancy between the transport pool and total cellular sodium could be explained by assuming that a large part of the cellular sodium is sequestered, for instance in the nuclei (see e.g. Mirsky and Allfrey, 1958; Zadunaisky and coworkers, 1968; Moore and Morrill, 1976). Furthermore, if only one single cell layer performs the main part of the sodium transport, the transport pool would be limited to this layer.

Some very recent observations indicate that another cellular compartment might be associated with active transport. During short-circuiting and with hydrostatic pressure applied on the inside of the frog skin the endoplasmatic reticulum is expanded, and in intimate connection with this system several vacuoles ('scalloped sacs') can be seen in the outer cells of str. granulosum. The formation of these vacuoles is dependent on the presence of sodium in the outer solution, they are not found in cases where choline is substituted for sodium. The number of vacuoles varies linearly with the magnitude of the short-circuit current. Whether the endoplasmic reticulum with the associated 'scalloped sacs' takes part in an auxiliary transport which could be active only during sodium overloading, or whether it is part of the main transport pathway cannot be decided (Voûte, Møllgaard and Ussing, 1975). But if some part of the cellular sodium is located in the endoplasmic reticulum the difference between the sodium transport pool and the total cellular sodium would find a natural explanation.

7. The K-selective Membrane

In the original model it was assumed that the inner membrane with the sodium pump should be the inward-facing membrane of the transporting cells, and that this barrier has a high permeability to potassium and a low permeability to sodium. These assumptions were based on experiments showing that the slope of the potential across the skin at different K concentrations of the serosal bathing medium approached that of a K-selective membrane. That the cells are very impermeable to sodium has now been shown. By direct measurements of sodium uptake from the serosal side it has been found that sodium distributes itself like mannitol (Biber, Aceves and Mandel, 1972a).

The Na pump was originally visualized as a one to one Na/K pump. However, it has not yet been possible to give any proof for a fixed coupling ratio between K uptake and Na transport (e.g. Curran and Cerejido, 1965). Electrogenity of the pump has been suggested (Bricker, Biber and Ussing, 1963), and a coupling ratio of 2–5 Na/K has recently been found (Biber, Aceves and Mandel, 1972a).

It is questionable though, whether it is possible to make a meaningful determination of the Na/K coupling ratio in a structure like that of the frog skin. If the main part of the sodium transport takes place in the outermost cell layer, it is the K exchange of these cells that should be correlated with the short-circuit current. As the intercellular space is very narrow at the level of these cells the chances are that potassium may recycle to an unknown extent

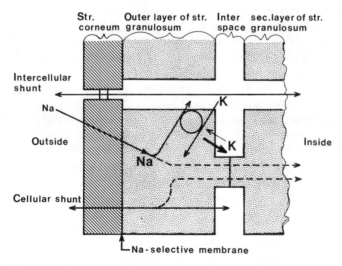

Figure 3 Schematic diagram of the outermost cell layers of frog skin epithelium with str. corneum, a cell from the first living cell layer, and a part of the next cell layer of str. granulosum. Features of the Na pathway, the cellular and the intercellular shunt pathways are indicated

before exchange with the inside solution. Furthermore the cells along the long winding interspace will function like an ion exchange column; as K concentration is low and Na concentration high in the Ringer solution a ^{42}K ion at one end of the interspace system is more likely to enter a cell than to pass further on.

Even if the coupling ratio of the pump is not established the pump might very well turn out to be partly electrogenic, but it seems improbable that it could be purely electrogenic. Thus the finding that an outward directed active K transport can be demonstrated when the outer membrane has been made permeable to potassium by amphotericin B (Nielsen, 1971) is quite in accordance with the original model; the pump appears as a pure sodium pump because the outer membrane is impermeable to potassium (Fig. 3).

8. Shunt Pathways

The magnitude of the spontaneous potential difference across frog skins shows a great biological variability due to the presence of a shunt from passively moving ions. Treatment of the outside of the skin with copper sulphate (10^{-5} mol l^{-1}) decreases chloride permeability considerably, and high potentials can be obtained (Koefoed-Johnsen and Ussing, 1958). The shunt can be both transcellular and intercellular as pointed out by Ussing and Windhager (1964).

Skins of freshly caught autumn frogs have a very high resistance which is maintained during winter, but in spring and summer the resistance can be extremely low. Moulting also gives rise to a transiently increased shunt conductance (Jørgensen, 1949; Larsen, 1971). Another factor which has a great influence on the shunt is the temperature at which the animals have been stored prior to the experiment. Thus it has turned out that skins of frogs which have been kept at 5 °C (i.e. in a cold room, the way frogs are usually stored nowadays) have high potentials and a rather low permeability to chloride which is not affected by treating the outside with copper ions. On the other hand if the frogs have been kept at room temperature for a few days before the experiment, the skins have low potentials and a high chloride permeability which can be markedly reduced by copper ions. Simultaneous determination of chloride and sulphate (or sucrose) permeabilities furthermore show that the permeability of sulphate or sucrose is quite unaffected by copper treatment, both in cold-adapted and in warm-adapted frogs, whereas chloride permeability is reduced only in warm-adapted frogs. Under the assumption that sulphate and sucrose permeabilities are mainly intercellular, the results seem to indicate the presence of two shunt pathways for chloride: (1) a cellular shunt pathway which is accesible only in warm-adapted frogs and which can be reduced by copper; and (2) a paracellular shunt pathway in common with sulphate and sucrose, and which cannot be affected by Cu treatment (Koefoed-Johnsen, Lyon and Ussing, 1973; Koefoed-Johnsen and Ussing, 1974).

The interspace of the frog skin is normally closed towards the outside by zonulae occludentes (Martinez-Palomo, Erlij and Bracho, 1971), and to what extent a paracellular shunt pathway through the 'tight' seals and/or a transcellular shunt are of importance for the overall function under normal conditions little is known.

When comparing shunt conductance with the total conductance of the skin of *Rana temporaria* it appears that the shunt conductance (which is proposed to be intercellular) may vary from about 15%—about 40% of the total conductance (Ussing and Windhager, 1964). Paracellular shunt paths also seem to dominate in the skin of *Rana pipiens* (Mandel and Curran, 1972a, 1972b). When poisoned by ouabain and under steady-state voltage clamping the fluxes of Na, K, Cl, and mannitol move in parallel with changes in urea movements. Furthermore, the fact that the ion fluxes respond to potential changes according to the constant field equation suggests that the path has the properties of a single barrier, and the authors conclude that the substances tested use an intercellular pathway in common with mannitol. In toad skin the major part of the passive sodium flux, and a minor part of the passive chloride flux are intercellular (Bruus, Kristensen and Larsen, 1975).

The existence of a transcellular pathway for sodium efflux in frog skin has also been proposed (Kirschner, 1955), and recently again on the basis of saturation kinetics for Na efflux (Biber and Mullen, 1976). As Na efflux increases after ADH treatment a cellular shunt path for sodium may also be present in toad skin, granted that ADH exerts its effect only on cell membranes (Ussing, 1973). In contrast ADH seems to have no effect on Na efflux in frog skin, an observation that has led to the conclusion that in- and efflux paths for sodium are not the same (Morel and Bastide, 1965; Bastide and Jard, 1968).

Although frog skin belongs to the category 'tight' epithelia, it can be reversibly transformed into a 'leaky' one under certain circumstances. If the osmotic pressure of the outside bathing solution is raised relative to that of the inside by addition of, e.g., urea the potential and resistance decreases, while the short-circuit current is more and less unaffected indicating creation of an intercellular shunt (Ussing and Windhager, 1964).

That this shunt path induced by hypertonicity of the outside medium is intercellular can be demonstrated directly. When the outside hyperosmotic bathing medium also contains Ba ions and the inside medium sulphate ions, a precipitate of $BaSO_4$ can be found in the interspaces, but not in the cells (Ussing, 1970). Lanthanum ions will also move into the skin under these circumstances, and are found in the interspaces (Erlij and Martinez-Palomo, 1972).

The outward directed osmotic gradient with the ensuing opening of the shunt is associated with an 'anomalous' transport inward of sucrose and a large variety of medium-sized hydrophilic molecules (not bigger than raffinose) when such substances are added in addition to the hypertonic agent

(Ussing, 1966; Franz and van Bruggen, 1967). The phenomenon is explained in various ways. Solute–solute interaction in the aqueous space is proposed by Franz and van Bruggen (1967), while Ussing (1969) describes it as a result of 'anomalous solvent drag', and in a model illustrates how such a transport could be brought about when it is coupled to local circulation of water. It is a prerequisite for the model that the osmotically active substance is so small that it can diffuse into the interspace, but on the other hand so large that it does not enter the cells as easily as water. Urea and thiourea have the most pronounced effect for a given osmolarity, but a series of other substances including D_2O are also effective (Lindley, Hoshiko and Leb, 1964). The model predicts that 'anomalous' transport should stop or reverse when suitable hydrostatic pressures are applied to the inside solution, and such a result has also been demonstrated (Ussing, 1969).

Regardless of the model, it is now evident that even 'tight' junctions in epithelia cannot be regarded simply as impermeable seals, but as barriers, the permeability of which can be regulated.

Thus the frog skin seems to possess a variable intercellular and cellular shunt which together with the active sodium transport mechanism determines the level of the frog skin potential.

9. Water Movement

It was demonstrated by Reid as early as 1892 that water is taken up through the isolated frog skin from isotonic solutions. Water permeability has originally been measured by the rate of osmosis resulting from an applied osmotic pressure gradient. Availability of isotopic tracers for water has provided another means of measuring water permeability, and thus made it possible to compare the two methods. Such a comparison has been made by Hevesy, Hofer and Krogh (1935) with the unexpected result that the permeability to water of the isolated frog skin measured by osmotic water uptake turns out to be 3–5 times greater than when measured by permeability to heavy water. The finding that the osmotic permeability coefficient, P_{os}, does not equal the diffusional permeability coefficient, P_{di}, must mean that water movement across this tissue cannot be described as simple diffusion according to Fick's first law. The ratio P_{os}/P_{di} has invariably been found bigger than unity in a variety of epithelia, and becomes even larger when ADH is present, for instance in frog and toad skins (Capraro and Bernini, 1952; Koefoed-Johnsen and Ussing, 1953). The disparity between P_{os} and P_{di}, and the finding that ADH increases P_{os} much more than P_{di} has been explained by assuming that net water flux involves water molecules moving in bulk flow through aqueous channels the diameters of which would be increased by ADH treatment. This is a consequence of the well known fact that diffusion is proportional to the second power of the radius of the aqueous pores, while bulk (viscous) flow increases with the fourth power of the radius.

An equation combining Fick's first law and Poiseuille's law would also allow an estimation of the effective pore radii in the membrane rate limiting to water (Koefoed-Johnsen and Ussing, 1953; Andersen and Ussing, 1957).

Thus in frog skin or even better in toad skin, when the outside solution is more dilute than the inside, ADH increases the permeability to water and water–water interaction can be observed. But the permeability to small hydrophilic substances is also increased in toad skin (Andersen and Ussing, 1957; Ussing, 1973) and in toad bladder (Leaf and Hays, 1962), and consequently water–solute interaction should also occur. The viscous flow of water through water-filled channels should disturb the unidirectional fluxes of small solutes passing through these channels, the rate of the molecules being enhanced by solvent drag in the same direction as the bulk flow and impeded in the opposite direction. Such a coupling of solute and solvent flows within aqueous channels has also been demonstrated by Andersen and Ussing (1957) who have shown that the net flux of thiourea and acetamide increases in proportion to the rate of osmotic flow in ADH treated toad skin. Solvent drag on urea, but not thiourea, has also been reported to take place in toad urinary bladder (Leaf and Hays, 1962).

In some cases ADH fails to increase the permeability to small solutes both in toad skin, but especially in toad bladder. In order to explain that difficulty Andersen and Ussing (1957), and Leaf and Hays (1962) have proposed that the outer epithelial surface is a composite dual barrier, consisting of a thin diffusional barrier which represents a restriction to the passage of small molecules, but allows water to move freely. In series is the second barrier, a porous membrane, and therefore rate limiting for osmotic water flow. The action of ADH, increase of pore size, is exerted on this membrane only.

Amphibian epithelia (skin, bladder) are composite structures and not homogenous membranes, and water moves across epithelial cell walls as well as in the intercellular spaces. Apart from the relative contributions of the transcellular and intercellular paths it is important to know the permeability to water of the apical and basal membranes of the cells in the transcellular path.

The permeability to water of the outer and inner boundary of the frog skin epithelium has been estimated by MacRobbie and Ussing (1961) by measuring the thickness of the epithelium in response to osmotic swelling and shrinkage. The inner border appears to be much more (20 times) permeable to water than the outer one. Thus the apical membrane is rate limiting to osmotic water flow as the presence of dilute media on the outside surface does not lead to swelling of the cells unless the skin is treated with ADH.

Unstirred layers of solution and/or unstirred regions within the tissue will reduce the size of the osmotic gradient across the epithelium itself below that of the apparent gradient between the bathing solutions (Dainty, 1963). Vigorous stirring of the solutions bathing the frog skin has been demonstrated to increase P_{di} twice (Dainty and House, 1966), while a tenfold increase has been obtained in toad bladder (Hays and Franki, 1970).

But comparison of P_{os}/P_{di} after correction for unstirred layers (skin, urinary bladder) still suggests that somewhere along the transepithelial route are channels that give rise to rapid water flow.

ADH does not increase P_{di} in frog skin even after vigorous stirring (Dainty and House, 1966). This is in contrast to the more than tenfold increase in P_{di} which has been found in urinary bladder after correction for unstirred layers and application of ADH (Hays and Franki, 1970). P_{di} may be even higher in this case if corrections had been made for barriers in series with the luminal membrane (cell cytoplasma, intracellular structures, etc.). Consequently it has been argued that ADH increases the number rather than the diameter of small aqueous pores in the toad urinary bladder (Hays, 1972b). Such an explanation would also account for the observation that penetration of the bladder by most small solutes does not increase in the presence of ADH (Hays and Franki, 1970; Hays, 1972a; Finkelstein, 1976).

Although the main resistance to osmotic water flow is the apical membrane of the outer epithelial cells both in toad urinary bladders and in frog and toad skins, water may move through the intercellular space, too, rather than by a transcellular route alone.

It seems that distension of intercellular spaces as a consequence of active solute and passive water transfer is a general feature of most transporting epithelia, for instance frog skin (Voute and Ussing, 1970), toad bladder (DiBona, Civan and Leaf, 1969; Grantham, Cuppage and Fanestil, 1971) and frog bladder (Jard and coworkers, 1971).

A phenomenon like solvent drag (which by the Dainty school is called an effect of unstirred layers within the tissue) is highly dependent on the length of the pathway. Such a drag effect is much more likely to occur in a multilayer epithelium (like e.g. frog and toad skins) where bulk flow through the long interspace system may produce a spectacular change in the transport kinetics for solutes through the epithelium. (For an excellent treatment of the problems concerning water transport, see House, 1974.)

B. The Isolated Urinary Bladder

The isolated urinary bladder of toad has many transport properties in common with frog skin.

Thus it was originally demonstrated by Leaf (1955) that the isolated bladder when bathed on both sides with Ringer solution generates a potential with the serosal surface electrically positive to the mucosal surface. Flux measurements of sodium by means of the double-labelling technique together with measurements of the short-circuit current have furthermore shown that the transepithelial net transport of sodium equals the short-circuit current (Leaf, Anderson and Page, 1958; Leaf, 1965).

Functionally the bladder may be described as one single layer of epithelial cells, connected by tight junctions, and supported by connective tissue, and histologically would seem to be a more simple preparation than the frog skin. However, the mucosa of the bladder contains four different cell types (Choi, 1963). The granular and the mitochondria-rich cells are probably the ones which are responsible for the transport characteristics, while the basal cells are young undifferentiated cells, and the goblet cells are mucus secreting.

The bladder has similar absorptive capacities as the distal convoluted tubule and collecting duct of the mammalian nephron, and furthermore it responds to the two major hormones regulating water and sodium metabolism, ADH and aldosterone. Understanding of the mechanisms of action of these hormones, and of many pharmaca, e.g. diuretics, has been greatly enhanced by their study in this biological preparation (see e.g. Sharp and Leaf, 1966; Leaf and Sharp, 1971; Crabbé, 1972; Edelman and Fimognari, 1969; Hays, 1972b; De Sousa, 1975; Urakabe and coworkers, 1976).

The conductance of the active pathway in the bladder of *Bufo marinus* (from Dominion Republic) is in the range 45–85% of the total conductance. Na^+ is the only actively moving ion. The passive movements of Na^+, K^+, and Cl^- seem to share common ionic pathways which do not discriminate between monovalent physiological ions on the basis of size, but which favour the transport of cations over that of anions (Saito, Lief and Essig, 1974). Low rates of Cl^-, K^+, and H^+ transports have been reported in bladders of Columbian toads, but not in those from the Dominican Republic (see e.g. Davies, Martin and Sharp, 1968; Frazier, 1974).

The active Na pathway of toad bladder consists of an apical passive permeability barrier in series with an Na/K activated ATPase (Na pump) at the baso-lateral membrane. The electrical profile is a two step process (Frazier, 1962).

Measurements of the intracellular sodium transport pool from analysis of whole tissue (Frazier, Dempsey and Leaf, 1962), and of scraped off cells (Macknight, Civan and Leaf, 1975a) have shown that passive entry of sodium can account for the initial step.

It has proved possible to scrape off epithelial cells from bladders after they have been previously mounted in a chamber and exposed to experimental solutions, and subsequently to analyse the cells for water and ion contents. Use of this technique has given the interesting result that in contrast to frog skin 80% of total cellular sodium equilibrates with ^{24}Na from the serosal side, while 20% equilibrates with ^{24}Na from the mucosal side. These findings suggest that there might be two discrete pools of sodium in bladder epithelial cells, one of mucosal and one of serosal origin, and raise the problem of the anatomical localization of these pools. One possibility would be compartmentalization within intracellular structure in the same cells, and an alternative explanation would be that different cell types might contain only sodium of mucosal or serosal origin. The mitochondria-rich cells might for instance under normal circumstances contain sodium of mucosal origin, while the more plentiful granular cells together with the basal cells equilibrated with serosal sodium. The stimulation of transepithelial sodium transport by ADH could then reflect an entry of mucosal sodium into the granular cells, as it has been shown that these cells are involved in water movements produced by this hormone (DiBona, Civan and Leaf, 1969). Such an explanation would also be in accordance with the claim (Mendoza, Handler and Orloff, 1970; Mendoza, 1972, 1973) that base-line Na transport and ADH-stimulated Na transport follow different pathways (Macknight, Civan and Leaf, 1975a).

The sodium pump has been claimed to be purely electrogenic on the grounds that, with the tip of a micropipette within the cell, a potential has been demonstrated across the serosal membrane even when the inner solution is sodium-free potassium Ringer (Frazier and Leaf, 1963). However, rigorous quantitative interpretation of microelectrode studies can be very difficult. It has recently been pointed out that the use of microelectrodes involves severe problems (Lindemann, 1975), and that many cells impaled with a microelectrode (filled with 3 mol l^{-1} KCl) swell spontaneously. The swelling is often shared by neighbouring cells showing that the cells are coupled or become coupled (Loewenstein, 1966) during swelling of the impaled cell (Ehrenfeld, Nelson and Lindemann, 1976b).

One of the inherent properties of a coupled Na/K pump, though, has been demonstrated in the toad bladder, viz. a drastic inhibition of transcellular Na transport when potassium is absent from the Ringer solution bathing the inside of the bladder (Essig and Leaf, 1963). The inhibition is less when all sodium is replaced by choline in the inner potassium-free solution (Essig, 1965) in which case Na transport is downhill.

Ouabain inhibits active sodium transport in toad bladder (Herrera, 1966) indicating that the Na/K activated ATPase is also involved in the energy-dependent step of Na transport. Inhibition of the toad bladder by ouabain is associated, as expected, with an increase in cellular sodium of mucosal origin and a concomitant loss of potassium into the serosal solution. Amiloride which blocks sodium entry into the cells (Bentley, 1968) prevents any gain of mucosal sodium in ouabain treated bladders (Macknight, Civan and Leaf, 1975b).

Attempts to establish the coupling ratio have also been made with varying results ranging from a coupling much less than one (Essig and Leaf, 1963) to a one to one ratio (Finn and Nellans, 1972). But, as discussed earlier in this chapter, the possibility of recycling of potassium makes measurements of coupling ratios very difficult, if not impossible.

Neither has it been possible to relate K content to Na transport from analyses of scraped off epithelial cells. The K content has been found to be the same in control cells (i.e. from hemibladder bathed in Ringer solution on both sides), and in cells where Na transport has been prevented either by substitution of choline for sodium in the mucosal solution or by inhibition with amiloride (Macknight, Civan and Leaf, 1975b). Consequently the authors conclude that transepithelial Na transport and Na/K exchange are dissociated processes, but that both can be ascribed to a ouabain-sensitive energy-dependent process. Furthermore they postulate the existence of another pump, ouabain-insensitive, and involved in volume regulation. The reason for this hypothetical pump is the finding that toad bladder epithelial cells scraped from hemibladders incubated with ouabain have the same water content as cells scraped from control hemibladders. Similar results have also been reported from experiments with isolated epithelial cells incubated *in vitro* (Lipton and Edelman, 1971; Macknight, Leaf and Civan, 1971; Handler, Preston and Orloff, 1972). Two sodium pumps with the same characteristics as the above mentioned have been postulated to be present also in frog bladder (Janáček, Rybová and Slaviková, 1972). However, the actual nature of this mechanism remains to be found out.

C. Interdependence of Active Transport and Metabolism

In an active transport process work is done at the expense of energy derived from metabolism.

The active transport of sodium in amphibian skin and urinary bladder is mainly energized by oxidative metabolism. Stoichiometry between sodium transported and net oxygen consumed (after subtraction of 'background', i.e. oxygen consumed when no transport takes place) has been established to be on an average 18 Na ions per molecule of O_2 in frog skin (Zerahn, 1956; Leaf and Renshaw, 1957). The same stoichiometry holds for the toad bladder (Leaf, Page and Anderson, 1959).

The ratio, transported equivalents of sodium/equivalent of oxygen, of 4–5 has been found to be the same in *Rana temporaria*, in *Rana esculenta*, and in *Bufo bufo*. Furthermore, the stoichiometry persists whether transport takes place against a concentration or potential gradient, or whether the transport is increased by stimulation from antidiuretic hormone (Zerahn, 1956).

Recent measurements of Na/O_2 ratio in *Rana pipiens* (Viera, Caplan and Essig, 1972) have shown a mean value of 15 equivalents of sodium per molecule of oxygen, not different from the value of 18 found earlier (Zerahn, 1956; Leaf and Renshaw, 1957), both numbers being means of a large number varying over a wide range. Thus the range in *R. temporaria* is between 14–28 (Zerahn, 1956) and 6–43 (Leaf and Renshaw, 1957), in *R. esculenta* 12–27 (Zerahn, 1956) and in *R. pipiens* 7–31 (Viera, Caplan and Essig, 1972). When net oxygen consumption and net sodium transport are measured in the same skin of *R. pipiens* under varying circumstances (varying spontaneous short-circuit current, stimulation by antidiuretic hormone) there is a linear relationship between net sodium uptake and net oxygen consumption giving an apparent stoichiometric ratio for each individual skin. As the Na/O_2 ratio vary within the range 7–31 in different skins, while the individual skins have their own stoichiometry, the authors claim that the assumption of a unique ratio is applicable to individual, but not necessarily to all skins (Viera, Caplan and Essig, 1972).

In studies of toad urinary bladders a direct relationship between the rate of sodium transport and net oxygen consumption has been found, but the ratio seems to be an increasing function of the rate of Na transport at low transport rates, and to saturate at higher rates at levels between 20–25 Na/O_2 (Nellans and Finn, 1974). In a very recent investigation of toad bladder the Na/O_2 ratio has been found to be constant for each tissue, but varying among different tissues. Only when the potential is varied does the ratio change with the rate of active transport. Under these circumstances the ratio is constant, but exceeds the one measured in the short-circuited state. This might suggest that coupling between transport and metabolism is not complete (Lang, Caplan and Essig, 1977).

The transport work for one equivalent of sodium in the short-circuited skin with Ringer on both sides and a Na flux ratio varying between 10 and 100 can be calculated to vary between 1340–2680 cal. With a net oxygen consumption of 1/4 equivalent per equivalent of transported Na the thermodynamic efficiency varies from 20%–50%. In terms of ATP this means that about 3 equivalents of sodium are transported per mole of ATP (Zerahn, 1956).

The direct proportionality between net oxygen consumption and the operation of the sodium pump must mean that active transport is a pacemaker of metabolism. The Na pump converts ATP to ADP during its action, and it is probably through its production of ADP that it regulates mitochondrial oxygen consumption, since ADP availability is a dominant factor in regulating respiration (Chance and Williams, 1956). When sodium transport is inhibited by ouabain, respiration is also inhibited (Coplon and Maffly, 1972).

The Na/K activated, ouabain-sensitive ATPase, which is part of or identical with the cation transport system, has been demonstrated to be present in frog skin and in toad urinary bladder, and Na/ATP ratio found to be 3.0 in frog skin and 2.5 in toad bladder (Bonting and Caravaggio, 1963).

A microsomal Na/K activated, ouabain-sensitive ATPase has also been found in the epithelium of *R. catesbiana*. Li cannot substitute for Na as an activator for the enzyme, but in high concentrations Li can partially substitute for K (Siegel, Tormay and Candia, 1975).

A highly active ATPase has been prepared from the isolated epithelium of *R. temporaria*. There seems to be much more enzyme available than that required to sustain the epithelial Na transport, a finding that would be in accordance with the view that only one single cell layer performs the main part of the active sodium transport (Ferreira, Ferreira and Lew, 1976).

D. Control Mechanisms at the Cellular Level

Modulation and control of ion permeabilities are of adaptive value for amphibians when changing surroundings, and passive permeability controlling downhill movements is obviously not a static property of membranes, but rather a dynamic one.

The observation (first made by Krogh, 1937) that frogs depleted of salt take up Na at a faster rate than non-depleted ones can be ascribed to an increase in the density of Na entry sites. Thus isolated skins from frogs, which have been depleted for a week in distilled water with spirolactone (for antagonizing the effect of endogenous aldosterone) can be shown to have a higher density of sodium entry sites than control skins, as measured by ^{14}C–amiloride binding (Cuthbert and Shum, 1976).

Conversely the effect of salinity adaptation on *Bufo viridis in vivo* (Katz, 1975a) and *in vitro* (Katz, 1975b) is a reduction of the number of Na-selective sites at the outer border as measured as a decrease in the amiloride-sensitive component of Na transport. Effects of prolonged salinity adaptation in *R. temporaria* and in *R. esculenta* can also be ascribed to a decreased permeability to sodium of the outer barrier when measured *in vitro* (Hornby and Thomas, 1969). A direct comparison of the acute effects of external Na concentration on short-circuit current and potential *in vivo* and *in vitro* has shown similar relations (Brown, 1962).

There is a variety of evidence showing that calcium ions are able to control the permeability of membranes to monovalent ions (see e.g. Cuthbert and Wong, 1971). Thus the inhibitory action of amiloride on Na transport is dependent on the presence of Ca^{2+} in the external medium (Cuthbert and Wong, 1972). Receptors controlling Na ion translocation in biological membranes have been discussed recently. The Na channels in excitable tissues and epithelia are compared, and some similarities between the initial entry stages for sodium with both types of channels pointed out (Cuthbert, 1976).

Some intracellular organelles, microtubules and microfilaments, seem to play an important role for the permeability change induced by vasopressine (and its mediator cyclic AMP). When added to the serosal bathing solution of toad bladder antimitotic agents like colchicine, vinblastine, podophyllotoxin or cytochalasin B, inhibit the hydro-osmotic, but not the natriferic response of vasopressin or cyclic-AMP (Taylor and coworkers, 1973). There is a well defined microfilament network at the apical surface of the granular cells, and microtubules in close relationship to secretion granula, which may be responsible for the permeability change (Masur and coworkers, 1971, 1972). Taylor and coworkers suggest that microtubules and filaments participate in the hydro-osmotic action of vasopressin through their involvement in the mechanism of releasing the secretion granula.

The mitochondria-rich cell type found in toad bladder (Choi, 1963) are also present in frog skin (Farquhar and Palade, 1965). The mitochondria-rich cells in frog skin have recently been found to be concerned with organic base transport by a mechanism which is dependent on sodium (Ehrenfeld, Masoni and Garcia-Romeu, 1976a). Several functions have been ascribed to these cells, e.g. the moulting process in frog skin (Voûte and coworkers, 1969), aldosterone controlled osmoregulation in frog skin (Voûte, Hänni and Amman, 1972), and also suggested for the urinary toad bladder (DiBona, Civan and Leaf, 1969). Furthermore a specific binding of aldosterone to isolated mitochondria-rich cells from toad bladder has been found (Sapirstein and Scott, 1975).

With the growing knowledge of cellular and intracellular structure it may be expected that future research on transport phenomena will be much concerned with relations between function and structure at the cellular level.

REFERENCES

Aceves, J., D. Erlij, and C. Edwards (1968). Sodium transport across the isolated skin of *Ambystoma mexicanum*. *Biochim. Biophys. Acta*, **150**, 744–6.

Aceves, J., D. Erlij, and G. Whittembury (1970). The role of the urinary bladder in water balance of *Ambystoma mexicanum*. *Comp. Biochem. Physiol.*, **33**, 39–42.

Ackrill, P., R. Hornby, and S. Thomas (1969). Responses of *Rana temporaria* and *Rana esculenta* to prolonged exposure to a saline environment. *Comp. Biochem. Physiol.*, **38**, 1317–29.

Adolph, E. F. (1933). Exchanges of water in frog. *Biol. Rev.*, **8**, 224–40.

Alvarado, R. H., and T. H. Dietz (1970a). Effect of salt depletion on hydromineral balance of larval Ambystoma gracile. I. Ionic composition. *Comp. Biochem. Physiol.*, **33**, 85–92.

Alvarado, R. H., and T. H. Dietz (1970b). Effect of salt depletion on hsdromineral balance in larval Ambystoma gracile. II. Kinetics of ion exchange. *Comp. Biochem. Physiol.*, **33**, 93–110.

Alvarado, R. H., T. H. Dietz, and T. L. Mullen (1975b). Chloride transport across isolated skin of *Rana pipiens*. *Am. J. Physiol.*, **229**, 869–76.

Alvarado, R. H., and S. R. Johnson (1965). The effects of arginine vasotocin and ocytocin on sodium and water balance in *Ambystoma*. *Comp. Biochem. Physiol.*, **16**, 531–46.

Alvarado, R. H., and S. R. Johnson (1966). The effects of neurohypophysial hormones on water and sodium balance in larval and adult bullfrogs (*Rana catesbiana*). *Comp. Biochem. Physiol.*, **18**, 549–61.

Alvarado, R. H., and L. B. Kirschner (1963). Osmotic and ionic regulation in *Ambystoma tigrinum*. *Comp. Biochem. Physiol.*, **10**, 55–67.

Alvarado, R. H., and L. B. Kirschner (1964). Effect of aldosterone on sodium fluxes in larval Ambystoma tigrinum. *Nature*, **202**, 922–23.

Alvarado, R. H., A. M. Poole, and T. L. Mullen (1975a). Chloride balance in *Rana pipiens*. *Am. J. Physiol.*, **229**, 861–8.

Alvarado, R. H., and D. F. Stiffler (1970). The transepithelial potential difference in intact larval and adult salamanders. *Comp. Biochem. Physiol.*, **33**, 209–212.

Anderson, B., and H. H. Ussing (1957). Solvent drag on non-electrolytes during osmotic flow through isolated toad skin, and its response to antidiuretic hormone. *Acta Physiol. Scand.*, **39**, 228–39.

Andreoli, T. E. (1973). On the anatomy of amphotericin B-cholesterol pores in lipid membranes. *Kidney International*, **4**, 337–45.

Andreoli, T. E., and M. Monahan (1968). The interaction of polyene antibiotics with thin lipid membranes. *J. Gen. Physiol.*, **52**, 300–25.

Baba, W. I., A. F. Lant, A. J. Smith, M. M. Townshend, and G. M. Wilson (1968).

Pharmacological effects in animals and normal human subjects of the diuretic amiloride hydrochloride (MK-870). *Clin. Pharmacol. Therapeut.*, **9**, 318–27.

Baldwin, R. A. (1974). The water balance response of the pelvic 'patch' of *Bufo punctatus* and *Bufo boreas*. *Comp. Biochem. Physiol.*, **47A**, 1285–95.

Balinsky, J. B., and E. Baldwin (1961). The mode of excretion of ammonia and urea in *Xenopus laevis*. *J. Exp. Biol.*, **38**, 695–705.

Balinsky, J. B., M. M. Cragg, and E. Baldwin (1961). The adaptation of amphibian waste nitrogen excretion to dehydration. *Comp. Biochem. Physiol.*, **3**, 236–44.

Balinsky, J. B., E. L. Choritz, C. G. L. Coe, and G. S. van der Schans (1967a). Urea cycle enzymes and urea excretion during the development and metamorphosis of *Xenopus laevis*. *Comp. Biochem. Physiol.*, **22**, 53–57.

Balinsky, J. B., E. L. Choritz, C. G. Coe, and G. S. van Der Schans (1967b). Aminoacid metabolism and urea synthesis in naturally aestivating *Xenopus laevis*. *Comp. Biochem. Physiol.*, **22**, 59–68.

Balinsky, J. B., T. L. Coetzer, and F. J. Matheyse (1972b). The effect of thyroxine and hypertonic environment on the enzymes of the urea cycle in *Xenopus laevis*. *Comp. Biochem. Physiol.*, **43B**, 83–95.

Balinsky, J. B., S. E. Dicker, and B. Elliott (1972a). The effect of long-term adaptation to different levels of salinity on urea synthesis and tissue aminoacid concentrations in *Rana cancrivora*. *Comp. Biochem. Physiol.*, **43B**, 71–82.

Barrington, E. J. W., and C. B. Jørgensen (Eds) (1968). *Perspectives in Endocrinology. Hormones in the Lives of Lower Vertebrates*, Academic Press, New York.

Bastide, F., and S. Jard (1968). Actions de la noradrénaline et de l'ocytocine sur le transport actif de sodium et la perméabilité à l'eau de la peau de grenouille. Rôle du 3′,5′-AMP cyclique. *Biochim. Biophys. Acta*, **150**, 113–23.

Bentley, P. J. (1963). Neurohypophyseal function in amphibians, reptiles, and birds. In H. Heller (Ed.) *Symp. Zool. Soc. Lond.*, No. 9, 141–52.

Bentley, P. J. (1966). The physiology of the urinary bladder of amphibia. *Biol. Rev.*, **41**, 275–316.

Bentley, P. J. (1968). Amiloride: a potent inhibitor of sodium transport across the toad bladder. *J. Physiol.*, **195**, 317–30.

Bentley, P. J. (1969). Neurohypophyseal function in amphibia: Hormone activity in the plasma. *J. Endocr.*, **43**, 359–69.

Bentley, P. J. (1971). *Endocrines and Osmoregulation. A Comparative Account of the Regulation of Water and Salt in Vertebrates*, Springer-Verlag, Berlin, Heidelberg, New York.

Bentley, P. J. (1973). Osmoregulation in the aquatic urodeles *Amphiuma means* (the Congo eel) and *Siren lacertina* (the mud eel). Effects of vasotocin. *Gen. Comp. Endocr.*, **20**, 386–391.

Bentley, P. J., and H. Heller (1964). The actions of neurohypophyseal hormones on the water and sodium metabolism of urodele amphibians. *J. Physiol.*, **171**, 434–53.

Bentley, P. J., and H. Heller (1965). The water retaining action of vasotocin on the fire salamander (*Salamandra maculosa*): The role of the urinary bladder. *J. Physiol.*, **181**, 124–9.

Bentley, P. J., A. K. Less, and A. R. Main (1958). Comparison of dehydration and hydration of the two genera of frogs (Heleioporus and Neobatrachus) that live in areas of varying aridity. *J. Exp. Biol.*, **35**, 677–84.

Bentley, P. J., and K. Schmidt-Nielsen (1971). Acute effects of sea-water on frogs (*Rana pipiens*). *Comp. Biochem. Physiol.*, **40A**, 547–8.

Biber, T. U. L. (1971). Effect of changes in transepithelial transport on the uptake of sodium across the outer surface of the frog skin. *J. Gen. Physiol.*, **58**, 131–44.

Biber, T. U. L., J. Aceves, and L. Mandel (1972a). Potassium uptake across serosal

surface of isolated frog skin epithelium. *Am. J. Physiol.*, **222**, 1366–73.
Biber, T. U. L., R. A. Chez, and P. F. Curran (1966). Na-transport across frog skin at low external sodium concentrations. *J. Gen. Physiol.*, **49**, 1161–1176.
Biber, T. U. L., L. Cruz, and P. F. Curran (1972b). Sodium influx at the outer surface of the frog skin. Evaluation of different extracellular markers. *J. Membr. Biol.*, **7**, 365–376.
Biber, T. U. L., and P. F. Curran (1970). Direct measurement of uptake of sodium at the outer surface of the frog skin. *J. Gen. Physiol.*, **56**, 83–99.
Biber, T. U. L., and T. L. Mullen (1976). Saturation kinetics of Na-efflux across isolated frog skins. *Am. J. Physiol.*, **231**, 995–1001.
Biber, T. U. L., and M. L. Sanders (1973). Influence of transepithelial potential differences on the Na-uptake at the outer surface of the isolated frog skin. *J. Gen. Physiol.*, **61**, 529–51.
Bonting, S. L., and L. L. Caravaggio (1963). Studies on the Na/K-activated ATPase. VI. Correlation of enzyme activity with cation flux in six tissues. *Arch. Biochim. Biophys.*, **101**, 37–46.
Bricker, N. S., T. U. L. Biber, and H. H. Ussing (1963). Exposure of the isolated frog skin to high potassium concentrations at the internal surface. I. Bioelectric phenomena and sodium transport. *J. Clin. Invest.*, **42**, 88–99.
Brown, A. C. (1962). Current and potential of frog skin *in vivo* and *in vitro*. *J. Cell. Comp. Physiol.*, **60**, 263–70.
Brunn, F. (1921). Beitrag zum kenntniss der wirkung von hypophysenextracten auf den wasserhaushalt des frosches. *Z. Ges. Exp. Med.*, **25**, 170.
Bruus, K., P. Kristensen, and E. H. Larsen (1976). Pathways for chloride and sodium transport across toad skin. *Acta Physiol. Scand.*, **97**, 31–47.
Capraro, V., and G. Bernini (1952). Mechanism of action of extracts of post-hypophysis on water transport through the skin of the frog. *Nature*, **169**, 454.
Cerejido, M., F. C. Herrera, J. W. Flanigan, and P. F. Curran (1964). The influence of Na-concentration on Na-transport across frog skin. *J. Gen. Physiol.*, **47**, 879–93.
Cerejido, M. and C. A. Rotunno (1967). Transport and distribution of sodium across frog skin. *J. Physiol.*, **190**, 481–97.
Cerejido, M. and C. A. Rotunno (1968). Fluxes and distribution of sodium in frog skin. A new model. *J. Physiol.*, **51**, 280–98.
Chance, B., and G. R. Williams (1956). The respiratory chain and oxidative phosphorylation. *Adv. Enzymol.*, **17**, 65–134.
Choi, J. K. (1963). The fine structure of the urinary bladder of the toad, *Bufo marinus*. *J. Cell. Biol.*, **16**, 53–72.
Christensen, C. U. (1974). Adaptations in the water economy of some anuran amphibia. *Comp. Biochem. Physiol.*, **47A**, 1035–49.
Cirne, B., and B. Lindemann (1975). Passive nature of Li- and Na-transport through the Na-selective membrane of frog skin. *Abstr. P-121, 5th Int. Biophys. Congr., Copenhagen 1975*.
Claussen, D. L. (1969). Studies on water loss and rehydration in anurans. *Physiol. Zool.*, **42**, 1–14.
Cohen, P. P. (1966). Biochemical aspects of metamorphosis: Transition from ammonotelism to ureotelism. *Harvey Lecture Ser.*, **60**, 119–54.
Cohen, P. P. (1970). Biochemical differentiation during amphibian metamorphosis. *Science*, **168**, 533–43.
Cohen, P. P. and G. W. Brown (1960). In M. Florkin and H. S. Mason (Eds) *Comparative Biochemistry*, Vol. 2, Academic Press, New York. pp. 161–244.
Colley, L., W. C. Rowe, A. K. Huggins, A. B. Elliott, and S. E. Dicker (1972). Effect

of short term external salinity changes on ornithine–urea cycle enzymes in *Rana cancrivora. Comp. Biochem. Physiol.*, **41B**, 307–22.
Coplon, N. S., and R. H. Maffly (1972). The effect of ouabain on sodium transport and metabolism of the toad bladder. *Biocheim. Biophys. Acta*, **282**, 250–4.
Coviello, A. (1970). Natriferic and hydro-osmotic effect of angiotensin II in the toad *Bufo paracnemis. Acta Physiol. Latinoam.*, **20**, 349–58.
Crabbé, J. (1961). Stimulation of active transport across the isolated toad bladder after injection of aldosterone to the animal. *Endocrinology*, **69**, 673–82.
Crabbé, J. (1972). Aldosterone: Mechanism of action on isolated sodium-transporting epithelia. *J. Steroid. Biochem.*, **3**, 557–66.
Crabbé, J., and E. N. Ehrlich (1968). Amiloride and the mode of action of aldosterone on Na-transport across toad bladder and skin. *Pflügers Arch.*, **304**, 284–96.
Cragg, M. M., J. B. Balinsky, and E. Baldwin (1961). A comparative study of nitrogen excretion in some amphibia and reptiles. *Comp. Biochem. Physiol.*, **3**, 227–35.
Curran, P. F., and M. Cerejido (1965). K-fluxes in frog skin. *J. Gen. Physiol.*, **48**, 1011–33.
Cuthbert, A. W. (1976). Receptors controlling sodium ion translocation in biological membranes. *J. Pharm. Pharmacol.*, **28**, 383–88.
Cuthbert, A. W., and W. K. Shum (1974). Amiloride and the sodium channel. *Arch. Pharmacol.*, **281**, 261–9.
Cuthbert, A. W., and W. K. Shum (1976). Induction of transporting sites in a sodium transporting epithelium. *J. Physiol.*, **260**, 223–35.
Cuthbert, A. W., and P. Y. D. Wong (1971). The effect of metal ions and antidiuretic hormone on oxygen consumption in toad bladder. *J. Physiol.*, **219**, 39–56.
Cuthbert, A. W., and P. Y. D. Wong (1972). The role of calcium ions in the interaction of amiloride with membrane receptors. *Molec. Pharmacol.*, **8**, 222–9.
Dainty, J. (1963). Water relations of plant cells. *Adv. Bot. Res.*, **1**, 279–326.
Dainty, J., and C. R. House (1966). Unstirred layers in frog skin. *J. Physiol.*, **182**, 66–78.
Davies, H. E. F., D. G. Martin, and G. W. G. Sharp (1968). Differences in physiological characteristics of bladders of toads from different geographical sources. *Biochim. Biophys. Acta*, **150**, 315–8.
De Sousa, R. C. (1975). Mécanismes de transport de l'eau et du sodium par les cellules des epithelia d'amphibiens et du tubule rénale isolé. *J. Physiologie*, **71**, 5A–71A.
Deyrup, I. (1964). Water balance and kidney. In J. A. Moore (Ed.), *Physiology of the Amphibia*, Academic Press, New York. pp. 251–328.
DiBona, D. R., M. M. Civan, and A. Leaf (1969). The cellular specificity of the effect of vasopressin on toad urinary bladder. *J. Membr. Biol.*, **1**, 79–91.
Dicker, S. E., and A. B. Elliott (1967). Water uptake by *Bufo melanostictus*, as affected by osmotic gradients, vasopressin and temperature. *J. Physiol.*, **190**, 359–70.
Dicker, S. E., and A. B. Elliott (1970). Water uptake by the crab-eating frog, *Rana cancrivora*, as affected by osmotic gradients and by neurohypophyseal hormones. *J. Physiol.*, **207**, 119–32.
Dietz, T., L. B. Kirschner, and D. Porter (1967). The roles of sodium transport and anion permeability in generating transepithelial potential differences in larval salamanders. *J. Exp. Biol.*, **46**, 85–96.
Du Bois Reymond, E. (1848). *Untersuchungen über die Tierische Elektricität*. Berlin 1848.
Edelman, I. S. and G. M. Fimognari (1969). On the biochemical mechanism of action of aldosterone. *Recent Prog. Horm. Res.*, **24**, 1–44.
Ehrenfeld, J., A. Masoni, and F. Garcia-Romeu (1976a). Mitochondria-rich cells of

frog skin in transport mechanisms: Morphological and kinetic studies on transepithelial excretion of methylene blue. *Am. J. Physiol.*, **231**, 120–6.

Ehrenfeld, J., D. J. Nelson, and B. Lindemann (1976b). Volume changes of epithelial cells of frog skin on impalement with a microelectrode. *Pflügers Arch. Ges. Physiol.*, **365**, R32.

Ehrlich, E. N., and J. Crabbé (1968). The mechanism of action of amipramizide. *Pflügers Arch. Ges Physiol.*, **302**, 79–96.

Eigler, J., J. Kelter, and E. Renner (1967). Wirkungscharacteristica eines neuen Acylguanidins-Amilorido-HCl (MK870) an der isolierten haut von amphibien. *Klin. Wschr.*, **45**, 737–8.

Engbaek, L., and T. Hoshiko (1957). Electrical potential gradients through frog skin. *Acta Physiol. Scand.*, **39**, 348–355.

Erlij, D. (1971). Salt transport across isolated frog skin. *Phil. Trans. R. Soc. Lond. B*, **262**, 153–61.

Erlij, D., and A. Martínez-Palomo (1972). Opening of the tight junctions in frog skin by hypertonic urea solutions. *J. Membr. Biol.*, **9**, 229–40.

Erlij, D., and M. W. Smith (1973). Sodium uptake by frog skin and its modification by inhibitors of transepithelial Na-transport. *J. Physiol.*, **228**, 221–39.

Essig, A. (1965). Active sodium transport in the toad bladder despite removal of serosal potassium. *Am. J. Physiol.*, **208**, 401–6.

Essig, A., and A. Leaf (1963). The role of potassium in active transport of sodium by the toad bladder. *J. Gen. Physiol.*, **46**, 505–15.

Ewer, R. F. (1952). The effect of pituitrin on fluid distribution in *Bufo regularis* Reuss. *J. Exp. Biol.*, **29**, 173–7.

Fanelli, G. M., and L. Goldstein (1964). Ammonium excretion in the neotenous newt *Necturus maculosus* (Refinesque). *Comp. Biochem. Physiol.*, **13**, 193–204.

Farquhar, M. G., and G. E. Palade (1964). Functional organization of amphibian skin. *Proc. Nat. Acad. Sci.*, **51**, 569–77.

Ferreira, K. G., H. G. Ferreira, and V. L. Lew (1976). On the amount of $(Na^+ + K^+)$-ATPase available for transepithelial sodium ion transport in the amphibian skin. *Biochim. Biophys. Acta*, **448**, 185–8.

Finkelstein, A. (1976). Nature of the water permeability increase induced by antidiuretic hormone (ADH) in toad urinary bladder and related tissues. *J. Gen. Physiol.*, **68**, 137–43.

Finn, A. L., and H. Nellans (1972). The kinetics and distribution of potassium in toad bladder. *J. Membr. Biol.*, **8**, 189–203.

Forster, R. P. (1938). The use of inulin and creatinine as glomerular measuring substances in the frog. *J. Cell. Comp. Physiol.*, **12**, 213–22.

Forster, R. P. (1954). Active cellular transport of urea by frog renal tubules. *Am. J. Physiol.*, **179**, 372–7.

Forster, R. P. (1970). Early history of renal clearance studies. Active tubular transport of urea and its role in environmental physiology. In B. Schmidt-Nielsen (Ed.), *Urea and the Kidney*, Excerpta Medica Foundation, Amsterdam, pp. 225–37.

Forster, R. P., B. Scmidt-Nielsen, and L. Goldstein (1963). Relation of renal tubular transport of urea to its biosynthesis in metamorphosing tadpoles. *J. Cell. Comp. Physiol.*, **61**, 239–47.

Francis, L. W. (1933). Output of electrical energy by frog skin. *Nature*, **131**, 805.

Franz, T. J., and J. T. van Bruggen (1967). Hyperosmolarity and the net transport of non-electrolytes in frog skin. *J. Gen. Physiol.*, **50**, 933–49.

Frazier, H. S. (1962). The electrical potential profile of the isolated toad bladder. *J. Gen. Physiol.*, **45**, 515–28.

Frazier, L. W. (1974). Interrelationship of H^+ excretion and Na^+ reabsorption in the toad urinary bladder. *J. Membr. Biol.*, **19**, 267–76.

Frazier, H. S., and A. Leaf (1963). The electrical characteristics of active sodium transport in the toad bladder. *J. Gen Physiol.*, **46**, 491–503.
Frazier, H. S., E. F. Dempsey, and A. Leaf (1962). Movement of sodium across the mucosal surface of the isolated toad bladder and its modification by vasopressin. *J. Gen. Physiol.*, **45**, 529–43.
Frieden, E. (1968). Biochemistry of amphibian metamorphosis. In *Metamorphosis: A Problem in Developmental Biology*, Appleton-Century Crofts, New York. pp. 349–98.
Fuchs, W., U. Gebhardt, and B. Lindemann (1972). Delayed voltage responses to fast changes of $(Na)_o$ at the outer surface of frog skin epithelium. In F. Kreutzer and J. F. G. Slegers (Eds), *Passive Permeability of Cell Membranes*, Plenum Press, New York. pp. 483–98.
Fuchs, W., E. H. Larsen, and B. Lindeman (1977). Current–voltage curve of sodium channels and concentration dependence of sodium permeability in the frog skin. *J. Physiol.*, **267**, 1–30.
Fukuda, T. R. (1942). Über die Bedingungen Für das Zustandekommen des Asymmetriepotentials der Froschhaut. *Jap. J. Med. Sci. III*, **8**, 123.
Galeotti, G. (1904). Über die elektromotorische Kräfte welche an der Oberfläche tierische Membrane bei der Berührung mit verschiedenen Elektrolyten zustande kommen. *Z. Physikal. Chemie*, **49**, 542.
Garcia-Romeu, F. (1971). Anionic and cationic exchange mechanisms in the skin of anurans, with special reference to Leptodactylidae in vivo. *Phil. Trans. R. Soc. Lond. B*, **262**, 163–74.
Garcia-Romeu, F., and J. Ehrenfeld (1972). The role of ionic exchangers and pumps in transepithelial sodium and chloride transport across frog skin. In L. Bolis (Ed.), *Roles of Membranes in Secretory Processes*, Elsevier, Amsterdam. pp. 264–83.
Garcia-Romeu, F., and J. Ehrenfeld (1975a). *In vivo* Na^+- and Cl^--independent transport across the skin of *Rana esculenta*. *Am. J. Physiol.*, **228**, 839–44.
Garcia-Romeu, F., and J. Ehrenfeld (1975b). Chloride transport through the non short-circuited isolated skin of *Rana esculenta*. *Am. J. Physiol.*, **228**, 845–9.
Garcia-Romeu, F., A. Salibían, and S. Pezzani-Hernandez (1969). The nature of the *in vivo* sodium and chloride uptake mechanisms through the epithelium of the Chilean frog, *Calyptocephallea gayi*. *J. Gen. Physiol.*, **53**, 816–35.
Gebhardt, U., W. Fuchs, and B. Lindemann (1972). Resistance response of frog skin to brief and long lasting changes of $(Na)_o$ and $(K)_o$. In L. Bollis (Ed.), *Role of Membranes in Secretory Processes*, Elsevier, Amsterdam. pp. 284–300.
Goldman, D. E. (1943). Potential, impedance and rectification in membranes. *J. Gen. Physiol.*, **27**, 37–60.
Gordon, M. S. (1962). Osmotic regulation in the green toad (*Bufo viridis*). *J. Exp. Biol.*, **39**, 261–70.
Gordon, M. S., and V. A. Tucker (1965). Osmotic regulation in the tadpoles of the crab-eating frog (*Rana cancrivora*). *J. Exp. Biol.*, **42**, 437–45.
Gordon, M. S., and V. A. Tucker (1968). Further observations on the physiology of salinity adaptation in the crab-eating frog (*Rana cancrivora*). *J. Exp. Biol.*, **49**, 185–93.
Gordon, M. S., K. Schmidt-Nielsen, and H. M. Kelly (1961). Osmotic regulation in the crab-eating frog (*Rana cancrivora*). *J. Exp. Biol.*, **38**, 659–78.
Grantham, J. J., F. E. Cuppage, and D. Fanestil (1971). Direct observation of toad bladder response to vasopressin. *J. Cell. Biol.*, **48**, 695–9.
Greenwald, L. (1971). Sodium balance in the leopard frog (*Rana pipiens*). *Physiol. Zool.*, **44**, 149–61.
Greenwald, L. (1972). Sodium balance in amphibians from different habitats. *Physiol. Zool.*, **45**, 229–37.

Handler, J. S., A. S. Preston, and J. Orloff (1972). Effect of ADH, aldosterone, ouabain, and amiloride on toad bladder epithelial cells. *Am. J. Physiol.*, **222**, 1071–4.

Hansen, H. H., and K. Zerahn (1964). Concentration of lithium, sodium and potassium in the epithelial cells of the isolated frog skin during active transport of lithium. *Acta Physiol. Scand.*, **60**, 189–96.

Hashida, K. (1922). Untersuchungen über das elektromotorische Verhalten der Froschhaut. I. Die Abhängigkeit des elektromotorischen Verhaltens der Froschhaut von den ableitenden Flüssigkeiten. *J. Biochem.*, **1**, 21–65.

Hays, R. M. (1972a). Independent pathways for water and solute movements across the cell membrane. *J. Membr. Biol.*, **10**, 367–71.

Hays, R. M. (1972b). The movement of water across vasopressin-sensitive epithelia. In F. Bronner and A. Kleinzeller (Eds) *Current Topics in Membranes and Transport*, Academic Press, New York.

Hays, R. M., and N. Franki (1970). The role of water diffusion in the action of vasopressin. *J. Membr. Biol.*, **2**, 263–76.

Heatwole, H., F. Torres, S. B. de Austin, and A. Heatwole (1969). Evaporative water loss in frog. *Comp. Biochem. Physiol.*, **28**, 245–69.

Heckmann, K. D., B. Lindemann, and J. Schnackenberg (1972). Current–voltage curves of porous membranes in the presence of pore-blocking ions. *Biophys. J.*, **12**, 683–702.

Heller, H. (1965). Osmoregulation in amphibia. *Arch. Anat. Microsc.*, **54**, 471–90.

Herrera, F. C. (1966). Action of ouabain on sodium transport in the toad urinary bladder. *Am. J. Physiol.*, **210**, 980–6.

Hevesey, G., E. Hofer, and A. Krogh (1935). The permeability of the skin of frogs to water as determined by D_2O and H_2O. *Skand. Arch. Physiol.*, **72**, 199–214.

Hodgkin, A. L., and B. Katz (1949). The effect of sodium ions on the electrical activity of the giant axon of the squid. *J. Physiol.*, **108**, 37–77.

Hornby, R., and S. Thomas (1969). Effect of prolonged saline exposure on sodium transport across frog skin. *J. Physiol.*, **200**, 321–344.

House, C. R. (1974). *Water transport in cells and tissues*. Edward Arnold, The Camelot Press, London.

Huf, E. (1935). Versuche über den Zusammenhang zwischen Stoffwechsel, Potentialbildung und Funktion der Froschhaut. *Pflügers Archs. Ges. Physiol.*, **235**, 655–73.

Janáček, K., R. Rybová, and M. Slaviková (1972). Sodium–potassium pump and cell volume regulation in frog bladder. *Biochim. Biophys. Acta*, **288**, 221–4.

Jard, S. (1966). Étude des effets de la vasotocine sur l'excretion de l'eau et des électrolytes par le rein de la granouille *Rana esculenta* L.: Analyse a l'aide d'analogues de l'hormone naturelle des caracteres structuraux requis pour son activité biologique. *J. Physiologies, extrait*, 124 pp.

Jard, S., and F. Morel (1963). Actions of vasotocin and some of its analogues on salt and water excretion. *Am. J. Physiol.*, **204**, 222–6.

Jard, S. J. Bourguet, P. Favard, and N. Carasso (1971). The role of intercellular channels in transepithelial transport of water and sodium in the frog urinary bladder. *J. Membr. Biol.*, **4**, 124–47.

Johnston, C. J., O. Davis, F. S. Wright, and S. S. Howards (1967). Effects of rennin and ACTH on adrenal steroid production in the American bullfrog. *Am. J. Physiol.*, **213**, 393–9.

Jørgensen, C. B. (1947). Influence of adenohypophysectomy on the transfer of salt across frog skin. *Nature*, **160**, 872.

Jørgensen, C. B. (1949). Permeability of the amphibian skin. II. Effect of moulting of

the skin of anurans on the permeability to water and electrolytes. *Acta Physiol. Scand.*, **18**, 171–80.
Jørgensen, C. B. (1950). The influence of salt loss on the osmotic regulation in anurans. *Acta Physiol. Scand.*, **20**, 56–61.
Jørgensen, C. B., and L. O. Larsen (1963). Effect of corticotrophin and growth hormone on survival in hypophysectomized toads. *Proc. Soc. Exp. Biol.* (*NY*), **113**, 94–6.
Jørgensen, C. B., and Larsen, L. O. (1967). Neuroendocrine mechanisms in lower vertebrates. In L. Martini and W. F. Ganong (Eds) *Neuroendocrinology*, Vol. 2, Academic Press, New York. pp. 485–528.
Jørgensen, C. B., H. Levi, and H. H. Ussing (1946). On the influence of the neurohypophyseal principles on the sodium metabolism in the axolotl (*Ambystoma mexicanum*). *Acta Physiol. Scand.*, **12**, 350–71.
Jørgensen, C. B., H. Levi, and K. Zerahn (1954). On active uptake of sodium and chloride ions in anurans. *Acta Physiol. Scand.*, **30**, 178–90.
Jørgensen, C. B., and P. Rosenkilde (1957). Chloride balance in hypophysectomized frogs. *Endocrinology*, **60**, 219–24.
Kalman, S. M., and H. H. Ussing (1955). Active sodium uptake by the toad and its response to the antidiuretic hormone. *J. Gen. Physiol.*, **38**, 361–70.
Katz, U. (1975a). NaCl adaptation in *Rana ridibunda* and a comparison with the euryhaline toad *Bufo viridis*. *J. Exp. Biol.*, **63**, 763–73.
Katz, U. (1975b). Salt induced changes in sodium transport across the skin of the euryhaline toad, *Bufo viridis*. *J. Physiol.*, **247**, 537–50.
Katzin, L. I. (1939). The ionic permeability of frog skin determined with the aid of radioactive indicators. *Biol. Bull.*, **77**, 302–3.
Kerstetter, T. H., and L. B. Kirschner (1971). The role of hypothalamoneurohypophyseal system in maintaining hydromineral balance in larval salamanders (*Ambystoma tigrinum*). *Comp. Biochem. Physiol.*, **40A**, 373–84.
Kirschner, L. B. (1955). On the mechanism of active sodium transport across the frog skin. *J. Cell. Comp. Physiol.*, **45**, 61–87.
Kirschner, L. B., L. Greenwald, and T. Kerstetter (1973). Effect of amiloride on sodium transport across body surfaces of freshwater animals. *Am. J. Physiol.*, **224**, 832–7.
Kirschner, L. B., T. Kerstetter, D. Porter, and R. H. Alvarado (1971). Adaptation of larval *Ambystoma tigrinum* to concentrated environments. *Am. J. Physiol.*, **220**, 1814–9.
Koefoed-Johnsen, V. (1954). Unpublished work, quoted by Ussing (1954). In A. T. Clarke and D. Nachmansohn (Eds) *Ion Transport Across Membranes*, Academic Press, New York. pp. 3–22.
Koefoed-Johnsen, V. (1957). The effect of g-strophanthin (ouabain) on the active transport of sodium through the isolated frog skin. *Acta Physiol. Scand.*, **42**, Suppl. 145, 87.
Koefoed-Johnsen, V. (1964). Unpublished work, quoted by Ussing and Windhager. In J. de Graeff and B. Leijnse (Eds) *Water and Electrolyte Metabolism II*, Elsevier, Amsterdam. pp. 3–19.
Koefoed-Johnsen, V., H. Levi, and H. H. Ussing (1952a). The mode of passage of chloride ions through the isolated frog skin. *Acta Physiol. Scand.*, **25**, 150–63.
Koefoed-Johnsen, V., I. Lyon, and H. H. Ussing (1973). Effect of Cu ion on permeability properties of isolated frog skin (*Rana temporaria*). *Acta Physiol. Scand. Suppl.*, **396**, 102.
Koefoed-Johnsen, V., and H. H. Ussing (1949). The influence of the corticotropic

hormone from ox on the active salt uptake in the axolotl. *Acta Physiol. Scand.*, **17**, 38–43.

Koefoed-Johnsen, V., and H. H. Ussing (1953). The contribution of diffusion and flow to the passage of D_2O through living membranes. *Acta Physiol. Scand.*, **28**, 60–76.

Koefoed-Johnsen, V., and H. H. Ussing (1958). The nature of the frog skin potential. *Acta Physiol. Scand.*, **42**, 298–308.

Koefoed-Johnsen, V., and H. H. Ussing (1971). Ion transport through biological membranes. *Proc. 8th Colloq. Int. Potash Inst.* International Potash Institute, Berne,

Koefoed-Johnsen, V., and H. H. Ussing (1974). Transport pathways in frog skin and their modification by copper ions. In N. A. Thorn and O. H. Petersen (Eds) *Secretory Mechanisms of Exocrine Glands*, Munksgård, Copenhagen, Academic Press, New York. pp. 411–22.

Koefoed-Johnsen, V., H. H. Ussing, and K. Zerahn (1952b). The origin of the short-circuited current in the adrenaline stimulated frog skin. *Acta Physiol. Scand.*, **27**, 38–48.

Kristensen, P. (1972). Chloride transport across isolated frog skin. *Acta Physiol. Scand.*, **84**, 338–46.

Krogh, A. (1937). Osmotic regulation in the frog (*Rana esculenta*) by active absorption of chloride ions. *Skand. Arch. Physiol.*, **76**, 60–74.

Krogh, A. (1938). The active absorption of ions in some freshwater animals. *Z. Vergleich, Physiol.*, **25**, 335–50.

Krogh, A. (1939). *Osmotic Regulation in Aquatic Animals*, Cambridge University Press, Cambridge.

Krough, A. (1946). The active and passive exchanges of inorganic ions through the surfaces of living cells and through living membranes generally. *Croon Lecture, Proc. R. Soc. B*, **133**, 140–200.

Lang, M. A., R. R. Caplan, and A. Essig (1977). Sodium transport and oxygen consumption in toad bladder. A thermodynamic approach. *Biochim. Biophys. Acta*, **464**, 557–82.

Larsen, E. H. (1971). The relative contributions of sodium and chloride ions to the conductance of toad skin in relation to shedding of the stratum corneum. *Acta Physiol. Scand.*, **81**, 254–63.

Leaf, A. (1955). Ion transport by the isolated bladder of the toad. *Proc. 3rd Int. Congr. on Biochemistry, Brussels.*

Leaf, A. (1965). Transepithelial transport and its hormonal control in the toad bladder. *Ergebn. Physiol.*, **56**, 216–63.

Leaf, A., and R. M. Hays (1962). Permeability of the isolated toad bladder to solutes and its modification by vasopressin. *J. Gen. Physiol.*, **45**, 921–32.

Leaf, A., and A. Renshaw (1957). Ion transport and respiration of isolated frog skin. *Biochem. J.*, **65**, 82–90.

Leaf, A., and G. W. G. Sharp (1971). The stimulation of sodium transport by aldosterone. *Phil. Trans. R. Soc.*, **262**, 323–32.

Leaf, A., J. Anderson, and L. B. Page (1958). Active sodium transport by the isolated toad bladder. *J. Gen. Physiol.*, **41**, 657–68.

Leaf, A., L. B. Page and J. Anderson (1959). Respiration and active sodium transport of isolated toad bladder. *J. Biol. Chem.*, **234**, 1625–8.

Leblanc, G. (1972). The mechanism of lithium accumulation in the isolated frog skin epithelium. *Pflügers Arch.*, **337**, 1–18.

Levinsky, N. G., and W. H. Sawyer (1953). Significance of the neurohypophysis in the regulation of fluid balance in the frog. *Proc. Soc. Exp. Biol. Med.*, **82**, 272–4.

Lichstenstein, N. S., and A. Leaf (1965). Effect of amphotericin B on the permeability of the toad bladder. *J. Clin. Invest.*, **44**, 1328–42.

Lindemann, B. (1975). Impalements artifacts in microelectrode recordings of epithelial membrane potentials. *Biophys. J.*, **15**, 1161–4.
Linderholm, H. (1952). Active transport of ions through frog skin with special reference to the action of certain diuretics. *Acta Physiol. Scand.*, **27**, *Suppl.* 97.
Lindley, D. B. (1969). Nerve stimulation and electrical properties of frog skin. *J. Gen. Physiol.*, **53**, 427–49.
Lindley, D. B., T. Hoshiko, and D. E. Leb (1964). Effects of D_2O and osmotic gradients on potential and resistance of the isolated frog skin. *J. Gen. Physiol.*, **47**, 773–93.
Lipton, P., and I. S. Edelman (1971). Effect of aldosterone and vasopressin on electrolytes of toad bladder epithelial cells. *Am. J. Physiol.*, **221**, 733–41.
Loewenstein, W. R. (1966). Permeability of membrane junctions. *Ann. NY Acad. Sci.*, **137**, 441–72.
Loewenstein, W. R., and Y. Kanno (1964). Studies on an epithelial (gland) cell junction. I. Modifications of surface membrane permeability. *J. Cell Biol.*, **22**, 565–86.
Lund, E. J., and P. Stapp (1947). Use of the iodine coulometer in measurement of bioelectrical energy and the efficiency of the bioelectrical process. In E. J. Lund (Ed.), *Bioelectric Fields and Growth*, University of Texas Press, Austin. pp. 235–54.
Machin, J. (1969). Water permeability of skin of Bufo. *Am. J. Physiol.*, **216**, 1562–8.
Macknight, A. D. C., M. M. Civan, and A. Leaf (1975a). The sodium transport pool in toad urinary bladder epithelial cells. *J. Membr. Biol.*, **20**, 365–86.
Macknight, A. D. C., M. M. Civan, and A. Leaf (1975b). Some effects of ouabain on cellular ions and water in epithelial cells of toad urinary bladder. *J. Membr. Biol.*, **20**, 387–401.
Macknight, A. D. G., A. Leaf, and M. M. Civan (1971). Effects of vasopressin on the water and ionic composition of toad bladder epithelial cells. *J. Membr. Biol.*, **6**, 127–37.
Maetz, J. (1963). Physiological aspects of neurohypophyseal functions in fishes with some reference to the amphibia. In H. Heller (Ed.) *Symp. Zool. Soc. No. 9*. pp. 107–40.
Maetz, J. (1968). Salt and water metabolism. In E. J. W. Barrington and C. B. Jørgensen (Eds), *Perspectives in Endocrinology*, Academic Press, New York.
Maetz, J., S. Jard, and F. Morel (1958). Action de l'aldosterone sur le transport actif de sodium de la peau de grenouille. *Compt. Rend.*, **247**, 516–8.
Maetz, J., F. Morel, and B. Race (1959). Mise en évidence dans neurohypophyse de *Rana esculenta* L. d'un factor hormonal nouveau stimulant de transport actif de sodium. *Biochim. Biophys. Acta*, **36**, 317–26.
Maffly, R. H., R. M. Hays, E. Lamdin, and A. Leaf (1960). The effect of neurohypophyseal hormones on the permeability of the toad bladder to urea. *J. Clin. Invest.*, **39**, 630–41.
Main, A. R., and P. J. Bentley (1964). Comparison of dehydration and hydration of burrowing desert frogs and tree frogs of the genus Hyla. *Ecology*, **45**, 379–82.
Mandel, L. J., and P. F. Curran (1972a). Response of the frog skin to steady-state voltage clamping. I. The shunt path. *J. Gen. Physiol.*, **59**, 503–18.
Mandel, L. J., and P. F. Curran (1972b). Chloride flux via a shunt pathway in frog skin: Apparent exchange diffusion. *Biochim. Biophys. Acta*, **282**, 258–64.
Maren, T. H. (1967). Carbonic anhydrase: chemistry, physiology and inhibition. *Physiol. Rev.*, **47**, 595–781.
Martin, D. W., and P. F. Curran (1966). Reverse potentials in frog skin. II. Active transport of chloride. *J. Cell. Comp. Physiol.*, **67**, 367–74.
Martínez-Palomo, A., D. Erlij, and H. Bracho (1971). Localization of permeability barriers in the frog skin epithelium. *J. Cell Biol.*, **50**, 277–87.

Masur, S. K., E. Holtzman, E. Schwartz, and R. Walther (1971). Correlation between pinocytosis and hydroosmosis induced by neurohypophyseal hormones and mediated by adenosine 3',5'-cyclic monophosphate. *J. Cell Biol.*, **49**, 582–94.

Masur, S. K., E. Holtzman, and R. Walther (1972). Hormone-stimulated exocytosis in the toad urinary bladder. Some possible implications for turn over of surface membranes. *J. Cell Biol.*, **52**, 211–9.

Mayhew, W. W. (1965). Adaptations of the amphibian, *Scaphiopus couchi*, to desert conditions. *Am. Midl. Naturalist*, **74**, 95–109.

McClanahan, L. (1967). Adaptations of the spadefoot toad, *Scaphiopus couchi*, to desert environments. *Comp. Biochem. Physiol.*, **20**, 73–99.

McClanahan, L., and R. Baldwin (1969). Rate of water uptake through the integument of the desert toad, *Bufo punctatus*. *Comp. Biochem. Physiol.*, **28**, 381–9.

McBean, R. L., and L. Goldstein (1967). Ornithine–urea cycle activity in *Xenopus laevis*: adaptation to saline. *Science*, **157**, 931–2.

McBean, R. L., and L. Goldstein (1970a). Renal function during osmotic stress in the aquatic toad *Xenopus laevis*. *Am. J. Physiol.*, **219**, 1115–23.

McBean, R. L., and L. Goldstein (1970b). Accelerated synthesis of urea in *Xenopus laevis* during osmotic stress. *Am. J. Physiol.*, **219**, 1124–30.

McRobbie, E. A. C., and H. H. Ussing (1961). Osmotic behaviour of the epithelial cells of frog skin. *Acta Physiol. Scand.*, **53**, 348–65.

Mendoza, S. A. (1972). Potassium dependence of base-line and ADH-stimulated sodium transport in toad bladder. *Am. J. Physiol.*, **223**, 120–4.

Mendoza, S. A. (1973). Sodium dependence of base-line and ADH-stimulated short-circuit current in toad bladder. *Am. J. Physiol.*, **225**, 476–80.

Mendoza, S. A., J. S. Handler, and J. Orloff (1970). Effect of inhibitors of sodium transport on response of toad bladder to ADH and cyclic AMP. *Am. J. Physiol.*, **219**, 1440–5.

Middler, S. A., C. R. Kleeman, and E. Edwards (1967). Neurohypophyseal function in the toad, *Bufo marinus*. *Gen. Comp. Endocrinol.*, **9**, 38–48.

Middler, S. A., C. R. Kleeman, and E. Edwards (1968). The role of the urinary bladder in salt and water metabolism of the toad, *Bufo marinus*. *Comp. Biochem. Physiol.*, **26**, 57–68.

Middler, S. A., C. R. Kleeman, E. Edwards, and D. Brody (1969). Effect of adenohypophysectomy on salt and water metabolism of the toad, *Bufo marinus*, studies in hormonal replacement. *Gen. Comp. Endocrinol.*, **12**, 290–304.

Mirsky, A. E., and V. Allfrey (1958). The role of the cell nucleus in development. In W. P. McElroy and B. Glass (Eds), *A Symposium on the Chemical Basis of Development*, The Johns Hopkins Press, Baltimore. pp. 94–102.

Moore, R. D., and G. A. Morill (1976). A possible mechanism for concentrating sodium and potassium in the cell nucleus. *Biophys. J.*, **16**, 527–33.

Morel, F., and F. Bastide (1965). Action de l'ocytocine sur la composante du transport de sodium par la peau de grenuille. *Biochim. Biophys. Acta*, **94**, 606–11.

Morel, F., and G. Leblanc (1973). Kinetics of sodium and lithium accumulation in isolated frog skin epithelium. In H. H. Ussing and N. A. Thorn (Eds), *Transport Mechanisms in Epithelia*, Munksgaard, Copenhagen, Academic Press, New York. pp. 73–85.

Motokawa, K. (1935a). Adsorption und bioelektrisches potential. *Jap. J. Med. Sci. III Biophys.*, **3**, 177–202.

Motokawa, K. (1935b). Über den einfluss der valenz un der konzentration der ionen auf die potentialdifferenz der froschhaut. *Jap. J. Med. Sci. III Biophys.*, **3**, 203–20.

Nagel, W., and A. Dörge (1970). Effect of amiloride on Na-transport of frog skin. I. Action on intracellular Na-content. *Pflügers Arch.*, **317**, 84–92.

Neill, W. T. (1958). The occurence of amphibians and reptiles in salt water areas, and a bibliography. *Bull. Mar. Sci. Gulf Carribean*, **8**, 1–97.
Nellans, H. N., and A. L. Finn (1974). Oxygen consumption and sodium transport in the toad urinary bladder. *Am. J. Physiol.*, **227**, 670–5.
Nielsen, R. (1971). Effect of amphotericin B on the frog skin *in vitro*. Evidence for active outward potassium transport across the epithelium. *Acta Physiol. Scand.*, **83**, 106–114.
Nielsen, R., and R. W. S. Tomlinson (1970). The effect of amiloride on sodium transport in the normal and the moulting frog skin. *Acta Physiol. Scand.*, **79**, 238–43.
Overton, E. (1904). Neununddreissig thesen über die wasserökonomie der amphibien und die osmotischen eigenschaften der amphibienhaut. *Verh. Phys.-Med. Ges. Würzburg*, **36**, 277–95.
Rawlins, F., L. Mateu, F. Fragachan, and G. Whittembury (1970). Isolated toad skin epithelium: transport characteristics. *Pflügers Arch.*, **316**, 64–80.
Reid, W. E. (1892). Reports on experiments upon 'absorption without osmosis'. *Br. Med. J.*, **13**, 323–6.
Rey, P. (1937). Recherches expérimentales sur l'economie de l'eau chez les Batrachiens. *Ann. Physiol. Physiochem. Biol.*, **13**, 1081–144.
Rosen, S., and J. Friedly (1973). Carbonic anhydrase activity in *Rana pipiens* skin: biochemical and histochemical analysis. *Histochemie*, **30**, 1–4.
Rosenberg, T. (1948). On accumulation and active transport in biological systems. I. Thermodynamical considerations. *Acta Chem. Scand.*, **2**, 14–133.
Rotunno, C. A., F. A. Vilallonga, M. Fernández, and M. Cerejido (1970). The penetration of sodium into the epithelium of the frog skin. *J. Gen. Physiol.*, **55**, 716–35.
Ruibal, R. (1962). The adaptive value of bladder water in the toad, *Bufo cognatus*. *Physiol. Zool.*, **35**, 218–23.
Saito, T., P. D. Lief, and A. Essig (1974). Conductance of active and passive pathways in the toad bladder. *Am. J. Physiol.*, **226**, 1265–71.
Salako, L. A., and A. J. Smith (1970a). Effect of amiloride on active Na-transport by the isolated frog skin: evidence concerning sites of action. *Br. J. Pharmacol.*, **38**, 702–18.
Salako, L. A., and A. J. Smith (1970b). Changes in Na-pool and kinetics of Na-transport in frog skin produced by amiloride. *Br. J. Pharmacol.*, **39**, 99–109.
Salibian, A., S. Pezzani-Hernandez, and F. Garcia-Romeu (1968). *In vivo* ionic exchange through the skin of the South American frog, *Leptodactylus ocellatus*. *Comp. Biochem. Physiol.*, **25**, 311–7.
Sapirstein, V. S., and W. N. Scott (1975). Binding of aldosterone by mitochondria-rich cells of the toad urinary bladder. *Nature*, **257**, 241–3.
Sawyer, W. H. (1951). Effect of posterior pituitary extract on permeability of frog skin to water. *Am. J. Physiol.*, **164**, 44–8.
Sawyer, W. H. (1957). Increased reabsorption of osmotically free water by the toad (*Bufo. marinus*) in response to neurohypophyseal hormones. *Am. J. Physiol.*, **189**, 564–8.
Sawyer, W. H. (1967). Evolution of antidiuretic hormones and their functions. *Am. J. Med.*, **42**, 478–86.
Sawyer, W. H., and R. M. Schisgall (1956). Increased permeability of the frog bladder to water in response to dehydration and neurohypophysial extracts. *Am. J. Physiol.*, **187**, 312–4.
Scheer, B. T., and R. P. Markel (1962). The effect of osmotic stress and hypophysectomy on blood and urine urea levels in frogs. *Comp. Biochem. Physiol.*, **7**, 289–97.

Schmid, W. D. (1965). Some aspects of the water economy of nine species of amphibians. *Ecology*, **46**, 261–9.

Schmid, W. D. (1968). Natural variations in nitrogen excretion of amphibians from different habitats. *Ecology*, **49**, 180–7.

Schmid, W. D., and R. E. Barden (1965). Water permeability and lipid content of amphibian skin. *Comp. Biochem. Physiol.*, **15**, 423–7.

Schmidt, K. P. (1957). Amphibians. *Mem. Geol. Soc. Am.*, **67**, 1211–2.

Schmidt-Nielsen, B., and R. P. Forster (1954). The effect of dehydration and low temperature on renal function in the bull frog. *J. Cell. Comp. Physiol.*, **44**, 233–46.

Schmidt-Nielsen, K., and P. Lee (1962). Kidney function in the crab-eating frog (*Rana cancrivora*). *J. Exp. Biol.*, **39**, 167–77.

Schoffeniels, E., and Tercafs, R. R. (1966). L'osmoregulation chez les batrachiens. *Ann. Soc. R. Zool. B.*, **96**, 23–39.

Sharp, G. W., and A. Leaf (1966). Mechanism of action of aldosterone. *Physiol. Rev.*, **46**, 593–633.

Shoemaker, V. H. (1964). The effects of dehydration on electrolyte concentration in a toad, *Bufo marinus*. *Comp. Biochem. Physiol.*, **13**, 261–71.

Shoemaker, V. H. (1965). The stimulus for the water balance responses to dehydration in toads. *Comp. Biochem. Physiol.*, **15**, 81–8.

Shoemaker, V. H., and H. Waring (1968). Effect of hypothalamic lesions on the water balance response of a toad, *Bufo marinus*. *Comp. Biochem. Physiol.*, **24**, 47–54.

Siegel, G. J., A. Tormay, and O. Candia (1975). Microsomal ($Na^+ + K^+$)-activated ATPase from frog skin epithelium. *Biochim. Biophys. Acta*, **389**, 557–66.

Singer, I., M. M. Civan, R. F. Baddour, and A. Leaf (1969). Interactions of amphotericin B, vasopressin and Ca^{++} in toad urinary bladder. *Am. J. Physiol.*, **217**, 938–45.

Smith, P. G. (1971). The low-frequency electrical impedance of the isolated frog skin. *Acta Physiol. Scand.*, **81**, 355–66.

Smith, C., W. D. Hughes, and E. G. Huf (1971). Movement of CO_2 and HCO_3^- across isolated frog skin. *Biochim. Biophys. Acta*, **225**, 77–88.

Taylor, A., M. Mamalak, E. Reaven, and R. Maffly (1973). Vasopressin: possible role of microtubules and microfilaments in its action. *Science*, **181**, 347–50.

Teorell, T. (1949). Membrane electrophoresis in relation to bio-electrical polarization effects. *Arch. Sci. Physiol.*, **3**, 205–19.

Tercafs, R. R., and E. Schoffeniels (1962). Adaptation of amphibians to salt water. *Life Sci.*, **1**, 19–23.

Thorsen, T. B. (1955). The relationship of water economy to terrestrialism in amphibians. *Ecology*, **36**, 100–16.

Thorsen, T. B., and A. Svihla (1943). Correlation of the habitats of amphibians with the ability to survive the loss of body water. *Ecology*, **24**, 374–81.

Urakabe, S., D. Shirai, S. Yuasa, G. Kimura, Y. Orita, and H. Abe (1976). Comparative studies of the effects of different diuretics on the permeability properties of the toad bladder. *Comp. Biochem. Physiol.*, **53C**, 115–9.

Ussing, H. H. (1948). The use of tracers in the study of active ion transport across animal membranes. *Cold Spring Harbor Symp. Quant. Biol.*, **8**, 193–200.

Ussing, H. H. (1949a). The active ion transport through the isolated frog skin in the light of tracer studies. *Acta Physiol. Scand.*, **17**, 1–37.

Ussing, H. H. (1949b). The distinction by means of tracers between active transport and diffusion. *Acta Physiol. Scand.*, **19**, 43–56.

Ussing, H. H. (1954). Ion transport across biological membranes. In A. T. Clarke and D. Nachmansohn (Eds), *Ion Transport across Membranes*, Academic Press, New York. pp. 3–22.

Ussing, H. H. (1966). Anomalous transport of electrolytes and sucrose through the isolated frog skin induced by hypertonicity of the outside bathing solution. *Ann. NY Acad. Sci.*, **137**, 543–55.
Ussing, H. H. (1969). The interpretation of tracer fluxes in terms of membrane structure. *Q. Rev. Biophys.*, **1**, 363–76.
Ussing, H. H. (1970). Tracer studies and membrane structure. In C. Crone and N. Lassen (Eds), *Capillary Permeability*, Munksgaard, Copenhagen, Academic Press, New York. pp. 653–6.
Ussing, H. H. (1972). The use of the flux ratio equation under steady-state conditions. In D. P. Agin (Ed.), *Perspectives in Membrane Biophysics*, Gordon and Breach, New York. pp. 211–7.
Ussing, H. H. (1973). Effects of ADH on transport paths in toad skin. In H. H. Ussing and N. A. Thorn (Eds), *Transport Mechanisms in Epithelia*, Munksgaard, Copenhagen, Academic Press, New York. pp. 11–16.
Ussing, H. H., and E. Windhager (1964). Nature of the shunt path and active sodium transport path through frog skin epithelium. *Acta Physiol. Scand.*, **61**, 484–504.
Ussing, H. H., and K. Zerahn (1951). Active transport of sodium as the source of electric current in the short-circuited isolated frog skin. *Acta Physiol. Scand.*, **23**, 110–27.
Viera, F. L., S. R. Caplan, and A. Essig (1972). Energetics of sodium transport in frog skin. I. Oxygen consumption in the short-circuited state. *J. Gen. Physiol.*, **59**, 60–76.
Voûte, C. L., and H. H. Ussing (1968). Some morphological aspects of active sodium transport. The epithelium of the frog skin. *J. Cell Biol.*, **36**, 625–38.
Voûte, C. L., R. Dirix, R. Nielsen, and H. H. Ussing (1969). The effect of aldosterone on the isolated frog skin epithelium (*R. temp.*). A morphological study. *Exp. Cell Res.*, **57**, 448–9.
Voûte, C. L., S. Hänni, and E. Ammann (1972). Aldosterone induced morphological changes in amphibian epithelia *in vivo*. *J. Steroid. Biochem.*, **3**, 161–5.
Voûte, C. L., K. Møllgaard, and H. H. Ussing (1975). Quantitative relationship between active sodium transport, expansion of endoplasmic reticulum, and specialized vacuoles ('scalloped saca') in the outermost living cell layer of the frog skin epithelium (*R. temp.*). *J. Membr. Biol.*, **21**, 273–89.
Voûte, C. L., and H. H. Ussing (1970a). The morphological aspects of shunt-paths in the epithelium of the frog skin (*R. temp.*). *Exp. Cell Res.*, **61**, 133–40.
Voûte, C. L., and H. H. Ussing (1970b). Quantitative relation between hydrostatic pressure gradient, extracellular volume, and active sodium transport in the epithelium of the frog skin (*R. temp.*). *Exp. Cell Res.*, **62**, 375–83.
Warburg, M. R. (1967). On thermal and water balance of three central Australian frogs. *Comp. Biochem. Physiol.*, **20**, 27–43.
Wixom, R. L., M. Kumudavalli Reddy, and P. Cohen (1972). A concerted response of the enzymes of urea biosynthesis during thyroxine-induced metamorphosis. *J. Biol. Chem.*, **247**, 3684–92.
Zadunaisky, J. A., O. A. Candia, and D. J. Chianrandini (1963). The origin of the short-circuit current in the isolated skin of South American frog, *Leptodactylus ocellatus*. *J. Gen. Physiol.*, **47**, 393–402.
Zadunaisky, J. A., J. Gennaro, N. Bashirelahi, and M. Hilton (1968). Intracellular redistribution of sodium and calcium during stimulation of sodium transport in epithelial cells. *J. Gen. Physiol.*, **51**, 290S–302S.
Zeiske, W., and B. Lindemann (1974). Chemical stimulation of Na^+-current through the outer surface of the frog skin epithelium. *Biochim. Biophys. Acta*, **352**, 323–6.
Zerahn, K. (1955). Studies on the active transport of lithium in the isolated frog skin. *Acta Physiol. Scand.*, **33**, 347–58.

Zerahn, K. (1956). Oxygen consumption and active sodium transport in the isolated and short-circuited frog skin. *Acta Physiol. Scand.*, **36**, 300–18.

Zerahn, K. (1969). Nature and localization of the sodium pool during active transport in the isolated frog skin. *Acta Physiol. Scand.*, **77**, 272–81.

Chapter 7

Control Mechanisms in Reptiles

W. A. Dunson

I. Introduction	273
II. Composition of the Extracellular Fluid	275
III. Regulation of Body Fluid Volumes	278
IV. Partitioning of Uptake and Loss	283
V. Achieving Salt and Water Balance	287
A. In terrestrial reptiles	287
B. In fresh-water reptiles	290
C. In estuarine reptiles	294
D. In marine reptiles	298
VI. Special Organs of Excretion and Uptake	301
A. Salt glands	301
B. The integument	310
C. Oral and cloacal membranes in aquatic reptiles	314
D. The kidney–bladder–gut complex in terrestrial reptiles	315
Acknowledgments	316
References	317

I. INTRODUCTION

Reptiles have evolved effective mechanisms of osmotic and ionic regulation that enable them to colonize a wide variety of habitats. They are common on land, including hostile deserts, in fresh-water, and even the sea. Thus they do not often seem to be ecologically limited by osmotic factors, as they are by low temperatures. The variety of osmoregulatory mechanisms in reptiles from different habitats, the relative importance of each in overall ion and water balance, and the physiological and cytological bases for ion and water transport in selected tissues and organs are the subjects of this review. Emphasis will be placed on those studies that contribute the most to our understanding of salt and water budgets of the whole animal, especially in aquatic reptiles.

Osmoregulation in reptiles is receiving an increasing amount of research attention and this is reflected in the number of recent review articles. Some aspects of renal and salt gland function have been thoroughly discussed by Dantzler and Holmes (1974), with an updated chapter by Dantzler (1976) on reptilian kidneys. Special aspects of salt and water balance are covered by Templeton (1972) for desert lizards and by Dunson (1975a) for sea snakes. Reptilian salt glands are the subject of reviews by Dunson (1969b, 1976) and Peaker and Linzell (1975). Bentley (1971) and Bradshaw (1972, 1975) have discussed the hormonal control of osmoregulation in reptiles. Minnich (1978) has prepared the first general review of the entire field of reptilian osmoregulation.

In considering the rates of uptake and loss of salts and water for an individual animal it is generally assumed that a steady state is in effect, and that balance is obtained by the interaction of several different processes. However research has generally been focused on a single organ such as the kidney, cloaca, salt gland or skin, and the broader concept of balance neglected. The analysis of balance is complicated since the relative importance of certain organs, such as the salt gland, is not a constant but changes with environmental factors related to such parameters as the internal state of hydration. Thus it is very difficult to construct a reliable balance sheet of routes of influx and efflux in any single species of reptile unless rather unnatural constraints are imposed. For example, the effect of activity on water loss is generally neglected. A greater degree of environmental relevance can be obtained by the use of free ranging animals for isotopic determination of turnover times, but these measurements provide little or no information on partitioning of function among different organs. Due to these problems it is unwise to generalize too freely about the importance of certain organs in overall ion and water balance until animals are tested under variations in environmental conditions normally encountered in different parts of their range. The time factor is also crucial since acclimation may occur slowly or in a seasonal fashion. Commercially supplied specimens normally should be carefully preconditioned before use or the natural conditions under which wild caught animals were living should be documented. The failure to demonstrate full salt gland function is often due to the absence of conditions necessary to stimulate it. The diamondback terrapin requires many months in sea-water to fully acclimate its salt glands (Dunson and Dunson, 1975); thus the temporal factor is crucial in assessing modes of osmoregulation.

Among the organs which contribute to osmotic and ionic regulation, the skin deserves special mention because its role has often been ignored. A common fault in the literature is the assumption that the horny skin of reptiles is impermeable to water. In terrestrial reptiles where the skin is considerably less permeable to water than in most amphibians, the integument is still the major source of water loss at moderate temperatures. Textbook assertions that 'the skin of reptiles is heavily keratinized and effectively impermeable to water and salts' (Lockwood, 1964), or that 'the integuments of all these animals are quite impermeable to water and solutes' (Gordon and coworkers, 1968), or 'the skin is also important in marine reptiles as a barrier against the loss of water by osmosis' (Bellairs, 1970) lack experimental proof. It appears that not only is the reptilian skin a major avenue of water loss, but that considerable interspecific differences in skin permeability occur. This may be related in part to habitat differences and to phylogenetic history. The driving force of this water movement across the skin of aquatic reptiles is not the difference in osmotic pressure between plasma and sea-water (which is about three-fold), but instead is the gradient in water concentration, which is very small. Thus there should be an influx of water into a marine reptile almost balancing the efflux, regardless of the overall permeability of the integument. Yet some modern textbooks continue to implicate osmotic gradients in control of water movements across membranes. It may well be that the primary adaptation of the integument of marine reptiles is for impermeability to Na, and only secondarily for reduction in bidirectional water flux.

The field of osmotic and ionic regulation in reptiles is currently in a vigorous phase of growth. Many important studies remain to be done and undoubtedly our concepts will continue to change for some time to come. I will attempt in this review to focus attention on areas in which further work would be especially desirable in order to elucidate basic mechanisms of transport. Some reptilian preparations are indeed favourable material for molecular studies of ion transport, and it is to be hoped that reptilian salt glands, cloaca, and skin may prove as useful to physiologists in the future as frog skin and toad bladders have in the past.

II. COMPOSITION OF THE EXTRACELLULAR FLUID

The extracellular fluid most commonly studied is the plasma, and a large number of measurements have been made on its composition. Dessauer (1970) should be consulted for a more complete list than that shown in Table 1. It is clear that the plasma of reptiles from a wide variety of habitats and of different taxonomic groups is similar to that of terrestrial vertebrates in general. Osmotic pressure is usually 300–400 mosmol kg^{-1} and NaCl is the major inorganic constituent. There may be a tendency for fresh-water turtles to have lower plasma osmotic pressures. Trends of this nature are difficult to substantiate unless the conditions under which the animals were maintained are known. Traditionally aquatic turtles are kept in water without adequate feeding, or at lower than preferred temperatures. Both of these conditions can lead to a gradual loss of body fluid electrolytes.

Reptiles are well able to tolerate substantial changes in plasma concentration, although they will regulate the level fairly closely if given the proper conditions. For example, the softshell turtle (*Trionyx spiniferus*) maintained plasma Na near 144 mmol l^{-1} when active and feeding; during hibernation there was a drop to 69–98 mmol l^{-1} (Dunson and Weymouth, 1965).

Table 1. Some representative electrolyte concentrations of reptilian plasma (after Dessauer, 1970, modified)

Species	Habitat[a]	Osmotic pressure (mosmol kg^{-1})	Electrolyte concentrations (mmol l^{-1})					
			Na	Cl	HCO$_3^-$	K	Ca	Mg
Crocodilians								
Alligator mississippiensis	F	284	141	112	19.8	3.8	5.1	2.9
Turtles								
Chelydra serpentina	F	315	132	76	48	3.2	3.8	2.7
Kinosternon subrubrum	F	288	121	98	30	4.2	3.5	1.0
Trionyx spiniferus	F	—	144	—	—	—	—	—
Terrapene carolina	T	345	130	108	—	4.7	1.3	3.5
Testudo hermanni	T	317	127	95	—	4.4	2.3	—
Caretta caretta	M	408	157	110	36	2.2	3.1	2.9
Lepidochelys olivacea	M	—	163	108	29	6.6	5.2	1.4
Chelonia mydas	M	—	158	—	33	1.5	—	—
Lizards								
Sauromalus obesus	T	—	169	127	—	4.9	—	—
Anolis carolinensis	T	—	157	127	15	4.6	2.9	—
Snakes								
Natrix rhombifera	F	359	155	139	7	4.0	3.9	—
Natrix sipedon	F	318	159	127	11	4.6	3.8	1.3
Thamnophis sirtalis	T	329	152	130	—	5.9	3.0	1.5
Coluber constrictor	T	375	151	101	14	4.1	3.2	1.5
Elaphe obsoleta	T	384	162	131	—	4.9	3.6	2.5
Laticauda semifasciata	M	320	159	—	—	—	—	—
Pelamis platurus[b]	M	—	210	167	—	8.1	—	—
Various sea snakes[b]	M	—	159	135	—	10.0	—	—

[a] F=fresh-water, T=terrestrial, M=marine.
[b] From Dunson (1975a).

Prolonged low temperatures led to a decline in plasma Na to lethal levels of 49–57 mmol l^{-1}. The effect of low temperature was ascribed to inhibition of the active uptake of Na from water. The decline in plasma Na of these animals was directly related to the observed drop in plasma osmotic pressure (which went from 280 in active animals to as low as 150 mosmol kg^{-1} in moribund ones; Dunson, 1967). Na losses alone accounted for 77% of the observed decrease in osmotic pressure.

Under dehydrating conditions the plasma of healthy reptiles may also become quite concentrated. A sample of 51 freshly caught sea snakes (*Pelamis platurus*) had a mean plasma Na of 210 mmol l^{-1} (range 200–230 mmol l^{-1}) (Dunson, Packer and Dunson, 1971). After starvation in the laboratory one *P. platurus* had a value as high as 307 mmol l^{-1} (Dunson, 1968). The ability to retain and tolerate excess electrolytes in the extracellular fluid is also found in terrestrial lizards such as *Amphibolurus maculosus* (Braysher, 1976) and *A. ornatus* (Bradshaw and Shoemaker, 1967). Bentley (1959) reported a similar elevation during the summer in the skink *Trachydosaurus rugosus*, although Braysher (1971) found little change in plasma Na of a coastal population of the same species possessing active salt glands.

The ability of marine reptiles to regulate plasma electrolytes when placed in fresh-water is a matter of some interest. The sea snake *P. platurus* has been maintained for six months in fresh-water feeding on goldfish, apparently without ill effect (Dunson and Ehlert, 1971). Similarly, young green sea turtles (*Chelonia mydas*) fed on shrimp grew well over a two month period in fresh-water; plasma Na dropped from 156 mmol l^{-1} in sea-water to 130 mmol l^{-1} in fresh-water (Holmes and McBean, 1964). Green sea turtles that are not fed undergo a continuing decline in plasma Na (88 mmol l^{-1} at 35 days; Kooistra and Evans, 1976). Estuarine diamondback terrapins (*Maclemys terrapin*) fed fish survive well in fresh-water; after 99 to 132 days the mean plasma Na was 131 mmol l^{-1}, similar to that of *C. mydas* after two months (Robinson and Dunson, 1976). Thus it is clear that marine reptiles may achieve a steady-state Na balance in fresh-water if they are feeding. The possible presence or absence of mechanisms allowing hyperosmotic regulation of fasted marine reptiles in fresh-water should be examined further. A most interesting natural case of this is the 'sea snake' *Hydrophis semperi* that lives in Lake Taal in the Philippine island of Luzon (Dunson and Dunson, 1974; Dunson, 1975a). This coastal lake is clearly fresh, yet the snake retains a functional salt secreting posterior sublingual gland. A solution to this enigma awaits further on-site research.

Although blood plasma is the body fluid most often analysed, it actually represents only about one-fourth of the total extracellular fluid. The remaining three-fourths, the interstitial fluid and the lymph, are ultrafiltrates of the plasma and are quite similar in electrolyte composition to it. In man, plasma is 142 mmol l^{-1} for Na and 103 mmol l^{-1} for Cl in comparison with values of

144 mmol l^{-1} for Na and 114 mmol l^{-1} for Cl in the interstitial fluid (Pitts, 1968). The only similar analysis that is available for a reptile is the interesting study made by Norris and Dawson (1964) on the lateral lymph sac fluid of the desert chuckwalla lizard. They found that the fluid would clot upon standing, had white blood cells but no erythrocytes in it, and had a Cl concentration of 122 to 136 mmol l^{-1}, almost identical to that of plasma.

III. REGULATION OF BODY FLUID VOLUMES

In man total body water (TBW, 50–70% body weight) is made up of intracellular fluid (ICF, 30–40% body weight), interstitial fluid (ISF, 16% body weight), plasma (PV, 4.5% body weight), and transcellular fluid (1–3% body weight) (Pitts, 1968). This basic scheme of classifying the components of body water is applicable to all vertebrates. Difficulties arise from the techniques used to estimate fluid volumes. All of the procedures that can be used *in vivo* are by simple dilution, either of a dye or a compound identifiable chemically or by radioisotopic labelling. Since no substance is confined solely to one of the major body water compartments, and metabolism and excretion of the substance will occur at varying rates, these measurements are far from being perfect. The best that can be achieved is to choose a substance that is the most suitable for the experiment in question, and provide comparisons with other commonly used techniques. In reptiles PV is usually measured with Evans Blue dye (T-1824), ECF by sucrose or thiocyanate space, and TBW by desiccation to a constant weight at 100 °C. Then by calculation (ECF − PV = ISF) and (TBW − ECF = ICF). For ECF, sucrose is preferable since thiocyanate is believed to enter the cells; it consistently gives higher values than sucrose. Thiocyanate may also be poisonous to some active transport systems (Epstein, Maetz and De Renzis, 1973), making it quite undesirable as a measure of body fluid volume.

Before any body fluid volume measurements were made on reptiles, the difference in water content of various tissues was recognized (Khalil and Abdel-Messeih, 1954). The adipose tissue, skeleton and skin contained only 11, 21, and 60% water respectively in the desert lizard *Uromastyx aegyptia* compared to TBW of 78–82% body weight. Another finding in agreement with that of earlier work on mammals was that younger animals had higher TBW than older ones. Correlation between tissue water content and functional status in osmoregulation is unclear and such measurements are currently of little importance.

Thorson (1968) is the only investigator to attempt to measure body water partitioning among a great variety of reptiles from different habitats. Some of his data are shown in Table 2. He found that TBW and ICF of marine and terrestrial species were lower than that of fresh-water reptiles, but ECF was larger (both PV and ISF). Marine forms showed this pattern in a more extreme form than terrestrial ones. Thorson (1968) pointed out that his tests

Table 2. Selected reptilian body fluid volumes (as % body weight); L = lizard, S = snake, T = turtle, C = crocodilian (ISF = sucrose space−plasma)

Habitat/Genus		Conditions	Body weight (g)	Total water	ECF				Blood	ICF	References
					Sucrose	Thio-cyanate	Plasma	ISF			
A. Terrestrial											
Dipsosaurus	(L)	No anaesth.	50	74.7	—	30.0	5.0	—	—	45.0	Minnich (1970a)
Iguana	(L)	Nembutal anaesth.	260–820	70.8	16.8	—	4.2	12.6	6.0	54.0	Thorson (1968)
Sauromalus	(L)	Ether anaesth., April	50–300	76.3	—	35.1	7.0	—	9.8	36.3	Nagy (1972)
Sauromalus	(L)	Ether anaesth., Oct.	50–300	73.0	—	32.9	6.1	—	8.6	40.0	Nagy (1972)
Constrictor	(S)	Nembutal anaesth.	460–3500	71.0	16.8	—	4.0	12.8	5.9	54.2	Thorson (1968)
Pituophis	(S)	Nembutal anaesth.	575–1050	70.0	16.7	—	4.1	12.6	6.0	53.3	Thorson (1968)
Gopherus	(T)	Nembutal anaesth.	2790–5740	69.9	16.8	—	4.1	12.7	6.0	53.1	Thorson (1968)
B. Fresh-water											
Chelydra	(T)	Nembutal anaesth.	2790–8510	72.9	14.9	—	3.3	11.6	4.7	58.0	Thorson (1968)
Alligator	(C)	Nembutal anaesth.	1860–2620	72.9	15.1	—	3.7	11.4	5.1	57.8	Thorson (1968)
C. Estuarine											
Malaclemys	(T)	Nembutal anaesth.	545–735	64.9	21.9	—	4.4	17.5	6.5	43.0	Thorson (1968)
Malaclemys	(T)	Fresh-water (FW)	1030	61.6	18.6	—	5.8	12.8	—	45.9	Robinson and Dunson (1976)
				(36.9)[a]	(11.6)[a]		(3.7)[a]	(8.0)[a]		(27.5)[a]	
Malaclemys	(T)	Sea-water	750	59.9	16.4	—	3.5	12.9	—	44.6	
				(27.5)[a]	(7.0)[a]		(1.5)[a]	(5.5)[a]		(20.5)[a]	
Malaclemys	(T)	FW and salt loads	1075	66.4	26.6	—	7.6	18.9	—	39.9	
				(44.7)[a]	(17.9)[a]		(5.2)[a]	(12.7)[a]		(26.8)[a]	
D. Marine											
Caretta	(T)	Nembutal anaesth.	77,270	64.0	19.3	—	4.4	14.9	6.7	44.7	Thorson (1968)
Caretta	(T)	Nembutal anaesth.	18,950	64.9	19.1	—	4.4	14.7	6.5	45.8	Thorson (1968)

[a] Values in parentheses are body fluid volumes expressed as ml cm^{-1} plastron length.

were static in nature and that interesting results would undoubtedly be obtained by transferring aquatic forms among different salinities. He particularly noted that the estuarine diamondback terrapin had the highest ECF (and ISF) and the lowest ICF of any species studied. This led him to propose that the ECF increased in terrapins placed in sea-water at the expense of the ICF.

As part of an overall study of Na and water balance in the diamondback terrapin, Robinson and Dunson (1976) measured body fluid volumes in sea-water and fresh-water. They discovered that terrapins are so impermeable to water, and especially Na, that placing them in sea-water has little immediate effect on them. The activation of the salt gland is very slow (Dunson and Dunson, 1975). The concept of 'acclimation' of a test animal has only a temporal importance in this context, and Thorson's (1968) measurements on terrapins placed in sea-water for 'a number of days' are certainly not indicative of real adjustment to sea-water conditions. Instead they reflect the unknown treatment of the previous several months. This osmotic insulation of the terrapin from its environment is a most interesting phenomenon, but it makes experimental studies extremely difficult since the physiological state of test animals cannot readily be synchronized by placing them in sea-water or fresh-water for a specified period. Another serious problem that arose over the prolonged periods of sea-water exposure necessary for the study was a small but significant loss of TBW, making body weight unreliable as a standard against which to express the size of body fluid compartments. To facilitate comparisons between animals of differing and unknown degrees of dehydration, a major skeletal character (mid-line plastron length) was used. Thus absolute values of body fluid volumes within a species can be expressed as ml fluids/cm plastron length (see Table 2).

Thorson's (1968) original postulate that the ECF increased in sea-water exposed terrapins at the expense of the ICF has not been confirmed. Instead Robinson and Dunson (1976) found that a pronounced loss of TBW in sea-water occurred from both ICF and ECF. The ICF decreased in proportion to the TBW loss (about 25%) whereas the ECF declined by 40% (based on absolute changes expressed as ml/cm plastron). Under these conditions PV declined more (60%) than the ISF (30%) did (Table 2). At this point some field observations made in the salt marsh habitat of the terrapin near Chincoteague, Virginia, provided an important clue to our understanding of the adaptive importance of changes in body fluid volumes. Freshly caught terrapins often have extremely baggy folds of skin around the legs which gradually disappear in captivity. My initial impression of this phenomenon was that it might be due to changes in ISF volumes, and that it could be related to storage of fluid for use during periods of dehydration in sea-water (Dunson, 1970). On interviewing a long-time terrapin trapper I was intrigued to discover that he referred to these swellings around the legs as 'water sacs'. Yet these sacs did not appear when sea-water exposed and dehydrated terrapins were put in fresh-water. A reasonable guess was that significant expansion of the ECF could not occur without a sufficient supply of NaCl, and that in nature the alternation in exposure to sea-water and then fresh-water provided access to salt. Possibly terrapins drink brackish water of appropriate salinity, but they do not drink full strength sea-water. To simulate this process in the laboratory terrapins were placed in fresh-water and injected with NaCl periodically (Robinson and Dunson, 1976). Under these conditions TBW and ECF increased dramatically, but ICF remained near the level observed in fresh-water (Table 2). PV and ISF both increased substantially. Thus it appears likely that terrapins can gradually utilize the fluid stored in the ECF during sea-water exposure. As water is lost, excess NaCl is at first stored and then eventually excreted by the salt gland as the gland is activated by plasma Na concentrations of 175 mmol l^{-1} and above (Dunson and Dunson, 1975).

It is indeed fortunate that some excellent studies by Gilles-Baillien (1970, 1973a, 1973b) on the plasma and body tissues of terrapins allow further insights into the mechanism of ECF and ICF volume regulation discussed above. In sea-water, terrapin plasma osmotic pressure increases considerably, due to NaCl and urea. In 50% sea-water the increase is mainly due to NaCl, with a major increase of urea in the change from 50% to 100% sea-water. The difference between sea-water and fresh-water plasma is least in July, due mainly to a drop in urea. In sea-water exposed terrapins plasma osmotic pressure began to increase in September and reached a maximum in April at the time of arousal from hibernation. When the animals began feeding regularly in May, a rather abrupt drop occurred. Plasma urea concentrations followed a similar pattern. An important observation was that there was no increase in plasma salts during hibernation, a good demonstration of the impermeability of the integument (see Robinson and Dunson, 1976). A sudden rise in plasma NaCl at arousal may indeed be due to drinking, but this is not certain. The accumulation of urea (as high as 115 mmol l^{-1}; Gilles-Baillien, 1970) in the ECF of terrapins exposed to sea-water has been interpreted as a mechanism to assist in regulation of water balance, although it may be simply due secondarily to an antidiuretic effect on urine flow. In fact it is doubtful if the 25% increase in osmotic pressure of the ECF observed would significantly affect the rate of water efflux which is primarily integumental. Instead the urea build-up may be more important in regulating the relative volumes of the ECF and the ICF. Gilles-Baillien (1973a) has clearly shown that the osmotic pressure of the ICF of many tissues is elevated, perhaps in association with the rise in urea in the ECF. Muscle ICF shows increased concentrations of ammonia, taurine, urea and to a lesser extent of amino acids (aspartate excepted). There is no change in inorganic ions. In other tissues there is some variation in this response, but the general pattern holds, except for increased K in the jejunal mucosa.

The really interesting question remains unanswered, as to the sequence of initiation of the above events and the controlling factors. There are so many different organs involved that the interactions are complicated. The possible roles of the salt gland and the urinary bladder are especially intriguing. Bentley, Bretz and Schmidt-Nielsen (1967) first reported the dramatic rise in osmotic pressure and K concentration of bladder urine of sea-water terrapins. Gilles-Baillien (1970) confirmed these findings and reported high concentrations also of urea and ammonia, with unidentified substances comprising an important remainder of the osmotic pressure. The mean bladder urine osmotic pressure for sea-water terrapins was 372 ± 94 mosmol l^{-1}, and although difficulties were encountered with the Cl titration, Cl was believed to be low in concentration. Robinson and Dunson (1976) measured even higher bladder urine osmotic pressures (550 ± 53 mosmol kg^{-1}) and K concentrations (89 ± 23 mmol l^{-1}); NaCl was very low. They also found that plasma Na concentration was directly related to bladder urine K, both of which were related to salt gland Na–K ATPase content (Dunson and Dunson, 1975). What remains unclear at present is the integration of the processes taking place in the bladder with changes in the ECF, ICF and the salt gland. The bladder fluid is a rather exotic mixture of K, urea, ammonia and some unknown probably nitrogenous compounds. The lack of NaCl strongly suggests the presence of an active reabsorption process, perhaps in exchange for K. The serosa is positive with respect to the mucosa and a respectable short-circuit current (40 μA cm^{-2}) has been demonstrated across the bladder of sea-water exposed animals (Robinson and Dunson, 1976). The possible paradox of the terrapin reabsorbing urinary Na, which may then be excreted by the salt gland, suggests that further study into water fluxes across the bladder might be fruitful. Gilles-Baillien (1976) has recently discussed the role of intestinal ion and water transport in the physiological response of the terrapin to sea-water exposure. There will be further discussion below of the overall pattern of Na and water balance of the terrapin, especially in relation to the important new finding that the skin is permeable to water but not to Na (Robinson and Dunson, 1976).

The only other reptile for which detailed information on body fluid volumes is available is the chuckwalla, a desert lizard (Nagy, 1972). Animals were collected over a six month period from April to October, including a drought in the summer. As the vegetation dried out, chuckwallas ceased feeding, became inactive and began to lose weight. By October TBW had decreased by 30%, whereas body weight declined by only 19%. Absolute changes in body fluid volumes were not precisely quantified against a static index (such as plastron length used for the terrapins). However it appears that the decline in TBW is attributable mainly to losses from the lymph sacs and the urinary bladder. The lymph sacs (Fig. 1) are structures first described by Norris and Dawson (1964) as possible storage areas for ECF. Filled during favourable periods, the fluid could then be used later to ameliorate the effects of dehydration. The lymph sacs of Nagy's chuckwallas were probably only partly filled, since he estimated a maximum volume of 2% body weight by palpation in comparison to Norris and Dawson's figure of 7.3% for this species. Nagy's data certainly show a preferential loss of fluid from the lymph sacs and the bladder. Yet he felt that the failure of the lymph sacs to fill completely in May, during a time of abundant water availability, indicated that they were not used as a fluid reserve during drought. This would certainly be an interesting topic for further research, since it is very unlikely that a structure capable of holding as much fluid as the entire blood vascular system

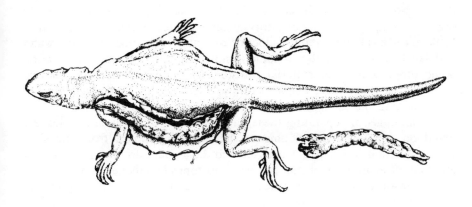

Figure 1 The lateral lymph sac system of the chuckwalla lizard (*Sauromalus hispidus*) from an island in the Gulf of California. There is a lower lobe extending behind the polyester cast shown *in situ* and additional bilateral sacs are found in the head and throat. From Norris and Dawson (1964); reproduced from *Copeia* by permission of the American Society of Ichthyologists and Herpetologists

is useless. As in the case of the 'water sacs' of terrapins, the availability of excess NaCl may well be crucial in allowing the lymph sacs to be filled by an expansion of the ECF.

IV. PARTITIONING OF UPTAKE AND LOSS

Physiologists have a tendency to work on one specific tissue of interest, whether it is the kidney, salt gland, skin, adrenals, etc., and to ignore other components of the osmoregulatory system. There is a great need for synoptic studies of water and electrolyte balance including investigations into the various routes of uptake and loss. Assumptions made about various parts of the system should be minimal; direct measurement of all parameters is the key to reliable results. In reptiles the following are major sources of materials transfer into and out of the body:

Influx	Efflux
(a) food	(a) faeces (gut)
(b) drinking water	(b) urine (cloaca, kidney and bladder)
(c) integumentary uptake (including oral and cloacal epithelial membranes)	(c) salt gland
	(d) respiratory tract
	(e) integumentary loss (including oral and cloacal epithelial membranes)

These categories may vary in importance from almost 0 to 100% of the total flux, depending on the species in question and the circumstances. Some organs may operate in an obligatory manner whereas others are facultative. Little enough is known about endocrine or neural mediation of overall water and salt balance to make this topic a most attractive one for future work. Some other sources of material transfer, which are considered minor at present, should also receive more scrutiny. These would include such topics as skin shedding and the secretions of cloacal, skin and venom glands.

The methods of separating these various routes of influx and efflux are based on a physical partitioning of the head and cloaca from the skin. Ingested food is weighed and analysed and the amount drunk is measured by gut content analysis of phenol red or an isotopically labelled substance not readily absorbed by the gut. Since no substance is ideal for these purposes (Diefenbach, 1973, has shown intestinal uptake of phenol red), drinking may also be estimated by the difference between total loss and uptake by the skin and in the food. In aquatic animals drinking can be reliably measured by confining the head in a chamber labelled with a radioisotope (assuming it can be shown that dermal or oral membrane transport of the substance is minimal). The cloacal component is usually measured by cannulation or by collecting cloacal fluid and faeces under oil. If a diuretic effect of cloacal cannulations is suspected, an additional method whereby the cloacal opening is sewed shut to measure the change from the unaltered condition is useful. In some cases it is possible to directly cannulate the ureters to separate the functions of the kidney and the cloaca–gut–bladder complex. The salt glands of snakes can readily be cannulated although the simpler measurement of extracloacal ion loss is usually a very good estimate of salt gland excretion (since the integument is impermeable). Losses from the respiratory tract must be measured by analysis of the water vapour in expired air, or more simply by the difference between total evaporative loss (by weight loss) and skin loss. In water, precise measurements of dermal flux can be made with tritiated water. All of these techniques are *in vivo* methods, which are strongly recommended over *in vitro* procedures when relevance to the normal functioning of the whole animal is important. Often it is necessary to study isolated skin, epithelial membranes, or renal tubules to establish unequivocally some special feature of ion or water transport. However great care is advised in generalizing such findings to the whole animal. Too often such studies become an end in themselves (e.g. biophysical studies on frog skin) and reveal little about ionic and osmotic regulation as an integrated physiological system. It is not uncommon to find that the permeability of isolated tissues changes drastically when removed from the normal neural, hormonal, and circulatory environment. For example isolated sea snake skin is much more permeable to water than it is *in vivo* (Dunson and Robinson, 1976). On the other hand the impermeability of the sea snake skin to Na, which is due to the outer keratin layers, is not affected by death or removal of the skin from the body since keratin is not a living tissue (Dunson and Robinson, 1976).

Almost any substance which is taken up by the body is also lost from it. The difference between influx and efflux is termed the net flux. This quantity is an ecologically significant measurement since it represents the new amount actually available to the body pool or lost from it under the conditions specified. Some studies on reptiles have not clearly separated unidirectional fluxes and net changes, leading to some confusion in the literature. For example Diefenbach (1973) claims to have established that significant amounts of water were not absorbed through the skin of caimans whose bodies were immersed in water while their heads were held out in the air. The movements of water were estimated by weight changes and obviously represent net water flux. Although it is undeniably true that net respiratory loss greatly exceeded net dermal uptake, there may well have been a large unobserved dermal influx accompanied by an almost equally large efflux. In this case it also seems likely that the experimental set-up (restraint of animals and low humidity of air) may have led to unnaturally high respiratory water loss.

A very useful means of estimating the rate of total water metabolism is the tritiated water turnover time. The rate of decline in plasma activity can be periodically monitored, even in animals released to the wild. Comparisons made with laboratory held controls provide an excellent estimate of the effects of normal activity patterns on water balance. A few values for reptiles are shown in Table 3. Turnover is much less in the two desert lizards studied than in the estuarine terrapin or the marine sea snake. There is a clear increase in turnover of wild chuckwallas over their laboratory brethren.

The relation between habitat and water balance may be seen more clearly if we look specifically at evaporative water losses, both respiratory and cutaneous (Table 4). These measurements were made at 23 °C. At higher temperatures respiratory losses generally increase more rapidly than cutaneous ones. As the habitat becomes more dehydrating, there is a decrease in both respiratory and cutaneous losses. There also seems to be a trend towards reduction in the relative contribution of the cutaneous component.

Even fresh-water reptiles have means of regulating water loss, as is clearly evident in a comparison of urinary and total evaporative loss in the course of dehydration of the alligator (Table 5). The animals were simply left in air without food or drinking water. There was an enormous decrease in urinary

Table 3. Body water turnover of some reptiles measured by tritiated water fluxes

Habitat/Species	Conditions	Body weight (g)	Total body water (% wt)	Turnover (ml kg^{-1} day^{-1})	$T_{1/2}$ (day)	Reference
A. Terrestrial						
Dipsosaurus dorsalis	Desert	10–65	—	15–65	—	Minnich and Shoemaker (1970)
Sauromalus obesus	Creosote scrub	50–300	73–77	24.5	14.5	Nagy (1972)
	Laboratory			18.1		Nagy (1972)
Uma scoparia	Desert	4–28	~74	12–24	—	Minnich and Shoemaker (1972)
B. Estuarine						
Malaclemys terrapin	Laboratory in sea-water	750–1030	60–62	43[a]	12[a]	Robinson and Dunson (1976)
C. Marine						
Pelamis platurus	Laboratory in sea-water	30–60	73.9	58[a]	9[a]	Dunson and Robinson (1976)

[a] Aquatic exchange only; minor aerial respiratory exchange not included.

Table 4. The relationship between habitat and evaporative water losses (measured as weight loss in dry air at 23 °C) in reptiles (after Bentley and Schmidt-Nielsen, 1966, modified)

Habitat/Species	Body weight (g)	Total loss (% wt day^{-1})	O$_2$ uptake (ml g^{-1} day^{-1})	Respiratory loss		Cutaneous loss	
				(mg g^{-1} day^{-1})	(mg ml^{-1} O$_2$)	(mg cm^{-2} day^{-1})	(% total)
Aquatic crocodilian:							
Caiman sclerops	124	11.5	1.8	9.6	4.9	32.9	87
Aquatic turtle:							
Chrysemys scripta	600	2.0	0.9	4.3	4.2	12.2	78
Terrestrial turtle:							
Terrapene carolina	305	0.9	0.6	2.6	4.2	5.3	76
Mesic lizard:							
Iguana iguana	124	0.8	2.6	3.4	0.9	4.8	72
Desert lizard:							
Sauromalus obesus	134	0.3	1.2	1.1	0.5	1.3	66

Table 5. Routes of water loss in fasted alligators with progressive dehydration at 28 °C and 50% relative humidity (after Coulson and Hernandez, 1964, modified)

Days of dehydration	Mean water loss (in % body weight/day)		
	Urinary loss	Evaporative loss (skin and respiratory)	$\dfrac{\text{Evaporative}}{\text{Total}} \times 100$
1	3.6	1.6	30.8
2	1.3	1.3	50.0
5	0.2	1.3	86.7
6	0.2	0.9	81.8
7	0.05	0.7	93.3

water loss and a lesser drop in evaporative loss. How much of the latter is due to changes in respiratory or cutaneous loss is unknown at present. The control of such a sequence of antidiuretic responses would be a most interesting topic for further study. From these data it is clear that the state of hydration of experimental animals must be carefully controlled for reliable comparisons between species from different habitats. For example, could the decline in evaporative water loss shown in Table 4 be at least partly due to the terrestrial animals being more dehydrated?

V. ACHIEVING SALT AND WATER BALANCE

A. In Terrestrial Reptiles

Desert lizards are certainly not typical of all terrestrial reptiles, but some excellent studies have been carried out on several species, illustrating extreme adaptations to an arid environment. The water and K budgets of the chuckwalla are shown in Fig. 2. K is illustrated because it is usually more abundant than Na in the diet of desert lizards. General aspects of the routes of exchange of water and K are similar to those of *Dipsosaurus dorsalis* (see Templeton, 1972; Minnich, 1976) and *Uma notata* (see Deavers, 1972; Minnich and Shoemaker, 1972). Chuckwallas probably do not drink (although other desert lizards may do so on occasion), so that preformed water in the food is the major source of water intake (Fig. 2(a)). Metabolic water from the oxidation of food constituents is less, but still important. Water losses, in decreasing order of importance, occur from the skin, in the faeces, from respiration, and in the urine. The relative contribution of cutaneous and respiratory losses to total evaporation is variable, depending on the temperature (Table 6). Respiratory water losses increase faster than cutaneous ones at higher temperatures in the chuckwalla and the desert tortoise. From Krakauer's (1970) studies it appears that this is not the case in water snakes, where little change occurs in the cutaneous contribution

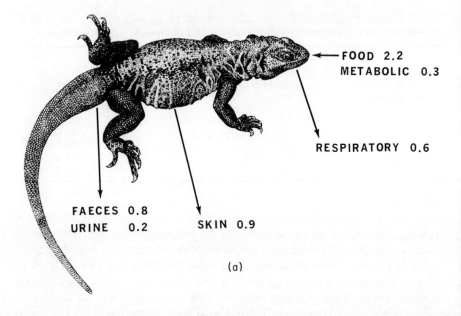

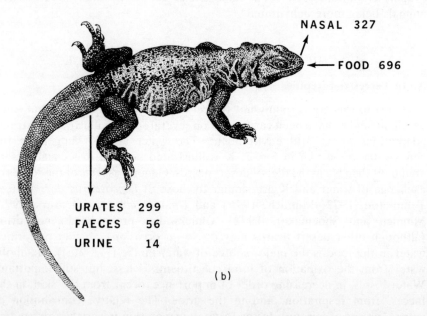

Figure 2 Daily water and K budgets in the desert chuckwalla lizard, *Sauromalus obesus*, at balance in the field (drawn by Richard Silvoy).
(a) Water budget (calculated from Nagy, 1972; Bentley and Schmidt-Nielsen, 1966). Values in ml $(100\ g)^{-1}\ d^{-1}$.
(b) K budget (calculated from Nagy, 1972). Values in μmol $(100\ g)^{-1}\ d^{-1}$

Table 6. The cutaneous contribution to total evaporative water loss of reptiles at various temperatures

Habitat/Species	Body weight (g)	Cutaneous loss as % of total evaporative loss				References
		15 °C	22–27 °C	30–35 °C	40 °C	
Desert lizard:						
Sauromalus obesus	134	—	76.5	—	43.6	Bentley and Schmidt-Nielson (1966)
Desert tortoise:						
Gopherus agassizii	725–2600	—	76	52	—	Schmidt-Nielsen and Bentley (1966)
Desert snakes:						
Spalerosophis cliffordi	218	—	~66	65	—	Dmi'el (1972)
Aspis cerastes	125	—	~66	56	—	Dmi'el (1972)
Crotalus atrox	123	—	30	—	—	Chew and Dammann (1961)
Crotalus scuetellatus	278	—	30	—	—	Chew and Dammann (1961)
Terrestrial snakes (mesic):						
Thamnophis sauritus	28	—	81.3	—	—	Krakauer (1970)
Pituophis catenifer	—	—	72.2	—	—	Krakauer (1970)
Coluber ravergieri	136	—	~66	79	—	Dmi'el (1972)
Vipera palaestinae	581	—	~66	64	—	Dmi'el (1972)
Aquatic snakes:						
Natrix fasciata pictiventris	152–278	86.5	92.2	90.3	—	Krakauer (1970)
Natrix fasciata compressicauda	135–192	91.4	80.4	90.2	—	Krakauer (1970)
Natrix cyclopion	—	—	96.8	—	—	Krakauer (1970)

between 15 and 32 °C (Table 6). One complicating factor is that there seem to be differences in total evaporative losses between taxonomic groups, regardless of the habitat. For example Dmi'el (1972) found that colubrid snakes lost much more water by evaporation than viperids. Yet within each group, desert representatives had lower evaporative losses than their mesic relatives (see also Table 4). Since evaporative water losses can be affected also by size (or surface area), activity, flow rate of air in the experimental system, and the state of shedding (Cohen, 1975), it is obvious that comparisons between different studies must be made with care. Further data illustrating these points may be found in Bogert and Cowles (1947), Templeton (1960), Schmidt-Nielsen (1964), Warburg (1965a, 1965b, 1966), Claussen (1967), Prange and Schmidt-Nielsen (1969), Gans, Krakauer and Paganelli (1968), Ernst (1968), Krakauer, Gans and Paganelli (1968), Dawson, Shoemaker and Licht (1966), Dmi'el and Zilber (1971), Minnich (1970b, 1978), and Snyder (1975). The most important finding to come out of the studies of the past ten years has been the previously unsuspected role of the skin in evaporative water loss. As recently as 1961, Chew and Dammann contended that 'the skin of at least certain reptiles is indeed nearly waterproof'. We now know that reptilian skin is permeable to water in both terrestrial and aquatic forms (see below).

The pattern of K regulation in the chuckwalla is quite different from that of water (Fig. 2(b)). Although the source of intake is the food, as is true also for water, routes of loss are through the salt gland and the cloaca. The nasal salt gland secretes intermittently, since it has the capacity to clear all K from the plasma in less than one hour (Templeton, 1972). The maximum secretory rate (Templeton, 1964) measured in one hour is 746 μmol $(100 \text{ g})^{-1}$ day^{-1}, more than twice that found over an entire day by Nagy (1972). The K and Na concentration of secreted fluid is variable. Templeton reports mean concentrations of 370 mmol l^{-1} for K and 121 mmol l^{-1} for Na in one experiment.

Almost equal in importance to the salt gland in K excretion is the cloaca (Fig. 2(b)). Large quantities of K are excreted as precipitated urate salts, with lesser amounts in the faeces and the liquid urine. Minnich (1970a) was the first to recognize the importance of this process, whereby desert iguanas can remove about 4.9 moles of K per litre of urine water lost. Total Na excretion by the chuckwalla is less than one tenth that of K and occurs mainly through the faeces and the salt gland.

B. In Fresh-water Reptiles

This subject has not attracted as much attention as the physiological problems of adaptation to desert life. Thus there is no single species which has been thoroughly studied. By utilizing data from Dunson (1967, unpublished), Dunson and Weymouth (1965), Seidel (1975), and Dantzler and Schmidt-Nielsen (1966), it is possible to piece together a picture of Na balance (Fig. 3)

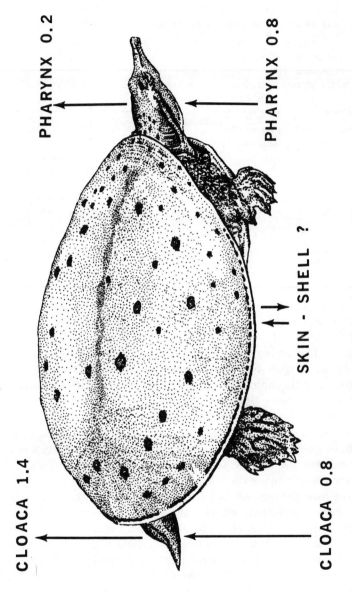

Figure 3 Sodium fluxes in the fasted fresh-water softshell turtle, *Trionyx spiniferus*, at balance. Values in μmol $(100\ g)^{-1}\ h^{-1}$ (calculated from Dunson, 1967, and unpublished data). Drawn by Richard Silvoy

which would be reasonably valid for the softshell turtle *Trionyx spiniferus* or the red-eared turtle *Chrysemys scripta*. For illustrative purposes the turtle in Fig. 3 is shown at balance since softshells have been proven capable of net uptake of Na from solutions as low as 5 μmol l^{-1} (Dunson and Weymouth, 1965). The only available flux data on softshells are shown in Table 7; efflux

Table 7. Na influx and efflux in the fresh-water softshell turtle, *Trionyx spiniferus* (Dunson, unpublished data). Fluxes in μmol (100 g)$^{-1}$ h^{-1}. Turtles maintained at 23 °C in tap water for one year on mealworm food. Fluxes measured by ^{22}NaCl uptake and loss (as in Dunson, 1967)

No.	Body weight (g)	Plasma (Na) (mmol l^{-1})	Total body Na (μmol g^{-1} wet wt)	Influx	Efflux
1	27.5	112	72.8	1.06	1.17
2	17.0	143	89.4	1.26	1.57
3	18.5	153	91.1	0.90	1.52
4	28.2	134	91.5	0.57	1.81
5	28.9	125	79.7	0.48	1.80
6	20.8	130	84.5	0.59	0.86
Mean	23.5	133	84.8	0.81	1.46
±SD	±5.3	±14	±7.4	±0.31	±0.37

exceeds influx, as would be expected in turtles that have not been stressed by fasting and 'wash-out' of Na by low temperatures or by repeated transfers to distilled water. The ability of these completely aquatic turtles to regulate hyperosmotically is shown by the plasma Na values (133 mmol l^{-1}) and especially the body Na content (about 85 μmol g^{-1}). The latter is higher than *C. scripta* (July, 71 μmol g^{-1}; September, 70 μmol g^{-1}; Dunson, 1967), even though *T. spiniferus* is much more specialized for aquatic life. The exchangeable Na pool was approximately equal to the total body Na in both species. Softshells have Na fluxes similar to those measured in *C. scripta* and it has been assumed that the relative contribution of the pharyngeal and cloacal membranes to Na uptake are the same (both about equal in *C. scripta*). The *in vivo* Na efflux from the pharynx and cloaca have been calculated from *in vitro* membrane studies showing the efflux in these two regions is roughly 25% and 50% of the influx respectively. The renal and gut contribution is then obtained by difference, but its order of magnitude was also confirmed by urine Na concentrations (about 20 mmol l^{-1}) from Seidel (1975) and urine flow rates (about 1.3 ml kg^{-1} h^{-1}) from Dantzler and Schmidt-Nielsen (1966). The Na permeabilities of the skin and shell are not known. However since *C. picta* skin is impermeable to Na in sea water (Robinson and Dunson, 1976), it is not unreasonable to postulate impermeability of *T. spiniferus* integument in fresh-water. Skins of fresh-water snakes also have a very low permeability to Na (Dunson, 1978). Although the data in Fig. 3 are thus a composite from two species, along with a few judicious guesses, they are considered a reasonably accurate portrayal of routes of Na regulation in fresh-water turtles. The most important feature of this system is the presence of active Na uptake in the pharynx, cloaca, and cloacal bursae (Dunson, 1967). Thus even when not feeding, turtles can maintain Na balance. However fresh-water turtles apparently lose this ability at low temperatures when the active transport mechanism is inhibited; plasma Na may drop to near 70 mmol l^{-1} during hibernation (Dunson and Weymouth, 1965). On removal from low temperatures plasma Na can be rapidly restored by active uptake alone. One softshell turtle kept at 8 °C had a plasma Na of 84 mmol l^{-1}. After being placed in 21 °C tapwater for 20 days, plasma Na was raised to 122 mmol l^{-1} without feeding. It is interesting that the cloacal route (primarily the urine) remains the largest source of Na loss. Yet there must be considerable reabsorption of Na, particularly when the body Na pool is depleted. Seidel (1975) reported cloacal fluid Na concentrations of about 20 mmol l^{-1} in softshells in fresh-water at 25 °C. Softshell turtles (some moribund) kept at 8 °C in simulated hibernation had much lower bladder urine concentrations (73–665 μmol l^{-1} for Na) and often had sizeable volumes of fluid in the bladder (Dunson, unpublished). Even at these low temperatures a large Na gradient between the ECF and the bladder was maintained. Further study of the mechanisms of Na retention in fresh-water turtles would certainly be useful. A preliminary study of *C. picta* showed no net uptake of Na or K from water (Trobec and Stanley, 1971). It would be interesting to know whether other aquatic reptiles possess a system of Na regulation similar to that described above, or whether it may be limited to certain turtles.

Too few data are available to construct a water budget for any fresh-water reptile. Dunson (1967) found that *Trionyx spiniferus* underwent an initial weight loss coincident with Na loss on placement in distilled water. The amount of weight loss (assumed to be water) was inversely related to body weight and varied from 1 to more than 5% per day in turtles of 20–280 g weight. It seems illogical that a net loss of water would occur when the water concentration of the softshell's body fluids is less than that of the medium. However the loss of water might be associated with volume regulation of the ECF. While the initial loss of Na into distilled water from a 184 g softshell turtle (Dunson, 1967) is only about 140 μmol at its peak, this represents approximately 4% of the ECF Na or 15% of plasma Na. Changes of this magnitude might well cause compensatory renal excretion of water.

A subsequent study by Bentley and Schmidt-Nielsen (1970) has suggested that the skin and/or the shell of *Trionyx spiniferus* are important sites for net water uptake in fresh-water and net water loss in 3.3% NaCl. The rate of cutaneous exchange in *Chrysemys scripta* was considerably less than in *T. spiniferus*. This is not too surprising in view of the large amount of cutaneous respiration that takes place in the softshell turtle (Dunson, 1960; Girgis, 1961).

Further evidence for the permeability to water of aquatic reptilian skin comes from a study by Bentley and Schmidt-Nielsen (1965) on caimans. They found that 70% of water uptake from water and 75% of the water loss in air occurs through the skin. Diefenbach (1973) has criticized these findings on the basis that the phenol red used to measuring drinking rates is partly absorbed by the gut and thus is not an accurate indicator of drinking. He believes that drinking was underestimated, and in his own experiments found no significant weight increase due to dermal intake of water. In Diefenbach's experiments, respiratory losses were large and apparently negated the effect of any net uptake of water across the skin. Body weight changes were used to measure water uptake; this method provides no information on bidirectional water fluxes. The question of water balance in fresh-water crocodilians needs to be re-examined with modern isotopic techniques.

Na permeability of the crocodilian skin is also a topic of some interest. It has generally been found that the skin of terrestrial, fresh-water, marine and estuarine reptiles is either impermeable or has an extremely low permeability to Na in both *in vivo* and *in vitro* tests (Robinson and Dunson, 1976; Dunson and Robinson, 1976; Dunson, 1978). A fresh-water turtle (*Chrysemys picta*) tested in sea-water had a Na influx as low as marine reptiles which have Na impermeable skins. Bentley and Schmidt-Nielsen (1965) measured a net loss of Na across the skin of caimans in distilled water of $0.01 \ \mu\text{mol cm}^{-2} \ \text{h}^{-1}$. This is quite comparable to Na fluxes measured between sea-water and Ringer's solution across the isolated skins of terrestrial and fresh-water snakes (Dunson, 1978). Evans and Ellis (1977) measured a Na efflux from *Crocodylus acutus* in fresh-water of $1.8 \ \mu\text{mol} \ (100 \ \text{g})^{-1} \ \text{h}^{-1}$, about 10–15% of which was assigned to the integumentary route. Note the similarity of this efflux value to those from fresh-water turtles (Fig. 3, Dunson, 1967). Further work on skin permeability is obviously necessary before a pattern emerges from these somewhat divergent findings.

C. In Estuarine Reptiles

There are very few truly estuarine reptiles that are not found mainly in fresh-water or in the sea. Those few whose distribution is known to be limited primarily to the estuary include the diamondback terrapin (*Malaclemys terrapin*), the salt marsh snakes (*Natrix fasciata clarki, N. f. taeniata, N. f. compressicauda, N. sipedon williamengelsi*) and the dog-faced water snake (*Cerberus rhynchops*). Others such as the loggerhead sea turtle (*Caretta caretta*) and the American crocodile (*Crocodylus acutus*) may fall into this category, but there is insufficient information on their natural history. Since the terrapin (*Malaclemys terrapin*) has been the most thoroughly studied of

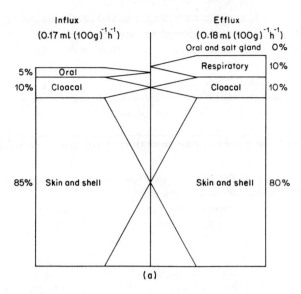

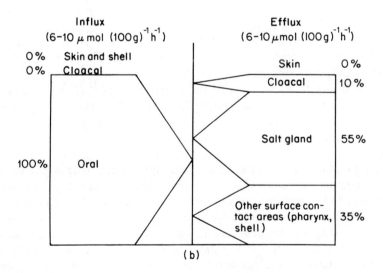

Figure 4 Water and Na budgets of the fasted estuarine diamondback terrapin, *Malaclemys terrapin*, in 100% sea-water. Routes of exchange are rounded off to the nearest 5%.
 (a) Water budget. The salt glands are not secreting.
 (b) Na budget. The salt glands are minimally active.
From Robinson and Dunson (1976); reproduced by permission of Springer-Verlag

these species, we will use it as an example of the physiological adaptations for osmoregulation that have developed in estuarine reptiles.

Mechanisms of body fluid volume regulation in the terrapin have already been discussed. We will focus here on the routes of uptake and loss of water and Na while the animals are fasting in sea-water (Fig. 4). A familiar picture emerges from an examination of water movements (Fig. 4(a)). Most of the influx and efflux are through the integument. Efflux is slightly higher than influx in fasted animals, so that a minute but significant dehydration occurs. The rate of net water loss is low enough that survival in sea-water without access to drinking water is possible for extremely long periods. It is not yet known if enough water can be obtained from the food to balance the net loss and enable indefinite survival in sea-water.

The pattern of Na exchange in the terrapin is completely different from that of water exchange (Fig. 4(b)). Uptake is completely oral; the total intake is little more than that of *Trionyx spiniferus* in fresh-water. Indeed the fresh-water turtle *Chrysemys picta* has a rate of Na influx in sea-water nearly identical to that of the terrapin. Na efflux in the terrapin is divided among a number of routes, but the salt gland is probably predominant. Even when the salt gland is 'inactive' there is a cephalic Na loss of about 3 μmol $(100 \text{ g})^{-1}$ h^{-1}. This can increase to over 70 (mean 43 μmol $(100 \text{ g})^{-1}$ h^{-1}) after salt loading. Thus excretory mechanisms seem more than adequate to cope with any possible Na increases in the ECF. However this is not true for water, where a gradual decline in TBW must occur unless feeding, or fresh- or brackish-water drinking occur. Robinson and Dunson (1976) have predicted from these data that the observed rise in Na concentration of the ECF is due almost entirely to water loss and not Na gain.

Studies currently in progress of the estuarine snakes *N. f. clarki* and *C. rhynchops* have yielded similar results (Dunson, 1978). The skin is permeable to water, but not to Na. Water turnover of fresh-water reptiles placed in sea-water is usually higher than in estuarine or marine forms. However efflux is only slightly greater than influx in any species. Fresh-water snakes also have a higher initial rate of Na influx in sea-water due mainly to uptake in the head region. It is questionable whether this cephalic uptake is due to voluntary drinking or to leakage into the mouth. It is believed that the resulting rise in ECF Na (and perhaps secondarily to net water loss) leads subsequently to massive drinking and eventual death. The ability of *Natrix fasciata clarki* to survive in sea-water, whereas *N. f. pictiventris* dies, has always been attributed solely to a behavioural reluctance to drink sea-water when it is dehydrated (Pettus, 1956, 1958, 1963). Although this may be partly true, it now appears that the differences in initial Na influx and water loss between the salt marsh and fresh-water races of *Natrix fasciata* might better explain sea-water tolerance in *N. f. clarki*. In other words *N. f. pictiventris* dies primarily because it becomes dehydrated much more rapidly and is then forced to drink sea-water. Pettus also claimed that the skins of both fresh-water and salt marsh subspecies were impermeable to water, which is certainly not true.

His technique of osmotically measuring net water movements was incapable of revealing the existence of almost equally balanced influxes and effluxes.

The total water turnover of *N. f. clarki* is very similar to that of *P. platurus* and *M. terrapin* in which the skin accounts for most of the water flux. Krakauer (1970) also reported that the skin of the mangrove snake (*N. f. compressicauda*) is permeable to water, although less so than that of *N. f. pictiventris*. At present it appears that *N. f. clarki* does not have a salt gland, but further tests will be required to definitely eliminate the possibility. This subspecies now appears to be one of the best examples known of the early stages of evolution of the ability to hypo-osmoregulate in sea-water. However the existence within the same species, in close geographical proximity, of races differing markedly in tolerance to sea-water seems without precedent. It would be most interesting to examine the physiological attributes of those individuals described by Pettus (1956) as intermediate in colour pattern (between the true salt marsh race and the true fresh-water race) which occur in intermediate brackish-water habitats. Physiological studies of another salt marsh snake, *Natrix sipedon williamengelsi* (presumably independently evolved) would also be useful. The importance of examining different evolutionary lines of marine reptiles is borne out by the recent exciting finding that a rear-fanged homalopsine snake (*Cerberus rhynchops*) from Philippine mangrove swamps has a new type of salt gland (Dunson and Dunson, unpublished). Reptiles have colonized the sea secondarily after a long terrestrial existence, and different mechanisms of osmoregulation have evolved in certain cases (i.e. different salt glands in lizards, turtles, and two of the three types of marine snakes), but not in others (i.e. all have skins virtually impermeable to Na). Of course only a few marine reptiles have been studied, and there may yet be unsuspected variation in many osmoregulatory mechanisms.

Seidel (1975) has recently reported on a population of softshell turtles living in a brackish-water habitat in the Pecos River in New Mexico (about 0.79% saline). Since sea-water is about 3.5% saline, the Pecos River could be considered the equivalent of about 23% sea-water, although the ionic composition is not the same. This does not represent a very severe osmoregulatory problem for *T. spiniferus* since the turtles are hyperosmotic to the water. The existence of the Nile softshell (*Trionyx triunguis*) along the Israeli sea coast is a more interesting case since the adults survive for long periods in sea-water (Neumann, unpublished). The plasma remains hypoosmotic to sea-water, despite the absence of a salt gland. The large size of these turtles, coupled with integumental impermeability to Na could account for the tolerance to sea water. However it would be most interesting to examine the water and Na budgets of these turtles while fasting in sea-water, since they are known to respire aquatically through the skin and pharynx (Girgis, 1961).

Several species of crocodiles appear to be primarily estuarine. *Crocodylus porosus* is the species most often encountered on isolated islands and reefs in the Western Pacific, but little is known about its osmoregulation. *C. acutus* lives in brackish mangrove swamps in the Americas and although it has been sighted in full-strength sea-water (as indeed have American alligators), it appears to be primarily a fresh-water and brackish-water species. A typical fresh-water crocodilian, the caiman, loses weight very rapidly in 3.3% NaCl (about 5% in 18 h) and drinks the saline water (about 4.4% body weight in 18 h) (Bentley and Schmidt-Nielsen, 1965). Serum Na rose rapidly from 134 to 160 mmol l^{-1} and the animals began to sicken. Young *C. acutus* lose weight less rapidly in 100% sea-water than the caiman (about 1.7% body weight per day; Dunson, 1970), but full strength sea-water is obviously not a healthy environment for them. Reptiles capable of tolerating sea-water immersion for long periods typically lose about 0.1–0.4% body weight per day, and larger animals are at an advantage in having a smaller relative surface area and a lower rate of weight loss (Dunson and Dunson, 1973). This may explain why a 3.4 kg *C. acutus* studied by Dunson (1970) was able to thrive in 100% sea-water on a diet of fresh-water fish. Dunson (1970) failed to find any evidence of functional salt glands in crocodiles, although further work was recommended on wild-caught crocodiles from highly saline areas.

Evans and Ellis (1977) have recently measured Na fluxes in small *C. acutus*. In fresh-water, Na efflux was 2.5 μmol (100 g)$^{-1}$ h^{-1}, very similar to values for fresh-water turtles (Dunson, 1967). This efflux was divided among the head and forelimbs (69%), integument posterior to the forelimbs (12%), and the cloaca (19%). No active uptake of Na was detected but it is not clear if the animals were depleted of body Na sufficiently to stimulate such a system. The essentially fresh-water physiology of young *C. acutus* was clearly demonstrated by their inability to tolerate salinities above 25% sea-water (about 125 mmol l^{-1} for Na which is hypo-osmotic to blood). The Na efflux remained low, about 5–14μmol (100 g)$^{-1}$ h^{-1}, of which 46% was from the head and forelimbs, 20% from the integument posterior to the forelimbs, and 34% from the cloaca.

D. In Marine Reptiles

Of all reptiles, the sea snakes show the most impressive adaptations to marine life. Many of them are completely divorced from their ancestral home, the land, unlike the sea turtles that must nest on shore. Sea snakes have only recently begun to receive the long overdue attention of experimental biologists (Dunson, 1975b). It is surprising that many aspects of the circulatory and respiratory physiology of sea snakes are similar to those of land snakes (Heatwole and Seymour, 1975). The major evolutionary changes necessary for colonization of the sea may have been related instead to physiological mechanisms of osmoregulation (Dunson, 1975a) and perhaps also to behaviour associated with feeding. In any case the sea snakes are superbly suited for hypo-osmotic regulation, despite their relatively small size and the lack of a kidney capable of excreting urine more concentrated than the plasma. The most obvious osmoregulatory organ of sea snakes is the posterior sublingual salt gland, which will be discussed later. In this section the routes of intake and loss of water and Na will be discussed, in comparison with data for the sea turtles *Caretta caretta* and *Chelonia mydas*.

A tentative water budget for fasting *Pelamis platurus* in sea-water is shown in Fig. 5. Both intake and loss are believed to be mainly dermal. The aquatic

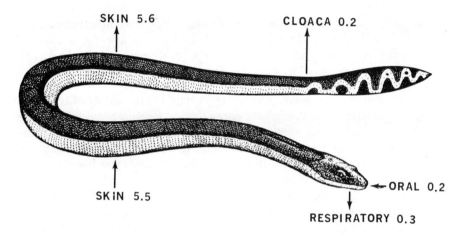

Figure 5 A diagrammatic representation of the daily water budget of the fasted yellow-bellied sea snake (*Pelamis platurus*) in sea-water. Data mainly from Dunson and Robinson (1976) and expressed as ml $(100 \text{ g})^{-1} \text{ d}^{-1}$

efflux of 5.8 ml $(100 \text{ g})^{-1}$ day^{-1} plus 0.3 ml $(100 \text{ g})^{-1}$ day^{-1} estimated respiratory loss yields a total efflux of 6.1. The respiratory loss was derived from a value of 4.5 ml $(100 \text{ g})^{-1}$ h^{-1} for aerial oxygen consumption for *P. platurus* while floating in water (Graham, 1974) and a water loss to oxygen ratio of about 2.8 mg ml^{-1} oxygen. The latter represents a figure chosen near the middle of the range of 0.5–4.9 mg ml^{-1} measured by Bentley and Schmidt-Nielsen (1966) in a variety of reptiles from arid to fresh-water habitats. The influx was calculated by subtracting the measured net loss of 0.4 (Dunson and Robinson, 1976) from the above total efflux of 6.1 ml $(100 \text{ g})^{-1}$ h^{-1}. The oral intake of water represents possible diffusion through oral membranes and a small amount of leakage around the labial scales. No real drinking is believed to take place in sea-water. A figure was derived by calculating the amount of sea-water that would have to be swallowed to account for the entire Na influx, which is completely oral. This probably gives an inflated value for water uptake since some Na movement must occur by diffusion across the oral epithelium. The dermal water efflux was calculated by subtracting a moderately low value for cloacal water loss of *P. platurus* (Dunson, 1968) from the aquatic efflux. Although these figures are tentative, the preponderant role of the skin in water balance is striking. This situation is of course very similar to that of the larger terrapin (Fig. 4(a)), which has a total efflux of about 4.3 ml $(100 \text{ g})^{-1}$ day^{-1}. Krakauer (1970) reported aquatic water effluxes in the mangrove water snake (*N. f. compressicauda*) of 4.85 ml $(100 \text{ g})^{-1}$ day^{-1}; the fresh-water banded water snake had a much greater efflux of 17.26 ml $(100 \text{ g})^{-1}$ day^{-1} in sea-water. An extensive survey of water flux rates in sea snakes (Dunson, 1978) shows however that several fully marine snakes have rates of water efflux (and influx) in excess of 17 ml $(100 \text{ g})^{-1}$ day^{-1}. The reason for this is not clear at present, although there might be an association with the extent of dermal gas transport. Thus the absolute magnitude of water fluxes does not predict how well a snake will survive in sea-water.

A factor which seems much more closely related to the ability of reptiles to tolerate sea-water is the dermal Na permeability. *Pelamis platurus*, as in the cases of several estuarine reptiles studied (i.e. *Malaclemys terrapin, Cerberus rhynchops*, and *Natrix fasciata clarki*), seems to have a Na impermeable skin (Dunson and Robinson, 1976). Both *in vivo* and *in vitro* experiments support this contention. When the head of a snake in ^{24}Na-labelled sea-water is held out of the solution, measurable uptake of ^{24}Na ceases. Isolated skin preparations, with few exceptions, show no transfer of ^{24}Na between sea-water on the outside and a Ringer's solution on the inside. Indeed this impermeability can be duplicated with a shed skin, showing that keratin is the barrier to Na movement. In fasting snakes Na influx $(8.2 \pm 7.1\ \mu\text{mol}\ (100 \text{ g})^{-1}\ \text{h}^{-1})$ balances efflux $(7.1 \pm 2.3\ \mu\text{mol}\ (100 \text{ g})^{-1}\ \text{h}^{-1})$ within the accuracy of the flux techniques used. The salt gland has an enormous capacity to excrete Na (up to 140 μmol $(100 \text{ g})^{-1}\ \text{h}^{-1}$) in comparison with the equilibrium fluxes. As in the terrapin (see Fig. 4(b)), *P. platurus* can readily maintain Na balance, but not water balance when fasting.

The concept advanced above of dermal impermeability to Na in those reptiles capable of surviving long periods in sea-water has been challenged by studies on sea turtles by Evans (1973) and Kooistra and Evans (1976). *Caretta caretta* in sea-water has a Na efflux of 300 μmol $(100 \text{ g})^{-1}\ \text{h}^{-1}$, which was calculated as 60% nasal, 35% dermal, and 5% cloacal. Evans clearly points out, however, that this calculation was based on several assumptions, including a crucial one that nasal salt gland excretion was equal to that measured by Holmes and McBean (1964) for *Chelonia mydas* (134 μmol $(100 \text{ g})^{-1}\ \text{h}^{-1}$). Since this is certainly not the case, we must dismiss Evan's speculation that the skin is permeable to Na in *C. caretta* until further experimental evidence is presented. The enormous Na efflux in comparison with those measured in *M. terrapin* and *P. platurus* is evidence that these *C. caretta* were not at equilibrium, and that the salt gland could account for most of the efflux. On placement in fresh-water, Na efflux in *C. caretta* dropped to 1 μmol (100 g)$^{-1}$ h^{-1} within 1–2 days, a dramatic demonstration of the ability of this species to adapt to fresh-water. *C. caretta* is often found in coastal areas and may be essentially estuarine in habits, at least at certain ages. In subsequent studies on the green sea turtle (*Chelonia mydas*) Kooistra and Evans (1976) directly partitioned Na efflux *in vivo*. They measured an efflux in sea water of 131 μmol (100 g)$^{-1}$ h^{-1}, again a high value for marine reptiles; 90% was ascribed to the head, 5% to the cloaca and 5% to the skin. Their conclusion was that a low epithelial Na permeability was one of the major mechanisms of hypo-osmoregulation. The disagreement over dermal Na permeability has thus been resolved into the question of whether there is merely a low permeability or essentially complete impermeability. Partitioning experiments are subject to the criticism that the animal is unnaturally restrained. For example Kooistra and Evans (1976) observed that the Na efflux in turtles in a chamber was less than half that of free swimming animals. The process of cannulation of the cloaca also blocks entry of ions and water from the medium, an important process in fresh-water forms (Dunson, 1967). Other serious difficulties arise from the possible permeability of the cannula balloon to the isotope under study, to leaks between the partitioned compartments, and to inadequate controls to determine if leaks have occurred from the injection site of the isotope. The most reliable way of confirming or denying the permeability of the sea turtle skin to Na would be by *in vitro* studies of the skin. Until such experiments are carried out, most evidence favours the hypothesis that the skin of all marine reptiles are virtually impermeable to Na.

VI. SPECIAL ORGANS OF EXCRETION AND UPTAKE

A. Salt Glands

The subject of reptilian salt glands has been recently reviewed by Dunson (1975a, 1976) and Peaker and Linzell (1975). These specialized transport organs have excited the attention of physiologists since their discovery in 1958 by Schmidt-Nielsen, Jörgensen and Osaki, and Schmidt-Nielsen and Fänge. Originally they were thought to secrete only highly concentrated solutions of NaCl (Schmidt-Nielsen, 1960), but glands in some terrestrial species are now known to excrete K and HCO_3 also.

Salt glands have evidently evolved many different times (Table 8, Fig. 6). Four non-homologous structures have developed in similar ways for electrolyte excretion; these are rectal, lachrymal, nasal and posterior sublingual glands. There is good evidence that a fifth, the premaxillary gland, is salt secreting in the marine homalopsine snake, *Cerberus rhynchops* (Dunson and Dunson, unpublished observations). Only two major groups, the lizards and the birds, have homologous nasal salt glands. This may be only a coincidence,

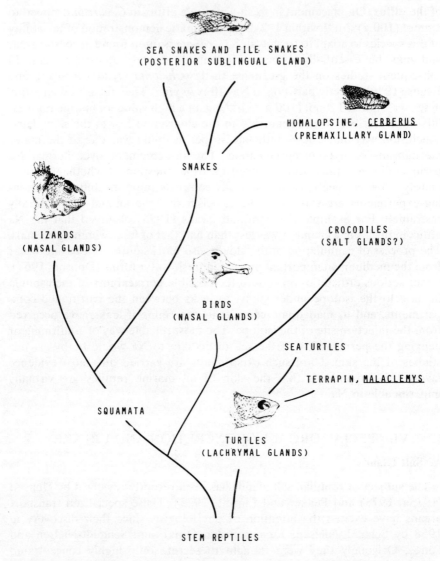

Figure 6 The independent evolution of salt glands in turtles, birds, lizards and snakes

rather than evidence for a common ancestor with nasal salt glands. However salt glands are common in terrestrial lizards, whereas they have not yet been found in terrestrial turtles or snakes. This indicates that the evolution of lachrymal salt glands in the sea turtles (Cheloniidae) and the diamondback terrapin (Emydidae) was independent. The same argument holds true for the posterior sublingual salt gland in the true sea snakes (Hydrophiidae) and the

Table 8. Evolution of salt glands in vertebrates

Taxonomic group	Location of salt glands	Genus	First published description
Elasmobranch fish	All rectal	*Squalus*	Burger and Hess (1960)
Turtles			
Cheloniidae	All lachrymal	*Caretta*	Schmidt-Nielsen and Fänge (1958)
Emydidae		*Malaclemys*	
Lizards	All nasal		
Iguanidae		*Amblyrhynchus*	Schmidt-Nielsen and Fänge (1958)
Agamidae		*Uromastyx*	Schmidt-Nielsen and coworkers (1963)
Scincidae		*Trachydosaurus*	Braysher (1971)
Xantusidae		*Xantusia*	Minnich, unpublished
Varanidae		*Varanus*	Green (1969)
Teiidae		*Ameiva*	Hillman and Pough (1976)
Lacertidae		*Acanthodactylus*	Duvdevani (1972)
Snakes	Two types:		
Hydrophiidae	Posterior sublingual	*Pelamis*	Dunson and coworkers (1971)
Acrochordidae	Posterior sublingual	*Acrochordus*	Dunson and Dunson (1973)
Homalopsidae	Premaxillary	*Cerberus*	Dunson and Dunson, unpublished
Crocodiles	Unknown	*Crocodylus*	Dunson (1970)
Birds	All nasal		Schmidt-Nielsen and coworkers (1958)

file snake (Acrochordidae). There has been some confusion in the literature regarding the possible presence of salt glands in the crocodilians. However there is absolutely no firm evidence for the presence of a functional salt gland (Dunson, 1970). Freshly captured *Crocodylus porosus* from a highly saline area need to be studied to insure that the process of acclimation of any possible salt glands is complete. Of the five known salt glands, one has developed in association with the rectum (elasmobranchs) one with the eyeball (turtles), one with the nasal passages (birds and lizards), one with the tongue sheath (sea and file snakes) and one with the oral cavity and snout (marine homalopsine snake). The convergence in ultrastructure between these five glands is indeed striking, with the exception that glands in marine birds develop basal invaginations not present in other forms. In the early stages of salt adaptation the gland of marine birds is essentially identical to that of reptiles. In terrestrial birds the basal infoldings probably never develop (Dunson, Dunson and Ohmart, 1976).

In evaluating the function of salt glands, the concentration of electrolytes and the overall rate of excretion are probably the two most important factors to consider (Table 9). Since most of the collections of salt gland fluid (with the exception of the sea snakes) are not from cannulated glands, caution in interpretation of the results is suggested. Lachrymal gland fluid collected from the surface of the cornea can be modified by evaporation. Nasal fluid is especially prone to modification in concentration by dissolution of salt encrustations and evaporation. It is also uncertain in many cases if the maximum obtainable rate of excretion has been reached by dehydrating or salt loading the animal over a brief or more prolonged period. Few if any species have been systematically tested for the lability of the gland response to injection or ingestion of different electrolytes. Our knowledge would be substantially advanced by initiation of such detailed studies on a few selected species. Comparative surveys are still recommended in those groups (i.e. the crocodilians and marine snakes) in which there is some reason to believe that an unknown gland may be present. Our current knowledge suggests that turtles and snakes have glands secreting mainly NaCl, at mean concentrations between 500 and 800 mmol l^{-1} (Table 9). Among the sea snakes there are two main groups of glands in relation to secretion concentration (Dunson and Dunson, 1974). One secretes near the concentration of sea-water and the other considerably above it. A most interesting aspect of this division in concentrating capacity is the lack of relation to the rate of secretion (and thereby the flow rate). For example the genus *Aipysurus* has a highly concentrated secretion, but some species have rates of excretion both higher and lower than those of forms with a secretory fluid considerably less concentrated. Gland weight, on the other hand, seems to be related to the total rate of excretion. However it is curious that the large variation in relative gland weight found among sea snakes has no obvious relation to their habits. They are all fully marine and almost all eat fish. It may be that such a situation

Table 9. Excretion of electrolytes by some reptilian salt glands

Species	Habitat[a]	Conditions	Fluid concentration (mmol l^{-1})			Mean maximum excretion rate (μmol (100 g body wt)$^{-1}$) h^{-1})			$\frac{Na}{K}$	Reference
			Cl	Na	K	Cl	Na	K		
Turtles										
Chelonia mydas	M	In sea-water	—	685	20.7	—	134	4.9	27.3–37.7	1
Malaclemys terrapin	E	In fresh-water	—	288	—	—	0.4	0.2	—	2
Malaclemys terrapin	E	In sea-water	—	682	32.4	—	43	—	24.0	2, 3
Lizards										
Dipsosaurus dorsalis	T	Various salt loads	—	494–1032	640–1387	—	0.01–12.5	0.6–18.7	0.02–3.0	4
Sauromalus obesus	T	Various salt loads	827	121–150	540–1102	—	0.7–3.25	12.5–31.1	0.04–0.5	5
Conolophus subcristatus	T	Various salt loads	486	692	214	—	16–26	1.6–5.0	3.2–15.9	6
Varanus semiremax	E, T	NaCl loads	745	686	57	—	33.9	2.7	2–21	7
Amblyrhynchus cristatus	M	Various salt loads	—	—	—	—	159–255	7.6–51	5–21	6
Snakes										
Acrochordus granulatus	M	NaCl loads	492	483	15	48	—	—	32	8
Pelamis platurus	M	NaCl loads	594	584	22.6	142	—	4.5	26	9
Hydrophis elegans	M	NaCl loads	520	509	20	35	28	0.9	25	10
Aipysurus fuscus	M	NaCl loads	749	—	—	24	—	—	—	10
Aipysurus laevis	M	NaCl loads	791	798	28	157	165	6	29	10
Aipysurus eydouxii	M	NaCl loads	749	703	34	222	—	—	21	10
Lapemis hardwickii	M	NaCl loads	704	676	23	162	158	5	29	10

[a] M = marine, E = estuarine, T = terrestrial.
1 Holmes and McBean (1964).
2 Dunson (1970).
3 Robinson and Dunson (1976).
4 Schmidt-Nielson and coworkers (1963); Templeton (1966); Minnich (1970); Shoemaker and coworkers (1972).
5 Norris and Dawson (1964); Nagy (1972); Templeton (1964).
6 Dunson (1969a).
7 Dunson (1974).
8 Dunson and Dunson (1973).
9 Dunson (1968).
10 Dunson and Dunson (1974).

is related to differing degrees of precision in regulation of ECF electrolyte concentration. A sea snake with a small gland capable of only a low rate of excretion might have to temporarily tolerate a larger rise in Na concentration after ingestion of salt.

Lizard nasal salt glands are especially fascinating because of their ability to excrete Na, K, Cl and HCO_3 (see Dunson, 1976). Although the data are very incomplete, it is quite likely that the ability to switch from transport of one ion to another is limited in a characteristic way for each species. The changeover occurs relatively slowly and seems well adapted for coping with shifts in dietary Na to K ratios in a desert herbivorous lizard such as *Dipsosaurus dorsalis*. Those lizards associated with coastal or marine habitats or with a more carnivorous diet may show a greater ability to excrete Na than their mainly herbivorous terrestrial relatives. However the herbivorous marine iguana holds the record for the greatest rate of Na or K excretion.

Shoemaker, Nagy and Bradshaw (1972) have experimentally verified that anionic ratios (Cl/HCO_3) also vary selectively with the salt injected into *Dipsosaurus dorsalis*. The question of control of this type of secretion is a most interesting topic that deserves a great deal more study. The classical view is of neural control by parasympathetic cholinergic fibres. Osmoreceptors might monitor the concentration of the body fluids. Sea snake salt glands respond to sucrose injections (Dunson, 1968). Those of the desert iguana do not (Shoemaker, Nagy and Bradshaw, 1972), and seem instead to be tied in with specific Na and K receptors. Since the change in Na to K ratio of the nasal fluid is slow, hormone mediation is suspected, perhaps from the adrenals.

The similarity in ultrastructure of the many independently evolved reptilian salt glands would suggest a similarity in basic secretory mechanism. We can consider the sea snake posterior sublingual gland as an example of the reptilian type. The general position of the gland in indicated in Fig. 7. It encloses the tongue sheath; multiple ducts enter the sheath and fluid is

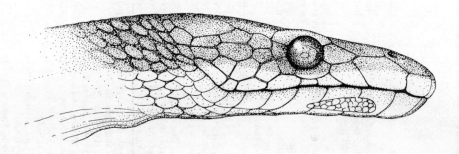

Figure 7 The head of the yellow-bellied sea snake, *Pelamis platurus*, showing the posterior sublingual salt gland in the lower jaw

expelled to the outside by extrusion on the tongue. In cannulation of the gland, a catheter is sewed into the oral opening of the sheath. The gland is composed primarily of a principal cell type (Fig. 8(a)), a mitochondria-packed cell arranged around a small central lumen. The lateral surfaces are enormously elaborated into folds which interdigitate with neighbouring cells across the

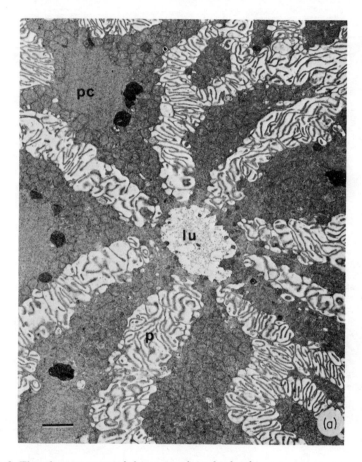

Figure 8 The ultrastructure of the sea snake salt gland.
(a) A cross section of a salt gland tubule from *Aipysurus duboisii*. Note the large intercellular spaces, the tiny apical area of the eight principal cells (pc) around the tubule lumen (lu), and the enormous numbers of lateral cell processes (p). Scale 2 μm. Micrograph by M. K. Dunson

interstitial space. An examination of the basal area of the principal cells (Fig. 8(b)) clearly reveals that they lack the pronounced basal invaginations of the salt-adapted marine bird nasal gland. In fact it appears that the actual basal surface of the cell is nearly as small as is the apical area. The elaborate lateral evaginations dominate the surface architecture of the cell. There are also two other less common cell types, one with abundant rough endoplasmic reticulum, and another that is apparently undifferentiated. In some species a small anterior portion of the gland also has a different structure.

The presumed function of the lateral processes of reptilian salt gland

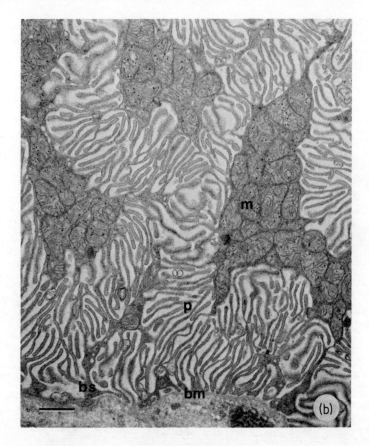

(b) The basal area of a salt gland tubule from *Aipysurus fuscus*. Note that the cells have numerous lateral evaginations, but no basal invaginations as in bird nasal salt glands. Scale 1 μm. Mitochondria (m), lateral cell processes (p), basal cell surface (bs) on the basement membrane (bm). From Dunson and Dunson (1974); reproduced by permission of the American Physiological Society

principal cells is absorption. Unfortunately it is not at all clear what, if anything, is passing through the cell into the lumen. Despite various elaborate theories, we are at a loss to propose a mechanism for active concentration of salts in the gland without some additional basic information on the structure

Table 10. Na–K ATPase content of fresh whole homogenates of reptilian salt glands at 37 °C. Means ±SD. (After Dunson and Dunson (1975), modified.)

Species	Conditions	Plasma Na (mmol l^{-1})	Na-K ATPase, (μmol PO$_4$ (mg wet wt)$^{-1}$ h^{-1})	No.
Terrapin (*Malaclemys terrapin*)	In fresh-water 54–130 days	131 ± 27	1.3 ± 0.6	7
	In sea-water 90–101 days	188 ± 9	2.5 ± 1.0	2
	In sea-water 45–127 days plus salt injections	201 ± 20	4.0 ± 1.0	11
	For terrapins with plasma	> 200	4.7 ± 1.0	5
Desert inguana (*Dipsosaurus dorsalis*)	Dehydrated	173 ± 19	4.6 ± 1.3	3
Sea snake (*Pelamis platurus*)	In sea-water or fresh-water	140–219	5.3 ± 0.7	11

and permeability of the tubule. Possibly the most important point that remains to be established is the nature of the apical junction between adjacent cells in the tubule. ECF can of course pass freely from the basal side up to the junctions, but can water and other small molecules diffuse through? No high resolution electron microscopy has provided a definite answer. A similar problem exists for bird nasal glands, where for years the assumption has been that the junctions were 'tight'. Yet Martin and Philpott (1973) believe that they are 'open' in the duck. As Peaker and Linzell (1975) have pointed out, essentially no physiological work has been done on this problem; an examination of possible inulin or sucrose secretion would be a worthwhile first step (see also Chapter 8). It is known that in resting sea snake glands, the ICF Na concentration is not elevated (Dunson and Dunson, 1974). The more difficult task is to determine the concentration of the ICF, luminal fluid and ECF in the channels between the cells during secretion. At present we do not even know whether the secreted fluid is isosmotic or hyperosmotic to plasma. This would be a fertile field for exploitation of microtechniques developed for the study of salivary glands.

Our biochemical knowledge of reptilian salt glands is limited to two studies on Na–K ATPase (Dunson and Dunson, 1974, 1975). This transport enzyme is present in high concentrations in the three species studied (Table 10). Is it coincidental that levels are quite similar in salt adapted lachrymal, nasal and sublingual salt glands (about 5 μmol (mg wet wt)$^{-1}$ h^{-1})? Differences in the concentrating capacity of sea snake glands were not related to Na–K ATPase levels. It is unfortunately extremely difficult to compare these data on Na–K ATPase content in reptilian glands with some of the many other studies on rectal and avian nasal glands. Differences in technique are quite common and they can cause large changes in the measured activity. A comparative study of Na–K ATPase in the four major types of salt glands would be quite interesting if the rates of gland electrolyte excretion were also measured and correlated with enzyme activity. The distinction in structure between the avian nasal gland (basal infoldings and some lateral processes) and the rectal and reptilian glands (more extensive lateral processes) might be associated with enzymatic differences.

B. The Integument

The skin and shell of reptiles probably comprise the largest combined organ of the body, yet they are one of the most neglected by physiologists. The fact that terrestrial reptiles have so obviously reduced their total water loss in comparison with amphibians led many scientists to believe that the reptilian integument was impermeable to water. This misconception has recently been laid to rest, yet the real role of the skin and shell in osmoregulation remains only partly understood.

Despite the paucity of physiological data on the integument, the morphology is fairly well known. In fact there are so many papers on the structure of the skin that it is possible to cite only the most recent ones that have come to my attention. Maderson and his coworkers have been primarily interested in light microscopic anatomy, embryonic development, the identification of specific cell types in the various layers of the skin, and the shedding process (Maderson, 1965a, 1965b, 1965c; Maderson and coworkers, 1972; Flaxman and Maderson, 1973). The types of keratin in reptilian epidermis have been

Figure 9 The 'scales' of the sea snake *Lapemis hardwickii*. Note the reduction from the typical terrestrial condition to a knobby configuration (photo by H. G. Cogger)

studied by Alexander and Parakkal (1969), Alexander (1970), Parakkal and Alexander (1972), and Baden and coworkers (1966, 1974). Roth and his associates have been especially interested in the ultrastructure, development, and enzymatic activity of the snake epidermis (Roth and Jones, 1967, 1970; Roth and Baden, 1967; Downing and Roth, 1974).

The skins of reptiles vary considerably in their external aspect, although most species are covered with some form of scales, or scale-like structures. In snakes, for example, the typical terrestrial type of scale is an overlapping imbricate type which seems highly effective as a flexible 'armour'. Aquatic species often show a reduction in the size of the scales, as can be seen in the sea snake *Lapemis hardwickii* (Fig. 9). Many sea snakes do retain more typical scale patterns, perhaps due to their habit of feeding in coral or rocks along the bottom. Perhaps the ultimate in reduction of shell scutes is seen in aquatic turtles such as the softshell *Trionyx spiniferus*, the peculiar plateless turtle *Carettochelys insculpta*, and the leatherback sea turtle *Demochelys*

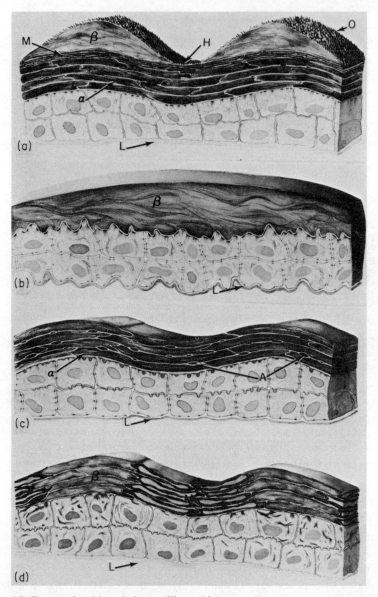

Figure 10 Types of epidermis in reptilian orders.
(a) Snakes and lizards have a series of layers. The *Oberhäutchen* (O) rests upon the β keratin which is thick except in hinge areas (H). The mesos (M) separates the β and α keratin, beneath which lies the living stratum over the basal lamina (L) or basement membrane.
(b) Turtle shell scutes have only a β keratin layer above the living cells.
(c) Turtle neck and leg skin in contrast has only an α keratin layer, with extracellular accumulations of PAS-positive material (A).
(d) Crocodilian epidermis has a variable composition. The centre of each scale has the appearance of β keratin. The hinge region seems to be a mesos layer (M) with characteristics of both α and β keratin.
From Alexander (1970); reproduced by permission of Springer-Verlag

coriacea. Although one might suspect that such species would automatically be more permeable to water and ions, such is probably not the case. Licht and Bennett (1972) and Bennett and Licht (1975) have studied water loss in two species of mutant scaleless snakes in which the skin is aberrantly thin. There was no effect on integumentary water loss, proving that the actual thickness of the skin is no reflection of its permeability.

The permeability of reptilian integument may be better understood by considering the composition of its major layers. Roth and Jones (1967) have classified the skin of the boa constrictor into five ultrastructurally distinct layers (from the inside to the outside): (1) the basal germinal cells that are relatively undifferentiated; (2) a differentiating layer of spinous cells and clear cells which may be in the process of maturation into α-layer cells; (3) the α layer made up of 70 Å filaments; (4) a mesos layer composed of amorphous material, and some filaments; and (5) the β layer made up of tightly packed filaments and amorphous material with a sculptured outer surface termed the *Oberhäutchen* (Fig. 10(a)). Only layers one and two are living. The others are generally termed keratin and presumably are the structures that are responsible for the special permeability characteristics of the skin. Little has been done to test this idea since it is not easy to examine the functions of each layer separately. However the characteristic cyclic shedding of squamates (especially snakes) provides an ideal opportunity to test the permeability of the shed keratin layers. The cast-off 'skin' is actually the old α and β keratin layers (Maderson, 1965a). Since these are non-living, there is no reason to suspect that their permeability would change with the shedding process. Indeed Dunson and Robinson (1976) found that the shed skin of the sea snake *Pelamis platurus* duplicated the Na impermeability of the entire skin. Another possible approach to this problem would be to strip off the outer keratin layers bit by bit with cellophane tape and to observe when abrupt changes in permeability occurred. This would probably allow a separation of the functions of the α and β layers.

It is important to note that all reptiles do not have the same types of arrangement of outer keratin layers (Parakkal and Alexander, 1972; Alexander and Parakkal, 1969; Alexander, 1970). Variations occur even between the shell and the skin of turtles (see Fig. 10). Snakes and lizards (squamates) have the general pattern described previously, except that the β layer becomes much thinner in the hinge regions between scales. The scutes of the turtle shell lack the α layer completely, whereas the skin in contrast has only an α layer. Crocodilian epidermis seems to consist of β keratin in the scales, with a mesos type layer (both α and β keratin) in between. Extensive studies of α type keratin show that it consists of 80 Å filaments embedded in a protein matrix. It is found in a wide variety of tissues such as mammalian epidermis and hair (wool is α keratin). β keratin is known mainly from bird feathers. The occurrence of α and β layers together in the squamate epidermis is unique (Alexander and Parakkal, 1969).

The permeability of the reptilian skin to Na has been discussed previously in relation to salt balance in estuarine and marine reptiles. Under the *in vitro* conditions used, terrapin and sea snake skin seem to be impermeable to Na with sea-water on the outside and a 200 mmol l^{-1} NaCl Ringer's solution on the inside (Robinson and Dunson, 1976; Dunson and Robinson, 1976). Tregear (1966a, 1966b) reports penetration rates (influx) for human skin of 0.0054 μmol cm^{-2} h^{-1} *in vitro* (155 mmol l^{-1} NaCl inside and out) and of 0.0096 μmol cm^{-2} h^{-1} *in vivo* (155 mmol l^{-1} NaCl placed on skin). These rates are only slightly less than Na influxes from sea-water across fresh-water and terrestrial snake skins (about 0.01–0.03 μmol cm^{-2} h^{-1}; Dunson, 1978). There is a need for further studies on the dermal salt fluxes of aquatic reptiles in relation to their life in fresh-water or sea-water.

Our concept of osmoregulation in reptiles has been radically changed by the demonstration that the skin is permeable to water, both in water and in air. Tercafs and Schoffeniels (1965) were probably the first to demonstrate this in reptiles placed in water. They found that the desert lizard *Uromastix acanthinurus* had a net water uptake through the skin of 2.2 μl cm^{-2} h^{-1}. Drinking was eliminated by the failure to find any of the fluorescein dye placed in the water in the gut. *In vitro* studies on skins of a variety of reptiles were carried out with Ringer's solution inside and one tenth Ringer's on the outside. The net uptake of water varied from 1.4 to 13.6 μl cm^{-2} h^{-1} (at 20–30 °C). A most interesting finding was that the fluxes across the skin were larger between two fluids than between a fluid (inside) and air (outside). Gans, Krakauer and Paganelli (1968) found dermal water effluxes (*in vivo*) of two species of terrestrial snakes immersed in deionized water to be 0.3–1.8 μl cm^{-2} h^{-1}. Efflux increased with temperature from 27 to 39 °C (Q_{10} = 2.2). Similarly Krakauer (1970) measured dermal water effluxes of water snakes (in either distilled water or sea-water) of 0.2–0.7 ml (100 g)$^{-1}$ h^{-1}. Sea snake skin (*P. platurus*) with sea-water outside and Ringer's solution inside had water influxes of 0.9–2.0 μl cm^{-2} h^{-1} and effluxes of 1.3–3.6 μl cm^{-2} h^{-1} (Dunson and Robinson, 1976, unpublished). The net water loss was equivalent to about 0.1 ml (100 g)$^{-1}$ h^{-1}. It is clear that there is considerable variation in skin water permeability, related to the aridity of the habitat. In emphasizing that reptilian skin in water and air is permeable to water, it should be realized that overall this permeability is not very large. For example the human stratum corneum is considered one of the most water impermeable of all biological tissues (Scheuplein and Blank, 1971). Yet the rates of water influx (0.3–3.9 μl cm^{-2} h^{-1}, depending on the region tested) are comparable to those measured on reptiles.

C. Oral and Cloacal Membranes in Aquatic Reptiles

Many aquatic turtles irrigate the mucous membranes of the mouth with water, probably for purposes of respiration (Dunson, 1960; Girgis, 1961).

This can be clearly observed in aquaria, even in marine species. Less commonly observed is the passage of water into the cloaca and cloacal bursae (Dunson, 1967). The intimate contact with environmental fluids, whether sea-water or fresh-water, must have profound effects on ion and water balance since these membranes are not protected by keratin layers. Dunson (1967) has examined the Na fluxes across the pharynx, cloacal bursa, and cloaca of the fresh-water turtle *Chrysemys picta*. All three tissues show a net inward Na flux of 0.2–1.2 μmol cm^{-2} h^{-1} in an Ussing cell. The total Na movement can account for the observed short-circuit current, suggesting that only Na is being taken up actively in a net fashion. Thus aquatic turtles possess a system similar to that of frog skin, except that it can be easily isolated from the environment by physical closure of the mouth and cloaca.

With Ringer's solution on both sides of the isolated turtle pharynx, the mucosa (outside) is negative with respect to the serosa (blood side). If distilled water is substituted on the mucosal side, the potential reverses and the short-circuit current almost disappears. The pharynx can then be 'titrated' by adding back known amounts of Na to the mucosal side. There is a resulting increase in potential and short circuit current (Dunson, 1969c). Above the outside Na concentration at which the potential is zero, net uptake of Na may occur. Thus it is interesting that the outside Na concentration of zero potential for the terrestrial box turtle was far higher than it was in the fresh-water *T. spiniferus* and *C. scripta*. This is a reflection of the specialized mechanism for Na uptake that has evolved in aquatic turtles. We do not know if such specialized transports exist in exposed tissues of other aquatic reptiles. It would be especially interesting to know what happens to ion transport in the pharynx of the estuarine terrapin as it moves from fresh-water to sea-water, and whether a salt excretory mechanism is present in the pharynx of marine turtles.

D. The Kidney–Bladder–Gut Complex in Terrestrial Reptiles

The kidney of reptiles plays a less conspicuous role in overall osmoregulation than in mammals. This is mainly due to the inability of the reptilian kidney to excrete a urine hyperosmotic to the plasma. However the kidney is extremely important as part of a system for excretion of nitrogenous compounds and K, and reabsorption of water. Urine elaborated by the kidney may or may not enter the bladder (assuming that the species in question has a bladder). Fluid passing into the cloaca might then be voided or held in the cloacal–intestinal area where modifications in concentration can occur. Regulation of ion and water transport can thus occur at the level of the kidney, the bladder, or the cloaca and gut. Only a brief summary of a few recent findings will be presented here. Dantzler (1976), Dantzler and Holmes (1974), and Minnich (1978) should be consulted for a detailed treatment of this subject.

Urate is the major excretory product of nitrogen metabolism in terrestrial reptiles. It has a low solubility in water in all forms and can thus precipitate out, in which form it contributes nothing to urine osmotic pressure. This allows for an efficient mechanism of water reabsorption in the large intestine and cloaca (Junqueira, Malnic and Monge, 1964, 1966; Minnich, 1972). Net water transport into the blood is believed to be due mainly to the higher colloid osmotic pressure of the plasma (Murrish and Schmidt-Nielsen, 1970). Most of the urate in urine is in the form of salts due to the high pH. Large amounts of salts (especially K) can be excreted in this fashion (see Fig. 2(b)). Minnich (1970a) calculated that the equivalent of about 5 mol l^{-1} Na + K was excreted in the urine of the desert iguana. This mechanism is widespread, but would be especially important in arid zone reptiles lacking salt glands.

A most extraordinary case of the effectiveness of the renal–cloacal–gut excretory system (in the absence of a salt gland) is the Australian agamid lizard *Amphibolurus maculosus*. This is a small animal (about 10 g) capable of producing a 'urine' hyperosmotic to plasma. Although Braysher (1976) refers to this as 'urine', the final process of concentration occurs in the cloaca and rectum, and a more accurate term might be 'cloacal fluid'. The diet is mainly ants, which are high in Na (about 270 mmol l^{-1}). Free water is available only periodically when it rains, and plasma ranged as high as 215 mmol l^{-1} Na and 445 mosmol kg^{-1} during dry periods. After salt loading the voided cloacal fluid had an osmotic pressure of 660–686 mosmol kg^{-1} and a Na concentration of 244–282 mmol l^{-1} (plasma was 452–463 mosmol kg^{-1} and 201–211 mmol l^{-1} Na). Ureteral urine remained hypo-osmotic at all times. It is indeed unfortunate that this species is so rare, since a detailed study of its excretory system would be most interesting. An understanding of the structure and physiology of the cloacal and rectal reabsorptive sites would be especially valuable. It has been suggested that nasal salt glands are necessary for the greatest efficiency in uric acid excretion and water conservation by the cloaca (Schmidt-Nielsen and coworkers, 1963). This hypothesis is based on the presumption that reabsorption of ions from the cloaca precedes the absorption of water and that the excess ions must be excreted by a salt gland. *Amphibolurus maculosus* may pose a contradiction to this idea, but more information is needed on partitioning of water and salt balance in comparison with species having salt glands. The general role of cloacal Na and K transport in the reabsorption of water needs to be examined further.

ACKNOWLEDGMENTS

Supported by NSF grants BMS75-18548 and DES74-24129 to WAD and OFS74-02888 and OFS74-01830 to the Scripps Institution of Oceanography for operation of the R/V Alpha Helix.

REFERENCES

Alexander, N. J. (1970). Comparison of α and β keratin in reptiles. *Z. Zellforsch.*, **110**, 153–65.
Alexander, N. J., and P. F. Parakkal (1969). Formation of α and β-type keratin in lizard epidermis during the molting cycle. *Z. Zellforsch.*, **101**, 72–87.
Baden, H. P., S. I. Roth, and L. C. Bonar (1966). Fibrous proteins of snake scale. *Nature*, **212**, 498–9.
Baden, H., S. Sviokla, and I. Roth (1974). The structural protein of reptilian scales. *J. Exp. Zool.*, **187**, 287–94.
Bellairs, A. (1970). *The Life of Reptiles*, Vol. 2, Universe Books, New York.
Bennett, A. F., and P. Licht (1975). Evaporative water loss in scaleless snakes. *Comp. Biochem. Physiol.*, **52A**, 213–5.
Bentley, P. J. (1959). Studies on the water and electrolyte metabolism of the lizard *Trachysaurus rugosus* (Gray). *J. Physiol.*, **145**, 37–47.
Bentley, P. J. (1971). *Endocrines and Osmoregulation*. Springer-Verlag, New York.
Bentley, P. J., and K. Schmidt-Nielsen (1965). Permeability to water and sodium of the crocodilian, *Caiman sclerops*. *J. Cell Comp. Physiol.*, **66**, 303–10.
Bentley, P. J., and K. Schmidt-Nielsen (1966). Cutaneous water loss in reptiles. *Science*, **151**, 1547–9.
Bentley, P. J., and K. Schmidt-Nielsen (1970). Comparison of water exchange in two aquatic turtles, *Trionyx spinifer* and *Pseudemys scripta*. *Comp. Biochem. Physiol.*, **32**, 363–5.
Bentley, P. J., W. L. Bretz, and K. Schmidt-Nielsen (1967). Osmoregulation in the diamondback terrapin, *Malaclemys terrapin centrata*. *J. Exp. Biol.*, **46**, 161–7.
Bogart, C. M., and R. B. Cowles (1947). Moisture loss in relation to habitat selection in some Floridian reptiles. *Am. Mus. Nov.*, No. 1358, 1–34.
Bradshaw, S. D. (1972). The endocrine control of water and electrolyte metabolism in desert reptiles. In W. S. Hoar and H. A. Bern (Eds), *Progress in Comparative Endocrinology, Proc. 6th Int. Symp. Comp. Endocrinol. Gen. Comp. Endocrinol. Suppl. 3*, Academic Press, New York.
Bradshaw, S. D. (1975). Osmoregulation and pituitary–adrenal function in desert reptiles. *Gen. Comp. Endocrinol.*, **25**, 230–48.
Bradshaw, S. D., and V. H. Shoemaker (1967). Aspects of water and electrolyte changes in a field population of *Amphibolurus* lizards. *Comp. Biochem. Physiol.*, **20**, 855–65.
Braysher, M. L. (1971). The structure and function of the nasal salt gland from the australian sleepy lizard *Trachydosaurus* (formerly *Tiliqua*) *rugosus*: family Scincidae. *Physiol. Zool.*, **44**, 129–36.
Braysher, M. L. (1976). The excretion of hyperosmotic urine and other aspects of the electrolyte balance of the lizard *Amphibolurus maculosus*. *Comp. Biochem. Physiol.*, **54A**, 341–5.
Burger, J. W., and W. N. Hess (1960). Function of the rectal gland in the spiny dogfish. *Science*, **131**, 670–1.
Chew, R. M., and A. E. Dammann (1961). Evaporative water loss of small vertebrates, as measured with an infrared analyzer. *Science*, **133**, 384–5.
Claussen, D. L. (1967). Studies of water loss in two species of lizards. *Comp. Biochem. Physiol.*, **20**, 115–30.
Cohen, A. C. (1975). Some factors affecting water economy in snakes. *Comp. Biochem. Physiol.*, **51A**, 361–8.
Coulson, R. A., and T. Hernandez (1964). *Biochemistry of the Alligator*. Louisiana State University Press, Baton Rouge.

Dantzler, W. H. (1976). Renal function (with special emphasis on nitrogen excretion). In C. Gans and W. R. Dawson (Eds) *Biology of the Reptilia, Physiology A*, Vol. 5, Academic Press, New York, pp. 447–503.

Dantzler, W. H., and W. N. Holmes (1974). Water and mineral metabolism in Reptilia. In M. Florkin and B. T. Scheer (Eds.) *Chemical Zoology*, Vol. 9, Academic Press, New York. pp. 277–336.

Dantzler, W. H., and B. Schmidt-Nielsen (1966). Excretion in fresh-water turtle (*Pseudemys scripta*) and desert tortoise (*Gopherus agassizii*). *Am. J. Physiol.*, **210**, 198–210.

Dawson, W. R., V. H. Shoemaker, and P. Licht (1966). Evaporative water losses of some small Australian lizards. *Ecology*, **47**, 589–94.

Deavers, D. R. (1972). Water and electrolyte metabolism in the arenicolous lizard *Uma notata notata*. *Copeia*, **1972**, 109–22.

Dessauer, H. C. (1970). Blood chemistry of reptiles: physiological and evolutionary aspects. In C. Gans and T. S. Parsons (Eds) *Biology of the Reptilia, Morphology C*, Vol. 3, Academic Press, New York. pp. 1–72.

Diefenbach, C. O. Da C. (1973). Integumentary permeability to water in *Caiman crocodilus* and *Crocodylus niloticus* (Crocodilia: Reptilia). *Physiol. Zool.*, **46**, 72–8.

Dmi'el, R. (1972). Effect of activity and temperature on metabolism and water loss in snakes. *Am. J. Physiol.*, **223**, 510–16.

Dmi'el, R., and B. Zilber (1971). Water balance in a desert snake. *Copeia*, **1971**, 754–5.

Downing, S. W., and S. I. Roth (1974). The derivation of the cells of the epidermal strata of the boa constrictor (*Constrictor constrictor*). *J. Invest. Derm.*, **62**, 450–7.

Dunson, M. K., and W. A. Dunson (1975). The relation between plasma Na concentration and salt gland Na–K ATPase content in the diamondback terrapin and the yellow-bellied sea snake. *J. Comp. Physiol.*, **101**, 89–97.

Dunson, W. A. (1960). Aquatic respuration in *Trionyx spinifer asper*. *Herpetologica*, **16**, 277–83.

Dunson, W. A. (1967). Sodium fluxes in fresh-water turtles. *J. Exp. Zool.*, **165**, 171–82.

Dunson, W. A. (1968). Salt gland secretion in the pelagic sea snake *Pelamis*. *Am. J. Physiol.*, **215**, 1512–7.

Dunson, W. A. (1969a). Electrolyte excretion by the salt gland of the Galápagos marine iguana. *Am. J. Physiol.*, **216**, 995–1002.

Dunson, W. A. (1969b). Reptilian salt glands. In S. Y. Bothelho, F. P. Brooks and W. B. Shelley (Eds) *Exocrine Glands*, University of Pennsylvania Press, Philadelphia. pp. 83–103.

Dunson, W. A. (1969c). Concentration of sodium by freshwater turtles. In D. J. Nelson and F. C. Evans (Eds) *Symposium on Radioecology*, Proc. 2nd Natl Symp. US AEC CONF-670503. pp. 191–7.

Dunson, W. A. (1970). Some aspects of electrolyte and water balance in three estuarine reptiles, the diamondback terrapin, American and 'salt water' crocodiles. *Comp. Biochem. Physiol*, **32**, 161–74.

Dunson, W. A. (1974). Salt gland secretion in a mangrove monitor lizard. *Comp. Biochem. Physiol.*, **47A**, 1245–55.

Dunson, W. A. (1975a). Salt and water balance in sea snakes. In W. A. Dunson (Ed.) *The Biology of Sea Snakes*, University Park Press, Baltimore. pp. 329–53.

Dunson, W. A. (Ed.) (1975b). *The Biology of Sea Snakes*. University Park Press, Baltimore.

Dunson, W. A. (1976). Salt glands in reptiles. In W. R. Dawson and C. Gans (Eds) *Biology of the Reptilia, Physiology A*, Vol. 5, Academic Press, New York. pp. 413–45.

Dunson, W. A. (1978). Role of the skin in sodium and water exchange of snakes placed in sea water. *Am. J. Physiol.* (in the press).
Dunson, W. A., and M. K. Dunson (1973). Covergent evolution of sublingual salt glands in the marine file snake and the true sea snakes. *J. Comp. Physiol.*, **86**, 193–208.
Dunson, W. A., and M. K. Dunson (1974). Interspecific differences in fluid concentration and secretion rate of sea snake salt glands. *Am. J. Physiol.*, **227**, 430–8.
Dunson, W. A., M. K. Dunson, and R. D. Ohmart (1976). Evidence for the presence of nasal salt glands in the roadrunner and the *Coturnix* quail. *J. Exp. Zool.*, **198**, 209–216.
Dunson, W. A., and G. W. Ehlert (1971). Effects of temperature, salinity and surface water flow on distribution of the sea snake *Pelamis. Limnol. Oceanogr.*, **16**, 845–53.
Dunson, W. A., and G. D. Robinson (1976). Sea snake skin: permeable to water but not to sodium. *J. Comp. Physiol.*, **108**, 303–11.
Dunson, W. A., and R. D. Weymouth (1965). Active uptake of sodium by softshell turtles (*Trionyx spinifer*). *Science*, **149**, 67–9.
Dunson, W. A., R. K. Packer, and M. K. Dunson (1971). Sea snakes: an unusual salt gland under the tongue. *Science*, **173**, 437–41.
Duvdevani, I. (1972). The anatomy and histology of the nasal cavities and the nasal salt gland in four species of fringed-toed lizards, *Acanthodactylus* (Lacertidae). *J. Morphol.*, **137**, 353–64.
Epstein, F. H., J. Maetz, and G. De Renzis (1973). Active transport of chloride by the teleost gill: inhibition by thiocyanate. *Am. J. Physiol.*, **224**, 1295–9.
Ernst, C. H. (1968). Evaporative water-loss relationships of turtles. *J. Herp.*, **2**, 159–61.
Evans, D. (1973). The sodium balance of the euryhaline marine loggerhead turtle, *Caretta caretta. J. Comp. Physiol.*, **83**, 179–85.
Evans, D. and T. Ellis (1977). Sodium balance in the hatchling American crocodile, *Crocodylus acutus. Comp. Biochem. Physiol.*, **58A**, 159–162.
Flaxman, B. A., and P. F. A. Maderson (1973). Relationship between pattern of cell migration from the germinal layer and changing patterns of differentiation in the lizard epidermis. *J. Exp. Zool.*, **183**, 209–16.
Gans, C., T. Krakauer, and C. V. Paganelli (1968). Water loss in snakes: interspecific and intraspecific variability. *Comp. Biochem. Physiol.*, **27**, 747–61.
Gilles-Baillien, M. (1970). Urea and osmoregulation in the diamondback terrapin *Malaclemys centrata centrata* (Latreille). *J. Exp. Biol.*, **52**, 691–7.
Gilles-Baillein, M. (1973a). Isomotic regulation in various tissues of the diamondback terrapin *Malaclemys centrata centrata* (Latreille). *J. Exp. Biol.*, **59**, 39–43.
Gilles-Baillien, M. (1973b). Hibernation and osmoregulation in the diamondback terrapin *Malaclemys centrata centrata* (Latreille). *J. Exp. Biol.*, **59**, 45–51.
Gilles-Baillien, M. (1976). Intestinal ion and water transport in the diamondback terrapin acclimatized either to sea or to fresh water. In J. W. L. Robinson (Ed.) *Intestinal Ion Transport*, University Park Press, Baltimore.
Girgis, S. (1961). Aquatic respiration in the common Nile turtle, *Trionyx triunguis* (Forskal). *Comp. Biochem. Physiol.*, **3**, 206–17.
Gordon, M. S., G. A. Bartholomew, A. D. Grinnell, C. B. Jørgensen, and F. N. White (1968). *Animal Function: Principles and Adaptations*, Macmillan, New York.
Graham, J. B. (1974). Aquatic respiration in the sea snake *Pelamis platurus. Resp. Physiol.*, **21**, 1–7.
Green, B. (1969). Water and electrolyte balance in the sand goanna *Varanus gouldii* (Gray). Ph.D. Thesis, University of Adelaide, Australia.
Heatwole, H., and R. Seymour (1975). Diving physiology. In W. A. Dunson (Ed.) *The Biology of Sea Snakes*, University Park Press, Baltimore. pp. 289–327.

Hillman, P. E., and F. H. Pough (1976). Salt excretion in a beach lizard (*Ameiva quadrilineata*, Teiidae). *J. Comp. Physiol.*, **109B**, 169–175.
Holmes, W. N., and R. L. McBean (1964). Some aspects of electrolyte excretion in the green turtle, *Chelonia mydas mydas*. *J. Exp. Biol.*, **41**, 81–90.
Junqueira, L. C. U., G. Malnic, and C. Monge (1964). Note on the function of the ophidian cloaca. *Anais Acad. Brasil. Ciênc.*, **36**, 311–2.
Junqueira, L. C. U., G. Malnic, and C. Monge (1966). Reabsorptive function of the ophidian cloaca and large intestine. *Physiol. Zool.*, **39**, 151–9.
Khalil, F., and G. Abdel-Messeih (1954). Water content of tissues of some desert reptiles and mammals. *J. Exp. Zool.*, **125**, 407–14.
Kooistra, T. A., and D. H. Evans (1976). Sodium balance in the green turtle, *Chelonia mydas* in seawater and freshwater. *J. Comp. Physiol.*, **107**, 229–40.
Krakauer, T. (1970). The ecological and physiological control of water loss in snakes. *Ph.D. Thesis*, University of Florida, Gainesville.
Kraukauer, T., C. Gans, and C. V. Paganelli (1968). Ecological correlation of water loss in burrowing reptiles. *Nature*, **218**, 659–60.
Licht, P., and A. F. Bennett (1972). A scaleless snake: tests of the role of reptilian scales in water loss and heat transfer. *Copeia*, **1972**, 702–7.
Lockwood, A. P. M. (1964). *Animal Body Fluids and Their Regulation*, Harvard University Press, Cambridge.
Maderson, P. F. A. (1965a). Histological changes in the epidermis of snakes during the sloughing cycle. *J. Zool.*, **146**, 98–113.
Maderson, P. F. A. (1965b). The structure and development of the squamate epidermis. In A. G. Lyne and B. F. Short (Eds) *Biology of the Skin and Hair Growth*, Elsevier, New York. pp. 129–53.
Maderson, P. F. A. (1965c). The embryonic development of the squamate integument. *Acta Zool.*, **46**, 275–95.
Maderson, P. F. A., B. A. Flaxman, S. I. Roth, and G. Szabo (1972). Ultrastructural contributions to the identification of cell types in the lizard epidermal generation. *J. Morph.*, **136**, 191–210.
Martin, B. J., and C. W. Philpott (1973). The adaptive response of the salt glands of adult mallard ducks to a salt water regime: an ultrastructural and tracer study. *J. Exp. Zool.*, **186**, 111–22.
Minnich, J. E. (1970a). Water and electrolyte balance of the desert iguana, *Dipsosaurus dorsalis*, in its natural habitat. *Comp. Biochem. Physiol.*, **35**, 921–33.
Minnich, J. E. (1970b). Evaporative water loss from the desert iguana, *Dipsosaurus dorsalis*. *Copeia*, **1970**, 575–8.
Minnich, J. E. (1972). Excretion of urate salts by reptiles. *Comp. Biochem. Physiol.*, **41A**, 535–49.
Minnich, J. E. (1976). Water procurement and conservation by desert reptiles in their natural environment. *Israel J. Med. Sci.*, **12**, 740–758.
Minnich, J. E. (1978). Osmotic and ionic regulation in reptiles. In G. M. O. Maloiy (Ed.) *Comparative Physiology of Osmoregulation in Animals*, Academic Press, New York, in the press.
Minnich, J. E., and V. H. Shoemaker (1970). Diet, behavior and water turnover in the desert iguana, *Dipsosaurus dorsalis*. *Am. Midl. Nat.*, **84**, 496–509.
Minnich, J. E., and V. H. Shoemaker (1972). Water and electrolyte turnover in a field population of the lizard, *Uma scoparia*. *Copeia*, **1972**, 650–9.
Murrish, D. E., and K. Schmidt-Nielsen (1970). Water transport in the cloaca of lizards: active or passive? *Science*, **170**, 324–6.
Nagy, K. A. (1972). Water and electrolyte budgets of a free-living desert lizard *Sauromalus obesus*. *J. Comp. Physiol.*, **79**, 39–62.

Norris, K. S., and W. R. Dawson (1964). Observations on the water economy and electrolyte excretion of chuckwallas (Lacertilia, *Sauromalus*). *Copeia*, **1964**, 638–46.
Parakkal, P. F., and N. J. Alexander (1972). *Keratinization*, Academic Press, New York.
Peaker, M. and J. L. Linzell (1975). *Salt Glands in Birds and Reptiles*, Cambridge University Press, Cambridge.
Pettus, D. (1956). Ecological barriers to gene exchange in the common water snake (*Natrix sipedon*). Ph.D. Thesis, University of Texas, Austin.
Pettus, D. (1958). Water relationships in *Natrix sipedon*. *Copeia*, **1958**, 207–11.
Pettus, D. (1963). Salinity and subspeciation in *Natrix sipedon*. *Copeia*, **1963**, 499–504.
Pitts, R. F. (1968). *Physiology of the Kidney and Body Fluids*, Year Book Medical Publishers, Chicago.
Prange, H. D., and K. Schmidt-Nielsen (1969). Evaporative water loss in snakes. *Comp. Biochem. Physiol.*, **28**, 973–5.
Robinson, G. D., and W. A. Dunson (1976). Water and sodium balance in the estaurine diamondback terrapin (*Malaclemys*). *J. Comp. Physiol.*, **105**, 129–52.
Roth, S. I., and H. P. Baden (1967). An autoradiographic study of the sites of protein synthesis in the epidermis of the indigo snake (*Drymarchon corais couperi*). *J. Exp. Zool.*, **165**, 345–54.
Roth, S. I., and W. A. Jones (1967). The ultrastructure and enzymatic activity of the boa constrictor (*Constrictor constrictor*) skin during the resting phase. *J. Ultra. Res.*, **18**, 304–23.
Roth, S. I., and W. A. Jones (1970). The ultrastructure of epidermal maturation in the skin of the boa constrictor (*Constrictor constrictor*). *J. Ultra. Res.*, **32**, 69–93.
Scheuplein, R. J., and I. H. Blank (1971). Permeability of the skin. *Physiol. Rev.*, **51**, 702–47.
Schmidt-Nielsen, K. (1960). The salt-secreting gland of marine birds. *Circulation*, **21**, 955–67.
Schmidt-Nielsen, K. (1964). *Desert Animals*. Oxford University Press, Oxford.
Schmidt-Nielsen, K., and P. J. Bentley (1966). Desert tortoise *Gopherus agassizii*: cutaneous water loss. *Science*, **154**, 911.
Schmidt-Nielsen, K., A. Borut, P. Lee, and E. Crawford Jr. (1963). Nasal salt excretion and the possible function of the cloaca in water conservation. *Science*, **142**, 1300–1.
Schmidt-Nielsen, K., and R. Fänge (1958). Salt glands in marine reptiles. *Nature*, **182**, 783–5.
Schmidt-Nielsen, K., C. B. Jørgensen, and H. Osaki (1958). Extrarenal salt excretion in birds. *Am. J. Physiol.*, **193**, 101–7.
Seidel, M. E. (1975). Osmoregulation in the turtle *Trionyx spiniferus* from brackish and freshwater. *Copeia*, **1975**, 124–8.
Shoemaker, V. H., K. A. Nagy, and S. D. Bradshaw (1972). Studies on the control of electrolyte excretion by the nasal gland of the lizard *Dipsosaurus dorsalis*. *Comp. Biochem. Physiol.*, **42A**, 749–57.
Snyder, G. K. (1975). Respiratory metabolism and evaporative water loss in a small tropical lizard. *J. Comp. Physiol.*, **104**, 13–18.
Templeton, J. R. (1960). Respiration and water loss at the higher temperatures in the desert iguana, *Dipsosaurus dorsalis*. *Physiol. Zool.*, **33**, 136–45.
Templeton, J. R. (1964). Nasal salt excretion in terrestrial lizards. *Comp. Biochem. Physiol.*, **11**, 223–9.
Templeton, J. R. (1966). Responses of the lizard nasal salt gland to chronic hypersalemia. *Comp. Biochem. Physiol.*, **18**, 563–72.

Templeton, J. R. (1972). Salt and water balance in desert lizards. In G. M. O. Maloiy (Ed.) *Comparative Physiology of Desert Animals*, *Symp. Zool. Soc. Lond.*, No. 31, 61–77.

Tercafs, R. R., and E. Schoffeniels (1965). Phénomeñes de perméabilité au niveau de la peau des reptiles. *Ann. R. Soc. Zool. Belg. Brux.*, **96**, 9–22.

Thorson, T. B. (1968). Body fluid partitioning in Reptilia. *Copeia*, **1968**, 592–601.

Tregear, R. T. (1966a). The permeability of mammalian skin to ions. *J. Invest. Derm.*, **46**, 16–23.

Tregear, R. T. (1966b). *Physical Functions of Skin*, Academic Press, New York.

Trobec, T. N. and J. G. Stanley (1971). Uptake of ions and water by the painted turtle *Chrysemys picta.*, *Copeia*, **1971**, 537–42.

Warburg, M. R. (1965a). The influence of ambient temperature and humidity on the body temperature and water loss from two Australian lizards, *Tiliqua rugosa* Gray (Scincidae) and *Amphibolurus barbatus* Cuvier (Agamidae). *Aust. J. Zool.*, **13**, 331–50.

Warburg, M. R. (1965b). Studies on the environmental physiology of some Australian lizards from arid and semi-arid habitats. *Aust. J. Zool.*, **13**, 563–75.

Warburg, M. R. (1966). On the water economy of several Australian geckos, agamids and skinks. *Copeia*, **1966**, 230–5.

Chapter 8

Control Mechanisms in Birds

M. Peaker

I. Introduction — 323
II. Kidney Function — 324
  A. General features — 324
  B. Kidney structure and function — 325
  C. Control of renal function — 331
III. Modification of Ureteral Urine — 332
IV. Salt Glands — 335
  A. General features — 335
  B. Secretory mechanism — 337
V. Other Systems — 343
References — 344

I. INTRODUCTION

It is often stressed that birds have successfully exploited a wide variety of ecological niches by virtue of their abilities for osmoregulation. Largely because of this diversity, studies of osmoregulation have often been more concerned with the ecological aspects rather than with the cellular mechanisms involved. In this article the main physiological and cellular processes will be stressed where these are known, and reference will be made to recent reviews wherever possible in order to enable the reader to delve more deeply into specific aspects.

As with so many aspects of avian physiology, some features of the excretory and osmoregulatory systems are basically reptilian, for example the presence of salt glands in some species and the modification of ureteral urine in the cloaca and rectum. The kidney however shows features which are normally regarded as being mammalian as well as features which are basically reptilian, and it is only in the past few years that we have begun to understand the workings of the avian kidney in terms other than as a 'black box'.

II. KIDNEY FUNCTION

A. General Features

In very general terms the kidneys of birds function like those of mammals and most other vertebrates, producing urine by processes of filtration, secretion and reabsorption. Compared with the mammalian kidney however there are three main and important differences, namely the presence of several types of nephron, the presence of a renal portal vascular supply and the excretion of uric acid or urate salts. Like mammals, but unlike extant reptiles, birds can produce urine hyperosmotic to plasma although the concentrating abilities are not nearly so marked.

In gross investigations of renal function in birds there are two complicating factors. The first of these is the secondary modification of urine in the cloaca and rectum and also therefore the contamination of voided urine by faeces. Difficulties arise when trying to assess the role of the kidneys *per se* in the overall process of excretion and osmoregulation. A number of methods have been devised to separate ureteral urine from faeces which rely either on the surgical diversion of the rectum or ureters directly to the exterior or, for short-term studies, the attachment of catheters to the ureteral openings in the urodaeum (see Sykes, 1971). While such techniques have yielded very useful information there are difficulties in interpretation because interference with the ureters may disturb their function and therefore kidney function as well. Furthermore the absence of modification of the urine in the cloaca and rectum may alter the salt and water balance of the whole bird and thereby affect renal excretion.

The second complicating factor is the formation of insoluble urate salts (p. 330) so that cations may be present in unionized as well as in ionized forms. Therefore if only the watery phase of urine is analysed the results of electrolyte balance studies will be misleading. Furthermore, the sequestration of cations makes calculations on possible mechanisms for ion transport operating in normal physiological conditions extremely hazardous.

The abilities of the kidneys in various species under different physiological circumstances have been reviewed by Skadhauge (1973) and by Willoughby and Peaker (1979). In general terms the rate of urine flow can change by up to twenty-fold depending on the state of hydration. In most species the maximum urine osmotic concentration that can be achieved is about 2–3 times that of plasma (i.e. $U/P = 2-3$). However in some birds from arid regions which lack salt glands and can survive without drinking fresh-water, the U/P ratio can approach 6. As well as being able to elaborate a concentrated urine, birds are also able to produce a dilute urine ($U/P = 0.2$ or lower). Therefore it is with this background that the micro-architecture of the kidney can be considered in relation to function.

B. Kidney Structure and Function

The avian kidney is a complex lobular structure and the ureter, as it runs along the ventral surface, is joined by secondary ureters from the numerous lobules. The general arrangement of a kidney lobule showing the medullary and cortical regions is shown in Fig. 1. The nephrons dicharge into collecting

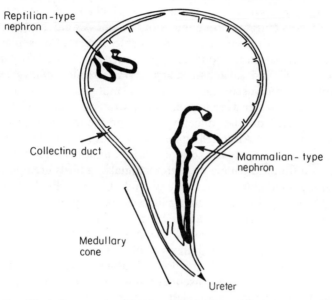

Figure 1 Simplified diagram showing the organization of the excretory elements in a single lobule of the avian kidney

ducts which enter medullary cones and then converge to form a secondary ureter at the apex of the cone. There are two main types of nephron—the reptilian type and the mammalian type. Reptilian-type nephrons have no loop of Henle and simply empty into the collecting ducts. By contrast, the mammalian type, which are fewer in number, do have a loop of Henle; these loops together with the associated vasa recta and the collecting ducts constitute the medullary cone (Fig. 1). Several cortical lobules may contribute to a medullary cone. For anatomical details of the avian kidney see Hüber (1971), Johnson (1968, 1974), Johnson and Mugaas (1970a, 1970b), Johnson, Phipps and Mugaas (1972), Johnson and Ohmart (1973), Braun and Dantzler (1972) and Siller (1971).

It is obvious that the parallel arrangement of the loops of Henle and collecting ducts in the medullary cones is analogous to the arrangement in the mammalian renal medulla, and it is generally believed that a counter-current multiplier system is responsible for the production of hyperosmotic urine in birds. Poulson (1965) found that the renal concentrating ability of a species is related to the number of medullary cones and therefore to the number of loops of Henle, and he suggested that the basis for the production of hyperosmotic urine in birds is similar to that in mammals. If this is so then one might expect a gradient of solutes along the medullary cone when a bird is producing hyperosmotic urine. Indeed Skadhauge and Schmidt-Nielsen (1967) found such a gradient in the domestic chicken and turkey during the formation of concentrated urine but not, as one would expect, during a diuresis induced by water-loading. More recently Amery, Poulson and Kinter (1972) have demonstrated medullary hypertonicity in the budgerigar (*Melopsittacus undulatus*) and savannah sparrow (*Passerculus sandwichensis rostratus*) using microcryoscopical methods; osmolalities as high as 1100–1800 mosmol kg^{-1} water were found, and autoradiographic studies indicated resorption of water from the collecting ducts and the transport of sodium out of the loops of Henle. Although some problems remain it does appear that the avian renal medulla acts as a counter-current multiplier. However, there is an important difference compared with mammals: in birds urea plays no part in establishing medullary hypertonicity and it seems likely that sodium chloride transport from the ascending limb of the loop of Henle is the main mechanism involved (see Skadhauge, 1973, 1975).

Skadhauge (1973) has also made the point that the ability of birds to produce a dilute urine during diuresis is also related to the presence of a counter-current multiplier system since a diluting segment in which sodium chloride resorption without a consequent movement of water may occur is a prerequisite for the operation of such a system.

Attempts have been made, as mentioned above, to correlate maximal osmotic U/P ratios with morphometric measurements in the avian kidney. In mammals a good correlation has been observed between the relative length of the renal papilla and concentrating ability. By contrast no such relationship is evident in different birds, although there is a positive correlation between maximal osmotic U/P and the relative number of medullary cones and therefore the number of loops of Henle (Poulson, 1965; Johnson and Mugaas, 1970b; Johnson, 1974). Skadhauge (1973) has suggested that a steeper gradient is built up by adding more loops of Henle relative to the number of collecting ducts. There are difficulties in applying morphometric methods because, as Braun and Dantzler (1972) have reiterated, medullary cones follow a tortuous path and one may be seen more than once in one section.

1. *Functional Significance of Different Types of Nephron*

Braun and Dantzler (1972, 1974) have studied the properties of the reptilian-type and mammalian-type nephrons in a species of desert quail, *Lophortyx gambelii*, that can attain an osmotic U/P ratio of about 2.5; about

10% of the nephrons are of the reptilian type. They used Hanssen's injection technique to estimate single nephron glomerular filtration rate (SNGFR) and found that the reptilian-type nephrons had a low GFR compared with the mammalian type, and this was associated with a low glomerular volume and only a few capillary loops in the glomerulus. By contrast the mammalian type

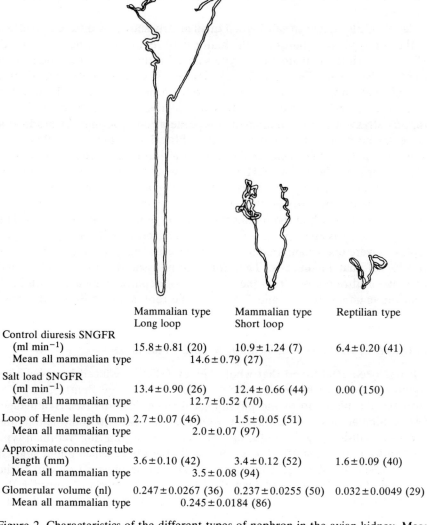

	Mammalian type Long loop	Mammalian type Short loop	Reptilian type
Control diuresis SNGFR (ml min^{-1})	15.8±0.81 (20)	10.9±1.24 (7)	6.4±0.20 (41)
Mean all mammalian type	14.6±0.79 (27)		
Salt load SNGFR (ml min^{-1})	13.4±0.90 (26)	12.4±0.66 (44)	0.00 (150)
Mean all mammalian type	12.7±0.52 (70)		
Loop of Henle length (mm)	2.7±0.07 (46)	1.5±0.05 (51)	
Mean all mammalian type	2.0±0.07 (97)		
Approximate connecting tube length (mm)	3.6±0.10 (42)	3.4±0.12 (52)	1.6±0.09 (40)
Mean all mammalian type	3.5±0.08 (94)		
Glomerular volume (nl)	0.247±0.0267 (36)	0.237±0.0255 (50)	0.032±0.0049 (29)
Mean all mammalian type	0.245±0.0184 (86)		

Figure 2 Characteristics of the different types of nephron in the avian kidney. Mean ± SE. From Braun and Dantzler (1972); reproduced by permission of the American Physiological Society

had a GFR as high as that of mammals (Fig. 2), a large glomerular volume and a more complex glomerular blood supply. Mammalian-type nephrons with a short loop of Henle were also studied; they were found to have a somewhat lower GFR and glomerular volume than the long-loop type. In view of these differences these workers investigated the effect of salt-loading on SNGFR in the different types of nephron. Under such circumstances in the chicken urine flow and total GFR fall, and evidence from para-aminohippurate secretion and glucose reabsorption studies indicated that the fall in GFR results from a decrease in the number of functioning nephrons (Dantzler, 1966).

The SNGFR studies in salt-loaded quail clearly showed a differential effect on the two types of nephron. In the majority of reptilian-type nephrons GFR fell to zero while the mammalian type showed little change or a tendency to decrease. Braun and Dantzler (1972) commented on the significance of their findings that since the small reptilian-type nephrons do not contribute to the concentrating ability it seems reasonable that they cease to function during osmotic stress, while the mammalian-type nephrons continue to produce a concentrated urine. The slight decrease in filtration by the mammalian-type nephrons could further increase the urine concentration by reducing the rate of flow through the loops of Henle which would enhance counter-current multiplication. Also leading to an increased concentration would be the reduced rate of flow through the collecting ducts in the medullary cones as a result of the decrease in the number of filtering reptilian-type nephrons. The overall effect was that in the salt-loaded quail only 16% of reptilian-type nephrons were open compared to 71% in diuresis, and whole kidney GFR fell to 40% of that in diuresis. Therefore during osmotic stress similar total amounts of filtrate arose from the two types of nephron compared with 25% from the mammalian type and 75% from the reptilian type during diuresis.

2. Effects of Arginine Vasotocin on Nephron Function

It has been established that whole-kidney GFR is reduced by exogenous arginine vasotocin (the antidiuretic hormone of birds, see Bentley, 1971) at probably the high end of the physiological range in birds (see Skadhauge, 1973). Similar doses in *Lophortyx gambelii* mimicked to some extent the effect of salt-loading in reducing the number of filtering reptilian-type nephrons although the effect was not exactly the same, which might suggest the involvement of other factors in addition to arginine vasotocin in the dehydrated bird (Braun and Dantzler, 1974). These authors consider this action on reptilian-type nephrons may be the main effect of antidiuretic hormone in birds and that an effect on the permeability of the distal tubule and collecting duct, which is normally regarded as the main site of action of vasopressin in mammals, may not even be present. However Skadhauge

(1973) found a decrease in urine flow and an increase in urine osmolality without a change in whole-kidney GFR or renal plasma flow with lower doses of arginine vasotocin. This would suggest that the overall effect of AVT under normal circumstances is on distal water reabsorption and on GFR as well as in severe dehydration. Braun and Dantzler suggested that cardiovascular changes are responsible for shunting blood away from the glomeruli of reptilian-type nephrons although the precise site of action remains unknown.

While it seems clear that whole-kidney GFR may be reduced in response to severe osmotic dehydration by a reduction in the number of filtering reptilian-type nephrons, the question remains as to whether the response is only a short-term emergency measure in times of extreme dehydration since a marked long-lasting reduction in GFR could lead to the accumulation of some substances, possibly toxic, within the body.

3. *Tubular Function*

Obviously the tubules are involved in reabsorption and secretion but the cellular mechanisms responsible, or whether they are similar to those in mammals, are unknown. Clearly the concentrations of sodium, potassium, chloride, etc. in urine must reflect relative dietary intake and the mechanisms must be sufficiently flexible to allow ionic balance to be maintained.

Shoemaker (1972) in his survey concluded that the fraction of filtered sodium and chloride that is excreted in urine may vary from less than 0.3 to 30%, and the fraction of filtered water from about 1 to 30%. This range results in urinary sodium and chloride concentrations from one-twentieth to twice the concentrations in plasma. There is also evidence that reabsorption of potassium can be varied, and some indication, as one might expect, for the tubular secretion of potassium (Sperber, 1960; Shoemaker, 1967).

The tubules must also be involved in regulating the excretion of other ions, calcium and magnesium, for example (see Willoughby and Peaker, 1977) and in the maintenance of acid–base balance (see Sykes, 1971).

4. *Nitrogenous Excretion*

The advantages of uricotelism as a means of water conservation have often been stressed since uric acid and its salts, being relatively insoluble, occupy little 'osmotic space' in urine. Although uric acid and urates are the major component of nitrogen excretion, ammonia (as NH_4) (and to a lesser extent urea) can be a major avenue. For example in fasted domestic chickens Sykes (1971) found that ammonia accounted for 23% of the total nitrogen in urine compared with 7% in fed birds. Similarly, Stewart, Holmes and Fletcher (1969) found ammonia accounted for approximately 30% in the domestic duck.

The physical chemistry of uric acid and urate salts has been excellently described by Sykes (1971). Obviously the proportions of acids and salts will vary with pH and the ionic composition of urine. Moreover the proportion of uric acid and urates present as a precipitate compared with that in solution or as a colloidal suspension varies. In dilute urine only 15–20% is present as a precipitate, whereas in concentrated urine over 90% is in the solid state (McNabb, 1974).

As mentioned earlier (p. 324) significant amounts of sodium and potassium can be associated with uric acid. McNabb, McNabb and Hinton (1973) found 3–75% of urinary sodium and 8–34% of potassium associated with precipitated urates, the higher proportions in birds on a high protein diet. Therefore not only does the excretion of uric acid simply contribute to water conservation by freeing water for reabsorption, it also sequesters cations thus freeing additional water for reabsorption. These authors consider that in addition to urate salt, soluble salts may be trapped between layers of insoluble uric acid. It seems likely that other cations are also associated with uric acid.

It is clear that, in contrast to mammals, the physical chemistry of bird urine is complex and cognisance of these matters is important in determining salt and water balance in birds from different environments in various physiological conditions and in attempting to determine the cellular mechanisms involved in osmoregulation, ionic regulation, nitrogenous excretion and acid–base balance.

Uric acid is synthesized in the liver and excreted by the kidney by filtration and tubular secretion. Within the physiological range of plasma uric acid concentrations tubular secretion, a saturable energy-dependent process, is the main route, the renal clearance in the chicken being 20 times that of inulin (see Sykes, 1971). Precipitation of uric acid begins in the collecting ducts and mucus secreted by the cells lining the ducts and ureteral branches is thought to prevent blockage of the ureteral system by acting as a lubricant. The precipitate consists of spheres, 2–8 μm in diameter (Folk, 1969), rather than the usual type of uric acid crystals. These spheres are apparently mainly uric acid with smaller amounts of soluble materials (Lonsdale and Sutor, 1971). They are unstable outside the body and gradually change into the usual type of uric acid crystal; the physicochemical processes involved in their formation intrarenally are unknown.

5. Renal Portal Blood Supply

The main point of interest as far as renal function is concerned is that renal portal vessels can deliver venous blood from the hind part of the bird to the renal tubules (but not the glomeruli). Therefore the renal portal supply contributes to the secretory capacity of the kidneys; indeed some consider the necessity for uric acid secretion to be the *raison d'être* for the retention of the supposedly primitive renal portal system in birds. In lower vertebrates the portal system may continuously supply blood to the kidneys but in birds the pattern of venous blood flow can be varied by operation of the renal portal valves and vasomotor control of the portal vessels. Therefore the kidneys may or may not be receiving portal blood at any one time. In fact the situation is even more complicated because not only can flow to the two kidneys be different but the anterior and posterior parts of a kidney may also receive different amounts of portal blood. The anatomy and control of venous blood flow in this region have been reviewed by Akester (1971) and Sykes (1971) but the functional significance is not understood. Changes may be more concerned with cardiovascular requirements rather than kidney function and it is not known whether renal portal blood flow is influenced by the osmoregulatory requirements of the bird.

6. *Kidney Function in Birds with Salt Glands*

In terms of renal physiology it is important to distinguish between short-term effects induced by salt-loading and the longer-term when input and output are virtually at steady state. In long-term studies on domestic ducks the main difference in renal excretion in birds drinking salt-water was the formation of hyperosmotic urine which was due to increases in sodium and chloride concentrations; the volume of urine produced did not change. Apart from a rise in calcium excretion, the concentrations of other substances, GFR and renal plasma flow were not significantly different compared with ducks drinking fresh-water. Therefore the significant feature is that the sodium concentration in urine is lower than in plasma which means that the tubular reabsorption of sodium is high (Fletcher and Holmes, 1968; Holmes, Fletcher and Stewart, 1968; Stewart and coworkers, 1969). It has been argued that tubular reabsorption of sodium and chloride in ducks on salt-water is maintained at high levels by some humoral factor (Holmes and Wright, 1969). The evidence, that sodium and chloride reabsorption is controlled under these conditions and not simply always high, is that while tubular water reabsorption is similar in birds drinking fresh-water or salt-water, sodium and water reabsorption are closely correlated in ducks on salt-water but not in those drinking fresh-water. Transient increases in cloacal sodium output have been reported in the short-term in gulls (Douglas, 1970) which would imply that when steady state is reached tubular sodium movements are being controlled while water movements are controlled by a different factor (see Peaker and Linzell, 1975).

The identity of the factors controlling renal function in birds with active salt glands is unknown but a number of hormones have been suggested as possible candidates (see Peaker and Linzell, 1975; Ensor, 1975). The case seems to be strongest for the involvement of prolactin, glucocorticoids and arginine vasotocin. The possibility of these and possibly other factors acting in concert must be considered in this complex situation. It should also be noted that in most studies ureteral urine has not been collected and a possible involvement of altered function in the cloaca and rectum should not be overlooked.

C. Control of Renal Function

Obviously the complexity of the avian kidney and its blood supply means that renal function is open to control by a variety of hormonal, other humoral or nervous factors no fewer in number and certainly no less complex than those which affect mammalian kidney function (see Chapter 9). Although a number of hormones affect urine output and composition including arginine vasotocin (see above), adrenocorticosteroids and prolactin (see Shoemaker, 1972; Ensor, 1975) the site and mechanism of action are usually unknown. For example, Holmes and Adams (1963) found exogenous corticosterone and aldosterone to decrease sodium excretion in fresh-water-loaded ducks; aldosterone also decreased potassium excretion, in contrast to mammals.

III. MODIFICATION OF URETERAL URINE

It has long been suspected that ureteral urine may be stored in the cloacal region, and in the past ten years direct evidence has been obtained that urine can pass back into the coprodaeum and large intestine. Several methods have been used to demonstrate this phenomenon including radiography during the renal excretion of X-ray dense material and the analysis of the inulin content of the urine–faeces mass in the intestine during intravenous inulin infusion (see Sykes, 1971; Skadhauge, 1973). Thus it is clear that the potential exists for the composition of urine to be modified during its sojourn in the large intestine, and studies by Skadhauge and his coworkers on the actual mechanisms involved in the movements of water and ions across the wall of the large intestine have added enormously to our knowledge of the processes that can occur. In most of their studies they perfused the lumen of the coprodaeum and large intestine with fluids of different composition in anaesthetized birds; the preparation they used is shown in Fig. 3 (Skadhauge, 1967; Bindslev and Skadhauge, 1971a, 1971b; Skadhauge, 1973).

In the chicken it was established that water could be absorbed from fluids up to 80 mosmol kg^{-1} water higher than plasma at perfusion rates of 5–9 ml h^{-1}. When perfusion rates were reduced to 0.8–1 ml kg^{-1} h^{-1}, similar to the rate of urine flow in dehydrated birds, it was estimated that water absorption could take place from perfusion fluids up to 175 mosmol kg^{-1} water higher than plasma. Therefore it was argued that osmotic dilution of the urine can be prevented during storage in the large intestine and that absorption of water may occur in the dehydrated bird.

These results showed that water can be absorbed against an osmotic gradient, which can be attributed to solute-linked water transport with the movements of water being coupled to the active transport of sodium out of the lumen. In chickens the solute-linked water flow was approximately 1 μl water/μmol sodium in normal birds and 1.5 μl μmol^{-1} during dehydration; the difference however was not statistically significant. Therefore the sodium concentration of the absorbate was 1250 mmol l^{-1}. In the dehydrated cockatoo, *Cacatua roseicapilla*, a species from arid regions of Australia,

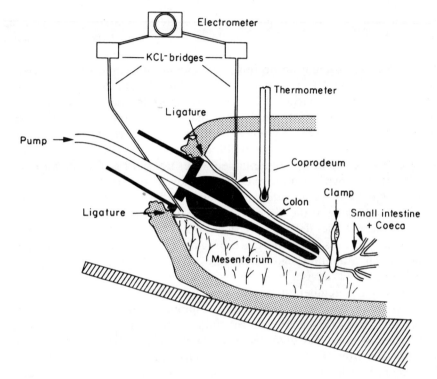

Figure 3 Preparation used by Skadhauge and his coworkers to study ion and water movements across the cloaca and large intestine in anaesthetized birds. From Bindslev and Skadhauge (1971a); reproduced by permission of the *Journal of Physiology*

Skadhauge (1974) found the solute-linked water transport to be 5 μl μmol^{-1}, i.e. an absorbate concentration of 385 mmol l^{-1}. The significance of this finding is that in this species relatively more water is carried across the wall and therefore the bird can store urine of a higher osmotic concentration than that of the dehydrated fowl without dilution of the urine occurring by osmotic water movements.

In addition it has been found that the rates of osmotic water movement are not equal in both directions, i.e. there is osmotic rectification. For example, in the chicken the osmotic permeability from blood to lumen (serosa to mucosa)

was only 55% of that in the opposite direction in normal birds and 36% during dehydration. Such differences would tend to ensure osmotic absorption of water from hypo-osmotic urine and retard osmotic dilution of hyperosmotic urine.

Related studies on ion movements have shown that sodium is actively absorbed from the large intestine and that potassium is secreted into the lumen. These movements are clearly under some form of systemic control since the absorption of sodium and chloride decreased markedly following the administration of sodium chloride to chickens. Local mechanisms also appear to be involved since, for example, high concentrations of potassium in the lumen decrease sodium absorption, and sodium movements are at least to some extent coupled to those of potassium in the reverse direction. Skadhauge (1973) concluded that the transport mechanisms in the large intestine are similar to those in mammals in that there is little passive movement of ions but that sodium and chloride are absorbed by a process that can be regulated according to physiological requirements, and potassium is secreted (see Chapter 9). Solute-linked water transport is also similar in that the absorbate is hyperosmotic so net water movements are determined by this process and by osmotic water movements in the same or in the opposite direction according to the osmolality of the fluid in the lumen.

While these studies demonstrate that ion and water movements can occur it is difficult to assess the importance of such processes under natural conditions. In an attempt to go some way in this direction, Skadhauge and Kristensen (1972) devised a computer simulation of the reabsorptive processes based on a number of assumptions and such variables as the rate of ureteral urine flow, urine osmolality, osmotic permeability, solute-linked water flow, etc. This mathematical model indicated that, in the chicken, hyperosmotic urine may enter the cloaca without resulting in further osmotic loss of water from the body and that reabsorption of up to 14% of the water can occur. To achieve this however a considerable amount of sodium must be reabsorbed. While this may be a disadvantage it may be important in sodium conservation since it has long been known that if urine is diverted directly to the exterior chickens become deficient in sodium (Hart and Essex, 1942). Skadhauge and Kristensen (1972) have also suggested that the reabsorption of sodium may be of benefit to desert birds which can tolerate an increased plasma osmolality, by helping to maintain plasma volume during dehydration. Another probable benefit, used in the past as an argument that post-renal modification of urine occurs, is that reabsorption of salts and water should permit the production of a sufficient volume of urine to eliminate uric acid in dehydrated birds; the uric acid can then be concentrated in the cloacal region by the reabsorption of water with further precipitation of urates leading to the freeing of more water for reabsorption (see Skadhauge, 1973).

Even though the simulation studies are valuable, there are even more complications in trying to achieve a full understanding because factors other than the actual transport mechanisms are involved. For example, knowledge of the extent to which urine flows into the large intestine and the length of time it is stored there is required. And of course it is not simply a problem related to urine because the large intestine is also involved in the absorption of salts and water from the material passing down the alimentary tract and so the overall process is one of mixture of the urine and intestinal contents with the walls of the large intestine (and also possibly the caeca) fulfilling two roles simultaneously. Therefore a knowledge of the volume and composition of the faeces is also needed as part of the overall view.

With these provisos in mind, Skadhauge's work shows that the transport mechanisms can be controlled and that there are significant adaptive differences between species with respect to the properties of these mechanisms. For example, in *Cacatua roseicapilla* Skadhauge (1974) has found that apart from the difference in solute-linked water transport described above, a high potassium concentration in the luminal fluid does not inhibit sodium absorption, as is the case in the chicken. During dehydration when the concentration of potassium in urine increases, sodium absorption can continue. Therefore sodium conservation may well be important in these species and it was estimated that 70% of the urine sodium can be reabsorbed without losing water osmotically to the stored urine. Further investigations on the cellular mechanisms involved in regulating the transport processes, on the transport mechanisms themselves and whether other solutes in urine, as well as sodium, are involved in water reabsorption (a possibility raised by Skadhauge, 1975) are clearly important in studies of osmoregulation at the cellular level.

Schmidt-Nielsen and coworkers (1963) suggested that salt-gland secretion may be required to take advantage of water reabsorption in the cloaca. Since water flow coupled to sodium reabsorption would lead to an excess of sodium in the body, extra renal excretion of sodium would be necessary for the efficient reabsorption of water from ureteral urine. However, it is now clear that some birds without salt glands can reabsorb sodium and water and that conservation of sodium is probably important in such birds.

Although there is some indication for the reabsorption of sodium chloride from ureteral urine in marine birds with active salt glands, this is an aspect of osmoregulation in these species which requires further study (see Peaker and Linzell, 1975).

IV. SALT GLANDS

A. General Features

As is now well known, following the studies of Knut Schmidt-Nielsen and his colleagues in the late 1950s, the nasal salt glands of marine birds excrete a markedly hyperosmotic fluid consisting mainly of sodium chloride and thus permit a bird to obtain osmotically-free water when eating marine invertebrates, with a high salt content, or drinking sea-water.

The nasal salt glands are located in or around the orbit and the secretion is formed by secretory tubules which radiate from the central canal or duct of each lobule. These central canals join to form secondary and then primary ducts which carry the secretion to the nasal cavity and exterior. Electron microscopical studies have shown that the secretory cells display signs of great

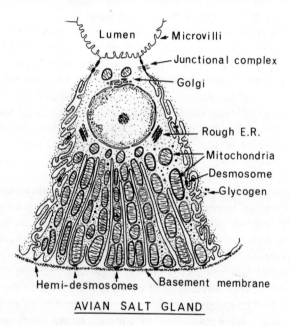

Figure 4 Diagram showing the structure of a salt gland secretory cell. From Peaker and Linzell (1975); reproduced by permission of Cambridge University Press

metabolic activity—as might be expected with secretory rates as high as 2 ml $(g\ tissue)^{-1}\ min^{-1}$; the cytoplasm is packed with mitochondria and the basal membrane greatly infolded (Fig. 4). The cells of the tubules at the periphery of the lobules, i.e. the distal or blind end, are not specialized and are believed to be the site of cell division. There is a gradient of cell types along the tubule, with the apparently more highly-specialized cells (more mitochondria and basal infolding) being situated adjacent to the central canal. The length of the tubules, the degree of specialization of the cells and the weight of the glands are all increased when birds drink salt-water (see p. 342). By contrast with the secretory cells of the tubule, the cells lining the duct system appear from electron micrographs not to be highly active and it can be argued that they are not involved in modifying the secretion but simply act as conduits to the exterior (see Peaker and Linzell, 1975).

It is not intended to consider the entire physiology of salt glands since this has been done recently in the monograph by Peaker and Linzell (1975). Instead, three main topics all concerned with cellular control will be stressed. However before dealing with these—the secretory mechanism, excitation–secretion coupling and adaptive changes in the gland—it is pertinent to point out the gross physiological control of secretion and other events that occur within the active gland.

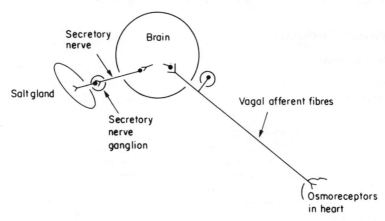

Figure 5 The reflex arc for stimulation of the salt gland. From Hanwell, Linzell and Peaker (1972); reproduced by permission of the *Journal of Physiology*

It is now well established that secretion is initiated and maintained by a nervous secretory reflex, the efferent limb being cholinergic (Fänge, Schmidt-Nielsen and Robinson, 1958). Hanwell, Linzell and Peaker (1972) concluded from a series of experiments involving infusion, perfusion and cross-circulation that the afferent pathway of the reflex runs in the vagus nerves and the receptors are located in or near the heart (Fig. 5). It is also well established that osmotic or tonicity receptors are responsible for detecting excess salt in the blood, rather than specific ion or blood volume receptors (see Peaker and Linzell, 1975).

Several hormones, for example, adrenocorticosteroids, prolactin, arginine vasotocin, are known to affect secretion or to be necessary for secretion to occur, but current evidence favours the view that they play a permissive or secondary role in modulating the nervous control of the gland (see Phillips and Ensor, 1972; Peaker and Linzell, 1975). Accompanying secretion is a fourteen-fold increase in the rate of blood flow through the gland. In domestic geese flows as high as 15 ml g^{-1} min^{-1} were obtained by Hanwell, Linzell and Peaker (1971). Even with such high rates the amounts of sodium and chloride extracted by the glands from arterial blood were also high. Calculations from blood flow and secretory rate using the Fick principle showed that as much as 70% of the chloride and 57% of the sodium were extracted from arterial plasma per unit time in geese with high secretory rates, although the median figures in this series of experiments were lower (chloride 21%, sodium 15%). The arrangement of the blood supply is such that the highly specialized secretory cells receive blood with the highest sodium chloride content. The capillaries then run along the tubules in a counter-current manner to the flow of secretion. Therefore blood passes the bases of many secretory cells, which could account for the high rates of extraction observed.

B. Secretory Mechanism

1. *Fluid Composition*

Sodium and chloride usually account for 97% of the osmolality of the secretion in marine birds (Table 1). The maximum concentrating ability

Table 1. Composition of nasal fluid in a gull (modified from Schmidt-Nielsen, 1960; from Peaker and Linzell, 1975, reproduced by premission of Cambridge University Press)

	Concentration (mmol l^{-1})
Sodium	718
Potassium	24
Calcium and magnesium	1
Chloride	720
Bicarbonate	13
Sulphate	0.35

clearly differs between species, the lower concentrations in species living in estuaries, the higher in pelagic, invertebrate-eating forms (Schmidt-Nielsen, 1960), the maximum sodium concentration recorded in a marine bird being 0.9–1.1 mol l^{-1} in a petrel. While the maximum concentration is probably characteristic of a species this full potential is not realised until a bird is accustomed to marine conditions and many studies have shown that the concentration and amount of secretion are increased when birds are drinking salt-water.

The potassium concentration of the secretion is usually low but Hughes (1970) has found that in a gull, *Larus glaucescens*, the concentration increases to a greater extent than that of sodium during adaptation to sea-water. This resulted in a decrease of the sodium:potassium ratio from 24:1 to 12:1.

2. Ion and Water Transport

The architecture of the tubular epithelium of the salt gland of a marine bird, showing the extensive basal infolding, the junctional complexes near the apex and the relatively flat apical or luminal cell membrane is shown in Fig. 4. Physiological evidence indicates that the passage of substances between the secretory cells by a paracellular pathway is not involved in the formation of the secretion, and therefore that the tight junctions connecting neighbouring epithelial cells are truly tight. It is generally considered that the passage of disaccharides or inulin across an epithelium suggests the presence of paracellular movements. Schmidt-Nielsen (1960) found that inulin did not appear in the secretion during intravenous infusion and similar results were obtained by Peaker and Hanwell (1974) with [^{14}C]sucrose. In the latter study only 0.4% of the concentration in plasma was detected in the secretion of geese after a 90 min infusion.

Frömter and Diamond (1972) presented evidence that greater potential, concentration and osmotic gradients are established across epithelia lacking a paracellular pathway than in ones with leaky junctions. In terms of these variables the salt-gland secretory tubule qualifies as a tight epithelium. This is particularly evident in the case of the osmotic gradient. In the leaky epithelia listed by Frömter and Diamond the ratio of the osmolarity of the transported fluid to that of the bathing solution does not exceed 1.0, but in the salt gland of marine birds this ratio is 3–7.5. Arguing teleonomically, truly tight junctions might be expected in the salt gland because passive paracellular movements would obviously militate against the formation of a hypertonic solution.

While physiological evidence favours the view that the secretory epithelium is tight and that substances secreted must traverse the two membranes of the cell, a morphological study suggests that horseradish peroxidase and lanthanum can penetrate some of the junctions in the domestic duck (Martin and Philpott, 1973). The reason for this discrepancy is not apparent but the possibility that fixation altered the structure of the junction and allowed large molecules to enter during this time must be considered.

The physiological, biochemical and histochemical evidence on the mechanism of ion transport has been considered in detail by Peaker (1971) and Peaker and Linzell (1975); the suggested scheme is shown in Fig. 6. The considerations on which this scheme is based will not be detailed again in this article except to mention that the salient features are:

(i) An electrogenic sodium-extruding pump on the luminal or apical membrane with chloride following passively; this pump is not coupled to inward potassium movements.
(ii) A sodium–potassium pump on the baso-lateral membrane serving to maintain a high intracellular potassium concentration and a low intracellular sodium during inactivity of the gland.

(iii) Na^+–H^+ and Cl^-–HCO_3^- exchange on the baso-lateral membrane, H^+ and HCO_3^- being derived from carbon dioxide. Therefore activity of the luminal pump, respiration and the exchange mechanisms are closely coupled during secretion.

(iv) Since potassium may under certain circumstances be increased (p. 338) there is the possibility either of the presence of a potassium-extruding pump on the luminal membrane or of the acceptance of potassium by the sodium pump.

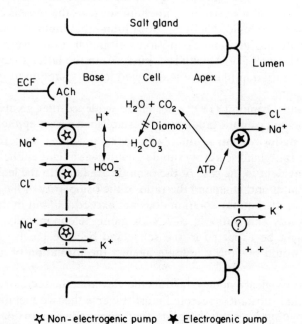

Figure 6 Suggested scheme for ion transport in the salt gland. From Peaker and Linzell (1975); reproduced by permission of Cambridge University Press

(v) The concentration gradient is established across the luminal membrane with water crossing either by solute-linked flow (solute drag) or osmotically; in either case the osmotic and passive ionic permeabilities of the luminal membrane must be low to permit the formation of hyperosmotic fluids.

(vi) The possibility of water and ion movements being influenced by the infolded basal membranes on the lines suggested by Diamond and Bossert (1968).

Since secretion is initiated and maintained by cholinergic nerves, acetylcholine must act at some point to stimulate these mechanisms. It has previously been argued that an increase in the intracellular sodium concentration brought about by an increase in the passive permeability of the basal membrane or activation of the sodium transport mechanisms on that membrane is not the intracellular stimulus for activation of the luminal pump. For example, Peaker and Stockley (1974) found that methacholine had an effect compatible with activation of the luminal pump on salt-gland slices in which the intracellular concentrations of sodium and chloride were increased by the addition of low concentrations of lithium to the incubation medium.

The question then arises of how acetylcholine, released from nerve terminals near the basal membrane, acts to stimulate the luminal pump, and it is on this problem that very recent evidence has been obtained (Peaker, 1977). In other exocrine glands such substances as cyclic nucleotides and calcium have been implicated as 'second messengers' acting intracellularly from a stimulus received at the cell membrane. In early experiments it was found that dibutyryl cyclic-AMP infused into a carotid artery did not initiate secretion in geese. However it is difficult to perform such experiments *in vivo* because agents given to a bird may affect secretion indirectly by stimulating or inhibiting nervous pathways or hormone release. Therefore the technique, first used in the salt gland by Van Rossum (1966), of loading slices with ^{24}Na and then studying the efflux has been used. As in Van Rossum's experiments methacholine temporarily stimulated sodium efflux. It was also established in slices loaded with both ^{24}Na and ^{36}Cl, that chloride efflux was similarly increased by methocholine. Therefore this strengthens the interpretation of Van Rossum that the increased efflux indicates activity of the luminal sodium pump with chloride following passively.

The effect of methacholine was almost completely inhibited by the omission of calcium (and the addition of EGTA to chelate residual calcium) from the fluid superfusing the slice. By contrast, the calcium ionophore A23187 stimulated sodium efflux in the absence of methacholine, and a similar but less prolonged response was obtained when the calcium concentration of the fluid was suddenly raised from 2.6 to 9.6 mmol l^{-1}. Therefore these results suggest that an elevated concentration of calcium inside the cell activates secretion and that acetylcholine *in vivo* alters the permeability of the basal membrane to permit calcium to enter. In other words, calcium appears to act as the second messenger, and preliminary experiments suggest that calcium fluxes are altered by cholinomimetics. The effects of monobutyryl and dibutyryl derivatives of cyclic-GMP and cyclic-AMP were also tested over a wide range of concentrations. Cyclic-GMP had no effect on sodium efflux while cyclic-AMP caused a very small increase, in no way comparable in magnitude to that obtained in response to methacholine. Therefore these data suggest that calcium movement is the major mechanism involved.

Clearly the mechanism by which calcium activates the luminal pump requires further study and with calcium playing such a vital role it follows that any agent which influences calcium transport across the cell or mitochondrial membranes or binding in the cell could modulate secretory activity induced by the secretory reflex; it is possible that hormones could act in this way.

3. *Adaptation to Environmental Salinity*

An important property of salt glands is their ability to adapt to the prevailing conditions of environmental salinity. Thus while the glands of marine or estuarine birds kept on fresh-water will start to secrete within minutes of the administration of salt-water, continued ingestion leads to an increase in their size and efficiency, in terms of the concentration of the secretion and the volume secreted per gland and per unit weight of tissue.

The time-course of the changes in weight, secretory ability and the accompanying cytological and biochemical changes have been studied, mainly in the domestic duck, by several groups of workers and comprehensive accounts are given by Holmes (1972) and Peaker and Linzell (1975). During adaptation the cells increase in size and they become more specialized in that the degree of basal infolding becomes more marked and the number of mitochondria increases. Therefore the proportion of such fully specialized cells in the tubule is increased (Ernst and Ellis, 1969). It has been suggested that the fully specialized cells can produce more fluid of a higher concentration than the relatively less specialized ones further along the tubule, which would account for the increase in secretory ability (Hanwell and coworkers, 1971). The structural changes are accompanied by increases in the total RNA content of the glands and in the RNA:DNA ratio, in the total protein content and, later, in the activity of ATPases, glycolytic and other enzymes.

Recently, the physiological mechanisms responsible for initiating adaptive hypertrophy have been investigated. The rise in the RNA content occurs during the first 24 h on salt-water. Therefore the 'trigger' for hypertrophy must operate shortly after a bird first ingests salt-water, and since the adaptive changes are reversed if birds are returned to fresh-water it would seem that a factor must maintain the glands in the adapted state. In order to investigate the importance of the nervous secretory reflex, Hanwell and Peaker (1973, 1975) denervated one gland of geese (a procedure which blocks secretion and vasodilatation) and found that after 24 h on salt-water the increase in RNA and RNA:DNA that occurred in the intact gland was not apparent in the denervated gland. This study showed that the nervous secretory reflex is vital for adaptive hypertrophy, and a similar conclusion was reached by Pittard and Hally (1973) who found no adaptive increase in salt-gland weight in ducklings with denervated glands after 10 days on salt-water.

It could be argued that without the increase in blood flow in the denervated glands, the failure of hormones, like corticosterone for which the gland has a high affinity (Bellamy and Phillips, 1966), to reach the gland in sufficient quantities could have been responsible for the difference observed. However, further experiments showed that hypertrophy depends on the nervous stimulation of secretion and not on the vasodilatation which accompanies secretion since atropine, which blocks secretion but not the increase in blood flow in response to a salt-load (Hanwell and coworkers, 1971), was also found to prevent hypertrophy (Hanwell and Peaker, 1975). Therefore it was suggested that the processes within the cell which are responsible for the initiation and maintenance of adaptive hypertrophy are obligatorily related to the process of secretion. Of interest will be whether the same or different intracellular messengers are responsible for initiating secretion and hypertrophy.

Further support for the view that nerves primarily control hypertrophy was obtained in experiments in ducks in which corticosterone and prolactin failed to induce hypertrophy in birds on fresh-water, even with very high does. However, the possibility remains that hormones may play some role in this process and it is interesting to note that prolactin while having no effect on salt-gland weight in ducks on fresh-water did cause a rise in the RNA concentration in the glands (Hanwell and Peaker, 1975).

Until recently it has not been clear whether the number of cells in the salt glands increases during adaptation or, in other words, whether there is hyperplasia as well as hypertrophy. Ellis and coworkers (1963) in a histological study suggested that the number of cells in each secretory tubule increased and that the mitotic rate must rise. In apparent confirmation, Holmes and Stewart (1968) found an increased total amount of DNA during adaptation in older ducks; this increase occurred during the first 24 h which would indicate the rapid onset of cell division at a high rate. In contrast, other workers have found no significant increase in DNA content after two days (Hanwell and Peaker, 1975) or four days (Ballantyne and Wood, 1969). In order to try to resolve this difference the incorporation of [^{3}H]thymidine into DNA has been studied. In ducks drinking fresh-water the incorporation of thymidine was very low and no change was apparent after 6 or 24 h on salt-water. By two days however there was a marked increase which coincided with the DNA content beginning to increase. By seven and fourteen days the DNA content was significantly raised but thymidine incorporation had fallen to a level little higher than that of birds on fresh-water. Therefore there appears to be a burst of cell division beginning approximately two days after a duck first drinks salt-water. From the rise in DNA it was calculated that the number of cells in the gland had increased by about 40% in the fourteen-day period—a figure similar to that obtained by Holmes and Stewart (1968). Whether the increase in the number of cells is due solely to an increase in the rate of cell division or to a decreased rate of cell loss as well remains to be determined, but clearly there is some factor, as yet unknown, which stimulates cell division. The cells at the blind end of the tubule appear to be the ones involved because cells in metaphase were observed in that region and in the ducts in birds given salt-water for two days and colchicine 3 h before killing (M. Peaker, unpublished).

4. Salt Glands in Non-marine Birds

Some terrestrial birds have salt glands. The most notable case is the ostrich, *Struthio camelus*, in which the fluid may contain high concentrations of potassium, sodium, calcium and chloride. In some samples the concentrations of sodium and potassium were similar whereas in others the potassium concentration exceeded that of sodium by five- to ten-fold (Schmidt-Nielsen and coworkers, 1963). Nothing is known of the mechanism of control of secretion in this species.

V. OTHER SYSTEMS

Other systems of the body are also involved in osmoregulation. Absorption from the intestine is an important but neglected feature particularly in marine birds in which salt and water must be absorbed and the sodium chloride then excreted by the salt glands at a higher concentration than that of the solution ingested. It is likely that marine birds resemble marine fish in that sodium and chloride are absorbed as a fluid isosmotic to the intestinal contents but hypertonic to plasma by processes of solute-linked water flow, i.e. water movements apparently against an osmotic gradient, coupled to active sodium transport (Douglas, 1970). The highest concentration of salt-water at which absorption can still occur is not known but it must, along with the abilities of the salt glands, be a main determinant of the salinity of drinking water to which a bird can successfully become adapted (Peaker and Linzell, 1975). Crocker and Holmes (1971) found, using everted sacs of small intestine from ducklings, that the rate of fluid uptake (from solutions approximately isosmotic to plasma) was much higher in birds given salt-water to drink than those on fresh-water. They also found that this adaptation occurred mainly during the first day on salt-water and that adrenocorticosteroids may be important in inducing the increase.

There is evidence that the rate of evaporative water loss may be controlled in the interests of water conservation. In birds for which data are available about half the water lost by evaporation occurs from the skin and the rest from the respiratory tract. It has been established that total evaporative loss falls in several species when the supply of drinking water is restricted (see Willoughby and Peaker, 1977), and Lee and Schmidt-Nielsen (1971) found that a decrease in the loss through the skin, rather than the respiratory tract, is involved.

The production of metabolic water may be an important feature of water balance in small birds. Skadhauge and Bradshaw (1974) have examined the water budget of the zebra finch (*Taeniopygia castanotis*) which can survive on a diet of seeds and 0.8 mol l^{-1} sodium chloride. The maximum urine concentration was only 1000 mosmol but it was found that only 10% of the water gained was from the salt-water; the remainder was from preformed water in the food, and the largest proportion from metabolic water. They suggested that salt-water may be consumed to maintain the volume of the extracellular space or to provide sodium chloride for the efficient reabsorption of water in the cloaca and rectum. The high rates of metabolic water production are a consequence of the relatively high metabolic rate of this small bird and it can be argued that osmoregulatory abilities are a function of body size in this case, rather than of special mechanisms of ion and water transport.

REFERENCES

Akester, A. R. (1971). The blood vascular system. In D. J. Bell and B. M. Freeman (Eds) *Physiology and Biochemistry of the Domestic Fowl*, Vol. 2, Academic Press, London. pp. 783–839.

Ballantyne, B., and W. G. Wood (1969). Mass and function of the avian nasal gland. *Cytobios*, **4**, 337–45.

Bellamy, D., and J. G. Phillips (1966). Effects of the administration of sodium chloride solutions on the concentration of radioactivity in the nasal gland of ducks (*Anas platyrhynchos*) injected with [^{3}H]corticosterone. *J. Endocrinol*, **36**, 97–8.

Bentley, P. J. (1971). *Endocrines and Osmoregulation*, Springer-Verlag, Berlin.

Bindslev, N., and E. Skadhauge (1971a). Salt and water permeability of the epithelium of the coprodeum and large intestine in the normal and dehydrated fowl (*Gallus domesticus*). *In vivo* perfusion studies. *J. Physiol. Lond.*, **216**, 735–51.

Bindslev, N., and E. Skadhauge (1971b). Sodium chloride absorption and solute-linked water flow across the epithelium of the coprodeum and large intestine in the normal and dehydrated fowl (*Gallus domesticus*). *In vivo* perfusion studies. *J. Physiol. Lond.*, **216**, 753–68.

Braun, E. J., and W. H. Dantzler (1972). Function of mammalian-type and reptilian-type nephrons in kidney of desert quail. *Am. J. Physiol.*, **222**, 617–29.

Braun, E. J., and W. H. Dantzler (1974). Effects of ADH on single-nephron glomerular filtration rates in the avian kidney. *Am. J. Physiol.*, **226**, 1–8.

Crocker, A. D., and W. N. Holmes (1971). Intestinal absorption in ducklings (*Anas platyrhynchos*) maintained on fresh water and hypertonic saline. *Comp. Biochem. Physiol.*, **40A**, 203–11.

Dantzler, W. H. (1966). Renal response of chickens to infusion of hyperosmotic sodium chloride solution. *Am. J. Physiol.*, **210**, 640–6.

Diamond, J., and W. H. Bossert (1968). Functional consequences of ultrastructural geometry in 'backwards' fluid-transporting epithelia. *J. Cell Biol.*, **37**, 694–702.

Douglas, D. S. (1970). Electrolyte excretion in seawater-loaded herring gulls. *Am. J. Physiol.*, **219**, 534–9.

Ellis, R. A., C. C. Goertemiller, R. A. DeLellis, and Y. H. Kablotsky (1963). The effect of a salt water regimen on the development of the salt glands of domestic ducklings. *Devl. Biol.*, **8**, 286–308.

Emery, N., T. L. Poulson, and W. B. Kinter (1972). Production of concentrated urine by avian kidneys. *Am. J. Physiol.*, **223**, 180–7.

Ensor, D. M. (1975). Prolactin and adaptation. In M. Peaker (Ed.) *Avian Physiology*, Symp. Zool. Soc. Lond. No. 35, Academic Press, London. pp. 129–48.

Ernst, S. A., and R. A. Ellis (1969). The development of surface specialization in the secretory epithelium of the avian salt gland in response to osmotic stress. *J. Cell Biol.*, **40**, 305–21.

Fänge, R., K. Schmidt-Nielsen, and M. Robinson (1958). Control of secretion from the avian salt gland. *Am. J. Physiol.*, **195**, 321–6.

Fletcher, G. L., and W. N. Holmes (1968). Observations on the intake of water and electrolytes by the duck (*Anas platyrhynchos*) maintained on fresh-water and on hypertonic saline. *J. Exp. Biol.*, **49**, 325–9.

Folk, R. L. (1969). Spherical urine in birds: petrography. *Science*, **166**, 1516–9.

Frömter, E., and J. Diamond (1972). Route of passive ion permeation in epithelia. *Nature, Lond.*, **235**, 9–13.

Hanwell, A., J. L. Linzell, and M. Peaker (1971). Salt-gland secretion and blood flow in the goose. *J. Physiol. Lond.*, **213**, 373–87.

Hanwell, A., J. L. Linzell, and M. Peaker (1972). Nature and location of the receptors for salt-gland secretion in the goose. *J. Physiol. Lond.*, **226**, 453–72.

Hanwell, A., and M. Peaker (1973). The effect of post-ganglionic denervation on functional hypertrophy in the salt gland of the goose during adaptation to salt-water. *J. Physiol. Lond.*, **234**, 78–80P.

Hanwell, A., and M. Peaker (1975). The control of adaptive hypertrophy in the salt-glands of geese and ducks. *J. Physiol. Lond.*, **248**, 193–205.

Hart, W. M., and H. E. Essex (1942). Water metabolism of the chicken with special reference to the role of the cloaca. *Am. J. Physiol.*, **136**, 657–68.

Holmes, W. N. (1972). Regulation of electrolyte balance in marine birds with special reference to the role of the pituitary-adrenal axis in the duck (*Anas platyrhynchos*). *Fed. Proc.*, **31**, 1587–98.

Holmes, W. N., G. L. Fletcher, and D. J. Stewart (1968). The patterns of renal electrolyte excretion in the duck (*Anas platyrhynchos*) maintained in freshwater and on hypertonic saline. *J. Exp. Biol.*, **48**, 487–508.

Holmes, W. N., and B. M. Adams (1963). Effects of adrenocortical and neurohypophysial hormones on the renal excretory pattern in the water-loaded duck. *Endocrinology*, **73**, 5–10.

Holmes, W. N., and D. J. Stewart (1968). Changes in nucleic acids and protein composition of the nasal glands from the duck (*Anas platyrhynchos*) during the period of adaptation to hypertonic saline. *J. Exp. Biol.*, **48**, 509–19.

Holmes, W. N., and A. Wright (1969). Some aspects of the control of osmoregulation and homeostasis in birds. In C. Gual and F. J. G. Ebling (Eds) *Progress in Endocrinology, Excerpta Medica Int. Congr. Ser.*, **184**, 237–48.

Hüber, G. C. (1917). On the morphology of the renal tubules of vertebrates. *Anat. Rec.*, **13**, 305–39.

Hughes, M. R. (1970). Cloacal and salt gland ion excretion in the seagull *Larus glaucescens* acclimated to increasing concentrations of sea water. *Comp. Biochem. Physiol.*, **32**, 315–25.

Johnson, O. W. (1968). Some morphological features of avian kidneys. *Auk*, **85**, 216–28.

Johnson, O. W. (1974). Relative thickness of the renal medulla in birds. *J. Morphol.*, **142**, 277–84.

Johnson, O. W., and J. N. Mugaas (1970a). Some histological features of avian kidneys. *Am. J. Anat.*, **127**, 423–36.

Johnson, O. W., and J. N. Mugaas (1970b). Quantitative and organizational features of the avian renal medulla. *Condor*, **72**, 288–92.

Johnson, O. W., and R. D. Ohmart (1973). Some features of water economy and kidney microstructure in the large-billed savannah sparrow (*Passerculus sandwichensis rostratus*). *Physiol. Zool.*, **46**, 276–84.

Johnson, O. W., G. L. Phipps, and J. N. Mugaas (1972). Injection studies of cortical and medullary organization in the avian kidney. *J. Morphol.*, **136**, 181–90.

Lee, P., and K. Schmidt-Nielsen (1971). Respiratory and cutaneous evaporation in the zebra finch: effect on water balance. *Am. J. Physiol.*, **220**, 1598–605.
Lonsdale, K., and D. J. Sutor (1971). Uric acid dihydrate in bird urine. *Science*, **172**, 958–9.
Martin, B. J., and C. W. Philpott (1973). The adaptive response of the salt glands of adult Mallard ducks to a salt water regime: an ultrastructural and tracer study. *J. Exp. Zool.*, **186**, 111–22.
McNabb, R. A. (1974). Urate and cation interactions in the liquid and precipitated fractions of avian urine, and speculations on their physico-chemical state. *Comp. Biochem. Physiol.*, **48A**, 45–54.
McNabb, R. A., F. M. A. McNabb, and A. P. Hinton (1973). The excretion of urate and cationic electrolytes by the kidney of the male domestic fowl (*Gallus domesticus*). *J. Comp. Physiol.*, **82**, 47–57.
Peaker, M. (1971). Avian salt glands. *Phil. Trans. R. Soc. B.*, **262**, 289–300.
Peaker, M. (1977). Paper in preparation.
Peaker, M., and A. Hanwell (1974). Transepithelial [^{14}C]sucrose movements the goose salt gland. Relation to the secretory mechanism. *Pflügers Arch.*, **352**, 365–6.
Peaker, M., and J. L. Linzell (1975). *Salt Glands in Birds and Reptiles*, Cambridge University Press, Cambridge.
Peaker, M., and S. J. Stockley (1974). The effects of lithium and methacholine on the intracellular ionic composition of goose salt gland slices: relation to sodium and chloride transport. *Experientia*, **30**, 158–9.
Phillips, J. G., and D. M. Ensor (1972). The significance of environmental factors in the hormone mediated changes of nasal (salt) gland activity in birds. *Gen. Comp. Endocrinol. Suppl.*, **3**, 393–404.
Pittard, J. B., and A. D. Hally (1973). The effect of denervation on genotypic and compensatory growth of the immature avian salt gland. *J. Anat.*, **114**, 303.
Poulson, T. L. (1965). Countercurrent multipliers in avian kidneys. *Science*, **148**, 389–91.
Schmidt-Nielsen, K. (1960). Salt-secreting gland of marine birds. *Circulation*, **21**, 955–67.
Schmidt-Nielsen, K., A. Borut, P. Lee, and E. Crawford (1963). Nasal salt excretion and the possible function of the cloaca in water conservation. *Science*, **142**, 1300–1.
Shoemaker, V. H. (1967). Renal function in the mourning dove. *Am. Zool.*, **7**, 736.
Shoemaker, V. H. (1972). Osmoregulation and excretion in birds. In D. S. Farner and J. R. King (Eds) *Avian Biology*, Vol. 2, Academic Press, New York. pp. 527–74.
Siller, W. G. (1971). Structure of the kidney. In D. J. Bell and B. M. Freeman (Eds) *Physiology and Biochemistry of the Domestic Fowl*, Vol. 1, Academic Press, London. pp. 197–231.
Skadhauge, E. (1967). *In vivo* perfusion studies of the cloacal water and electrolyte resportion in the fowl (*Gallus domesticus*). *Comp. Biochem. Physiol.*, **23**, 483–501.
Skadhauge, E. (1973). Renal and cloacal salt and water transport in the fowl (*Gallus domesticus*). *Danish Med. Bull.*, **20**, Suppl. 1, 1–82.
Skadhauge, E. (1974). Cloacal resorption of salt and water in the galah (*Cacatua roseicapilla*). *J. Physiol. Lond.*, **240**, 763–73.
Skadhauge, E. (1975). Renal and cloacal transport of salt and water. In M. Peaker (Ed.) *Avian Physiology, Symp. Zool. Soc. Lond.*, No. 35, Academic Press, London. pp. 97–106.
Skadhauge, E., and S. D. Bradshaw (1974). Saline drinking, and cloacal excretion of salt and water in the Australian zebra finch (*Taeniopygia castanotis*). *Am. J. Physiol.*, **227**, 1263–7.
Skadhauge, E., and K. Kristensen (1972). An analogue computer simulation of cloacal

resorption of salt and water from ureteral urine in birds. *J. Theor. Biol.*, **35,** 473–87.

Skadhauge, E., and B. Schmidt-Nielsen (1967). Renal medullary electrolyte and urea gradient in chickens and turkeys. *Am. J. Physiol.*, **212,** 1313–8.

Sperber, I. (1960). Excretion. In A. J. Marshall (Ed.) *Biology and Comparative Physiology of Birds*, Vol. 1, Academic Press, London. pp. 469–92.

Stewart, D. J., W. N. Holmes, and G. L. Fletcher (1969). The renal excretion of nitrogenous compounds by the duck (*Anas platyrhynchos*) maintained on freshwater and on hypertonic saline. *J. Exp. Biol.*, **50,** 527–9.

Sykes, A. H. (1971). Formation and composition of urine. In D. J. Bell and B. M. Freeman (Eds) *Physiology and Biochemistry of the Domestic Fowl*, Vol. 1, Academic Press, London. pp. 233–78.

Van Rossum, G. D. V. (1966). Movements of Na^+ and K^+ in slices of herring-gull salt gland. *Biochim. Biophys. Acta*, **126,** 338–49.

Willoughby, E. J., and M. Peaker (1979). Osmotic and ionic regulation in birds. In G. M. O. Maloiy (Ed.) *Comparative Physiology of Osmoregulation in Animals*, Academic Press, London. In the press.

Chapter 9

Control Mechanisms in Mammals

M. ABRAMOW

I. INTRODUCTION	349
II. THE KIDNEY AND THE CONTROL OF LOSSES	350
A. Glomerular filtration	351
B. Tubular handling of sodium chloride and water	354
C. The concentrating mechanism and the role of urea	393
III. CONCLUDING REMARKS	402
ACKNOWLEDGMENTS	402
REFERENCES	403

I. INTRODUCTION

In the course of evolution, mammals, as other terrestrial animals, were faced with a new osmotic challenge; the acquisition of freedom from the aquatic environment means that the osmotic problem which it generates is exchanged for the need for fighting against water losses through evaporation in the air. Moreover, in mammals, some of these losses are incurred through thermoregulation. Thus, because water is not continuously available for intake, mammals must be able to excrete small volumes of urine when necessary. In addition, mammals are the only fully terrestrial vertebrates which are ureotelic, thus lacking the ability to excrete nitrogenous waste in isosmotic urine with a minimum of water. In order to comply with the double need to conserve water and to excrete urea, mammals must be able to form a concentrated urine. The mammalian kidney has evolved as an organ which can meet these challenges very adequately and the development of medullary counter-current mechanisms is a key feature of the kidney which allows it to play its unique role in osmoregulation. In fact, the mammalian kidney has assumed most of the osmoregulatory functions which are normally distributed among skin, gills, salt gland and kidney in other vertebrates.

In mammals because the availability of water is usually separate from that of solutes, both inputs are regulated independently through mechanisms for thirst and salt appetite under neurohumoral control. When access to water becomes limited, the maintenance of constant osmotic pressure of the body fluids requires that the kidneys excrete relatively large amounts of salt and urea in small volumes of water.

Accounts of these intake control mechanisms are given in Chapters 10 and 14; our discussion will thus be centred around the physiology of the mammalian kidney, with particular emphasis on the molecular basis of the intrarenal mechanisms which are involved in osmoregulation.

The solutes which contribute most importantly to the osmotic pressure of the blood (NaCl) and the urine (NaCl and urea) are handled in the mammalian kidney by the processes of glomerular filtration and net tubular reabsorption with some participation of net secretion in discrete regions of the nephron. The terms reabsorption and secretion are used here in a broad sense, namely transfer from tubular urine to renal interstitium or blood (reabsorption) or in the reverse direction (secretion) without implications as to the mechanisms underlying transport. Water transfer is either osmotically linked to solute transport or driven through a number of interfaces by hydrostatic and/or osmotic forces.

Although it is customary, or simply logical, to describe kidney function starting from the glomerulus and travelling down the nephron with the bulk of urine, the author wishes to stress that from the standpoint of osmoregulation, the overwhelming superiority of the terminal portions of the nephrons would dictate that we work backwards from the collecting duct to the glomerulus. For the sake of clarity, however, we will follow the traditional approach.

II. THE KIDNEY AND THE CONTROL OF LOSSES

A. Glomerular Filtration

Mammals filter each day at their glomeruli the equivalent of 15 to 100 times the volume of their extracellular fluid. Since fluid intake amounts to only a small fraction of the latter, one understands that the process of glomerular filtration would endanger life if it was not associated with the simultaneous tubular reabsorption of more than 95% of the filtrate formed.

1. *Mechanisms of Glomerular Filtration*

The process of glomerular filtration is of a purely physical nature. According to the Ludwig–Starling hypothesis (Ludwig, 1843; Starling, 1896, 1899) it results from the imbalance between hydrostatic and osmotic pressures exerted across the complex glomerular barrier. In this process, which is essentially that of an ultrafiltration, about one fifth of the plasma flowing through the glomerular capillary is filtered giving way in the Bowman's capsule to a glomerular fluid, essentially devoid of proteins but otherwise with a similar composition to that of plasma. The latter proposition holds if correction is made for the Donnan distribution of diffusible electrolytes and allowance is made for those solutes which are substantially bound to proteins.

The simplest equation describing the process of ultrafiltration would be:

$$Q_f = K_f(\Delta P - \Pi) \tag{1}$$

where Q_f is the rate of formation of the filtrate (ml min^{-1} or cm^3 s^{-1}); K_f is the filtration permeability coefficient (an intrinsic property of the glomerular barrier), with the dimensions of ml min^{-1} per mm Hg effective filtration pressure; ΔP is the difference between the hydrostatic pressure within the capillary and that in the Bowman's capsule; and Π is the colloid osmotic pressure of the plasma proteins (which is essentially equal to the colloid osmotic pressure difference across the glomerular barrier since in the Bowman's capsule the osmotic pressure is virtually zero).

Because the protein free filtrate is formed continuously along the axis of the glomerular capillary, there is a progressive increase in Π which can equal ΔP at or before the end of the capillary. At this point, filtrate formation should cease. Such a phenomenon has been taken into account in recent models of the ultrafiltering glomerulus (Brenner, 1974) and forms the basis of the theory of plasma flow dependence of glomerular filtration rate (GFR) in the rat.

One will note that for the phenomenon of glomerular filtration to proceed no energy is required on the part of the kidney if one neglects the metabolism of the glomerular cells which is necessary to sustain their life and maintain the physical integrity of their membranes. The driving force for the passage of fluid from the plasma to the tubular urine is imparted to the blood by the beat of the heart, as recognized more than one century ago (Ludwig, 1843).

A great deal of work has been devoted to the study of the passage through the glomerular capillary wall of solutes with ranges of molecular sizes intermediate between those of molecules freely filtered (such as inulin, MW 5000) and the largest macromolecules, which are totally impermeant (Renkin and Gilmore, 1973).

Since it was observed that restriction to the passage of these solutes increases with the molecular size, a behaviour similar to that of porous artificial membranes, the pore concept of glomerular filtration has received wide acceptance (Lambert, Gregoire and Malmendier, 1957; Pappenheimer, 1953; Wallenius, 1954).

The development and application of the equations describing the sieving nature of the glomeruli and the derivation of hydrodynamic parameters of the glomerulus from the sieving data of macromolecules (Lambert and coworkers, 1971) are beyond the scope of the present discussion. The interested reader should consult excellent recent reviews on the subject (Brenner and Deen, 1974; Renkin and Gilmore, 1973).

2. The Measurement of Glomerular Filtration Rate

Several methods have been used to measure GFR. The most practical is indirect and based on the clearance of a glomerular marker. The rate of appearance of the marker in the final urine is divided by its concentration in the plasma. The glomerular marker must be ideal in that the solute used should be completely filterable at the glomerular barrier and neither reabsorbed nor secreted by the tubules. Inulin certainly qualifies as such an ideal marker despite arguments to the contrary (Scott and coworkers, 1964) which were later disproved (Gutman, Gottschalk and Lassiter, 1965; Marsh and Frasier, 1965; Maude and coworkers, 1965).

The clearance techniques permit an assessment of the GFR of the whole kidney. However, the nephron population is heterogeneous (Jamison, 1973) and for a number of reasons it is desirable to measure GFR at the single nephron level (SN GFR). This is feasible by two different approaches, one of which requires micropuncture. In this case a tubule segment appearing at the surface of the kidney (most often in the proximal convoluted tubule) and a complete timed collection of the fluid flowing along that tubule is made. The amount of inulin recovered per minute divided by its concentration in the arterial plasma is a measure of the SN GFR. SN GFR is valid provided that the oil drop inserted distal to the site of puncture to ensure quantitative collection of the filtered fluid or provided that the suction of the tubular fluid through the pipette, have not themselves interfered with the normal pattern of glomerular hydrodynamics by altering the intratubular pressure (Schnermann, Horster and Levine, 1968; Wright and Giebisch, 1972). This method is applicable to most mammalian species in which capsular puncture cannot be made (Horster and Thurau, 1968) because the glomeruli do not appear at the surface (Lewy and Windhager, 1968; Liebau, Levine and Thurau, 1968).

More elaborate data obtained directly on single nephron parameters determining glomerular filtration rate in the mammal have required the breeding and use of the so-called Munich strain of rats (Brenner and Deen, 1974) in which a substantial number of glomeruli can be seen below the renal capsule and punctured. The possibility of a redistribution of GFR within the kidney under various conditions has been explored by comparing the whole kidney GFR with that of the SN GFR of the superficial nephrons. More appropriately SN GFR of superficial and deep nephrons can be compared directly in species where the juxtamedullary nephrons become accessible at the papilla such as *Psammomys obesus* (de Rouffignac and Morel, 1969) or the young rat (Jamison, 1970). In the adult rat part of the renal cortex has to be excised to expose the papilla (Horster and Thurau, 1968).

Methods for estimating GFR of the superficial, deep as well as intermediate nephrons and not requiring micropuncture have been developed (Coelho, Chien and Bradley, 1972; de Rouffignac, Deiss and Bonvalet, 1970; Hanssen, 1961, 1963). They combine the perfusion of the glomerular marker sodium ferrocyanide with maceration and dissection of the tubules. The amount of ferrocyanide recovered 'within' the tubular lumen (counted as ^{14}C ferrocyanide or revealed *in situ* as Prussian blue) serves as an index of SN GFR in various populations of nephrons.

A summary of SN GFR and of total kidney GFR in the kidneys of various mammals is given in Table 1.

Table 1. Glomerular filtration rate in the kidney of different species

Species	Total kidney GFR (ml min^{-1})	SNGFR (nl min^{-1})		References
		Superficial and (intermediate)	Juxtamedullary	
Rat	2.43	27.9 (33.3)	38.5	de Rouffignac, Bonvalet and Menard (1974)
Psammomys obesus	1.34	8 (16)	29	Baines and de Rouffignac (1969)
Rabbit	4–8	25	31	Forster (1947)
Dog	43	60	72	Bruns and coworkers (1974)

3. *Control of GFR in relation to Salt and Water Balance*

It is apparent from Table 1 that SN GFR increases from superficial to deep nephrons. Since it was also demonstrated that renin content of the juxtaglomerular apparatuses decreases from superficial to deep glomeruli (de Rouffignac, Bonvalent and Menard, 1974; Flamenbaum and Hamburger, 1974) and since it was proposed that the composition and flow of tubular fluid at the macula densa exerts a feedback control on the GFR of the same nephron (Thurau and Mason, 1974) there have been intensive searches in recent years for correlations between SN GFR and changes in renin content of the glomeruli in a variety of experimental conditions. The results so far do not permit a definitive answer to that important question (Bonvalet and de Rouffignac, 1975).

Acute Na loading, which is known to expand the extracellular space brings about a slight increase in total GFR. Results concerning a possible redistribution of SN GFR in this experimental situation have been largely equivocal (Bonvalet and de Rouffignac, 1975).

Chronic expansion of the extracellular fluid volume obtained by feeding rats with a high Na diet has been reported to elevate slightly total GFR while considerably altering the pattern of intrarenal distribution of SN GFR (Horster and Thurau, 1968). Rats on a low Na diet had an overall mean

SN GFR of 31.4 nl min^{-1} (calculated by dividing total kidney GFR by the total glomerular count) while superficial cortical glomeruli and juxtamedullary glomeruli filtered 23.5 and 58.2 nl min^{-1} respectively (tubular micropuncture data). Rats on a high Na intake had filtration rates of 33.7, 38.1 and 16.5 nl min^{-1} respectively (Horster and Thurau, 1968). Thus there was a shift of filtrate formation from deep to superficial nephrons. The latter are shown to filter more and assumed to reabsorb less Na. If true, the situation would represent a very appropriate renal adaptation for the excretion of sodium in the mammal. Subsequent studies however, while confirming a slight increase in GFR with chronic Na-loading failed to reproduce the intrarenal shift in SN GFR (Bonvalet and de Rouffignac, 1975).

While several factors intrinsic to the kidney (with the widely described but not well understood phenomenon of autoregulation) or extrinsic to it, including adrenergic and cholinergic influences, doubtless participate in the control of GFR (Renkin and Gilmore, 1973), it seems safe to conclude at this juncture that changes in GFR do not play a central role in the control of Na and fluid balance in the mammal.

The same considerations apply to the phenomenon of glomerular recruitment. While recruitment of glomerular activity has been unequivocally demonstrated in vertebrates other than mammals with clear-cut instances where the needs for conserving water are met through the action of neurohypophysial peptides (which decrease urine flow by reducing GFR) (Jard, 1966; Lahlou, 1966) such a mechanism seems to be lacking in mammals (Renkin and Gilmore, 1973).

Clearly the mammalian kidney does not primarily regulate sodium and water balance through GFR. An understanding of how it does so requires knowledge of the mechanisms which control the powerful tubular reabsorptive processes according to the needs of the organism.

B. Tubular Handling of Sodium Chloride and Water

As emphasized earlier, the daily urinary excretion of water and electrolyte is only a very small fraction of the filtered load. Thus in man, dog and rat the tubular epithelium reabsorbs more than 99% of the filtrate. It is apparent that small variations of that reabsorptive capacity can greatly affect daily outputs. Consequently, the regulation of the volume and composition of the body fluids requires very accurate adjustments of tubular reabsorption.*

---

* It is convenient to describe tubular transport in the nephron by subdividing it into successive anatomical and functional segments, namely, the proximal convoluted and straight tubules, the thin descending and thin ascending limbs of Henle's loop, the thick ascending limb, the distal convoluted tubule, the cortical collecting tubule and the papillary collecting ducts. A detailed anatomical account of the mammalian kidney is outside the scope of the present discussion.

1. Proximal Tubule

The proximal tubule is the site where the largest portion of the filtered Na, Cl, bicarbonate and fluid is reabsorbed as evidenced by data from a number of micropuncture studies in laboratory animals (Gottschalk, 1961; Walker, 1941; Windhager and Giebisch, 1961).

a. *Proximal Convoluted Tubule* Measurements of the fractional reabsorption of the filtered load have been carried out in so-called free-flow micropuncture experiments. The anaesthetized animal is perfused with inulin and enough saline 'to replace surgical losses'. The tubular samples are

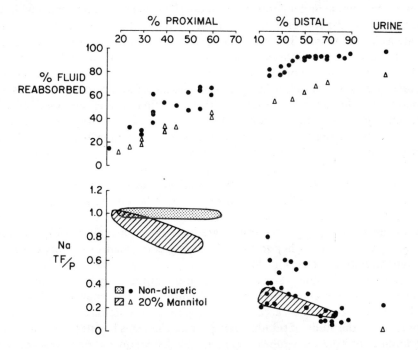

Figure 1 Micropuncture study of sodium and water handling by rat kidney.

Fractional fluid reabsorption and ratio of tubular urine to plasma concentrations of sodium under non-diuretic conditions and under mannitol diuresis. In the upper portion of the graph the fractional fluid reabsorption, calculated from TF/P inulin ratios, is plotted as a function of fractional distance along the proximal or distal convoluted tubule and compared to the value in the ureteral urine.

In the lower portion, the tubular fluid to plasma concentration ratio of Na is represented either as individual values or as areas covering the range of values. The sodium concentration in the proximal tubule remains identical to that of plasma except when a poorly reabsorbed solute retards fluid absorption and allows the Na concentration in the proximal tubule to decrease below the plasma level as it normally does in the diluting portions of the nephron. From Giebisch and Windhager (1964); reproduced by permission of the *American Journal of Medicine* analysed for inulin and wherever appropriate, osmolality, Na, KCl, etc. the linear increase in the tubule fluid to plasma concentration ratio (TF/P) of inulin along the length of the accessible portions of the proximal convoluted tubule quantitatively reflects the fraction of the glomerular fluid absorbed. For example a TF/P inulin of 2 indicates that, at the site of collection of the sample, fifty per cent of the filtrate has been reabsorbed. Fig. 1 (Giebisch and Windhager, 1964) shows a plot of fractional fluid absorption along cortical segments of the nephron together with the TF/P ratios of sodium measured in the same studies. Conditions of antidiuresis and mannitol diuresis are compared.

It is apparent that large amounts of fluid are reabsorbed in the proximal convoluted tubule under control conditions. This phenomenon occurs without any change in the concentration of Na in the luminal fluid, which remains the same as in the plasma. The proximal reabsorption of fluid is always isosmotic (Gottschalk, 1961; Gottschalk and Mylle, 1959; Walker and coworkers, 1941; Walker and coworkers, 1937; Wirz, 1956). When a poorly reabsorbed solute such as mannitol is infused, it produces an osmotic diuresis. Under such conditions, clearance studies show that the final urine, obtained during maximal diuresis is still isosmotic (Wesson and Anslow, 1948) but its sodium concentration drops to very low levels. Wesson and Anslow hypothesized that during such a strong diuresis the final urine was mainly composed of proximal tubular fluid that had escaped reabsorption and had not been greatly altered during its rapid transit through the distal portions of the nephron. They postulated that the sodium concentration must be lowered in the proximal tubule and that proximal tubule reabsorption was an active process, proceeding against the chemical potential gradient (Wesson and Anslow, 1948). Wesson and Anslow's hypothesis was amply confirmed in subsequent micropuncture studies (see Fig. 1).

In addition to studies using micropuncture methodology which contributed immensely to our knowledge on the magnitude and mechanisms of sodium chloride reabsorption in accessible segments of the nephron, a new direct approach to the study of tubular transport mechanisms has stemmed from the techniques of isolating nephron segments by dissection and perfusing them *in vitro*. The isolated tubule technique of Burg, Grantham, Abramow and Orloff (Burg and coworkers, 1966) are depicted schematically in Fig. 2.

These techniques allow the study of virtually all the nephron segments including such portions as the pars recta of the proximal tubule, the thick ascending limb of Henle's loop and the cortical collecting tubule which are not reached by the micropuncture approach.

In addition, the rate of perfusion as well as the composition of the peritubular and luminal fluids can be controlled. Long term measurement of transepithelial potential differences and electrical resistance are easier to perform than in *in vivo* preparations.

The only mammalian species studied so far with these techniques has been the rabbit although it has also been possible to investigate some properties of the isolated human collecting duct (Abramow and Dratwa, 1974).

Figure 3 summarizes data on net fluid absorption in the isolated proximal tubule of the rabbit. It is apparent also that a transepithelial voltage of about −5 mV (lumen negative) exists under control conditions. Ouabain in the medium bathing the tubule induces a reversible decrease of both the voltage and the fluid absorption to almost zero.

There are several lines of evidence establishing that the reabsorption of fluid in the proximal tubule is a result of active net reabsorption of Na. First there are instances in which, as stated above, the sodium concentration is

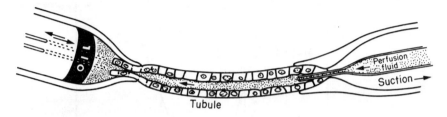

Figure 2 Arrangement for perfusing isolated kidney tubules. After dissection, one tubule end is aspirated towards the constriction of a holding pipette (right) and cannulated with a perfusing pipette. The perfused fluid accumulates at the other end and can be removed with calibrated pipettes for measurements of flow rates and analysis. Not shown are pipettes containing the liquid dielectric Sylgard 184 which come over the holding pipettes and provide good seals at the ends of the epithelium. From Burg and Orloff (1968); reproduced by permission of the Rockefeller University Press

20–50 mmol l^{-1} lower than in the peritubular space when a poorly reabsorbed solute such as mannitol or raffinose is present in the lumen (see Fig. 1). This also occurs in isolated proximal tubules perfused with solutions containing raffinose (Kokko, Burg and Orloff, 1971). Second, there is a transepithelial electrical potential which is so oriented that the lumen is negative relative to the interstitium (Fig. 3). Thus sodium is 'pumped' against an electrochemical gradient. Third, inhibitors known to interfere with active sodium transport in other tissues such as ouabain (which inhibits Na–K-activated ATPase) strongly decrease fluid absorption and cause the potential difference to drop to zero (Fig. 3). Removal of potassium has the same effects, presumably through a similar mechanism. Luminal Na is thought to enter the cells through the apical membrane under favourable electrical and chemical gradients and to be actively pumped across the baso-lateral membranes (Giebisch and Windhager, 1973).

Water reabsorption is believed to be coupled to sodium by osmosis although no osmotic gradient has been detected across the epithelium. This has been explained by postulating that there is a region within the epithelium where sodium pumping creates local hypertonicity. This constitutes a driving force for the movement of water by osmosis. Theoretical models for isotonic fluid transport coupled to active solute transfer have been developed (Curran, 1965; Curran and MacIntosh, 1962). They incorporate a middle compartment interposed between the two other compartments (lumenal and peritubular) available for sampling. Diamond and Bossert (1967) have adapted the

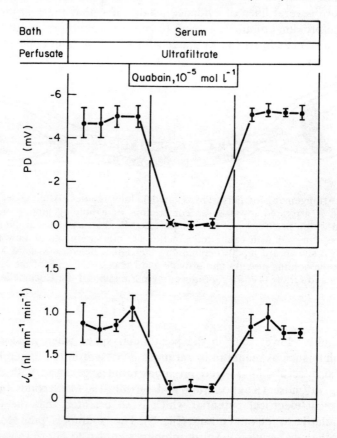

Figure 3 Effect of ouabain on the rate of fluid absorption (J_v) and the transepithelial voltage (PD) across isolated rabbit proximal convoluted tubules. See text for documentation. From Cardinal and coworkers (1975); reproduced by permission of *Kidney International*

model of Curran and MacIntosh (1962) to transport in epithelia by postulating that the intercellular spaces and basal infolding could serve as the 'middle compartment'. In this view, sodium is transported actively from the cells into the interspaces. Since these are long and narrow channels, mixing is incomplete and hypertonicity drives water from the cells to the interspaces where hydrostatic pressure increases, thus forcing the fluid across the basement membrane and the peritubular capillary wall. The osmotic gradient can be dissipated progressively along the channel through osmotic transfer of water from cell to interspace. Thus the intercellular compartment would be hypertonic in the region close to the tight junction and isotonic toward the basal portion of the channel. This is the so-called 'standing gradient model' (Diamond and Bossert, 1967). Modified versions of this model have been proposed recently by Sackin and Boulpaep (Sackin and Boulpaep, 1975) to account for proximal tubule Na reabsorption. They include different boundary conditions and provide for a permeable tight junction and electrical as well as chemical driving forces for ion flows. In these models the interspaces remain hyperosmotic with uniform salt concentration profiles and no standing gradient was found. However the degree of hypertonicity of the fluid emerging from the interspace may be small enough to be experimentally indistinguishable from isotonic transport. Finally, whatever the model considered, the reabsorbate is driven through the basement membrane and the capillary wall by the net balance between hydrostatic and colloid osmotic pressure differences across these boundaries, the so-called Starling forces.

In addition to providing an explanation for isotonic fluid transfer driven by active sodium reabsorption, the models of Diamond and Bossert and of Sackin and Boulpaep emphasize the importance of the intercellular or so-called paracellular 'shunt' pathway (Ussing and Windhager, 1964) for a valid description of transepithelial transport of water and solute in the proximal tubule. This description takes into account a number of experimental observations. Electrophysiological studies have suggested that this shunt pathway is a low resistance route from luminal to peritubular fluids (Boulpaep, 1976). The total effective transepithelial resistance is much lower than the sum of the resistances of the single cell boundaries. Thus this shunt determines the passive permeabilities of the proximal tubule epithelium. The proximal tubule certainly ranks among the 'leaky' epithelia, with a low transepithelial resistance and a high permeability to sodium (see Table 2 and also Fig. 5) (Ussing, Erlij and Lassen, 1974).

Several morphological observations suggest that one component of the intercellular pathway, namely the tight junction or *zonula occludens* plays an important role in the regulation of fluid and solute transfer in the proximal tubule as well as in other nephron segments or non-renal epithelia. These studies indicate that 'leaky' epithelia such as the proximal tubule or the gall bladder have shallow tight junctions with few junctional strands whereas 'tight' epithelia such as frog skin, toad bladder, collecting duct have deeper and denser tight junctions (Claude and Goodenough, 1973; Farquhar and Palade, 1963, 1964; Humbert and coworkers, 1977; Martinez-Palomo and Erlij, 1973; Tisher and Yarger, 1973; Whittembury, Rawlins and Boulpaep, 1973).

The shunt pathways with the *zonula occludens* is believed to be the site and

Table 2. Permeability characteristics and electrical parameters in the proximal tubule

Segment	Species	Preparation	Hydraulic conductivity, L_p ($\times 10^{-5}$ cm^3 cm^{-2} s^{-1} atm^{-1})	Diffusional permeability, P ($\times 10^{-5}$ cm s^{-1})				Potential difference, PD (mV)	Specific transverse resistance (Ω cm^2)	References
				water	urea	Na	Cl			
Proximal convoluted tubule (PCT)	rat	*in vivo*	18		11.1					Ullrich, Rumrich and Fuchs (1964)
	rat	*in vivo*	15.5	560						Persson and Ulfendahl (1970)
	rat	*in vivo*	8.3							
	rat	*in vivo* early						−1.5		Fromter and Gessner (1974)
	rat	*in vivo* intermediate and late						+1.8		Fromter and Gessner (1974)
	rat	*in vivo* early							11.6	Seely (1973a)
	rat	*in vivo* intermediate and late							5.6	Seely (1973a)
	dog	auto perfused *in vivo*						−2	5.6	Boulpaep and Seely (1971)
	rabbit	isolated perfused	2.9–6.3							Kokko, Burg and Orloff (1971)
	rabbit	isolated perfused				9.3				Kokko (1972)
	rabbit	isolated perfused			5.3		3.8			Rocha and Kokko (1974)
	rabbit	isolated perfused					9			Stoner, Burg and Orloff (1974)
	rabbit	isolated perfused						−4 to −6		Burg and Orloff (1970); Kokko and Rector (1971)
	rabbit	isolated perfused							7.0	Lutz, Cardinal and Burg (1973)
Proximal straight tubule (PST)	rabbit	isolated perfused				2.6	5.6			Schafer, Troutman and Andreoli (1974)
	rabbit	isolated perfused superficial						−1.3 to −2		Lutz, Cardinal and Burg (1973); Schafer, Troutman and Andreoli (1974)
	rabbit	isolated perfused juxtamedullary					2.1	−1.8		Kawamura and Kokko (1976)
	rabbit	isolated perfused superficial							8.2	Lutz, Cardinal and Burg (1973)

modulator of backflux of sodium from peritubular to luminal compartments (Bank, Yarger and Aynedjian, 1971; Boulpaep, 1976; Green, Windhager and Giebsch, 1974). Thus experimentally induced changes in net sodium and fluid reabsorption may reflect alterations in the rate of passive intercellular backflux rather than an action on the sodium pumping mechanism. Such changes have been demonstrated to occur as responses to the so-called Starling forces (Giebisch and Windhager, 1973). Physical factors such as the peritubular osmotic or hydrostatic pressure are determinants of a sizeable moiety of net fluid absorption both in amphibian (Boulpaep, 1976) and mammalian (Grantham, Qualizza and Welling, 1972; Hayslett, 1973; Imai and Kokko, 1972; Seely, 1973b) proximal tubules. It is believed that, by affecting the rate of removal of the reabsorbate out of the intercellular spaces, the peritubular physical forces act upon the resistance of the shunt path through pressure induced changes in the permeability of the zonula occludens (Windhager and Giebisch, 1976). Increasing the hydrostatic pressure or decreasing the osmotic pressure of the peritubular environment was shown to reduce the electrical resistance of the paracellular route (Boulpaep, 1972; Grandchamp and Boulpaep, 1974; Seely, 1973b).

Decreased electrical resistance presumably reflects increased backflux into the lumen, of sodium which had been pumped, thus decreasing the efficiency of net sodium transport in the proximal tubule.

An effect of peritubular protein oncotic pressure has been demonstrated in several micropuncture studies including free-flow experiments where proximal tubular reabsorption was directly correlated with postglomerular protein concentration under infusions of saline, isoncotic or hyperoncotic solutions (Brenner and coworkers, 1969; Brenner and coworkers, 1973). In other studies the microenvironment of the punctured tubules was selectively altered and the same qualitative findings were obtained. The well known inhibition of proximal tubular fluid reabsorption during acute expansion of the extracellular fluid volume with saline infusions was reversed when the protein concentration in peritubular capillaries was increased back to normal by perfusing the capillaries with solutions isoncotic to normal plasma (Brenner and coworkers, 1973). The latter findings suggest that at least some of the inhibition of the proximal fluid reabsorption during volume expansion with saline infusions is a consequence of a decrease in the protein concentration in the postglomerular capillary bed.

Under physiological conditions, the main source of variation of the postglomerular oncotic pressure is the rate of glomerular filtration. As GFR increases, the concentration of proteins in the efferent arteriole and peritubular capillary network increases proportionally. As a consequence, net tubular reabsorption of Na and fluid increases thus providing a simple control mechanism which adapts the rate of proximal tubular reabsorption to the changing filtered load. This adaptation is of prime importance since in its absence, even relatively small changes in the filtered load presumably would overwhelm the regulatory capacity of the distal nephron segments and result in large alterations in the excretion of sodium chloride and water with an increased risk of jeopardizing overall osmoregulation. Whatever the mechanisms underlying the proximal tubular adaptive changes, the phenomenon has been termed 'glomerulo-tubular balance'. Although more elaborate theories have been proposed to explain the phenomenon, they have been generally less convincing (Gertz and Boylan, 1973).

A series of interesting observations derive from studies with isolated perfused proximal convoluted tubule of the rabbit. In this experimental situation, the capillary network is eliminated. Yet increasing the protein concentration in the medium bathing the tubules is associated with increased net fluid absorption (Burg and coworkers, 1976; Grantham, Qualizza and Welling, 1972; Imai and Kokko, 1972). That the protein effect was, at least in part, due to changes in oncotic pressure is suggested by the findings than in tubules bathed in rabbit serum and perfused with an isosmolal ultrafiltrate of the same serum, the addition of 3.5 g dl^{-1} of polyvinylpyrrolidone (PVP) to the bath produced a 67% increase in the fluid reabsorption (Imai and Kokko, 1974a). The transtubular electrical potential difference was not affected by the changes in oncotic pressure. When ouabain had inhibited net fluid absorption, PVP still exerted a stimulating effect. The absolute oncotic effect was decreased. However the fractional increase in reabsorption was the same as in the experiments without the inhibitor of active transport. Since it is not certain that ouabain eliminated all of the active transport under these conditions, the significance of the osmotic effect in inhibited tubules remains unclear. Imai and Kokko concluded that active transport pathways are coupled to pathways which are responsive to oncotic forces (Imai and Kokko, 1974a). Evidence for the involvement of the 'shunt' pathway derives from the demonstration that removal of protein from the bath is associated with increased bath to lumen flux of ^{14}C sucrose, a marker which is assumed not to penetrate the cells (Imai and Kokko, 1972).

While it is possible that the capillary wall and the geometry of the interstitial spaces are important mediators of the oncotic effects *in vivo*, another site of action of protein must be searched to explain the effect *in vitro*. Welling and Grantham (1972) showed that albumin exerted an oncotic effect across the isolated basement membrane. In addition, however, the basement membrane is sufficiently permeable to albumin (Tisher and Kokko, 1974; Welling and Grantham, 1972) to permit it to exert its effect at sites within the lateral spaces. An ultrastructural study of isolated proximal convoluted tubules of the rabbit by Tisher and Kokko (1974) showed widening of lateral and basilar intercellular spaces when the tubules were exposed to hyperoncotic or isoncotic bathing solution at the time of fixation. Stated another way, shrinking and irregularity of the interspaces was observed in tubules bathed in hypo-oncotic protein-free fluids.

A general feature of the oncotic and hydrostatic effects in the proximal tubule is the asymmetrical behaviour of the epithelium in response to the

Starling forces. Thus, introduction of hyperoncotic solutions into the lumen produces little or no changes in net water absorption (Burg and coworkers, 1973; Giebisch and coworkers, 1964; Imai and Kokko, 1974a; Kashgarian and coworkers, 1964).

It has been assumed previously, on the basis of a number clearance (Bresler, 1976; Malvin and coworkers, 1958; Sonnenberg, 1973; Vander and coworkers, 1958; Vereerstraeten and Demyttenaere, 1968) and micropuncture (Brenner and coworkers, 1969; Brenner and Troy, 1971; Persson, Agerup and Schnermann, 1972; Spitzer and Windhager, 1970) studies demonstrating the oncotic effect, that the oncotic gradient was a direct driving force for passive sodium and fluid reabsorption (Bresler, 1976; Vander and coworkers, 1958; Vereerstraeten and Demyttenaere, 1968). The old concept of Carl Ludwig (Ludwig, 1843) which was that renal tubular reabsorption is the passive result of physical forces was thus used as an explanation for the oncotic effect. The concept had to be abandoned for two main reasons.

First, the hydraulic conductivity (L_p) of the proximal tubule appears too low to permit water flux to proceed at the high rates observed when measuring net fluid absorption (Ullrich, 1964). Second, the asymmetrical effects of the oncotic force required another explanation such as that offered by the 'peritubular control theory' (Lewy and Windhager, 1968).

The coefficient of hydraulic permeability can be measured in perfused tubules, *in vivo* or *in vitro*, according to the following formula:

$$J_v = L_p \text{ (osmotic) } \Delta \Pi + L_p \text{ (hydrostatic) } \Delta P \qquad (2)$$

where J_v is the transtubular volume flux ($cm^3 \, cm^{-2} \, s^{-1}$) induced by the imposed osmotic ($\Delta \Pi$) or hydrostatic (ΔP) pressure difference (in atm or ml H_2O) across the tubule wall; $\Delta \Pi$ is the effective osmotic pressure gradient, equal to $\sigma RT \Delta c$, where σ is the reflection coefficient of the solute used to generate the osmotic pressure difference, Δc is the mean difference in concentration of the solute between the lumen and peritubular fluid, R and T have their usual meaning.

Because of water transfer, $\Delta \Pi$ varies along the length of the tubule segments. When σ is unity an exact mean integrated $\Delta \Pi$ can be computed (Dubois, Verniory and Abramow, 1976). Otherwise, an arithmetic (Grantham and Orloff, 1968) or logarithmic (Abramow, 1974; Kokko, 1970) mean osmotic gradient is empirically used.

From equation (2) L_p can be calculated. Usually one measures the osmotic L_p since much larger pressure gradients can be produced by experimentally changing osmotic rather than hydrostatic pressures. L_p values are given for the proximal tubule in Table 2.

Recently, the concept of peritubular oncotic and hydrostatic pressures acting as direct driving forces for fluid absorption was again proposed by Schnermann and his colleagues (Persson and coworkers, 1975). They found hydraulic conductances 3–8 times higher than previously reported, when measurements of L_p in rat proximal tubule were obtained with albumin and PVP as the impermeant molecules. Green, Windhager and Giebisch (1974) reinvestigated the problem by doubly perfusing rat proximal tubules with their peritubular capillaries. According to the concept of the direct driving force, colloid effects should be produced whether or not active transport of sodium is present.

When the sodium pump was poisoned with 4 mmol l^{-1} Na cyanide, albumin in the lumen had no effect and peritubular albumin had only a minor effect (Green, Windhager and Giebisch, 1974) whereas the peritubular effect was marked in the presence of normal active Na transfer. The hydraulic conductances were low (8.26×10^{-5} cm^3 cm^{-2} s^{-1} atm^{-1}) and essentially the same whether raffinose or albumin was used for generating osmotic gradients.

The reasons for the disagreement between both groups of studies are not clear.

However, since the results of Green and coworkers are in accord with free-flow experiments *in vitro* there is a tendency to accept them as valid. According to these authors, the magnitude of the fraction of the filtrate which could be directly reabsorbed through the osmotic effect of the plasma proteins would be only 8%. Thus these studies confirm many others in demonstrating that the protein effect is mainly to remove fluid that has been reabsorbed secondary to active Na transport. Interestingly, a recent study by Horster and Larsson (1976) indicates that the hydrostatic hydraulic conductance L_p, obtained by measuring J_v under appropriate hydrostatic pressure gradients across isolated rabbit proximal convoluted tubules amounts to 8.62×10^{-5} cm^3 cm^{-2} s^{-1} atm^{-1}. This value is the same as those (see above and Table 2) measured with osmotic gradients, further validating the latter estimates. In contrast, tubules from 2–6 day old rabbits had a hydrostatic L_p seven times larger, suggesting that isotonic fluid absorption in the neonatal proximal tubule may be more dependent on transepithelial pressure gradients than in the mature tubule.

Further insight into the mechanisms of fluid transport in proximal convoluted tubules has been derived from recent studies by Burg and coworkers on the ionic requirements for (Burg and Green, 1976) and the role of organic solutes in (Burg and coworkers, 1976) net fluid absorption and transepithelial voltage. Net fluid transport stops and voltage is abolished when Na is removed from the lumen and peritubular bathing fluids, whether the cation is replaced by Li, choline or tetramethylammonium (Burg and Green, 1976). As expected K is also essential for fluid absorption and voltage. Omission of bicarbonate slows absorption by 33%. Thus the effect is less marked than in the case of removal of Na or K. In contrast, replacement of all the chloride with nitrate or perchlorate has no effect on fluid absorption or potential difference (Burg and Green, 1976).

In addition the presence in the lumen of organic solutes such as glucose, alanine, lactate or citrate all enhanced fluid absorption (Burg and coworkers, 1976). Sugars and amino acid presumably are co-transported with Na. Their presence stimulates fluid transport and increases luminal negativity because of their own transport, not because they act as fuels for the Na pump. This is inferred from the fact that a-methyl-D-glucoside and cycloleucine, which are transported but not metabolized by the tubular cells, both stimulate fluid absorption when added to the perfusate (Burg and coworkers, 1976).

The above *in vitro* studies bear directly on the *in vivo* observations that several organic solutes including glucose and aminoacids are extensively reabsorbed in the first few millimetres after the glomerulus so that their concentration in the tubular fluid drops rapidly to very low levels along the proximal convolution. This contrasts with the sodium and osmolality in the tubular fluid, which remain the same as in plasma. Similarly the concentration of bicarbonate falls, due to the secretion of hydrogen ions into the lumen. There is a reciprocal rise in the concentration of chloride ions. The mechanisms by which the chloride rises remain unclear. For several reasons, among others, because chloride in the tubular cells is higher than predicted for a Donnan equilibrium (Abramow, Burg and Orloff, 1967) and because in isotonic $NaHCO_3$ droplets injected into proximal tubules the steady-state chloride concentration rose to levels 40% greater than plasma levels (Malnic, Mello-Aires and Vieira, 1970) a mechanism for the exchange of HCO_3 for Cl has been proposed in relation to the process of proximal tubular acidification (Rector, 1976). Whatever the mechanisms responsible for the development of a tubular to plasma chloride gradient in the late proximal convoluted tubule, this gradient constitutes a driving force for passive reabsorption of Na ions. Preferential bicarbonate reabsorption also augments Na reabsorption by solvent drag because of the difference in reflection coefficients of sodium chloride and sodium bicarbonate. In addition the voltage changes polarity along the proximal convoluted tubule, becoming lumen positive in the late portions (see Table 2). The explanation for the change in polarity is the following: in the early proximal tubule, the concentrations of glucose and alanine are sufficient to produce a negative voltage (Kokko, 1973). As their concentration decreases, so does the voltage. Additionally a positive potential is generated because of the establishment of both a chloride and a bicarbonate transepithelial gradient (bi-ionic potential). Since chloride is more permeant than bicarbonate, the chloride gradient is responsible for the positive voltage. The higher effective osmotic pressure difference generated by the bicarbonate as compared to the chloride gradient tends to favour flow of water out of the lumen. Solvent drag might be an additional driving force for passive sodium efflux from lumen to peritubular space.

Thus, although the basic active mechanism for Na reabsorption remains the same in early and late proximal convoluted tubules, the importance of the driving forces for passive Na efflux may change along this segment of the nephron.

With respect to the relative importance of the active and passive components of Na reabsorption, it should be stressed that conclusions derived from the phenomenological description of proximal transport (Fromter, Rumrich and Ullrich, 1973) differ from those which are based on direct experimental manipulations both *in vivo* and *in vitro* (Windhager and Giebisch, 1976).

With the thermodynamic treatment of the ion fluxes, the transepithelial voltage and net fluid absorption, only one-third of the reabsorbed Na is predicted to be active; one-half of the passive component (which amounts to two-thirds of the total Na efflux) is driven by the voltage while the other half is absorbed by solvent drag (Fromter, Rumrich and Ullrich, 1973).

In contrast, in isolated rabbit proximal tubules, bicarbonate and chloride concentration differences *per se* did not cause changes in fluid absorption (Cardinal and coworkers, 1975). Qualitatively similar results were obtained *in vivo* in the rat using simultaneous tubular and capillary microperfusion with artificial solution (Green and Giebisch, 1975a, 1975b).

The phenomenological approach treats the tubule as a single homogenous membrane. Because of the number of boundaries which actually exist in the proximal tubule and the lack of experimental evaluation of their permeability properties and the real driving forces, the limitations of the thermodynamic treatment have been emphasized and its conclusions have been challenged (Windhager and Giebisch, 1976).

Despite the amount of experimental work on the function of the proximal convoluted tubule, and although the definite inhibitory effect of acute expansion of the extracellular fluid volume on net sodium and fluid reabsorption has been a universal finding with emphasis on the role of the peritubular Starling forces in mediating this effect, it is still unclear as to what extent the proximal tubular epithelium plays a regulatory role in the day to day sodium excretion (*vide infra*).

b. *Proximal Straight Tubule* This portion of the proximal tubule has a rectilinear shape, at least in the superficial nephrons, and courses down to the outer medulla where it gives way to the thin descending limb of Henle's loop. Since most of it is not accessible from the surface of the kidney it could not be explored directly by micropuncture techniques. Micropuncture of the last accessible portions of the proximal convoluted tubule and of the early distal convoluted tubule of the same superficial nephrons permits extrapolation to the intervening segments called 'short loops' of Henle. However these are a rather heterogenous mixture of late proximal convoluted, proximal straight tubule, thin descending and ascending limb, thick ascending limb of Henle's loop and early distal convoluted tubule. The technique of *in vitro* perfusion of isolated segments of the nephron (Burg and coworkers, 1966) has clarified the function of these important portions of the tubular epithelium.

In isolated proximal straight tubules perfused with ultrafiltrate of the same rabbit serum used as bathing fluid, the net fluid absorption per millimetre length was found to be half as great as in the convoluted portion (Burg and

Orloff, 1968). Active sodium transport is the same basic mechanism for reabsorbing fluid as in the convoluted portion. Chloride absorption is passive, driven by the electrical potential difference which is lumen negative (see Table 2) (Kawamura and coworkers, 1975; Schafer, Troutman and Andreoli, 1974).

There are several important functional differences between convoluted and straight portions however. Glucose transport is negligible in the straight portion (Tune and Burg, 1971). On the other hand PAH secretory capacity is nearly three times larger in the straight tubule (Tune, Berg and Patlak, 1969). Glucose or amino acids in the perfusate do not alter the rate of fluid absorption or the transepithelial voltage. The absence of a glucose co-transport system may be an important determinant of the intrinsically lower capacity of the straight segment to reabsorb fluid and sodium.

When proximal straight tubules were perfused with a fluid simulating that which leaves the proximal convoluted tubule (having a lower concentration of bicarbonate and a higher concentration of chloride than in the bathing medium), a positive bi-ionic potential was found, as in the convoluted portion. There were differences, however, with respect to the mechanism of fluid transport. Ouabain stops fluid transport in straight tubules perfused with ultrafiltrates. However, when the chloride in the perfusate replaced the bicarbonate, one-third of the fluid absorption persisted in the face of high concentrations of ouabain. Thus in this segment as opposed to the convoluted portion, there is definite experimental evidence for a passive component of fluid transfer driven by the opposing concentration gradients of anions with different reflection coefficients (Burg, 1976a).

The high salt permeability and low effective resistance (see Table 2) suggest an important paracellular shunt pathway, as in the proximal convoluted tubule.

Heterogeneity between superficial and deep proximal straight tubules has been found with respect to the relative sodium and chloride permeabilities. The straight tubules of the deep nephrons are more permeable to Na than to Cl. The reverse was true for the superficial straight tubules (see Table 2).

Acidification clearly occurs in the proximal straight tubule since the total CO_2 in the collected fluid decreases considerably in tubules perfused at slow rates with HCO_3 25 mmol l^{-1} in the lumen and the bath (McKinney and Burg, 1976; Warnock and Burg, 1977).

2. Thin Limbs of Henle's Loop

The transport characteristics of the loop of Henle are complex and play a key role in the concentrating process since they form the basis for the counter-current multiplication system. This system is responsible for the progressive accumulation of solutes from outer medulla to the papillary tip.

Only a sketchy outline of the functional properties of the loop will be given here while some aspects will be treated in more detail in the section on the counter-current multiplier.

The transition from the *pars recta* of the proximal tubule to the descending thin limb of Henle's loop forms the boundary between outer and inner stripes of the outer medulla. The descending limb from intermediate and juxtamedullary nephrons penetrates the inner medulla and forms a hairpin turn at the beginning of the thin ascending limb of the loop. The level of the hairpin turn is variable. Thus, the environment of the loops may vary according to the type of nephrons from which they originate. The long loops traverse regions of increasing osmolalities.

It has been concluded from micropuncture studies that approximately 25% of the filtered Na is reabsorbed in the 'loop', i.e. between the last accessible convolution of the proximal tubule and the first accessible portion of the distal tubule. One must examine the reabsorptive capacity of each of the intervening segments of the 'loop' as defined in micropuncture work. The straight portion reabsorbs fluid at a rate less than half that of the convoluted tubule (Schafer, Troutman and Andreoli, 1974; Tune, Berg and Patlak, 1969). In addition no evidence for active transport of solute (Kokko, 1970) or of sodium (Abramow, unpublished) was obtained in isolated perfused thin descending limbs of Henle's loop. No evidence for active sodium efflux from the thin ascending limb has been obtained (Imai, 1977; Imai and Kokko, 1974b). Thus most of the salt which is reabsorbed between the end of the proximal convoluted tubule and the early distal tubule must be transported out of the thick ascending limb of Henle's loop.

A number of different events have been shown to occur as the tubular fluid in the loop enters and subsequently leaves the medulla.

It has been known for a long time, from the remarkable observation by Wirz, Hargitay and Kuhn (1951) that *in situ* frozen kidneys from hydropenic rats displayed melting at the same temperature of the ice in all the tubular and vascular structures taken from the same level in the medulla, suggesting a tendency towards osmotic equilibration of the loop and collecting duct fluid with the hypertonic environment. Using a similar approach Bray (1960) later noted that the content of ascending limb fluid was less hypertonic than the contents of all the other structures. The mechanism by which the fluid in the descending limb (which is initially isosmotic with systemic plasma when leaving the *pars recta* of the proximal tubule) progressively equilibrates with the hypertonic medulla has been found to be basically different when the results of micropuncture studies in *Psammomys obesus* (de Rouffignac and Morel, 1969) were compared with *in vitro* perfusion of thin descending limbs isolated from the rabbit kidney (Kokko, 1970). In the former studies, sodium chloride in the interstitium exerted little or no osmotic pressure and fluid within the loop became hypertonic through difusion of solute from peritubular fluid into the urine with no water efflux (de Rouffignac and Morel, 1969). In contrast isolated rabbit descending limbs equilibrated with the hypertonic bath more than 95% by water removal from the lumen (Kokko, 1970; Abramow, unpublished). Results of micropuncture studies in other rodents were characterized by a combination of water extraction and solute addition, the main solute diffusing into the loop being urea (Jamison, Buerkert and Lacy, 1973; Pennell, Lacy and Jamison, 1974). The various mechanisms are explained by the fact that the thin limbs are endowed with very special permeability characteristics for both water and solutes (see Table 3). However, it is still not clear whether the differing results reflect differences between the experimental approaches or truly represent species differences. The question is of some importance since it bears on the

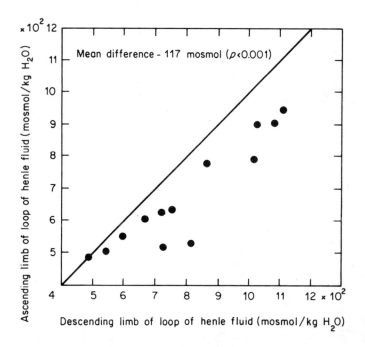

Figure 4 Comparison of osmolalities of fluid from thin ascending limbs and adjacent descending limbs of the loop of Henle in the inner medulla.
The dilution of the fluid in the ascending limb presumably reflects countercurrent multiplication by the thin loops. The process is not necessarily active in nature. From Jamison and coworkers (1967); reproduced by permission of the American Physiological Society

nature of the countercurrent multiplication system in the inner medulla (*vide infra*).

At the bend of the loop the permeability characteristics of the epithelium change drastically. The thin ascending limb is a water impermeable segment (see Table 3). There are also ultrastructural differences between both limbs

Table 3. Permeability characteristics and electrical parameters of segments of the loop of Henle

Segment	Species	Preparation	Hydraulic conductivity, L_P ($\times 10^{-5}$ cm^3 cm^{-2} s^{-1} atm^{-1})	Permeability, P ($\times 10^{-5}$ cm s^{-1}) water	urea	Na	Cl	Reflection coefficient, σ NaCl	urea	Potential difference, PD (mV)	Specific transverse resistance (Ω cm^2)	References
Thin descending limb	rat	in vitro papilla	39	119	13			<0.5	<0.5			Morgan and Berliner (1968)
	rat	in vitro papilla	16			47						Morgan (1974)
	rat	in vitro isolated perfused	17									Imai (1977)
	rabbit	in vitro isolated perfused			1.5			0.95				Kokko (1972)
	rabbit	in vitro isolated perfused				1.6		0.96		0		Kokko (1973)
	rabbit	isolated perfused								0	582	Abramow and Burg, unpublished; Imai (1977)
	hamster	in vitro isolated perfused	37.9									
	hamster	in vivo								-3		Windhager (1964)
Thin ascending limb	rat	in vitro papilla	0.3	50	14	7		1	<0.5			Morgan and Berliner (1969)
	rat	in vitro papilla	0.18		23	80	184					Morgan (1974)
	rat	in vitro isolated perfused	0.09		6.7	25.5	117			0		Imai (1977); Imai and Kokko (1974b)
	hamster	in vivo								-9 to -11		Marsh and Solomon (1965); Windhager (1964)
	hamster	in vitro isolated perfused	0.21		19	88	196					Imai (1977)
Thick ascending limb	rabbit	in vitro isolated perfused	0.08			2.8	1.4			+3.5 to 6.4	21.5 to 24.6	Burg and Green (1973c)
	rabbit	in vitro isolated perfused	0		0.9	6.3	1.1			+6.7		Rocha and Kokko (1973); Rocha and Kokko (1974)

(Schwartz and Venkatachalam, 1974), although a precise morphofunctional correlation remains to be established.

As shown in Fig. 4, micropuncture studies in the exposed renal papilla of the rat revealed that the osmolalities of the fluid from the thin ascending limb were about 100 mosmol lower than those in samples from adjacent descending limbs (Jamison, Bennett and Berliner, 1967). Thus, in this segment, solutes (mainly NaCl) are reabsorbed across a barrier impermeable to solvent, and water lags behind. The authors attributed their findings to the existence of active transport of sodium out of this segment although passive mechanisms have later been shown to be compatible with such findings (Kokko and Rector, 1972) (*vide infra*).

3. Thick Ascending Limb of Henle's Loop or Diluting Segment

On the basis of indirect evidence from micropuncture studies showing that samples collected from the early distal tubule are always hypo-osmotic relative to systemic plasma, with Na concentrations reaching low values, it has been concluded that the major segment responsible for the elaboration of dilute urine (also called 'diluting segment') was the thick ascending limb of Henle's loop (Gottschalk and Mylle, 1959; Walker and coworkers, 1941; Wirz, 1956). In addition, by extrapolation from findings in proximal tubules or other epithelial membranes, it had been universally assumed in the absence of precise knowledge of the driving forces for movement of solutes across the thick ascending limb of Henle's loop that active sodium transport was responsible for the reabsorption of salt. When diluting segments of cortical (Burg and Green, 1973c) or medullary origin (Rocha and Kokko, 1973) could be studied in isolation using the *in vitro* perfusion technique (Burg and coworkers, 1966) it was discovered that an active chloride rather than sodium pump was causing salt reabsorption against a chemical gradient. This stems from several experimental observations. First, in segments bathed and perfused with identical solutions a transepithelial electrical potential difference which is lumen positive (see Table 3) develops while the osmolality and Na and Cl concentrations in the luminal fluid decrease. Since transport of chloride is against an electrochemical potential, it must be active. Second, when chloride in the perfusate was replaced by nitrate or sulfate, the voltage dropped to zero (Burg and Green, 1973c). Conversely, replacement of Na with choline did not eliminate the voltage. This is in sharp contrast with the transport characteristics in the proximal tubule where Na is essential for the negative voltage to develop and net transport to proceed.

The permeability to sodium of the thick ascending limb is quite high with a relatively low electrical resistance (compare Tables 2 and 3 and see Fig. 5). Because of the high permeability to Na the small positive voltage has been calculated to be sufficient to drive all of the Na out of isolated thick ascending limbs. Thus sodium reabsorption is likely to be entirely passive. Ouabain in the bath inhibits chloride transport and decreases the positive voltage. This is disturbing since the known molecular actions of ouabain are to inhibit Na–K-activated ATPase which is present in thick ascending limb as well as in the proximal tubule (Schmidt and Dubach, 1969). Na–K-activated ATPase is believed to be an integral part of an Na for K exchange pump and there is no known chloride-activated ATPase. Since thick ascending limbs of Henle's loop have a high K content with a rapid turnover (Burg and Abramow, 1966) it is conceivable that the Na–K-activated ATPase in this segment is related to

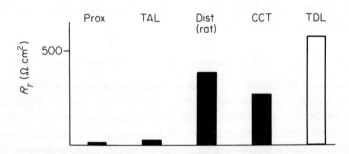

Figure 5 Comparison of the specific transverse resistance of isolated perfused tubules. The values were obtained from the rabbit tubules *in vitro* except in the case of the distal convoluted tubule of the rat *in vivo*. References in parentheses.
Prox: proximal convoluted tubule (Lutz and coworkers, 1973);
TAL: thick ascending limb of Henle's loop (Burg and Green, 1973e);
Dist: Distal convoluted tubule (Malnic, 1972);
CCT: cortical collecting tubule (Stoner and coworkers, 1974);
TDL: thin descending limb of Henle's loop.
Prox and TAL segments rank among 'leaky' epithelia' The other segments are electrically 'tight'. From Abramow and Burg, unpublished

this cellular K transport which, in turn would be necessary for the transepithelial chloride pump to work properly.

A more direct and perhaps more specific pharmocological action on the chloride transport mechanism in the thick ascending limb is that demonstrated by Burg in studies on the so-called 'loop diuretics' (Burg, 1976b). Low concentrations of furosemide in the perfusate very rapidly and reversibly inhibited active chloride transport while much larger concentrations had little effect when applied to the peritubular surface. The effect is specific for the thick ascending limb of Henle's loop in the sense that the drug does not greatly affect transport properties in other segments of the nephron (Burg and coworkers, 1973).

The same observations with respect to inhibition of the active chloride pump apply to the mercurial diuretics (Burg and Green, 1973a) and ethacrynic acid (Burg and Green, 1973b). Ethacrynic acid in addition may affect water reabsorption in the terminal parts of the nephron since it strongly inhibits the hydro-osmotic effect of vasopressin in the isolated rabbit collecting tubule (Abramow, 1974).

The sodium chloride reabsorptive rate in the thick ascending limb of Henle's loop exhibits a rather large reserve capacity *in vivo*. About 25–40% of the filtered sodium is reabsorbed in the entire short loops in rodents, the proportion being less in dog or monkey (Giebisch and Windhager, 1973). Moreover, salt reabsorption varies with the load which has escaped reabsorption in the proximal tubule. Thus, when proximal tubular reabsorption decreases as a consequence of saline infusions (Landwehr, Klose and Giebisch, 1967; Stein and coworkers, 1973) loop reabsorption increases. Conversely, reducing glomerular filtration rate increases proximal tubular reabsorption of salt with a reciprocal decrease in loop reabsorption of sodium chloride (Anagnostopoulos, Kinney and Windhager, 1971; Landwehr and coworkers, 1968). The mechanism of the adjustment is not entirely clear but it must be different from that of the glomerulo-tubular balance in the proximal tubule (*vide supra*). Part of the transport mechanism in the thick ascending limb is probably flow dependent since the drop in salt concentration along the segment may impose a limit to the capacity of the chloride pump. Increasing the flow rate through the loop would tend to elevate the salt concentration in late portions of the thick ascending limb and thus allow increased reabsorption. However this explanation remains speculative since the available evidence *in vivo* rests on determinations of NaCl concentrations in the early distal tubule. An *in vivo* microperfusion of Morgan and Berliner (1969) in the rat showed that the relationship between flow rate in the loop and Na concentration in the early distal tubule held for perfusion rates above 15 nl min^{-1}. Sodium concentration increased again as flow rate declined below 15 nl min^{-1}. The findings are in conflict with the more direct observations on isolated thick ascending limbs from cortex or outer medulla in the rabbit in which the lowest concentrations of Na or total solutes were noted at the slowest perfusion rates (below 1 nl min^{-1}) (Burg and Green, 1973c; Rocha and Kokko, 1973). The reasons for the differences are not apparent.

Humoral, possibly hormonal regulation of the salt transport in the thick ascending limb has not been demonstrated directly although there are grounds to suspect some variations in the function of the loop with the state of hydration (de Rouffignac and Imbert, 1975). In addition, measurements of basal and ADH stimulated adenylate cyclase activity in single dissected rabbit tubules after digestion with collagenase, by Imbert and coworkers (1975) revealed the existence of an ADH-stimulated activity that was not restricted to the various components of the collecting system known to be physiological targets of the hormone. These authors found that ADH-activated adenylate cyclase was also contained in both thin and thick ascending limbs of Henle's loop. However, much larger concentrations of the hormone were needed to elicit a response in these segments as compared to the segments in the collecting system. The biological significance of these findings remains to be established but the authors believe that, given the absence of a clear-cut effect of ADH on water permeability in *in vitro* perfused ascending limbs (Rocha and Kokko, 1974) the hormone may well stimulate active salt reabsorption in the thick ascending limb (de Rouffignac and Imbert, 1975). The postulated effect would somehow explain the dual effects of ADH observed when perfusing the intact rat during water diuresis at different dosage levels (Atherton, Green and Thomas, 1971). While low dosage levels of the hormone cause the expected fall in urine flow and free water clearance, which clearly results from ADH effect on water permeability in the collecting system (see below), larger doses induce natriuresis and increased accumulation of sodium in the medulla. One must emphasize that the transient natriuresis seen under some conditions with exogenous vasopressin does not constitute proof of a physiological effect of the hormone and that the role of ADH in Na homeostasis as opposed to water metabolism remains unclear in mammalian species. It should be recalled that the interest focused upon the thick ascending limb stems from its pivotal role not only as a diluting segment but also as the main, if not exclusive, site where metabolic energy is expended for producing an hypertonic medulla and a concentrated urine (*vide infra*).

4. Distal Convoluted Tubule

The distal convoluted tubule is defined as that segment of the nephron which begins at the *macula densa* and ends at the junction with another distal convoluted tubule to form the beginning of the cortical collecting tubule. This is an arbitrary definition which has been convenient for micropuncture experiments but has led to some confusion. Indeed, the distal convoluted tubule has been demonstrated to be rather heterogeneous on both anatomical and functional grounds (Morel, Chabardes and Imbert, 1976; Woodhall and Tisher, 1973).

Micropuncture studies disclose that fluid sampled from the early distal convoluted segments of the superficial nephrons is dilute with respect to plasma. This reflects the diluting properties of the preceding segment, the thick ascending limb of Henle's loop. However, the dilution process continues along the length of the distal convoluted tubule, at least in some of the mammalian species studied, and during water diuresis, in all of them. This indicates that salt is reabsorbed in excess of water in the latter segment also. The basic mechanism for salt reabsorption in the distal tubule must be different from that in the thick ascending limb of Henle's loop since the voltage is now oriented lumen negative. Thus net active reabsorption of sodium against both a chemical and an electrical potential gradient is the driving force for salt transfer. Chloride is driven passively by the favourable voltage gradient.

The amplitude of the voltage increases along the length of the distal convoluted tubule and reaches values as high as -52 mV in rats in antidiuresis (Giebisch and Windhager, 1973) or -62 mV in *in vitro* perfused distal convoluted segments of the rabbit (Gross, Imai and Kokko, 1975). High transepithelial electrical gradients are typical for 'tight' epithelia (Ussing, Erlij and Lassen, 1974) and can be sustained only when ion backflux through the shunt pathway is sufficiently low. The same consideration applies to the large concentration gradient for Na. Thus the effective transepithelial resistance in this segment is high, averaging $500 \Omega \text{cm}^2$ (Boulpaep, 1976; Boulpaep and Seely, 1971) a value about two orders of magnitude higher than in the proximal convoluted tubule (see also Fig. 5).

In some instances the voltage is small and positive in the very early part of the distal tubule in rats, presumably indicating that these segments belong functionally to the thick ascending limb of Henle's loop.

The capacity of the Na transport system, i.e. the rate at which Na is reabsorbed in the distal convoluted tubule, is much less than that in the proximal tubule averaging about a quarter of the latter (Giebisch and Windhager, 1973).

In the presence of ADH, distal fluid reabsorption represents about 10% of the filtered load. The state of hydration is important here since L_p in micropuncture experiments was found to increase from 1.6 to 6.3×10^{-5} cm³ cm⁻² s⁻¹ atm⁻¹ when switching rats from water diuresis to antidiuresis (Persson, 1970). During water diuresis, the fluid along the distal convoluted tubule remains hypotonic in all species studied. However, during antidiuresis, late distal tubular fluid regains isotonicity with respect to plasma in rats, at least in some strains (Wirz and Dirix, 1973). In other strains of rats, in the dog, the monkey and Meriones, the late distal tubular fluid remains hypotonic in the face of adequate circulating levels of ADH (Wirz and Dirix, 1973). The reasons for the differing results, which raise the question of ADH action in so-called 'distal convoluted tubule', have been explained by the heterogeneity of this arbitrarily defined segment (Woodhall and Tisher, 1973).

Woodhall and Tisher (1973) on the basis of morphological studies, showed that, in some instances late segments of superficial distal tubules reaching the renal capsule, actually had distinct morphological features of the cortical collecting tubules and moreover responded to ADH infusion by cell swelling and enlargment of intercellular spaces typical of the segments of the collecting system (Abramow and Dratwa, 1974; Ganote and coworkers, 1968; Grantham and coworkers, 1969; see also Fig. 8). When the superficial loops of the distal tubule had ultrastructural features characteristic of the distal convoluted tubule, no morphological effects of ADH were detected (Woodhall and Tisher, 1973). It is tempting to conclude that the 'distal convoluted tubule' is, in fact heterogenous and that the target of ADH action on water reabsorption does not extend proximal to the collecting system.

In confirmation of these findings isolated *in vitro* perfused distal convoluted tubules of the rabbit had L_p values which could not be distinguished experimentally from zero. Addition of 200 μU ml^{-1} ADH to the bath produced no significant net water absorption along an osmotic gradient (Gross, Imai and Kokko, 1975).

Evidence for a considerable morphological as well as functional heterogeneity of the so-called 'distal convoluted tubule' has been obtained by Morel, Chabardes and Imbert (1976) using an entirely different approach. Stereomicroscopic observations of rabbit distal convoluted tubules (DCT) microdissected from collagenase treated kidneys allowed up to four portions starting from the *macula densa* to be distinguished. The first portion, very short (DCTa) was of the same appearance as the thick ascending limb preceding the *macula densa*; the second (DCTb) had a bright appearance; the third portion (DCTg) had a granular appearance and most often was connected to a portion of the cortical collecting tubule having the same appearance (CCTg). A fourth portion (DCTl) was found in many DCT appearing below the renal capsule and branching to form an initial cortical collecting tubule (CCTl) of the same 'light' appearance. Despite the rather crude method used for defining these segments anatomically, they proved to behave in a quite distinctive manner with respect to hormone stimulated adenylate cyclase activity. Maximal concentrations of arginine–vasopressin parathyroid hormone and isoproterenol were tested. Vasopressin activated adenylate cyclase only in DCTl and CCTl, whereas significant stimulation by parathyroid hormone was obtained in DCTa, DCTg and CCTg and by isoproterenol in DCTg, CCTg and CCTl. These results establish the functional heterogeneity of the distal convoluted tubule. They suggest that the very first portion of the segment (DCTa) is functionally related to the thick ascending limb while the terminal portions of DCT in the superficial cortex which are usually of the DCTl type are functionally related to the cortical collecting tubule. DCTb is a well defined portion truly belonging to the distal convoluted tubule. At the present time, the physiological significance of the hormone responsiveness of the various portions of DCT is apparent only with respect to vasopressin, since data pertaining directly to hormonally controlled transport characteristics in this region of the nephron are not available for the other hormones which were tested for activation of adenylate cyclase. It is confirmed that in the distal convoluted tubule, receptors for ADH are lacking, further suggesting that this region remains poorly permeable to water, whatever the state of hydration of the organism. It remains an exciting challenge for future investigation to discover which hormonally controlled functions are hidden behind the puzzling mosaïc of the distal convoluted tubule.

Aside from these recent findings an important hormonal control with respect to sodium balance has been known for several years to be located in at least some portion of the 'distal convoluted tubule'. This hormonal control is exerted by some adrenal steroids. Only a brief account of the effect of mineralocorticoids will be given here since hormonal control of osmoregulation is dealt with in a separate chapter. Adrenalectomy in rats has been reported to reduce the rate of Na reabsorption along the distal convoluted tubule and to decrease the steady-state Na concentration gradient acorss the distal tubular wall.

Aldosterone in low doses injected at least 60 min prior to micropuncture reversed the sodium transport defect (Wiederholt and coworkers, 1966). Actinomycin D, a powerful inhibitor of protein synthesis specifically blocks that fraction of distal Na transport which is under the control of aldosterone. This observation and the lag period for aldosterone action suggest that a specific transport protein necessary for a fraction of Na translocation across the distal tubular wall is induced by the mineralocorticoids (Wiederholt and coworkers, 1966). Experiments on isolated rabbit distal tubules lead to different conclusions (Gross, Imai and Kokko, 1975). The negative voltage in the isolated distal convoluted tubule is insensitive to variations in the level of mineralocorticoids whereas a distinctive effect of a low Na, high K diet plus DOCA administration on the voltage in the isolated collecting tubule is observed, with the lumen becoming some 30 mV more negative than in rabbits kept on a regular diet (Gross, Imai and Kokko, 1975). Additional studies including measurements of ionic permeabilities and transport rates are necessary to confirm these findings. They suggest however that the main target for mineralocorticoids is the collecting tubule rather than the distal convoluted tubule. Given the problems raised by the heterogeneity of the distal tubular elements and especially by their identification in the course of micropuncture experiments and by possible species differences, the question of the exact area responsive to aldosterone in the distal nephron of mammals remains unsettled.

Whatever the physiological basis for the multiple hormonal responsiveness in the distal convoluted tubule, this segment of the nephron does not seem to play a central role in the regulation of Na excretion. Micropuncture studies of segmental sodium reabsorption along the nephron disclose little changes in sodium transport along the distal tubule when large experimental variations of sodium excretion are brought about by saline infusion (Morgan and Berliner, 1969), blood volume expansion (Sonnenberg, 1972) or salt depletion (Stein and coworkers, 1974).

In contrast to the modest part played by the distal tubule in the control of sodium reabsorption, changes in potassium metabolism are accompanied by conspicuous changes in the rate of net potassium secretion, especially in the 'late' distal tubule (Giebisch and Windhager, 1973). Mechanisms of potassium transport in this segment are complex and include an active transport step from lumen to cell and a component of net secretion which becomes important as the voltage increases in the late part of the segment (Giebisch and Windhager, 1973). Whatever the mechanisms involved, potassium transport in the distal nephron (including the collecting system) is critical for potassium homeostasis since filtered potassium is extensively reabsorbed up to the distal tubule, irrespective of the needs. Thus, urinary potassium essentially derives from 'distal' tubular secretion.

5. Collecting System

The collecting system includes the cortical collecting tubule and the papillary collecting duct. Its function has been studied by micropuncture comparing the composition of the pelvic urine with that of samples obtained from the last accessible part of distal tubules necessarily belonging to superficial nephron. Such an approach has been repeatedly presented as possibly leading to erroneous conclusions. Evidently, final urine is a mixture of urine from all the nephrons while superficial distal urine may differ considerably in composition from that in the distal convolution of deep nephrons, which is unknown (Giebisch and Windhager, 1964; Sonnenberg, 1972; Stein and coworkers, 1973).

More direct evaluation of the collecting system has been possible by the development of the *in vitro* microperfusion techniques of isolated tubules (Abramow, 1974, 1975; Abramow and Dratwa, 1974; Burg and coworkers, 1966, 1970; Grantham and Burg, 1966; Grantham and Orloff, 1968; Grantham and Burg, 1971; Schafer and Andreoli, 1972a, 1972b), by catheterization of the Bellini ducts or direct puncture of the exposed medulla (Giebisch and Windhager, 1973; Jamison, Buerkert and Lacy, 1971; Morgan and Berliner, 1968; Morgan, Sakai and Berliner, 1968).

Water conservation essentially depends on the function of the collecting system which is poorly permeable to water in the absence of vasopressin leading to water diuresis. During hydropenia the same membranes under vasopressin become much more permeable to water which is allowed to back-diffuse along an osmotic gradient, explaining the formation of a small volume of concentrated urine.

Independently of being the site of the vasopressin mediated regulation of water reabsorption, the collecting system has been recognized as playing an important role in the regulation of sodium excretion.

a. *Cortical Collecting Tubule. Water permeation.* With the development of the *in vitro* perfusion of isolated tubules (Burg and coworkers, 1966) the permeability to water of this segment has been studied extensively (Abramow, 1974, 1975, 1976; Abramow and Dratwa, 1971; Andreoli and Schafer, 1976; Grantham and Burg, 1966; Grantham and Orloff, 1968; Helman, Grantham and Burg, 1971) both with respect to mechanisms of water permeation and to molecular events involved in ADH action. In the absence of ADH, L_p of the cortical collecting tubule is distinctly low but by no means negligible (see Table 4). A low L_p can be measured with some degree of accuracy provided a sufficiently large area for water permeation (long tubules) is used and some other technical precautions are taken (Abramow, 1975).

In Fig. 6 a frequency histogram of L_p values of 88 collecting tubules incubated in the absence of vasopressin is presented. The mean value is 2.63×10^{-6} ml cm^{-2} s^{-1} atm^{-1} which is very significantly different from zero. Thus, albeit low, the osmotic water permeability permits some degree of back

Table 4.
Permeability characteristics in the collecting sysem

Segment	Species	Preparation	Hydraulic conductivity, L_P ($\times 10^{-6}$ cm^3 cm^{-2} s^{-1} atm^{-1})	Diffusional permeability ($\times 10^{-5}$ cm s^{-1}) water	Na	urea	References
Cortical collecting tubule no ADH	rabbit	isolated in vitro perfused	4	38		0.1	Burg and coworkers (1970); Grantham and Burg (1966)
			2.6	44		0	Abramow (1975); Schafer and Andreoli (1972a)
			1	47	0.06		Frindt and Burg (1972)
					0.08		Stoner, Burg and Orloff (1974); Abramow and Dratwa (1974)
	man	isolated in vitro perfused	1.7				
Cortical collecting tubule with ADH	rabbit	isolated	16	97		0.1	Burg and coworkers (1970); Grantham and Burg (1966)
		in vitro perfused	14	142		0	Schager and Andreoli (1972a); Abramow (1975)
		isolated	12	145			Abramow and Dratwa (1974)
	man	in vitro perfused	8.2				
Papillary collecting duct no ADH	rabbit	isolated in vitro perfused	4.8	40		2.4	Rocha and Kokko (1974)
with ADH			3	57		2.2	Rocha and Kokko (1974)
no ADH	rat	in vitro		45		20	Morgan and Berliner (1968); Morgan, Sakai and Berliner (1968)
		papilla					Morgan (1974)
with ADH			20	87	5.4	30	Morgan and Berliner (1968); Morgan, Sakai and Berliner (1968)
					5.9		Morgan (1974)

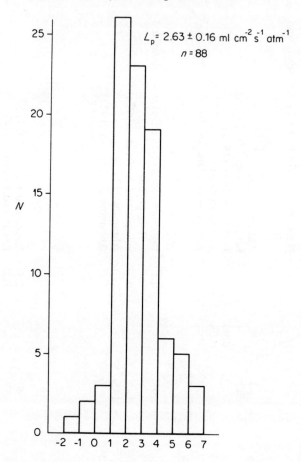

Figure 6 Osmotic water permeability of isolated cortical collecting tubules *in vitro* (vasopressin absent). Range of L_p values ($\times 10^{-6}$ ml cm^{-2} s^{-1} atm^{-1}) in abcissa. Number of tubules according to the class intervals in ordinate. Only 3 out of 88 tubules had L_p at or below zero ('negative absorption'). Despite its very low mean value, L_p can be determined experimentally (see text for explanation). After Abramow (1975); modified

diffusion of water in the cortex, given the large osmotic gradient created by continuous active salt absorption. Although the exact fraction of filtered water reabsorbed in the cortical portion of the collecting system during water diuresis is not known it is an important determinant of the volume of water ultimately delivered to the medulla in deeper segments of the collecting system. That L_p is not negligibly small is attested by the production of hypertonic urine in the absence of ADH when urine flow rate is experimentally lowered (Berliner and Davidson, 1957).

Addition of vasopressin to the medium bathing the tubules as in Fig. 7 and Table 4 produces a dramatic increase in L_p. The phenomenon, which is reversible, represents an unique hormonally controlled membrane property. It is also the most important step in the efferent pathway of the osmotic control of body fluids.

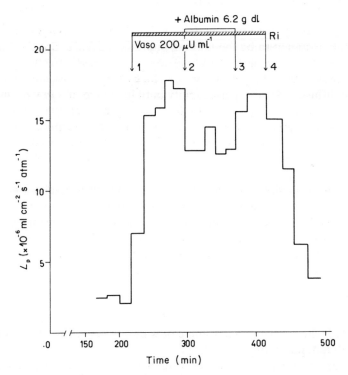

Figure 7 Effect of vasopressin on osmotic water permeability of an isolated perfused rabbit collecting tubule. Time in abcissa. Hydraulic conductivity in ordinate.
Arrows:
(1) addition of ADH (vaso) to the medium is attended by a switch from low to high L_p;
(2) addition of albumin to the bath does not accelerate water flow but rather depresses L_p;
(3) removal of protein restores L_p to its previous high value under ADH;
(4) removal of ADH is immediately followed by a progressive drop of L_p towards low values typical of the absence of the hormone.
From Abramow, unpublished

The hydraulic conductivity L_p increases four- to tenfold, depending on the experimental conditions (see Table 4).

The permeability of tubules to water can also be measured in the absence of an osmotic gradient provided a tracer for water diffusion is used. Tritiated water (THO) is added to the perfusion fluid and the rate at which it disappears from the lumen is noted. The diffusional permeability to tagged water, P_{Dw}, is calculated according to the following equation (Grantham and Burg, 1966):

$$P_{Dw} = \frac{V}{A} \ln \frac{c_o}{c_i} \tag{3}$$

where V is the perfusion rate (cm^3 s^{-1}); A is the area of the luminal membrane, assumed to be the limiting interface; c_o and c_i are the concentrations of THO in the collected and perfused fluids, respectively.

The dimensions of P_{Dw} are those of velocity (cm s^{-1}). Results of measurements of diffusional water permeability both in *in vivo* and *in vitro* microperfusion experiments are given in Tables 1 to 5. Although they may have less physiological meaning than L_p in terms of water reabsorption secondary to the existing driving forces in the kidney, they have been used as a tool for evaluating the membrane pathways and mechanisms involved in ADH controlled water permeation.

Indeed it is possible to compare osmotic and diffusional water permeabilities in the same segments under the same experimental conditions (with or without ADH). To this end, the units of L_p have to be transformed into the same dimensions of velocity as P_{Dw}. A permeability coefficient for flow, P_f (cm s^{-1}) is calculated according to the following equation (Andreoli and Schafer, 1976):

$$P_f = \frac{L_p RT}{\bar{V}_w} \tag{4}$$

where $\bar{V}_w$ is the partial molar volume of water, R is the gas constant and T is the absolute temperature.

If osmotic water flow is the result of net water diffusion secondary to a transmembrane difference in water activity then P_f and P_{Dw} should be equal. The experimental evidence in a number of epithelia (Andreoli and Schafer, 1976) and in the cortical collecting tubule in particular (Table 5) indicates that P_{Dw} and P_f values are not the same, P_{Dw} usually being substantially less than P_f.

The discrepancy between P_f and P_{Dw} has led to the suggestion that mechanisms other than simple diffusion underlie water transfer and its acceleration by ADH. Koefoed-Johnsen and Ussing (1953) found a P_f/P_{Dw} ratio of 5.3 in toad skin. ADH disporportionately increased P_f relative to P_{Dw}, and P_f/P_{Dw} rose to 33. Since P_{Dw} appeared too low to account for osmosis, Koefoed-Johnsen and Ussing postulated the existence of aqueous channels or pores in the membrane. Osmosis would proceed by bulk water flow through the aqueous pores and ADH would increase the size of the pores (Koefoed-Johnsen and Ussing, 1953).

The effective radii of pores in the membrane can be computed from the combination of Fick's first law and Poiseuille's law into an equation of the form (Andreoli and Schafer, 1976):

$$\frac{P_f}{P_{Dw}} = 1 + \frac{r^2RT}{8\eta \bar{V}_w D_w^\circ} \tag{5}$$

where r is the effective pore radius, D_w° is the free diffusion coefficient for water and η is the viscosity.

The apparent pore radius in a number of ADH-sensitive and ADH-insensitive epithelia has been calculated (Andreoli and Schafer, 1976). We

Table 5. Apparent 'pore' radius (r_p) in the isolated collecting tubule

ADH concentration (μU ml^{-1})	P_f (cm s^{-1})	$P_{Dw} \times 10^{-4}$	P_f/P_{Dw}	r_p (Å)	References
0	6	4.7	1.27	1.9	Schafer and Andreoli
250	186	14.2	13.1	12.5	(1972b)
0	33.8	4.37	7.73	9.42	Abramow (1975)
2.5	131.7	8.20	16.06	13.5	
250	157	14.48	10.82	11.15	

will limit our discussion to the case of the mammalian collecting system. Table 5 shows computations of apparent pore radii in isolated cortical collecting tubules. In the experiments reported by Schafer and Andreoli, the pore radius is only 1.9 Å in the absence of vasopressin and rises to 12.5 Å with vasopressin (Schafer and Andreoli, 1972b). In contrast our measurements indicate (Abramow, 1975) that the mean radius is relatively large in the absence of ADH (9.4 Å) and does not increase substantially when vasopressin 2.5 and 250 μU ml^{-1} is added. Since both the diffusional and osmotic permeabilities increase when ADH concentration is brought from 0 to 2.5 and from 2.5 to 250 μU ml^{-1}, an increase in total number of 'pores' rather than mean pore radius is inferred. The different conclusion reflects the unusually low L_p found by Schafer and Andreoli (1972b) in the absence of ADH as compared to our values (see Table 5 and Fig. 6) and those of Grantham and Orloff (1968). Taken at face value, the calculated pore radius (1.9 Å) in the first row of Table 5 is dangerously close to the molecular radius of water.

In any event, the pore theory has been challenged on a number of grounds and Schafer and Andreoli do not believe in the existence of membrane pores (Schafer and Andreoli, 1972b; Andreoli and Schafer, 1976). It has been proposed by several authors (reviewed in Andreoli and Schafer, 1976) that the discrepancy between P_{Dw} and P_f would not result from bulk flow through aqueous channels but rather would arise from unstirred layers in series with the membrane. The unstirred layers are expected to lower the apparent P_{Dw}

(retarding diffusion) without much affecting P_f. Vigorous stirring of the bulk phase bathing the membrane and removal of stromal elements did much to reduce the P_f/P_{Dw} discrepancy by increasing apparent P_{Dw}.

Schafer and Andreoli (1972b) presented evidence that bulk phase unstirred layers were negligible in cortical collecting tubules and pointed to the cellular layer as the site of constraint to diffusion. In their view ADH acts on water permeability by increasing the diffusion of water and lipophilic solutes across the luminal membrane. The water 'diffusibility' would increase through an increase in the partition coefficient between aqueous and membrane phases (Andreoli and Schafer, 1976).

Despite the impressive analysis presented by Andreoli and Schafer (1976) further work appears necessary to exclude totally the participation of aqueous channels as a partial explanation for the 'abnormal' P_f/P_{Dw} ratio and to establish diffusion as the sole mode of osmotic water reabsorption in the collecting system.

Striking morphological changes can be seen to occur in collecting tubules a few minutes after exposure to vasopressin (Ganote and coworkers, 1968; Grantham and coworkers, 1969; see also Fig. 8). They include cell swelling and dilatation of intercellular spaces and suggest that the rate limiting barrier for water permeation is located in the luminal membrane.

A sizeable fraction, perhaps 25%, of the water flowing across the epithelium takes the intercellular route when going from lumen to the peritubular space. The driving force for water flow from cell to lateral interspace is provided by the continuous diffusion of salt from peritubular space to lateral space through the basement membrane and the basilar slit (Ganote and coworkers, 1968; Grantham and coworkers, 1969). In contrast to the situation in other epithelia, where fluid flows in the interspaces are coupled to salt pumps, the mechanism here is purely a consequence of the transtubular osmotic gradient. No morphological changes are detected with vasopressin in the absence of an osmotic gradient although the hormone is shown to be active in that it accelerates diffusion of THO (Ganote and coworkers, 1968; Grantham and coworkers, 1969). Moreover, inhibition of active transport with ouabain does not suppress osmotic swelling of the cells and widening of the interspaces under vasopressin and an osmotic gradient. The apical tight junction is believed to remain 'tight' and bulk water gains access to the intercellular space by crossing the lateral membranes. Under these circumstances, the hydrostatic pressure in the interspace is estimated to become a few centimetres of water greater than that of the lumen on external bath (Grantham, 1974).

In an attempt to determine if pressure build-up in the cells and lateral spaces may retard overall osmotic water flow, oncotic pressure was increased substantially in the peritubular medium (see Fig. 7). It is apparent that an oncotic pressure of about 30 cm H_2O (presumably exerted partly across the basement membrane and partly across the baso-lateral membrane) failed to

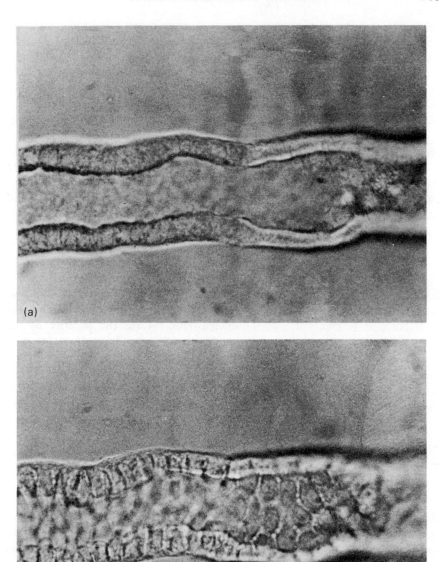

Figure 8 Photomicrograph of an isolated human collecting duct perfused with a solution which is hypotonic relative to the bath. After addition of vasopressin (b) cell swelling and dilatation of the intercellular spaces become obvious. From Abramow and Dratwa (1974); reproduced by permission of McMillan Journals Ltd.

enhance L_p when osmotic water flow was maximally increased with vasopressin. Thus, in contrast to the proximal tubule where peritubular oncotic pressure increases fluid absorption secondary to active salt transport, the same experimental manipulation does not lead to accelerated water flow under an osmotic gradient in the collecting tubule. Actually, a decrease of L_p (which was reversible after removal of protein) was observed (see Fig. 7). Of course osmotic effects are even less likely to be a direct driving force than in proximal tubule, given the low 'overall' L_p in the collecting tubule (Abramow, unpublished).

Electrolyte transport. In isolated collecting tubules in the absence of ADH the sodium concentration decreases and the potassium concentration increases along the length of the segments. The largest transepithelial gradients for the ions are obtained with the largest contact times (Grantham, Burg and Orloff, 1970). There is active sodium reabsorption with a (lumen negative) voltage of about -35 mV (Grantham, Burg and Orloff, 1970; Stoner, Burg and Orloff, 1974). In some instances, active potassium secretion also occurs (Grantham, Burg and Orloff, 1970). Both transport processes are linked although in a loose fashion. Removal of sodium from perfusate and bath, omission of K from bath, addition of amiloride or ouabaïn all inhibited both Na and K transport. These manoeuvres reverse the polarity of the voltage which becomes lumen positive (Stoner, Burg and Orloff, 1974). The positive voltage is attributed to the unmasking of the hydrogen ion secretion, since it decreases upon addition of acetazolamide. Under 'normal' conditions, acetazolamide increases the negative voltage (Stoner, Burg and Orloff, 1974). The ratio of the net sodium reabsorption to the net potassium secretion rate varies with the flow rate, being 4 to 1 at fast perfusion rates and closer to one at slow (<1 nl min^{-1}) perfusion rates.

Chloride absorption is passive, driven by the voltage, with an important component of exchange diffusion (Stoner, Burg and Orloff, 1974).

The collecting tubule is a 'tight' epithelium (Ussing, Erlij and Lassen, 1974) with effective transverse resistance in the range of 266 Ω cm^2 (Stoner, Burg and Orloff, 1974) (see also Fig. 5) to 900 Ω cm^2 (Helman, Grantham and Burg, 1971). Consistent with these findings, the tracer permeability to sodium is very low (see Table 4) and the unidirectional sodium flux from peritubular fluid to lumen is only 15% of that from lumen to peritubular fluid (Stoner, Burg and Orloff, 1974). ADH produces transient increase of the negative voltage with short-lived increase in lumen to bath flux (Frindt and Burg, 1972) and no effect on the electrical resistance (Helman, Grantham and Burg, 1971). Obviously the pathways for Na and water transfer are separate and so is their ADH dependence. The physiological significance of the short-lived effect of ADH on sodium transfer is not clear. Thus there is no clear-cut experimental evidence in the mammals for an osmoregulatory role of ADH other than on promoting water conservation.

b. *Papillary Collecting Duct* The behaviour of this segment with respect to water permeability and response to vasopressin is qualitatively similar to that of the cortical portions (see Table 4). The environment here however is hypertonic relative to the systemic blood. The luminal fluid is hypotonic with respect to the blood in water diuresis, with further dilution by active sodium reabsorption along the whole collecting system (Jamison, Buerkert and Lacy, 1971; Jamison and Lacy, 1972).

During hydropenia, L_p increases in the whole collecting system and the urine comes into osmotic equilibrium with the peritubular environment. Thus the urine becomes isotonic in the cortical portion of the collecting system and progressively hypertonic along the papillary portion of the system.

It should be emphasized that in spite of the fact that the volume of the final urine by definition is much smaller during antidiuresis, than during water diuresis, the absolute rate of water reabsorption in the permeable papillary collecting duct is actually less than in water diuresis. By combining the results of micropuncture studies in rats and *in vitro* microperfusion of rabbit tubules, Jamison (1974) calculated that the fraction of filtered water reabsorbed in the cortex and outer medulla increases from 81% in water diuresis to 95% in antidiuresis. In contrast, the fraction of filtered water reabsorbed more distally actually decreases from 5.6% to 3.9% in the inner medulla and from 3.2% to 1.7% in the papilla when switching from water diuresis to antidiuresis (Jamison, 1974).

The apparent paradox vanishes when it is realized that in antidiuresis the volume of fluid entering the medullary portion of the collecting system has been considerably reduced because of an important increase in reabsorption in the cortical region of the system. Moreover, the driving force for water reabsorption in the inner medulla (the osmotic pressure gradient) is also smaller in antidiuresis. It is apparent that reduced rate of delivery and reduced driving force more than compensate for the increased L_p. Thus osmotic equilibration in the most hypertonic regions occurs with a minimum of water reabsorption whereas the largest volume of water is returned to the body in the cortex where it can be disposed thanks to the large blood flow. This device helps to protect the counter-current system in the inner medulla against excessive wash-out of the interstitial solutes. Otherwise, because of the relatively small blood flow, the effectiveness of the urinary concentrating mechanisms would be diminished by the diluting effect of the reabsorbate.

It has become increasingly clear in recent years that the collecting system plays an important role in the regulation of sodium excretion despite the modest part taken by the collecting ducts in terms of the fractional reabsorption of the filtered sodium (Sonnenberg, 1974). Great changes in Na reabsorption with the state of the animal are found however when Na handling by the collecting duct is examined as the fraction of the Na load which is offered to the duct. For example, the papillary collecting duct of control rats reabsorbs 51% of the sodium load it receives. After salt deprivation, the reabsorbed fraction increased to 81% whereas in salt-loaded rats it dropped to 28% directly influencing urinary Na excretion (Diezi, and coworkers, 1973). Other instances where collecting duct transport has been stimulated by volume depletion or suppressed by extracellular volume expansion (Sonnenberg, 1972) or salt-loading in the so-called DOCA-escaped state (Sonnenberg, 1973) have been documented and reviewed recently (Stein and Reineck, 1974).

The nature of the control mechanisms involved in the regulation of Na transport in the collecting duct is not known. Although aldosterone definitely influences Na transport in the papillary collecting duct (Uhlich, Baldamus and Ullrich, 1969) as well as in the more cortical portion of the collecting system, other factors are probably operative. It would not be surprising to discover in the future, among the heterogenous population of the so-called 'natriuretic factors' (Dirks, Seely and Levy, 1976) at least one which would act more or less specifically on the Na pump or the passive backflux in the collecting system.

c. *Action of Vasopressin at the Cellular Level* A large body of evidence has accumulated in recent years in support of the concept that the effect of vasopressin to increase water permeability of the luminal plasma membrane in the collecting tubule cells depends on stimulation of an adenylate cyclase located in the baso-lateral membrane. The resulting increase in the level of intracellular cyclic adenosine 3',5'-monophosphate (cyclic-AMP) is ultimately responsible for the changes in permeability. The concept has been fathered by physiologists working with ADH-responsive amphibian epithelia (Grantham, 1974; Handler and Orloff, 1973). The model was later found to apply to the rabbit collecting tubule studied in isolation (Abramow, 1974; Grantham and Burg, 1966; Grantham and Orloff, 1968). The experimental backbone of the concept is that water permeability of the collecting system is increased not only by vasopressin but also by exogenous cyclic-AMP or analogues of cyclic-AMP which penetrate the cells more easily (Abramow, 1974; Grantham and Burg, 1966) and by theophylline (Abramow, 1974; Grantham and Orloff, 1968) an agent which retards the degradation of intracellular cyclic-AMP through inhibition of cyclic nucleotide phosphodiesterases. The degradation product of cyclic-AMP, 5'-AMP was demonstrated to be inactive (Grantham and Burg, 1966).

Biochemists have done much to refine and complete the model. They studied ADH-activated membrane bound adenylate cyclase in the renal medulla of several mammalian species and also the characteristics of the stereospecific receptor for vasopressin binding (reviewed in Dousa and Valtin, 1976; Grantham, 1974). Moreover cyclic-AMP content of renal medulla has been found to increase under stimulation by vasopressin (Dousa and Valtin, 1976). More recently attention has been focused on the cyclic-AMP-dependent protein kinase. The hypothesis which brings us one step further in the chain of events linking the nucleotide level and the permeability changes is the following: the water permeability properties of the luminal membrane would depend on whether or not some specific membrane protein is phosphorylated. The membrane protein is serving as a substrate for cyclic-AMP-dependent protein kinase. Phosphorylation of the protein would change membrane structure in such a way as to render it more water permeable. Reversal to the water-impermeable condition would depend on removal of phosphate from the membrane protein, catalysed by a protein phosphatase.

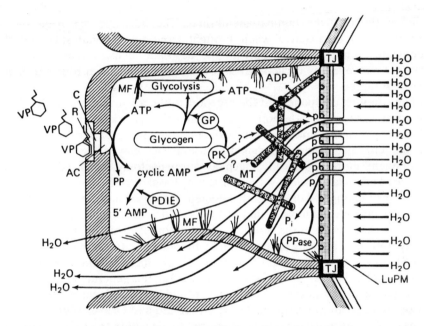

Figure 9 Schematic representation of the cellular action of vasopressin.
AC, adenylate cyclase; ADP, adenosine 5′-diphosphate; 5′AMP, adenosine 5′-monophosphate; ATP, adenosine 5′-triphosphate; C, coupling component of adenylate cyclase; GP, glycogen phosphorylase; MF, microfilaments; MT, microtubules; LuPM, luminal plasma membrane; p, phosphate attached to protein; cyclic AMP, cyclic adenosine 3′5′-monophosphate; PDIE, cyclic AMP phosphodiesterase; PK, cyclic AMP-dependent protein kinase; P_1, inorganic phosphorus; PP, pyrophosphate; PPase, protein phosphatase; R, receptor for vasopressin; TJ, tight junction; VP, vasopressin molecule.
The connection of MT and MF with the plasma membranes is uncertain (see text for explanations). From Dousa and Valtin (1976); reproduced by permission of *Kidney International*

Although all of the pieces of the puzzle have been found in the renal medulla, several questions remain to be answered such as the exact location (cytosolic or in luminal plasma membrane) of the protein kinase, the subcellular location of the protein which is phosphorylated and of the specific protein phosphatase (Dousa and Valtin, 1976).

Furthermore it is not certain that the enzyme activities which were found beyond cyclic-AMP generation represent obligatory steps linking VP binding to the target tissue and its physiological response or are mere byproducts of the complex metabolic response. The model as it stands now is depicted schematically in Fig. 9. It includes hypothetical events together with more firmly established aspects of ADH action which should be kept in mind. The complex receptor-coupling component to adenylate cyclase is located in the basal (and lateral) membrane whereas the final changes occur at the other pole of the cell, the luminal membrane. The hormone has been shown repeatedly to be inactive when brought from the luminal side and the adenylate cyclase is found only in the baso-lateral membrane. Thus, the message conveyed by cyclic-AMP has to travel by unknown routes to reach the final target.

The coupling component of the receptor to the adenylate cyclase has been kinetically described as a non-linear relationship (Bockaert and coworkers, 1973) between the binding of the hormone to the receptor and the activation of the enzyme. Some experimental conditions may affect the coupling component more specifically; such as adrenalectomy which reduced VP-activated adenylate cyclase without affecting basal or NaF-stimulated activity. Hormone binding was only slightly affected (Rajerison and coworkers, 1974). Another possible example is that of PGE_1, which inhibits the hydro-osmotic effect of vasopressin in the isolated collecting tubule while slightly increasing L_p by itself with a strong potentiation by theophylline (Grantham and Orloff, 1968). Although PGE_2 rather than PGE_1, is produced naturally by the renal medulla it has been hypothesized that prostaglandins generated in the kidney in response to the hormonal challenge would serve as local modulators of vasopressin action (Grantham, 1974; Grantham and Orloff, 1968).

Another exogenous compound supposed to act within the receptor–adenylate cyclase complex is the diuretic ethacrynic acid which strongly inhibits the hydro-osmotic action of vasopressin in the isolated collecting tubule but has no effect on L_p by itself and does not affect the permeability response to theophylline or dibutyryl-cyclic-AMP (Abramow, 1974). The drug may interact with the coupling component in the membrane although an effect on the binding step of the hormone to the receptor cannot be ruled out (Abramow, 1974).

One possible effector of cyclic-AMP in the collecting tubule cell (whether directly or via a protein kinase activation) is the microtubular system (see Fig. 9). Microtubule disrupting agents such as colchicine interfere with ADH action in toad bladder and on urinary concentration in rats (Dousa and Valtin, 1976). Figure 10 shows that preincubation of an isolated collecting tubule with colchicine prevents the subsequent vasopressin effect. The onset of ADH response is delayed for about two hours and is obtained more than thirty minutes after removal of the microtubule disrupting drug (Abramow, 1976).

The microfilaments (MF) have also been implicated in the action of vasopressin since cytochalasin B, which disrupts MF, was shown to block the hydro-osmotic action of vasopressin in toad bladder (reviewed in Dousa and Valtin, 1976). It should be emphasized however that this effect could not be reproduced in the mammals *in vivo*. Moreover, in the isolated rabbit collecting tubule we showed that vasopressin effect on L_p was potentiated, not blocked (Abramow, 1976) (see Fig. 11). The drug did not affect basal L_p nor did it alter the effect of vasopressin on the permeability to tritiated water

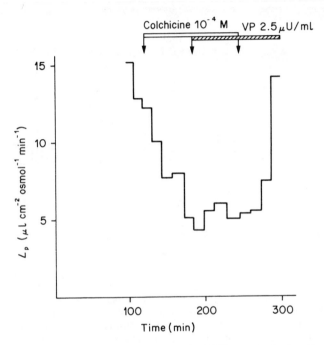

Figure 10 Effect of colchicine and vasopressin (VP) on the hydraulic conductivity (L_p) of an isolated perfused collecting tubule. Time in minutes after sacrifice. The initial decline in L_p is a common finding in this preparation and is not dependent on the preincubation with colchicine. Compare time course of vasopressin response with that in Fig. 7. See text for explanation. From Abramow (1976); reproduced by permission of MTP Press Ltd.

in the absence of an osmotic gradient (Abramow, 1976). In the presence of an osmotic gradient and ADH, cytochalasin B produced giant vacuolization of the cells suggesting that disruption of MF had altered the pathways for bulk water flow within the epithelium without fundamentally disturbing the permeability barrier (Abramow, 1976).

The reason for the difference between the results on toad bladder and the isolated mammalian tubule are not apparent but it is obvious here again that generalizing from findings in one species or tissue may be hazardous. At this juncture it is not entirely clear as to how the microtubular system is implicated in the action of vasopressin (Dousa and Valtin, 1976). Further studies are also needed before one can draw conclusions about the effects of the cytochalasins and whether a role exists for microfilaments in the changes in permeability.

Despite the biochemical refinements in identifying the steps leading to the physiological response to antidiuretic hormone the very nature of the

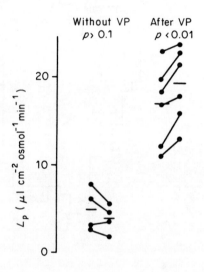

Figure 11 Effects of cytochalasin B (2–5 γ ml^{-1}) on osmotic water permeability in the absence (left) or presence (right) of vasopressin. Values connected by lines belong to the same tubule, the first point representing a control period and the second a period with cytochalasin B. From Abramow (1976); reproduced by permission of MTP Press Ltd.

permeability changes remains elusive. Perhaps more knowledge of the physico-chemical basis of membrane transport properties is required to gain better understanding of the mechanisms underlying the switch from low to high L_p under vasopressin. A better look at the ultrastructure of the pertinent membranes such as afforded by the freeze–fracture approach in isolated tubules (Humbert and coworkers, 1977) may also be of some help in view of the information gained with the technique in amphibian bladders (Wade, Kachadorian and Discala, 1977). Through this approach, some light is expected to be shed on the intimate changes related to the demonstrated increase in luminal membrane deformability (Grantham, 1974).

C. The Concentrating Mechanism and the Role of Urea

It has been known for decades that the acquisition in mammals and birds of the ability to produce a concentrated urine was somehow related to the development of the loop structures in the medulla. A correlation between the length of the loops of Henle and the degree to which the urine is concentrated has been found (Sperber, 1944). More specifically there was a close correlation between the thickness of the medulla and the ability to concentrate electrolytes in the urine (Schmidt-Nielsen and O'Dell, 1961).

The beaver, with 100% short-looped nephrons and a poorly developed inner medulla, has the lowest maximal urinary osmolality (about 450 mosmol $(kg\ H_2O)^{-1}$) the rabbit with 44% long-looped nephron concentrates the urine to about 1200 mosmol $(kg\ H_2O)^{-1}$, whereas the desert rodent Psammomys with 100% long-looped nephron produces urines averaging 3000 mosmol $(kg\ H_2O)^{-1}$ (Schmidt-Nielsen and O'Dell, 1961). In fact, the relative proportion of long- and short-looped nephrons appeared less important than the relative medullary thickness (Schmidt-Nielsen and O'Dell, 1961; Sperber, 1944). Desert rodents are among the mammals with the longest papilla, sometimes protruding deeply into the renal pelvis (Schmidt-Nielsen and O'Dell, 1961). Notable exceptions to the relationship between the development of the inner medulla and the maximum urinary concentrating ability have been observed in rhesus and other macaque monkeys (Tisher, 1971; Tisher, Schrier and McNeil, 1972). In these species the inner medulla is very poorly developed, with a paucity of long-looped nephrons, yet the maximum attainable urinary osmolality is of the order of 1500 mosmol $(kg\ H_2O)^{-1}$.

The latter studies cast doubt on the necessity, for at least some of the mammalian kidneys, to possess long loops in order to concentrate the urine adequately. Their conclusions however do not detract from the necessity for the concentrating kidney to possess a diluting segment together with looped structures in the outer medulla. These anatomical and functional features are the basic ingredients of the so-called countercurrent multiplier.

1. *The Countercurrent Multiplier*

The first model suggesting that the loops of Henle might function as a countercurrent multiplier was that of Kuhn and Ryfell (1942). In this and subsequent versions, three channels are juxtaposed in such a way that counter-flow of fluid occurs within the channels, two of them being connected at one end, by a 'U turn'. A source of energy in the form of a concentrated solution flowing through the third channel or of a small hydrostatic pressure between channels is provided.

Whatever the details of the models, as long as the components have the right sets of permeabilities to water and/or solutes, they predict that a small 'single effect', requiring energy and created transversally in the form of a relatively slight osmotic pressure difference between the limbs of the loop, can result in a considerable amplification of the effect in the longitudinal axis of the loop. Although the initial model (Kuhn and Ryffel, 1942) was not intended to mimic the concentrating medulla in all respects, for the first time if allowed the understanding of how the large corticopapillary gradient of solutes could be generated at a relatively low energetic cost. Besides, some wisdom rather than *'caprice de la nature'* was found in the curious double hairpin turn of the urine on its travel along the nephron.

The model of Kuhn and Ryffel did not attract the attention of physiologists until after the publication of the work of Wirz, Hargitay and Kuhn (1951) referred to above.

It became apparent from both micropuncture (Gottschalk and Mylle, 1959; Wirz, 1956; Wirz and Dirix, 1973) and *in vitro* microperfusion experiments (Burg and Green, 1973c; Imai and Kokko, 1974b; Rocha and Kokko, 1973) that the source of energy for the 'single effect' was the active salt reabsorption (later identified as a chloride pump) in the water-impermeable thick ascending limb of Henle's loop or diluting segment in the outer medulla. It was not clear however how this effect could be the source of a concentration profile of interstitial solutes progressing into the deeper portions of the inner medulla where only thin limbs make up the loops. Because of the apparent morphological 'simplicity' of the cells of the thin ascending limbs of Henle's loop, physiologists were reluctant to see them as engaged in active solute transport. At the same time mathematicians warned that all the models which involved passive transport of a single solute out of the thin ascending limb could not account for the rising concentration profile of the interstitial solutes in the inner medulla (Jamison, 1974). This concentration profile was demonstrated experimentally in a number of studies involving the measurement of solute content in slices of the medulla (Jamison, 1974). Attempts to build models ascribing the source of energy in the inner medulla to the collecting duct were not successful (Jamison, 1974).

That the loop rather than the collecting duct contributed to the increased sodium concentration in the inner medulla was suggested by the finding of Jamison, Bennett and Berliner (1967). By micropuncturing the papilla of the rat after removal of the overlying cortex, they established that dilution clearly occurred in the thin ascending limb of Henle's loop (please refer to Fig. 4). At the time, it was believed that the basic mechanism was active salt transport with a counter-current multiplication system analogous to that in the outer medullary thick ascending limb despite the considerable structural differences between the thick and thin portions of the limb.

Finally, the available models and data offered no convincing explanation for the corticopapillary accumulation of urea together with sodium and for the role of urea in concentrating non-urea solutes in the urine (Gamble and coworkers, 1934; Jaenike, 1960; Levinsky and Berliner, 1959; Pennel and coworkers, 1975; Roch-Ramel, Chomety and Peters, 1968; Wirz and Dirix, 1973).

Recent models of the countercurrent multiplication system were constructed on the assumption that the dilution of the fluid in the thin ascending limb of Henle's loop is of a purely passive nature (Kokko and Rector, 1972; Stephenson, 1972). These models work provided the components of the long loops have the required permeability characteristics and that a second solute besides NaCl, namely urea is incorporated into the models. The models, which were formulated independently by Kokko and Rector (1972) and by Stephenson, (1972) have clarified the understanding of the countercurrent multiplier in the inner medulla in a major way.

According to the Kokko and Rector model, depicted in Fig. 12, the events are the following. Fluid flowing down the descending limb of Henle's loop progressively equilibrates with the hypertonic interstitium through water abstraction. Since the interstitium contains more urea than the luminal fluid and since the epithelium is poorly permeable to both Na and urea (σ_{NaCl} and σ_{urea} near 1), the concentration of NaCl in the lumen is raised to a level higher than in the interstitium. At the hairpin turn, the permeabilities of the loop change drastically: the epithelium of the thin ascending limb is now very permeable to NaCl and urea but impermeable to water, allowing NaCl to leave the lumen down its chemical potential. As a result, the interstitium of the inner medulla becomes enriched with NaCl. Urea influx is also expected to occur, but, because the permeability to urea is less than to Na, the rate of urea influx is smaller than that of NaCl efflux so that there is effective dilution of the fluid in the thin ascending limbs by purely passive means. The system requires osmotic work for the maintenance of the high urea in the inner medulla. Energy for this is provided by the chloride pump in the outer medulla. The pump achieves dilution. Urea in the fluid downstream will be concentrated as water escapes under the influence of ADH. The distal tubule and cortical and outer medullary collecting tubule are indeed poorly permeable to urea. Urea is then deposited in the inner medullary interstitium through the urea-permeable papillary collecting duct under the combined forces of solvent drag and an appropriate chemical potential (Kokko and Rector, 1972).

The central core model of Stephenson (1972) is of a more general applicability than that of Kokko and Rector since it allows the thin ascending limb to function either on a passive mode, as in the Kokko–Rector model (the solute mixing model with the concentrated urea used to generate the Na gradient), or on an active mode (through active transport of a single solute out of the thin ascending limb). The central core is the system of the vasa recta which is exchanging freely with the interstitium. Transport and exchange is possible between the core and each tubule segment (collecting duct, descending and ascending thin limbs). The active and passive mode in Stephenson's model are not mutually exclusive.

The experimental evidence supporting the recent models is impressive (see Tables 3 and 4). There are however a few notable exceptions deserving careful consideration.

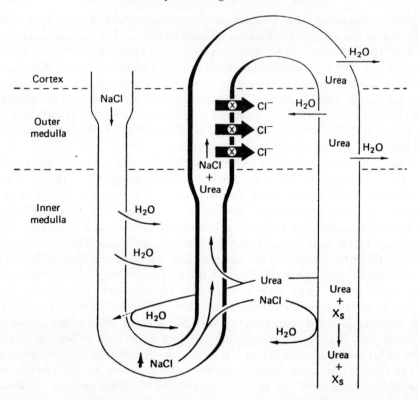

Figure 12 Passive model for the counter-current multiplication system in the inner medulla. The thin ascending limb functions as a purely passive equilibrating segment. The osmotic work created by the chloride pump in the outer medulla allows to concentrate urea in the fluid of the water permeable collecting system and is transmitted to the inner medulla in the form of a high unterstitial urea (provided by the urea permeable papillary collecting duct). Hypertonic urea, in turn, extracts water from the thin descending limb with resultant increase in loop NaCl. Hypertonic NaCl flows into the inner medulla through the salt-permeable, water-impermeable thin ascending limb and extracts water from the papillary collecting duct. From Kokko and Tisher (1976); reproduced by permission of *Kidney International*

The Kokko–Rector model is entirely consistent with the experimental data obtained in *in vitro* microperfused rabbit tubules (Imai, 1977; Imai and Kokko, 1974b; Kokko, 1970, 1972; Rocha and Kokko, 1974; see also Fig. 13 and Tables 3 and 4). The descending limb of the rabbit is indeed very permeable to water and osmotic equilibration in an hypertonic milieu is calculated to be 96% by water abstraction and only 4% by urea entry (Kokko, 1972). Passive permeability to Na and urea is low (see Table 3). The isolated thin ascending limb of Henle's loop was shown by Imai and Kokko

(1974b) to dilute luminal fluid by passive diffusion of solutes. When the perfusate was a 600 mosmol kg^{-1} solution containing predominantly NaCl and the bath was a 600 mosmol kg^{-1} buffer where urea contributed 50% to the osmotic pressure, the collected fluid osmolality decreased by about 300 mosmol (kg H$_2$O)$^{-1}$ (mm tubular length)$^{-1}$.

The passive permeabilities to Na and Cl of thin ascending limbs were very high (see Table 3) and even higher than in the proximal tubule (compare with Table 2). There was no voltage in the absence of salt asymmetry and no evidence for active solute transport in these segments (Imai and Kokko, 1974b).

In contrast to the findings in the rabbit, the results of micropuncture studies in *Psammomys obesus* by de Rouffignac and Morel (1969) are clearly in contradiction with the Kokko and Rector model. These authors showed beyond doubt that fluid in the descending limb of *Psammomys* is concentrated almost entirely by solute addition. NaCl entry accounted for 85% of the increase in osmolality (de Rouffignac and Morel, 1969). With such a behaviour of the descending limb, it would not be possible to account for the hypertonicity of the papilla in *Psammomys*, were the thin ascending limb devoid of an active process for diluting its content.

In the diabetes insipidus Brattleboro rat *in vivo*, Jamison and coworkers studied the contribution of water removal and solute addition under conditions of water diuresis and antidiuresis induced by exogenous ADH (Jamison, Buerkert and Lacy, 1973). At the bend of the loop they found a rise in TF/P inulin and a fall in flow rate during antidiuresis. The authors calculated that water extraction was the predominant mode by which fluid in the descending limb was concentrated during water diuresis while solute addition and water removal contributed about equally during antidiuresis. Thus the state of hydration seems to influence the process of osmotic equilibration in the descending limb although it is not certain that the intrinsic properties of the loop are altered. As far as urea entry is concerned the findings of Jamison and coworkers are more consistent with the permeability measurements of Morgan and Berliner (1968) in the rat papilla *in vitro* than with the values obtained in rabbits (see Table 3). However they are still compatible with a passive model for dilution in the thin ascending limb.

In the hamster kidney, Marsh (1970) also found entry of urea and some water abstraction but no evidence for NaCl addition.

Given the differences between the results with the isolated rabbit descending limbs and those obtained in other mammalian species one may wonder if the behaviour of the limb may not change along its length even in the same species under the same experimental conditions. Kokko perfused descending limbs with the end of the *pars recta* attached to the holding pipette. Thus, presumably the first two millimetres of the limb were so studied. Recently, we (Abramow and Burg, to be published) reinvestigated the problem by perfusing segments of thin descending limbs belonging to long loops and located closer to the bend of the loops. ADH was added to the bath so as to mimic the antidiuretic state shown to be associated with a larger participation of solute entry (Jamison, Buerkert and Lacy, 1973). The results suggested that this segment had a very poor ionic permeability as attested by the high transepithelial electrical resistance we found (see Table 3 and Fig. 5). When incubated with an hypertonic medium where the difference is osmotic pressure relative to the luminal fluid was due to NaCl and urea in equal proportions, osmotic equilibration was complete and essentially obtained by

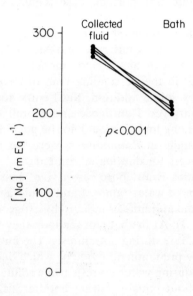

Figure 13 Passive generation of a sodium gradient between fluid in the inner medullary thin descending limb of Henle's loop and peritubular fluid. When tubules were bathed in a medium containing NaCl 220 mEq l^{-1} and urea 150 mmol l^{-1}, the Na concentration in the lumen rose from 150 mEq l^{-1} to 280 mEq l^{-1} with a resultant Na gradient of about 70 mEq l^{-1} between the collected fluid and the bath. The lines connect the experimental results from the same tubules. From Abramow and Burg, unpublished

water withdrawal. Under these conditions, Na and Cl concentrations increased to the same extent as did osmolality, confirming the low permeability to these ions and the negligible contribution of urea entry. Thus a considerable inside to outside Na gradient was generated (see Fig. 13). Our studies validate further the passive model of Kokko and Rector as applied to the rabbit although they certainly do not permit us to rule out some active pump in the thin ascending limb which would be assisted by a favourable Na gradient.

Imai was able recently to measure the permeability characteristics of thin ascending limbs of Henle's loop in rats and hamsters (Imai, 1977). The results agree with the previous findings of Imai and Kokko on the rabbit thin ascending limb (Imai and Kokko, 1974b).

This segment again is virtually impermeable to water and very permeable to Na and to Cl, with a significant although less important permeability to urea (see Table 3). The voltage was zero and the previously reported value of -11 mV in the hamster *in vivo* by Windhager (1964) rather than representing an active transport potential (Windhager, 1964) is claimed by Imai (1977) to depend largely on a liquid junction potential. It is not certain that some weak active transport process could not go on unnoticed with the approach used. However Imai does not believe in this hypothetical weak pump because such a mechanism would not be capable of net uphill Na transport in the face of the high permeability of the membrane to NaCl which he measured.

Thus one is inclined to conclude that a passive countercurrent multiplication system is a distinct possibility in several mammalian species although important questions remain to be clarified. One question is that of the very existence of the favourable inside to outside gradient for Na which should extend from the bend of the loop up to the active diluting segment (the thick ascending limb).

In the hamster Marsh and Azen (1975) found no difference between tubule fluid and ascending *vasa recta* Na, except when flow rate in the loop was slowed by simultaneous collection of fluid upstream in the loop. In this case an outside to inside gradient was demonstrated which Marsh and Azen believed indicative of active transport out of the loop (Marsh and Azen; 1975). However the significance of the results was questioned by Imai (1977) who pointed out that the blood in descending rather than ascending *vasa recta* is expected to reflect more closely the composition of the interstitial fluid in contact with the thin ascending limb. Imai also suggested that an artefactually lowered Na concentration in the loop fluid by retrograde contamination could not be ruled out in the experiments of Marsh and Azen (Marsh and Azen, (1975).

More recently, Johnston and coworkers (1977) detected in young Munich-Wistar rats a sodium concentration difference of about 60 mEq l^{-1} between loop fluid and *vasa recta* plasma in adjacent sites. When corrected for plasma water the difference was still 40 mEq l^{-1} indicating the presence *in vivo* of a key requirement of the passive model.

Another requirement of a purely passive model is that urea concentration should be sufficiently low in the fluid entering the thin ascending limb so as to conserve an outside to inside gradient of urea up to the thick ascending limb, a prerequisite for the simultaneous passive outflux of sodium.

The available evidence points to a concentration of urea in the loop fluid which is considerably in excess of that assumed to exist according to the Kokko and Rector model (Kokko and Rector, 1972).

2. Medullary Recycling of Urea

Urea is handled by the kidney in a rather complex fashion. At the end of the proximal a relatively fixed amount (about 50%) of filtered urea is reabsorbed (Lassiter, Gottschalk and Mylle, 1961) irrespective of the urine osmolality (Jamison, 1974). Urea is clearly added by net secretion in the 'loop' somewhere between the end of the accessible proximal tubule and the beginning of the accessible distal tubule of the superficial nephrons since 110% of the filtered urea is now recovered in the early distal tubule (Lassiter, Gottschalk and Mylle, 1961). Thus about 60% of the filtered load has been added to the tubular fluid by secretion. It was later shown (see Jamison, 1974) that the amount of urea apparently secreted was related to the urine flow. As urine flow is reduced (antidiuresis) the fraction of the 'filtered' urea added to short loops increases while the fraction which is reabsorbed by the collecting duct increases in a strikingly parallel fashion and amounts to nearly 60% in maximal antidiuresis (Lassiter, Gottschalk and Mylle, 1961). The fraction reabsorbed in the 'collecting duct' is the difference between the fraction recovered in the urine and that remaining at the end of the superficial distal tubule.

These findings are the main evidence for the so-called recycling of urea. While there is agreement as to the finding of urea recycling from collecting duct to loop of Henle, the site or sites in the loop where urea enters has been a matter of debate. From the above discussion and the measured P_{urea} and σ_{urea} in Tables 2 to 4, it is apparent that a significant amount of urea may enter the thin descending limb in the rat but not in the rabbit, where the thin ascending limb is a more likely site. The thin ascending limb is also a more appropriate site for urea entry if the passive model of Kokko and Rector is to work efficiently. Any urea entering the nephron or remaining in the tubular fluid at sites proximal to the bend of the loop is expected to decrease the efficiency of water withdrawal. A recent study with isolated rabbit tubules by Kawamura and Kokko (1976) gave support to the possibility (not tested directly in the rabbit) that loop urea concentration may indeed be more important than suspected on the basis of the low P_{urea} of the descending limb. They found a component of active urea secretion in the proximal straight tubule of superficial and juxtamedullary nephrons. Thus urea accumulated in the medulla may recirculate by two routes: the thin ascending limb of Henle's loop, and the *pars racta* as it courses through the outer medulla. The authors recognize that the latter process decreases the efficiency of the passive concentrating mechanism but state that as long as the amount of urea entry remains within the range observed in the micropuncture studies the model remains valid (Lassiter, Gottschalk and Mylle, 1961, 1966; Pennell, Lacy and Jamison, 1974; Pennell and coworkers, 1975; Roch-Ramel, Chomety and Peters, 1968).

Additional micropuncture studies of urea-loaded animals have confirmed the role of urea in extracting water from the descending limb. Pennel and coworkers (1975) showed that urea-infused rats had a considerable increase in TF/P inulin at the end of the descending limb. At the same time, the secretion of urea in the loop was also considerably increased since 300% of the filtered urea remained at the end of the descending limb. At first sight it seems difficult to combine urea induced water abstraction and urea permeation in one single process.

Pennel and coworkers, using computer simulation of the rat descending limb with the published P_{urea} values of rat and rabbit (see Table 3) concluded that the same segment could carry out both functions, the first part being chiefly involved in water abstraction and the later part serving as the main site of urea influx (Pennell and coworkers, 1975).

Even *Psammomys*, when urea loaded, show an increased TF/P inulin at the bend of the loop if compared to salt-loaded animals (Imbert and de Rouffignac, 1976). However this finding is hardly sufficient to explain the concentrating mechanism in this species by the passive models. *Psammomys* is indeed an unusual species which not only lives in dry habitat but also feeds on succulent plants which are very rich in NaCl (Schmidt-Nielsen and O'Dell, 1961). The kidneys in this species are as much involved in eliminating large Na loads as in conserving water. This behaviour may somehow explain why urea plays such a small role in their concentrating process. Of interest is the finding that urea-loaded *Psammomys* excreted a urine which was less concentrated than similarly NaCl-loaded animals (Imbert and de Rouffignac, 1976).

With the exception of *Psammomys*, the passive models make it clearer how the 'economy of water referable to urea' (Gamble and coworkers, 1934) can be understood. Increasing urea excretion makes more urea available for trapping or recycling in the medulla thus balancing the higher urea concentration in the collecting duct and preventing the obligation of extra water for the excretion of the increased urea load. In addition, the ability of urea to increase the concentration of non-urea solutes in the urine is obvious given the role which is assigned to urea in the passive model.

It appears that the largest discrepancy in the literature over the concentrating process concerns the results from studies on *Psammomys in vivo* and rabbit tubules *in vitro*. In order to clarify the origin of the discrepancy it might be of help to study isolated perfused thin limbs of *Psammomys*. Such an approach would indicate whether their intrinsic permeability characteristics differ so drastically from those of the rabbit.

3. *Medullary Circulation*

The medullary circulation, with the *vasa recta* functioning as countercurrent exchangers is functionally arranged so as to prevent the dissipation of the corticopapillary solute gradient (Jamison, 1974; Wirz and Dirix, 1973).

The function of the countercurrent exchanger is to maximize exchange between the limbs of the vascular loop (arterial and venous capillaries) and to minimize the axial flow. Thus, solute losses from medulla and water entry into the medulla are kept to a minimum. It should be recalled that the great difference between the exchanger and the multiplier is that the exchanger does not require an internal source of energy. It seems that the physiological work on the medullary circulation has been less rewarding than its detailed anatomical description for the understanding of the exchange with the tubular fluid compartments and for the mapping of the route taken by the solutes in the recycling phenomenon (Jamison, 1974). What is still poorly understood is how the *vasa recta* can help to keep the solutes in the medulla and at the same time to dispose of the excess water extracted from the descending limb and the collecting duct.

III. CONCLUDING REMARKS

In this chapter we mainly focused interest on some of the efferent pathways involved in osmoregulation in mammals. Certainly, afferent signals indicating whether to conserve or excrete water or solutes are also of prime importance. However, if one excepts the case of osmoregulation of vasopressin secretion, the relative importance of several afferent pathways in the control of osmotic pressure and volume of the body fluids remains to be established despite the amount of work devoted to the subject (Dirks, Seely and Levy, 1976).

Even the hormonal control of Na excretion is ill understood with the exception of some aspects of the renin–angiotensin–aldosterone system, which are dealt with in another chapter of this book.

The factors regulating the day to day balance of sodium, the main osmotically active solute of the extracellular fluid, are essentially unknown despite the great effort made to identifying them (Dirks, Seely and Levy, 1976). It is possible that most of the models used to study the phenomenon are not entirely appropriate. An example is acute loading with isotonic saline. Perhaps the intrarenal physical changes which are involved in the altered sodium excretion and the secretion of the still elusive natriuretic hormone which is expected to be triggered by these manipulations, have little to do with the subtle adjustments of osmoregulation of volume control.

Clearly there is a need for a better understanding of the coordinated changes which occur in the whole organism.

At the same time continuing efforts are necessary to gain more insight into the ways the mammalian kidney admirably performs its multiple tasks.

ACKNOWLEDGMENTS

The work reported in this section was supported by grant No. 3.4546.75 of the Fonds de la Recherche Scientifique Médicale, Belgium. The technical assistance of Mrs S. Foulon and the secretarial help of Mrs M. N. Doutreluingne are gratefully acknowledged.

REFERENCES

Abramow, M. (1974). Effects of ethacrynic acid on the isolated collecting tubule. *J. Clin. Invest.*, **53**, 796–804.
Abramow, M. (1975). La perméabilité à l'eau du tubule collecteur isolé et perfusé. Effets de l'hormone antidiurétique et de certains diurétiques. *Thèse d'agrégation de l'enseignement supérieur*, Université Libre de Bruxelles.
Abramow, M. (1976). Effect of vasopressin on water transport in the kidney: Possible role of microtubules and microfilaments. In J. W. L. Robinson (Ed.) *Intestinal Ion Transport*, M.T.P. Press, Lancaster. pp. 173–82, 187–8.
Abramow, M., M. Burg, and J. Orloff (1967). Chloride flux in rabbit kidney tubule *in vitro*. *Am. J. Physiol.*, **213**, 1249–53.
Abramow, M., and M. Dratwa (1974). Effect of vasopressin on the isolated human collecting duct. *Nature*, **250**, 492–3.
Anagnostopoulos, T., M. Kinney, and E. Windhager (1971). Salt and water reabsorption by short loops of Henle during renal vein constriction. *Am. J. Physiol.*, **22**, 1060–70.
Andreoli, T. E., and J. A. Schafer (1976). Mass transport across cell membranes: The effects of antidiuretic hormone on water and solute flows in epithelia. *A. Rev. Physiol.*, **38**, 451–500.
Atherton, J. C., R. Green, and S. Thomas (1971). Influence of lysine-vasopressin dosage of the time course of changes in renal tissue and urinary composition in the conscious rat. *J. Physiol.*, **213**, 291–309.
Baines, A. D., and C. de Rouffignac (1969). Functional heterogeneity of nephrons. II. Filtration rates, intraluminal flow velocities and fractional water reabsorption. *Arch. Ges. Physiol.*, **308**, 274–84.
Bank, N., W. E. Yarger, and H. S. Aynedjian (1971). A microperfusion study of sucrose movement across the rat proximal tubule during renal vein constriction. *J. Clin. Invest.*, **50**, 294–302.
Berliner, R. W., and P. G. Davidson (1957). Production of hypertonic urine in the absence of pituitary antidiuretic hormone. *J. Clin. Invest.*, **36**, 1416–27.
Bockaert, J., C. Roy, R. Rajerison, and S. Jard (1973). Specific binding of receptor of ^{3}H lysine-vasopressin to pig kidney plasma membranes. Relationship of receptor occupancy to adenylate cyclase activation. *J. Biol. Chem.*, **248**, 5922–31.
Bonvalet, J. P., and C. de Rouffignac (1975). Hétérogénéité fonctionnelle des néphrons. *J. Physiol., Paris*, **71**, 73A–121A.
Boulpaep, E. L. (1972). Permeability changes of the proximal tubule of Necturus during saline loading. *Am. J. Physiol.*, **222**, 517–31.
Boulpaep, E. L. (1976). Recent advances in electrophysiology of the nephron. *A. Rev. Physiol.*, **38**, 20–36.
Boulpaep, E. L., and J. F. Seely (1971). Electrophysiology of proximal and distal tubules in the perfused dog kidney. *Am. J. Physiol.*, **221**, 1084–96.
Bray, G. A. (1960). Freezing point depression of rat kidney slices during water diuresis and antidiuresis. *Am. J. Physiol.*, **199**, 915.
Brenner, B. M., and W. M. Deen (1974). The physiological basis of glomerular ultrafiltration. In K. Thurau (Ed.) *Kidney and Urinary Tract Physiology: MTP International Review of Science*, Butterworths, University Park Press, London. p. 335.
Brenner, B. M., K. H. Falchuk, R. I. Keimowitz, and R. W. Berliner (1969). The

relationship between peritubular capillary protein concentration and fluid reabsorption by the renal proximal tubule. *J. Clin. Invest.*, **48**, 1519–31.

Brenner, B. M., and J. L. Troy (1971). Postglomerular vascular protein concentration: evidence for a causal role in governing fluid reabsorption and glomerulotubular balance by the renal proximal tubule. *J. Clin. Invest.*, **50**, 336–49.

Brenner, B. M., J. L. Troy, T. M. Daugherty, and R. M. McInnes (1973). Quantitative importance of changes in postglomerular colloid osmotic pressure in mediating glomerulotubular balance in the rat. *J. Clin. Invest.*, **52**, 190–7.

Bresler, E. H. (1976). Ludwig's theory of tubular reabsorption: the role of physical factors in tubular reabsorption. *Kidney Int.*, **9**, 313–22.

Bruns, F. J., E. A. Alexander, A. L. Riley, and N. G. Levinsky (1974). Superficial and juxta-medullary nephron function during saline loading in the dog. *J. Clin. Invest.*, **53**, 971–9.

Burg, M. B. (1976a). The renal handling of sodium chloride. In B. M. Brenner and F. C. Rector (Eds), *The Kidney*, Saunders, Philadelphia. pp. 272–98.

Burg, M. B. (1976b). Tubular chloride transport and the mode of action of some diuretics. *Kidney Int.*, **9**, 189–97.

Burg, M. B., and M. Abramow (1966). Localization of tissue sodium and potassium compartments in rabbit renal cortex. *Am. J. Physiol.*, **211**, 1011–17.

Burg, M. B., J. Grantham, M. Abramow, and J. Orloff (1966). Preparation and study of fragments of single rabbit nephrons. *Am. J. Physiol.*, **210**, 1293–8.

Burg, M. B., and N. Green (1973a). Effect of mersalyl on the thick ascending limb of Henle's loop. *Kidney Int.*, **4**, 245–51.

Burg, M. B., and N. Green (1973b). Effect of ethacrynic acid on the thick ascending limb of Henle's loop. *Kidney Int.*, **4**, 301–8.

Burg, M., and N. Green (1973c). Function in the thick ascending limb of Henle's loop. *Am. J. Physiol.*, **224**, 659–68.

Burg, M. B., and N. Green (1976). Role of monovalent ions in the reabsorption of fluid by isolated perfused proximal renal tubules of the rabbit. *Kidney Int.*, **10**, 221–8.

Burg, M. B., S. Helman, J. Grantham, and J. Orloff (1970). Effect of vasopressin on the permeability of isolated rabbit cortical collecting tubules to urea, acetamide and thiourea. In B. Schmidt-Nielsen and D. W. S. Kerr (Eds), *Urea and the Kidney*, Excerpta Medica Foundation, Amsterdam. pp. 193–9.

Burg, M., L. Isaacson, J. Grantham, and J. Orloff (1968). Electrical properties of isolated rabbit renal tubules. *Am. J. Physiol.*, **215**, 788–94.

Burg, M. B., and J. Orloff (1968). Control of fluid absorption in the renal proximal tubule. *J. Clin. Invest.*, **47**, 2016–24.

Burg, M. B., and J. Orloff (1970). Electrical potential difference across proximal convoluted tubules. *Am. J. Physiol.*, **219**, 1714–16.

Burg, M., C. Patlak, N. Green, and D. Villey (1976). The role of organic solutes in fluid absorption by proximal convoluted tubules. *Am. J. Physiol.*, **231**, 627–37.

Burg, M., L. Stoner, J. Cardinal, and N. Green (1973). Furosemide effect on isolated perfused tubules. *Am. J. Physiol.*, **225**, 119–24.

Cardinal, J., M. D. Lutz, M. D. Burg, and J. Orloff (1975). Lack of relationship of potential difference to fluid absorption in the proximal renal tubule. *Kidney Int.*, **7**, 94–102.

Claude, P., and D. A. Goodenough (1973). Fracture faces of zonulae occludentes from 'tight' and 'leaky' epithelia. *J. Cell Biol.*, **58**, 390–400.

Coelho, J. B., K. C. H. Chien, and S. E. Bradley (1972). Measurement of single nephron glomerular filtration rate without micropuncture. *Am. J. Physiol.*, **223**, 832–9.

Curran, P. F. (1965). Ion transport in intestine and its coupling to other transport processes. *Fed. Proc.*, **24**, 993–9.

Curran, P. F., and J. R. MacIntosh (1962). A model system for biological water transport. *Nature*, **193**, 347–8.

Diamond, J. M., and W. H. Bossert (1967). Standing-gradient osmotic flow. A mechanism for coupling of water and solute transport in epithelia. *J. Gen. Physiol.*, **50**, 2061–83.

Diezi, J., P. Michoud, J. Aceves, and G. Giebisch (1973). Micropuncture study of electolyte transport across papillary collecting duct of the rat. *Am. J. Physiol.*, **224**, 623–34.

Dirks, J. H., J. F. Seely, and M. Levy (1976). Control of extracellular fluid volume and the pathophysiology of edema formation. In B. M. Brenner and F. C. Rector (Eds), *The Kidney*, Saunders, Philadelphia. pp. 495–552.

Dousa, T. P., and H. Valtin (1976). Cellular actions of vasopressin in the mammalian kidney. *Kidney Int.*, **10**, 46–63.

Dubois, R., A. Verniory, and M. Abramow (1976). Computation of the osmotic water permeability of perfused tubule segments. *Kidney Int.*, **10**, 478–9.

Farquhar, M. G., and G. E. Palade (1963). Junctional complexes in various epithelia. *J. Cell Biol.*, **17**, 375–412.

Farquhar, M. G., and G. E. Palade (1964). Functional organization of amphibian skin. *Proc. Natl Acad. Sci. USA*, **51**, 569–77.

Flamenbaum, W. R., and J. Hamburger (1974). Superficial and deep juxtaglomerular apparatus renin activity of the rat kidney. *J. Clin. Invest.*, **54**, 1373–81.

Forster, R. P. (1947). An examination of some factors which alter glomerular activity in the rabbit kidney. *Am. J. Physiol.*, **150**, 523–33.

Frindt, G., and M. B. Burg (1972). Effect of vasopressin on sodium transport in renal cortical collecting tubules. *Kidney Int.*, **1**, 224–31.

Fromter, E., and K. Gessner (1974). Free flow potential profile along rat kidney proximal tubule. *Pflüger Arch.*, **351**, 69–83.

Fromter, E., G. Rumrich, and K. J. Ullrich (1973). Phenomenologic description of Na^+, Cl^- and HCO_3^- absorption from proximal tubules of the rat kidney. *Pflügers Arch. Ges. Physiol.*, **343**, 189–220.

Gamble, J. L., C. F. McKhann, A. M. Butler, and E. Tuthill (1934). An economy of water in renal function referable to urea. *Am. J. Physiol.*, **109**, 139.

Ganote, C. E., J. J. Grantham, H. L. Moses, M. B. Burg, and J. Orloff (1968). Ultrastructural studies of vasopressin effect on isolated perfused renal collecting tubules of the rabbit. *J. Cell Biol.*, **36**, 355–67.

Gertz, K. H., and J. W. Boylan (1973). Glomerular-tubular balance. In J. Orloff and R. W. Berliner (Eds), *Handbook of Physiology, Section 8: Renal Physiology*, Washington, American Physiological Society. pp. 763–90.

Giebisch, G., R. M. Klose, G. Malnic, W. J. Sullivan, and E. E. Windhager (1964). Sodium movement across single perfused proximal tubules of rat kidneys. *J. Gen. Physiol.*, **47**, 1175–94.

Giebisch, G., and E. E. Windhager (1964). Renal tubular transfer of sodium, chloride and potassium. *Am. J. Med.*, **36**, 643–69.

Giebisch, G., and E. E. Windhager (1973). Electrolyte transport across renal tubular membranes. In J. Orloff and R. W. Berliner (Eds), *Handbook of Physiology, Section 8: Renal Physiology*, Washington, American Physiological Society. pp. 315–76.

Gottschalk, C. W. (1961). Micropuncture studies of tubular function in the mammalian kidney. *Physiologist*, **4**, 35–55.

Gottschalk, C. W., and M. Mylle (1959). Micropuncture study of the mammalian urinary concentrating mechanism: evidence for the counter-current hypothesis. *Am. J. Med.*, **196**, 927–36.

Grandchamp, A., and E. L. Boulpaep (1974). Pressure control of sodium reabsorption and intracellular backflux across proximal kidney tubule. *J. Clin. Invest.*, **54**, 69–84.

Grantham, J. J. (1970). Vasopressin: effect on deformability of urinary surface of collecting duct cells. *Science*, **168**, 1093–5.

Grantham, J. J. (1974). In K. Thurau (Ed.), *Kidney and Urinary Tract Physiology*, Butterworths, University Park Press, London. pp. 247–72.

Grantham, J. J., and M. B. Burg (1966). Effect of vasopressin and cyclic AMP on permeability of isolated collecting tubules. *Am. J. Physiol.*, **211**, 255–9.

Grantham, J. J., M. B. Burg, and J. Orloff (1970). The nature of transtubular Na and K transport in isolated rabbit renal collecting tubules. *J. Clin. Invest.*, **49**, 1815–26.

Grantham, J. J., C. E. Ganote, M. B. Burg, and J. Orloff (1969). Paths of transtubular water flow in isolated renal collecting tubules. *J. Cell Biol.*, **41**, 562–76.

Grantham, J. J., and J. Orloff (1968). Effect of prostaglandin E_1 on the permeability response of the isolated collecting tubule to vasopressin, adenosine 3'5'-monophosphate and theophylline. *J. Clin. Invest.*, **47**, 1154–61.

Grantham, J. J., P. B. Qualizza, and L. W. Welling (1972). Influence of serum proteins on net fluid reabsorption of isolated proximal tubules. *Kidney Int.*, **2**, 66–75.

Green, R., and G. Giebisch (1975a). Ionic requirements of proximal tubular sodium transport. I. Bicarbonate and chloride. *Am. J. Physiol.*, **229**, 1205–15.

Green, R., and G. Giebisch (1975b). Ionic requirements of proximal tubular sodium transport. II. Hydrogen ion. *Am. J. Physiol.*, **229**, 1216–26.

Green, R., E. E. Windhager, and G. Giebisch (1974). Protein osmotic pressure effects on proximal tubular fluid movement in the rat. *Am. J. Physiol.*, **226**, 265–76.

Gross, J. B., M. Imai, and J. P. Kokko (1975). A functional comparison of the cortical collecting tubule and the distal convoluted tubule. *J. Clin. Invest.*, **55**, 1284–94.

Gutman, Y., C. Gottschalk, and W. E. Lassiter (1965). Micropuncture study of inulin absorption in the rat kidney. *Science*, **47**, 753–4.

Handler, J. S., and J. Orloff (1973). The mechanism of action of antidiuretic hormone. In J. Orloff and R. W. Berliner (Eds), *Handbook of Physiology, Section 8: Renal Physiology*, Washington, American Physiological Society. pp. 791–814.

Hanssen, O. E. (1961). The relationship between glomerular filtration and length of the proximal convoluted tubules in mice. *Acta Pathol. Microbiol. Scand.*, **53**, 265–79.

Hanssen, O. E. (1963). Method for comparison of glomerular filtration in individual rat nephrons. *2nd Int. Congr. on Nephrology, Prague (Abstract)*, Excerpta Medica, Amsterdam. p. 527.

Hayslett, J. P. (1973). Effects of changes in hydrostatic pressure in peritubular capillaries on the permeability of the proximal tubule. *J. Clin. Invest.*, **52**, 1314–19.

Helman, S. I., J. J. Grantham, and M. B. Burg (1971). Effect of vasopressin on electrical resistance of renal cortical collecting tubules. *Am. J. Physiol.*, **220**, 1825–32.

Horster, M., and L. Larsson (1976). Mechanisms of fluid absorption during proximal tubule development. *Kidney Int.*, **10**, 348–63.

Horster, M., and K. Thurau (1968). Micropuncture studies on the filtration rate of single superficial and juxtamedullary glomeruli in the rat kidney. *Arch. Ges. Physiol.*, **301**, 162–81.

Humbert, F., M. Abramow, A. Perrelet, and L. Orci (1977). Feasibility of freeze–fracturning single isolated renal tubules. *Kidney Int.*, **12**, 66–71.

Imai, M. (1977). Function of the thin ascending limb of Henle of rats and hamsters perfused *in vitro*. *Am. J. Physiol.*, **232**, F201–9.

Imai, M., and J. P. Kokko (1972). Effect of peritubular protein on reabsorption of sodium and water on isolated perfused proximal tubules. *J. Clin. Invest.*, **51**, 314–25.
Imai, M., and J. P. Kokko (1974a). Transtubular oncotic pressure gradients and net fluid transport in isolated proximal tubules. *Kidney Int.*, **6**, 138–45.
Imai, M., and J. P. Kokko (1974b). Sodium chloride urea and water transport in the thin ascending limb of Henle. Generation of osmotic gradients by passive diffusion of solutes. *J. Clin. Invest.*, **53**, 393–402.
Imbert, M., D. Chabardes, D. Montegut, A. Clique, and F. Morel (1975a). Vasopressin dependent adenylate cyclase in single segments of rabbit kidney tubule. *Pflügers Arch.*, **357**, 173–86.
Imbert, M., D. Chabardes, M. Montegut, A. Clique, and F. Morel (1975b). Adenylate cyclase activity along the rabbit nephron as measured in single isolated segments. *Pflügers Arch.*, **354**, 213–28.
Imbert, M., and C. de Rouffignac (1976). Role of sodium and urea in the renal concentration mechanism in *Psammomys obesus*. *Pflügers Arch.*, **361**, 107–14.
Jaenike, J. R. (1960). Urea enhancement of water reabsorption in the renal medula. *Am. J. Physiol.*, **199**, 1205.
Jamison, R. L. (1970). Micropuncture study of superficial and juxtamedullary nephrons in the rat. *Am. J. Physiol.*, **218**, 46.
Jamison, R. L. (1973). Intrarenal heterogeneity: the case for two functionally dissimilar populations of nephrons in the mammalian kidney. *Am. J. Med.*, **54**, 281–9.
Jamison, R. L. (1974). Countercurrent systems. In K. Thurau (Ed.), *Kidney and Urinary Tract Physiology*, Butterworths, University Park Press, London. pp. 199–245.
Jamison, R. L., C. M. Bennett, and R. W. Berliner (1967). Countercurrent multiplication by the thin loops of Henle. *Am. J. Physiol.*, **212**, 357–66.
Jamison, R. L., J. Buerkert, and F. B. Lacy (1971). A micropuncture study of collecting tubule function in rats with hereditary diabetes insipidus. *J. Clin. Invest.*, **50**, 2444–52.
Jamison, R. L., J. Buerkert, and F. Lacy (1973). A micropuncture study of Henle's thin loop in Brattleboro rats. *Am. J. Physiol.*, **224**, 180–5.
Jamison, R., and F. Lacy (1972). Evidence for urinary dilution by the collecting tubule. *Am. J. Physiol.*, **223**, 898–902.
Jard, S. (1966). Etude des effets de la vasotocine sur l'excrétion de l'eau et des électrolytes par le rein de la grenouille. *Rana esculenta L*: analyse à l'aide d'analogues artificiels de l'hormone naturelle des caractères structuraux requis pour son activité biologique. *J. Physiol., Paris*, **58**, *Suppl. 15*, 1–24.
Johnston, P. A., C. A. Battilana, F. B. Lacy, and R. L. Jamison (1977). Evidence for a concentration gradient favoring outward movement of sodium from the thin loop of Henle. *J. Clin. Invest.*, **59**, 234–40.
Kashgarian, M. H., Y. Warren, R. L. Mitchell, and F. H. Epstein (1964). Effect of protein in tubular fluid upon proximal tubular absorption. *Proc. Soc. Exp. Biol. Med.*, **117**, 848–50.
Kawamura, S., M. Imai, D. W. Seldin, and J. P. Kokko (1975). Characteristics of salt and water transport in superficial and juxtamedullary straight segments of proximal tubules. *J. Clin. Invest.*, **55**, 1269–77.
Kawamura, S., and J. P. Kokko (1976). Urea secretion by the straight segment of the proximal tubule. *J. Clin. Invest.*, **58**, 604–12.
Koefoed-Johnsen, V., and H. H. Ussing (1953). The contribution of diffusion and flow to the passage of D_2O through living membrane. *Acta Physiol. Scand.*, **28**, 60–76.

Kokko, J. P. (1972). Urea transport in the proximal tubule and the descending loop of Henle. *J. Clin. Invest.*, **51**, 1999–2008.

Kokko, J. P. (1973). Proximal tubule potential difference. Dependence on glucose, HCO_3 and amino acids. *J. Clin. Invest.*, **52**, 1362–7.

Kokko, J. P. (1973). Sodium chloride and water transport in the descending limb of Henle. *J. Clin. Invest.*, **49**, 1838–46.

Kokko, J. P., M. B. Burg, and J. Orloff (1971). Characteristics of NaCl and water transport in the renal proximal tubule. *J. Clin. Invest.*, **50**, 69–76.

Kokko, J. P., and F. C. Rector (1971). Flow dependence of transtubular potential difference in isolated perfused segments of rabbit proximal convoluted tubule. *J. Clin. Invest.*, **50**, 2745–50.

Kokko, J. P., and F. C. Rector, Jr (1972). Countercurrent multiplication system without active transport in inner medulla. *Kidney Int.*, **2**, 214–23.

Kokko, J. P., and C. Tisher (1976). Water movement across nephron segments involved with the countercurrent multiplication system. *Kidney Int.*, **10**, 64–81.

Kuhn, W., and K. Ryffel (1942). Herstellung konzentrierter Lösungen aus verdünnten durch blosse Membranwirkung. Ein Modellversuch zur Funktion der Neiere. *Hoppe-Seylers Z. Physiol. Chem.*, **276**, 145–78.

Lahlou, B. (1966). Mise en évidence d'un 'recrutement glomérulaire' dans le rein des Téléostéens d'aprés la mesure du Tm glucose. *Compt. Rend.*, **262**, 1356–8.

Lambert, P. P., J. P. Gassee, A. Verniory, and P. Ficheroulle, P. (1971). Measurement of glomerular filtration pressure from sieving data for macromolecules. *Arch. Ges. Physiol.*, **329**, 34–58.

Lambert, P. P., F. Gregoire, and C. Malmendier (1957). La perméabilité glomérulaire aux substances protidiques (Editorial). *Rev. Franç. Etudes Clin. Biol.*, **2**, 15–21.

Landwehr, D., R. Klose, and G. Giebisch (1967). Renal tubular sodium and water reabsorption in the isotonic sodium chloride-loaded rat. *Am. J. Physiol.*, **212**, 1327–33.

Landwehr, D., J. Schnermann, R. Klose, and G. Giebisch (1968). Effect of reduction of filtration rate on renal tubular sodium and water reabsorption. *Am. J. Physiol.*, **215**, 687–95.

Lassiter, W. E., C. W. Gottschalk, and M. Mylle (1961). Micropuncture study of net transtubular movement of water and urea in nondiuretic mammalian kidney. *Am. J. Physiol.*, **200**, 1139–47.

Lassiter, W. E., C. W. Gottschalk, and M. Mylle (1966). Micropuncture study of urea transport in rat renal medulla. *Am. J. Physiol.*, **210**, 965–70.

Levinsky, N. G., and R. W. Berliner (1959). The role of urea in the urine concentrating mechanism. *J. Clin. Invest.*, **38**, 741–48.

Lewy, J. E., and E. E. Windhager (1968). Peritubular fluid reabsorption in the rat kidney. *Am. J. Physiol.*, **214**, 943–54.

Liebau, G., D. Z. Levine, and K. Thurau (1968). Micropuncture studies on the dog kidney. I. The response of the proximal tubule to changes in systemic blood pressure within and below the autoregulatory range. *Arch. Ges. Physiol.*, **304**, 57–68.

Ludwig, C. (1843). *Beitrage zur Lehre vom Mechanismsus der Harnsecretion*, N. G. Elwert'sche Universitäts und Verlagsbuchhandlung, Marburg.

Lutz, M. D., J. Cardinal, and M. B. Burg (1973). Electrical resistance of renal proximal tubule perfused *in vitro*. *Am. J. Physiol.*, **225**, 729–34.

Malnic, G. (1972). Some electrical properties of distal tubular epithelium in the rat. *Am. J. Physiol.*, **223**, 797–808.

Malnic, G., M. Mello-Aires, and F. Vieira (1970). Chloride excretion in nephrons of the rat during alterations of acid–base equilibrium. *Am. J. Physiol.*, **218**, 20–26.

Malvin, R. L., W. S. Wilde, A. J. Vander, and L. P. Sullivan (1958). Localization and characterisation of sodium transport among the renal tubule. *Am. J. Physiol.*, **195**, 549–57.
Marsh, D. J. (1970). Solute and salt flows in thin limbs of Henle's loop in the hamster kidney. *Am. J. Physiol.*, **218**, 824–31.
Marsh, D. J., and J. P. Azen (1975). Mechanisms of NaCl reabsorption by hamster thin ascending limb of Henle's loop. *Am. J. Physiol.*, **228**, 71–9.
Marsh, D. J., and C. Frasier (1965). Reliability of inulin for determining volume flow in rat renal cortical tubules. *Am. J. Physiol.*, **209**, 283–6.
Marsh, D. J., and S. Solomon (1965). Analysis of electrolyte movement in thin Henle's loop of hamster papilla. *Am. J. Physiol.*, **208**, 1119–28.
Martinez-Palomo, A., and D. Erlij (1973). The distribution of lanthanum in tight junctions of the kidney tubule. *Pflügers Arch.*, **343**, 267–72.
Maude, D. L., W. N. Scott, I. Shehadeh, and A. K. Solomon (1965). Further studies on the behaviour of inulin and serum albumin in rat kidney tubules. *Arch. Ges. Physiol.*, **285**, 313–6.
McKinney, T. D., and M. Burg (1976). Effect of acetazolamide on absorption of bicarbonate and fluid in rabbit proximal straight tubules. *Kidney Int.*, **1**, 592.
Morel, F., D. Chabardes, and M. Imbert (1976). Functional segmentation of the rabbit distal tubule by microdetermination of hormone-dependent adenylate cyclase activity. *Kidney Int.*, **9**, 264–77.
Morgan, T. (1974). A microperfusion study in the rat of the permeability of the papillary segments of the nephron to ^{24}Na. *Clin. Exp. Pharmacol. Physiol.*, **1**, 23.
Morgan, T., and R. W. Berliner (1968). Permeability of the loop of Henle, *vasa recta*, and collecting duct to water, urea and sodium. *Am. J. Physiol.*, **215**, 108–15.
Morgan, T., and R. W. Berliner (1969). A study by continuous microperfusion of water and electrolyte movements in the loop of Henle distal and tubule of the rat. *Nephron*, **6**, 388–405.
Morgan, T., F. Sakai, and R. W. Berliner (1968). *In vitro* permeability of medullary collecting ducts to water and urea. *Am. J. Physiol.*, **214**, 574–81.
Pappenheimer, J. R. (1953). Passage of molecules through capillary walls. *Physiol. Rev.*, **33**, 387–423.
Pennell, J. P., F. B. Lacy, and R. L. Jamison (1974). An *in vivo* study of the concentrating process in the descending limb of Henle's. *Kidney Int.*, **5**, 337–47.
Pennell, J. P., V. Sanjana, N. R. Frey, R. L. Jamison (1975). The effect of urea infusion on the urinary concentrating mechanism in protein-depleted rats. *J. Clin. Invest.*, **55**, 399–409.
Person, E. (1970). Water permeability in rat distal tubules. *Acta Physiol. Scand.*, **78**, 364–75.
Persson, A. E. G., B. Agerup, and J. Schnermann (1972). The effect of luminal application of colloids on rat proximal tubular net fluid flux. *Kidney Int.*, **2**, 203–213.
Persson, A. E. G., J. Schermann, B. Agerup, and N. E. Eriksson (1975). The hydraulic conductivity of the rat proximal tubular wall determined with colloidal solutions. *Pflügers Arch.*, **360**, 25–44.
Persson, E., and H. R. Ulfendahl (1970). Water permeability in rat proximal tubules. *Acta Physiol. Scand.*, **78**, 353–63.
Rajerison, R., J. Marchetti, C. Roy, J. Bockaert, and S. Jard (1974). The vasopressin-sensitive adenylate cyclase of the rat kidney. Effect of adrenalectomy and corticosteroids on hormonal receptor-enzyme coupling. *J. Biol. Chem.*, **249**, 6390–400.
Rector, F. C., Jr. (1976). Renal acidification and ammonia production; Chemistry of

weak acids and bases; Buffer mechanisms. In B. M. Brenner and F. C. Rector (Eds), *The Kidney*, Saunders, Philadelphia. Chap. 9.

Renkin, E. M., and J. P. Gilmore (1973). Glomerular filtration. In J. Orloff and R. W. Berliner (Eds), *Handbook of Physiology, Section 8: Renal Physiology*. Washington, American Physiological Society. pp. 185–248.

Rocha, A. S., and J. P. Kokko (1973). Sodium chloride and water transport in the medullary thick ascending limb of Henle: Evidence for active chloride transport. *J. Clin. Invest.*, **52**, 612–23.

Rocha, A. S., and J. P. Kokko (1974). Permeability of medullary nephron segments to urea and water: Effect of vasopressin. *Kidney Int.*, **6**, 379–87.

Roch-Ramel, F., F. Chomety, and G. Peters (1968). Urea concentration in tubular fluid and in renal tissue of nondiuretic rats. *Am. J. Physiol.*, **215**, 429–38.

de Rouffignac, C., J. P. Bonvalet, and J. Menard (1974). Renin content in superficial and deep glomeruli of normal and salt loaded rats. *Am. J. Physiol.*, **226**, 150–4.

de Rouffignac, C., S. Deiss, and J. P. Bonvalet (1970). Détermination du taux individuel de filtration glomérulaire des néphrons accessibles et inaccessibles à la microponction. *Pflügers Arch.*, **315**, 273–99.

de Rouffignac, C., and M. Imbert (1975). Données récentes sur les mécanismes de concentration et de dilution de l'urine. *J. Physiol. (Paris)*, **71**, 183A–255A.

de Rouffignac, C., and F. Morel (1969). Micropuncture study of water, electrolytes and urea movements along the loop of Henle in *Psammomys*. *J. Clin. Invest.*, **48**, 474–86.

Sackin, H., and E. L. Boulpaep (1975). Models for coupling of salt and water transport. Proximal tubular reabsorption. In: Necturus kidney. *J. Gen. Physiol.*, **66**, 671–733.

Schafer, J. A., and T. E. Andreoli (1972a). Cellular constraints to diffusion. The effect of antidiuretic hormone on water flows in isolated mammalian collecting tubules. *J. Clin. Invest.*, **51**, 1264–78.

Schafer, J. A., and T. E. Andreoli (1972b). The effect of antidiuretic hormone on solute flows in mammalian collecting tubules. *J. Clin. Invest.*, **51**, 1279–86.

Schafer, J. A., S. L. Troutman, and T. E. Andreoli (1974). Volume reabsorption, transepithelial potential differences, and ionic permeability properties in mammalian superficial proximal straight tubules. *J. Gen. Physiol.*, **64**, 582–607.

Schmidt, U., and U. C. Dubach (1969). Activity of (Na^+K^+)-stimulated adenosinetriphosphatase in the rat nephron. *Arch. Ges. Physiol.*, **306**, 351–65.

Schmidt-Nielsen, B., and R. O'Dell (1961). Structure and concentrating mechanism in the mammalian kidney. *Am. J. Physiol.*, **200**, 1119–24.

Schnermann, J., J. M. Horster, and D. Z. Levine (1968). The influence of sampling technique on the micropuncture determination of GFR and reabsorptive characteristics of single rat proximal tubules. *Arch. Ges. Physiol.*, **309**, 48–58.

Schwartz, M. M., and M. A. Venkatachalam (1974). Structural differences in thin limbs of Henle: physiological implications. *Kidney Int.*, **6**, 193–208.

Scott, W. N., D. L. Maude, I. Shehadeh, and A. K. Solomon (1964). Inulin and albumin absorption from the proximal tubules in Necturus kidney. *Science*, **146**, 1588–90.

Seely, J. F. (1973a). Variation in electrical resistance along length of rat proximal convoluted tubule. *Am. J. Physiol.*, **225**, 48–57.

Seely, J. E. (1973b). Effect of peritubular oncotic pressure on rat proximal tubule electrical resistance. *Kidney Int.*, **4**, 28–35.

Sonnenberg, H. (1972). Renal response to blood volume expansion: distal tubular function and urinary excretion. *Am. J. Physiol.*, **223**, 916–24.

Sonnenberg, H. (1973). Proximal and distal tubular function in salt-loaded deoxycortisone acetate-escaped rats. *J. Clin. Invest.*, **52**, 263–72.
Sonnenberg, H. (1974). Medullary collecting duct function in antidiuretic and in salt—or water—diuretic rats. *Am. J. Physiol.*, **226**, 501–6.
Sperber, I. (1944). Studies on the mammalian kidney. *Zool. Bidrag. Fran. Uppsala*, **22**, 249–431.
Spitzer, A., and E. E. Windhager (1970). Effect of peritubular oncotic pressure changes on proximal tubular fluid reabsorption. *Am. J. Physiol.*, **218**, 1188–93.
Starling, E. H. (1896). On the absorption of fluids from the connective tissue space. *J. Physiol., Lond.*, **19**, 312–26.
Starling, E. H. (1899). The glomerular functions of the kidney. *J. Physiol., Lond*, **24**, 317–30.
Stein, J., R. Osgood, S. Boonjarern, S. Cox, and T. Ferris (1974). Segmental sodium reabsorption in rats with mild and severe volume depletion. *Am. J. Physiol.*, **227**, 351–9.
Stein, J., R. Osgood, S. Boonjarern, and T. Ferris (1973). A comparison of the segmental analysis of sodium reabsorption during Ringer's and hyperoncotic albumin infusion in the rat. *J. Clin. Invest.*, **52**, 2313–23.
Stein, J. H., and H. J. Reineck (1974). The role of collecting duct in the regulation of excretion of sodium and other electrolytes. *Kidney Int.*, **6**, 1–9.
Stephenson, J. L. (1972). Central core model of the renal counterflow system. *Kidney Int.*, **2**, 85–94.
Stoner, L. C., M. B. Burg, and J. Orloff (1972). Ion transport in cortical collecting tubule; effect of amiloride. *Am. J. Physiol.*, **227**, 453–9.
Thurau, K., and J. Mason (1974). The intrarenal function of the juxtaglomerular apparatus. In K. Thurau (Ed.), *Kidney and Urinary Tract Physiology: MTP International Review of Science*, Butterworths, University Park Press, London. pp. 357–389.
Tisher, C. C. (1971). Relationship between renal structure and concentrating ability in the rhesus monkey. *Am. J. Physiol.*, **220**, 1100–6.
Tisher, C. C., and J. P. Kokko (1974). Relationship between peritubular oncotic pressure gradients and morphology in isolated proximal tubules. *Kidney Int.*, **6**, 146–56.
Tisher, C. C., R. W. Schrier, and J. S. McNeil (1972). Nature of urine concentrating mechanism in the macaque monkey. *Am. J. Physiol.*, **223**, 1128–37.
Tisher, C. C., and W. E. Yarger (1973). Lanthanum permeability of the tight junction (zonula occludens) in the renal tubule of the rat. *Kidney Int.*, **3**, 238–50.
Tune, B., and B. Burg (1971). Glucose transport by proximal tubules. *Am. J. Physiol.*, **221**, 580–5.
Tune, B., M. Burg, and C. Patlak (1969). Characteristics of p-aminohippurate transport in proximal renal tubules. *Am. J. Physiol.*, **217**, 1057–63.
Uhlich, E., C. A. Baldamus, and K. J. Ullrich (1969). Einfluss von Aldosteron auf den Sammelrohren der Säugetierniere. *Pflügers Arch.*, **308**, 111–26.
Ullrich, K. J. (1973). Permeability characteristics of the mammalian nephron. In J. Orloff and R. W. Berliner (Eds), *Handbook of Physiology, Section 8: Renal Physiology*, Washington, American Physiological Society. pp. 377–98.
Ullrich, K. J., G. Rumrich, and G. Fuchs (1964). Wasser permeabilität und transtubulärer Wasserfluss corticaler Nephronabschnitte bei verschiedenen Diuresezu tanden. *Pflügers Arch. Ges. Physiol.*, **280**, 99–107.
Ussing, H. H., D. Erlij, and U. Lassen (1974). Transport pathways in biological membranes. *A. Rev. Physiol.*, **36**, 17–49.

Ussing, H. H., and E. E. Windhager (1964). Nature of shunt path and active sodium transport path through frog skin epithelium. *Acta Physiol. Scand.*, **61**, 484–604.
Vander, A. J., R. L. Malvin, W. S. Wilde, and L. P. Sullivan (1958). Re-examination of salt and water retention in congestive heart failure. *Am. J. Med.*, **25**, 497–502.
Vereerstraeten, P., and M. Demyttenaere (1968). Effect of raising the transtubular oncotic gradient on sodium excretion in the dog. *Arch. Ges. Physiol.*, **302**, 1–12.
Wade, J. B., W. A. Kachadorian, and V. A. Discala (1977). Freeze–fracture electron microscopy: relationship of membrane structural features to transport physiology. *Am. J. Physiol.*, **232**, F77–83.
Walker, A., P. Bott, J. Oliver, and M. MacDowell (1941). The collection and analysis of fluid from single nephrons of the mammalian kidney. *Am. J. Physiol.*, **134**, 580–95.
Walker, A., C. Hudson, T. Findley, and A. N. Richards (1937). The total molecular concentration and NaCl concentration of fluid from different segments of the renal tubule of amphibia. *Am. J. Physiol.*, **118**, 121–9.
Wallenius, G. (1954). Renal clearance of dextran as measure of glomerular permeability. *Acta Soc. Med. Uppsala*, **59**, 1–91.
Warnock, D. G., and M. B. Burg (1977). Urinary acidification: CO_2 transport by the rabbit proximal straight tubule. *Am. J. Physiol.*, **232**, F20–5.
Welling, L. W., and J. J. Grantham (1972). Physical properties of isolated perfused renal tubules and tubular basement membrane. *J. Clin. Invest.*, **51**, 1063–70.
Wesson, L. G. Jr., and Anslow Jr. (1948). Excretion of sodium and water during osmotic diuresis in the dog. *Am. J. Physiol.*, **153**, 465–474.
Whittembury, G., F. A. Rawlins, and E. L. Boulpaep (1973). Paracellular pathway in kidney tubules: Electrophysiological and morphological evidence. In H. H. Ussing and N. A. Thorn (Eds), *Transport Mechanisms in Epithelia*, Academic Press, New York. pp. 577–88.
Wiederholt, M., H. Stolte, J. P. Brecht, and K. Hierholzer (1966). Mikropunktions untersuchungen über den Einfluss von Aldosteron, Cortison und Dexamethason auf die renale Natriumresorption adrenalektomierter Ratten. *Pflügers Arch.*, **291**, 43–63.
Windhager, E. E. (1964). Electrophysiological study of renal papilla of golden hamsters. *Am. J. Physiol.*, **206**, 694–700.
Windhager, E. E., and G. Giebisch (1961). Micropuncture study of renal tubular transfer of sodium chloride in the rat. *Am. J. Physiol.*, **200**, 581–90.
Windhager, E. E., and G. Giebisch (1976). Proximal sodium and fluid transport. *Kidney Int.*, **9**, 121–33.
Wirz, H. (1956). Der osmotische Druck in den corticalen Tubuli der Rattenniere. *Helv. Physiol. Pharmacol. Acta*, **14**, 354–62.
Wirz, H., and R. Dirix (1973). Urinary concentration and dilution. In J. Orloff and R. W. Berliner (Eds), *Handbook of Physiology, Section 8, Renal Physiology*, Washington, American Physiological Society. p. 415.
Wirz, H., B. Hargitay, and W. Kuhn (1951). Localization des Konzentrierungsprocesses in der Niere durch direkte Krioskopie. *Helv. Physiol. Pharmacol. Acta*, **9**, 196–207.
Woodhall, P. B., and C. C. Tisher (1973). Response of the distal tubule and cortical collecting duct to vasopressin in the rat. *J. Clin. Invest.*, **52**, 3095–108.
Wright, F. S., and G. Giebisch (1972). Glomerular filtration in single nephrons. *Kidney Int.*, **1**, 201–9.

Chapter 10

Hormones and Osmoregulation in the Vertebrates

W. N. HOLMES AND R. B. PEARCE

I. Introduction	413
II. The Neurohypophysial Hormones	415
A. Chemical composition	416
B. Site of production and release	418
C. Effects in the organism	419
D. Effects on the cells of responsive tissues	435
E. Molecular basis of hormonal action	452
III. The Adrenocortical Steroids	463
A. Chemical composition	464
B. Site of production and release	467
C. Effects in the organism	471
D. The cellular and molecular effects in response tissues	488
E. The possible roles of the proteins	504
Acknowledgments	507
References	508

I. INTRODUCTION

The maintenance of steady-state homeostatic conditions in all vertebrate organisms requires that the daily intakes of water and electrolytes by an organism are matched by their daily rates of loss from the body. Thus, the volume of water ingested via the gastrointestinal tract (I) in all terrestrial air-breathing vertebrates, and in vertebrates that breathe by a gill mechanism, must equal the sum of the volumes lost via renal excretory pathway (V), the faeces (F), the extrarenal excretory pathway (E), the respiratory surface (R) and the integument (S). Among the fresh-water vertebrates, however, a continuous osmotic influx of water often occurs across the integument and, if they have gills, a similar influx may also occur the respiratory epithelium. In these species, therefore, the fluxes of water across the respiratory (R) and the integumentary (S) surfaces will supplement the volume ingested via the gastrointestinal tract (I). Therefore, a general relationship for all vertebrates may be expressed as follows:

$$I = V + F + E \pm R \pm S \tag{1}$$

Using a similar rationale, the daily intakes of any osmotically active substance (x) must balance the sum of the net losses of this substance via all pathways. This relationship is expressed most simply for air-breathing vertebrates where the volume of water lost via the respiratory pathway does not involve the simultaneous loss of dissolved electrolytes.

Thus

$$i_x I = u_x V + f_x F + e_x F + e_x E + s_x S \tag{2}$$

where i_x represents the concentration of substance x in the volume of water ingested (I) and u_x, f_x, e_x and s_x represent the concentrations of x in the volumes of water (V, F, E and S) discharged from the organism.

The factors affecting the osmotic balance of organisms living in fresh-water or sea-water media are slightly more complex and the factors representing the movement of ions across the gill epithelium ($r_x R$) and the skin ($s_x S$) may have to be assigned either positive or negative values dependent upon whether the organisms live in fresh-water or marine environments. Thus, a generalized relationship for the electrolyte balance in all vertebrates may be stated as follows:

$$i_x I = u_x V + f_x F + e_x E \pm r_x R \pm s_x S \tag{3}$$

All vertebrate organisms live in continuously changing environments, particularly with respect to those factors influencing their osmotic balance. From time to time, therefore, some degree of obligatory change must occur in some of the factors represented in equations (1), (2) and (3). These facultative responses involve changes in the permeability and active transport properties of epithelial cells in the intestinal mucosa, the integument, the kidney tubules and the various extrarenal excretory organs that have evolved in the different vertebrate classes. Furthermore, the passage of water and solutes through these cells may be regulated by one or more hormones acting either separately or in combination on each of the target organs.

The principal hormones involved in these facultative regulatory processes are the neurohypophysial peptides and the adrenocortical steroids. We have attempted to review briefly the gross effects of these hormones on the responsive target tissues and their role in contributing to the overall osmotic balance. We have also tried to summarize the cellular and molecular bases of these hormonal effects in terms of their known actions *in vitro* in some responsive tissues.

II. THE NEUROHYPOPHYSIAL HORMONES

In mammals the neurohypophysial peptides cause increases in arterial blood pressure (vasopressor effect), reductions in the rate of urine flow (antidiuretic or hydro-osmotic effects), increases in the rate of urinary Na^+ excretion (natriuretic effect), contraction of smooth muscle cells in the uterine myometrium (oxytocic effect) and contraction of myoepithelial cells surrounding the mammary alveoli (milk-ejection or 'let-down' effect).

In non-mammalian vertebrates these hormones have been shown to have various effects on vascular and non-vascular smooth muscle and the movement of water and electrolytes across the skin and several renal and non-renal epithelial tissues.

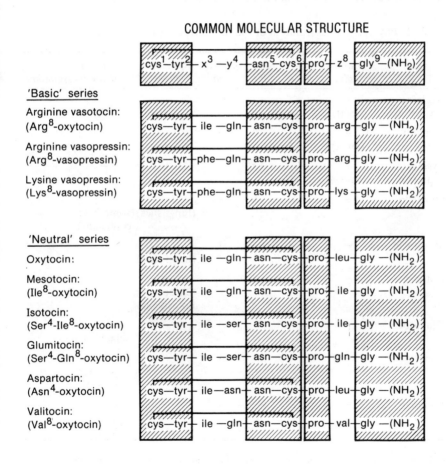

Figure 1 The structure of nine naturally occurring neurohypophysial peptides. The shaded portions represent the stable segments that are common to the known peptides

A. Chemical Composition

The molecular structures of the neurohypophysial peptides have been defined in many vertebrate species and in each case they consist of a pentapeptide ring with a tripeptide amide side chain. The pentapeptide ring contains 20 atoms and is joined by a disulphide bond between two half-*cistine* molecules situated at the 1 and 6 positions (Fig. 1). The specific amino acids located at the 3, 4 and 8 positions are variable and they bestow upon these molecules their particular physiological and pharmacological properties; the individual amino acids found at the 1, 2, 5, 6, 7 and 9 positions are the same in all known neurohypophysial hormones (Fig. 1). The phylogenetic distribution of these neurohypophysial hormones and their relative biological activities in several bioassay systems is summarized in Tables 1 and 2.

Table 1. Phylogenetic distribution of the neurohypophysial hormones in vertebrates

	Mammalia	Aves	Reptilia	Amphibia	Osteichthyes	Chondrichthyes	Agnatha
Arginine-vasotocin	+ (fetal)	+	+	+	+ (all)	+ (all)	+
Arginine-vasopressin	+ (most)	−	−	−	−	−	−
Lysine-pressin	+ (suina)	−	−	−	−	−	−
Oxytocin	+	(+)	(+)	(+)	(+) (lungfishes)	(+) (some holocephalans)	−
Mesotocin	−	+	+	+	+ (lungfishes)	−	−
Isotocin	−	−	−	−	+ (Teleosts & holosteans)	−	−
Glumitocin	−	−	−	−	−	+ (selachians)	−
Aspartocin	−	−	−	−	−	+ (some sharks)	−
Valitocin	−	−	−	−	−	+ (some sharks)	−

(+) = uncertain identity. In the absence of specific chemical identification the discrete identification of either oxytocin, mesotocin or both peptides is difficult due to the close similarity between the potencies of these peptides when they are bioassayed in the systems that are commonly used (cf. Table 2).

Table 2. Comparison of relative potencies of neurohypophysial hormones in various biological assays

	Biological activities (international units μmol^{-1})*								
	Rat anti-diuresis	Rat vaso-pressor	Rat uterus contraction	Rabbit milk ejection	Fowl vaso-depressor	Fowl oviduct contraction	Frog bladder 'natriferic' effect	Frog bladder 'water-balance' effect	Eel vaso-pressor (ventral aorta)
Oxytocin	5[a]	5[a]	450[a]	450[a]	450[a]	29[b]	450[c]	450[c]	450[i]
Mesotocin	1[a]	6[d]	289[d]	328[d]	489[d]	—	405[e]	450[k]	—†
Isotocin	0.18[f]	0.06[f]	150[f]	300[f]	320[f]	—	7[g]	9[g]	1350[i]
Arginine-vasotocin	250[h]	245[h]	115[h]	210[h]	285[h]	1640[b]	6750[c]	16000[c]	120[i]
Arginine-vasopressin	400[a]	400[a]	20[a]	70[a]	60[a]	320[b]	20[i]	21[c]	30[i]
Lysine-vasopressin	250[a]	270[a]	5[a]	60[a]	40[a]	29[a]	8[i]	5[c]	<1[i]

The letters denote the references in which the data are reported: [a]Boissonas and coworkers (1961); [b]Heller and Pickering (1961); [c]Bourguet and Maetz (1961); [d]Berde and Konzett (1960); [e]Guttmann and coworkers (1962); [f]Heller and coworkers (1961); [g]Follett and Heller (1964a, 1964b); [h]Berde and coworkers (1962); [i]Jard and Morel (1963); [j]Chan and Chester Jones (1969); [k]Bentley (1969b).

* The international unit for oxytocin and vasopressin is the amount of pituitary hormone contained in 0.5 mg of acetone-dried posterior pituitary powder. One milligram of pure synthetic oxytocin and vasopressin has an activity of 450 IU, or 1 IU of oxytocic potency is contained in 2.2 μg of synthetic oxytocin. One milligram of synthetic arginine-vasopressin has a pressor potency of 400 IU, or 1 IU represents either pressor or antidiuretic potency of 2.5 μg of arginine-vasopressin. All assays are based on these standards.

† Mesotocin causes a similar vasopressor response in the Asiatic eel, *A. Japonica*, and is slightly more potent than oxytocin in this species (D. K. O. Chan, unpublished results).

B. Site of Production and Release

The neurohypophysial hormones are produced by neurones whose cell bodies are concentrated in the single pre-optic nucleus of fishes and amphibians and in the paired supraoptic and the paraventricular nuclei of reptiles, birds and mammals. The axones of these cells run through the infundibular stalk to terminate in the neurohypophysis where the hormones are released into the perivascular space. The neurohypophysial hormones are probably synthesized in the cell bodies of the neurones and transported to the neurohypophysis in membrane-bound secretory granules. The membrane-bound granules contain the neurohypophysial hormones bound to their respective polypeptide carriers (neurophysins) and they are stored in this form in the swollen terminals of the axones situated in the neurohypophysis. Action potentials passing along the axones cause a depletion of granules in both the axones and their terminals and this depletion is believed to be

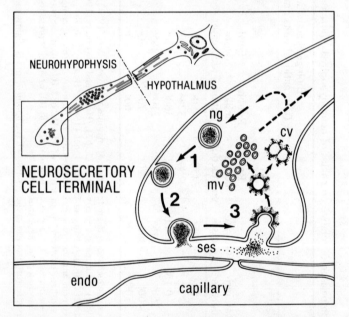

Figure 2 Schema showing the origin of microvesicles (mv) and the release of hormone, together with its associated carrier-protein, in the neurohypophysial cell. The neurosecretory granules (ng) migrate to the plasma membrane in the cell terminal. This is followed by the fusion of the membrane around the granule with the plasma membrane (1) and then the release of its contents by exocytosis (2) into the subendothelial space (ses). The membrane is retrieved through the formation of coated pits from the emptied vesicles (3). The coated vesicles (cv) then shed their coats to form microvesicles (mv) which may be either removed by lysosomal digestion or may be recycled to form once again vesicles containing neurosecretory granules (dashed line). This sequence of events has been termed 'exocytosis-vesiculation' (Douglas and Nagasawa, 1971)

associated with the release of neurohypophysial hormones from the terminals (Dyball and Koizumi, 1969; Douglas, 1973). The release of hormone is believed to occur by a process described as exocytosis-vesiculation (Douglas, 1973, 1974). This process is outlined diagrammatically in Fig. 2.

In the mammal, arginine-vasopressin is released from the neurohypophysis in response to a number of physiological changes. Verney (1947) showed that a 2% increase above the normal plasma osmolality was sufficient to reduce urine flow by 90% in the hydrated dog and he noted that the response was greater when the hypertonic saline was infused into the carotid artery rather than the malleolar vein. He therefore proposed that osmoreceptors were situated along the course of the carotid artery. Subsequent work suggested that osmoregulators might also be located in the supraoptic nuclei and certain cells of the liver (Brooks, Ushiyama and Lange, 1962; Niijima, 1969). Baroreceptors, and either stretch or volume receptors may also be implicated in the control of neurohypophysial hormone release (Share, 1961; Claybaugh and Share, 1973).

Although neurohypophysial peptides may be liberated in response to a variety of stimuli, the concensus suggests that a fine control of plasma osmolality is provided by atrial stretch receptors which continuously monitor blood volume. However, when large decreases in blood volume occur, the body may also be served by baroreceptors. In addition to these monitoring systems, osmoreceptors in the carotid artery, and perhaps the supraoptic nuclei, may respond to changes in the osmotic concentration of extracellular fluid. Thus, each of these receptors may transmit afferent impulses to the central nervous system to cause degranulation of the neurosecretory cells of the hypothalamo-neurohypophysial system.

C. Effects in the Organism

1. *Fishes*

Most fresh-water fishes must rid themselves continuously of the water that passively diffuses into their bodies by excreting a copious and dilute urine. It is unlikely, therefore, that the classical antidiuretic and hydro-osmotic actions on the kidney and the integument are of great importance to the survival of fresh-water fishes. On the other hand, these fishes must conserve electrolytes by maintaining low concentrations and by actively extracting essential ions such as Na^+ and Cl^- from their environments. It is not surprising, therefore, that unlike Amphibia, a water balance effect has not been seen following the injection of neurohypophysial peptides into fresh-water fishes. The fresh-water lamprey (*Lampetra fluviatilis*), for example, does not increase in weight after the injection of arginine-vasotocin, although this hormone may induce a net loss of Na^+ by increasing the urinary Na^+ concentration without causing concomitant change in the rate of urine flow (Bentley and Follett, 1962, 1963; Heller and Bentley, 1965).

In fresh-water fishes the turnover of Na^+ is low compared to the total mass of Na^+ within the organism. A net uptake of Na^+ and Cl^- from the environment ($-r_{Na}R$ and $-r_{Cl}R$ in equation (3)) occurs because the active uptake of these ions (f_{in}) across the branchial epithelium is 20–50% higher than their rates of passive efflux (f_{out}). Krogh (1937, 1939) postulated that this uptake of Na^+ and Cl^- by the gill epithelium of fresh-water teleosts was at least in part dependent on the simultaneous exchange of NH_4^+ and HCO_3^-. This was later confirmed by Garcia Romeu and Maetz (1964) and Maetz (1972).

Measurements of simultaneous influx and efflux of either Na^+ or Cl^- across the branchial epithelium of the intact goldfish (*Carassius auratus*) have suggested that arginine-vasotocin and isotocin have quite distinct effects on the overall electrolyte balance of the fish; isotocin predominantly stimulating the active uptake of Na^+ from the environment and arginine-vasotocin causing an increase in the net excretion of Na^+ that is probably mediated via the kidneys (cf. Maetz, 1956; Sawyer, 1970; Maetz and Lahlou, 1974).

It has been argued that since fresh-water teleosts may never experience the need to restrict urine flow, attempts to demonstrate an antidiuretic effect of neurohypophysial peptides in fresh-water fishes are meaningless (Maetz and Lahlou, 1974). In apparent support of this belief several studies have shown that high doses and, in some instances, quite low doses of neurohypophysial peptides may cause either an increase in urine flow, an increase in the glomerular filtration rate or both of these responses when injected into fresh-water teleostean and holostean fishes (Sexton, 1955; Holmes and McBean, 1963; Maetz and coworkers, 1964; Sawyer, 1966, 1970; Chester Jones and coworkers, 1971). Also, even though very low urine flow rates are characteristic of the marine teleost, the pituitaries from these fish contain a diuretic substance that increases glomerular filtration rate when injected into a fresh-water fish.

One of the most thorough studies on the regulation of kidney function in a freshwater teleost has been conducted on the African lungfish, *Protopterus aethiopicus* (Sawyer, 1970). In this study increases in glomerular filtration rate, the rate of urine flow and the rate of Na^+ excretion were observed following the injection of several doses of arginine-vasotocin; in contrast mesotocin had no significant effect on any of these parameters. Similar responses to 5 ng arginine-vasotocin per kg body weight have been recorded in the North American lungfish, *Amia calva* (Chester Jones and coworkers, 1971).

The possible role of neurohypophysial peptides in the regulation of renal excretion in teleosts becomes even more confusing when one considers the fact that a profound antidiuresis, due mainly to a reduction in glomerular filtration rate, occurs in several euryhaline teleosts during the period immediately *preceding* migration to the sea and *following* abrupt transfer to sea-water in the laboratory (R. M. Holmes, 1961; Holmes and McBean, 1963; Sharratt, Chester Jones and Bellamy, 1964; Holmes and Stainer, 1966; Lahlou, 1967; Babiker and Rankin, 1973). In the trout, at least, it is difficult to reconcile the experimentally observed diuretic effects of neurohypophysial peptides with the adaptive changes known to occur both in kidney function and the neurosecretory activity of the hypothalamo-hypophysial system during transfer from fresh-water to sea-water (Carlson and Holmes, 1962; Holmes and McBean, 1963; Lederis, 1964).

The results of some recent studies on the control of renal function in fresh-water and marine fishes go some way towards resolving this paradox. In the fresh-water European eel (*Anguilla anguilla*) very small intravenous doses of arginine-vasotocin, isotocin and oxytocin ranging from 4×10^{-13} to 10^{-8} g per kg body weight cause dose-dependent *antidiuretic* responses, while at higher doses they each induce a dose-dependent diuretic response. The dose-response curve for each hormone shows a marked inflexion at a dose of about 10^{-8} g per kg body weight (Babiker and Rankin, 1973). In a similar study on the fresh-water eel, diuretic and antidiuretic doses of arginine-vasotocin have been shown to cause commensurate increases and decreases respectively in the rate of urine flow, the glomerular filtration rate and the tubular glucose reabsorption maximum (Tm_G) values (Wales, Henderson and Chester Jones, 1973). These investigators found no evidence to suggest that arginine-vasotocin altered the permeability of the renal tubule and they concluded that the diuretic and antidiuretic responses reflected changes in the number of actively functioning glomeruli. When injected into the sea-water-adapted eel, the minimally active doses of each of the neurohypophysial hormones are in the range that causes diuresis and the normally antidiuretic doses are without any detectable effect on renal function (Babiker and Rankin, 1973). Thus, normally circulating levels of neurohypophysial hormones in the sea-water-adapted fish are presumed to have a sustained antidiuretic effect and the addition of more hormone to the circulation may raise the plasma concentrations into the range that has been noted to cause diuresis (Babiker and Rankin, 1973; Wales, Henderson and Chester Jones, 1973).

In contrast to fishes living in fresh-water, most marine fishes tend to dehydrate continuously while electrolytes in the surrounding sea-water diffuse into the animal through their highly vascularized branchial epithelia. The difficulty that this situation presents is further compounded by the intestinal uptake of ingested sea-water, an action which tends to lead to even further salt loading of the organism. Mechanisms designed to conserve water and, at the same time, provide for the continuous extrusion of these excess salts are clearly necessary to ensure the survival of the marine teleost. When living in sea-water, most teleosts seem to conserve some water by maintaining low urine flow rates. As we have seen, this antidiuresis may in some instances be attributed to the effects of neurohypophysial hormones. But, since the nephron of the fish kidney is unable to produce hypertonic urine, the kidneys play only a minor role in the relief of the osmotic burden incurred by marine fishes. That is to say, the factor $U_{Na}V$ in equation (3) is small and if the organism is to maintain osmotic independence from its environment, another factor to the right-hand side of this equation must increase.

In 1931, Ancel Keys showed that an isolated heart-gill preparation from the sea-water-adapted eel (*Anguilla anguilla*) could actively secrete Cl^- into the sea-water medium bathing the cells (Keys, 1931a, 1931b). Histological comparisons of the gill epithelium at the bases of the gill lamellae in fresh-water and marine teleosts have revealed the presence of cells that are rich in mitochondria; these cells which seem to be more abundant in species that have a need to excrete large quantities of salt extrarenally have been described as the 'chloride-secreting cells' (Keys and Wilmer, 1932; Morris, 1965). Keys and Bateman (1932) showed that the extrusion of Cl^- stopped when vasodilation was induced by adrenaline in perfused gills that were bathed in sea-water. Further anatomical studies of the branchial vasculature have revealed that the 'chloride-secreting cells' at the bases of the lamellae are supplied by blood flowing through the central compartments of the gill and that these vessels are distinct from those supplying the respiratory regions in the gill lamellae. These observations have led to the suggestion that the pattern of blood flow in different parts of the branchial apparatus may be controlled by the neurohypophysial hormones and that under some circumstances more blood may be routed via the mitochondria-rich 'chloride-secreting cells' (Steen and Kruysse, 1964; Maetz and Rankin, 1969; Richards and Fromm, 1969). Thus, it has been proposed that under conditions of continuous salt loading, such as those which occur in the marine teleost and in the euryhaline fresh-water teleost migrating from fresh-water to sea-water, osmoreceptors will sense the salt-loaded condition of the fish and the resulting release of vasoactive neurohypophysial hormone will cause an increase in blood flow through the central compartment of the gill and the 'chloride-secreting cells' will be stimulated to secrete Na^+; that is to say, the factor $+r_{Na}R$ in equation (3) will increase.

In fish, as in mammals, the earliest efforts to identify the actions of neurohypophysial hormones relied upon the injection of large doses of mammalian neurohypophysial extracts. The response to these extracts was similar to that observed in mammals, the primary effect being an elevation of blood pressure (cf. Magnus and Schaffer, 1901; MacKay, 1931; Nelson, 1934). However, whereas the increase in blood pressure that was observed in mammals is now considered to have been a pharmacological response, in fishes the primary physiological role of their native neurohypophysial hormones may involve some type of haemodynamic regulation (Sawyer, 1967; Rankin and Maetz, 1971). Detailed dose-dependent changes in ventral aortic pressures have been recorded in the fresh-water eel (*Anguilla anguilla*) given several doses of either isotocin, oxytocin, arginine-vasotocin, arginine-vasopressin or lysine-vasopressin (Chan and Chester Jones, 1969). The vasopressor effects of low doses of native peptide hormones have also been recorded in other species of boney fishes as well as cyclostomes and elasmobranches (see reviews by Bentley, 1971; Maetz and Lahlou, 1974).

Only indirect methods have been used to assess the hypothesis that naturally occurring neurohypophysial hormones in marine and euryhaline fishes may act primarily to shift the pattern of blood flow through the branchial apparatus when the organism needs to excrete a burden of excess electrolytes. Using the method developed by Chan and Chester Jones (1969), Maetz and Rankin (1969) were able to measure alternately the blood pressures in the dorsal and ventral aortae in the sea-water-adapted European eel (*Anguilla anguilla*). Their results suggest that arginine-vasotocin may increase the supply of blood to the regions at the base of the gill lamellae and that this in turn may facilitate any enhanced secretion of Na^+; in contrast, the presence of adrenaline may reduce blood supply to this region and cause the activities of the 'chloride-secreting cells' to cease. Also, using an isolated perfused gill preparation, Rankin and Maetz (1971) showed that concentrations of 10^{-11} to 10^{-9} mol l^{-1} vasotocin and 10^{-13} mol l^{-1} isotocin were able to elicit changes in branchial blood flow.

2. Amphibians

When living in a fresh-water environment amphibians tend to gain water continuously through a sustained osmotic influx of water across the skin and their primary physiological response to this situation is to get rid of the excess water by excreting an equivalent volume of dilute urine (equation (1)). Some anuran amphibians, however, show a marked capacity to accelerate the rate of water uptake across their skin and this ability enables them to restore their body water composition to normal following periods of dehydration. Thus, when a terrestrial frog or toad is transferred to fresh-water, an increase in body weight occurs during the next few hours and in 1921 Brunn showed that this increase was accelerated when the animals were injected with extracts of the mammalian neurohypophyses. These findings were later confirmed by Novelli (1936). This water balance response, or 'Brunn effect', seems to be better developed in species that spend more of their time in the terrestrial habitat (Table 2).

In those anuran amphibians that show a pronounced 'Brunn effect', the primary effect of the neurohypophysial extract is to promote an accelerated influx of water through the skin (hydro-osmotic effect), accompanied by a stimulated increase in the active uptake of Na^+ from the environmental water (natriferic effect). Studies on dehydrated desert toads, *Bufo punctatus*, have indicated that the skin is not uniformly permeable to water; skin in the region of the ventral pelvis absorbing water at over $400\,\mu l\,cm^{-2}\,h^{-1}$ whereas absorption through the skin from the ventral thorax is too small to measure (McClanahan and Baldwin, 1969). Similar differences in permeability have been recorded between samples of skin from the dorsal and ventral surfaces of the toad, *Bufo marinus*, and four species of tree frogs, *Hylidae* (Bentley and Yorio, 1976; Yorio and Bentley, 1977). These differential permeabilities are, at least in the tree frogs, associated with correspondingly different responses to arginine-vasotocin (Yorio and Bentley, 1977).

The physiological action of neurohypophysial extract may not be restricted to the skin. Collection of urine during the period immediately following the administration of extract indicates that the stimulated increase in water uptake across the skin is accompanied by a reduced discharge of urine from the cloaca. To this end the neurohypophysial extract may act simultaneously at three distinct sites. As in the mammals, an antidiuresis may be induced through the action of the neurohypophysial hormones on the permeability of water through the distal portions of the nephron. In addition, the hormones may also cause constriction of the afferent glomerular arterioles and so reduce the rate of glomerular filtration. This latter effect may act either to reduce temporarily the number of active glomeruli in the kidneys or to diminish uniformly the rate of filtration in all glomeruli.

During the time that urine remains in the bladder, however, the neurohypophysial extract may act to cause an even further decrease in volume. This is caused by a dual effect of the extract on the mucosal cells lining the urinary bladder; first of all, hormonal peptides in the extract may act to increase the passive osmotic flux of water between the hypotonic urine in the bladder and the more concentrated extracellular fluid on the serosal side of the cells lining the bladder (hydro-osmotic effect), and secondly, the active uptake of Na^+ from urine (natriferic effect) may induce the movement of even more water from the urine into the extracellular compartment.

Similar neurhypophysial extracts may not cause the same degree of response in the skin, the nephron and the urinary bladder of all amphibian species and each tissue from the same species may respond quite differently to individual native peptides contained in a neurohypophysial extract. Therefore, we must examine in more detail the response of each tissue to the different peptides and attempt to evaluate the role that each hormone might play in ensuring adequate osmoregulatory control in those species that have been studied.

In 1941, Heller compared the biological activities of extracts derived from frog pituitaries with those of the commercial preparations, Pitressin and Pitocin, derived from mammalian tissues. He found that although the frog extracts contained significant antidiuretic activity when tested in rats, the mammalian preparations caused little or no increase in body weight when injected into frogs whereas extracts of frog pituitaries were extremely potent in this regard. Convinced that the frog pituitaries contained something that was chemically distinct from the antidiuretic material extracted from mammalian pituitaries, Heller called the substance 'amphibian water balance principle.' Nearly 20 years later, his scientific judgement was vindicated when the biological activity of vasotocin was compared to that of neurohypophysial extracts known to contain 'amphibian water balance principle'; in a series of several bioassay systems their pharmacological spectra of relative activities were congruent and in particular vasotocin was a potent stimulant of water uptake across the amphibian skin (Table 2 and Pickering and Heller, 1959). Thus, the existence of a neurohypophysial peptide was predicted on the basis of the distinct biological properties contained in the tissue extracts and this same peptide was independently synthesized *in vitro* before it was chemically isolated from the neurohypophysial tissue (cf. Heller, 1941; Katsoyannis and du Vigneaud, 1958).

A substance with oxytocic properties has also been identified in amphibian neurohypophysial extracts. This substance was shown by Follett and Heller (1964a, 1964b) to be pharmacologically similar to 8-isoleucine oxytocin and was chemically identified as this compound and given the name mesotocin by Acher and coworkers (1964).

A comparison of the biological activities of hormones present in amphibian neurohypophysial extracts has shown that the water balance effect on the whole organism as well as the individual natriferic and hydro-osmotic effects on the urinary bladder are each stimulated most effectively by vasotocin; the doses of mesotocin and oxytocin necessary to elicit the same response being at least one order of magnitude greater than the corresponding dose of vasotocin (Bentley, 1969a, 1969b).

The increase in body weight induced by these hormones, however, reflect their net simultaneous effects on the kidney tubules as well as the skin and the urinary bladder. But, since each of these physiological responses is directed toward water retention, it may be reasonable to assume that these hormones, particularly vasotocin, has a similar water-conserving effect on the nephron. Direct measurements of the effects of neurohypophysial hormones on renal function in amphibians have shown that a variety of neurohypophysial peptides cause a reduction in urine flow rates in both anurans and urodeles; these responses, however, seem to be greater in terrestrial and amphibious species that they are in more exclusively aquatic forms such as *Xenopus laevis* and *Necturus maculosus*. Although mammalian peptides may produce antidiuretic effects in several amphibious and terrestrial species, amphibian neurohypophysial extract and vasotocin are the most active in this regard. In the bullfrog (*Rana catesbiana*) and the American toad (*Bufo marinus*), the injection of neurohypophysial extract equivalent to as little as one-hundredth of a gland will cause antidiuresis and in the bullfrog and the European frog (*Rana esculenta*) does of vasotocin estimated to produce plasma concentrations between 10^{-11} and 10^{-10} mol l^{-1} have been shown to have similar effects; these plasma concentrations of vasotocin are similar to those found in the plasma of three species of toads and frogs, *Bufo marinus, Rana cates-*

beiana and *Xenopus laevis*, following dehydration (Sawyer, 1957a, 1957b; Jard, Maetz and Morel, 1960; Uranga and Sawyer, 1960; Jard and Morel, 1963; Bentley, 1969a). The reduced rates of urine flow are caused by a reduction in glomerular filtration rate and an increase in the reabsorption of free water from the distal tubular fluid. The decreases in glomerular filtration caused by vasotocin are probably due to a temporary decrease in the number of actively functioning glomeruli (Richards and Schmidt, 1924; Sawyer, 1951; Jard, 1966). Vasotocin may induce changes in the distal tubular reabsorption of water, however, without causing concomitant changes in glomerular filtration rate (Sawyer, 1957a, 1957b; Jard, Maetz and Morel, 1960). Since the tubular fluid in the distal nephron is hypo-osmotic with respect to plasma, an induced change in the osmotic permeability of the mucosal surface of the distal tubular cells will result in an efflux of water similar to that described for the epithelial cells of the skin and the bladder (Walker and coworkers, 1937; Whittembury, Sugino and Solomon, 1960; Whittembury, 1962).

When present in high concentrations both mesotocin and oxytocin may act to antagonize the antidiuretic effects of vasotocin by causing increase in urine flow (Uranga and Sawyer, 1960; Morel and Jard, 1963; Jard, 1966). However, neither oxytocin nor mestotocin seems to affect water reabsorption in the distal tubule and the diuresis appears to be caused through a direct vasoactive effect of the hormones on the glomerular circulation.

It is clear, therefore, that recovery from the brief periods of dehydration experienced by some amphibious and terrestrial anurans is dependent upon their return to fresh-water. If these species are exposed indefinitely to hypertonic saline, however, they will continue to dehydrate and will eventually die. In contrast, *Rana cancrivora*, the crab-eating frog of Southeast Asia, can live indefinitely in sea-water (Gordon, Schmidt-Nielsen and Kelly, 1961; Gordon and Tucker, 1965; Dicker and Elliott, 1970a).

During adaptation to sea-water, *Rana cancrivora*, like the marine elasmobranchs, develops a degree of physiological uremia that is sufficient to maintain the osmolality of its extracellular fluid slightly greater than that of the surrounding sea-water (Gordon, Schmidt-Nielsen and Kelly, 1961). In this way, the organism avoids dehydration by sustaining a continuous osmotic influx of water across the skin. High plasma urea concentrations are maintained by increasing the rate of urea synthesis in proportion to the increase in the osmotic concentration of the surrounding sea-water and by increasing the reabsorption of urea from urine entering the bladder (Balinsky, Dicker and Elliott, 1972; Chew, Elliott and Wong, 1972; Colley and coworkers, 1972).

Although the plasma vasotocin concentration in *Rana cancrivora* increases after brief exposure to sea-water, the permeability of the skin, unlike the skin of other anurans, does not increase in response to the presence of this hormone (Elliott and Chew in Elliott, 1977). When exposed to neurohypophysial hormones, however, the permeability of the isolated bladder from *Rana cancrivora* increases with respect to both water and urea (Dicker and Elliott, 1970b; Chew, Elliott and Wong, 1972). With respect to water permeability, dose-response studies indicate that this tissue is at least two orders of magnitude more sensitive to arginine-vasotocin than either oxytocin or arginine-vasopressin (Dicker and Elliott, 1970b). Thus, Elliott and her coworkers have proposed that the survival of *Rana cancrivora* during the early stages of adaptation to sea-water is due mainly to increased rates of water and urea reabsorption from urine in the bladder and that this increase is induced by released neurohypophysial hormones. Experiments with cystectomized *Rana cancrivora* have tended to support this hypothesis, for when the urinary bladders are removed prior to exposure of the frogs to sea-water, the physiological uremia does not develop and the animals perish. In contrast, cystectomized *Rana cancrivora* survive indefinitely when maintained in fresh-water. Thus, the early stages of adaptation may depend entirely on hormonally induced water reabsorption and a concomitant uptake of urea from the urinary bladder (Elliott, 1977).

3. *Reptiles*

The osmoregulatory organs of reptiles include the kidney, the urinary bladder, the cloaca and the salt glands. The neurohypophysial hormones that have been isolated from reptilian neurohypophysial tissues are arginine-vasotocin, mesotocin and oxytocin and their physiological actions seem to be confined to the kidney and possibly the cloaca.

Individuals of some species of reptiles show a wide range of urine flow rates under a variety of environmental and physiological conditions. For example, the urine flow rate in the Australian lizard, *Trachysaurus rugosus*, may increase almost fifty-fold under conditions of water loading and yet when water intake is restricted, the volume of urine excreted may be too small to measure accurately (Bentley, 1959). Such variations in urine flow rates may be caused by changes in the glomerular filtration rate and the rate of water reabsorption from filtrate in the distal portions of the tubule. As in the amphibia, modification of the glomerular filtration rate may be attributed either to increases and decreases in the pressure of blood perfusing the glomeruli or to changes in the actual number of actively functioning glomeruli. The latter process has been shown to occur in the snake, *Natrix sipedon*, and the turtle, *Pseudemys scripta* (Lebrie and Sutherland, 1962; Dantzler and Schmidt-Nielsen, 1966; Dantzler, 1967a).

The amount of water reabsorbed from the glomerular filtrate may also vary between species; in the well-hydrated horned toad 45% of the filtrate is reabsorbed, whereas even under conditions of dehydration this value may rise to only 60% (Roberts and Schmidt-Nielsen, 1966). In contrast, the reabsorption of water in the renal tubular fluid of the Australian lizard may increase from 40% to more than 95% of the volume filtered under conditions of high and low urine flow rates respectively (Shoemaker, Licht and Dawson, 1966). Tubular reabsorption processes have been examined by 'stop-flow' analysis in *Natrix sipedon* and the results suggest that during antidiuresis additional water is reabsorbed from the filtrate in the distal nephron (Dantzler, 1967b).

The neurohypophysial peptides have an antidiuretic action on the reptilian kidney and this may be attributed to either reduced glomerular filtration, increased reabsorption of water from the distal tubule, or both of these mechanisms working in concert (Bentley, 1971). However, direct evidence that an antidiuretic hormone of hypothalamic origin may influence the pattern of renal excretion in reptiles has been derived from studies on only one species, the Australian lizard, *Amphibolurus ornatus* (Bradshaw, 1975). Although severing the hypothalamo-hypophysial tracts in this species does not cause any change in the rate of urine flow as long as the animals are well hydrated, an antidiuresis does fail to develop when they are loaded with saline instead of distilled water. The antidiuresis that occurs when intact lizards are given saline appears to be entirely attributable to a reduction in glomerular filtration rate and the administration of arginine-vasotocin to the lizards with severed hypothalamic tracts causes a reduction in glomerular filtration rate and a commensurate decrease in the rate of urine flow.

The renal effects of purified arginine-vasotocin have been most extensively studied in only one species of fresh-water snake, *Natrix sipedon* (Dantzler, 1967a). Low pressor doses of arginine-vasotocin cause reductions in the glomerular filtration rate without any discernible effect on the systemic circulation. Even when somewhat higher doses or arginine-vasotocin are given, the $C_{H_2O}:C_{inulin}$ ratio continues to be less than unity for some time after glomerular filtration rate has returned to the control level. These observations suggest that the effect of the hormone on tubular water reabsorption persists longer than its effects on the glomerulus. Indeed, the hormone-induced alterations in tubular water reabsorption may make progressively larger contributions to the overall reduction in urine flow as the kidney is exposed to higher and higher concentrations of vasotocin (Dantzler, 1967a). Other reptiles may respond differently. In the fresh-water turtle (*Pseudemys scripta*) doses of neurohypophysial extract equivalent to 2.9 mU vasopressor activity per kg body weight have a tubular effect and when the dosage is increased both glomerular and tubular responses occur (Dantzler and Schmidt-Nielsen, 1966). A similar pattern of response is seen in the fresh-water turtle subjected to dehydration or salt loading; the increase in tubular permeability to water preceding any decrease in glomerular filtration rate and persisting after the glomerular filtration rate has returned to normal (Dantzler and Schmidt-Nielsen, 1966).

If arginine-vasotocin has the ability to reduce the glomerular filtration rate at doses which do not affect the systemic circulation, then the hormonal effect must be confined to vasoconstriction of the afferent arteriole. However, it has been suggested that the hormone may also alter the number of functioning nephrons but the manner in which a hormone may cause the intermittent functioning of some nephrons is not known (LeBrie and Sutherland, 1962; Dantzler and Schmidt-Nielsen, 1966).

There is little direct evidence to suggest that neurohypophysial hormones influence the uptake of Na^+ across the tubules of the reptilian kidney. The $C_{Na}:C_{inulin}$ ratio decreases significantly following the injection of arginine-vasotocin into the fresh-water snake, *Natrix sipedon* (Dantzler, 1967a).

Uptake of Na^+ and water may also occur across the mucosal epithelia of the cloaca and urinary bladder. However, because of their anatomical association with each other and with the large intestine, studies on cloaca and urinary bladder of reptiles are difficult. Braysher and Green (1970) have shown that in the Australian lizard, *Varanus gouldii*, arginine-vasotocin may cause a doubling of the rates of Na^+ and water reabsorption when isosmotic Ringer solution is placed in the cloaca; water uptake from the reptilian cloaca, however, may be secondary to the active transport of Na^+ from the mucosal to serosal sides of the epithelial cells lining the organ (Bentley and Schmidt-Nielsen, 1965; Jungueira, Malnic and Monge, 1966).

The urinary bladder present in some species of chelonian and lacertilian reptiles can serve to actively accumulate Na^+ (Brodsky and Schlib, 1960; Bentley, 1962). In this respect, the reptilian bladder resembles the urinary bladder in toads and frogs. But, in contrast to the amphibian tissues, no reports of enhanced Na^+ transport and water reabsorption have been reported following treatment with neurohypophysial peptides. Although extrarenal sites for salt extrusion are important in the osmoregulation of many species of *Chelonia* and *Squamata*, they do not appear to be under the control of neurohypophysial hormones (Dantzler and Holmes, 1974).

4. Birds

In an early series of experiments, Burgess, Harvey and Marshall (1933) showed that the injection of mammalian neurohypophysial extracts into normal hydrated domestic fowl caused a decrease in urine flow. This effect was attributed to a probable decrease in glomerular filtration rate as well as a stimulation of water reabsorption across the distal tubule. High doses of vasopressin (Pitressin) and oxytocin (Pitocin) have similar antidiuretic effects in the intact fresh-water-loaded Pekin duck; but, while low doses of Pitressin had no antidiuretic effect, low doses of oxytocin significantly increased the rate of urine flow (Holmes and Adams, 1963). The neurohypophysis in birds, however, contains arginine-vasotocin and either oxytocin or mesotocin and when tested in the hydrated chicken, arginine-vasotocin is a much more potent antidiuretic agent than the mammalian antidiuretic hormone, vasopressin, and only high doses of oxytocin elicit a significant antidiuresis (Munsick, Sawyer and van Dyke, 1960).

When small amounts of hormone are injected into the renal portal vein, the renal tubules are exposed to the hormone before it passes into the general circulation and returns to the glomerulus at a much lower plasma concentration. Under these circumstances, arginine-vasotocin causes a decrease in urine flow without causing any simultaneous change in either the glomerular filtration rate or the rate of renal blood flow; the antidiuresis is therefore attributed primarily to an increase in the tubular reabsorption of water (Skadhauge, 1964). Much higher doses of vasotocin were required to increase the rates of glomerular filtration and renal blood flow.

There is an abundance of evidence to indicate, however, large doses of neurohypophysial peptides do cause decreases in the blood pressure of a variety of bird species, including pigeons, penguins, emus, cormorants and chickens (Waring, Morris and Stephens, 1956; Woolley, 1959). Although such decreases in blood pressure could cause reduced rates of renal blood flow and glomerular filtration, there is no direct evidence to support the view that the antidiuresis induced by arginine-vasotocin is induced by this mechanism.

The effects of neurohypophysectomy on the renal function in birds has only been studied in the domestic fowl and the Pekin duck (Shirley and Nalbandov, 1956; Wright and coworkers, 1967; Bradley, Holmes and Wright, 1971). Neurohypophysectomy in both species produces an immediate and profound polydipsia and polyuria. During the first few days after removal of the neurohypophysis from the duck, the rates of water intake and urine flow rise to values that are many times greater than normal. But, one week after surgery these rates start to diminish and at fourteen days after surgery they have stabilized at levels which are approximately three times higher than those found in intact birds (Bradley, Homes and Wright, 1971). A similar polydipsic response occurs in the neurohypophysectomized chicken but the decline is slower than that observed in the duck (cf. Shirley and Nalbandov, 1956; Bradley, Holmes and Wright, 1971). A marked polydipsia has also been observed in the chicken following ablation of portions of the rostral hypothalamus (Ralph, 1960). In contrast to the immediate effects of neurohypophysectomy, however, less than normal quantities of water are consumed for a few days following placement of electrolytic lesions, but once the polydipsia is established, it continues unabated. This observation in birds reinforces the earlier observation in mammals that destruction of the neurosecretory neurones in the hypothalamus produces a condition that is more reminiscent of *diabetes inspidus* than that which is produced by removal of the neurohypophysis (cf. Verney, 1947; Ralph, 1960).

The intramuscular administration of vasopressin to neurohypophysectomized chickens and ducks causes an immediate but short-lived decrease in urine flow (Shirley and Nalbandov, 1956; Bradley, Holmes and Wright, 1971). The injected hormone seems to be metabolized quite rapidly and in order to sustain an antidiuresis, successive doses of hormone must be injected at frequent intervals (cf. Holmes and Adams, 1963). These observations are consistent with the low biological half-lives ($T_{1/2}$) that have been recorded for oxytocin ($T_{1/2} = 9$ min) and arginine-vasopressin ($T_{1/2} = 13$ min) in the chicken (Hasan and Heller, 1968). A comparison of the effects of different neurohypophysial hormone preparations injected into the neurohypophysectomized duck reveals that the native antidiuretic hormone, arginine-vasotocin, is at least one order of magnitude more potent than vasopressin (Pitressin) when judged on the basis of its ability to diminish the rate of urine flow to the level found in intact birds. Also, doses of vasopressin that are sufficient to reduce the urine flow rates to normal cause changes in the rates of Na^+ excretion whereas no such effects are observed with similarly effective doses of arginine-vasotocin. In contrast, even high doses of oxytocin have no significant effect on either the rate of urine flow or the rate of Na^+ excretion in the neurohypophysectomized duck (Bradley, Holmes and Wright, 1971).

5. Mammals

The antidiuretic effect of the neurohypophysial hormones in mammals is restricted to a hydro-osmotic action on the cells of the collecting tubules in the kidney. This action is mediated through the release of either arginine- or lysine-vasopressin and the resulting reduction in urine flow may be accompanied by a proportionate increase in urine osmolality. The pressor actions of the neurohypophysial hormones are seen only after the injection of relatively high doses. These effects, unlike those recorded in fish, are considered to be pharmacological and the changes in blood pressure and renal blood flow that may be brought about by these large doses of hormone are not believed to be involved in the normal modification of urine production.

The occurrence of an antidiuretic response following the injection of neurohypophysial extracts was first reported, quite independently, by Farini (1913) and von den Velden (1913). Both of these physicians noted a frequent coincidence between the occurrence of high rates of urine flow and the presence of pathological lesions in the neurohypophysis. Following the injection of neurohypophysial extracts into patients suffering *diabetes insipidus*, they observed a reduction in urine flow, an increase in the specific gravity of the urine and an alleviation of thirst.

These observations were later confirmed in many other clinical and experimental studies (see historical review by Heller, 1974) but they did not resolve the question as to whether the high rate of urine flow or the high rate of drinking was the primary cause of the syndrome identified as *diabetes insipidus*. In 1924, Starling and Verney were able to separate the physiological functions of urine flow and thirst in the dog by measuring the production of urine from an isolated kidney that was perfused by a heart–lung preparation. They showed that such a perfused kidney normally excreted a copious and dilute urine but, when an extract of the neurohypophysis was perfused through the isolated kidney, the rate of urine flow was diminished and the concentration of Cl⁻ increased. Under conditions of these experiments, thirst could not have possibly played a role in the control of urine production.

It was recognized quite early that removal of the neurohypophysis from experimental animals often resulted in just a transient high rate of urine flow and studies on the cat showed that an experimental syndrome reminiscent of persistent *diabetes insipidus* could be produced only after destructive bilateral lesions were placed in the ventral hypothalamus (Fisher, Ingram and Ranson, 1938). Immediately after section of the supraoptico-hypophysial tracts, the animal goes through a period of polydipsia and polyuria; this is followed by a brief period when the fluid exchange is within normal limits and finally a permanent phase of polydipsia and polyurea ensues (O'Connor, 1962). In contrast, removal of the neurohypophysis may cause only a transient period of profound polydipsia and polyuria, and only slightly higher than normal states persist indefinitely. These results were interpreted to indicate that neurohypophysial function depended upon intact innervation of the neurohypophysis by the supraoptico-hypophysial tracts. The true relationship between the hypothalamus and the hormonal secretions of the neurohypophysis, however, did not become clear until much later when the principles of neuroendocrine secretions were established (Fig. 2).

Among all of the vertebrates, only the kidneys of birds and mammals can secrete urine that is hypertonic to plasma. In these two classes, this ability is due to the presence of the loop of Henle, the geometrical arrangement of which permits two counter-current exchange systems to operate; one of them operates continuously between the ascending and descending limbs of the loop of Henle and the other operates intermittently and facultatively between the ascending limb and the collecting duct. The first counter-current mechanism is a multiplier system that, according to one theory, creates a pool of high solute concentrations in the medullary and papillary interstitium by actively transporting electrolytes out of the ascending limb into adjacent compartments of the descending limb (Wirz, Hargitay and Kuhn, 1951; Gottschalk and Mylle, 1959; Burg and Green, 1973). As a result, hypotonic fluid emerges from the ascending limb and enters the distal convoluted tubule and a concentrated sink of osmotically active material is created in the hairpin bend of the loop of Henle and in the surrounding interstitial fluid of the medulla and papilla. The second counter-current system functions to a variable extent depending upon the permeability of the cells of the collecting duct. Thus, in the absence of vasopressin, although the rate of solute excretion is proportional to the rate of urine flow, the urine is always hypo-osmotic with respect to plasma; in contrast, at the low urine flow rates made possible through the presence of neurohypophysial peptides and the attendant increase in the permeability of the collecting duct, a consistently hyperosmotic urine is produced. Any increase in the permeability of these cells to water will promote some osmotic equilibration between the hyperosmotic fluid in the medullary interstitium and the luminal fluid passing down the collecting duct. Such changes in permeability are stimulated by vasopressin and the resulting antidiuretic response is both immediate and graduated according to the extent of the permeability changes evoked by the hormone.

Samples collected by micropuncture have revealed that fluid entering the distal convoluted tubule in laboratory rats is osmotically dilute both in the presence and absence of arginine-vasopressin (Fig. 3). Indeed, the concentration of the distal tubular fluid does not equal that of plasma until it has travelled at least half way along the length of the distal tubule and in dogs and rhesus monkeys the fluid remains hypotonic until it reaches the collecting ducts (Walker and coworkers, 1941; Wirz, 1956, 1957; Gottschalk and Mylle, 1959; Clapp and Robinson, 1966; Bennett, Brenner and Berliner,

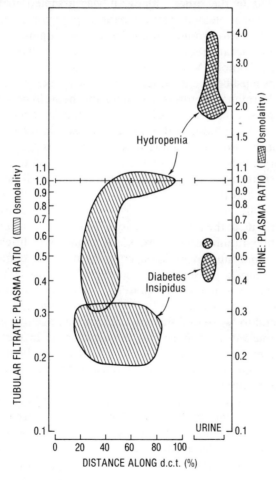

Figure 3 A comparison of the osmolality of distal tubular fluid and the urine: plasma osmolality ratio under conditions of high and low urine flow rates in the rat. After Gottschalk (1961); modified

1968). Thus, the facultative reabsorption of water leading to the excretion of hypertonic urine in mammals must take place primarily in the collecting ducts, although in some species vasopressin may also influence the permeability of some of the cells situated towards the end of the distal convoluted tubule (Fig. 3).

Berliner and Davidson (1957) have shown that the mechanism of urine concentration which accompanies the antidiuretic response to vasopressin is separable from the effect of the hormone on free water absorption. They found that when glomerular filtration was abruptly reduced by compression of one of the renal arteries in a dog undergoing water diuresis, the urine coming from the kidney supplied by this vessel became hypertonic; since diuresis persisted in the other kidney, it was assumed that no arginine-vasopressin had been released from the neurohypophysis. Similarly, rats with a hereditary inability to produce biologically active arginine-vasopressin can nevertheless concentrate their urine when the glomerular filtration rate is reduced by dehydration (Valtin, 1967). The conservation of water under these conditions must be due to either an intrinsic property of the kidney or to the action of hormonal factors other than the neurohypophysial peptides. Arginine-vasopressin can, however, lead to a further concentration of the urine beyond the level induced through a reduction in renal blood flow.

The fact that the osmotic pressure of kidney tissue is greatest in the region of the papillary tip has been known since the turn of the century but its physiological significance was not fully appreciated until nearly fifty years later (cf. Filehne and Biberfeld, 1902; Hirokawa, 1908; Wirz, Hargitay and Kuhn, 1951). Solutes other than Na^+, K^+, HCO_3^- and Cl^-, however, are now known to contribute to the total osmolarity of renal tissue and in the medullary and papillary interstitium, urea is particularly abundant (Ullrich and Jarausch, 1956). The idea that neurohypophysial peptides may help to maintain the high medullary and papillary osmotic pressure gradient has commanded some recent attention. A sustained water diuresis such as may occur in animals that either suffer from genetic diabetes insipidus or are subjected to chronic water loading may tend to 'wash out' the corticomedullary osmotic gradient. Single doses of arginine-vasopressin do not inhibit diuresis in these animals and antidiuretic responses may occur only after the gradient has been re-established through the continued administration of arginine-vasopressin (Harrington and Valtin, 1968).

The presence of urea in the inner kidney has long been regarded as an important factor in the production of concentrated urine (Gamble and coworkers, 1934). The counter-current exchange mechanism of the *vasa recta* tends to trap urea in the medullary and papillary interstitial fluids and although the amount of urea in the extracellular space of the inner kidney may depend upon many factors, a high protein diet certainly seems to be associated with an increase ability to concentrate urine (Epstein and coworkers, 1957).

Several studies have shown that the permeation of urea through the cells of the collecting ducts as they pass through the region of the papilla increases in the presence of vasopressin and the resulting increase in the localized concentration of extracellular urea may add to the concentrating ability of the kidney (Jaenike, 1961; Lassiter, Gottschalk and Mylle, 1961; Gardner and Maffly, 1964; Lassiter, Mylle and Gottschalk, 1966; Morgan and Berliner, 1968; Morgan, Sakai and Berliner, 1968; Rocha and Kokko, 1974). For example, in desert rodents where the need for water retention is great, large amounts of urea are recycled between the collecting ducts and the ascending limb of the loop of Henle and in *Psammomys obesus*, the urea titre in the fluid entering the distal convoluted tubule can be three to four times the filtered load (Morel and de Rouffignac, 1970). Thus, large quantities of urea may be added to the filtrate in the loop of Henle. Finally, as mentioned earlier, vasopressin appears to facilitate the extrusion of urea from the terminal segments of the mammalian collecting ducts (Morgan, Sakai and Berliner, 1968; Rocha and Kokko, 1974). Urea may be actively extruded from the collecting ducts, a process of greatest importance when dietary intake of protein is low (Clapp, 1966; Lassiter, Mylle and Gottschalk, 1966).

D. Effects on the Cells of Responsive Tissues

Single naturally occurring membranes are difficult to manipulate experimentally and the biological barriers we must use to study cellular transport mechanisms are willy-nilly larger and more complex than simple single membranes. A typical and frequently used responsive tissue is amphibian skin; when mounted *in vitro* as a dividing partition in a chamber containing physiological saline, the tissue can support a transepithelial electrical potential and this type of preparation has been used extensively to examine the influence of hormones on ion and water fluxes. The urinary bladder of the toad also possesses these properties and may represent an even better tissue for this type of study. The mucosa of the bladder consists of a single layer of epithelial cells that is supported by a basal layer of connective tissue with interdigitated smooth muscle fibres. Like the frog skin, the bladder may be mounted in an incubation chamber to form separate compartments on the mucosal and serosal sides. The toad bladder, however, is a paired organ and one half may be used as a control preparation while the other may be used experimentally. Another epithelium that is known to respond to the presence of neurohypophysial peptides is that lining the most distal portions of the mammalian nephron. As we have seen, the action of vasopressin on the kidney is primarily to increase water reabsorption from the tubular filtrate.

Before we consider in detail the cellular effects of neurohypophysial peptides in responsive tissue we should be aware that many commercial preparations of neurohypophysial peptides contain chlorobutanol as a preservative. This substance is vasoactive and may modify or mask the responses of an organism or tissue to the peptide being studied. This may be a particularly important factor in experiments where very large doses of the mammalian peptide preparations are used to observe the hydro-osmotic and natriferic effects in toad bladder and frog skin preparations. These tissues are notoriously insensitive to heterologous mammalian peptides and relatively large doses must be used to elicit responses (Table 2). Furthermore, commercial preparations of neurohypophysial hormones are frequently impure. For example, the commonly used preparation, Pitressin, in addition to being contaminated by a preservative, consists of a mixture of arginine- and lysine-vasopressin. Therefore, some scepticism may be in order when extremely precise inferences are drawn from experiments using high doses of these impure mixtures of heterologous hormones on amphibian tissues.

1. Na^+ Transport

Ussing and Zerahn (1951) showed that a potential difference as great as 100 mV (serosal side positive) could be maintained *in vitro* across the two sides of a frog skin preparation bathed in Ringer solution. The electrical current necessary to reduce this generated potential to zero was defined as the short-circuit current (SCC) and these investigators were able to show that the SCC was proportional to the current generated by the active Na^+ flux across the skin. It was also revealed that when extracts of the neurohypophysis were added to the chamber on the serosal side of the preparation, the electrical potential difference was heightened and an increase in the SCC was necessary to reduce the potential to zero (Fuhrman and Ussing, 1951; Ussing and Zerahn, 1951). Thus, direct evidence was obtained for the stimulation of active Na^+ transport mechanisms by a neurohypophysial principle. Later, using the same techniques for studies on the toad bladder, Leaf and Dempsey (1960) showed that arginine-vasopressin had the same effect on the urinary bladders of *Bufo marinus* and *Bufo bufo*. Furthermore, in contrast to the effect of aldosterone on Na^+ transport through this tissue, the development of a SCC following exposure to arginine-vasopressin is immediate and is not impaired by prior exposure of the tissue to Actinomycin D (Fig. 4 and Crabbé and De Weer, 1964). Although the active Na^+ transport mechanisms in these preparations are aerobic, arginine-vasopressin does not directly stimulate cell metabolism. Since the oxygen consumption is not affected if the bladders are stimulated while bathed in Na^+-free Ringer's solution, the increase in oxygen utilization must be secondary to the hormonal stimulation of Na^+ transport (Table 3).

Studies *in vitro* on the mechanisms of hormonally stimulated Na^+ transport across amphibian skin and bladder tissues suggest that Na^+ enters the cell passively through the apical surface and that this influx is driven by the chemical gradient of Na^+ existing between the fluid bathing the mucosa and the intracellular environment (Koefoed-Johnsen and Ussing, 1958; Frazier,

Dempsey and Leaf, 1962; Cereijido and Curran, 1965; Leb and coworkers, 1965; Civan, Kedem and Leaf, 1966; Civan and Frazier 1968; Gatzy and Berndt, 1968; Civan, 1970; Macknight, Leaf and Civan, 1970; Rawlins and coworkers, 1970; Handler and Orloff, 1971; Yonath and Civan, 1971).

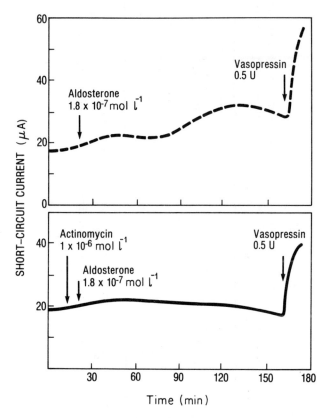

Figure 4 A comparison of the effects of arginine-vasopressin and aldosterone on Na^+ transport across the isolated toad bladder. In the upper graph the immediate response to arginine-vasopressin is compared to the delayed response to aldosterone. Exposure of the tissue to Actinomycin D inhibits the effect of aldosterone and suggests that the aldosterone-induced Na^+ transport may depend on the stimulation of some form of specific protein synthesis (lower graph); in contrast the effect of arginine-vasopressin is not impaired by Actinomycin D. Redrawn from Sharp and Leaf (1966); modified

Some estimates of the total intracellular Na^+ concentrations in anuran tissues seem too high to permit the passive entry of Na^+ (Gatzy and Berndt, 1968). However, the determination of intracellular Na^+ levels is difficult and the methods often yield spuriously high values (see below). Having entered the cell, the Na^+ leaves through the plasma membrane on the basal, and possibly the lateral, surfaces by an active extrusion mechanisms that may be

Table 3. A comparison of the Na$^+$ fluxes and respiration rates (Q_{O_2}) occurring in isolated toad bladders incubated in the presence and absence of neurohypophysial hormone. The toad bladders were bathed in either Na$^+$ Ringer's solution or Na$^+$-free choline Ringer's solution. (Data from Leaf and Dempsey (1960); modified)

	Na$^+$ flux[a] (μA cm^{-2})		Respiration rate (Q_{O_2}) in Na$^+$-Ringer[b] (μl (mg dry tissue)$^{-1}$ h^{-1})				Respiration rate (Q_{O_2}) in choline-Ringer[c] (μl (mg dry tissue)$^{-1}$ h^{-1})			
	M–S	S–M	hour 1	hour 2	hour 3	p value	hour 1	hour 2	hour 3	p value
Control	24.4	3.1	1.20	1.22	1.06	n.s.	1.05	0.93	0.90	n.s.[d]
Hormone	34.8	3.2	1.23	1.72	1.76	0.001	1.13	1.08	1.07	n.s.[d]

[a] Figures are the mean values obtained during two successive 30 min periods before and two 30 min periods after the addition of 1.0 unit of commercial vasopressin.
[b] Hormone added after the first hour; 0.2 unit of oxytocin used in eight experiments and 0.085 unit of arginine-vasopressin in two experiments.
[c] Hormone added after the first hour; either 0.2 or 0.4 unit of oxytocin used.
[d] n.s. = not significant.

coupled to the simultaneous uptake of K^+. The possibility exists, therefore, that neurohypophysial hormones stimulate the movement of Na^+ through the transporting epithelial cells either by increasing the rate of passive Na^+ influx at the apical surface or by augmenting the active extrusion of Na^+ through the membrane on the serosal side of the cell. However, an increase in the passive influx of Na^+ that is not accompanied or followed by an increase in active extrusion would soon eliminate whatever gradient is necessary to maintain passive influx. Conversely, the enhancement of active extrusion through the basal membrane might, unless the rate of passive influx was increased, soon deplete the transportable Na^+ pool within the cell. Thus, as the most recent studies indicate, the passive and active movements of Na^+ through the apical and basal membranes must be coupled and both processes seem at least to be partly, but not necessarily directly, under the influence of the neurohypophysial hormones.

The suggestion that Na^+ may enter the cell passively through the mucosal surface and leave by an active process in the basal membrane was first outlined by Koefoed-Johnsen and Ussing in 1958. However, the intracellular fluids of both frog epidermal and toad bladder mucosal cells bathed in Ringer's solution are electrochemically positive with respect to the fluid bathing their mucoasl surfaces and electrically negative with respect to their serosal fluids. Such differences in electrical potential would tend to oppose both the influx and the efflux of positively charged particles moving through the cell (Engbaek and Hoshiko, 1957). Therefore, the chemical gradient for the transportable pool of Na^+ must be at least great enough to compensate for this unfavourable electrical potential; in other words, the intracellular Na^+ concentration in the transportable pool must be low enough to allow passive diffusion of Na^+ across the apical border. Several groups have shown that in the short-circuited state, when no net Na^+ transport is occurring, the cell interior is negative with respect to the mucosal solution (Frazier, 1962; Cereijido and Curran, 1965; Rawlins and coworkers, 1970).

Assuming that a favourable chemical gradient exists between the interior of the cell and the fluid bathing the mucosal surface, then any hormonal stimulation that acts primarily to facilitate the passive entry of ions at the apical border would be expected to cause an increase in the intracellular Na^+ concentration. Conversely, if the hormones act principally at the site of active Na^+ transport along the basal or serosal border of the cell, a decrease in the intracellular concentration of Na^+ might be expected. Clearly, an accurate measurement of intracellular Na^+ concentration is the touchstone for confirmation or rejection to this hypothesis. Unfortunately this parameter cannot be measured satisfactorily; even the best methods available yield equivocal and non-reproducible results. The problem is further complicated by the fact that Na^+ is not distributed homogenously within the tissue and only a small fraction of the total intracellular Na^+ may participate in active transport mechanism (Cereijido, Reisin and Rotunno, 1968; Finn, 1971, 1975; Finn and Rockoff, 1971). In an effort to circumvent these problems, the effect of arginine-vasopressin on the entry of NaCl into epithelial cells has been studied in isolated mucosal cells of the toad bladder maintained *in vitro* (Macknight, Leaf and Civan, 1970). The 'extracellular space' of the culture was measured by determining the concentration of known amounts of labelled inulin added to the medium. Samples of these cells taken after the addition of arginine-vasopressin to the medium revealed increases in the total

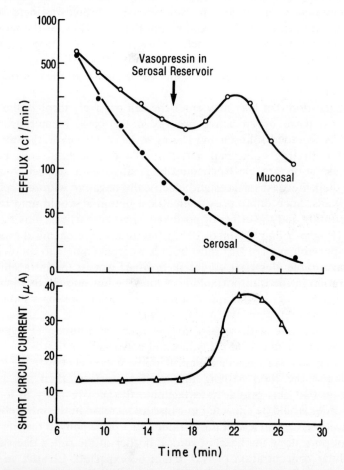

Figure 5 The effect of vasopressin on the rate of loss of ^{24}Na through the mucosal and serosal surfaces of toad bladder cells that had been previously equilibrated in a physiological medium containing ^{24}NaCl. Ringers solution made from unlabelled salts was used to bathe both the mucosal and serosal surfaces. At the time indicated by the arrow, 5 units vasopressin were added to the 250 ml Ringers solution in the serosal chamber. The rate of active Na$^+$ transport across the bladder was measured simultaneously by short-circuit current and is illustrated in the lower graph. Redrawn from Frazier and Hammer (1963); modified

Table 4. The effects of vasopressin on the composition to toad bladder tissue. Tissue was incubated for 30 min and then exposed to vasopressin (50–100 mmol l^{-1} ml^{-1}) for 5 to 10 min. Control tissue was incubated in Ringer's solution throughout. Twelve separate observations were made on tissue pooled from 36 toads and values are expressed as mean ±S.E. (Data from Macknight, Leaf and Civan (1970); modified)

	Intracellular water (kg (kg dry matter)$^{-1}$)	Tissue composition					
		Intracellular contents (mmol l^{-1} (kg dry matter)$^{-1}$)			Intracellular concentration (mmol l^{-1} (kg intracellular water)$^{-1}$)		
		Na$^+$	K$^+$	Cl$^-$	Na$^+$	K$^+$	Cl$^-$
Ringer's solution	3.34 ±0.06	167 ±6	443 ±7	226 ±7	50 ±2	133 ±3	68 ±1
Ringer's solution + vasopressin	3.74[a] ±0.08	225[a] ±6	437[c] ±9	285[a] ±4	61[b] ±2	117[b] ±3	76[a] ±1

[a] $P < 0.001$.
[b] $P < 0.01$.
[c] Not significant.

intracellular Na^+, Cl^- and water contents but, although its intracellular concentration declined due to the influx of water, no significant change was observed in the amount of intracellular K^+ (Table 4). These findings favour the hypothesis that neurohypophysial peptides act to stimulate the influx of Na^+.

We may, however, obtain some idea as to the extent and locations of Na^+ fluxes by observing the movements of radioactive sodium ions through cells that have been equilibrated for some time in Ringer's solution labelled with $^{24}NaCl$. The effect of neurohypophysial hormones on the subsequent movements of $^{24}Na^+Cl$ into unlabelled Ringer's solution on the mucosal and serosal sides of the tissue have at least suggested that one possible locus of sensitivity to the hormone resides in the mucosal membrane (Fig. 5 and Frazier and Hammer, 1963). However, other results from similar experiments have suggested that the neurohypophysial hormones may also act to stimulate an active efflux of ^{24}Na into the serosal chamber (Finn, 1971).

Many groups have emphasized the possible direct stimulation of active Na^+ transport by neurohypophysial peptides (Morel and Bastide, 1965; Janáček and Rybová, 1967, 1970; Finn, 1968; Lipton and Edelman, 1969). Some workers have tested this hypothesis through the use of Amphotericin B in isolated toad bladder preparations. This compound is believed to disrupt the organization of the cell membrane and render it more permeable to Na^+ (Marty and Finkelstein, 1975). The addition of Amphotericin B to the mucosal solution appears to increase the permeability of the mucosal membrane to Na^+ (Lichtenstein and Leaf, 1965; Finn, 1968). If, however, vasopressin is added subsequent to the addition of Amphotericin B, the net transport of Na^+ is stimulated (Finn, 1968). Thus, a net efflux of Na^+ is increased by vasopressin even though the permeability of Na^+ through the mucosal membranes has already become maximal following treatment with Amphotericin B.

These results, however, contradict the earlier evidence that Amphotericin B and arginine-vasopressin can independently enhance net Na^+ flux (Fig. 6). When arginine-vasopressin is added to a toad bladder preparation that has just been exposed to Amphotericin B, the effects of the two stimulants do *not* summate (Fig. 6). These findings have been interpreted to suggest that Amphotericin B and vasopressin act in a similar fashion to increase the passive flux of Na^+ at the mucosal border of anuran epithelia (Lichtenstein and Leaf, 1965).

Studies on the mammalian kidney are no less equivocal. Frindt and Burg (1972) support a possible role for arginine-vasopressin on active Na^+ uptake at the surface of the rabbit cortical collecting tubule. This conclusion is based on perfusion studies which show no change in electrical resistance across the epithelium exposed to arginine-vasopressin, but do show a transient increase in the electrical potential difference and net Na^+ flux (Helman, Grantham and Burg, 1971; Frindt and Burg, 1972). On the other hand, Ullrich and his associates concluded on the basis of their micropuncture studies of the rat collecting tubule *in situ* that antidiuretic hormone (Pitressin) affects only the passive entry of Na^+ into the cell through the apical or the luminal surface (Ullrich and coworkers, 1969).

The fact that changes in the serosal K^+ concentration can alter both the electrical potential difference and the active transport of Na^+ across the basal membrane of frog skin, supports the belief that a $Na^+:K^+$ exchange mechanism exists in this membrane (Huf, 1955; Koefoed-Johnsen and

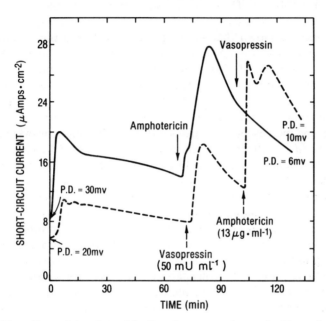

Figure 6 The effect of Amphotericin B and vasopressin on the short-circuit current across the isolated toad bladder. The addition of Amphotericin B to the medium bathing the mucosal surface caused a large stimulation of the short-circuit current and this was not augmented by the subsequent addition of vasopressin (solid line). When vasopressin was added to another tissue sample, it caused a small rise in short-circuit current and the subsequent addition of Amphotericin B to the chamber caused a large increase to follow (dotted line). The differences between the initial and final potential differences (PD) across the bladder samples indicate that these changes were accompanied by decreases in electrical resistance. Redrawn from Lichtenstein and Leaf; modified

Ussing, 1958). Cytochemical evidence points to the presence of a Na^+- and K^+-dependent ATPase in the basal and basilateral membranes of the cells in the amphibian epidermis (Farquahar and Palade, 1966). Such a Na^+ pump mechanism located along the serosal margin of the cells would be an energy-consuming system and the presence of inhibitors might be expected to cause an elevation in the intracellular Na^+ concentration. Crabbé and De

Weer (1969). however, were unable to detect any increase in the Na^+ concentration of toad bladder cells following the addition of either ouabain or CN^- to the chamber on the serosal side of the preparation. However, ouabain has been shown to have a dual effect on Na^+ entry into the cell (Finn, 1975). In addition to reducing Na^+ efflux through the basal membrane, the drug also reduces the passive entry of Na^+ at the apical site; this is especially so when the driving force for entry is made less favourable by either the lowering of the mucosal Na^+ concentration or by clamping the transepithelial potential difference at $+100$ mV. Thus, these results offer a possible explanation for the failure of ouabain to increase the intracellular Na^+ transport pool.

Other work has shown that the increase in Na^+ transport stimulated by antidiuretic hormone was not significantly impaired by ouabain (Mendoza, Handler and Orloff, 1970). Furthermore, the addition of ouabain (5×10^{-5} mol l^{-1}) to the serosal chamber was shown to reduce SCC and the transepithelial potential across toad bladders that were not depleted of metabolic substrate, but it did not affect the ability of the tissue to respond to vasopressin by increasing Na^+ transport (Crabbé and coworkers, 1974). Thus, the natriferic effect of vasopressin is not affected by changes in the activity of the ouabain-sensitive ATPase system in the serosal membranes of toad bladder cells and in the presence of ouabain the reduction in Na^+ transport across toad bladder cells exposed to vasopressin is only diminished by an amount corresponding to the reduction of basal transport observed in the ouabain-treated, but hormonally unstimulated, cells; in other words, the hormonally induced increment in Na^+ transport through the apical membrane persists even in the presence of ouabain (Crabbé and coworkers, 1974).

These data require us to consider that the primary action of arginine-vasopressin on these tissues is to increase the passive permeability of Na^+ through the membrane at the apical border and perhaps to a lesser extent stimulate the active movement of Na^+ through the membrane along the serosal margin.

2. *Water Transport*

Whereas the passage of Na^+ through epithelial cells in frog skin, toad bladder and the mammalian kidney depends in part on active transport, the movement of water through these cells is entirely passive and, in general, a net flux only occurs down an osmotic gradient. The permeability of water through the epithelial cells is believed to be controlled principally at the apical barrier and much of the experimental evidence suggests that this process is separate and distinct from that controlling the movement of Na^+ through the apical membrane.

The possibility exists that the transport of water and Na^+ through anuran epithelial cells occurs via separate pathways (Civan and DiBona, 1974; Scott, Sapirstein and Yoder, 1974). For example, although it has been shown that in the presence of neurohypophysial hormone both prostaglandin E_1 and Amphotericin B seem to enhance the permeability of the toad bladder to Na^+, they decrease and increase respectively the hormone-induced flow of water through this tissue (Lichtenstein and Leaf, 1965; Finn, 1968; Lipson and Sharp, 1971; Yuasa and coworkers, 1975). Conversely, Na^+ uptake by the toad bladder mucosa is inhibited by several agents which do not affect the vasopressin-induced transport of water and, although water permeation through the toad bladder may be reduced by either excess Ca^{2+} or colchicine in the bathing medium, neither agent seems to affect the transport of Na^+ (Bentley, 1960, 1966; Peterson and Edelman, 1964; Taylor and coworkers, 1973; Yuasa and coworkers, 1975). Furthermore, water permeability through the skin of the desert toad, *Scaphiopus couchi*, increases in response to both arginine-vasopressin and arginine-vasotocin only on the day after emergence from dormacy, when recent rains have created pools suitable for breeding, but the natriferic response to these hormones is present throughout the year (Hillyard, 1976).

Thus, *in vitro* and *in vivo* studies suggest that separate pathways may exist for the hydro-osmotic and natriferic responses to neurohypophysial peptides but it is not clear whether these responses are controlled by separate pathways within the same cell or by separate cells specialized to perform each function. Two cell types do exist in the toad bladder mucosa and the collecting ducts of the rat kidney (Choi, 1963; Ganote and coworkers, 1968). While the majority of the cells in these tissues have granular cytoplasm, there are some cells which do not show this granulation but do contain abundant mitochondria. Profiles of the luminal epithelium, as viewed by freeze etching and scanning electron microscope techniques, have shown that the surface characteristics of the granular cells, but not the mitochondria-rich cells, are altered during vasopressin-induced water transport (Chevalier, Bourguet and Hugon, 1974a; Davis and coworkers, 1974; Spinelli, Grosso and de Sousa, 1975). Various other histological observations tend to support this suggestion (Chevalier and coworkers, 1974b; Civan and DiBona, 1974).

Attempts have been made to separate the two types of cells in the toad bladder mucosa by density gradient centrifugation (Scott, Sapirstein and Yoder, 1974). Judged by their ability to produce cyclic-AMP (see below), only the mitochondria-rich cells may respond to hormone stimulation. However, Handler and Preston (1976) have questioned whether cells subjected to this isolation procedure (which requires the removal of Ca^{2+}) are still viable and using an alternative method, they have shown that both types of cell possess the enzymatic potential to form and metabolize cyclic-AMP.

The morphological heterogeneity of neurohypophysial-sensitive tissues may or may not reflect the dissociability of the Na^+ and water transport mechanisms. Although cell-specific changes can be induced by vasopressin, it would be premature to assign a particular role to each cell type. Hormonal stimulation of a net water flux, which in frog and toad bladders may increase by as much as 200-fold, must occur by reducing the resistance to flow at the restricting barrier (Fig. 7 and Hays and Leaf, 1962a). The suggestion that this is accomplished by increasing either the number or the diameter of cylindrical pores through a water-retarding barrier at the apical pole of the cell was made by Koefoed-Johnsen and Ussing (1953).

If we assume that vasopressin acts to increase water permeability at the mucosal surfaces of transporting epithelia, then stimulation should cause an initial increase in cell water and a swelling at the apex of the cell. Also, in the absence of the hormone, the membranes on the basal and lateral surfaces of

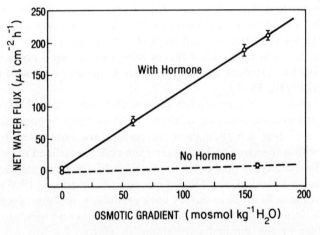

Figure 7 The results of a series of experiments designed to show the relationship that exists between the osmotic gradient maintained between the mucosal and the serosal sides of a toad bladder preparation and the net water flux occurring in the presence and absence of neurohypophysial hormone. After Hays and Leaf (1962a); modified

the cell should be more permeable to water than the membrane along the apical surface. Some of the electron micrographs of vasopressin-stimulated cells derived from frog bladder and mammalian collecting duct tissue show convincing evidence of apical swelling of the mucosal cells (Carasso and coworkers, 1966; Ganote and coworkers, 1968). When the hormone is absent, however, it does not matter whether a transepithelial osmotic gradient favouring net water uptake exists or not, the cells are essentially impermeable to water. Likewise, if the hormone is present but no osmotic gradient exists, there is no net movement of water and the histological appearance of the cells remains unchanged. Furthermore, when the serosal surface of toad bladder is exposed to Ringer's solution labelled with tritiated water, the specific activity of the tissue becomes more nearly equal to that of the fluid on the serosal side than it does if the mucosal side is exposed to the labelled Ringer's solution (Hays and Leaf, 1962a). These results, together with those from a very large number of experiments on the permeability of water through the cells of frog skin, toad bladder and mammalian collecting tubule tend to confirm the hypothesis that the changes in water permeability that are induced by neurohypophysial hormones occur at the apical surface of the responding cell (Whittembury, 1962; Morel, Mylle and Gottschalk, 1965).

As we have stated earlier, a fundamental prerequisite of a net passive water movement through these epithelial cells is the presence of an osmotic gradient between the mucosal and serosal sides of the tissue. Within the physiological limits that may define the range of this gradient, the rate of water permeability through the cells should be directly proportional to the difference between the osmotic concentrations on the two sides of the barrier. Experiments on both frog and toad urinary bladders, on frog skin and on the distal portions of the mammalian nephron have each confirmed the validity of this belief (Sawyer, 1960; Hays and Leaf, 1962a; Berliner and Bennet, 1967; Schafer and Andreoli, 1972a). The results of a typical experiment designed to measure the effect of a neurohypophysial hormone on the net water flux across a toad bladder exposed to different osmotic gradients is illustrated in Fig. 7.

The apical surface of responsive epithelia such as the skin of fresh-water frogs, the mucosal lining of the urinary bladder in a toad or the inner surfaces of the distal convoluted tubules and the collecting ducts in the kidneys of mammalian and non-mammalian vertebrates are always exposed *in vivo* to fluids that contain less osmotically active solute than do the extracellular fluids bathing their respective basal or serosal surfaces. However, experiments *in vitro* have shown that if the osmotic gradient across a toad bladder is reversed, the presence of vasopressin will stimulate an accelerated net flow of water from the serosal to the mucosal side (Bentley, 1961). It has been shown, however, that under conditions of reversed osmotic gradients and in the absence of hormone, water and solutes favour passage between rather than through the cells of the toad bladder (Civan and DiBona, 1974). This is in contrast to the transcellular route taken by water during osmotic flow down a normal concentration gradient. Provided hormone is present a transcellular route also predominates during osmotic flow under the condition of a reversed osmotic gradient (Civan and DiBona, 1974). Under normal conditions where water flows from the hypotonic solution bathing the mucosal surface to the hypertonic medium bathing the serosal side, the accelerated flow of water induced by vasopressin may not be due to a stimulation of active Na^+ transport because an increased flow of water can still be induced even when the hypotonic saline medium is replaced by a mannitol solution of similar osmolarity (Bentley, 1961). This study seems to dispense with the possibility that the accelerated movement of water in response to neurohypophysial stimulation is dependent upon the prior movement of Na^+ across the membrane. It is also worthwhile recalling that the nature of the mechanism seems to be passive and at least the acute responses to neurohypophysial stimulation seem to be largely independent of metabolic processes in the tissue. Therefore, we may be quite justified in attempting to seek primarily physical explanations for the net increases in water flux that have been observed in tissues exposed to neurohypophysial hormones *in vitro*.

The early work of Hevesy, Hofer and Krogh (1935) clearly indicated that the net efflux of water through the skin of the frog was dependent upon the water activities of each side of the membrane; that is to say the osmotic gradient between the extracellular fluid and the environmental water in which the animal was maintained. This net water flux, however, was much greater than could be predicted either from their observed rates of free diffusion through animals maintained in isotonic saline or their estimations of the rate of free diffusion of water in water. Koefoed-Johnsen and Ussing (1953) later found that powdered neurohypophysial extracts nearly doubled the net osmotic flux of water through frog skin while the unidirectional influx of heavy water increased by only 10%. They postulated, therefore, the existence of pores through which water could flow in bulk.

Since diffusion depends on the total pore area available (i.e. $N\pi r^2$, where N = number of pores and r = pore radius), then if water moves through the epithelium solely by diffusion, any increase in the rate of water transport should be proportional to the square of the pore radius. The studies of Hevesy, Hofer and Krogh (1935) and Koefoed-Johnsen and Ussing (1953) using D_2O showed that this is not so and that the net water flux may be many times greater than the total surface area of the theoretical pores in the barrier would allow. The model of Koefoed-Johnsen and Ussing states that water flows through pores in bulk and that the frictional forces between water molecules and the membrane and between water molecules and other molecules in the system will be reduced under these conditions. Laminar flow of this type is described by the Poiseuille law which relates the flux rate as a direct function of the hydrostatic pressure difference (ΔP) and the fourth power of the pore radius (r^4) and indirectly to the viscosity (η) of the fluid inside the pore and the length of the pore channel (l). Thus:

$$F_w = \frac{1}{8} \cdot \frac{\pi r^4}{\bar{V}_w} \cdot \frac{1}{\eta} \cdot N \cdot \frac{\Delta P}{l}$$

where N is the number of pores per unit area and $\bar{V}_w$ is the partial molar volume of water.

Hays and Leaf (1962a) showed that in the presence of vasopressin, the ratio of the osmotic water permeability coefficient (P_f) to the diffusional water permeability coefficient (P_{DW}) in the toad bladder was very much greater than unity ($P_f : P_{DW} = 185 \times 10^{-4}$ cm s^{-1} : 1.7×10^{-4} cm s^{-1} = 109). This indicated that osmotic water flow through this tissue must occur by a mechanism other than simple diffusion, possibly by bulk flow. On the basis of Poiseuille's law and the $P_f : P_{DW}$ ratio, a theoretical value can be obtained for the relative pore diameter in the presence and absence of hormone. In toad bladder, it has been estimated that the pore diameter would increase from 16.8 Å to 80 Å in the presence of vasopressin (Hays and Leaf, 1962a). These investigators also revealed that vasopressin decreased the activation energy for the diffusion of tritiated water to a level comparable to that for the diffusion of water in water (Hays and Leaf, 1962b). This finding was integrated into the pore theory by assuming that the initial constraints to water diffusion were due to the presence of structured water lining the inside of the pores; vasopressin supposedly acted by 'melting' the more organized water in the pore and thus led to an increase in the effective pore diameter.

The pore theory also seemed to account for the enhanced uptake of some solutes by anuran epithelial tissues exposed to neurohypophysial hormones. For example, Andersen and Ussing (1957) showed that the uptake of small molecules such as urea, thiourea and acetamide was enhanced in direct proportion to, and in the same direction as, the osmotic water flux. Not only did this suggest a common pathway for water and amide permeation, but it also suggested that solute movement was coupled to the flow of solvent through the pores. This phenomenon, which was termed 'solvent drag', also seemed to apply in the toad bladder for when an increase in water flow was stimulated by vasopressin, the influx of urea increased proportionately (Leaf and Hays, 1962).

The fact that these epithelial cells showed some selectivity towards the permeation of Na^+, urea, thiourea, acetamide, K^+, Cl^- and other electrolytes was integrated into a modified pore model (Andersen and Ussing, 1957; Hays and Leaf, 1962a, 1962b; Leaf and Hays, 1962). This model proposed that an outer selective barrier occurred in series with an inner vasopressin-sensitive porous barrier. The outer barrier would restrict the passage of most solutes but would allow the selective entry of some solutes and the free entry of water.

The development of the dual barrier hypothesis was primarily based on the following types of evidence: the $P_f:P_{DW}$ ratios in responsive tissues were found to be greater than unity, the apparent occurrence of 'solvent drag' during hormone-stimulated water transport, and the observation that the activation energy for water diffusion through vasopressin-treated cells of the toad bladder was comparable to that for the diffusion of water in free solution, such as might occur in open pores filled with water.

Attention has been focused recently on the possibility that unstirred layers of water overlying the mucosal surface may retard the free diffusion of water through isolated tissue preparations; this may lead to an underestimation of P_{DW} values in *in vitro* preparations (Dainty, 1963; House, 1974). For example, it was reported that the P_{DW} for tritiated water in the vasopressin-stimulated bladder was increased almost three-fold when the fluid in the mucosal chamber was stirred (Hays and Franki, 1970). From these and other data the apparent 'thickness' of the unstirred layer was estimated to range from 0.2 to 1 mm (Hays and Franki, 1970; Andreoli and Schafer, 1976; Pietras and Wright, 1975). In the nephron where unstirred layers probably do not occur, the P_{DW} values for labelled water are comparable to those observed for toad bladder preparations maintained in well-stirred chambers (Grantham and Orloff, 1968; Hays and Franki, 1970; Schafer and Andreoli, 1972a; Rocha and Kokko, 1974). In cortical collecting tubules from the rabbit and the human fetal kidney and in isolated toad bladder cells, P_{DW} values of 14.2×10^{-4}, 11.98×10^{-4} and 10.8×10^{-4} cm s^{-1} respectively have been recorded in the presence of vasopressin (Hays and Franki, 1970; Frindt and Burg, 1972; Schafer and Andreoli, 1972a; Abramow and Dratwa, 1974). Although these P_{DW} values are higher than those originally reported for toad bladder, the P_f:P_{DW} ratios are still substantially greater than one. This disparity may be due to the diffusional constraints imposed by portions of the epithelial cells other than the luminal membrane (Andreoli and Schafer, 1972a). Thus, only for the case of water transport across lipid bilayers exposed to vigorously stirred aqueous phases has a P_f:P_{DW} ratio equal to one been reported (Cass and Finkelstein, 1967).

By comparing the diffusion of small lipiphilic solutes (P_{Di}) through artificial lipid bilayers and natural membranes we may gain some insight into the possible nature of the diffusional constraints in the cell. Studies on lipid bilayers have shown that the diffusion of lipophilic solute is certainly increased when fluid in the bathing chamber is stirred (Holtz and Finkelstein, 1970). In some natural membranes, such as the mammalian collecting duct epithelium, there may be little or no unstirred layer but the magnitude of the P_{Di} value for a lipophilic solute suggests that the P_f:P_{DW} ratio may still be greater than unity and cellular resistance to diffusion may still exist both in the presence and absence of the hormone. Furthermore, the magnitude of the cellular resistance seems to be proportionate to the extent to which the diffusion of lipophilic solute is retarded during passage through the cellular barrier (Schafer and Andreoli, 1972b). Thus, the very high P_f:P_{DW} ratios originally assigned to toad bladder and collecting tubule cells were more likely due to an attenuation of diffusion due to the presence of either unstirred layers of water on the mucosal surface or other undefined cellular constraints rather than to an augmentation of water influx by bulk passage through enlarged pores.

The apparent effect of arginine-vasopressin on the activation energy of tritiated water transport across toad bladder cells may also be re-evaluated in terms of the effects of unstirred layers of water. When Hays and his associates (1971) repeated their original experiments, they found that with high rates of stirring, vasopressin did little to lower the activation energy for water diffusion: 10.7 versus 9.3 kcal mol^{-1} (Hays, Franki and Soberman, 1971). Furthermore, these values are quite similar to the activation energy for water diffusion across synthetic lipid bilayers (Price and Thompson, 1969).

The fact that vasopressin can increase the permeation of small lipophilic solutes through cortical collecting tubule membranes supports the belief that the hormone may act to modify the physical characteristics of the apical membrane (Schafer and Andreoli, 1972b; Pietras and Wright, 1974). This is further supported by studies showing that vasopressin can increase the diffusional permeability of branched lipophilic solutes to which the membrane is normally impermeable (Al-Zahid, Schafer and Andreoli, 1975). In the presence of arginine-vasopressin, the permeability of isobutyramide through toad bladder cells is increased from $16 \pm 1 \times 10^{-7}$ cm s^{-1} to $35 \pm 3 \times 10^{-7}$ cm s^{-1} (Pietras and Wright, 1974). In contrast, the permeability of n-butyramide is increased by only 13%.

Recent studies in a variety of related fields suggest that neurohypophysial hormones may initiate osmotic water flow by increasing the relative fluidity of the membranes through a reduction in the quantity, or changes in the organization, of structural elements such as proteins, sterols and saturated and unsaturated fatty acids in the membrane (Finkelstein and Cass, 1967; Träuble, 1971; Graziani and Livne, 1972; de Kruyff and coworkers, 1973). Low concentrations of arginine-vasopressin, but not oxytocin, may also act *directly* on synthetic membranes to augment osmotic water flow to an extent comparable to that seen in the toad bladder exposed to similar doses of the hormone (Graziani and Livne, 1971). Other similarities between synthetic and natural membranes include the ability of Ca^{2+} and prostaglandin E_1 to inhibit the vasopressin-induced hydro-osmotic effect in lipid bilayers by presumably competing for the electrostatic and hydrophobic sites occupied by the hormone.

Recent freeze-fracture studies on the toad bladder have shown a significant correlation between the aggregation of apical membrane particles, possibly proteins, and hormone-induced osmotic water flow (Chevalier, Bourguet and Hugon, 1974a; Kachadorian, Wade and Discala, 1975). Such changes in the organization of the membrane may reflect an increase in membrane fluidity as postulated by the solubility diffusion model. Pietras, Naujokatis and Szego (1975) sought to characterize the dynamics of the aggregation of membrane proteins using lectin probes. They were able to show that a vasopressin-dependent increase of haemaglutinability occurred in isolated epithelial cells from the frog and toad bladders. Since haemagglutination depends upon the redistribution of Concanavalin-A binding sites, rather than a change in the total number of binding sites, the effect is probably attributable to lateral movement of the binding sites within the membrane; this in turn may be due to the ability of arginine-vasopressin to cause a phase change in the membrane and thus create a more fluid state.

Since the permeability of water may be accounted for without reference to pores in the luminal or apical membrane, we must re-examine and explain the phenomenon described as solvent drag. Theoretical and experimental considerations show that coupling of solute to solvent *can* occur under appropriate conditions. Indeed, Amphotericin B apparently interacts with membranes to form apparent thoroughfares equivalent to pores approximately 8 Å in diameter (Holtz and Finkelstein, 1970; Andreoli, Schafer and Troutman, 1971; Andreoli, 1973; Marty and Finkelstein, 1975). The coupling of glycerol and mesoerythritol to the osmotic water flow is evident in Amphotericin-B-treated lipid bilayers (Andreoli, Schafer and Troutman, 1971). However, when acetamide fluxes are measured across toad urinary bladder mounted in well-stirred chambers, no solvent drag occurs; that is to say, the influx:outflux ratio approaches unity during vasopressin-induced osmotic water flow from the mucosal to the serosal chamber (Hays, 1972). Even though solvent drag may occur across lipid bilayers where the molecular radius of the coupled solute is less than the pore radius, it does not seem to apply to epithelial cells that are responsive to neurohypophysial hormones.

Some investigators have attempted to explain the vasopressin-dependent increase in urea and small amide permeation through the urinary bladder epithelium in terms of a carrier-mediated process (Levine, Franki and Hays, 1973a, 1973b). The influx of radioactively labelled urea, acetamide and proprionamide is inhibited by phloretin (10^{-4} mol l^{-1}). This compound, which has been found to inhibit other transport processes, does not affect the vasopressin-induced water transfer across the toad bladder. Also, the rate of influx of acetamide becomes maximal above concentrations of 150 mmol l^{-1} in the mucosal chamber and this suggests that a carrier process may be involved (Levine, Franki and Hays, 1973b). Finally, through competition with unlabelled acetamide the uptake of labelled urea and acetamide by the toad bladder is inhibited. There are some indications that a surface glycoprotein may be an essential component of the proposed urea transport mechanism in the toad bladder (Rubin, 1975).

E. Molecular Basis of Hormonal Action

Since increases in Na^+ and water transport occur immediately following exposure of the responsive tissues to neurohypophysial hormones, it seems improbable that the synthesis of new protein must precede the physiological manifestations of these hormone actions (Fig. 4). Attention has therefore been focused on the possibility that the plasma membrane, rather than an intracellular locus, is the first site of neurohypophysial hormone action. During studies on other hormones believed to act at the cell surface, Sutherland and Rall (1957) noted that $3',5'$-cyclic adenosine-monophosphate (c-AMP), was liberated from the debris of liver cell homogenates that had been exposed to epinephrine. This compund stimulated the conversion of inactive phosphorylase from liver cells into an active phosphorylase enzyme and as a result, glucose-l-phosphate was formed from glycogen (Fig. 8). The role of c-AMP in glycogenolysis is now fairly well understood and the scheme of this reaction serves as a general paradigm for the possible mode of action of other surface active hormones where the participation of c-AMP has been implicated.

Lysine-vasopressin was also found to have a glycogenolytic effect in liver cells similar to that mediated by c-AMP in liver cells that had been exposed to epinephrine. Furthermore, when lysine-vasopressin was injected into the adrenal artery of the dog, it stimulated the synthesis and release of corticosteroids. This action was similar to that of adrenocorticotrophic hormone

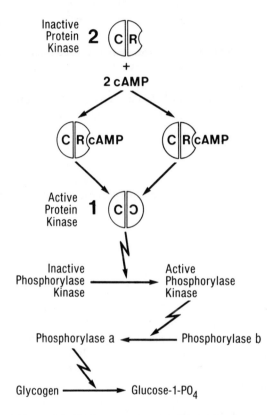

Figure 8 A schematic representation of the possible molecular mechanisms involved in the glycogenolytic effect of cyclic-AMP. The inactive protein kinase consists of a catalytic (C) and a regulatory (R) subunit and upon dissociation of two inactive molecules the two catalytic subunits combine to form an active molecule of protein kinase. Redrawn from Garren, Gill and Walton (1971a); modified

(ACTH), yet another hormone believed to act at the cell surface to stimulate enhanced intracellular synthesis of c-AMP (Hilton and coworkers, 1959). On the basis of these tenuous, and possibly unrelated observations, Orloff and Handler (1961) suggested that c-AMP may play some role in the antidiuretic action of vasopressin. The studies which followed confirmed that c-AMP alone could cause increases in both Na^+ and water transport across the toad bladder and frog skin (Handler and coworkers, 1965; Baba, Smith and Townshend, 1967; Bastide and Jard, 1968; Ganote and coworkers, 1968).

Furthermore, the movement of water out of the collecting tubules in the mammalian kidney could be stimulated by c-AMP in the *absence* of vasopressin (Grantham and Burg, 1966; Grantham and Orloff, 1968; Barraclough and Jones, 1970). Exposure of responsive amphibian epithelial cells and cells from the mammalian kidney to vasopressin was shown to cause an increase in the intracellular concentrations of c-AMP (Brown and coworkers, 1963; Handler and coworkers, 1965; Orloff and Handler, 1967; Ferguson and Price, 1972; Stoff, Handler and Orloff, 1972).

The formation of c-AMP within these cells requires the activation of adenyl cyclase through the binding of vasopressin to a receptor in the membranes of the responding epithelial cells. The enzyme, as well as the receptors, resides in the plasma membrane and the induced activity is directly proportional to the amount of hormone bound to the membrane receptors (Davoren and Sutherland, 1963; Bär and coworkers, 1970; Bockaert and coworkers, 1973; Rajerison and coworkers, 1974). The receptors show an apparent preferential affinity to bind with native peptides. For example, in the pig kidney, c-AMP production is highest following stimulation with lysine-vasopressin and lowest following stimulation with oxytocin; arginine-vasopressin being intermediate in this regard (Dousa and coworkers, 1971). The greater responses of lower vertebrate tissues to the homologous antidiuretic hormone, arginine-vasotocin, compared to their responses to heterologous hormones such as arginine-vasopressin, may also suggest a preferential affinity of these receptors in the plasma membranes of their respective responsive tissues (Table 2). Control of the intracellular concentrations of c-AMP may occur through regulation of either the activating enzyme, adenyl cyclase, or the degrading enzyme, phosphodiesterase. Thus, theophylline, which is an inhibitor of the phosphodiesterase, can act synergistically with vasopressin to cause an enhancement of the antidiuretic response by elevating the c-AMP concentrations in the responding cells (Orloff and Handler, 1962; Handler and coworkers, 1965).

Theophylline may only increase the SCC indirectly by having its principle action on Cl^- permeability rather than a direct inhibition of the phosphodisterase enzyme (Cuthbert and Painter, 1968). Thus, when Cl^- transport is eliminated by the substitution of isothionate ion for Cl^- in the mucosal bath, theophylline has no effect on the Na^+ transport whereas vasopressin does increase Na^+ transport even though no Cl^- is transported. However, when this work was repeated substituting SO_4^{2-} for Cl^-, theophylline did cause an increase in SCC (Kristensen, 1970). Until the mode of action of theophylline on frog skin is resolved more satisfactorily, the suggestion that its pharmacological properties are due solely to the inhibition of the phosphodiesterase activity in responding cells must be regarded critically.

Through a pursuit of the analogy between the classic scheme developed to explain the role of c-AMP in the stimulation of liver glycogenolysis by epinephrine, there have been many reports of the presence of c-

AMP-dependent protein kinase in a variety of tissues known to respond to neurohypophysial peptides (Kuo and Greengard, 1969). When a c-AMP-dependent protein kinase was isolated from frog bladder epithelium, 30% of the activity was found in the microsomal fraction of the cells and the remainder was present in the cytosolic fraction (Jard and Bastide, 1970). This

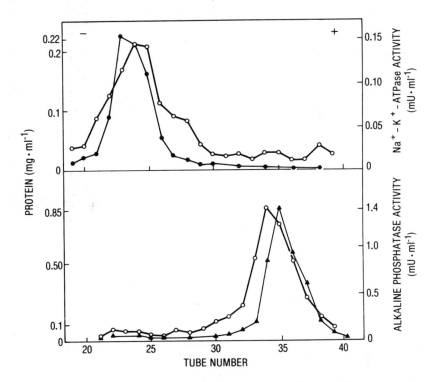

Figure 9 The isolation by free-flow electrophoresis of two different membrane fractions isolated from cells of the proximal convoluted tubule in rate kidney cortex. The membrane fraction containing a high Na–K dependent ATPase activity (—●—, upper graph) possessed junctional complexes and resembled plasma membranes from the contraluminal (basal) side of the cell. The other fraction was rich in alkaline phosphatase activity (—▲—, lower graph) and contained only the microvilli that are characteristic of the brush border (luminal surface) of the proximal tubular cells. In each instance the enzyme activities were coincident with the distribution of cell protein (—O—). After Heidrich and coworkers (1972); modified

finding suggested that the kinase activity present in the microsomal fraction may be related to the vasopressin-induced changes in membrane permeability. This belief was reinforced by the results of other studies which showed that the plasma membranes of the hog kidney contained a c-AMP-dependent protein kinase (Forte and coworkers, 1972). The localization of kinase activity associated with the plasma membrane was a novel finding. It immediately suggested that c-AMP may have the potential to alter the level of membrane phosphorylation. Dousa, Sands and Hechter (1972) have isolated a c-AMP-dependent protein kinase from the rat kidney medulla that was capable of phosphorylating isolated plasma membrane preparations. Specifically, it was shown that the phosphorylation was reversible and the fact that

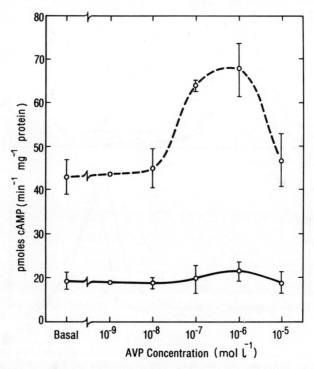

Figure 10 Effect of arginine-vasopressin on the adenylate cyclase activity of luminal membranes (———) and contraluminal membranes (— — —) isolated from bovine collecting duct cells. Redrawn from Schwartz and coworkers (1974); modified

digestion by pronase, but not Phospholipase C, would cause the release of inorganic phosphate suggested that the kinase had induced phosphorylation of a protein and not a phospholipid moiety with the cell membrane. In contrast, other workers reported a *decrease* in the phosphorylation level of a specific protein ('Protein D') in toad bladder membranes that had been exposed to either c-AMP or arginine-vasopressin (De Lorenzo and coworkers, 1973). They also suggested that the effect of c-AMP may be a direct one upon a 'Protein D' phosphatase.

Perhaps the best evidence to indicate that a c-AMP-activated protein kinase can phosphorylate the membrane fraction of a tissue that is sensitive to arginine-vasopressin was demonstrated by Schwartz and coworkers (1974). In this and in a previous study, these workers were able to separate the mucosal and serosal membranes from bovine collecting duct cells and proximal convoluted tubules in the rat kidney cortex (Heidrich and coworkers, 1972). The purity of the fractions was verified cytochemically by identifying Mg^{2+} and Na^+/K^+-dependent ATPase activities in the serosal

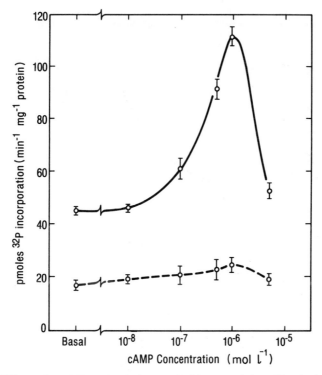

Figure 11 Effect of c-AMP on intrinsic protein kinase activity of luminal membranes (———) and contraluminal membranes (— — —) isolated from bovine collecting duct cells. Redrawn from Schwartz and coworkers (1974); modified

(contraluminal) membranes and either HCO_3^--Na^+/K^+-dependent ATPase or alkaline phosphatase activities in the mucosal (luminal) brush border membranes (Fig. 9). In these membrane preparations they found that the adenyl cyclase activity which was sensitive to arginine-vasopressin was associated with the serosal (contraluminal) membrane (Fig. 10) and that a c-AMP-dependent protein kinase was associated with the mucosal or luminal surface membranes (Fig. 11). Also, following the addition of c-AMP and a protein kinase preparation, the membranes derived from the mucosal surface incorporate inorganic phosphate to a greater extent than those derived from the serosal (contraluminal) surfaces (Fig. 12).

The stimulatory effect of c-AMP and vasopressin on cellular metabolism appears to be secondary to active Na+ transport (Handler, 1964). Arginine-vasotocin and arginine-vasopressin do not stimulate glycolysis and O_2 consumption by toad bladder tissue when there is no Na+ available in the mucosal solution but, under these conditions, there is some activation of the glycolytic phosphorylase enzyme system (Leaf and Dempsey, 1960; Handler

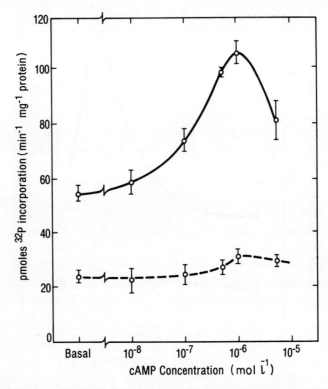

Figure 12 Effect of c-AMP on the phosphorylation of luminal membranes (———) and contraluminal membranes (— — —) isolated from bovine collecting duct cells in the presence of papillary cytosolic protein kinase. Redrawn from Schwartz and coworkers (1974); modified

and Orloff, 1963; Handler, Preston and Rogulski, 1968). According to Goodfriend and Kirkpatrick (1963), neurohypophysial peptides can increase the rate of glucose and pyruvate utilization even when the rate of Na+ transport is minimal. However, the demonstration of a primary effect of vasopressin on cellular metabolism has been difficult and the role of c-AMP on metabolic processes in responsive epithelial cells is still unresolved.

While there is no question that the synthesis of c-AMP increases following exposure of some tissues to the neurohypophysial peptides, this cyclic nucleotide may not be directly involved in all of the physiological responses induced by these hormones. The application of a single neurohypophysial peptide to the serosal membranes of amphibian skin and bladder and mammalian collecting tubules elicits several distinct responses. The suggestion that one type of cell may control hydro-osmosis while another type may control Na^+ transport has been considered above. There are, however, other possible interpretations of these data. One may be that two adenyl cyclase-receptor systems exist within a single cell and another is that the tissue may respond differently to low and high doses of vasopressin (cf. Peterson and Edelman, 1964; Bourguet and Morel, 1967; Morel and Jard, 1968; Wright and Snart, 1971).

The possible involvement of the microtubules and microfilaments in the hydro-osmotic response of some vertebrate epithelia to neurohypophysial peptides has been recognized only recently. Under circumstances where water transfer is known to be occurring, cytoplasmic vesicles appear to become fused with the apical membranes of the granular cells in the toad bladder mucosa and electron microscopic examinations of these cells have suggested some functional relationship may exist between the microfilaments, the peripheral vesicles and the movement of water (Masur, Holtzman and Walter, 1972). It has been suggested that the transport of these vesicles to the plasma membrane may depend upon the assembly of a microtubule or microfilament complex and that hydro-osmotic effects of a neurohypophysial peptide may depend upon the hormonal stimulation of the assembly of these complexes (Carasso, Favard and Bourguet, 1973; Taylor and coworkers, 1973; Yuasa and coworkers, 1975).

The use of drugs that inhibit the assembly of microtubules has indicated that microtubules are a necessary component in the sequence of events leading to the stimulation of hydro-osmosis by vasopressin. Thus, a reversible inhibition of enhanced water transport has been observed when vasopressin-stimulated cells of the toad bladder were exposed to 2×10^{-5} mol l^{-1} colchicine for three hours. In the absence of vasopressin the normal slow downhill osmotic flow of water continues in the presence of colchicine concentrations ranging from 2×10^{-7} mol l^{-1} to 2×10^{-4} mol l^{-1}. Other inhibiting agents, such as vinblastine and podophyllotoxin, in concentrations ranging from 2×10^{-7} mol l^{-1} to 2×10^{-5} mol l^{-1} have also been shown to have similar effects on the vasopressin-stimulated movement of water through these cells (Table 5).

Furthermore, the time required for the inhibitory effect of colchicine to be observed in terms of reduced hydro-osmosis corresponds well with the time needed for colchicine to bind to tubulin *in vitro* (cf. Wilson and Friedkin, 1967). The fact that colchicine does not alter basal levels of water flux through the cells is in agreement with the observation that this drug inhibits the assembly of new microtubules rather than disrupts the structure of existing tubules (Wilson and Bryan, 1974). Exposure to relatively high concentrations (2×10^{-5} mol l^{-1}) of lumicolchicine, a compound that is structurally similar to colchicine, does not affect water transport and does not inhibit microtubule assembly by binding to tubulin (Wilson and Friedkin, 1967).

Studies on the mammalian kidney have shown that agents that disrupt the assembly of microtubules also inhibit the antidiuretic effects of arginine-vasopressin in rabbits (Dousa and Barnes, 1947a, 1947b). Colchicine and

Table 5. The effects of colchicine, lumicolchicine, vinblastine and podophyllotoxin on the net osmotic water movement through the toad bladder in response to vasopressin and cyclic AMP. All agents were added to the bathing medium of the experimental hemibladders 4 hours prior to the addition of vasopressin or cyclic AMP. (Data from Taylor and coworkers (1973); modified)

Test agent	N	Mean weight loss (mg min^{-1})[a]		Per cent change	P value[b]
		Control	Experimental		
Vasopressin, 20 mu ml^{-1}					
2×10^{-5} mol l^{-1} colchicine	7	30.7 ± 3.8	16.2 ± 2.3	−47.9 ± 3.0	<0.001
2×10^{-5} mol l^{-1} lumicolchicine	6	35.5 ± 3.1	37.2 ± 1.8	+9.4 ± 9.8	n.s.[c]
2×10^{-6} mol l^{-1} vinblastine	8	36.2 ± 3.9	18.7 ± 0.9	−44.7 ± 5.4	<0.001
1.5×10^{-5} mol l^{-1} podophyllotoxin	6	30.1 ± 5.1	7.7 ± 1.7	−74.3 ± 3.7	<0.001
Cyclic AMP, 2 mmol l^{-1}					
2×10^{-5} mol l^{-1} colchicine	6	35.8 ± 4.8	13.5 ± 3.9	−65.0 ± 5.8	<0.001

[a] Measured over the 2-hour period following addition of either vasopressin or cyclic AMP.
[b] Calculated for paired experiments.
[c] n.s. = not significant.

vinblastine do not interfere with the activity of endogenous arginine-vasopressin. Furthermore, lumicolchicine has no effect either on the enhancement of protein kinase and protein phosphatase activities by c-AMP or on the antidiuretic effect in response to arginine-vasopressin. There is also some evidence that the stimulation of a microtubule-associated protein kinase and the phosphorylation of microtubules is dependent upon the presence of c-AMP (Murray and Froscio, 1971). However, the existence of such a microtubule-associated protein which possesses protein kinase activity has been questioned and the significance of microtubule phosphorylation is unknown (Rappaport, Leterrier and Nunez, 1975; Soilfer and coworkers, 1975).

Changes in water transport that occur subsequent to hormone stimulation require that the microtubule assembly mechanism is intact. The transport of vesicles to the apical surface may also be dependent on 'functional' microtubules. Exocytosis may confer upon the apical membrane the properties necessary to permit enhanced water permeation. In support of this proposal is the observation purporting to show Golgi-derived vesicles fusing with the plasma membrane and in so doing inserting portions of membrane with different permeability characteristics (Hicks, 1966).

Together, these studies suggest that microtubules may provide a link between hormonal stimulation of the responsive tissue and the subsequent increase in water permeability (Taylor and coworkers, 1973; Dousa and Barnes, 1974a, 1974b). Microtubules and microfilaments, in addition to their association with apical vesicles, amy also be involved in the lateral movement of some membrane proteins (Taylor and coworkers, 1971; Edidin and Weiss, 1972; de Petris, 1974, 1975; Pietras, Naujokatis and Szego, 1975). Thus, when conditions are unfavourable to either microtubule assembly or microfilament stability, the changes associated with enhanced mobility of surface-active proteins are not observed. For example, the ability of arginine-vasopressin to induce the aggregation of lectin binding sites in toad and frog bladder epithelial cells is depressed when the assay is performed at 4 °C, a temperature at which most microtubules are unstable (Olmsted and Borisy, 1973; Pietras, Naujokatis and Szego, 1975).

The precise interrelationship between the movement of membrane proteins, the organization of microtubule and microfilaments complexes, the occurrence of vesicles in the granular cells and the ultimate manifestation of the hydro-osmotic response to neurohypophysial peptides in the toad bladder is at present not clear. Evidence obtained from other cells, however, suggests that membrane proteins may be displaced laterally into areas of the plasma membrane which are distal to the region where either exocytosis, micropinocytosis or phagocytosis is occurring (Ukena and Berlin, 1972). The observation that the hydro-osmotic response in frog and toad bladder cells is associated with the aggregation of proteins in the mucosal membrane tends to support the belief that the neurohypophysial peptides act by inducing exocytosis and possibly micropinocytosis in this membrane (Masur, Holtzman and Walter, 1972).

In summary, the initial action of the neurohypophysial peptides in the induction of both hydro-osmotic and natriferic effects is to bind with a specific receptor situated in the serosal membrane of the responsive cells and to interact with an adenyl cyclase system in this membrane. This results in an increase in the cellular content of a second messenger, c-AMP, which may independently control water and Na^+ transport through these cells by influencing one or a combination of the several possible mechanisms outlined in Fig. 13.

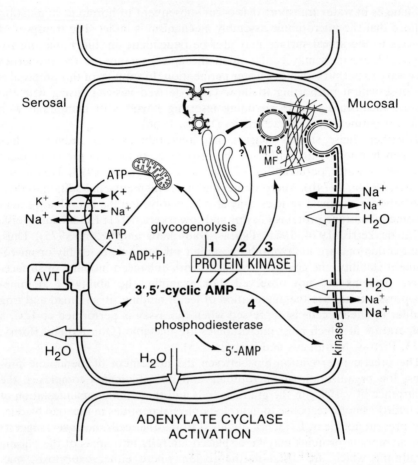

Figure 13 A summary showing the possible sites of neurohypophysial peptide action in a responsive epithelial cell. The membrane-bound adenyl cyclase is activated by the binding of a homologous hormone to receptors situated in the serosal membrane. Activation of the adenyl cyclase results in the production and intracellular release of c-AMP which, in turn, stimulates either a soluble or a membrane-bound protein kinase (cf. Fig. 8). The activated protein kinase is presumed to mediate either Na^+ or water transport by one or a combination of the following four cellular mechanisms: (1) stimulation of glycogenolysis and the consequent production of energy to stimulate active Na^+ transport; (2) changes in the Na^+ and water premeability of the membranes of Golgi-derived vesicles which then fuse with the plasma membrane at the mucosal surface; (3) phosphorylation of the microtubules that play an obligatory role in the hydro-osmotic response and also may regulate the transport of vesicles to the mucosal surface of the cell. Alternatively, the microtubules may induce the aggregation of membrane proteins within the apical membrane (not shown); (4) stimulation of a membrane-bound protein kinase which may change the level of phosphorylation and affect the permeability of the membrane to either water and Na^+ or the Na^+ alone

III. THE ADRENOCORTICAL STEROIDS

In a paper presented to the South London Medical Society in 1849, Thomas Addison described some possible connections between the clinical symptoms he had observed in three patients and the evidence of diseased 'suprarenal capsules' that he found post-mortem. Six years later he published what is now recognized to be the first description of the effects of adrenocortical insufficiency in humans (addison, 1885). The complex of symptoms he described was later named 'Addison's diseased and they included the development of irregular patches of bronze pigmentation in the skin, a general muscular weakness, hypotension and electrolyte imbalance.

Soon afterwards, Brown-Séquard (1856, 1857, 1858) reported that removal of the adrenal glands from dogs, cats, rabbits and guinea pigs caused them to die sooner than they did after removal of the kidneys. He paid little attention, however, to the fact that the animals from which only one adrenal had been removed survived but a few hours longer than bilaterally adrenalectomized animals. In retrospect, therefore, it seems that Brown-Séquard's experimental animals may have died from the effects of surgical trauma rather than adrenal insufficiency. Nevertheless, he concluded correctly, albeit on the basis of dubious and possible misinterpreted data, that the adrenal glands were essential for life.

None of these studies established whether the mortality among pateients with Addison's disease and among animals from which the adrenals had been removed was due primarily to the loss of either the medullary or the cortical portion of the adrenal gland. In spite of the fact that the lives of Addisonian patients and adrenalectomized animals were not prolonged through treatment with medullary extracts, the belief nevertheless persisted that the fatal effect of adrenal insufficiency was due to the absence of a factor produced by the adrenal medulla (Boinet, 1903; Biedl, 1913). This notion was resurrected periodically for another 25 years and in this regard it is interesting to note the early and definitive work of Biedl (1899, reported in Biedl, 1913) was largely ignored. Using elasmobranch fishes, in which the adrenocortical and adrenomedullary tissues are anatomically separate, Biedl showed that only when the cortical tissue was removed did the animals succumb for reasons other than surgical trauma. This finding was later confirmed in dogs (Crowe and Wislocki, 1914; Houssay and Lewis, 1923) and subsequently, as the techniques improved, bilateral adrenalectomy was shown to be consistently fatal in all of the vertebrate species examined (see Chester Jones, 1957).

Following publication of Addison's monograph on adrenal insufficiency, more than 70 years elapsed before the essential life-sustaining factor in the adrenal gland was shown to be present in the lipid fraction extracted from cortical tissue. Thus, although the first extracts prepared from beef adrenals sustained adrenalectomized dogs for only about 10 days, later preparations contained much more biological activity and adrenalectomized cats were maintained in good health for over 200 days (Rogoff and Stewart, 1928; Pfiffner and Swingle, 1929, 1931; Hartman, Brownell and Hartman, 1930; Swingle and Pfiffner, 1930a, 1930b, 1930c, 1931a, 1931b). These successes heralded a period of intensive research on the chemical nature of the physiologically active principles contained in the Swingle–Pfiffner extracts of adrenocortical tissues.

A. Chemical Composition

Between 1935 and 1942, Kendall and Mason at the Mayo Foundation, Cleveland, Wintersteiner and Pfiffner at Columbia University, New York, and Reichstein and his associates in Zurich, Switzerland, isolated 27 different steroids from the lipid extracts of beef adrenal glands but only six of them possessed the properties necessary to alleviate some of the symptoms of adrenocortical insufficiency. These compounds belonged to the pregnane series of steroids but each of them differed from progesterone (4-pregnene-3,20-dione) by having a hydroxyl substitution at the C-21 position (Fig. 14). Indeed, this C-21 hydroxylated derivative represents the minimal molecular structure that will ensure the survival of an adrenalectomized organism and it has been assigned the trivial name of *deoxycorticosterone* (21-hydroxy-4-pregnene-3, 20-dione). The other five steriod hormones may be derived by addition and deletion of hydroxyl and ketone groups at the C-21 and C-17 positions. These modifications of the basic molecule yield *corticosterone* (11β,21-dihydroxy-4-pregnene-3,20-dione), *cortisol* (11β,17a,21-trihydroxy-4-pregnene-3,20-dione), 11-*deoxycortisol* (17a,21-dihydroxy-4-pregnene-3,20-dione), *cortisone* (17a,21-dihydroxy-4-pregnene-3,11,20-trione) and 11-*dehydrocorticosterone* (21-hydroxy-4-pregnene-3,11,20-trione) and each is accompanied by a change in biological potency (cf. Fig. 14 and Table 6).

Following adrenalectomy in mammals two distinct patterns of physiological change may occur. The first is identified during the first 24 hours after removal of the adrenals when the Na$^+$ reabsorptive and K$^+$ secretory properties of the cells in renal tubules become impaired. The electrolyte balance that ensues may be substantially corrected through the addition of NaCl to the drinking water. But, adrenalectomized animals maintained in this manner are still extremely vulnerable to abrupt environmental change, fasting and the metabolic demands that accompany prolonged muscular activity. Each of these circumstances is considered 'stressful' and the inability of the adrenalectomized organism to cope with such 'stresses' is also attributable to absence of adrenocortical steroids. These implied actions of the adrenocortical steroids are mediated largely through a stimulation of gluconeogenesis. Although the adrenocortical hormones may influence both patterns of physiological change, they are not all equally potent with respect to their relative abilities to influence the pattern of renal electrolyte excretion and to modify carbohydrate, protein and fat metabolism. In mammals, some of these

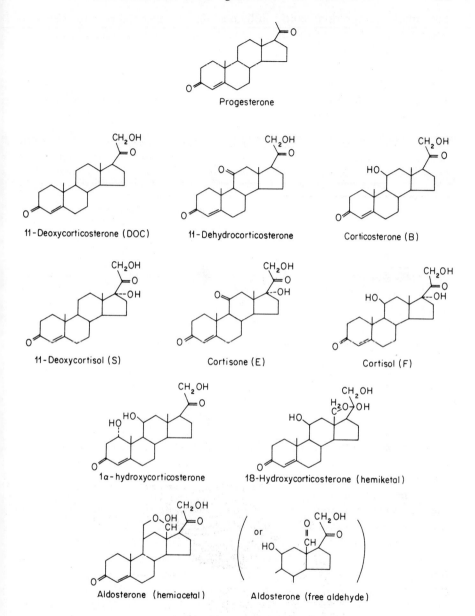

Figure 14 The structure of progesterone and the physiologically active corticosteroids that are derived from this compound and have been isolated from adrenocortical tissues and blood of vertebrate species

hormones may predominantly influence electrolyte balance and they have been collectively called the 'mineralocorticoids' whereas others are particularly active in stimulating glyconeogenesis and they have been termed 'glucocorticoids' (Table 6).

Following the isolation of *deoxycorticosterone, corticosterone, cortisol, 11-deoxycortisol, cortisone* and *11-dehydrocorticosterone* from beef adrenals tissue, the accrued 'mother liquors' which resisted further crystallization were

Table 6. The comparative potencies of some hormones expressed in terms of the minimum dose necessary to sustain adrenalectomized rats and to correct some aspects of their impaired electrolyte and carbohydrate metabolism. (Data assembled from numerous sources by Freiden and Lipner (1971); modified)

	Minimally effective dose (mg (100 g body wt)$^{-1}$)		
	Life maintenance[a]	Na$^+$ retention[b]	Liver glycogen deposition[c]
Aldosterone	0.025	0.007	0.33
Deoxycorticosterone	0.50	0.20	10.0
Cortisol	2.00	2.00	0.10
Cortisone	2.00	2.90	0.14
Corticosterone	2.70	1.25	0.33

[a] Based on the survival of adrenalectomized rats maintained on a limited salt and water diet.
[b] Based on the renal excretion of Na$^+$ by adrenalectomized rats maintained on controlled salt and water intake.
[c] Based on the deposition of liver glycogen in fasting adrenalectomized rats.

far more potent than any of the chemically defined hormones with respect to correcting the electrolyte balance of adrenalectomized rats. Attempts to identify the active principle in this amorphous fraction were unsuccessful until 1952 when Simpson and Tait in London succeeded in partially characterizing the hormone. Collaboration between these investigators and a team in Switzerland led to the structural definition of a compound which was called *aldosterone*, 18,11-hemiacetal of 11β,21-dihydroxy-20-oxopregn-4-en-18-ol-3-one (Simpson and coworkers, 1954; Simpson and Tait, 1955). This hormone, as the name implies, has an aldehyde substitution at the C-19 position, a configuration that is readily masked by reversible interaction with the hydroxyl group at the C-11 position to form a hemiacetal (Fig. 14). Aldosterone is the most potent mineralocorticoid that has been isolated from the adrenal cortex and, in contrast to deoxycorticosterone, it also has significant glucocorticoid action (Table 6).

The most recent additions to the list of probable hormonal corticosteroids are 18-hydroxycorticosterone (11β,18,21-Trihydroxy-4-pregnene-3,20-dione), a secretory companion of aldosterone, and 1*α-hydroxycortico-*

sterone (1α,11β,21-trihydroxy-4-pregnene-3,20-dione) that seems to be a major constituent of interrenal tissue of elasmobranch fishes (Idler and Truscott, 1967).

The identification of a biologically active corticosteroid in either the extracts of adrenal tissue or the medium in which adrenocortical slices have been incubated *in vitro* does not imply that the compound circulates in the blood and acts hormonally *in vivo*. The presence of some compounds may represent an accumulation of biosynthetic intermediates and in closed incubation media high concentrations of some steroids may result from abnormal precursor–product relationships (Sandor and Idler, 1972). Using the minimal criteria for the presumptive identification of a steroid hormone, the formation derived from the limited number of species that have been examined suggests that all vertebrate classes have a collective potential to synthesize and secrete not more than eight corticosteroids (Table 7). Studies on mammals indicate that corticosteroids with a C-11 keto group are metabolites of secreted glucocorticoids. Therefore, the *cortisone* identified in peripheral plasma from some cyclostomes, teleosts and chondrostean fishes may have been derived from *cortisol* subsequent to its secretion by the interrenal tissues.

B. Site of Production and Release

Tissues that produce steroid hormones are derived from the embryonic mesenchymal cells of the dorsal mesentery in an area anterior to the mesenophros. The differentiation of the cells leads to the formation of cell types that may, even in the same tissue, synthesize a variety of steroid hormones. Such is the case in the adrenal cortex of eutherian mammals where the cells of the zona glomerulosa may produce mostly aldosterone, whereas the cells of the zona fasciculata may synthesize the other corticosteroids that are characteristic of the species. Recent evidence has suggested that the interrenal tissue of amphibians and birds may also be organized into functionally separate regions (Pehleman and Hanke, 1968; for a review, see Holmes and Phillips, 1976).

The biosynthesis of corticosteroids is controlled primarily through the action of adrenocorticotrophic hormone (ACTH), although angiotensin II may also play an essential role in sustaining the biosynthesis of aldosterone. The initial steps in this biosynthesis are common to all steroidogenic tissues and involve the conversion of cholesterol to pregnenolone. In adrenocortical cells this reaction is stimulated following their interaction with ACTH. The cellular interaction between angiotensin II and the cells of the zona glomerulosa in mammals is not clear but ACTH certainly plays a permissive role in the synthesis of aldosterone.

In 1957, Sutherland and Rall demonstrated that ACTH stimulates the production of c-AMP in mammalian adrenal cell homogenates. In the same

Table 7. The adrenocortical steroids that have been identified in plasma from some species in each vertebrate class

Class	Deoxycorticosterone	Corticosterone	Cortisol	Cortisone	1α hydroxycorticosterone	Aldosterone
Agnatha (cyclostomes)		✓	✓ᵃ	✓		
Chondrichthyes Holocephali (ratfish)						
Elasmobranchi (sharks and rays)					✓	
Osteichthyes Teleosti (boney fishes)		✓	✓	✓		
Holostei (gar and bowfish)			✓			
Chondrostei (sturgeons, paddlefish and bichir)		✓	✓	✓		
Dipnoi (lungfish)		✓	✓ᵃ			✓
Amphibia (frogs, toads and salamanders)		✓				✓
Reptilia (turtles, lizards and crocodiles)	(✓)	(✓*)				(✓)
Aves (birds)		✓ᵇ				✓
Mammalia Prototheria (duckbilled platypus and spiny anteaters)		✓	✓			✓
Metatheria (marsupials)		✓	✓			✓
Eutheria	✓ᵇ	✓ᵇ	✓ᵃ			✓

ᵃ The presence of an 11-deoxygenated form of this hormone (i.e. 11-deoxycortisol) has also been detected in plasma from some species.

ᵇ The presence of an 18-oxygenated form of these hormones (18-hydroxycorticosterone and 18-hydroxy-11-deoxycorticosterone) may also be present in the plasma of some species.

(✓) and (✓*) The methods used to derive the available data on corticosteroid concentrations in samples of reptilian plasma are inadequate; the presence of these compounds in plasma is tentatively presumed on the basis of these incomplete identifications and positive identifications derived from *in vitro* adrenal incubation studies (see Sandor, 1972).

year, Haynes and Berthet also demonstrated that the activity of a glycogen phosphorylase in adrenocortical cells was enhanced by ACTH and they suggested that c-AMP may act to stimulate glycogenolysis in these cells (Haynes and Berthet, 1957). The increased availability of glucose-1-phosphate was viewed as the first major step in the hexose monophosphate shunt pathway leading to the production of d-ribulose 5-phosphate and cytoplasmic reducing power in the form of NADPH. Since the reductive biosynthesis of steroids from cholesterol requires NADPH, it was postulated that ACTH could promote steroidogenesis indirectly by controlling the availability of the hydrogen carrier. The likelihood that the production and release of adrenal steroids is controlled solely in terms of the Haynes and Berthet hypothesis has, however, diminished in recent years.

The production of c-AMP in response to ACTH stimulation precedes corticosteroidogenesis and does not appear to require the synthesis of new protein. Steroidogenesis, however, does appear to require the translation of a stable mRNA (Ferguson, 1963; Grahame-Smith and coworkers, 1967). Puromycin, cycloheximide and chloramphenicol, but not actinomycin D, each curtail the synthesis of corticosteroids in response to amino acid uptake or synthesis of new protein (Ferguson, 1963; Farese, 1964; Garren, Ney and Davis, 1965; Ney, Davis and Garren, 1966). The adrenocortical cell is fully capable, therefore, of increasing its output of steroid hormone in the relatively short time required for the translation of a pre-existing stable mRNA.

These observations, together with the isolation of an ACTH receptor protein and an ACTH-sensitive adenyl cyclase from adrenocortical cell membrane fragments from the basis of the belief that ACTH exerts its effect on the cell by reacting with the outer surface of the membrane (Fig. 15; Grahame-Smith and coworkers, 1967; Lefkowitz, Roth and Pasten, 1970; Finn, Widnell and Hofmann, 1972). Furthermore, the addition of c-AMP to adrenocortical cells *in vitro* can mimic the effect of ACTH on corticosteroidogenesis (Roberts and coworkers, 1964; Creange and Roberts, 1965a, 1965b).

The c-AMP produced as a result of ACTH stimulation is believed to become associated with a binding protein in the cytosolic fraction of the cell and this in turn leads to the stimulation of a protein kinase (Fig. 8). The protein kinase may then act in several ways to enhance the synthesis of pregnenolone from cholesterol; these may be summarized as follows:

(1) The uptake of cholesterol from the plasma into the adrenocortical cells may be stimulated (Dexter and coworkers, 1967; Dexter, Fishman and Ney, 1970).

(2) The cholesterol esterase activity may be enhanced to accelerate the hydrolysis of cholesterol esters in the lipid droplet and so increase the amount of free cholesterol substrate available to the biosynthetic pathway (Simpson and coworkers, 1972a; Simpson and coworkers, 1972b).

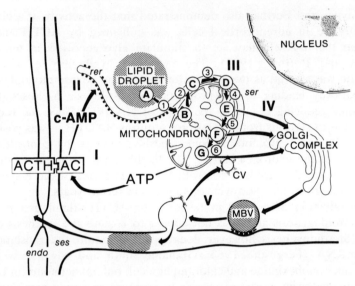

Figure 15 A diagrammatic representation of the synthesis of corticosteroids and their release from adrenocortical cells. The possible relationships between various steps in this biosynthetic process and the various organelles are indicated. ACTH binds to a specific receptor situated in the plasma membrane and there it activates a membrane-bound adenylate cyclase. This enzyme is responsible for the formation of 3′,5′-cyclic-AMP from ATP (I). Operating through a protein kinase, the cyclic-AMP may initiate the synthesis of a proteinaceous factor capable of promoting steroidogenesis (II). The reaction sequence shows the probable location within the adrenocortical cell of the enzyme systems responsible for the formation of some typical vertebrate corticosteroids; some steps in the biosynthesis occur within the mitochondria while others require the precursors and products to move out of the mitochondria and before the synthesis of yet other hormones can be completed, some steroid must re-enter the mitochondria (III). The final hormonal product of this synthesis may then be 'packaged' in the Golgi apparatus (IV). Recent studies have suggested that steroids may be released from the adrenocortical cell by a process of exocytosis coupled to the re-uptake of membrane by coated vesicles (V). Abbreviations used in the diagram are as follows: MBV, membrane-bound vesicles; endo, endothelial cell of a capillary; rer, rough endoplasmic reticulum; ser, smooth endoplasmic reticulum; ses, subendothelial space; (A) cholesterol esters; (1) prognenolone synthetase complex; (B) pregnolone; (2) 3β-hydroxydehydrogenase and Δ^5-Δ^4 isomerase; (C) progesterone; (3) 21-hydroxylase; (D) deoxycorticosterone, (4) 11β-hydroxylase; (E) corticosterone; (5) 18-hydroxylase; (F) 18-hydroxycorticosterone; (6) 18-hydroxysteroid dehydrogenase; (G) aldosterone

(3) A ribosomal protein may be phosphorylated; this may lead to the translation of a stable mRNA and ultimately to the synthesis of new protein (Garren and coworkers, 1971b). Two functions have been postulated for this new protein. First, it may serve to bind with cholesterol and so increase the availability of cholesterol to the mitochondria and secondly, it may act to stimulate directly the side-cleavage of cholesterol within the mitochondria

(Farese, 1967; Ichii, Ikeda and Izawa, 1970; Simpson and coworkers, 1972b; Kan and coworkers, 1972).

In addition to the above effects ACTH may, in the course of activating the adenyl cyclase, cause Ca^{2+} to be released from the inner surface of the plasma membrane (Rubin, 1974). The Ca^{2+} may then be actively taken up by the mitochondria and, since Ca^{2+} has been implicated in both the production of NADPH and in the side chain cleavage of cholesterol, the synthesis of progesterone may be further facilitated (Carofoli and Lehninger, 1971; Simpson and Williams-Smith, 1975).

Upon leaving the mitochondrion, the pregnenolone molecule is converted to progesterone through the actions of 3β hydroxysteroid dehydrogenase and $\Delta^5-\Delta^4$ isomerase located in the endoplasmic reticulum (Fig. 15). The progesterone molecule is then further hydroxylated at either the C-21 position to produce deoxycorticosterone, or at the C-21 and C-17 positions to produce 11-deoxycortisol (Figs 14 and 15). In some species some of these compounds may leave the adrenocortical cell to function as corticosteroid hormones. On the other hand, some of the deoxycorticosterone and 11-deoxycortisol may re-enter the mitochondria and become hydroxylated to form corticosterone, cortisol and 18-OH corticosterone. Again, these compounds may leave the mitochondria to enter the circulation in some species (Table 7). Some of the 18-OH corticosterone, however, may remain in the mitochondria to undergo further dehydrogenation to form aldosterone (Marusic, White and Aedo, 1973).

The mechanism by which steroids are released from the adrenocortical cell is not fully understood. Although the hormonal products of steroidogenesis have been frequently presumed to leave the cell by diffusion, this assumption does not explain why the intermediates in corticosteriod synthesis do not leave the cell with equal rapidity. Under some experimental conditions, low concentrations of Ca^{2+} have been shown to impair the release but not the production of steroid hormones and this suggests that the two processes may be distinct (Jaanus, Rosenstein and Rubin, 1970; Rubin, Jaanus and Carchman, 1972). Preliminary biochemical and ultrastructural evidence suggests that corticosteroids may be released from the adrenocortical cells by a process of exocytosis (Rubin and coworkers, 1974; Pearce, Cronshaw and Holmes, 1977).

C. Effects in the Organism

1. *Fishes*

Removal of the pituitary gland from both fresh-water and marine teleosts is followed by characteristic changes in their water and electrolyte balance. In general, the composition of the body fluids tends towards equilibrium with the environmental water. Although these compositional changes may be attributed to the absence of several hormones, such as prolactin, thyroxin and perhaps the neurohypophysial hormones, there is also good evidence to suggest that diminished adrenocortical function is responsible for some of the osmoregulatory dysfunction (see review by Henderson and Chester Jones, 1972). Studies on a variety of fresh-water and sea-water teleosts have provided both direct and indirect evidence that adrenocortical steroids may influence the osmoregulatory roles of the kidneys, the gill epithelium and the intestinal mucosa.

The possibility that the gross physiological effects of adrenocortical hormones in teleosts may not conform to those observed in mammals is suggested by the results of studies on fresh-water trout (Chester Jones, 1956; Holmes and Butler, 1963). In these studies, fresh-water trout (*Salmo trutta* and *Salmo gairdneri*) were injected with either deoxycorticosterone, corticosterone, cortisol or aldosterone and in each instance both the Na^+ concentration and the $Na^+:K^+$ ratio in plasma declined. The overall effects of these hormones, therefore, appeared to be opposite to those that might be expected in similarly treated mammals, where their characteristic antinatriuretic and kaliuretic effects would cause increases in the plasma Na^+ concentration and the $Na^+:K^+$ ratio. Since the effects of corticosteroids on the compartmental volumes in this species are not known, it is impossible to determine whether the observations reflect either an actual net loss of Na^+ from the body or merely the redistribution of this ion between the extracellular, intracellular and transcellular fluids.

This dilemma was partially resolved by the observation that the total amount of Na^+ excreted by the saline-loaded fresh-water rainbow trout increased following the injection of either cortisol or deoxycorticosterone (Holmes, 1959; Holmes, Phillips and Chester Jones, 1963). The increase in the total excretion of Na^+, however, was due to simultaneous but opposite effects of these corticosteroids on the renal and extrarenal excretory pathways (Holmes, 1959). Thus, although the hormones seem to have typical antinatriuretic effect on the kidney, this effect is much less than the increment which they stimulate in the extrarenal excretion of this ion; the overall effect of each hormone, therefore, is to stimulate a net increase in the excretion of Na^+. Since corticosterone may significantly reduce the glomerular filtration rate in this species, the apparent antinatriuretic effect may not be entirely attributable to the stimulation of increased tubular reabsorption of Na^+ (Holmes and McBean, 1963). This type of overall response has also been observed in the fresh-water European eel, *Anguilla anguilla* (Chan and coworkers, 1967; Chan, Rankin and Chester Jones, 1969). Once again, injections of cortisol into the intact fish caused a significant increase in the net *loss* of Na^+ from the organism. Aldosterone and deoxycorticosterone, however, had no effect on these parameters in the eel.

Progress on the further elucidation of the true osmoregulatory roles of the adrenocortical hormones in the lives of fresh-water and marine teleost fishes has been aided considerably through studies on the European eel (*Anguilla anguilla*). This species is euryhaline and therefore possesses the necessary regulatory mechanisms that enable it to alternately extract Na^+ from the fresh-water environment and excrete the excess Na^+ that continuously diffuses into the organism living in sea-water. The shape of the fish also permits it to be maintained in an inverted U-tube so that the influx and efflux of ions across the branchial epithelium may be monitored separately from the quantities being excreted via the kidneys (Bellamy and Chester Jones, 1961). Finally, the European eel is the only known teleost species that has been successfully adrenalectomized (Chester Jones, Henderson and Moseley, 1964).

Following removal of the adrenal glands from Na^+-depleted fresh-water eels, the rates of Na^+ uptake and K^+ loss across the branchial epithelium become severely impaired and only after treatment of these adrenalectomized fish with aldosterone, and possibly low doses of cortisol, are the rates of Na^+ influx and K^+ efflux restored to normal (Henderson and Chester Jones, 1967). By simultaneously injecting the synthetic 'glucocorticoid', betamethasone, and the 11β hydroxylase inhibitor, metyrapone, the secretion of ACTH and the endogenous production of deoxycorticosterone and 11-deoxycortisol may be diminished significantly and a condition similar to surgical adrenalectomy ensues. Under these circumstances, the extrarenal influx of Na^+ is also depressed significantly (Henderson and Chester Jones, 1967).

Injection of aldactone, the anti-aldosterone drug, into intact Na^+-depleted eels also significantly reduces the net influx of Na^+ and the net efflux of K^+ across the branchial epithelium (Chester Jones and Bellamy, 1964; Henderson and Chester Jones, 1967). Since this compound is believed to compete with and displace endogenously produced aldosterone molecules from their specific receptors in target tissues, one may presume tentatively that either aldosterone or some similarly sensitive hormone molecule acts directly on cells in the gill epithelium of fresh-water eels to control the influx and efflux of Na^+ and K^+.

When the circulating concentrations of adrenal steroids in plasma of fresh-water eels are increased by either subjecting them to stress or injecting them with cortisol, the Na^+ flux across the branchial epithelium is reversed and the fishes consistently show net losses of Na^+. In contrast, the adrenalectomized eels show no change in their pattern of extrarenal electrolyte excretion following exposure to stress (Henderson and Chester Jones, 1967). But, when 'unstressed' eels are injected with aldosterone, the net effect of the hormone is opposite to that of cortisol and an *increase* in the rate of Na^+ uptake is a constant feature of the response (Henderson and Chester Jones, 1967).

The enhanced extrarenal loss of Na^+ that occurs when both the eel and the trout are treated with cortisol suggests that the action of this hormone on the branchial epithelium may be crucial to the survival of euryhaline species during the early period of adaptation to the marine habitat. To test this hypothesis, the extrarenal Na^+ efflux has been measured in intact and adrenalectomized eels during the period immediately following their transfer to sea-water (Mayer and coworkers, 1967). During adaptation to sea-water the turnover of exchangeable Na^+ was found to be greatly reduced following removal of the interrenal tissue but it could be readily restored to normal by treating the adrenalectomized eels with comparatively small doses (50 μg per 100 g body weight) of cortisol. The diminished rates of turnover of exchangeable Na^+ in the adrenalectomized eels, and its restoration to normal following treatment with cortisol, probably reflect the corresponding changes which occur in the rates of Na^+ efflux across the branchial epithelium. Indeed, an examination of the temporal development of this enhanced Na^+ efflux during the early phase of adaptation of sea-water has shown that immediately following transfer the increase in Na^+ efflux is slow but after two to three hours, the rate accelerates until after twelve hours it has increased to between one-half and two-thirds of the rate of Na^+ efflux found in the fully adapted fish. In contrast, the rate of Na^+ efflux in the adrenalectomized eel does not change significantly for at least five hours after transfer to sea-water and thereafter the rate of increase is small compared to that found in the intact fish.

The consensus emerging from this important series of experiments on the eel is that the extrarenal uptake of Na^+ by the fresh-water fish is controlled by aldosterone, and that the net extrarenal efflux of Na^+ occurring in the sea-water-adapted form is controlled by either corticosterone or cortisol. Of course this hypothesis is predicated in part on the belief that aldosterone is a natural secretory product of the teleost interrenal. If it is not a natural secretory product of the teleost interrenal, then some other corticosteroid with similar biological properties must be postulated. An early analysis of blood collected from the sockeye salmon (*Onchorhynchus nerka*) during their upstream migration in fresh-water, suggested the presence of aldosterone (Phillips, Holmes and Bondy, 1959). However, although a more recent analysis has positively identified this hormone in the plasma of a marine teleost, the Atlantic herring (*Clupea harengus*), none has confirmed its presence in a teleost living in fresh-water (Idler and Truscott, 1969; Idler, 1971). Identification of aldosterone in the plasma of fresh-water teleosts, however, is the keystone to this hypothesis; the Na^+-conserving properties of this hormone are not consistent with life in sea-water. Furthermore, if the renin–angiotensin II system in teleosts has a regulatory role similar to that in mammals, then the hypernatremic state of the marine teleost would mitigate against the release of aldosterone (Sokabe, Mizogami and Sato, 1968; Sokabe and coworkers, 1969; Sokabe and Nakajima, 1972). Indeed, during the early phases of adaptation to sea-water, when hypernatremic conditions are most pronounced, the plasma cortisol concentrations increase in both the European (*A. anguilla*) and the Japanese (*A. japonica*) eels (Hirano, 1969; Ball and coworkers, 1971; Hirano and Utida, 1971).

During the course of migration into the ocean, the eel commences to drink sea-water and after excreting the excess electrolyte via the extrarenal mechanism in the gill, the fish gains sufficient osmotically free water to compensate for water lost by diffusion (Smith, 1930; Keys, 1931a, 1931b, 1933; Keys and Willmer, 1932). As a consequence of drinking sea-water, one of the first adaptive responses in these migrating fish, and presumably other euryhaline teleosts, is to increase the electrolyte water reabsorbing properties of the small intestine (Oide and Utida, 1967, 1968; Maetz and Skadhauge, 1968; Skadhauge, 1969). There is also some evidence to suggest that the need to transfer more electrolyte and water across the mucosa may be anticipated before the fish actually enter sea-water, for these adaptive changes in the intestine have been observed in the Japanese eel as it migrates down-river during the autumn (Oide and Utida, 1967; Utida and coworkers, 1967).

Ingested sea-water is generally believed to be diluted by the passive diffusion of water into the gastrointestinal tract and upon entering the small intestine, the active uptake of ions from an isosmotic flud facilitates the solute-linked uptake of water. It is clear, therefore, that during migration the rate of ion uptake must increase as the concentration of the ingested drinking water becomes greater. Furthermore, as the concentration of the ingested sea-water increases, the rate of mucosal water transfer per unit mass or unit area of small intestine must increase commensurately in order to yield the same volume of water to the organism (for review see Hirano and coworkers, 1975).

During the first two hours after transfer of the Japanese eel from fresh-water to sea-water, the plasma cortisol concentration doubles and this is followed by a significant increase in the rate of mucosal water transfer (Hirano, 1967, 1969; Oide and Utida, 1967). A similar increase in the rate of mucosal Na^+ and water transfer also occurs in isolated intestinal sacs prepared from *fresh-water* eels that have been treated with either ACTH or cortisol; this is a dose-dependent response that occurs between ten and twenty-four hours after treatment with the hormone (Hirano and Utida, 1968; Utida and coworkers, 1972; Ando, 1974). The delay between initial exposure to cortisol and the recognition of a physiological response may be due to the hormone acting on functionally undifferentiated cells in the crypts of the villi; only when these stimulated cells have migrated to the functional regions of the villi will their enhanced Na^+ and water transfer properties become apparent (Crocker and Munday, 1967a, 1967b). In contrast to the effects of cortisol and ACTH, treatment of fresh-water eels with aldosterone, deoxycorticosterone and corticosterone has little effect on mucosal transfer.

In a study on adrenalectomized European eels, Gaitskell and Chester Jones (1970) were able to show a direct relationship between the circulating levels of cortisol and the rates of mucosal water transfer in the intestine; the mucosa in the posterior regions of the intestine showing a greater sensitivity to the presence of hormone than that of the anterior intestine.

The results of studies on intestinal transfer in the goldfish (*Carassius auratus*), a stenohaline fresh-water teleost, essentially confirm those obtained from studies on the eel. The uptake of Na^+ and water by the intestinal mucosa is stimulated by the injection of cortisol, depressed following removal of the pituitary and restored in hypophysectomized individuals after treatment with cortisol; as in the eel, aldosterone, deoxycorticosterone and corticosterone exert no effect in goldfish (Ellory, Lahlou and Smith, 1971, 1972; Porthe-Nibellé and Lahlou, 1975).

The cortisol-induced increase in mucosal water transfer in the eel is accompanied by an increase in the electrical negativity of fluid on the serosal side of the mucosa (Ando, 1974; Ando, Utida and Nagahama, 1975). This is believed to reflect the development of an increased capacity for mucosal cells to actively transport Cl^- out of the intestinal lumen and thus permit the passive uptake of Na^+ and water. The precise locus of such a hormonally induced mechanism is not known.

2. *Amphibians*

In addition to the kidney, the skin and the urinary bladder of many *Amphibia* also constitute important osmoregulatory organs and each, like the kidney, may respond to the presence of corticosteroids. But, since there have been few attempts to remove the adrenals from amphibians, the available data are sparse and do not yield much insight into the role of adrenal steroids in the class of vertebrates. The earliest attempts to adrenalectomize an amphibian were made on the toad, *Bufo arenarum*, and a general loss of salt was reported (Marenzi and Fustinoni, 1938). Later, destruction of the adrenal cortical tissue in the frog, *Rana temporaria*, was shown to have little recognizable effect in individuals that were adapted to winter conditions but when performed on frogs adapted to summer conditions, they died within two days of the operation; these frogs appeared to have accumulated K^+ and suffered large losses of Na^+ in a manner similar to the classic mammalian pattern of adrenocortical insufficiency (Fowler and Chester Jones, 1955). When the adrenalectomized 'summer' frogs were placed in isosmotic saline rather than tap-water, the loss of tissue Na^+ was partially arrested and they survived much longer.

A progressive loss of Na^+ and Cl^- has been observed in *Rana temporaria, Rana esculenta* and *Bufo marinus* following hypophysectomy but the loss seems to be much slower than that observed following adrenalectomy (Jørgensen, 1947; Jørgensen and Rosenkilde, 1957; Middler and coworkers, 1969). In some instances the excessive loss of Na^+ may be successfully prevented, and the survival of hypophysectomized individuals prolonged, by injecting mammalian corticotrophin (see Jørgensen and Larsen, 1963; Middler and coworkers, 1969). The loss of Na^+ in the adenohypophysectomized anurans, however, may be due to an accelerated efflux of Na^+ through the skin rather than an absence of the nitriferic action of the endogenous corticosteroids. No changes in the pattern of renal Na^+ excretion have been identified following hypophysectomy but this may be due to the continued presence of aldosterone in these frogs. Nevertheless, aldosterone has not been shown to act definitively on the renal tubules of any amphibian (Mayer, 1963, 1969; Middler and coworkers, 1969). Immersion of the intact bullfrog (*Bufo marinus*) in distilled water leads to Na^+ depletion and under these circumstances the circulating levels of aldosterone increase significantly, while the plasma concentrations of corticosterone remain almost unchanged. Such an increase in aldosterone secretion is accompanied by an increase in the rates of Na^+ uptake across the skin, the urinary bladder and the colon (Crabbé, 1961, 1966). The simultaneous injection of mammalian corticotrophin and angiotensin II has also been observed to cause increases in the rates of Na^+ transport across these membranes and Crabbé has suggested that this is the result of an increase in the endogenous rates of aldosterone secretion (Crabbé, 1966).

3. Reptiles

Attempts to adrenalectomize reptiles frequently leads to persistent haemorrhage and the operation has been successfully accomplished in only a few species. Among the lizards, *Trachysaurus rugosus* and *Dipsosaurus dorsalis* have been reported to die within a few days of the operation and in *Trachysaurus rugosus*, death is preceded by hyperkalemia, although the plasma Na^+ concentration remains unchanged (Bentley, 1959; Templeton and coworkers, 1968). The response of the painted turtle, *Chrysemys picta*, is more classical, albeit slow; the plasma Na^+ concentration decreases while the K^+ concentration rises two weeks after removal of the adrenal glands (Butler and Knox, 1970). By tying off the blood supply to the adrenal glands of water snakes (*Natrix cyclopion*), the adrenal tissue can be caused to become atrophic (Elizondo and LeBrie, 1969). Under these conditions of presumed adrenocortical insufficiency, urine flow and glomerular filtration rates remain unchanged but the renal excretory rates of Na^+, Cl^- and total osmotically active material increase significantly; these changes probably reflect impaired reabsorptive properties in the proximal tubule (LeBrie and Elizondo, 1969; Elizondo and LeBrie, 1969).

Hypophysectomy of the lizard, *Agama agama*, and the grass snake, *Natrix natrix*, also causes atrophy of the adrenocortical tissue, but this is not associated with any significant change in the plasma and muscle electrolyte concentrations (Wright and Chester Jones, 1957). However, removal of the pituitary from the iguanid lizard, *Dipsosaurus dorsalis*, does cause a rise in the plasma water and Na^+ concentrations, an increase in the total and intracellular water content of muscle, an increase in the intracellular muscle Na^+ concentration and a decrease in the intracellular muscle K^+ concentration; each of these changes can be almost completely rectified by injections with corticosterone (Chan, Callard and Chester Jones, 1970).

Studies on the effects of adrenal steroids in intact reptiles have revealed very little about their role in the control of renal function. The urine composition of well-hydrated intact water snakes, *Natrix cyclopion*, does not change in response to injections of either aldosterone, corticosterone or spironolactone (Elizondo and LeBrie, 1969). High doses of aldosterone, however, cause significant increases in the rates at which water, Na^+ and total osmotically active materials are reabsorbed across the proximal renal tubules of snakes that have received intraperitoneal loads of NaCl. Using much smaller doses of aldosterone (10 μg per 100 g body weight per day), Bradshaw, Shoemaker and Nagy (1972) were unable to detect any effect of aldosterone on kidney function in intact saline-loaded lizards (*Dipsosaurus dorsalis*).

In contrast to the reported effects of aldosterone in intact reptiles, corticosterone causes a *decrease* in the fractional reabsorption of filtered Na^+, thereby increasing the rate of urinary Na^+ excretion by agamid lizards (*Amphibolurus ornatus* and *Dipsosaurus dorsalis*) given saline loads (Bradshaw, 1972; Bradshaw, Shoemaker and Nagy, 1972). These investigators also noted that saline loading in both species led to an increase in the rate of corticosteroidogenesis that was accompanied by a decrease in the tubular reabsorption of Na^+. Furthermore, the creation of adrenal insufficiency, by either hypophysectomy or dexamethasone blockade, effectively abolished these responses. These results are consistent with those obtained from studies on *Dipsosaurus dorsalis*, where hypophysectomy was found to cause an accumulation of Na^+ and water in the tissues (Chan, Callard and Chester Jones, 1970).

Attempts to block the effects of corticosteroids on target organs in reptiles have failed to show any effect on either electrolyte concentrations in plasma or the rates of electrolyte excretion via the kidney (Elizondo and LeBrie, 1969; Butler and Knox, 1970; Bradshaw, Shoemaker and Nagy, 1972). These studies have frequently involved the use of spironolactone and it has been widely presumed that this drug competes with only aldosterone for binding sites in target organs. This presumption is based solely on the action of the drug in mammals and fails to recognize that many osmoregulatory tissues in lower vertebrates are controlled by cortisol and corticosterone and not by aldosterone. The effects of spironolactone are certainly not limited to an inhibition of aldosterone action (Crocker and Holmes, 1971, 1976). An examination of the effects of this drug in relation to the changes known to occur in the rate of corticosterone synthesis in the saline-loaded agamid lizard may yield some interesting insights into the control of kidney function in these reptiles.

The cloaca and the large intestine of the snakes and turtles cannot be readily separated but some active Na^+ transport does appear to occur in this region of the gut (Baillien and Schoffeniels, 1961; Bentley, 1962; Junqueira, Malnic and Monge, 1966). In the turtle, *Testudo graeca*, this process is not affected *in vitro* by the presence of aldosterone. The urinary bladder of the chelonians, *Pseudemys scripta* and *Testudo graeca*, also possesses the ability to actively transport Na^+ from the mucosal to the serosal surfaces (Brodsky and Schlib, 1960; Bentley, 1962). In *Testudo graeca*, aldosterone increases the rate of Na^+ transport across the bladder and if individuals are pretreated with spironolactone, the transport of Na^+ is reduced.

The extrarenal excretory organs among extant reptiles have been collectively described as 'salt glands' and they fall into two functional classes; the glands in marine reptiles tending to secrete a fluid containing more Na^+ than K^+ and the terrestrial reptiles from arid environments tending to secrete a fluid containing relatively more K^+. The functional characteristics of these salt glands have been reviewed by Dantzler and Holmes (1974).

In marine reptiles, 75-90% of the total Na^+ excreted is discharged via the extrarenal pathway (Holmes and McBean, 1964; Dunson, 1968). Under natural conditions the salt glands seem to function intermittently, presumably in response to feeding and the ingestion of sea-water. After feeding, the salt glands of the green turtle, *Chelonia mydas*, start to secrete Na^+ and K^+ but if they are given Amphenone B, a compound known to inhibit corticosteroidogenesis, immediately following the meal, the extrarenal response is blocked (Holmes and McBean, 1964). An intramuscular saline load will elicit a similar extrarenal response in these turtles and this can also be blocked by pretreatment with Amphenone B; but, the administration of corticosterone prior to treatment with Amphenone B permits a normal pattern of Na^+ excretion to occur. Thus, we have some indirect evidence to suggest that the chelonian salt gland is partially controlled, like the nasal glands in marine birds, through the action of adrenocortical hormones.

In contrast, the action of adrenal steroid hormones on the salt gland in reptiles inhabiting the arid terrestrial environment may be directed towards Na^+ retention. Adrenalectomy of the desert iguana, *Dipsosaurus dorsalis*, results in the copious secretion of a salt gland fluid containing relatively large amounts of Na^+ and the administration of aldosterone, cortisol and corticosterone to intact individuals either reduces or abolishes secretion (Templeton and coworkers, 1968, 1972; Shoemaker, Nagy and Bradshaw, 1972).

Although extrarenal excretory pathways have been identified in the yellow-bellied sea snake (*Pelamis platurus*) and the banded sea snake (*Laticauda semifasciata*) and some form of extrarenal excretion is presumed to occur in crocodiles inhabiting coastal waters, nothing is known concerning the control of these excretory organs (see review, Dantzler and Holmes, 1974).

4. Birds

The post-operative survival of adrenalectomized birds is a short time compared with most mammals and this probably reflects the high turnover rate of adrenocortical steroids in birds (Bradley and Holmes, 1971; Chan, Bradley and Holmes, 1972; Holmes, Broock and Devlin, 1974; Thomas and Phillips, 1975a). There have been only a few studies conducted on the changes occurring in the salt and water balance of birds following removal of the adrenal glands and these have been restricted to a series of studies on the duck (Thomas and Phillips, 1975b, 1975c, 1975d). Although no change occurs in the total body water content of this species following removal of the adrenal glands, the daily turnover of water is reduced to almost one half the rate observed in sham-operated birds. At the same time, the rate of urinary water loss does not decline and we must therefore presume that following removal of the adrenals, the rate of water loss via a pathway other than the kidneys is diminished; this is probably associated with a decline in metabolism and a consequent reduction in the rate of respiratory water loss (equation (1)).

The urine flow rate in adrenalectomized ducks declines significantly during the first three days after adrenalectomy and, despite a 75% reduction in intake, the birds enter a state of negative Na^+ balance (Thomas and Phillips, 1975b, 1975d). The initial effects of adrenalectomy on K^+ balance also appear to conform to the classic mammalian pattern when drinking water is available *ad libitum*. Three to five days after adrenalectomy, the urinary concentrations of Na^+ and K^+ return to normal, but the urine flow rate remains low compared to sham-operated birds. This apparent restoration of the urinary concentrating mechanism, however, does not restore the electrolyte balance of the birds and lowered plasma $Na^+:K^+$ concentration ratios (20.7 versus 30.4) persists until death at approximately one week after surgery (Thomas and Phillips, 1975b).

The renal responses in intact ducks given exogenous corticosteroids have also been examined (Holmes and Adams, 1963). Their response to a single dose of hormone, however, is short-lived and injections must be given frequently in order to sustain an identifiable effect throughout a prolonged period of urine collection. This is further evidence for the rapid metabolism of corticosteroids in birds. Corticosterone causes a mild diuresis in the constantly water-loaded duck and this is accompanied by typical kaliuretic and antinatriuretic responses. Such effects are also consistent with the responses observed in mammals. It is noteworthy, however, that when a high dose of either corticosterone or cortisol is given, the antinatriuresis is not accompanied by a kaliuretic effect, whereas a low dose of either hormone stimulates the excretion of K^+ as well as the reabsorption of Na^+ (Holmes and Adams, 1963). When given aldosterone, however, the total excretion of both Na^+ and K^+ by the intact bird decreases; this anomalous response to aldosterone is consistent with the kaliuresis observed in water-loaded adrenalectomized ducks (cf. Holmes and Adams, 1963; Thomas and Phillips, 1975d).

The bird, unlike other non-mammalian vertebrates, can produce hypertonic urine and this ability is probably associated with a counter-current multiplier arrangement of the nephrons (Poulson, 1965). The fresh-water-maintained duck, however, does not produce hypertonic urine even under conditions of extreme antidiuresis (Holmes and Adams, 1963; Holmes, Fletcher and Stewart, 1968; Bradley, Holmes and Wright, 1971). But, when the birds are given hypertonic saline drinking water or they become adapted to the marine environment, their pattern of renal excretion changes and the urine becomes hyperosmotic with respect to plasma (Holmes, Fletcher and Stewart, 1968; Stewart, Holmes and Fletcher, 1969). Nevertheless, the available osmotic space in the urine is such that the limiting isorrheic concentration of Na^+ is always less than that necessary to permit all of the ingested Na^+ to be excreted via the kidneys. To maintain homeostasis, therefore, an extrarenal excretory mechanism must compliment the limited excretory abilities of the kidneys (equation (3)).

The extrarenal excretory organs in marine birds are developed from paired nasal glands situated in the orbit. When stimulated by an osmotic load, these glands discharge a concentrated solution of electrolytes into the anterior nasal cavity and in this way a marine bird may excrete more than 90% of the Na^+ and Cl^- and 70% of the K^+ in its diet (Fletcher and Holmes, 1968).

In response to the uptake of osmotically active material, the stimulation of a receptor associated with some part of the extracellular compartment elicits a sensory input to the central nervous system and a sequence of events leading to the onset of nasal gland secretion is initiated. Two motor pathways seem to be necessary for sustained nasal gland function. Visceral motor impulses stimulate the glands directly and efferent impulses terminating in the median eminence activate the adenohypophysial-adrenocortical axis (for review, see Holmes and Phillips, 1976).

When mallard ducks that are adapted to either fresh-water of hypertonic saline drinking water are given hypertonic saline loads, they show immediate four- to five-fold increases in their plasma corticosterone concentrations and these levels are sustained for two to three hours (Allen, Abel and Takemoto, 1975). Similar increases in adrenocortical activity have been observed in Pekin ducks loaded with hypertonic saline (Donaldson and Holmes, 1965; Macchi, Phillips and Brown, 1967). Removal of the adenohypophysis prevents this increase in plasma corticosterone concentration and the secretion of nasal gland fluid in response to the saline load is reduced to only 5% of normal. Conversely, treatment of the adenohypophysectomized birds with ACTH increases their peripheral plasma corticosterone concentrations and restores a normal extrarenal excretory response (Wright, Phillips and Huang, 1966; Bradley and Holmes, 1972; Holmes, Lockwood and Bradley, 1972). The imposition of partial adrenocortical insufficiency through either acute discontinuance of ACTH replacement therapy in adenohypophysectomized birds or by unilateral adrenalectomy on the previous day causes the extra-renal response to a hypertonic saline load to decrease significantly (Holmes, Lockwood and Bradley, 1972). Furthermore, either the removal of both adrenal glands or the treatment of birds with metyrapone almost totally abolishes the secretion of nasal gland fluid and a normal response can be restored by corticosteroid replacement therapy (Phillips, Holmes and Butler, 1961; Thomas and Phillips, 1975d; Cheeseman and Phillips in Holmes and Phillips, 1976). Augmentation of the endogenously produced corticosteroids by injection of the hypertonic saline-loaded bird with cortisol, deoxycorticosterone, 18-OH-corticosterone and ACTH significantly enhances the extra-renal excretion of nasal gland fluid in respone to a hypertonic saline load (Holmes, Phillips and Butler, 1961; Phillips and Bellamy, 1962; Lanthier and Sandor, 1973). Aldosterone is without any effect on nasal gland secretion; this is not surprising, however, since the hypernatraemia that precedes nasal gland secretion is not consistent with the stimulation of aldosterone release (Phillips and Bellamy, 1962).

The stimulatory effects of corticosterone on nasal gland tissue appear to be mediated by a direct intracellular action. The intravenous administration of labelled corticosteroids at the time the birds are loaded with hypertonic saline results in an intracellular accumulation of radioactivity in the nasal gland cells but not in the cells of the adjacent Harderian gland (Phillips and Bellamy, 1967). Evidence that corticosteroid hormones interact with specific cytosolic and nuclear receptors in the nasal gland secretory cells will be discussed later. In view of these observations, however, any indirect stimulation of nasal gland function through the gluconeogenic action of corticosterone would appear to be minor (cf. Peaker and coworkers, 1971; Holmes, Lockwood and Bradley, 1972).

As in the case of euryhaline teleost fishes, the adaptation of some species of birds to the coastal marine environment involves changes in the absorptive properties of the mucosa in the small intestine. Thus, within a few hours of being given hypertonic saline drinking water, ducklings that have been raised on fresh drinking water show increases in the rates of water and Na^+ uptake across the intestinal mucosa. Within 48 hours the rate becomes maximal along the whole length of the small intestine and this high rate is sustained as long as the birds are fed the hypertonic drinking water. Upon return to fresh drinking water, the mucosal transfer rates soon start to decline and within three days they have returned to normal (Crocker and Holmes, 1971). However, if the birds are treated with either spironolactone or metyrapone before they are given hypertonic drinking water, the increases in mucosal transfer do not occur. Furthermore, increases similar to those occurring in the birds given hypertonic saline can be induced in *fresh-water* ducklings by treating them with corticosterone; as in the case of birds given hypertonic saline, this response can be effectively blocked by an oral dose of spironolactone prior to injection of the corticosterone (Crocker and Holmes, 1971, 1976).

Development of increased rates of mucosal water transfer seem to be essential for sustained nasal gland function in birds that are continuously exposed to hypertonic drinking water; if the development is blocked by treatment with either spironolactone or metyrapone, then the nasal glands do not appear to secrete fluid in response to the ingestion of sea-water (Crocker and Holmes, 1971). Thus, the initiation and continuation of full nasal gland function in coastal marine birds may depend on the establishment of increased absorptive properties in the intestinal mucosa and the development of these properties seems to depend on an adrenocortical hormone, probably corticosterone (Crocker and Holmes, 1976).

5. Mammals

The adrenocortical insufficiency that follows removal of the adrenal glands is always fatal but two distinct patterns of physiological change have been recognized during the post-operative period. Some individuals show a sudden fall in blood pressure, that may be accompanied by hypoglycaemia, and death is probably attributable to shock in the absence of glucocorticoids. Most adrenalectomized individuals, however, appear fairly normal during the first few days after surgery but then their condition gradually deteriorates and they die over the next one to two weeks. In the absence of some replacement therapy the actual period of survival following removal of the adrenal glands varies from species to species and between individuals of the same species. In general, however, death is due to a disruption of electrolyte balance and is characterized by negative Na^+ balance and an unduly high accumulation of K^+ within the body. The early literature dealing with these effects in several mammalian species has been reviewed by Gaunt and Chart (1962).

The electrolyte imbalance that occurs following removal of adrenal glands is characterized by changes in the distribution of electrolytes within both the extracellular and the intracellular compartments. In a variety of mammalian species, including man, the plasma Na^+ concentrations decrease while in each case the plasma K^+ concentration increases. At the same time the distribution of Na^+ and K^+ within most tissues may change to reflect the generalized loss of Na^+ and an accumulation of K^+.

Although corticosteroids may influence the distribution of ions and water between the extracellular and intracellular compartments, the persistent loss of Na^+ in the urine and the failure of cells in the nephron to secrete adequate amounts of K^+ are the primary reasons for the electrolyte imbalance that follows adrenalectomy. Thus, when an adrenalectomized rat is given aldosterone, the rate of reabsorption of Na^+ (antinatriuresis) and the rate of K^+ secretion (kaliuresis) are increased and the electrolyte imbalance is corrected (Fig. 16).

Stop-flow analysis and free-flow micropuncture sampling techniques have indicated that aldosterone may modify the composition of urine in adrenalectomized animals by stimulating the Na^+ reabsorptive properties of cells in the distal convoluted tubule (Vander and coworkers, 1958; Hierholtzer and coworkers, 1965; Murayama and coworkers, 1968; Wright and coworkers, 1969). Although there is some evidence to suggest that Na^+ reabsorptive properties of the proximal tubule in adrenalectomized individuals may be impaired, the available evidence that corticosteroids may affect Na^+ reabsorption in the proximal portions of the nephron is rather equivocal (see the review of Forman and Mulrow, 1975). A large fraction of the filtered K^+ is reabsorbed by the cells of the proximal tubule and the loop of Henle and, in the normal animal, 60 to 90% of the K^+ appearing in the urine is secreted by cells of the distal nephron. Adrenalectomy causes a large reduction in this secretory activity but replacement therapy with aldosterone or deoxycorticosterone quickly restores the pattern of K^+ excretion to normal (Fig. 16).

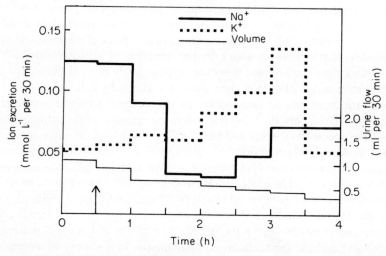

Figure 16 The effect of a single dose of aldosterone (0.1 μg) on the urinary excretion of water, Na^+ and K^+ in adrenalectomized male rats. The arrow indicates the time of injection. After Morris, Berek and Davis (1973); modified

For many years, investigators have been aware of the possibility that the antinatriuretic and the kaliuretic effects of aldosterone may be independent phenomena (Barger, Berlin and Tulenko, 1958). The separate nature of these mechanisms, however, has not always been recognized because most of the experiments have been conducted on adrenalectomized animals and the separate effects of aldosterone are not always apparent under conditions of extreme adrenal insufficiency. Nevertheless, one may now conclude even from the results of experiments on adrenalectomized rats, that the Na^+-

retaining and the K^+-secreting mechanisms may be largely independent of one another; the kaliuretic effect of aldosterone being delayed for at least two hours after injection whereas the antinatriuretic effect is maximal after only one hour (Fig. 16). But, when the effects of aldosterone are examined in intact animals, the two mechanisms can be easily identified as independent phenomena. For example, in the intact dog, a single injection of aldosterone influences only the secretion of K^+ and a similar response to a dose of aldosterone is seen in intact rats maintained on diets containing normal amounts of electrolytes (Fig. 17 and Barger, Berlin and Tulenko, 1958). In

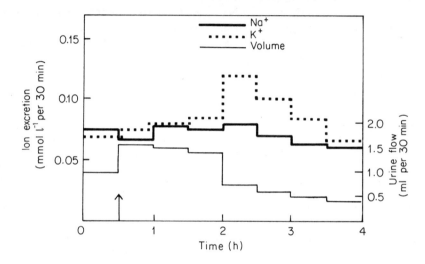

Figure 17 The effect of a single dose of aldosterone (0.1 μg) on the urinary excretion of water, Na^+ and K^+ in intact male rats. The arrow indicates the time of injection.
After Morris, Berek and Davis (1973); modified

some instances, however, an antinatriuretic response to aldosterone does occur in the intact rat and such an experimental instance provided conclusive evidence that the antinatriuretic and kaliuretic effects of aldosterone were separate mechanisms (Williamson, 1963). In his experiments, Williamson showed that although Actinomycin D alone had no effect on the pattern of renal excretion in intact rats, it completely blocked the antinatriuretic effect in rats treated with aldosterone. At the same time, however, it had no effect on the rate of K^+ excretion. These results were later confirmed in adrenalectomized dogs, although Actinomycin D did have a slight kaliuretic effect in this species (Table 8).

Thus, experiments of this type not only confirm the distinct nature of the Na$^+$ reabsorptive and K$^+$ secretory mechanisms in the distal nephron but also suggest that whereas the reabsorption of Na$^+$ may depend on the synthesis of new protein, the secretion of K$^+$ may not be induced by the same type of mechanism (Fimognari, Fanestil and Edelman, 1967).

Some of the earliest work on the effects of adrenocortical steroids in mammals showed that the intestinal absorption of Na$^+$ and Cl$^-$ were reduced in both the dog and the rat following adrenalectomy (Clark, 1939; Dennis and Wood, 1940; Stein and Wertheimer, 1941). Furthermore, the administration of deoxycorticosterone to the adrenalectomized rat was shown to restore these intestinal uptake rates to normal. Later, aldosterone was found to induce increases in the rates of Na$^+$ transfer across the intestinal mucosa of several mammalian species (Glouston, Harrison and Skyring, 1963; Sulya and coworkers, 1963; Crocker and Munday, 1967a, 1967b). Using changes in electrical potential and short-circuit current as indices of change in active Na$^+$ transport, Edmonds and Marriott (1967, 1968, 1969, 1970) showed that the

Table 8. The effect of Actinomycin D and aldosterone on electrolyte excretion in conscious adrenalectomized dogs. Urine was collected for a one-hour control period prior to the administration of either aldosterone or Actinomycin D. At the end of the control period the dogs were injected i.v. with either Actinomycin D (200–300 μg (kg body weight)$^{-1}$), Actinomycin D followed by aldosterone (100 μg kg^{-1}) 0.5 h later, or aldosterone. Urine was collected for four more one-hour periods and the period during which maximal change was observed is recorded as the 'peak period'. (Data from Lifschitz, Schrier and Edelman (1973); modified)

Treatment	Rate of excretion (μEq min^{-1})					
	Na$^+$		K$^+$		H$^+$	
	Control period	Peak period	Control period	Peak period	Control period	Peak period
Control	114 ±36	135 ±35	41 ±11	36 ±11	56 ±7	59 ±12
Aldosterone	108 ±47	33[a] ±12	40 ±5	78[a] ±8	56 ±15	69[a] ±11
Actinomycin D + aldosterone	110 ±37	109 ±43	37 ±8	71[a] ±13	47 ±14	73[a] ±21
Control	157 ±48	124 ±35	51 ±11	49 ±10	50 ±22	68 ±44
Actinomycin D	126 ±19	124 ±32	55 ±13	74[a] ±14	40 ±11	41 ±13

[a] $P < 0.05$ when compared to the corresponding value for control dogs.

rate of Na^+ absorption in tissues derived from rats treated with aldosterone was elevated in a dose-dependent manner.

The colon may also be a site of hormonally-regulated ion transfer. Studies on humans have indicated that a highly significant correlation may exist between the plasma aldosterone concentration and the potential difference found between the lumen of the colon and the extracellular fluid surrounding this portion of the intestine (Efstratopoulos, Peart and Wilson, 1974; Efstratopoulos and Peart, 1975). Clinical studies employing faecal dialysis have shown that aldosterone causes a decrease in Na^+ concentration and an increase in K^+ concentration within the lumen and that treatment with spironolactone causes the concentrations of these ions to return to normal; indeed, changes in the $Na^+:K^+$ ratio of faecal dialysates may be used as a reliable indicator of endogenous mineralocorticoid activity (Richards, 1969; Charron and coworkers, 1969). Some clinical studies, however, have yielded equivocal data concerning the action of aldosterone on Na^+ and K^+ transfer out of and into the colonic lumen. While on series of studies has shown that the transfer of both Na^+ and K^+ are increased in patients with primary aldosteronism, another study has suggested that Na^+ absorption is the primary function of the human colon (cf. Levitan and Ingelfinger, 1965; Shields, Mulholland and Elmslie, 1966; Shields, Miles and Gilbertson, 1968). The effect of aldosterone on ion movement through the colonic mucosa, however, may vary along the length of the organ. In the rat, for example, no response to aldosterone occurs in the ascending region of the colon whereas the most distal region shows a marked response to the hormone (Edmonds and Marriott, 1967, 1968, 1969, 1970; Thompson and Edmonds, 1971, 1974). Furthermore, the effects of aldosterone on Na^+ and K^+ transfer across the intestinal mucosa may not be stoichiometrically linked in a one-to-one fashion; indeed, this was suggested some years ago by Berger, Kanzaki and Steele (1960) who noted that, while deoxycorticosterone-induced increases in both Na^+ and K^+ transfer across the large intestine of the dog, the experiments which showed the largest transfer of Na^+ were not the same experiments as those where the largest transfer of K^+ was observed.

There have been only a few studies in which the effects of aldosterone on ion transfer across the rectal mucosa have been measured. Changes in potential difference across this mucosa suggest that high levels of Na^+ transfer occur in patients suffering from hyperaldosteroidism and similar increases in potential may be induced in normal patients following the injection of 9α-fluorohydrocortisol (Edmonds and Richards, 1970).

D. The Cellular and Molecular Effects in Response Tissues

Much of the early work that contributed to our understanding of corticosteroid action on osmoregulary tissues was conducted on amphibian skin and urinary bladder epithelia but more recently studies on the cellular and molecular basis of these actions have been conducted on the mammalian kidney. There have also been a few studies on the cellular actions of corticosterone and aldosterone in responsive cells of the avian salt gland and the intestinal mucosa of birds and mammals; the results of these studies, however, have contributed little that is new to the current theories on hormone action.

In an attempt to avoid the confusion that has frequently arisen when the results of experiments on one of these tissues have been used to explain a phenomenon observed in another, we have considered the cellular responses of each tissue separately. Following the discussion of these hormonal effects in the individual tissues, some of the generalizations that may be derived from their common biophysical and biochemical effects in the various tissues are presented in support of each of the major hypotheses.

1. *The Amphibian Skin and Urinary Bladder*

Many of the pioneering techniques used by investigators to study the effects of vasopressin on ion transport through the cells of anuran skin and urinary bladder epithelia were later applied to the study of the cellular action of aldosterone. Particularly useful was the SCC method devised by Ussing and Zerhan (1951) to measure the active transport of Na^+ across anuran skin and bladder preparations (Fig. 24). Using this technique, Crabbé (1961, 1963) observed that Na^+ transport was enhanced when aldosterone was added to the medium bathing either the serosal or the mucosal side of the toad bladder. In contrast to the very rapid effect of neurohypophysial hormones on Na^+ transport, however, the effect of aldosterone was only observed after a lag of 60 to 90 minutes (Fig. 4; Crabbé, 1963; Edelman, Bogoroch and Porter, 1963). Since aldosterone enters the cells very quickly, it was assumed that the latent period of the hormonal action reflected the stimulation of one or more intermediate steps. Also, since more than 90% of the ^3H-aldosterone added to the preparation could be recovered after 5 hours of incubation, the toad bladder tissue did not appear to be an active site of aldosterone metabolism (Crabbé, 1963; Edelman, Bogoroch and Porter, 1963).

In an effort to further resolve the nature of the intracellular action of the hormone, the effects of long- and short-term exposure of tissue to puromycin and Actinomycin D were examined. Results showed that both of these agents substantially inhibited the normal response to aldosterone and typical responses in the presence and absence of Actinomycin D are shown in Fig. 4. The drugs do not, however, interfere with the basal level of Na^+ transport and the tissue responds quite normally to the addition of either vasopressin or glucose (cf. Fig. 4 and Edelman, Bogoroch and Porter, 1963). On the basis of these results it was suggested that the latent period was necessary to complete the *de novo* synthesis of a protein or proteins responsible for the enhancement of Na$^+$ transport. Furthermore, the reduction in Na$^+$ transport which occurred in the presence of transcriptional inhibitors supported the suggestion that hormonal activity was the result of genetic regulation.

Autoradiographic studies revealed that labelled aldosterone, but not labelled progesterone, became closely associated with the nuclear compartments of stimulated toad bladder cells (Edelman, Bogoroch and Porter, 1963). In an effort to explore the notion that aldosterone acts to promote the transcription of mRNA, Edelman and his coworkers followed the incorporation of ^{3}H-uridine into responding cells. They found that aldosterone significantly increases the total water-soluble RNA content of the hormone-treated bladders and at the same time increases the number of autoradiographic silver grains in the vicinity of the nuclei (Table 9 and Porter, Bogoroch

Table 9. Effect of aldosterone on RNA synthesis and Na$^+$ transport (short-circuit current) in separated epithelial cells of the toad bladder. The preparations were pulse-labelled with tritiated uridine one hour before the tissues were removed for analysis. Zero time was set as the time of addition of aldosterone (7×10^{-7} mol l^{-1} final concentration) to the serosal chamber of the experimental hemi-bladder. (Data from Porter, Bogoroch and Edelman (1964); modified)

Time (h)	Aldosterone-treated: control ratio	
	RNA specific activity	Short-circuit current
0.75	1.12 ± 0.10	0.97 ± 0.02
1.50	1.28^a ± 0.08	0.98 ± 0.03
3.00	1.43^a ± 0.21	1.30^a ± 0.06
6.00	1.30^a ± 0.13	1.74^a ± 0.20

^{a} Significant difference with respect to corresponding ratio at time zero.

and Edelman, 1964). Also, the inhibition of polymerase I activity was shown to block the increased rate of Na$^+$ transport that normally follows exposure of the tissue to aldosterone; since this enzyme apparently regulates the polyadenylation of mRNA precursor, and also governs transcription of the genes coding for rRNA, there appears to be a close link between protein synthetic mechanisms regulated by the nucleus and the ultimate manifestation of the physiological response to the hormone (Chu and Edelman, 1972). It has been further demonstrated that aldosterone, but neither its stero-isomer, 17a-isoaldosterone nor spironolactone, will cause an increase in the incorporation of ^{3}H-uridine into specific 9–12 S RNA species (Rossier, Wilce and Edelman, 1974). This discrete class of RNA is most likely the mRNA species that is responsible for the Actinomycin-D-sensitive response to aldosterone (Edelman, 1975).

2. The Avian Nasal Gland

Labelled corticosterone is taken up preferentially by cells in the nasal glands of both normal and salt-loaded ducks (Phillips and Bellamy, 1962, 1967; Sandor and Fazekas, 1973; Allen, Abel and Takemoto, 1975). In contrast, no such selective accumulation occurs in the cells of the adjacent Harderian gland. The secretory cells of the nasal glands, therefore, appear to be specific targets for the direct action of circulating corticosteroids (Table 10). This conjecture has been substantiated through the identification of specific receptors in both cytoplasm and the nuclei of nasal gland cells (Sandor and Fazekas, 1973; Allen, Abel and Takemoto, 1975; Sandor, Mehdi and Fazekas, 1977). Studies *in vitro* have shown that both high affinity–low capacity and low affinity–high capacity receptors are present in the cytosolic fraction of nasal gland cells. Each type of receptor has a greater affinity to bind corticosterone than either aldosterone or deoxycorticosterone and incubation of the cytosolic fraction of nasal gland tissue with trypsin causes an almost total loss of the binding properties (Sandor and Fazekas, 1973; Allen, Abel and Takemoto, 1975). The high affinity receptors in the cytosolic fraction have an apparent dissociation constant (k_D) of 10^{-9} mol l^{-1} and the number of binding sites in both active and quiescent tissue is equivalent to approximately 10^{-12} mol mg^{-1} protein (Sandor, Mehdi and Fazekas, 1977). In common with receptors isolated from other target tissues, isolated cytosolic receptors from nasal gland cells will only bind efficiently with labelled aldosterone at 0 °C and at 25 °C, only 10% of the initial binding at 0 °C is observed. The physiological significance of receptors that have much greater affinity for the hormonal ligand at such low temperatures is not clear, particularly when the receptors have been isolated from responsive tissues in homeothermic vertebrates. Studies *in vitro* have shown that when labelled corticosterone is added to the cytosolic fraction from nasal gland cells, it becomes associated with a heavy (9–11 S) and a light (3–4 S) protein fraction. The peak of heavy protein is sensitive to high ionic concentrations and in 0.4 mol l^{-1} KCl shifts to the 4 S region when isolated by centrifugation in a linear sucrose gradient. Recombination studies show that when a cytosolic fraction containing labelled corticosterone complexes is incubated in the presence of nuclei that have not been exposed to labelled corticosterone, the receptors in the cytosolic fraction become depleted of labelled material and the ligand becomes associated with the nuclei (Table 11).

In the nuclei, the labelled material becomes associated with a tris-soluble macromolecule and with a tris-insoluble macromolecule that is soluble in 0.4 mol l^{-1} KCl; this latter macromolecule is believed to be associated with chromatin in the nuclei (Table 11). It is important to note, however, that most

Table 10. The distribution of receptor-bound radioactivity in the cellular fractions of the different tissues from the duck 20 minutes after the injection of tritiated corticosterone. (Data from Sandor, Mehdi and Fazekas (1977); modified)

	Tritiated corticosteroid concentration (fmol (mg^{-1} protein)$^{-1}$)								
	Cytosolic complex			Tris-soluble nuclear complex			Chromatin-bound nuclear complex		
	Total^a	Compound A^b	Compound B^c	Total	Compound A	Compound B	Total	Compound A	Compound B
Nasal gland	2.91	1.43	0.12	6.31	3.01	0.12	14.2	5.37	0.35
Harderian gland	0.48	0.11	0.17	2.18	0.52	0.52	1.05	0.26	0.36
Liver	2.85	0.02	0.04	0.40	0.02	0.05			
Lung	5.84	0.08	0.36						

^{a} Total = total bound tritiated activity, corticosterone.
^{b} Compound A = tritiated 11-dehydrocorticosterone.
^{c} Compound B = tritiated corticosterone.

Table 11. The distribution of receptor-bound radioactivity following a cell-free recombination of labelled cytosolic and unlabelled nuclear fractions from salt-activated nasal glands. The cytosol was labelled with [^{3}H]corticosterone and incubated with crude nuclei at 25 °C. (Data from Sandor, Mehdi and Fazekas (1977); modified)

Cell fraction	Tritiated 11-dehydrocorticosterone concentration present in fractions after various periods of incubation (fmol (mg protein)$^{-1}$)			
	Incubation time (min)			
	0	5	15	25
Labelled cytosolic fraction alone	46.7	—	—	41.1
Labelled cytosolic fraction plus nuclei	46.7	—	8.0	6.3
Tris-soluble nuclear complex derived from nuclei	0	17.3	24.8	21.0
Chromatin-bound steroid complex derived from nuclei (0.4 mol l^{-1} KCl soluble)	0	49.9	61.2	58.6

of the tritiated material associated with these nuclear receptors is in the form of tritiated 11-dehydrocorticosterone (Table 10). This conversion of labelled hormone presumably occurs in the cytoplasm prior to transfer of the labelled ligand to the nuclear receptors. Thus, after incubating a cytosolic fraction from actively secreting nasal gland cells for 120 minutes in the presence of ^{3}H-corticosterone, 95% of the original radioligand is converted to ^{3}H-11-dehydrocorticosterone. Cytosolic and nuclear fractions isolated from the adjacent Harderian gland and from more remote tissues such as lung and liver do not possess the same abilities either to bind or to metabolize significant amounts of labelled corticosterone (Table 10).

As similar metabolic conversion of labelled corticosterone appears to occur *in vivo*, for 20 minutes following the intravenous injection of ^{3}H-corticosterone, the ratio of labelled 11-dehydrocorticosterone: labelled corticosterone in the cytosolic fraction of nasal tissue is about 11:1 (Sandor, Mehdi and Fazekas, 1977). Although the relative amount of labelled metabolite is consistently greater in both *in vivo* and *in vitro* experiments, there nevertheless remains a small quantity of unaltered ^{3}H-corticosterone bound to the cytosolic receptors (Table 10). One cannot overlook the possibility, therefore, that either one or both of these corticosteroids may play a crucial role in the physiologic response of the nasal gland cells in birds drinking sea-water.

3. *The Intestinal Mucosa*

Protein receptors which are specific for aldosterone and have molecular weights ranging from 50 000 to 150 000 have been isolated from the mucosal cytoplasm of the duodenum, jejunum, ileum and colon in the rat (Swaneck, Highland and Edelman, 1969; Feldman, 1975; Pressley and Funder, 1975). The variation in the amounts of aldosterone that become bound to the mucosal cells from different portions of the intestine seems to suggest that the relative abundance of these receptors may vary along the entire length of the intestine (Fig. 18). Through the use of techniques which permit the rapid separation of free and complexed steroid, these intestinal cytosolic extracts have been shown to have a dissociation constant (k_D) for aldosterone of $6.6 \pm 3.8 \times 10^{-8}$ mol l^{-1} (Watts and Wheldrake, 1976).

Receptors have also been isolated from the nuclei of rat and human colon (Alberti and Sharp, 1970). These corticosteroid receptor complexes in the nuclei are much more stable *in vitro* than similar complexes isolated from the toad bladder. However, the aldosterone receptors from the human colon show a high affinity but a low capacity to bind the endogenous hormone whereas receptors isolated from rat colon show a considerable affinity for other steroids (Postel-Vinay and coworkers, 1974).

As in the case of other corticosteroid-responsive cells, the presence of steroid hormone in the nucleus suggests that it may be involved in the initiation of RNA synthesis. However, unlike most other tissues, an extremely long interval elapses between the initial exposure of an intestinal tissue to aldosterone and the recognition of a physiological response such as Na^+ reabsorption; for example, intervals as long as 27 hours may elapse between the treatment of rats with aldosterone and the occurrence of increased mucosal transfer in the jejunum (Crocker and Munday, 1969). The effect of aldosterone can be inhibited by Actinomycin D and this has been interpreted

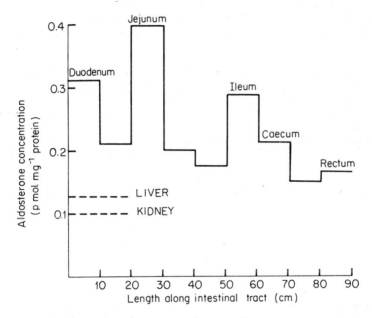

Figure 18 Localization of aldosterone along the gastrointestinal tract of the rat. The rats were sacrificed 30 min after a single injection of ^{3}H-aldosterone (10 μCi, 2 nmol (100 g body weight)$^{-1}$) and the amount of hormone present in successive 10 cm lengths of intestine was determined. The amounts of labelled hormone found in samples of liver and kidney are also indicated. After Watts and Wheldrake (1976); modified

to support the hypothesis that protein synthesis is involved in the final responses of the intestinal mucosa to aldosterone (Fimognari, Fanestil and Edelman, 1967; Crocker, 1969). It has been suggested, however, that the long delay observed in the response of the rat jejunum may be due to the aldosterone becoming associated with receptors in cells situated in the crypts of Lieberkühn. This hypothesis does not discount the possibility of RNA induction and the subsequent synthesis of new specific protein because cells stimulated at this locus would then have to migrate to the tops of the intestinal villi before their enhanced functional activity would become apparent; an interval of 24 to 27 hours may well be necessary to accomplish this migration (Crocker and Munday, 1967a, 1967b).

Further evidence in support of the belief that aldosterone may induce the synthesis of new protein is derived from studies showing that while adrenalectomy reduces the incorporation of ^{14}C-formate into RNA fractions of both the large and small intestine, pretreatment of the rats with aldosterone restores the rate of ^{14}C-formate incorporation to normal (Watts and Wheldrake, 1976). Efforts to follow the incorporation of ^{3}H-uridine and ^{3}H-orotate into mucosal cells, however, have failed to detect any stimulatory effect of aldosterone. Thus, there is only indirect evidence to suggest that

Table 12. The effects of spironolactone, metyrapone and corticosterone on the rates of mucosal water transfer in fresh-water- and sea-water-adapted ducklings (*Anas platyrhynchos*). The mucosal water transfer rates were measured *in vitro* in a series of everted sacs prepared from the small intestine. Sea-water-maintained birds were given either 5 mg spironolactone orally or 80 mg kg^{-1} metyrapone intraperitoneally 24 hours before the experiment. Fresh-water-maintained birds were given either 0.5 mg corticosterone (i.m.) only or 0.5 mg corticosterone (i.m.) followed by 5 mg spironolactone orally 24 hours before the experiment. (Data from Crocker and Holmes (1976); modified)

Treatment	Mucosal water transfer[a] (ml (g wet weight)$^{-1}$ h^{-1})				
	I	II	III	IV	V
Sea-water-maintained					
Untreated controls	1.63 ±0.07	1.80 ±0.07	2.09 ±0.08	1.79 ±0.17	1.70 ±0.13
Spironolactone-treated	1.11[b] ±0.03	1.03[b] ±0.08	1.36[b] ±0.04	1.16[d] ±0.11	0.99[c] ±0.06
Metyrapone-treated	0.87[b] ±0.03	1.03[b] ±0.07	1.20[b] ±0.07	1.25[d] ±0.12	1.04[d] ±0.14
Fresh-water-maintained					
Untreated controls	1.09 ±0.05	1.05 ±0.07	1.20 ±0.08	0.99 ±0.06	1.01 ±0.05
Corticosterone-treated	1.80[b] ±0.11	1.91[b] ±0.05	1.90[b] ±0.07	1.73[b] ±0.12	1.38[c] ±0.09
Spironolactone- and corticosterone-treated	1.16[e] ±0.08	1.10[e] ±0.06	1.28[e] ±0.08	1.25[e] ±0.07	1.17[e] ±0.06

[a] Staring 10 cm posterior to the entry of the bile duct and ending at the junction of the caecal pouches, the intestine was divided in five equal segments; these segments were numbered I through V from the anterior to the posterior ends.
[b] $P < 0.001$.
[c] $P < 0.01$.
[d] $P < 0.05$.
[e] Not significant with respect to the corresponding values for untreated controls.

aldosterone operates to stimulate the transcription of RNA species from DNA. Nevertheless, the physiologic response of intestinal mucosal cells is believed to be mediated through the induction of *de novo* synthesis of specific proteins.

The presumed occupation of mucosal binding sites by spironolactone and the resulting failure of the small intestine in marine teleosts and birds to respond to the ingestion of sea-water suggest the presence of somewhat similar stimulatory pathways in these animals (see Table 12; also Sections III-C-1 and III-C-4). The fact that these tissues in lower vertebrates respond to either corticosterone or cortisol and not to aldosterone suggests that perhaps interesting qualitative differences may exist between the receptors isolated from mammalian tissues and those yet to be isolated from fish and birds.

4. The Mammalian Nephron

a. *Receptors* Penetration of the cells of the target tissue by the steroid hormone must be the first step in the sequence of events leading to a physiologic response. The discovery that the nuclear fraction of kidney homogenates from adrenalectomized rats contained specific aldosterone-binding substances was therefore fundamental to our understanding of the nature of hormone action in these cells (Fanestil and Edelman, 1966). Later, complexes between aldosterone and macromolecules were isolated from both nuclear and cytosolic fractions of rat kidneys (Herman, Fimognari and Edelman, 1968). Since these early reports, however, evidence has accumulated that aldosterone binds specifically to three macromolecular fractions within the cells of the renal tubule; these fractions are the specific cytosolic receptors, the nuclear receptors and the chromatin in the nuclei of the responding cells.

Based on their susceptibility to cleavage by a range of enzymes, it has been demonstrated that the macromolecules in the cytosolic and nuclear fractions that form specific complexes with aldosterone are primarily proteins (Fanestil and Edelman, 1966; Herman, Fimognari and Edelman, 1968; Swaneck, Chu and Edelman, 1970; Ludens and Fanestil, 1971). Thus, assuming that extensive cleavage of these receptors would result in a rapid release of the ligand, it has been shown that whereas DNase, RNase, lipase, phospholipase D and neuraminidase fail to accelerate the release of tritiated aldosterone *in vitro*, some of the proteolytic enzymes are quite effective in this regard. Further studies have revealed that the complexes are relatively resistant to hydrolysis by trypsin and collagenase, suggesting that binding proteins are not histones containing large amounts of either arginine or lysine and also that they are not rich in proline residues (Fanestil and Edelman, 1966; Edelman and Fimognari, 1968).

Cytosolic fractions prepared from kidney tissue appear to contain three separate classes of binding sites or receptors. Comparisons of the relative affinities of these binding sites for certain steroids has led to the belief that the three classes of cytosolic receptors represent both the mineralocorticoid type I receptors and the glucocorticoid types II and III receptors (Rosseau and coworkers, 1972; Funder, Feldman and Edelman, 1973; Agarwal, 1975, 1976a, 1976b). The type I sites have a high affinity for aldosterone and a somewhat lower affinity for corticosterone; conversely, corticosterone has less than 2% of the affinity of aldosterone for type I sites (Rosseau and coworkers, 1972). The higher affinity *in vitro* of type I sites for mineralocorticoids may not necessarily reflect the actual *in vivo* relationship between these receptors and the circulating corticosteroids. The plasma concentrations of corticosterone and deoxycorticosterone are each far greater than those of aldosterone. They may, therefore, 'override' by mass action the higher affinity of the type I sites for aldosterone and under some *in vivo* conditions they may become the principle hormones occupying type I receptors. The tendency for this to occur, however, may be somewhat offset by the relative amount of each hormone that is bound to plasma protein (Funder, Feldman and Edelman, 1973). Since the proportions of the total circulating corticosterone and deoxycorticosterone that are bound to plasma proteins are in each case much greater than the proportion of bound aldosterone, the concentration of each steroid in the interstitial fluid, and thus the concentration available to the responsive cell, may be such that the sites maintain their relative affinities for the different hormones (see Funder, Feldman and Edelman, 1973). Furthermore, since more than 90% of the total blood flowing through the kidney is circulated through the glomeruli, this is the most probable pathway whereby the adrenal steroids are transported to their target cells in the kidney; also, the proximal reabsorption of glomerular filtrate may further alter the concentrations of hormone presented to the lumenal surfaces in the distal nephron.

The aldosterone-binding macromolecules in the cytosolic fraction may exist *in vitro* in at least two forms (Ludens and Fanestil, 1971). When one sample of kidney tissue is homogenized in 0.25 mol l^{-1} sucrose and another is homogenized in 0.3 mol l^{-1} KCl, the protein peaks containing the aldosterone-macromolecular complexes from the two cytosolic preparations do not appear in the same fractions eluted from Sephadex G-100 columns. The two forms of the complexes appear to be related, for they can be converted from one to the other by the addition or removal of KCl (Fig. 19). Further characterization of these receptors by density gradient analysis has indicated sedimentation values of 4.5 S in high salt media (0.4 mol l^{-1} KCl) and 8.5 S in low salt media. As in the case of the macromolecular complexes isolated by column chromatography, the addition of 0.4 mol l^{-1} KCl to cytosolic fractions that have been labelled *in vitro* with ^{3}H-aldosterone causes the 8.5 S peak to shift to a 4.5S peak (Marver, Goodman and Edelman, 1972). These results suggest that the 8.5 S and the 4.5 S forms of the cytosolic receptors found in low salt gradients are respectively aggregated and deaggregated forms of the same species. The observations are also in agreement with the allosteric-equilibrium model for steroid action (Fig. 20 and Edelman, 1975). According to this model, two forms of the cytoplasmic receptor exist, the inactive form which does not bind steroid, and the active form which does form a receptor–ligand complex with aldosterone.

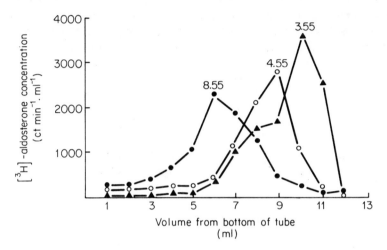

Figure 19 The effects of high salt and Ca^{2+} concentrations on density gradient patterns found in cytosolic fractions isolated from adrenalectomized rat kidney tissue. Kidneys were homogenized in 0.1 mol l^{-1} Tris-HCl, pH 7.2, and the cytosolic fraction was prepared by centrifugation at $19000 \times g$ for 15 min. The fraction was labelled by incubation with 1×10^{-8} mol l^{-1} tritiated aldosterone for 30 min at 0 °C. Aliquots were layered on a 13–34% glycerol gradient containing 0.1 mol l^{-1} Tris-HCl at pH 7.2, 1.6×10^{-9} mol l^{-1} tritiated aldosterone, and either (a) 1.5 mmol l^{-1} EDTA (—●—), (b) 1.5 mmol l^{-1} EDTA and 0.4 mol l^{-1} KCl (—O—), or (c) 1.5 mmol l^{-1} EDTA, 0.4 mol l^{-1} KCl and 6 mmol l^{-1} $CaCl_2$ (—▲—). One millilitre fractions were collected and filtered through G-50 Sephadex to remove free tritiated aldosterone. After Marver, Goodman and Edelman (1972); modified

It is most likely that the active form of cytosolic receptor is transported into the nucleus. Nuclear receptors, however, have unique chemical characteristics which indicate that either a modification of the cytoplasmic receptor occurs prior to its entry into the nucleus or that the steroid is transferred from the cytoplasm to a nuclear receptor. Aldosterone-binding nuclear proteins were first isolated and purified from the kidneys of adrenalectomized rats that had been injected with tritiated aldosterone (Herman, Fimognari and Edelman, 1968). The nuclear receptors are also tris-soluble and have a sedimentation value of 3 S that, unlike the cytoplasmic receptors, is stable over a range of ionic strengths (Marver, Goodman and Edelman, 1972).

The aldosterone-receptor complex associated with chromatin extracted from homogenates of kidney tissue from adrenalectomized male rats that have received a single dose of labelled aldosterone has a sedimentation value of 4 S (Swaneck, Chu and Edelman, 1970). In these experiments 55% of the total nuclear content of ^{3}H-aldosterone was recovered with the chromatin and, after passage through a Sephadex G-50 column, 76% of the chromatin-bound steroid was bound to soluble non-histone protein (Table 13).

In summary, therefore, three classes of active receptors for aldosterone have been isolated from homogenates of rat kidney tissue; these are a 4.5 S (high salt) cytosolic species which may, in a low salt medium, aggregate to

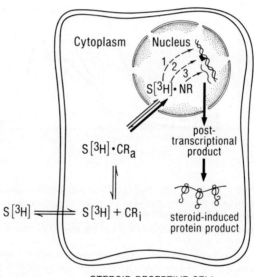

Figure 20 A diagrammatic representation of the possible mode of action of corticosteroids on responsive cells. Through the use of radioactively labelled hormone (S[^{3}H]), the movement of a corticosteroid can be traced from the plasma and interstitial fluid to the nucleus. The hormone is believed to pass freely through the plasma membrane into the intracellular compartment of the target tissue. In the cytoplasm of these cells, inactive corticosteroid receptors (CR$_i$) may be either activated to form an active receptor (CR$_a$) or may react directly with the labelled hormone and in so doing form active corticosteroid–receptor complexes (S[^{3}H]CR$_a$). At some point, the activated cytoplasmic receptor is modified and assumes the characteristics of a nuclear receptor (NR) which in turn gives rise to a nuclear corticosteroid–receptor complex (S[^{3}H]NR). Although the precise mechanism of interaction between this corticosteroid–receptor complex and the genetic material in the nucleus is not clear, this complex may modulate gene expression in one of several ways: (1) it may interact directly with the DNA of the chromatin; (2) it may become associated with a chromatin protein such as a histone or non-histone acidic protein; and (3) the steroid–receptor complex may interact with a regulatory gene product to derepress the induction of a specific protein product

d-aldosterone > 9α-fluorocortisol > deoxycorticosterone acetate > 6α-methylprenesolone > estradiol-17β = progesterone = 17α-isoaldosterone = no steroid (Herman, Fimognari and Edelman, 1968; Swaneck, Chu and Edelman, 1970; Ludens and Fanestil, 1971; Chu and Edelman, 1972; Feldman, Funder and Edelman, 1972; Marver, Goodman and Edelman, 1972; Liew, Suria and Gornall, 1973). The minimal effect of 17α-isoaldosterone is of considerable interest because this molecular form of

Table 13. The distribution of labelled aldosterone in various fractions derived from isolated nuclei of rat kidney tissue. The nuclear fractions were isolated following incubation of unlabelled nuclei in the presence of tritiated aldosterone receptor complexes prepared from the cytosolic fractions of kidney tissues. Both the nuclei and the cytosolic receptor complexes were derived from the kidneys of adrenalectomized male rats. (Data from Swaneck and coworkers, 1970); modified)

Fraction	Intranuclear distribution of ^{3}H-aldosterone	
	Total (%)	Bound (%)
Tris-CaCl$_2$ extract	27.6 ± 0.9	63[b]
Chromatin	55.1 ± 2.1	76[c]
Residual	17.3 ± 0.3	

[a] Means ± S.E.M. Bound refers to the proportion of the total ^{3}H-aldosterone in the particular fraction that was bound to a macromolecule.
[b] Determined by precipitation with 50% saturated $(NH_4)_2SO_4$ and G-50 Sephadex gel filtration.
[c] The chromatin was sheared and the resultant soluble-nucleohistone was passed through G-50 Sephadex columns.

form an active 8.5 S complex, a 3 S tris-soluble nuclear species and a 4 S (high salt) chromatin-bound species. The cytosolic, nuclear and chromatin-binding receptors have all been found to be stereospecific for aldosterone and related mineralocorticoids (Swaneck, Chu and Edelman, 1970). Assuming an adequate concentration in the interstitial fluid or the renal tubular filtrate, this stereospecificity ensures that only mineralocorticoids will bind to these target tissue receptors to induce the physiological response; the extent of the physiological response being proportional both to the number of cytoplasmic receptors and to the amounts of hormone actively bound to these specific receptors (Feldman, Funder and Edelman, 1972; Funder, Feldman and Edelman, 1973). Competition studies show that the presence of competing steroids may impair the formation of the aldosterone-receptor complexes in direct proportion to their potencies as mineralocorticoids. The relative effectiveness of several steroids to compete with aldosterone is as follows: aldosterone is the stereoisomer of d-aldosterone and the inability of this molecule to displace the d-aldosterone from the macromolecular receptor emphasizes the very high stereospecificity of the receptors for aldosterone and other potent mineralocorticoids.

Spironolactone has been shown to block the action of aldosterone on Na^+ transport. This was first believed to be due to competitive binding with chromatin but later studies have suggested it may occupy aldosterone-receptors in the cytoplasm (cf. Swaneck, Chu and Edelman, 1970; Marver and coworkers, 1974). The cytoplasmic receptors appear to form complexes with spironolactone that are unable to transfer the ligand to the chromatin sites within the nucleus. It has been suggested, therefore, that the competitive binding of spironolactone to specific cytoplasmic receptor sites inhibits the

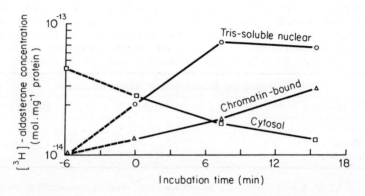

Figure 21 The transfer of tritiated aldosterone from cytosolic fractions of kidney tissue containing tritiated aldosterone–receptor complexes to nuclear fractions from previously unlabelled nuclei. The cytosolic fractions were labelled by incubation with tritiated aldosterone (1.3×10^{-8} mol l^{-1}) for 30 min at 0 °C. The labelled cytosolic fractions were then mixed with unlabelled nuclei and incubated for various times at 25 °C. The 'zero time' values represent measurements obtained on mixtures of these fractions when they were processed immediately after mixing at 25 °C. An extrapolated 'true zero' point at -6 min is included to account for the binding that may have developed during the 6 min required to process the mixture. After Morris and Davis (1973); modified

activation of the ligand–macromolecular complex and its subsequent transfer into the nucleus.

According to one theory, aldosterone moves through the cytoplasm of the responding cells and into their nuclei, where it finally becomes associated with the chromatin. There are at least three ways in which this transfer could take place: (1) aldosterone could be transferred from a cytoplasmic receptor to a nuclear receptor; (2) the cytoplasmic receptor could actually carry the aldosterone into the nucleus by migrating through the nuclear envelope; and (3) unbound aldosterone in the cytoplasm could diffuse into the nucleus and become bound to a nuclear receptor.

Reconstitution experiments have shown that when cytosolic receptors associated with labelled aldosterone are mixed and incubated with nuclear receptors from kidney cells of adrenalectomized animals that have not been exposed to labelled hormone, the labelled material associated with the cytosolic receptors decreases. At the same time that this decline is occurring, the labelled material associated with the tris-soluble nuclear receptors increases and later the specific activity of the chromatin-bound receptors increases (Fig. 21). Density gradient analysis of the complexes isolated after incubation has revealed that the sequential appearance of labelled material in the nucleus is first associated with the 3 S tris-soluble fraction and later with the 4 S chromatin-bound species.

The cytoplasmic steroid-receptor complexes have been found to be heat labile *in vitro*. Raising the temperature of the steroid-receptor complex from 0 °C to 37 °C causes the receptor to liberate the bound steroid and this prevents the uptake of labelled material by the nuclear receptors (Marver, Goodman and Edelman, 1972). Results from reconstitution experiments of this type have been interpreted to indicate that the generation of the nuclear receptors may depend on the presence of receptor complexes in the cytosolic fraction and that these receptors may actually go into the nucleus to form the nuclear receptor species. However, an alternative interpretation of these data suggests that the changes in the ratios of nuclear and cytosolic binding with time after injection of aldosterone reflects the different decay rates for the isolated nuclear and cytosolic binding proteins (Mills, Wheldrake and Feltham, 1970). Nevertheless, the finding that the unbound steroid released by heating the cytosolic fraction is not able to enter the nucleus also casts doubt on the possibility of direct action between free steroid in the cytoplasm and receptors in the nuclei.

b. *Evidence for the Regulation of Gene Expression* The reactions between the tris-soluble aldosterone-receptor complexes and chromatin and the mechanism whereby gene expression is modulated as a result of this reaction is poorly defined. The DNA of chromatin may be directly involved in the aldosterone-acceptor activity of the chromatin. Reconstitution experiments with kidney cytosol and chromatin fractions have shown that the uptake of ^{3}H-aldosterone by chromatin is reduced by 75% after destruction of half of the DNA by DNase (Marver and Edelman, 1975). Also, pretreatment of nuclei with either ethidium bromide or proflavine sulphate, both of which intercalate between DNA base pairs so that portions of the molecules project into the major groove of the DNA helix, reduces chromatin acceptor activity by 80 to 90%. In contrast, Actinomycin D, which intercalates specifically between the GpC base pairs with its peptide subunits projecting into the minor groove of the DNA helix, impairs acceptor activity by only 20%. Similar reductions can be obtained by exposing nuclei to netropsin, an ApT-specific reagent that elongates the contour length of the helix and binds primarily in the minor groove. All of these results indicate that either the local DNA sites or the proteins associated with the DNA sites, probably in the major groove, directly affect the binding of aldosterone to chromatin.

The delay between exposure of a responsive tissue to aldosterone and the recognition of the active movement of Na^+ has been interpreted as the time necessary to stimulate the synthesis of RNA which in turn leads to the synthesis of specific protein(s). The fact that Actinomycin D and other inhibitors of RNA and protein synthesis have the ability to block the movement of Na^+ through the cells lends further support for a hypothesis involving the induction of specific protein synthesis.

Quite small doses of aldosterone can stimulate increases in the quantity of total RNA in subcellular fractions of kidneys and the incorporation of ^3H-uridine-5'-monophosphate into RNA isolated from kidneys of adrenalectomized rats (Castles and Williamson, 1965). These effects on RNA synthesis are totally blocked by pretreatment with Actinomycin D. Furthermore, soon after the administration of aldosterone *in vivo*, homogenates of kidney tissues show enhanced rates of orotate-^3H incorporation into RNA in nuclear and cytosolic fractions of the tissue (Fimognari, Fanestil and Edelman, 1967; Forte and Landon, 1968). In some *in vitro* studies, it has also been shown that Actinomycin D inhibits approximately 65% of the orotate-^3H incorporation into nuclear RNA. This observation suggests that most of this incorporation was DNA-dependent rather than an increase in the turnover rate of the terminal cytidines of tRNA because Actinomycin D blocks RNA synthesis by interacting with DNA itself (Fimognari, Fanestil and Edelman, 1967).

Aldosterone may also stimulate nuclear RNA polymerase activity in the kidney tissue of adrenalectomized rats (Chu and Edelman, 1972; Liew, Suria and Gornall, 1973). In these studies evidence is presented to suggest that distinct RNA polymerases may exist in kidney tissue; a polymerase I located in the nucleolus may regulate transcription of genes coded for rRNA precursor, while a polymerase II may regulate nucleoplasmic mRNA synthesis. In renal nuclear fractions the RNA polymerase I activity increases relative to the polymerase II activity when the adrenalectomized rats are treated with aldosterone. This implies a shift in the polymerase I: II ratio in favour of nucleolar RNA polymerase activity and in increase in rRNA synthesis.

In contrast, other investigators have found no evidence that aldosterone has a direct effect on polymerase activity (Mishra and Feltham, 1975). They have suggested that it is the binding of the aldosterone-receptor complex directly to chromatin that results in enhanced template activity. These findings imply that aldosterone augments ribosomal RNA synthesis and that ribosomal or mRNA synthesis or both are required for the expression of mineralocorticoid activity. Although there have been demonstrations that mRNA synthesis can be specifically increased due to aldosterone treatment in other tissues such as the toad bladders, there has yet to be any demonstration of this sort in the kidney.

Attempts to demonstrate increased template activity of purified renal chromatin from aldosterone-treated adrenalectomized rats have been unsuccessful (Trachewsky and Cheah, 1971). However, there is indirect evidence derived from observations on the binding of labelled Actinomycin D to DNA that aldosterone treatment may change the physical or chemical state of the chromatin. In these studies, 1.2% of the guanine–cytosine nucleotide pairs of the chromatin from control rats bound ^{14}C-Actinomycin D, whereas only 0.6% of the guanine–cytosine base pairs of aldosterone-treated animals did so (Trachewsky and Cheah, 1971).

The diminished binding of Actinomycin D-^{14}C to the renal chromatin from aldosterone-treated rats may reflect an increase in the amount of histones or acidic chromosomal proteins that 'mask' or 'protect' more genes than in the chromatin from control rats. It would require only a small change in the positioning of the chromosomal proteins to change the pattern of active and inactive genes (Trachewsky and Cheah, 1971). Thus, there is some indirect evidence to suggest that aldosterone-receptor complexes may directly affect the genome. Following the administration of aldosterone to adrenalectomized rats, an increase in acetylation, followed by phosphorylation of chromosomal proteins has been observed (Liew, Suria and Gornall, 1973). The incorporation of ^{3}H-acetate and ^{32}P-phosphoric acid into kidney histones and nuclear acidic proteins also increased to maximum rates at 40 and 150 minutes respectively. The increase of acetylation and phosphorylation of the chromosomal proteins may be responsible for the change in the pattern of active and inactive genes.

Although the mechanism is unclear, it is generally believed that aldosterone induces a net increase in various RNA species. The next step in the sequence of events leading to an increase in protein synthesis involves the translation of the genetic material in the ribosomes of renal cells. The actual aldosterone-induced proteins (AIP) have not been identified but indirect evidence suggests that protein synthesis is increased by this hormone. For example, a significant increase has been observed in the *in vivo* incorporation of ^{3}H-leucine into the proteins of the microsomal fractions of kidney homogenates from aldosterone-treated rats (Fimognari, Fanestil and Edelman, 1967). In contrast, however, some recent experiments have shown no effect on the incorporation of ^{3}H-leucine into the *total* ribosomal protein content *in vivo* following aldosterone stimulation (Fanestil, 1969; Trachewsky and Yang, 1975). Some recent experiments have shown no alteration of the rate of protein biosynthesis *in vivo* following aldosterone administration (Fanestil, 1969; Rosseau and Crabbé, 1972). Trachewsky and Yang (1975) were also unable to measure any change of the radioactive leucine content of the total ribosomal population in incubated kidney slices. They did, however, detect a relative decrease in the labelling of specific ribosomal protein(s) with molecular weight(s) between 57 000 and 62 000 daltons.

Modification of the ribosomal protein(s) could be one manner in which changes in the structure and function of the ribosomes may be induced. It has also been shown that aldosterone treatment alters the phosphorylation and acetylation of specific renal cortical ribosomes (Hill and Trachewsky, 1974; Trachewsky, 1974). The phosphorylation of individual proteins of ribosomes was compared in aldosterone-treated and control rats by a double-labelling technique using ^{32}P and ^{33}P. Slices of kidney cortex from aldosterone-treated animals were labelled with ^{32}P and the effect of aldosterone on ribosomal protein phosphorylation could be estimated from the $^{32}P:^{33}P$ ratios in different proteins of the ribosome. Phosphorylation of some high molecular weight proteins increased while the phosphorylation of other lower molecular weight ribosomal proteins decreased (Hill and Trachewsky, 1974). A decrease in the acetylation of some ribosomal proteins has also been shown in kidney tissue following treatment with aldosterone (Trachewsky, 1974). Although the functional significance of ribosomal protein phosphorylation and acetylation is not known, these results do at least suggest that aldosterone may induce some changes in the structure of ribosomal protein. Aldosterone, by inducing structural changes in ribosomal proteins, may affect the synthesis of protein(s) responsible for the physiologic response of the kidney to the hormone.

Thus, the mechanism of action of aldosterone, as is the case for other steroids, seems to involve the interaction between the DNA template and DNA-dependent RNA polymerases followed by increased RNA synthesis and alterations of ribosomal proteins. Specific receptors for the hormone have been isolated from the cytosol and nuclei of various responsive tissues, and they appear to function as carriers, ferrying the hormone from the plasma to the nucleus where it becomes associated with the chromatin. The identification of the specific mineralocorticoid-induced proteins produced by the association of mRNA with the ribosomes has not been achieved. The evidence which currently favours the role of aldosterone-induced proteins (AIP) in antinatriuresis has been derived from the different experimental tissues, principally urinary bladder and kidney. It is from a consideration of the many physiological parallels of the responses to corticosteroids by these two tissues that the discussion of the precise role of corticosterone-induced proteins is made possible.

E. The Possible Roles of the Corticosteroid-induced Proteins

The accumulated experimental evidence suggests that one or more corticosteroids enter the responsive cells in a variety of osmoregulatory tissues in vertebrate organisms and that they bind stereospecifically with high affinity receptors in the cytoplasm of these cells. A temperature-sensitive activation of the corticosteroid-receptor complex then occurs and the complex either passes through the nuclear envelope into the nucleus or it facilitates the transfer of the hormonal ligand to another macromolecular receptor within the nucleus. In the nucleus the active complex becomes attached to chromatin and the transcription of specific RNA is induced and this in turn leads to the syntheis of new protein (Fig. 22). The corticosteroid-induced protein then stimulates directly or indirectly the movement of ions across the epithelia of responsive cells in the target tissue.

Thus, underlying the physiological response of the various tissues to corticosteroids is an apparent requirement for *de novo* protein synthesis. Although the aldosterone-induced proteins have not been identified, there

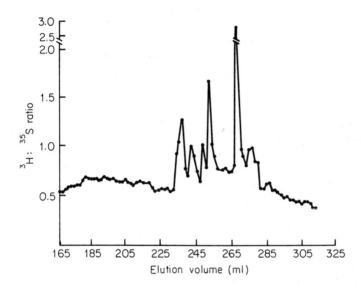

Figure 22 The chromatographic (Sephadex G-75) isolation of aldosterone-induced proteins from mitochondria-rich cells of the toad urinary bladder. The synthesis of induced protein was determined by measuring the simultaneous incorporation of ^{3}H-methionine into hemi-bladders exposed to aldosterone (2×10^{-8} mol l^{-1}) and ^{35}S-methionine into hemi-bladders that had not been exposed to this hormone. Protein synthesis is expressed as the ^{3}H: ^{35}S ratio occurring in the corresponding eluate fractions derived from the hormone-stimulated and control hemi-bladder extracts.
After Scott and Sapirstein (1975); modified

is convincing evidence of their existence. Ribosomes isolated from aldosterone-treated toad bladder cells have been shown to cause more tritiated phenylalanine to be incorporated into nascent polypeptide chains than ribosomes from non-treated cells (Wilce, Rossier and Edelman, 1976). Also, analysis of the isolated mitochondria-rich cells from toad bladder tissue that has been stimulated by aldosterone has revealed the presence of several new proteins with molecular weights from 17 000 to 38 000 daltons (Fig. 22).

The manner in which these proteins influence the polar movements of ions through the responding cells in not known but several theories have been proposed. One of the most widely supported theories concerns the belief that under aerobic conditions aldosterone-induced protein may enhance the production of glycolytic substrates leading to the oxidative phosphorylation of ADP to ATP (Spooner and Edelman, 1976a). But, the observation that a hormonally-induced increase in SCC occurs in toad bladders maintained under anaerobic conditions appears to make this hypothesis less tenable (Handler, Preston and Orloff, 1969). However, to counter this objection, Edelman and his associates, have pointed out that the SCC observed under anaerobic conditions may not be an accurate measure of increased Na$^+$ flux since, in the absence of oxygen, the $J_{net}^{Na^+}$, when measured by tracer kinetics, exceeds the SCC and does not change in the presence of aldosterone (Spooner and Edelman, 1976b). The question as to whether the metabolic effects of the hormone are secondary to the transport of Na$^+$ through the tissue is the keystone of this hypothesis. Studies involving the inhibition of basal Na$^+$ efflux with ouabain show that a *decrease* in lactate formation occurs following the addition of aldosterone (Handler, Preston and Orloff, 1969). According to the metabolic theory, aldosterone should cause an increase in lactate production in the absence of net Na$^+$ transport. Other investigators have pointed out that ouabain inhibits the natriferic response of the toad bladder to aldosterone only after glucose is added to the medium (Crabbé and coworkers, 1974). Furthermore, in substrate-depleted tissues, ouabain reduces the hormone-stimulated increase in SCC only by an amount that only represents the *basal* activity of the ouabain-sensitive Na$^+$/K$^+$ ATPase. That is to say, the tissue still responded to aldosterone and Na$^+$ transport was still increased, even though the ouabain-sensitive Na$^+$/K$^+$ ATPase activity was reduced. Since both decamethasone and corticosterone have been shown to stimulate Na$^+$/K$^+$ ATPase activity, observations suggesting that aldosterone may directly stimulate the Na$^+$/K$^+$ ATPase could possibly reflect the glucocorticoid activity of this hormone (Skou, 1964; Landon, Jazab and Forte, 1966; Katz and Epstein, 1967; Jørgensen, 1968, 1969; Hendler and coworkers, 1972).

The possibility has also been suggested that aldosterone-induced proteins may alter the permeability characteristics of the apical (luminal) surfaces of responsive cells warrants careful consideration. This has been called the 'permease' theory and it postulates that this induced protein may act to stimulate the influx of Na$^+$ through the apical (luminal) membrane. Stimulation of Na$^+$ transport by aldosterone is accompanied by a decrease in the transepithelial electrical resistance of toad bladders that are depleted of metabolic substrates. This may indicate a change in the permeability of the cell but since pyruvate does not alter the apical resistance of these cells, it has been suggested that the aldosterone-induced protein acts primarily on the apical surfaces of the transporting cells (Hoffmann and Civan, 1971). The most serious objection to the belief that aldosterone acts solely to facilitate the entry of Na^+ through the apical membrane is the observation that aldosterone can independently increase the activity of some of the enzymes of the tricarboxylic acid cycle (Kirsten and coworkers, 1968). However, it has also been pointed out that utilization of tricarboxylic acid substrates and cellular respiration depend on the availability of transportable Na^+ and, therefore, the activity of these enzymes may not be directly increased by aldosterone.

Finally, aldosterone-induced proteins have been implicated in the stimulation of Na^+ transport by inducing increases in the carbon anhydrase activities of the mitochondria-rich cells of frog skin (Voûte, Thummel and Brenner, 1975). Histochemical studies have indicated that this enzyme is secreted by the mitochondria-rich cells in the skin and deposited along the apical (mucosal) border of adjacent granular cells. For some time it has been known that the Na^+ permeability of this border is dependent on H^+ concentration (Bentley, 1966). Thus, the enhanced carbonic anhydrase activity along the apical border of the granular cells may lead to an acidification of the apical environment and a subsequent increase in the Na^+ permeability of this membrane. Thus, only the mitochondria-rich cell is sensitive to the hormone but its effect on enzyme activity in the cells may alter the Na^+ transport properties of the neighbouring granular cells.

The controversy over the possible intracellular mode of action of corticosteroid will not be resolved until corticosteroid-induced proteins are characterized functionally as well as structurally. It is entirely possible that aldosterone has a dual effect on Na^+ transport, facilitating both the entry of Na^+ into the cell and stimulating the metabolic enzymes controlling the production of ATP. A similar unification of these theories has been proposed to explain the disparate findings of studies on the natriferic actions of the neurohypophysial peptides (Finn, 1975).

ACKNOWLEDGEMENTS

The cost of preparing the manuscript for this chapter was defrayed by funds from the University of California and the National Science Foundation, Washington, DC. The bibliographic search was completed in part through a grant to Professor M. Marcus, Institute for the Interdisciplinary Applications of Algebra and Combinatorics, from the Office of Science Information Services of the National Science Foundation, Washington, DC (grant No. DSI 76-09080). We also wish to express our appreciation of the assistance we have received from Miss Jane Gorsline and Miss Virginia Hui-Fang Young in reviewing the literature in some of the areas covered by this chapter. Finally, we are deeply grateful to Mrs Ilene Hames who has maintained the bibliographic records and compiled the final list of references and typed all the drafts, as well as the final copy of the manuscript.

REFERENCES

Abramow, M., and M. Dratwa (1974). Effect of vasopressin on isolated human collecting duct. *Nature, Lond.*, **250**, 492–3.

Acher, R. R., J. Chauvet, M. T. Chauvet, and D. Crepy (1964). Phylogénie des peptides neurohypophysaires: isolement de la mésotocine (Ileu$_8$-ocytocine) de la grenouille, intermédiare entre la Ser$_4$-Ileu$_8$-ocytocine des poissons osseaux et 1-ocytocine des mammifères. *Biochim. Biophys. Acta*, **90**, 613–4.

Addison, T. (1885). On the constitutional and local effects of disease of the suprarenal capsules, Highley, London. Also in *Med. Classics*, **2**, 244–80 (1937).

Agarwal, M. K. (1975). Chromatographic demonstration of mineralocorticoid-specific receptors in rat kidney. *Nature, Lond.*, **254**, 623–5.

Agarwal, M. K. (1976a). Demonstration of steroid specific hormone receptors by chromatography. *Febs. Lett. (Fed. Eur. Biochem. Soc.)*, **62**, 25–9.

Agarwal, M. K. (1976b). Identification and properties of renal mineralocorticoid binders in rat liver and kidney. *Biochem. J.*, **154**, 567–5.

Alberti, K. G. M. M., and G. W. G. Sharp (1969). The isolation of bound aldosterone from the nuclei of human, toad and rat colon. *Ann. Endocrinol.*, **31**, 777–83.

Allen, J. C., J. H. Abel, and D. J. Takemoto (1975). Effect of osmotic stress on serum corticoid and plasma glucose levels in the duck (*Anas platyrhynchos*). *Gen. Comp. Endocrinol.*, **26**, 209–16.

Al-Zahid, G., J. A. Schafer, and T. E. Andreoli (1975). ADH action in isolated cortical collecting tubules: Evidence for increases in luminal membrane fluidity. *The Physiologist (Abstr.)*, **18**, 120.

Andersen, B., and H. H. Ussing (1957). Solvent drag on non-electrolytes during osmotic flow through isolated toad skin and its response to antidiuretic hormone. *Acta Physiol. Scand.*, **39**, 228–39.

Ando, M. (1974). Effects of cortisol on water transport across the eel intestine. *Endocrinol. Jap.*, **21**, 539–46.

Ando, M., S. Utida, and H. Nagahama (1975). Active transport of chloride in eel intestine with special reference to sea water adaptation. *Comp. Biochem. Physiol.*, **51A**, 27–32.

Andreoli, T. E. (1973). On the anatomy of amphotericin B-cholesterol pores in lipid bilayer membranes. *Kidney Int.*, **4**, 337–45.

Andreoli, T. E., and J. A. Schafer (1976). Mass transport across all membranes: the effects of antidiuretic hormone on water and solute flows in epithelia. *A. Rev. Physiol.*, **38**, 451–500.

Andreoli, T. E., J. A. Schafer, and S. L. Troutman (1971). Coupling of solute and solvent flows in porous lipid bilayer membranes. *H. Gen. Physiol.*, **57**, 479–93.

Baba, W. I., A. J. Smith, and M. W. Townshend (1967). The effects of vasopressin, theophylline and cyclic 3′,5′-adenosine monophosphate (cyclic AMP) on sodium transport across the frog skin. *Q. J. Exp. Physiol.*, **52**, 416–21.

Babiker, M. M., and C. J. Rankin (1973). Effect of neurohypophysial hormones on renal function in the freshwater- and seawater-adapted eel (*Anguilla anguilla* L.). *Proc. Soc. Endocrinol., J. Endocrinol.*, **57**, xi–xii.

Baillien, M., and E. Schoffeneils (1961). Origin of the potential difference in the intestinal epithelium of the turtle. *Nature, Lond.*, **190**, 1107–8.

Balinsky, J. B., S. E. Dicker, and A. B. Elliott (1972). The effect of long-term adaptation to different levels of salinity on urea synthesis and tissue amino acid concentrations in *Rana cancrivora*. *Comp. Biochem. Physiol.*, **43B**, 71–82.

Ball, J. N., I. Chester Jones, M. E. Forster, G. Hargreaves, E. F. Hawkins, and K. P. Milne (1971). Measurement of plasma cortisol levels in the eel *Anguilla anguilla* in relation to osmotic adjustments. *J. Endocrinol.*, **50**, 75–96.

Bär, H. P., O. Hechter, I. L. Schwartz, R. Walter, and W. Roderich (1970). Neurohypophyseal hormone-sensitive adenyl-cyclase of toad urinary bladder. *Proc. Natl Acad. Sci. USA*, **67**, 7–12.

Barger, A. C., R. D. Berlin, and J. F. Tulenko (1958). Infusion of aldosterone, 9α-fluorohydrocortisone and antidiuretic hormone into the renal artery of normal and adrenalectomized, unanesthetized dogs: Effect on electrolyte and water excretion. *Endocrinology*, **62**, 804–15.

Barraclough, M. A., and N. F. Jones (1970). Effects of adenosine-3′,5′-monophosphate on renal function in the rabbit. *Br. J. Pharmacol. Chemother.*, **40**, 334–41.

Bastide, F., and S. Jard (1968). Action of noradrenaline and oxytocin on the active transport of Na^+ and the permeability of frog skin to water. Role of cyclic AMP. *Biochim. Biophys. Acta*, **150**, 113–23.

Bellamy, D., and I. Chester Jones (1961). An apparatus for studying the gain and loss of electrolytes by the eel. *Comp. Biochem. Physiol.*, **3**, 223–6.

Bennett, C. M., B. M. Brenner, and R. W. Berliner (1968). Micropuncture study of nephron function in the rhesus monkey. *J. Clin. Invest.*, **47**, 203–16.

Bentley, P. J. (1959). Studies on the water and electrolyte metabolism of the lizard *Trachysaurus rugosus* (Gray). *J. Physiol., Lond.*, **145**, 37–47.

Bentley, P. J. (1960). The effects of vasopressin on the short-circuit current across the wall of the isolated bladder of the toad, *B. marinus*. *J. Endocrinol.*, **21**, 161–70.

Bentley, P. J. (1961). Directional differences in the permeability to water of the isolated urinary bladder of the toad, *Bufo marinus*. *J. Endocrinol.*, **22**, 95–100.

Bentley, P. J. (1962). Studies on the permeability of the large intestine and urinary bladder of the tortoise (*Testudo graeca*) with special reference to the effects of neuro-hypophysial and adrenocortical hormones. *Gen. Comp. Endocrinol.*, **2**, 323–8.

Bentley, P. J. (1966). The physiology of the urinary bladder of amphibia. *Biol. Rev.*, **41**, 275–316.

Bentley, P. J. (1969a). Neurohypophysial function in amphibia: hormone activity in the plasma. *J. Endocrinol.*, **43**, 359–69.

Bentley, P. J. (1969b). Neurohypophyseal hormones in amphibia: A comparison of their actions and storage. *Gen. Comp. Endocrinol.*, **13**, 39–44.

Bentley, P. J. (1971). *Endocrines and Osmoregulation. Zoophysiology and Ecology*, Vol. 1, Springer-Verlag, New York.

Bentley, P. J., and B. K. Follett (1962). The action of neurohypophysial and adrenocortical hormones on sodium balance in the cyclostome, *Lampetra fluviatilis*. *Gen. Comp. Endocrinol.*, **2**, 329–35.

Bentley, P. J., and B. K. Follett (1963). Kidney function in a primitive vertebrate, the cyclostome *Lampetra fluviatilis*. *J. Physiol., Lond.*, **169**, 902–18.

Bentley, P. J., and K. Scmidt-Nielsen (1965), Permeability of water and sodium of the crocodilian *Caiman sclerops*. *J. Cell. Comp. Physiol.*, **66**, 303–9.

Bentley, P. J., and T. Yorio (1976). The passive permeability of the skin of anuran amphibia: A comparison of frogs (*Rana pipiens*) and toads (*Bufo marinus*). *J. Physiol., Lond.*, **261**, 603–15.

Berde, B., and H. Konzett (1960). Isoleucyl[8]-oxytocine, ein biologisch hochwirksames Polypeptid. *Med. Exp.*, **2**, 317–22.

Berde, B., R. Huguenin, and E. Stürmer (1962). The biological activities of arginine-vasotocin obtained by a new synthesis. *Experientia*, **18**, 444–5.

Berliner, R. W., and G. M. Bennet (1967). Concentration of urine in the mammalian kidney. *Am. J. Med.*, **42**, 777–89.

Berliner, R. W., and D. G. Davidson (1957). Production of hypertonic urine in the absence of pituitary antidiuretic hormone. *J. Clin. Invest.*, **36**, 1416–27.

Berger, E. Y., G. Kanzaki, and J. M. Steele (1960). The effect of deoxycorticosterone on the unidirectional transfers of sodium and potassium into and out of the dog intestine. *J. Physiol., Lond.*, **151**, 352–62.
Biedl, A. (1913). *The Internal Secretory Organs*. W. Wood and Co., New York.
Bockaert, J., C. Roy, R. Rajerison, and S. Jard (1973). Specific binding of [^{3}H]lysine-vasopressin to pig kidney plasma membranes. Relationship of receptor occupancy to adenylate cyclase activation. *J. Biol. Chem.*, **248**, 5922–31.
Boinet, E. (1903). Dangers de l'adrénaline dans certains cas de maladie bronzée d'Addison. *C. R. Séanc. Soc. Biol., Paris*, **55**, 1471–3.
Boissonas, R. A., S. T. Guttman, B. Berde, and H. Konzett (1961). Relationships between the chemical structures and the biological properties of the posterior pituitary hormones and their synthetic analogues. *Experientia*, **17**, 377–90.
Bourguet, J., and J. Maetz (1961). Arguments en faveur de l'indépendence des mécanismes d'action de divers peptides neurohypophysaires sur de flux osmotique d'eau et sur le transport actif de sodium au sein d'un même récapteur; études sur las vessie et la peau de *Rana esculenta*. *Biochim. Biophys. Acta*, **52**, 552–65.
Bourguet, J., and F. Morel (1967). Independance des variations de permeabilité a l'eau et au sodium produites par les hormones neurohypophysaires sur la vessie de grenouille. *Biochim. Biophys. Acta*, **135**, 693–700.
Bradley, E. L., and W. N. Holmes (1971). The effects of hypophysectomy on adrenocortical function in the duck (*Anas platyrhynchos*). *J. Endocrinol.*, **49**, 437–57.
Bradley, E. L., and W. N. Holmes (1972). The role of the nasal glands in the survival of ducks (*Anas platyrhynchos*) exposed to hypertonic saline drinking water. *Can. J. Zool.*, **50**, 611–17.
Bradley, E. L., W. N. Holmes, and A. Wright (1971). The effects of neurohypophysectomy on the pattern of renal excretion in the duck *Anas platyrhynchos*. *J. Endocrinol.*, **51**, 57–65.
Bradshaw, S. D. (1972). The endocrine control of water and electrolyte metabolism in desert reptiles. *Gen. Comp. Endocrinol., Suppl.*, **3**, 360–73.
Bradshaw, S. D. (1975). Osmoregulation and pituitary-adrenal function in desert reptiles. *Gen. Comp. Endocrinol.*, **25**, 230–48.
Bradshaw, S. D., V. H. Shoemaker, and K. A. Nagy (1972). The role of adrenal corticosteroids in the regulation of kidney function in the desert lizard *Dipsosaurus dorsalis*. *Comp. Biochem. Physiol.*, **43A**, 621–35.
Braysher, M., and B. Green (1970). Absorption of water and electrolytes from the cloaca of an Australian lizard, *Varanus gouldii* (Gray). *Comp. Biochem. Physiol.*, **35**, 607–14.
Brodsky, W. A., and T. P. Schilb (1960). Electrical and osmotic characteristics of the isolated turtle bladder. *J. Clin Invest.*, **39**, 974–5.
Brooks, C. M., J. Ushiyama, and G. Lange (1962). Reactions of neurons on or near the supraoptic nuclei. *Am. J. Physiol.*, **202**, 487–90.
Brown, E., D. L. Clarke, V. Roux, and G. H. Sherman (1963). The stimulation of adenosine 3′,5′-monophosphate production by antidiuretic factors. *J. Biol. Chem.*, **238**, 852–3.
Brown-Séquard, C. E. (1856). La physiologie et la pathologie des capsules surrénales. *Arch. Gén. Méd., Sér.* 5, **8**, 385.
Brown-Séquard, C. E. (1857). Nouvelles recherches sur les capsules surrenales. *C. R. Hebd. Séanc. Acad. Sci., Paris*, **45**, 1036–9.
Brown-Séquard, C. E. (1858). Nouvelles recherches sur l'importance des fonctions des capsules surrénales. *J. Physiol. Path. Gén.*, **1**, 160.
Burg, M. B., and N. Green (1973). Function of the thick ascending limb of Henle's loop. *Am. J. Physiol.*, **224**, 659–68.

Burgess, W. W., A. M. Harvey, and E. K. Marshall (1933). The site of the antidiuretic action of pituitary extract. *J. Pharmacol Exp. Ther.*, **49**, 237–48.
Butler, D. G., and W. H. Knox (1970). Adrenalectomy of the painted turtle (*Chrysemys picta belli*): Effect on ion regulation and tissue glycogen. *Gen. Comp. Endocrinol.*, **14**, 551–6.
Carasso, N., P. Favard, and J. Bourguet (1973). Action de la cytochalasine B sur la réponse hydrosmotique et l'ultrastructure de lá vessie urinaire de la grenouille. *J. Microscopie*, **18**, 383–400.
Carasso, N., P. Favard, J. Bourguet, and S. Jard (1966). Rôle du flux net d'eau dans les modifications ultrastructurales de la vessie de Grenouille stimulée par l'ocytocine. *J. Microscopie*, **5**, 519–22.
Carlson, I. H., and W. N. Holmes (1962). Changes in the hormone content of the hypothalamo-hypophysial system of the rainbow trout (*Salmo gairdneri*). *J. Endocrinol*, **24**, 23–32.
Carofoli, E., and A. Lehninger (1971). A survey of the interaction of calcium ions with mitochondria from different tissues and species. *Biochem. J.*, **122**, 681–90.
Cass, A., and A. Finkelstein (1967). Water permeability of thin lipid membrane. *Am. J. Physiol.*, **50**, 1765–84.
Castle, T. R., and H. E. Williamson (1965). Stimulation *in vivo* of renal RNA-synthesis by aldosterone. *Proc. Soc. Exp. Biol. Med.*, **119**, 308–11.
Cereijido, M., and P. F. Curran (1965). Intracellular electrical potentials in frog skin. *J. Gen. Physiol.*, **48**, 543–57.
Cereijido, M., I. Reisin, and C. A. Rotunno (1968). The effect of sodium concentration on the content and distribution of sodium in the frog skin. *J. Physiol., Lond.*, **196**, 237–53.
Chan, D. K. O., and I. Chester Jones (1969). Pressor effects of neurohypophysial peptides in the eel, *Anguilla anguilla* L., with some reference to their interaction with adrenergic and cholinergic receptors. *J. Endocrinol.*, **45**, 161–74.
Can, D. K. O., I. Chester Jones, I. W. Henderson, and J. C. Rankin (1967). Studies on the experimental alteration of water and electrolyte composition of the eel (*Anguilla anguilla* L.). *J. Endocrinol.*, **37**, 297–317.
Chan, D. K. O., I. P. Callard, and I. Chester Jones (1970). Observations on the water and electrolyte composition of the Iguanid lizard *Dipsosaurus dorsalis* (Baird and Girard), with special reference to the control by the pituitary gland and the adrenal cortex. *Gen. Comp. Endocrinol.*, **15**, 374–87.
Chan, D. K. O., J. C. Rankin, and I. Chester Jones (1969). Influences of the adrenal cortex and the corpuscles of Stannius on osmoregulation in the European eel (*Anguilla anguilla* L.) adapted to freshwater. *Gen. Comp. Endocr., Suppl.*, **2**, 342–53.
Chan, M. Y., E. L. Bradley, and W. N. Holmes (1972). The effects of hypophysectomy on the metabolism of adrenal steroids in the pigeon (*Columba livia*). *J. Endocrinol.*, **52**, 435–50.
Charron, R. C., C. E. Leme, D. R. Wilson, T. S. Ing, and O. M. Wrong (1969). The effect of adrenal steroids on stool composition, as revealed by *in vivo* dialysis of faeces. *Clin. Sci.*, **37**, 151–67.
Chester Jones, I. (1956). The role of the adrenal cortex in the control of water and salt-electrolyte metabolism in vertebrates. In I. Chester Jones and P. Eckstein (Eds) *The Comparative Endocrinology of Vertebrates. II. The Hormonal Control of Water and Salt-Electrolyte Metabolism in Vertebrates: Mem. Soc. Endocrinol.*, **5**, Cambridge University Press, Cambridge. pp. 102–24.
Chester Jones, I. (1957). *The Adrenal Cortex*. Cambridge University Press, Cambridge.
Chester Jones, I., and D. Bellamy (1964). Hormonal mechanisms in the homeostatic

regulation of the vertebrate body with special reference to the adrenal cortex. *Symp. Soc. Exp. Biol.*, **18**, 195–236.

Chester Jones, I., I. W. Henderson, and W. Moseley (1964). Methods for the adrenalectomy of the European eel (*Anguilla anguilla* L.). *J. Endocrinol.*, **30**, 155–6.

Chester Jones, I., I. W. Henderson, N. A. M. Wales, and H. O. Garland (1971). Effects of arginine vasotocin in fish and amphibia. In G. E. W. Wolstenholme and J. Birch (Eds), *Neurohypophysial Hormones*, Ciba Foundation Study Group No. 39, The Williams and Wilkins Co., Baltimore. pp. 29–41.

Chevalier, J., J. Bourguet, and J. S. Hugon (1974a). Membrane associated particles: distribution in frog urinary bladder epithelium at rest and after oxytocin treatment. *Cell Tiss. Res.*, **152**, 129–40.

Chevalier, J., P. Ripoche, M. Pisam, J. Bourguet, and J. S. Hugon (1974b). A time course study of water permeability and morphological alterations induced by mucosal hyperosmolarity in frog urinary bladder. *Cell Tiss. Res.*, **154**, 345–56.

Chew, M.-M., A. B. Elliott, and H. Y. Wong (1972). Permeability of urinary bladder of *Rana cancrivora* to urea in the presence of oxytocin. *J. Physiol., Lond.*, **223**, 757–72.

Choi, J. K. (1963). The fine structure of the urinary bladder of the toad, *Bufo marinus*. *J. Cell Biol.*, **16**, 53–72.

Chu, L. L. H., and I. S. Edelman (1972). Cordycepin and α-amantin: inhibitors of transcription as probes of aldosterone action. *J. Membr. Biol.*, **10**, 291–310.

Civan, M. M. (1970). Path of bulk water movement through the urinary bladder of the toad. *J. Theor. Biol.*, **27**, 387–91.

Civan, M. M., and D. R. Di Bona (1974). Pathways for movement of ions and water across toad urinary bladder. II. Site and mode of action of vasopressin. *J. Membr. Biol.*, **19**, 195–220.

Civan, M. M., and H. S. Frazier (1968). The site of the stimulating action of vasopressin on sodium transport in toad bladder. *J. Gen. Physiol.*, **51**, 589–605.

Civan, M. M., O. Kedem, and A. Leaf (1966). Effect of vasopressin on toad bladder under conditions of zero net sodium transport. *Am. J. Physiol.*, **211**, 569–75.

Clapp, J. R. (1966). Renal tubular reabsorption of urea in normal and protein-depleted rats. *Am. J. Physiol.*, **210**, 1304–8.

Clapp, J. R., and R. R. Robinson (1966). Osmolality of distal tubular fluid in the dog. *J. Clin Invest.*, **45**, 1847–53.

Clark, W. G. (1939). Effect of adrenalectomy upon intestinal absorption of sodium chloride. *Proc. Soc. Exp. Biol. Med.*, **40**, 468–70.

Claybaugh, J. R., and L. Share (1973). Vasopressin, renin, and cardiovascular responses to continuous slow hemorrhage. *Am. J. Physiol.*, **224**, 519–23.

Colley, L., W. C. Rowe, A. K. Huggins, A. B. Elliott, and S. E. Dicker (1972). The effect of short-term changes in the external salinity on the levels of the non-protein nitrogenous compounds and the ornithine-urea cycle enzymes in *Rana cancrivora*. *Comp. Biochem. Physiol.*, **41B**, 307–22.

Crabbé, J. (1961). Stimulation of active sodium transport across the isolated toad bladder after injection of aldosterone to the animal. *Endocrinology*, **69**, 673–82.

Crabbé, J. (1963). Site of action of aldosterone on the bladder of the toad. *Nature, Lond.*, **200**, 787–8.

Crabbé, J. (1966). La régulation de la sécrétion d'aldostérone chez *Bufo marinus*. *Ann. Endocrinol.*, **27**, 501–5.

Crabbé, J., and P. De Weer (1964). Action of aldosterone on the bladder and skin of the toad. *Nature, Lond.*, **202**, 298–9.

Crabbé, J., and P. De Weer (1969). Relevance of sodium transport pool measurements in toad bladder tissue for the elucidation of the mechanism whereby hormones stimulate active sodium transport. *Arch. Ges. Physiol.*, **313**, 197–221.

Crabbé, J. D. D. Fanestil, M. Pelletier, and G. A. Porter (1974). Effect of ouabain on sodium transport across hormone-stimulated toad bladder and skin. *Pflügers Arch. Ges. Physiol.*, **347**, 275–96.

Creange, J. E., and S. Roberts (1965a). Stimulation of steroid C-11β and C-18 hydroxylation in rat adrenal homogenates by adenosine 3′,5′-phosphate via a mechanism not requiring endogenous precursor, glycogen phosphorylation or NADPH generation. *Steroids Suppl.*, **II**, 13–28.

Creange, J. E., and S. Roberts (1965b). Studies on the mechanism of action of cyclic 3′,5′-adenosine monophosphate on steroid hydroxylations in adrenal homogenates. *Biochim. Biophys. Res. Commun.*, **19**, 73–8.

Crocker, A. D. (1969). The effect of aldosterone and angiotensin on the intestinal transport of Na^+ and water in the rat. *J. Endocrinol.*, **44**, ii–iii.

Crocker, A. D., and W. N. Holmes (1971). Intestinal absorption in ducklings (*Anas platyrhynchos*) maintained on freshwater and hypertonic saline. *Comp. Biochem. Physiol.*, **40A**, 203–11.

Crocker, A. D., and W. N. Holmes (1976). Factors affecting intestinal absorption in ducklings (*Anas platyrhynchos*). *Proc. Soc. Endocrinol*, May, 1976.

Crocker, A. D., and K. A. Munday (1967a). Aldosterone and angiotensin action on water absorption in the rat jejunum. *J. Physiol., Lond.*, **192**, 36P–8P.

Crocker, A. D., and K. A. Munday (1967b). Effect of aldosterone on sodium and water absorption from rat jejunum. *J. Endocrinol*, **38**, xxv–xxvi.

Crocker, A. D., and K. A. Munday (1969). Factors affecting mucosal water and sodium transfer in everted sacs of rat jejunum. *J. Physiol., Lond.*, **202**, 329–338.

Crowe, S. J., and G. B. Wislocki (1914). Experimental observations on the suprarenal glands with especial reference to the functions of their interrenal portions. *Bull. Johns Hopkins Hosp.*, **25**, 287–304.

Cuthbert, A. W., and E. Painter (1968). Independent action of antidiuretic hormone theophylline and cyclic 3′,5′-adenosine monophosphate on cell membrane permeability in frog skin. *J. Physiol., Lond.*, **199**, 593–612.

Dainty, J. (1963). Water relations of plant cells. *Adv. Bot. Res.*, **1**, 279–326.

Dantzler, W. H. (1967a). Glomerular and tubular effects of arginine-vasotocin in water snakes (*Natrix sipedon*). *Am. J. Physiol.*, **212**, 83–91.

Dantzler, W. H. (1967b). Stop-flow study of renal function in conscious water snakes (*Natrix sipedon*). *Comp. Biochem. Physiol.*, **22**, 131–40.

Dantzler, W. H., and W. N. Holmes (1974). Water and mineral metabolism in reptilia. In *Chemical Zoology*, Vol. IX, Academic Press, New York. pp. 277–336.

Dantzler, W. H., and B. Schmidt-Nielsen (1966). Excretion in fresh-water turtle (*Pseudomys scripta*) and desert tortoise (*Gopherus agassizii*). *Am. J. Physiol.*, **210**, 198–210.

Davis, W. L., D. B. P. Goodman, J. H. Martin, J. L. Matthews, and H. Rasmussen (1974). Vasopressin-induced changes in the toad urinary bladder epithelial surface. *J. Cell Biol.*, **61**, 544–7.

Davoren, P. R., and E. W. Sutherland (1963). The cellular location of adenyl cyclase in the pigeon erythrocyte. *J. Biol. Chem.*, **238**, 3016–23.

DeLorenzo, R. J., K. G. Walton, P. F. Curran, and P. Greengard (1973). Regulation of phosphorylation of a specific protein in toad-bladder membrane by antidiuretic hormone and cyclic AMP, and its possible relationship to membrane permeability changes. *Proc. Natl Acad. Sci., USA*, **70**, 880–4.

Dennis, C., and E. H. Wood (1940). Intestinal absorption in adrenalectomized dog. *Am. J. Physiol.*, **129**, 182.

Dexter, R. N., L. M. Fishman, and R. L. Ney (1970). Stimulation of adrenal cholesterol uptake from plasma by adrenocorticotrophin. *Endocrinology*, **87**, 836–46.

Dexter, R. N., L. M. Fishman, R. L. Ney, and G. W. Liddle (1967). An effect of adrenocorticotrophic hormone on adrenal cholesterol accumulation. *Endocrinology*, **81**, 1185–7.

Dicker, S. E., and A. B. Elliott (1970a). Water uptake by the crab-eating frog, *Rana cancrivora*, as affected by osmotic gradients and by neurohypophysial hormones. *J. Physiol., Lond.*, **207**, 119–32.

Dicker, S. E., and A. B. Elliott (1970b). Effect of neurohypophysial hormones on fluid movement across isolated bladder of *Rana cancrivora, Rana temporaria* and *Bufo melanostictus*. *J. Physiol., Lond.*, **210**, 137–49.

Donaldson, E. M., and W. N. Holmes (1965). Corticosteroidogenesis in the freshwater and saline-maintained duck (*Anas platyrhynchos*). *J. Endocrinol.*, **32**, 329–36.

Douglas, W. W. (1973). How do neurones secrete peptides? Exocytosis and its consequences, including synaptic vesicle formation, in the hypothalamoneurohypophyseal system. *Prog. Brain Res.*, **39**, 21–38.

Douglas, W. W. (1974). Mechanism of release of neurohypophysial hormones: stimulus-secretion coupling. In R. O. Greep and E. B. Astwood (Eds) *Handbook of Physiology, Section 7, Endocrinology*, Vol. IV, American Physiological Society, Washington. pp. 191–224.

Douglas, W. W., and J. Nagasawa (1971). Membrane vesiculation at sites of exocytosis in the neurohypophysis, adenohypophysis and adrenal medulla: a device for membrane conservation. *J. Physiol., Lond.*, **218**, 94P–5P.

Dousa, T. P., and L. D. Barnes (1974a). Effects of colchicine (CLC), vinblastine (VBL) and cytochalasin B (CB) on the renal response to vasopressin. *Fed. Proc. Fedn Am. Soc. Exp. Biol.*, **33**, 388.

Dousa, T. P., and L. D. Barnes (1974b). Effects of colchicine and vinblastine on the cellular action of vasopressin in mammalian kidney. A possible role of microtubules. *J. Clin. Invest.*, **54**, 252–62.

Dousa, T. P., O. Hechter, I. L. Schwartz, and R. Walter (1971). Neurohypophyseal hormone-responsive adenylate cyclase from mammalian kidney. *Proc. Natl Acad. Sci., USA*, **68**, 1693–7.

Dousa, T. P., H. Sands, and O. Hechter (1972). Cyclic AMP-dependent reversible phosphorylation of renal medullary plasma membrane protein. *Endocrinology*, **91**, 757–63.

Dunson, W. A. (1968). Salt gland secretion in the pelagic sea snake *Pelamis*. *Am. J. Physiol.*, **215**, 1512–7.

Dyball, R. E. J., and K. Koizumi (1969). Electrical activity in the supraoptic and paraventricular nuclei associated with neurohypophysial hormone release. *J. Physiol., Lond.*, **201**, 711–22.

Edelman, I. S. (1975). Mechanism of action of steroid hormones. *J. Steroid Biochem.*, **6**, 147–59.

Edelman, I. S., and G. N. Fimognari (1968). On the biochemical mechanisms of action of aldosterone. *Recent Prog. Horm. Res.*, **24**, 1–44.

Edelman, I. S., R. Bogoroch, and G. A. Porter (1963). On the mechanism of action of aldosterone on sodium transport: the role of protein synthesis. *Proc. Natl Acad. Sci., USA*, **50**, 1169–77.

Edelman, I. S., R. Bogoroch, and G. A. Porter (1964). Specific action of aldosterone on RNA synthesis. *Trans. Ass. Am. Physns*, **77**, 307–16.

Edidin, M., and A. Weiss (1972). Antigen cap formation in cultured fibroblasts: a reflection of membrane fluidity and cell motility. *Proc. Natl Acad. Sci., USA*, **69**, 2456–9.

Edmonds, C. J., and J. C. Marriott (1967). The effect of aldosterone and adrenalectomy on the electrical potential difference of rat colon and on the transport of sodium, patoassium chloride and bicarbonate. *J. Endocrinol.*, **39**, 517–31.

Edmonds, C. J., and J. C. Marriott (1968). Factors influencing the electrical potential across the mucosa of rat colon. *J. Physiol., Lond.*, **194**, 457–78.

Edmonds, C. J., and J. C. Marriott (1969). The effect of aldosterone on the electrical activity of rat colon. *J. Endocrinol.*, **44**, 363–77.

Edmonds, C. J., and J. C. Marriott (1970). Sodium transport and short-circuit current in rat colon *in vivo* and the effect of aldosterone. *J. Physiol., Lond.*, **210**, 1021–39.

Edmonds, C. J., and P. Richards (1970). Measurement of rectal electrical potential difference as an instant screening-test for hyperaldosteronism. *Lancet*, **2**, 624–7.

Efstratopoulos, A. D., and W. S. Peart (1975). Effect of single and combined infusions of angiotensin II and aldosterone on colonic potential difference, blood pressure and renal function, in patients with adrenal deficiency. *Clin. Sci. Molec Med.*, **48**, 219–26.

Efstratopoulos, A. D., W. S. Peart, and G. A. Wilson (1974). The effect of aldosterone on colonic potential difference and renal electrolyte excretion in normal man. *Clin. Sci. Molec. Med.*, **46**, 489–99.

Elizondo, R. S., and S. J. LeBrie (1969). Adrenal-renal function in water snakes (*Natrix cyclopion*). *Am. J. Physiol.*, **217**, 419–25.

Elliott, A. B. (1977). Hormonal control of plasma osmolality in the S. E. Asian euryhaline frog *Rana cancrivora*. In B. Lofts and S. T. H. Chan (Eds) *Biological Research in Southeast Asia*, Occasional Publication No. 3, Department of Zoology, University of Hong Kong Press, Hong Kong. pp. 57–68.

Ellory, J. C., B. Lahlou, and M. H. Smith (1971). Intestinal transport of sodium by goldish adapted to a saline environment. *J. Physiol., Lond.*, **216**, 46P–7P.

Ellory, J. C., B. Lahlou, and M. W. Smith (1972). Changes in the intestinal transport of sodium induced by exposure of goldfish to a saline environment. *J. Physiol., Lond.*, **222**, 497–509.

Engbaek, L., and T. Hoshiko (1957). Electrical potential gradients through frog skin. *Acta Physiol. Scand.*, **39**, 348–55.

Epstein, F. H., C. R. Kleeman, S. Pursel, and A. Hendrikx (1957). The effect of feeding protein and urea on the renal concentrating process. *J. Clin Invest.*, **36**, 635–41.

Fanestil, D. D. (1969). Mechanism of action of aldosterone. *A. Rev. Med.*, **20**, 223–32.

Fanestil, D. D., and I. S. Edelman (1966). Characteristics of the renal nuclear receptors for aldosterone. *Proc. Natl Acad. Sci., USA*, **56**, 872–9.

Farese, R. V. (1964). Inhibition of the steroidogenic effect of ACTH and incorporation of amino acid into rat adrenal protein *in vitro* by chloramphenicol. *Biochim. Biophys. Acta*, **87**, 699–701.

Farese, R. V. (1967). Adrenocorticotrophin-induced changes in the steroidogenic activity of adrenal cell-free preparations. *Biochemistry, Wash. DC*, **6**, 2052–65.

Farini, F. (1913). Diabete insipido éd opoterapia. *Gazz. Osped. Clin.*, **34**, 1135–9.

Farquahar, M. G., and G. E. Palade (1966). Adenosine triphosphatase localization in amphibian epidermis. *J. Cell Biol.*, **30**, 359–88.

Feldman, D. (1975). The role of hormone receptors in the action of adrenal steroids. *A. Rev. Med.*, **26**, 83–90.

Feldman, D., J. W. Funder, and I. S. Edelman (1972). Subcellular mechanisms in the action of adrenal steroid. *Am. J. Med.*, **53**, 545–60.

Ferguson, D. R., and R. H. Price (1972). The actions of cyclic nucleotides on the toad bladder. *Adv. Cyclic Nucleotide Res.*, **1**, 113–9.

Ferguson, J. J., Jr. (1963). Protein synthesis and adrenocorticotrophin responsiveness. *J. Biol. Chem.*, **238**, 2754–9.

Filehne, W., and Biberfeld, H. (1902). Beiträge zur Diurese. *Arch. Ges. Physiol.*, **91**, 569–73.

Fimognari, G. M., D. D. Fanestil, and I. S. Edelman (1967). Induction of RNA and protein synthesis in the action of aldosterone in the rat. *Am. J. Physiol.*, **213**, 954–62.

Finkelstein, A., and A. Cass (1967). Effect of cholesterol on the water permeability of thin lipid membranes. *Nature, Lond.*, **216**, 717–8.

Finn, A. L. (1968). Separate effects of sodium and vasopressin on the sodium pump in toad bladder. *Am. J. Physiol.*, **215**, 849–56.

Finn, A. L. (1971). The kinetics of sodium transport in the toad bladder. II. Dual effects of vasopressin. *J. Gen. Physiol.*, **57**, 349j062.

Finn, A. L. (1975). Action of ouabain on sodium transport in toad urinary bladder. Evidence for two pathways for sodium entry. *J. Gen. Physiol.*, **65**, 503–14.

Finn, A. L., and M. L. Rockoff (1971). The kinetics of sodium transport in the toad bladder. I. Determination of the transport pool. *J. Gen. Physiol.*, **57**, 326–48.

Finn, F. M., C. C. Widnell, and K. Hofmann (1972). Localization of an adrenocorticotropic hormone receptor on bovine adrenal cortical membranes. *J. Biol. Chem.*, **247**, 5695–702.

Fisher, C., W. R. Ingram, and S. W. Ranson (1938). *Diabetes Insipidus and the Neurohormonal Control of Water Balance*. Edwards, Ann Arbor, Michigan.

Fletcher, G. L., and W. N. Holmes (1968). Observations on the intake of water and electrolytes by the duck (*Anas platyrhynchos*) maintained on freshwater and on hypertonic saline. *J. Exp. Biol.*, **49**, 325–39.

Follett, B. K., and H. Heller (1964a). The neurohypophysial hormones of lungfishes and amphibians. *J. Physiol., Lond.*, **172**, 92–106.

Follett, B. K., and H. Heller (1964b). The neurohypophysial hormones of bony fishes and cyclostomes. *J. Physiol., Lond.*, **172**, 72–91.

Forman, B. H., and P. J. Mulrow (1975). Effect of corticosteroids on water and electrolyte metabolism. In R. O. Greep and E. B. Astwood (Eds) *Handbook of Physiology, Section 7, Endocrinology*, Vol. VI, American Physiological Society, Washington. pp. 179–89.

Forte, L., and E. J. Landon (1968). Aldosterone-induced RNA synthesis in the adrenalectomized rat kidney. *Biochim. Biophys. Acta*, **157**, 303–9.

Forte, L. R., W. T. H. Chao, R. J. Walkenbach, and K. H. Byington (1972). Kidney membrane cyclic AMP receptor and cyclic AMP-dependent protein kinase activities: comparison of plasma membrane and cytoplasmic fractions. *Biochim. Biophys. Res. Commun.*, **49**, 1510–7.

Fowler, M. A., and I. Chester Jones (1955). The adrenal cortex in the forg, *Rana temporaria*, and its relation to water and salt electrolyte metabolism. *J. Endocrinol.*, **13**, vi–vii.

Frazier, H. S. (1962). The electrical potential profile of the isolated toad bladder. *J. Gen. Physiol.*, **45**, 515–28.

Frazier, H. S., and E. I. Hammer (1963). Efflux of sodium from isolated toad bladder. *Am. J. Physiol.*, **205**, 718–22.

Frazier, H. S., E. F. Dempsey, and A. Leaf (1962). Movement of sodium across the mucosal surface of the isolated toad bladder and its modification by vasopressin. *J. Gen. Physiol.*, **45**, 529–43.

Frieden, E., and H. Lipner (1971). *Biochemical Endocrinology of the Vertebrates*. Prentice-Hall, Englewood Cliffs, New Jersey.

Frindt, G., and M. B. Burg (1972). Effect of vasopressin on sodium transport in renal cortical collecting tubules. *Kidney Int.*, **1**, 224–31.

Fuhrman, F. A., and H. H. Ussing (1951). A characteristic response of the isolated frog skin potential to neurohypophysial principles and its relation to the transport of sodium and water. *J. Cell. Comp. Physiol.*, **38**, 109–30.

Funder, J. W., D. Feldman, and I. S. Edelman (1973). The roles of plasma binding and receptor specificity in the mineralocorticoid action of aldosterone. *Endocrinology*, **92**, 944–1004.

Gaitskell, R. E., and I. Chester Jones (1970). Effects of adrenalectomy and cortisol injection on the *in vitro* movement of water by the intestine of the freshwater European eel (*Anguilla anguilla* L.). *Gen. Comp. Endocrinol.*, **15**, 491–3.

Gamble, J. L., C. F. McKhann, A. M. Butler, and E. Tuthill (1934). An economy of water in renal function referable to urea. *Am. J. Physiol.*, **109**, 139–54.

Ganote, C. E., J. J. Grantham, H. L. Moses, M. B. Burg, and J. Orloff (1968). Ultrastructural studies of vasopressin effect on isolated perfused renal collecting tubules of the rabbit. *J. Cell Biol.*, **36**, 355–67.

Garcia Romeu, F., and J. Maetz (1964). The mechanism of sodium and chloride uptake by the gills of a fresh-water fish, *Carassius auratus*. *J. Gen. Physiol.*, **47**, 1195–207.

Gardner, K. D., Jr., and R. H. Maffly (1964). An *in vitro* demonstration of increased collecting tubular permeability to urea in the presence of vasopressin. *J. Clin. Invest.*, **43**, 1968–75.

Garren, L. D., G. N., and G. N. Walton (1971a). The isolation of a receptor for adenosine 3′,5′-cyclic AMP (cAMP) and the adrenal cortex: The role of the receptor on the mechanism of action of c-AMP. *Ann. NY Acad. Sci.*, **185**, 210–26.

Garren, L. D., G. N. Gill, H. Masui, and G. M. Walton (1971b). On the mechanism of action of ACTH. *Recent Prog. Horm. Res.*, **27**, 433–78.

Garren, L. D., R. L. Ney, and W. W. Davis (1965). Studies on the role of protein synthesis in the regulation of corticosterone production by adrenocorticotropic hormone *in vivo*. *Proc. Natl Acad. Sci., USA*, **53**, 1443–50.

Gatzy, J. T., and W. O. Berndt (1968). Isolated epithelial cells of the toad bladder. Their preparation, oxygen consumption and electrolyte content. *J. Gen. Physiol.*, **51**, 770–84.

Gaunt, R., and J. J. Chart (1962). Mineralocorticoid action of adrenocortical hormones. In H. Wendler Deane (Ed.) *Handbuch der Experimentallen Pharmakologie*, Vol. XIX, No. 1, *The Adrenocortical Hormones, Their Origin, Chemistry, Physiology and Pharmacology*, Springer-Verlag, Berlin.

Goodfield, T., and J. Kirkpatrick (1963). Effects of neurohypophysial hormones on oxidative metabolism of the toad bladder *in vitro*. *Endocrinology*, **72**, 742–8.

Gordon, M. S., and V. A. Tucker (1965). Osmotic regulation in the tadpoles of the crab-eating frog (*Rana cancrivora*). *J. Exp. Biol.*, **42**, 437–45.

Gordon, M. S., K. Schmidt-Nielsen, and H. M. Kelly (1961). Osmotic regulation on the crab-eating frog (*Rana cancrivora*). *J. Exp. Biol.*, **38**, 659–78.

Gottschalk, C. W. (1961). Micropuncture studies of tubular function in the mammalian kidney. *Physiologist*, **4**, 35–55.

Gottschalk, C. W., and M. Mylle (1959). Micropuncture study of the mammalian urinary concentrating mechanism: evidence for the counter-current hypothesis. *Am. J. Physiol.*, **196**, 927–36.

Goulston, K., D. D. Harrison, and A. P. Skyring (1963). Effect of mineralocorticoids on the sodium/potassium ratio of human ileostomy fluid. *Lancet*, **2**, 541–2.

Grahame-Smith, D. H., R. W. Butcher, R. L. Ney, and L. W. Sutherland (1967). Adenosine 3′,5′-monophosphate as the intracellular mediator of the action of adrenocorticotropic hormone on the adrenal cortex. *J. Biol. Chem.*, **242**, 5535–41.

Grantham, J. J., and M. B. Burg (1966). Effect of vasopressin and cyclic-AMP on permeability of isolated collecting tubules. *Am. J. Physiol.*, **211**, 255–9.

Grantham, J. J., and J. Orloff (1968). Effect of prostaglandin E_1 on the permeability response of the isolated collecting tubule to vasopressin, adenosine 3′,5′-monophosphate and theophylline. *J. Clin. Invest.*, **47**, 1154–61.

Graziani, Y., and A. Livne (1971). Vasopressin and water permeability of artificial lipid membranes. *Biochim. Biophys. Res. Commun.*, **45**, 321–6.

Graziani, Y., and A. Livne (1972). Water permeability of bilayer lipid membranes: sterol-lipid interactions. *J. Membr. Biol.*, **7**, 275–84.

Guttman, S., B. Berde, and E. Stürmer (1962). The synthesis and some pharmacological effects of serine[4]-isoleucine[8]-oxytocin, a probable neurohypophysial hormone. *Experientia*, **18**, 445–6.

Handler, J. S. (1964). Metabolic effects of antidiuretic hormone. In J. Metcoff (Ed.), *Renal Metabolism and Epidemiology of Some Renal Diseases. Proceedings of the Fifteenth Annual Conference on the Kidney*, Maple Press, York, Pa. pp. 159–74.

Handler, J. S., and J. Orloff (1963). Activation of phosphorylase in toad bladder and mammalian kidney by antidiuretic hormone. *Am. J. Physiol.*, **205**, 298–302.

Handler, J. S., and J. Orloff (1971). Factors involved in the action of cyclic AMP on the permeability of mammalian kidney and toad urinary bladder. *Ann. NY Acad. Sci.*, **185**, 345–50.

Handler, J. S., and A. S. Preston (1976). Study of enzymes regulating vasopressin-stimulated cyclic AMP metabolism in separated mitochondria-rich and granular epithelial cells of toad urinary bladder. *J. Membr. Biol.*, **26**, 43–50.

Handler, J. S., R. W. Butcher, E. W. Sutherland, and J. Orloff (1965). The effect of vasopressin and of theophylline on the concentration of adenosine-3′,5′-phosphate in the urinary bladder of the toad. *J. Biol. Chem.*, **240**, 4524–6.

Handler, J. S., A. S. Preston, and J. Orloff (1969). The effect of aldosterone on glycolysis in the urinary bladder of the toad. *J. Biol. Chem.*, **244**, 3194–9.

Handler, J. S., A. S. Preston, and J. Rogulski (1968). Control of glycogenolysis in the toad's urinary bladder. The effect of anaerobiosis, sodium transport, and arginine vasotocin. *J. Biol. Chem.*, **243**, 1376–83.

Harrington, A. R., and H. Valtin (1968). Impaired urinary concentration after vasopressin and its gradual correction in hypothalamic diabetes insipidus. *J. Clin. Invest.*, **47**, 502–10.

Hartman, F. A., K. A. Brownell, and W. E. Hartman (1930). A further study of the hormone of the adrenal cortex. *Am. J. Physiol.*, **95**, 670–80.

Hasan, S. H., and H. Heller (1968). The clearance of neurohypophysial hormones from the circulation of non-mammalian vertebrates. *Br. J. Pharmacol. Chemother.*, **33**, 523–30.

Haynes, R. C., and L. Berthet (1957). Studies on the mechanism of action of the adrenocorticotrophic hormone. *J. Biol. Chem.*, **225**, 115–24.

Hays, R. M. (1972). The movement of water across vasopressin-sensitive epithelium. In F. Bronner and A. Kleinzeller (Eds) *Current Topics in Membranes and Transport*, Vol. 3, Academic Press, New York. pp. 339–66.

Hays, R. M., and N. Franki (1970). The role of water diffusion in the action of vasopressin. *J. Membr. Biol.*, **2**, 263–76.

Hays, R. M., and A. Leaf (1962a). Studies on the movement of water through the isolated toad bladder and its modification by vasopressin. *J. Gen. Physiol.*, **45**, 905–19.

Hays, R. M., and A. Leaf (1962b). The state of water in the isolated toad bladder in the presence and absence of vasopressin. *J. Gen. Physiol.*, **45**, 933–48.

Hays, R. M., N. Franki, and R. Soberman (1971). Activation energy for water diffusion across the toad bladder: evidence against the pore enlargement hypothesis. *J. Clin. Invest.*, **50**, 1016–18.

Heidrich, H. G., R. Kinne, E. Kinne-Saffran, and K. Hannig (1972). The polarity of the proximal tubule cell in rat kidney. Different surface charges for the brush-border microvilli and plasma membranes from the basal infoldings. *J. Cell Biol.*, **54**, 233–45.

Heller, H. (1941). Differentiation of an (amphibian) water balance principle from the antidiuretic principle of the posterior pituitary gland. *J. Physiol., Lond.*, **100**, 125–41.

Heller, H. (1974). History of neurohypophysial research. In R. O. Greep and E. B. Astwood (Eds) *Handbook of Physiology, Section 7, Endocrinology*, Vol. VI, American Physiological Society, Washington. pp. 103–17.

Heller, H., and P. J. Bentley (1965). Phylogenetic distribution of the effects of neurohypophysial hormones on water and sodium metabolism. *Gen. Comp. Endocrinol.*, **5**, 96–108.

Heller, H., and B. T. Pickering (1961). Neurohypophysial hormones of non-mammalian vertebrates. *J. Physiol., Lond.*, **155**, 98–144.

Heller, H., B. T. Pickering, J. Maetz, and F. Morel (1961). Pharmacological characteristics of oxytocic peptides in the pituitary of a marine telest (*Polachius virens*). *Nature, Lond.*, **191**, 670–1.

Helman, S. I., J. J. Grantham, and M. B. Burg (1971). Effect of vasoiressin on electrical resistance of renal cortical collecting tubules. *Am. J. Physiol.*, **220**, 1825–32.

Henderson, I. W., and I. Chester Jones (1967). Endocrine influences on the net extracellular fluxes of sodium and potassium in the European eel, *Anguilla anguilla* L., *J. Endocrinol.*, **37**, 319–25.

enderson, I. W., and I. Chester Jones (1972). Hormones and osmoregulation in fishes. *Ann. Michel Pacha*, **5**, 69–235.

Hendler, E. D., J. Torretti, L. Kupor, and F. H. Epstein (1972). Effects of adrenalectomy and hormone replacement on Na-K-ATPase in renal tissue. *Am. J. Physiol.*, **222**, 754–60.

Herman, T. S., G. M. Fimognari, and I. S. Edelman (1968). Studies on renal aldosterone-binding proteins. *J. Biol. Chem.*, **243**, 3849–56.

Hevesy, G., E. Hofer, and A. Krogh (1935). The permeability of the skin of frogs to water as determined by D_2O and H_2O. *Skand. Arch. Physiol.*, **72**, 199–214.

Hicks, R. M. (1966). The function of the Golgi apparatus in transitional epithelium. Synthesis of the cell membrane. *J. Cell Biol.*, **30**, 623–43.

Hierholzer, K., W. Wiederholt, H. Holzgreve, G. Giebisch, R. M. Klose, and E. E. Windhager (1965). Micropuncture study of renal transtubular concentration gradients of sodium and potassium in adrenalectomized rats. *Arch. Ges. Physiol.*, **285**, 193–210.

Hill, A. M., and D. J. Trachewsky (1974). Effect of aldosterone on renal ribosomal protein phosphorylation. *J. Steroid Biochem.*, **5**, 561–8.

Hillyard, S. D. (1976). Variation in the effects of antidiuretic hormone on the isolated skin of the toad, *Scaphiopus couchi*. *J. Exp. Zool.*, **195**, 199–206.

Hiltin, J. G., L. F. Scian, C. D. Westerman, and O. R. Kruesi (1959). The effect of synthetic lysine vasopressin on adrenocortical secretion. *Science, NY*, **129**, 971.

Hirano, T. (1967). Effect of hypophysectomy on water transport in isolated intestine of the eel, *Anguilla japonica*. *Proc. Japan Acad.*, **43**, 793–6.

Hirano, T. (1969). Effects of hypophysectomy and salinity change on plasma cortisol concentration in the Japanese eel (*Anguilla japonica* L.). *Endocrinol. Jap.*, **16**, 557–60.

Hirano, T., and S. Utida (1968). Effects of ACTH and cortisol on water movement in isolated intestine of the eel, *Anguilla japonica. Gen. Comp. Endocrinol.*, **11**, 373–80.

Hirano, T., and S. Utida (1971). Plasma cortisol concentration and the rate of intestinal water absorption in the eel, *Anguilla japonica. Endocrinol. Jap.*, **18**, 47.

Hirano, T., M. Morisawa, M. Ando, and S. Utida (1975). Adaptive changes in ion and water transport mechanism in the eel intestine. In J. W. L. Robinson (Ed.) *Intestine Ion Transport*, Medical and Technical Publishing Co., Lancaster, England. pp. 301–17.

Hirokawa, W. (1908). Ueber den osmotischen Druck des Nierenparenchyms. Hofmeisters *Beitr. Physiol. Pathol.*, **11**, 458–78.

Hoffman, R. E., and M. M. Civan (1971). Effect of aldosterone on electrical resistance of toad bladder. *Am. J. Physiol.*, **220**, 324–8.

Holmes, R. M. (1961). Kidney function in migrating salmonids. *Rep. Challenger Soc., Camb.*, **3**, No. 13, 23.

Holmes, W. N. (1959). Studies on the hormonal control of sodium metabolism in the rainbow trout (*Salmo gairdneri*). *Acta Endocrinol., Copnh.*, **31**, 587–602.

Holmes, W. N., and B. M. Adams (1963). Effects of adrenocortical and neurohypophysial hormones on the renal excretory pattern of the water-loaded duck (*Anas platyrhynchos*). *Endocrinology*, **73**, 5–10.

Holmes, W. N., and D. G. Butler (1963). The effect of adrenocortical steroids on the tissue electrolyte composition of the fresh water rainbow trout (*Salmo gairdneri*). *J. Exp. Biol.*, **40**, 457–64.

Holmes, W. N., and R. L. McBean (1963). Studies on the glomerular filtration rate of rainbow trout (*Salmo gairdneri*). *J. Exp. Biol.*, **40**, 335–41.

Holmes, W. N., and R. L. McBean (1964). Some aspects of electrolyte excretion in the green turtle, *Chelonia mydas mydas. J. Exp. Biol.*, **41**, 81–90.

Holmes, W. N., and J. G. Phillips (1976). The adrenal cortex of birds. In I. Chester Jones and I. W. Henderson (Eds) *General Comparative and Clinical Endocrinology of the Adrenal Cortex*, Academic Press, London. pp. 293–420.

Holmes, W. N., and I. M. Stainer (1966). Studies on the renal excretion of electrolytes by the trout (*Salmo gairdneri*). *J. Exp. Biol.*, **44**, 33–46.

Holmes, W. N., R. L. Broock, and J. Devlin (1974). Tritiated corticosteroid metabolism in intact and adenohypophysectomized ducks (*Anas platyrhynchos*). *Gen. Comp. Endocrinol.*, **22**, 417–27.

Holmes, W. N., G. L. Fletcher, and D. J. Stewart (1968). The patterns of renal electrolyte excretion in the duck (*Anas platyrhynchos*) maintained on freshwater and on hypertonic saline. *J. Exp. Biol.*, **48**, 487–508.

Holmes, W. N., L. N. Lockwood, and E. L. Bradley (1972). Adenohypophysial control of extrarenal excretion in the duck (*Anas platyrhynchos*). *Gen. Comp. Endocrinol.*, **18**, 59–68.

Holmes, W. N., J. G. Phillips, and D. G. Butler (1961). The effect of adrenocortical steroids on the renal and extra-renal responses of the domestic duck (*Anas platyrhynchos*) after hypertonic saline loading. *Endocrinology*, **69**, 483–95.

Holmes, W. N., J. G. Phillips, and I. Chester Jones (1963). Adrenocortical factors associated with adaptation of vertebrates to marine environments. *Recent Prog. Horm. Res.*, **19**, 619–72.

Holtz, R., and A. Finkelstein (1970). The water and nonelectrolyte permeability induced in thin lipid membranes by the polyone antibiotics nystatin and amphotericin B. *J. Gen. Physiol.*, **56**, 125–45.

House, C. R. (1974). *Water Transport in Cells and Tissues*. Edward Arnold, London.

Houssay, B. A., and J. T. Lewis (1923). The relative importance to life of the cortex and medulla of the adrenal glands. *Am. J. Physiol.*, **64**, 512–21.

Huf, E. G. (1955). Ion transport and ion exchange in frog skin. In A. M. Shanes (Ed.) *Electrolytes in Biological Systems*, American Physiological Society, Washington. pp. 205–38.
Ichii, S., A. Ikeda, and M. Izawa (1970). Short- and long-term effect of ACTH *in vivo* on the incorporation of ^{3}H-leucine into the submitochondrial fractions from rat adrenal glands. *Endocrinol. Jap.*, **17**, 365–8.
Idler, D. R. (1971). Some comparative aspects of corticosteroid metabolism. *Proc. 3rd Int. Congr. on Hormonal Steroids*, Hamburg, 1970, *Excerpta Med. Found. Int. Congr. Ser.*, **219**, 14–28.
Idler, D. R., and B. Truscott (1967). 1a-hydroxycorticosterone: synthesis *in vitro* and properties of an interrenal steroid in the blood of cartilaginous fish (Genus *Raja*). *Steroids*, **9**, 457–77.
Idler, D. R., and B. Truscott (1969). Production of 1a-hydroxycorticosterone *in vivo* and *in vitro* by elasmobranchs. *Gen. Comp. Endocrinol.*, Suppl., **2**, 325–30.
Jaanus, S. D., M. J. Rosenstein, and R. P. Rubin (1970). On the mode of action of ACTH on the isolated perfused adrenal gland. *J. Physiol., Lond.*, **209**, 539–56.
Jaenike, J. R. (1961). The influence of vasopressin on the permeability of the mammalian collecting duct to urea. *J. Clin. Invest.*, **40**, 144–51.
Janáček, K., and R. Rybová (1967). Stimulation of the sodium pump in frog bladder by oxytocin. *Nature, Lond.*, **215**, 992–3.
Janáček, K., and R. Rybová (1970). Nonpolarized frog bladder preparation. The effects of oxytocin. *Pflügers Arch. Ges. Physiol.*, **318**, 294–304.
Jard, S. (1966). Etude des effets de la vasotocine sur l'excrétion de l'eau et des électrolytes par la rein de la grenouille *Rana esculenta* L.: analyse à l'aide d'analogues artificiels de l'hormone naturelle des caractères structuraux requis pour son activité biologique. *J. Physiol., Paris*, Suppl. 15, **58**, 1–124.
Jard, S., and F. Bastide (1970). A cyclic AMP-dependent protein kinase from frog bladder epithelial cells. *Biochim. Biophys. Res. Commun.*, **39**, 559–66.
Jard, S., and F. Morel (1963). Actions of vasotocin and some of its analogues on salt and water excretion by the frog. *Am. J. Physiol.*, **204**, 222–6.
Jard, S., J. Maetz, and F. Morel (1960). Action de quelques analogues de l'ocytocine sur différents récepteurs intervenant dans l'osmorégulation de *Rana esculenta*. *C. R. hebd. Séanc. Acad. Sci., Paris*, **251**, 788–90.
Jørgensen, C. B. (1947). Influence of adenohypophysectomy on the transfer of salt across the frog skin. *Nature, Lond.*, **160**, 872.
Jørgensen, C. B., and L. O. Larsen (1963). Neuro-adenohypophysial relationships. *Symp. Zool. Soc. Lond.*, **9**, 59–82.
Jørgensen, C. B., and P. Rosenkilde (1957). Chloride balance in hypophysectomized frogs. *Endocrinology*, **60**, 219–24.
Jørgensen, P. O. (1968). Regulation of the ($Na^+ + K^+$)-activated ATP hydrolyzing enzyme system in rat kidney. I. The effect of adrenalectomy and the supply of sodium on the enzyme system. *Biochim. Biophys. Acta*, **151**, 212–24.
Jørgensen, P. O. (1969). Regulation of the ($Na^+ + K^+$)-activated ATP hydrolyzing enzyme system in rat kidney. II. The effect of aldosterone on the activity in kidneys of adrenalectomized rats. *Biochim. Biophys. Acta*, **192**, 326–34.
Junqueira, L. C. U., G. Malnic, and C. Monge (1966). Reabsorptive function of the ophidian cloaca and large intestine. *Physiol. Zoöl.*, **29**, 151–9.
Kachadorian, W. A., J. B. Wade, and V. A. Discala (1975). Vasopressin: induced structural change in toad bladder luminal membrane. *Science, NY*, **190**, 67–9.
Kan, K. W., M. C. Ritter, F. Ungar, and M. E. Dempsey (1972). The role of a carrier protein in cholesterol and steroid hormone synthesis by adrenal enzymes. *Biochim. Biophys. Res. Commun.*, **48**, 423–9.

Katsoyannis, P. G., and V. du Vigneaud (1958). Arginine-vasotocin, a synthetic analogue of the posterior pituitary hormones containing the ring of oxytocin and the side chain of vasopressin. *J. Biol. Chem.*, **233**, 1352–4.

Katz, A. L., and F. H. Epstein (1967). The role of sodium-potassium-activated adenosine triphosphatase in the reabsorption of sodium by the kidney. *J. Clin. Invest.*, **46**, 1999–2011.

Keys, A. B. (1931a). The heart gill preparation of the eel and its perfusion for the study of a natural membrane *in situ*. *Z. Vergl. Physiol.*, **15**, 352–63.

Keys, A. B. (1931b). Chloride and water secretion and absorption by the gills of the eel. *Z. Vergl. Physiol.*, **15**, 364–88.

Keys, A. B. (1933). The mechanism of adaptation to varying salinities in the common eel, and the general problem of osmotic regulation in fishes. *Proc. R. Soc.*, **112**, 184–99.

Keys, A. B., and J. B. Bateman (1932). Branchial response to adrenaline and to pitressin in the eel. *Biol. Bull. Mar. Biol. Lab.*, *Woods Hole*, **63**, 327–36.

Keys, A. B., and E. N. Willmer (1932). 'Chloride secreting' cells in the gills of fishes with special reference to the common eel. *J. Physiol., Lond.*, **76**, 368–78.

Kirsten, E., R. Kirsten, A. Leaf, and G. W. G. Sharp (1968). Increased activity of enzymes of tricarboxylic acid cycle in response to aldosterone in toad bladder. *Pflügers Arch. Ges. Physiol.*, **300**, 213–25.

Koefoed-Johnsen, V., and H. H. Ussing (1953). The contributions of diffusion and flow to the passage of D_2O through living membranes. *Acta Physiol. Scand.*, **28**, 60–76.

Koefoed-Johnsen, V., and H. H. Ussing (1958). The nature of the frog skin potential. *Acta Physiol. Scand.*, **42**, 298–308.

Kristensen, P. (1970). The action of theophylline on the isolated skin of the frog (*Rana temporaria*). *Biochim. Biophys. Acta*, **203**, 579–82.

Krogh, A. (1937). Osmotic regulation in the frog (*R. esculenta*) by active absorption of chloride ions. *Skand. Arch. Physiol.*, **76**, 60–74.

Krogh, A. (1939). *Osmotic Regulation in Aquatic Animals*. Cambridge University Press, Cambridge.

de Kruyff, B., W. J. de Greef, R. V. W. van Eyk, R. A. Demel, and L. L. M. van Deenen (1973). The effect of different fatty acids and sterol composition on the erythritol flux through the cell membrane of *Acholesplasma laidlawii*. *Biochim. Biophys. Acta*, **298**, 479–99.

Kuo, J. F., and P. Greengard (1969). Cyclic nucleotide-dependent protein kinases. IV. Widespread occurrence of adenosine 3′,5′-monophosphate-dependent protein kinase in various tissues and phyla of the animal kingdom. *Proc. Natl Acad. Sci. USA*, **64**, 1349–55.

Lahlou, B. (1967). Excrétion rénale chez un poisson euryhaline, le flet (*Platichythys flesus* L.): Caractéristiques de l'urine normale en eau douce et en eau de mer et effets des changements de milieu. *Comp. Biochem. Physiol.*, **20**, 925–38.

Landon, E. J., N. Jazab, and L. Forte (1966). Aldosterone and sodium–potassium-dependent ATPase activity of rat kidney membranes. *Am. J. Physiol.*, **211**, 1050–6.

Lanthier, A., and T. Sandor (1973). The effect of 18-hydrocorticosterone on the salt-excreting gland of the duck (*Anas platyrhynchos*). *Can. J. Physiol. Pharmacol.*, **51**, 776–8.

Lassiter, W. E., C. W. Gottschalk, M. Mylle (1961). Micropuncture study of net transtubular movement of water and urea in nondiuretic mammalian kidney. *Am. J. Physiol.*, **200**, 1139–47.

Lassiter, W. E., M. Mylle, and C. W. Gottschalk (1966). Micropuncture study of urea transport in rat renal medulla. *Am. J. Physiol.*, **210**, 965–70.

Leaf, A., and E. Dempsey (1960). Some effects of mammalian neurohypophyseal hormones on metabolism and active transport of sodium by the isolated toad bladder. *J. Biol. Chem.*, **235**, 2160–3.
Leaf, A., and R. M. Hays (1962). Permeability of the isolated toad bladder to solutes and its modification by vasopressin. *J. Gen. Physiol.*, **45**, 921–32.
Leb, D. E., C. Edwards, B. D. Lindley, and T. Hoshiko (1965). Interaction between the effects of inside and outside Na^+ and K^+ on bullfrog skin potential. *J. Gen. Physiol.*, **49**, 309–20.
LeBrie, S. J., and R. S. Elizondo (1969). Saline-loading and aldosterone in water snakes, *Natrix cyclopion*. *Am. J. Physiol.*, **217**, 426–30.
LeBrie, S. J., and I. D. W. Sutherland (1962). Renal function in water snakes. *Am. J. Physiol.*, **203**, 995–1000.
Lederis, K. (1964). Fine structure and hormone content of the hypothalamoneurohypophysial system of the rainbow trout (*Salmo irideus*) exposed to sea water. *Gen. Comp. Endocrinol.*, **4**, 638–61.
Lefkowitz, R. J., J. Roth, and I. Pasten (1970). Effects of calcium on ACTH stimulation of the adrenal: Separation of hormone binding from adenyl cyclase activation. *Nature, Lond.*, **228**, 864–6.
Levine, S., N. Franki, and R. M. Hays (1973a). Effect of phloretin on water and solute movement in the toad bladder. *J. Clin. Invest.*, **52**, 1435–42.
Levine, S., N. Franki, and R. M. Hays (1973b). A saturable vasopressin-sensitive carrier for urea and acetamide in the toad bladder epithelial cell. *J. Clin. Invest.*, **52**, 2083–6.
Levitan, R., and F. J. Ingelfinger (1965). Effect of *d*-aldosterone on salt and water absorption from the intact human colon. *J. Clin. Invest.*, **44**, 801–8.
Lichtenstein, N. S., and A. Leaf (1965). Effect of amphotericin B on the permeability of the toad bladder. *J. Clin. Invest.*, **44**, 1328–42.
Liew, C. C., D. Suria, and A. G. Gornall (1973). Effects of aldosterone on acetylation and phosphorylation of chromosomal proteins. *Endocrinology*, **93**, 1025–34.
Lifschitz, M. D., R. W. Schrier, and I. S. Edelman (1973). Effect of actinomycin D on aldosterone-mediated changes in electrolyte excretion. *Am. J. Physiol.*, **224**, 376–80.
Lipson, L. C., and G. W. G. Sharp (1971). Effect of prostaglandin E_1 on sodium transport and osmotic water flow in the toad bladder. *Am. J. Physiol.*, **220**, 1046–52.
Lipton, P., and I. S. Edelman (1969). Effect of regulatory hormones on intracellular Na^+ and K^+ of toad bladder epithelial cells. *Abstr. Biophys. Soc. 13th Ann. Meetings, Los Angeles, Biophys. J.*, **9**, A164.
Ludens, J. H., and D. D. Fanestil (1971). Studies on cytosol aldosterone binding macromolecules. *Biochim. Biophys. Acta*, **244**, 360–71.
Macchi, I. A., J. G. Phillips, and P. Brown (1967). Relationship between the concentration of corticosteroids in avian plasma and nasal gland function. *J. Endocrinol.*, **38**, 319–29.
MacKay, M. E. (1931). The action of some hormones and hormone-like substances on the circulation of the skate. *Contr. Can. Biol. Fish.*, **7**, 19–28.
Macknight, A. D. C., A. Leaf, and M. M. Civan (1970). Vasopressin: evidence for the cellular site of the induced permeability change. *Biochim. Biophys. Acta*, **222**, 560–3.
Maetz, J. (1956). Les échanges de sodium chez le poisson *Carassius auratus* L. Action d'un inhibiteur de l'anhydrase carbonique. *J. Physiol., Paris*, **48**, 1085–99.
Maetz, J. (1972). Interaction of salt and ammonia transport in aquatic organisms. In J. W. Campbell and L. Goldstein (Eds) *Nitrogen Metabolism and the Environment*, Academic Press, New York. pp. 105–54.

Maetz, J., and B. Lahlou (1974). Actions of neurohypophysial hormones in fishes. In R. O. Greep and E. B. Astwood (Eds) *Handbook of Physiology, Section 7, Endocrinology*, Vol. IV, American Physiological Society, Washington. pp. 521–44.

Maetz, J., and J. C. Rankin (1969). Quelques aspects du rôle biologique des hormones neurohypophysaires chez les poissons. *Colloq. Intn. Centre Natl Rech. Sci.*, Paris, **177**, 45–54.

Maetz, J., and E. Skadhauge (1968). Drinking rates and gill ionic turnover in relation to external salinities in the eel. *Nature, Lond.*, **217**, 371–2.

Maetz, J., J. Bourguet, B. Lahlou, and J. Hourdry (1964). Peptides neurohypophysaires et osmorégulation chez *Carassius auratus*. *Gen. Comp. Endocrinol.*, **4**, 508–22.

Magnus, R., and E. A. Schafer (1901). The action of pituitary extracts upon the kidney. *J. Physiol., Lond.*, **27**, ix–x.

Marenzi, A. D., and O. Fustinoni (1938). El potasio de la sangre y de los tejidos de los sapos suprarenoprivos. *Rev. Soc. Arg. Biol.*, **14**, 118–22.

Marty, A., and A. Finkelstein (1975). Pores formed in lipid bilayer membranes by Nystatin. Differences in its one-sided and two-sided action. *J. Gen. Physiol.*, **65**, 515–26.

Marusic, E. T., A. White, and A. R. Aedo (1973). Oxidative reactions in the formation of an aldehyde group in the biosynthesis of aldosterone. *Arch. Biochem. Biophys.*, **157**, 320–1.

Marver, D., and I. S. Edelman (1975). A technique for differential extraction of nuclear receptors. *Meth. Enzym.*, **36A**, 286–92.

Marver, D., D. Goodman, and I. S. Edelman (1972). Relationships between renal cytoplasmic and nuclear aldosterone-receptors. *Kidney Int.*, **1**, 210–23.

Marver, D., J. Steward, J. W. Funder, D. Feldman, and I. S. Edelman (1974). Renal aldosterone receptors: studies with [^{3}H]aldosterone and the anti-mineralocorticoid [^{3}H]spirolactone (SC-26304). *Proc. Natl Acad. Sci. USA*, **71**, 1431–5.

Masur, S. K., E. Holtzman, and R. Walter (1972). Hormone-stimulated exocytosis in the toad urinary bladder. Some possible implications for turnover of surface membranes. *J. Cell Biol.*, **52**, 211–19.

Mayer, N. (1963). Nouvelles recherches sur l'adaption des grenouilles vertes *Rana esculenta* à des milieux de salinité variée. Etude spéciale de l'excrétion rénale de l'eau et des électrolytes. *Diplôme d'Etudes Supérieures*, Paris.

Mayer, N. (1969). Adaption de *Rana esculenta* à des milieux variés. Etude spéciale de l'excrétion rénale de l'eau et des electrolytes au cours de changements de milieux. *Comp. Biochem. Physiol.*, **29**, 27–50.

Mayer, N., J. Maetz, D. K. O. Chan, M. Forster, and I. Chester Jones (1967). Cortisol, a dosium excreting factor in the eel (*Anguilla anguilla* L.) adapted to sea water. *Nature, Lond.*, **214**, 1118–20.

McClanahan, L., and R. Baldwin (1969). Rate of water uptake through the integument of the desert toad, *Bufo cognatus*. *Comp. Biochem. Physiol.*, **28**, 381–9.

Mendoza, S. A., J. S. Handler, and J. Orloff (1970). Effect of inhibitors of sodium transport on response of toad bladder to ADH and cyclic AMP. *Am. J. Physiol.*, **219**, 1440–5.

Middler, S. A., C. R. Kleeman, E. Edwards, and D. Brody (1969). Effect of adenohypophysectomy on salt and water metabolism of the toad *Bufo marinus*: studies in hormonal replacement. *Gen. Comp. Endocrinol.*, **12**, 290–304.

Mills, A. J., J. F. Wheldrake, and L. A. W. Feltham (1970). Properties of renal aldosterone-binding proteins. *Biochem. J.*, **120**, 23P–4P.

Mishra, R. K., and L. A. W. Feltham (1975). RNA polymerase stimulation: effect of aldosterone and other adrenocorticoids on RNA turnover in rat kidney. *Can. J. Biochem.*, **53**, 70–8.

Morel, F., and F. Bastide (1965). Action de l'ocytocine sur la composante active du transport de sodium par la peau de grenouille. *Biochim. Biophys. Acta.*, **94,** 609–11.
Morel, F., and S. Jard (1963). Inhibition of frog (*Rana esculenta*) antidiuretic action of vasotocin by some analogues *Am. J. Physiol.*, **304,** 227–32.
Morel, F., and S. Jard (1968). Actions and functions of the neurohypophyseal hormones and related peptides in lower vertebrates. In B. Berde (Ed.) *Handbuch der Experimentallen Pharmacologie*, Vol. 23, Springer-Verlag, New York. pp. 655–716.
Morel, F., and C. de Rouffignac (1970). Micropuncture study of urea medullary recycling in desert rodents. In B. Schmidt-Nielsen (Ed.) *Urea and the Kidney*, Excerpta Medica Foundation, Amsterdam. pp. 401–13.
Morel, F., M. Mylle, and C. W. Gottschalk (1965). Tracer microinjection studies of the effect of ADH on renal tubular diffusion of water. *Am. J. Physiol.*, **209,** 179–87.
Morgan, T., and R. W. Berliner (1968). Permeability of the loop of Henle, vasa recta, and collecting duct to water, urea, and sodium. *Am. J. Physiol.*, **215,** 108–15.
Morgan, T., F. Sakai, and R. W. Berliner (1968). *In vitro* permeability of medullary collecting ducts to water and urea. *Am. J. Physiol.*, **214,** 574–81.
Morris, D. J., and R. P. Davis (1973). Complex formation by ^{3}H-aldosterone in rat kidney and liver. *Steroids*, **21,** 383–96.
Morris, D. J., J. S. Berck, and R. P. Davis (1973). The physiological response of aldosterone in adrenalectomized and intact rats and its sex dependence. *Endocrinology*, **92,** 989–93.
Morris, R. (1965). Studies on salt and water balance in *Myxine glutinosa* L. *J. Exp. Biol.*, **42,** 359–71.
Munsick, R. A., W. H. Sawyer, and H. D. Van Dyke (1960). Avian neurohypophyseal hormones. Pharmacological properties and tentative identification. *Endocrinology*, **66,** 860–71.
Murayama, Y., A. Suzuki, M. Tadokoro, and F. Sakal (1968). Microperfusion of Henle's loop in the kidney of the adrenalectomized rat. *Jap. J. Pharmacol.*, **18,** 518–9.
Murray, A. W., and M. Froscio (1971). Cyclic nucleotide-dependent phosphorylation of microtubular protein by protein kinases. *Proc. Aust. Biochem. Soc.*, **4,** 8.
Nelson, E. E. (1934). The diuretic effect of posterior pituitary extract in the anaesthetized animal. *J. Pharmacol. Exp. Ther.*, **52,** 184–95.
Ney, R. L., W. W. Davis, and L. D. Garren (1966). Heterogeneity of template RNA in adrenal glands. *Science, NY*, **153,** 896–7.
Niijima, A. (1969). Afferent discharges from osmoreceptors in the liver of the guinea pig. *Science, NY*, **166,** 1519–20.
Novelli, A. (1936). Lobulo posterior de hipofisis e imbibicion de los batrachios. II. Mecanismo de su accion. *Rev. Soc. Argent. Biol.*, **12,** 163–4.
O'Connor, W. J. (1962). *Renal Function*. Edward Arnold, London.
Oide, H., and S. Utida (1968). Changes in intestinal absorption and renal excretion of water during adaptation to seawater in the Japanese eel. *Marine Biol.*, **1,** 172–7.
Oide, M., and S. Utida (1967). Changes in water and ion transport in isolated intestine of the eel during salt-adaptation and migration. *Marine Biol.*, **1,** 102–6.
Olmsted, J. B., and G. G. Borisy (1973). Characterization and microtubule assembly in porcine brain extracts by viscometry. *Biochem. NY*, **12,** 4282–9.
Orloff, J., and J. S. Handler (1961). Vasopressin-like effects of adenosine 3′,5′-phosphate (cyclic 3′,5′-AMP) and theophylline in the toad bladder. *Biochim. Biophys. Res. Commun.*, **5,** 63–6.
Orloff, J., and J. S. Handler (1962). The similarity of effects of vasopressin adenosine-3′5′-monophosphate (cyclic AMP) and theophylline on the toad bladder. *J. Clin. Invest.*, **41,** 702–9.

Orloff, J., and J. S. Handler (1967). The role of adenosine 3′,5′-phosphate in the action of antidiuretic hormone. *Am. J. Med.*, **42**, 757–68.

Peaker, M., S. J. Peaker, J. G. Phillips, and A. Wright (1971). The effects of corticotrophin, glucose and potassium chloride on secretion by the nasal salt gland of the duck (*Anas platyrhynchos*). *J. Endocrinol.*, **50**, 293–9.

Pearce, R. B., J. Cronshaw, and W. N. Holmes (1977). The fine structure of the interrenal cells of the duck (*Anas platyrhynchos*) with evidence for the possible exocytotic release of steroids. *Cell Tiss. Res.*, **183**, 203–220.

Pehleman, F. W., and W. Hanke (1968). Funktionomorphologie des Interrenalorgans von *Rana temporaria* L. *Z. Zellforsch. Mikrosk. Anat.*, **89**, 281–302.

Petersen, M. J., and I. S. Edelman (1964). Calcium inhibition of the action of vasopressin on the urinary bladder of the toad. *J. Clin. Invest.*, **43**, 583–94.

de Petris, S. (1974). Inhibition and reversal of capping by cytochalasin B, vinblastine and colchicine. *Nature, Lond.*, **250**, 54–6.

de Petris, S. (1975). Concanavalin-A receptors, immunoglobulins and antigens of the lymphocyte surface. Interactions with concavavalin-A and with cytoplasmic structures. *J. Cell Biol.*, **65**, 123–46.

Pfiffner, J. J., and W. W. Swingle (1929). The preparation of an active extract of the suprarenal cortex. *Anat. Rec.*, **44**, 225.

Pfiffner, J. J., and W. W. Swingle (1931). Studies on the adrenal cortex. III. The revival of cats prostrate from adrenal insufficiency with an aqueous extract of the cortex. *Am. J. Physiol.*, **96**, 180–90.

Phillips, J. G., and D. Bellamy (1962). Aspects of the hormonal control of nasal gland secretion in birds. *J. Endocrinol.*, **24**, vi–vii.

Phillips, J. G., and D. Bellamy (1967). The control of nasal gland function with special reference to the role of adrenocorticosteroids. *Proc. 2nd Int. Congr. on Hormonal Steroids, Milan, 1966, Excerpta Medi. Found. Int. Congr. Ser.*, **132**, 1065–9.

Phillips, J. G., W. N. Holmes, and P. K. Bondy (1959). Adrenocorticosteroids in salmon plasma (*Oncorhynchus nerka*). *Endocrinology*, **65**, 811–8.

Phillips, J. G., W. N. Holmes, and D. G. Butler (1961). The effect of total and subtotal adrenalectomy on the renal and extra-renal response of the domestic duck (*Anas platyrhynchos*). *Endocrinology*, **69**, 958–69.

Pickering, B. T., and H. Heller (1959). Chromatographic and biological characteristics of fish and frog neurohypophysial extracts. *Nature, Lond.*, **184**, 1463–4.

Pietras, R. J., and E. M. Wright (1974). Non-electrolyte probes of membrane structure in ADH-treated toad urinary bladder. *Nature, Lond.*, **247**, 222–4.

Pietras, R. J., and E. M. Wright (1975). The membrane action of antidiuretic hormone (ADH) on toad urinary bladder. *J. Membr. Bio.*, **22**, 107–23.

Pietras, R. J., P. J. Naujokatis, and C. M. Szego (1975). Surface modifications evoked by antidiuretic hormone in isolated epithelial cells: Evidence from lectin probes. *J. Supramol. Struct.*, **3**, 391–400.

Porter, G. A., R. Bogoroch, and I. S. Edelman (1964). On the mechanism of action of aldosterone on sodium transport: the role of RNA synthesis. *Proc. Natl Acad. Sci. USA*, **52**, 1326–33.

Porthe-Nibellé, J., and B. Lahlou (1975). Effects of corticosteroid hormones and inhibitors of steroids on sodium and water transport by goldfish intestine. *Comp. Biochem. Physiol.*, **50A**, 801–5.

Postel-Vinay, M. C., G. M. Alberti, C. Ricour, J. M. Limal, R. Rappaport, and P. Royer (1974). Pseudohypoaldosteronism: persistence of hyperaldosteronism and evidence for renal tubular and intestinal responsiveness to endogenous aldosterone. *J. Clin. Endocrinol. Metab.*, **39**, 1038–44.

Poulson, T. L. (1965). Countercurrent multipliers in avian kidneys. *Science, NY*, **148**, 389–91.

Pressley, L., and J. W. Funder (1975). Glucocorticoid and mineralocortical receptors in gut mucosa. *Endocrinology*, **97,** 588.
Price, H. D., and T. E. Thompson (1969). Properties of lipid bilayer membranes separating two aqueous phases: temperature dependence of water permeability. *J. Molec. Biol.*, **41,** 443–57.
Rajerison, R., J. Marchetti, C. Roy, J. Bockaert, and S. Jard (1974). The vasopressin-sensitive adenylate cyclase of the rat kidney. *J. Biol. Chem.*, **249,** 6390–400.
Ralph, C. L. (1960). Polydipsia in the hen following lesions in the supraoptic hypothalamus. *Am. J. Physiol.*, **198,** 528–30.
Rankin, J. C., and J. Maetz (1971). A perfused teleostean gill preparation: vascular actions of neurohypophysial hormones and catecholamines. *J. Endocrinol.*, **51,** 621–35.
Rappaport, L., J. F. Leterrier, and J. Nunez (1975). Protein kinase activity *in vitro*. Phosphorylation and polymerization of purified tubulin. *Ann. NY Acad. Sci.*, **253,** 611–29.
Rawlins, F., L. Mateu, F. Fragachan, and G. Whittenbury (1970). Isolated toad skin epithelium: transport characteristics. *Pflügers Arch. Ges. Physiol.*, **316,** 64–80.
Richards, A. N., and C. F. Scmidt (1924). A description of the glomerular circulation in the frog's kidney and observations concerning the action of adrenalin and other substances upon it. *Am. J. Physiol.*, **71,** 178–208.
Richards, B. D., and P. O. Fromm (1969). Patterns of blood flow through filaments and lamellae of isolated-perfused rainbow trout (*Salmo gairdnerii*). *Comp. Biochem. Physiol.*, **29,** 1063–70.
Richards, P. (1969). Clinical investigation of the effects of adrenal corticosteroid excess on the colon. *Lancet, Mar.* **1,** 437–42.
Roberts, J. S., and B. Schmidt-Nielsen (1966). Renal ultrastructure and excretion of salt and water by three terrestrial lizards. *Am. J. Physiol.*, **211,** 476–86.
Roberts, K. D., L. Bandi, H. I. Calvin, W. Drucker, and S. Lieberman (1964). Evidence that steroid sulfates serve as biosynthetic intermediates. IV. Conversion of cholesterol sulfate *in vivo* to urinary C_{19} and C_{21} steroidal sulfates. *Biochemistry, Washington*, **3,** 1983–8.
Rocha, A. S., and J. P. Kokko (1974). Permeability of medullary nephron segments to urea and water: effect of vasopressin. *Kidney Int.*, **6,** 379–87.
Rogoff, J. M., and G. N. Stewart (1928). Studies on adrenal insufficiency in dogs. V. The influence of adrenal extracts on the survival period of adrenalectomized dogs. *Am. J. Physiol.*, **84,** 660–74.
Rousseau, G., and J. Crabbé (1972). Effects of aldosterone on RNA and protein synthesis on the toad bladder. *Eur. J. Biochem.*, **25,** 550–9.
Rousseau, G., J. D. Baxter, J. W. Funder, I. S. Edelman, and G. M. Tomkins (1972). Glucocorticoid and mineralocortical receptors for aldosterone. *J. Steroid Biochem.*, **3,** 219–27.
Rossier, B. C., P. A. Wilce, and I. S. Edelman (1974). Kinetics of RNA labeling in toad bladder epithelium: Effects of aldosterone and related steroids. *Proc. Natl Acad. Sci. USA*, **71,** 3101–5.
Rubin, M. S. (1975). Chemical modification of vasopressin (ADH)-induced urea transport across toad bladder. *Fedn Proc. Fedn Am. Soc. Exp. Biol.*, **34,** 327.
Rubin, R. P. (1974). *Calcium and the Secretory Process.* Academic Press, New York.
Rubin, R. P., S. D. Jaanus, and R. A. Carchman (1972). The role of calcium and adenosine cyclic 3′,5′-phosphate in action of adrenocorticotropin. *Nature New Biol.*, **240,** 150–2.
Rubin, R. P., B. Shield, R. McCauley, and S. G. Laychock (1974). ACTH-induced protein release from the perfused cat adrenal gland: Evidence for exocytosis? *Endocrinology*, **95,** 370–8.

Sandor, T. (1972). Corticosteroids in Amphibia, Reptilia, and Aves. In D. R. Idler (Ed.) *Steroids in Nonmammalian Vertebrates*, Academic Press, New York. pp. 253–327.
Sandor, T., and A. G. Fazekas (1973). Corticosteroid binding macromolecules in the nasal gland of the domesticated duck. *Proc. 7th Conf. Eur. on Comparative Endocrinology, Budapest, Abstr.*, **71**.
Sandor, T., and D. R. Idler (1972). Steroid methodology. D. R. Idler (Ed.) *Steroids in Nonmammalian Vertebrates*, Academic Press, New York. pp. 6–36.
Sandor, T., A. Z. Mehdi, and A. G. Fazekas (1977). Corticosteroid-binding macromolecules in the salt-activated nasal gland of the domestic duck (*Anas platyrhynchos*). *Gen. Comp. Endocrinol.*, **32**, 348–59.
Sawyer, W. H. (1951). Effect of posterior pituitary extracts on urine formation and glomerular circulation in the frog. *Am. J. Physiol.*, **164**, 457–66.
Sawyer, W. H. (1957a). Increased renal reabsorption of osmotic free water by the toad (*Bufo marinus*) in response to neurohypophysial hormones. *Am. J. Physiol.*, **189**, 564–8.
Sawyer, W. H. (1957b). The antidiuretic action of neurohypophysial hormones in amphibia. In *Colston Papers*, Vol. VIII, *Proc. 8th Symp. of the Colston Research Society*, Butterworths, London. pp. 171–9.
Sawyer, W. H. (1960). Increased water permeability of the bullfrog (*Rana catesbeiana*) bladder *in vitro* in response to synthetic oxytocin and arginine vasotocin and to neurohypophysial extracts from nonmammalian vertebrates. *Endocrinology*, **66**, 112–20.
Sawyer, W. H. (1966). Diuretic and natriuretic responses of lungfish (*Protopterus aethiopicus*) to arginine vasotocin. *Am. J. Physiol.*, **210**, 191–7.
Sawyer, W. H. (1967). Evolution of antidiuretic hormones and their functions. *Am. J. Med.*, **42**, 678–86.
Sawyer, W. H. (1970). Vasopressor, diuretic and natriuretic responses by lungfish to arginine vasotocin. *Am. J. Physiol.*, **218**, 1789–94.
Schafer, J. A., and T. E. Andreoli (1972a). Cellular constraints to diffusion. The effect of antidiuretic hormone on water flows in isolated mammalian collecting tubules. *J. Clin. Invest.*, **51**, 1264–78.
Schafer, J. A., and T. E. Andreoli (1972b). The effect of antidiuretic hormone on solute flows in isolated mammalian collecting tubules. *J. Clin. Invest.*, **51**, 1279–86.
Schwartz, I. L., L. J. Shlatz, E. Kinne-Saffran, and R. Kinne (1974). Target cell polarity and membrane phosphorylation in relation to the mechanism of action of antidiuretic hormone. *Proc. Natl Acad. Sci. USA*, **71**, 2595–9.
Scott, W. N., and V. S. Sapirstein (1975). Identification of aldosterone-induced proteins in the toad's urinary bladder. *Proc. Natl Acad. Sci. USA*, **72**, 4056–60.
Scott, W. N., V. S. Sapirstein, and M. J. Yoder (1974). Partition of tissue fractions in epithelia: Localization of enzymes in 'mitochondria-rich' cells of toad urinary bladder. *Science, NY*, **184**, 797–9.
Sexton, A. W. (1955). Factors influencing the uptake of sodium against a diffusion gradient in the goldfish gill. *Diss. Abstr.*, **15**, 2270–1.
Share, L. (1961). Acute reduction in extracellular fluid volume and the concentration of antidiuretic hormone in blood. *Endocrinology*, **26**, 925–33.
Sharp, G. W. G., and A. Leaf (1966). Mechanism of action of aldosterone. *Physiol. Rev.*, **46**, 593–631.
Sharratt, B. M., I. Chester Jones, and D. Bellamy (1964). Water and electrolyte composition of the body and renal function of the eel (*Anguilla anguilla* L.). *Comp. Biochem. Physiol.*, **11**, 9–18.

Shields, R., J. B. Miles, and C. Gilbertson (1968). Absorption and secretion of water and electrolytes by the intact colon in a patient with primary aldosteroidism. *Br. Med. J.*, **1**, 93–6.

Shields, R., A. T. Mulholland, and R. G. Elmslie (1966). Action of aldosterone upon the intestinal transport of potassium, sodium, and water. *Gut*, **7**, 686–96.

Shirley, H. V., and A. V. Nalbandov (1956). Effects of neurohypophysectomy in domestic chickens. *Endocrinology*, **58**, 477–83.

Shoemaker, V. H., P. Licht, and W. R. Dawson (1966). Effects of temperature on kidney function in the lizard *Tiliqua rugosa*. *Physiol. Zoöl.*, **39**, 244–52.

Shoemaker, V. H., K. A. Nagy, and S. D. Bradshaw (1972). Studies on the control of electrolyte excretion by the nasal gland of the lizard *Dipsosaurus dorsalis*. *Comp. Biochem. Physiol.*, **42A**, 749–57.

Simpson, E. R., and D. L. Williams-Smith (1975). Effect of calcium (ion) uptake by rat adrenal mitochondria on pregnenolone formation and spectral properties of cytochrome P-450. *Biochim. Biophys. Acta*, **404**, 309–20.

Simpson, E. R., C. R. Jefcoate, A. C. Brownie, and G. S. Boyd (1972a). The effect of ether anaesthesia stress on cholesterol side-chain cleavage and cytochrome P 450 in rat adrenal mitochondria. *Eur. J. Biochem.*, **28**, 442–50.

Simpson, E. R., W. H. Trzeciak, J. L. McCarthy, C. R. Jefcoate, and G. S. Boyd (1972b). Factors affecting cholesterol esterase and cholesterol side-chain cleavage activities in rat adrenal. *Biochem. J.*, **129**, 10P–11P.

Simpson, S. A., and J. F. Tait (1955). Recent progress in methods of isolation, chemistry and physiology of aldosterone. *Recent Prog. Horm. Res.*, **11**, 183–210.

Simpson, S. A., J. F. Tait, A. Wettstein, R. Neher, J. von Euw, O. Schindler, and T. Reichstein (1954). Aldosteron-Isolierung und Eigenschaften über Bestandteile der Nebennierenrinde und verwandte Stoffe. *Helv. Chim. Acta*, **37**, 1163–200.

Skadhauge, E. (1964). Effects of unilateral infusion of arginine-vasotocin into the portal circulation of the avian kidney. *Acta Endocrinol., Copnh.*, **47**, 312–30.

Skadhauge, E. (1969). The mechanism of salt and water absorption in the intestine of the eel (*Anguilla anguilla*) adapted to waters of various salinities. *J. Physiol., Lond.*, **204**, 135–58.

Skou, J. C. (1964). Enzymatic aspects of active linked transport of Na^+ and K^+ through the cell membrane. *Prog. Biophys. Molec. Biol.*, **14**, 133–66.

Smith, H. W. (1930). The absorption and excretion of water and salts by marine teleosts. *Am. J. Physiol.*, **93**, 480–505.

Soifer, D., A. Laszlo, K. Mach, J. Scotto, and L. Siconolfi (1975). The association of a cyclic AMP-dependent protein kinase activity with microtubule protein. *Ann. NY Acad. Sci.*, **253**, 598–610.

Sokabe, H., and T. Nakajima (1972). Chemical structure and role of angiotensins in the vertebrates. *Gen. Comp. Endocrinol., Suppl.*, **3**, 382–92.

Sokabe, H., S. Mizogami, and A. Sato (1968). Role of renin in adaptation to sea water in euryhaline fishes. *Jap. J. Pharmacol.*, **18**, 332–43.

Sokabe, H., M. Ogawa, M. Oguri, and H. Nishimura (1969). Evolution of the juxtaglomerular apparatus in the vertebrate kidneys. *Tex. Rep. Biol. Med.*, **27**, 867–85.

Spinelli, F., A. Grosso, and R. C. de Sousa (1975). The hydrosmotic effect of vasopressin: A scanning electron-microscope study. *J. Membr. Biol.*, **23**, 139–56.

Spooner, P. M., and I. S. Edelman (1976a). Effects of aldosterone on Na^+ transport in the toad bladder. I. Glycolysis and lactate production under aerobic conditions. *Biochim. Biophys. Acta*, **444**, 653–62.

Spooner, P. M., and I. S. Edelman (1976b). Effects of aldosterone on Na^+ transport in the toad bladder. II. The anaerobic response. *Biochim. Biophys. Acta*, **444**, 663–73.

Starling, E. H., and E. B. Verney (1924). The secretion of urine as studied on the isolated kidney. *Proc. R. Soc., Lond.* B., **97**, 321–63.

Steen, J. B., and A. Kruysse (1964). The respiratory function of teleostean gills. *Comp. Biochem. Physiol.*, **12**, 127–42.

Stein, L., and E. Wertheimer (1941). Effect of adrenalectomy on intestinal absorption involving osmotic work in rats. *Proc. Soc. Exp. Biol. Med.*, **46**, 172.

Stewart, D. J., W. N. Holmes, and G. L. Fletcher (1969). The renal excretion of nitrogenous compounds in the duck (*Anas platyrhynchos*) maintained on freshwater and hypertonic saline. *J. Exp. Biol.*, **50**, 527–39.

Stoff, J. S., J. S. Handler, and J. Orloff (1972). The effect of aldosterone on the accumulation of adenosine 3'-5'-cyclic monophosphate in toad bladder epithelial cells in response to vasopressin and theophylline. *Proc. Natl Acad. Sci. USA*, **69**, 805–8.

Sulya, L. L., C. S. McCaa, V. H. Read, and D. Bomer (1963). Uptake of tritiated aldosterone by rat tissues. *Nature, Lond.*, **200**, 788–9.

Sutherland, E. W., and T. W. Rall (1957). The properties of an adenine ribonucleotide produced with cellular particles, ATP, Mg^{++}, and epinephrine or glucagon. *J. Am. Chem. Soc.*, **79**, 3608.

Swaneck, G. E., L. L. H. Chu, and I. S. Edelman (1970). Stereospecific binding of aldosterone to renal chromatin. *J. Biol. Chem.*, **245**, 5382–9.

Swaneck, G. E., E. Highland, and I. S. Edelman (1969). Stereospecific nuclear and cytosol aldosterone-binding proteins of various tissues. *Nephron*, **6**, 297–316.

Swingle, W. W., and J. J. Pfiffner (1930a). An aqueous extract of the suprarenal cortex which maintains the life of bilaterally adrenalectomized cats. *Science, NY*, **71**, 321–2.

Swingle, W. W., and J. H. Pfiffner (1930b). Further observations on adrenalectomized cats treated with aqueous extract of the suprarenal cortex. *Science, NY*, **71**, 489–90.

Swingle, W. W., and J. J. Pfiffner (1930c). The revival of comatose adrenalectomized cats with an extract of the suprarenal cortex. *Science, NY*, **72**, 75–6.

Swingle, W. W., and J. J. Pfiffner (1931a). Studies on the adrenal cortex. I. The effect of a lipid fraction upon the life-span of adrenalectomized cats. *Am. J. Physiol.*, **96**, 153–63.

Swingle, W. W., and J. J. Pfiffner (1931b). Studies on the adrenal cortex. II. An aqueous extract of the adrenal cortex which maintains the life of bilaterally adrenalectomized cats. *Am. J. Physiol.*, **96**, 164–79.

Templeton, J. R., D. E. Murrish, E. M. Randall, and J. N. Mugaas (1968). The effect of aldosterone and adrenalectomy on nasal salt excretion of the desert iguana, *Dipsosaurus dorsalis*. *Am. Zool.*, **8**, 818–9.

Templeton, J. R., D. E. Murrish, E. M. Randall, and J. N. Mugaas (1972). Salt and water balance in the desert iguana, *Dipsosaurus dorsalis*. II. The effect of aldosterone and adrenalectomy. *Z. Vergl. Physiol.*, **76**, 255–69.

Thomas, D. H., and J. G. Phillips (1975a). Studies in avian adrenal steroid function. V. Hormone kinetics and the differentiation of mineralocorticoid and glucocorticoid effects. *Gen. Comp. Endocrinol.*, **26**, 440–50.

Thomas, D. H., and J. G. Phillips (1975b). Studies in avian adrenal steroid function. I. Survival and mineral balance following adrenalectomy in domestic ducks (*Anas Platyrhynchos* L.). *Gen. Comp. Endocrinol.*, **26**, 394–403.

Thomas, D. H., and J. G. Phillips (1975c). Studies in avian adrenal steroid function. II. Chronic adrenalectomy and the turnover of $[^3H]_2O$ in domestic ducks (*Anas platyrhynchos* L.). *Gen. Comp. Endocrinol.*, **26**, 404–11.

Thomas, D. H., and J. G. Phillips (1975d). Studies in avian adrenal steroid function. III. Adrenalectomy and the renal-cloacal response in water-loaded domestic ducks (*Anas platyrhynchos* L.). *Gen. Comp. Endocrinol.*, **26**, 412–9.

Thomas, D. H., and J. G. Phillips (1975e). Studies in avian adrenal steroid function. IV. Adrenalectomy and the response of domestic ducks (*Anas platyrhynchos* L.) to hypertonic NaCl loading. *Gen. Comp. Endocrinol.*, **26**, 427–39.

Thompson, B. D., and C. J. Edmonds (1971). Comparison of effects of prolonged aldosterone administration on rat colon and renal electrolyte excretion. *J. Endocrinol.*, **50**, 163–9.

Thompson, B. D., and C. J. Edmonds (1974). Aldosterone, sodium depletion and hypothyroidism on the ATPase activity of rat colonic epithelium. *J. Endocrinol.*, **62**, 489–96.

Trachewsky, D. (1974). Effects of aldosterone on acetylation of ribosomal proteins in outer and inner zones of kidney. *Proc. Soc. Exp. Biol. Med.*, **147**, 396–8.

Trachewsky, D., and A. M. Cheah (1971). Modification of rat renal cortical chromatin by aldosterone treatment. *Can. J. Biochem.*, **49**, 496–500.

Trachewsky, D., and H. Yang (1975). Alteration in the labelling of renal ribosomal protein by aldosterone. *J. Steroid Biochem.*, **6**, 1157–64.

Träuble, H. (1971). The movement of molecules across lipid membranes: A molecular theory. *J. Membr. Biol.*, **4**, 193–208.

Taylor, A., M. Mamelak, F. Reaven, and R. Maffly (1973). Vasopressin: possible role of microtubules and microfilaments in its action. *Science, NY*, **181**, 347–50.

Taylor, R. B., W. P. H. Duffus, M. C. Raff, and S. de Petris (1971). Redistribution and pinocytosis of lymphocyte surface immunoglobulin molecules induced by anti-immunoglobulin antibody. *Nature New Biol.*, **233**, 225–9.

Ukena, T. E., and R. D. Berlin (1972). Effect of colchicine and vinblastine on the topographical separation of membrane functions. *J. Exp. Med.*, **136**, 1–7.

Ullrich, K. J., and K. H. Jarausch (1956). Untersuchungen zum. Problem der Harnkonzentrierung und Harnverdünnung. Uber die Verteilung von Elektrolyten (Na, K, Ca, Mg, Cl, anorganischen Phosphat). Harnstoff, Aminosäuren und exogenem Kreatinin in Rinde und Mark der Hundeniere bei verschiedenen Diuresezuständon. *Arch. Ges. Physiol.*, **262**, 537–50.

Ullrich, K. J., C. A. Baldamus, E. Uhlich, and G. Rumrich (1969). Influence of ionic calcium and antidiuretic hormone on transtubular sodium transport in the rat kidney. *Arch. Ges. Physiol.*, **310**, 369–76.

Uranga, J., and W. H. Sawyer (1960). Renal responses of the bullfrog to oxytocin, arginine-vasotocin and frog neurohypophysial extract. *Am. J. Physiol.*, **198**, 1287–90.

Ussing, H. H., and K. Zerahn (1951). Active transport of sodium as the source of electric current in the short-circuited isolated frog skin. *Acta Physiol. Scand.*, **23**, 110–27.

Utida, S., T. Hirano, H. Oide, M. Ando, D. W. Johnson, and H. A. Bern (1972). Hormonal control of the intestine and urinary bladder in teleost osmoregulation. *Gen. Comp. Endocrinol., Suppl.*, **3**, 317–24.

Utida, S., M. Oide, S. Saishu, and M. Kamiya (1967). Préétablissement du mécanisme d'adaptation à l'eau de mer dans l'intestine et les branchies isolées de l'Anguille argentée au cours de sa migration catadrome. *C. R. Séanc. Soc. Biol.*, **161**, 1201–4.

Valtin, H. (1967). Hereditary hypothalamic diabetes insipidus in rats (Brattleboro strain). *Am. J. Med.*, **42**, 814–27.

Vander, A. J., R. L. Malvin, W. S. Wilde, J. Lapides, L. P. Sullivan, and V. M. McMurray (1958). Effects of adrenalectomy and aldosterone on proximal and distal tubular sodium reabsorption (24338). *Proc. Soc. Exp. Biol. Med.*, **99**, 323–5.

von den Velden, R. (1913). Beiträge zur Wirkung von Hypophsenestrakten. *Berl. Klin. Wschr.*, **50**, 1969.

Verney, E. B. (1947). Croonian lecture: The antidiuretic hormone and the factors which determine its release. *Proc. R. Soc., Lond.* B, **135**, 25–105.

Voûte, C. L., J. Thummel, and M. Brenner (1975). Aldosterone effect in the epithelium of the frog skin—a new story about an old enzyme. *J. Steroid Biochem.*, **6**, 1175–9.

Wales, N. A. M., I. W. Henderson, and I. Chester Jones (1973). Renal tubular maxima studies of glucose in the freshwater eel, *Anguilla anguilla* L., with special reference to the action of arginine-vasotocin. *Proc. Soc. Endocrinol., J. Endocrinol.*, **57**, xv.

Walker, A. M., P. A. Bott, J. Oliver, and M. C. MacDowell (1941). The collection and analysis of fluid from single nephrons of the mammalian kidney. *Am. J. Physiol.*, **134**, 580–95.

Walker, A. M., C. L. Hudson, T. J. Findley, and A. N. Richards (1937). The total molecular concentration and the chloride concentration of fluid from different segments of the renal tubule of amphibia. The site of chloride reabsorption. *Am. J. Physiol.*, **118**, 121–9.

Waring, H., L. Morris, and G. Stephens (1956). The effect of pituitary posterior lobe extracts on the blood pressure of the pigeon. *Aust. J. Exp. Biol. Med. Sci.*, **34**, 235–8.

Watts, R. W., and J. F. Wheldrake (1976). Aldosterone induced changes in RNA synthesis in rate intestine. *J. Steroid Biochem.*, **7**, 263–6.

Whittembury, G. (1962). Action of antidiuretic hormone on the equivalent pore radius at both surfaces of the epithelium of the isolated frog skin. *J. Gen. Physiol.*, **46**, 117–30.

Whittembury, G., N. Sugino, and A. K. Solomon (1960). Effect of antidiuretic hormone and calcium on the equivalent pore radius of kidney slices from *Necturus*. *Nature, Lond.*, **20**, 700–1.

Wilce, P. A., B. C. Rossier, and I. S. Edelman (1976). Actions of aldosterone on polyadenylated ribonucleic-acid and Na^+ transport in toad bladder. *Biochem. J.*, **15**, 4279–85.

Williamson, H. E. (1963). Mechanism of the antinatriuretic action of aldosterone. *Biochem. Pharmacol.*, **12**, 1449–50.

Wilson, L., and J. Byran (1974). Biochemical and pharmacological properties of microtubules. *Abstr. Cell Molec. Biol.*, **3**, 21–72.

Wilson, L., and M. Freidkin (1967). The biochemical events of mitosis. II. The *in vivo* and *in vitro* binding of colchicine in grasshopper embryos and its possible relation to inhibition of mitosis. *Biochemistry*, **6**, 3126–35.

Wirz, H. (1956). Der osmotische Druck in den corticalen Tubuli der Rattenniere. *Helv. Physiol. Pharmacol. Acta*, **14**, 353–62.

Wirz, H. (1957). The location of the antidiuretic action in the mammalian kidney. In H. Heller (Ed.) *The Neurohypophysis*, Butterworths, London. pp. 157–66.

Wirz, H., B. Hargitay, and W. Kuhn (1951). Lokalisation des Konzentrierungsprozesses in der Niere durch direkte Kryoskopie. *Helv. Physiol. Pharmacol. Acta*, **9**, 196–207.

Wooley, P. (1959). The effect of posterior lobe pituitary extracts on blood pressure in several vertebrate classes. *J. Exp. Biol.*, **36**, 453–8.

Wright, A., and I. Chester Jones (1957). The adrenal glands in lizards and snakes. *J. Endocrinol.*, **15**, 83–99.

Wright, A., J. G. Phillips, and D. P. Huang (1966). The effect of adenohypophysectomy on the extrarenal and renal excretion of the saline-loaded duck (*Anas platyrhynchos*). *J. Endocrinol.*, **36**, 249–56.

Wright, A., J. G. Phillips, M. Peaker, and S. J. Peaker (1967). Some aspects of the endocrine control of water and salt-electrolytes in the duck (*Anas Platyrhynchos*).

Proc. 3rd Asia and Oceania Congr. on Endocrinology, Manilla, Phillippines, pp. 322–7.

Wright, D., and R. S. Snart (1971). Simultaneous measurement of the effect of vasopressin on sodium and water transport across toad bladder. *Life Sci.*, **10,** 301–8.

Wright, F. S., F. G. Knox, S. S. Howards, and R. W. Berliner (1969). Reduced sodium reabsorption by the proximal tubule of DOCA-escaped dogs. *Am. J. Physiol.*, **216,** 869–75.

Yonath, J., and M. M. Civan (1971). Determination of the driving force of the Na^+ pump in toad bladder by means of vasopressin. *J. Membr. Biol.*, **5,** 366–85.

Yorio, T., and P. J. Bentley (1977). Asymmetrical permeability of the integument of tree frogs (*Hylidae*). *J. Exp. Biol.*, **67,** 197–204.

Yuasa, S., S. Urakabe, G. Kimura, D. Shirai, Y. Takamitsu, Y. Orita, and H. Abe (1975). Effect of colchicine on the osmotic water flow across the toad urinary bladder. *Biochim. Biophys. Acta*, **413,** 277–82.

Part D

Osmoregulation and Ecology in the Aquatic Environment

Chapter 11

Temperature and Osmoregulation in Aquatic Species

F. J. VERNBERG AND S. U. SILVERTHORN

I. Introduction	537
II. Whole Animal Responses	537
A. Temperature and osmoregulation	537
B. Survival studies	539
C. Larval survival and growth	542
III. Physiological Responses	543
A. Composition of body fluids	543
B. Other responses	547
IV. Effects of Pollutants	548
V. Mechanisms of Adaptation	548
References	554

I. INTRODUCTION

Since temperature influences most biological phenomena, it is not surprising that the osmoregulation of aquatic organisms may be responsive to thermal effects. Temperature not only is an important physical factor in determining basic osmotic characteristics of fluids, but also it may exert additional effects on living systems by influencing water movement across cell membranes (House, 1974) or differential effects on active ion uptake relative to ion loss (Lockwood, 1962), to cite a few examples (see also Chapters 1, 2). The principal purpose of this review is to emphasize the ecological significance of temperature–osmoregulatory interactions on aquatic organisms by selectively citing some published papers; space limitation restricts a more detailed analysis. A general account of osmoregulation in aquatic arthropods is found in papers by Schoffeniels and Gilles (1970) and by Gilles (1975) (see also chapters 4 and 5 of this volume).

II. WHOLE ANIMAL RESPONSES

A. Temperature and Osmoregulation

The comparative effect of temperature on the osmoregulatory ability of two species of shrimp (genus *Crangon*) common to the North Sea was investigated by Spaargaren (1971). These species vary in their migratory behaviour: *Crangon crangon* migrates to coastal and inland waters in the spring and with the approach of winter it returns to the North Sea, while in contrast *C. allmanni* typically remains in deeper waters throughout the year. The rate of accommodation of the blood concentration to a new salinity was temperature dependent, i.e. the rate increased as the temperature increased. Both species appeared to be completely adapted within 5 days. However, Spaargaren generalized that marine species show a quicker adaptation rate than estuarine species. The salinity at which maximal regulation in *C. crangon* occurs decreases with high temperature as was also shown by Weber and Spaargaren (1970). This shifting of the regulation range to reduced salinities at elevated temperatures appeared to be correlated with the summer migration of this species to brackish water. Geographical differences in response of animals were observed in that Flügel (1963) and Grimm (1969), working with species of *Crangon* from the Baltic Sea and Scotland respectively, observed a different response pattern (Weber and Spaargaren, 1970). In contrast to the osmoregulatory ability of *C. crangon*, Spaargaren (1971) reported that *C. allmanni* showed a very high degree of osmoconformity, and no difference in blood concentration was observed at different temperatures (5–15 °C). Temperature and salinity did interact to influence the lethal limits of this species. For example, at 5–7 °C a salinity as low as 10‰ could be tolerated, but at 21 °C salinities below 25‰ were lethal. This response is the reverse of *C. crangon* where lower salinities were tolerated at higher temperatures. However, Binyon (1961) found that temperature had no effect on the inability of the starfish, *Asterias rubens*, to regulate its weight in dilute sea-water.

Spaargaren (1972) further investigated the comparative effects of temperature and salinity on the osmoregulatory ability of two species of prawns from the Bay of Naples, a region of high salinity throughout the year. At low temperature and high salinity *Palaemon serratus* is a strong regulator of total osmotic concentration in the body fluids while at low salinities conformity is noted. At high temperature the regulation range is shifted towards lower salinities, a response contrary to that of the blood. Electrolyte concentration shows a similar shift, but the response of chloride concentration is not clear. *Lysmata seticaudata* shows conformity in respect to total osmotic concentration, but at higher temperatures some regulation in electrolyte concentration occurs. When comparing the responses of these two species with the two species of *Crangon* previously cited, Spaargaren concluded that *L. seticaudata* and *C. allmanni* show similar osmoregulatory responses except that *C. allmanni* can withstand lower salinities better at low temperatures while *L. seticaudata* tolerates reduced salinities better at higher temperatures. This response pattern correlates well with conditions normally encountered in their respective habitats. This can be further correlated with the less effective cellular electrolyte regulation of *Lysmata seticaudata* at lower temperatures. Unlike *C. crangon* a causal connection between osmoregulation patterns and migration was not noted for *P. serratus*. Spaargaren suggested that the effect of temperature on osmoregulation is a secondary adaptation to changing seasonal conditions.

In contrast, the differential effects of temperature on the osmoregulatory ability of two closely related species of shrimp influence their seasonal distribution. Both species can hyperosmoregulate in reduced salinity at those temperatures normally encountered during the summer, and they may be found in estuaries. Although at low temperature this ability to osmoregulate is impaired in both species, *Penaeus duorarum* is less affected and can overwinter in the cold inshore waters of the Carolina coast (Williams, 1960).

Segal and Burbanck (1963) found distinct differences in osmoregulatory abilities of two populations of an isopod (*Cyathura polita*) in response to temperature and salinity. As indicated previously *Crangon crangon* populations from the North Sea appear to have different osmoregulatory responses to temperature than populations from Scotland and the Baltic Sea (Flügel, 1963; Grimm, 1969; Weber and Spaargaren, 1970). Whether the differences were due to genetic differences or differences in environmental factors were not known.

B. Survival Studies

Many studies on the interactions of temperature and salinity have concerned themselves with the effects of these two factors on survival. Although not always stated, it is frequently implied that mortality results from osmoregulatory failure.

One trend which is apparent from the studies of thermal effects on osmoregulation is that high temperatures narrow the salinity tolerance of the animal (*Lytechinus variegatus*, sea urchin, Lawrence, 1975; *Turbanella ocellata*, gastrotrich, Hummon, 1975). The gastropod, *Nassarius reticulatus*, survives at 25 °C in 20–30 ‰, but at 5 °C its salinity range is widened to 10–40 ‰ (Eriksson and Tallmark, 1974). The narrowing of salinity tolerance with increasing temperature in this case is an example of combined temperature–salinity shock, as the snails were acclimated to 7 °C prior to testing at 25 °C. However, all animals do not behave in this manner. For example, the shrimp, *Crangon septemspinosa*, shows maximum tolerance to a wider range of salinities at higher temperatures (Haefner, 1969) and Panikkar (1940) suggested that tropical animals can withstand low salinities better at high temperatures thereby making it easier for them to colonize estuaries and fresh-water (also see Kinne, 1970). At the tissue level Vernberg, Schlieper and Schneider (1963) found that isolated pieces of gill tissue from two intertidal bivalves (the oyster, *Crassostrea virginica*, and the ribbed mussel, *Modiolus demissus*) survived low salinity better at higher temperature, a condition which they are more likely to encounter in their estuarine habitat than a subtidal bivalve, the scallop, *Aequipecten irradians*.

The ability of a species to survive a given combination of salinity and temperature may be correlated with the ecology of the species. For example, Kenny (1969) reported that the polychaete, *Clymenella torquata*, showed a greater tolerance to low salinity at lower temperatures. This species is abundant in an estuary where the lowest salinities typically are found during the winter.

Many other aquatic organisms survive low salinity best at colder temperatures as demonstrated in the protozoan, *Paramecium caudatum* (Poljansky and Sukhanova, 1967), the sublittoral scallop, *Aequipecten irradians*, the European mussel, *Modiolus modiolus* (Vernberg, Schlieper and Schneider, 1963), and the gastrotrich, *Turbanella ocellata* (Hummon, 1975). The cod, *Gadus morhua*, survived 8‰ well at 0 and 2 °C but showed nervous pathologies at 10 °C (Jones and Scholes, 1974).

Although a number of organisms survive low salinities best in the lower part of their temperature range, survival in cold extremes is generally enhanced by higher salinities. Bivalve molluscs have been widely studied for their response to cold since isolated pieces of gill tissue are easily observed for ciliary activity. Bivalves living in 15‰ are less resistant to cold than those in 32‰, but they can enhance their cold tolerance by acclimation to higher salinities (Theede, 1965; Schlieper, Flügel and Theede, 1967; Theede, 1969). Williams (1970) acclimated *Mytilus edulis* to 150% sea-water and increased the tolerance of the animal to temperatures below −10 °C. However, exposure of isolated gill tissue to 150% sea-water had no influence on resistance to freezing which suggests that a systemic response enhances tolerance in the intact animal. Murphy and Pierce (1975) conducted a similar study on *Modiolus demissus* and found that acclimation to high salinity increased freezing resistance by lowering the amount of tissue water which froze.

Mires, Shak and Shilo (1974), working on mullet fry, *Mugil capito*, reported that the transfer of animals acclimated to 17 °C from 50% sea-water to fresh-water resulted in 41% mortality while fresh-water animals acclimated to 17 °C survived the thermal shock of being exposed to 13 °C. However, the combination of the temperature and salinity shocks mentioned above increased mortality to 68%. In contrast, animals moved from fresh-water to sea-water with the 4 °C drop in temperature only suffered 2% mortality. These data suggest that it is more difficult for fish to acquire salts from dilute media than it is for them to secrete salts in hyperosmotic media.

The same conclusion was reached by Garside and Chin-Yuen-Kee (1972) in their study of osmotic stress on the upper lethal temperature in *Fundulus heteroclitus*. They found that over a range of acclimation temperatures the upper lethal temperature was higher at 32‰ than at 0‰ with highest resistance at the isosmotic point.

The salinity at which fish were isosmotic with the environment (about 12‰) also provided maximum protection for a number of fresh-water species (Strawn and Dunn, 1967; Garside and Jordan, 1968; Jordan and Garside, 1972). Embryos of the spiny dogfish, *Squalus acanthias*, survive heat shock of

20 °C best at their isosmotic salinity of 26‰ (Jones and Price, 1974). A study of temperature effects on flatfish, *Limanda limanda*, in hypersaline water (60–70‰) indicated that high temperatures have a strong contributory effect on mortality with short term exposure (Lowthion, 1974).

A variety of fresh-water fish show increasing heat resistance in water with some ionic content as compared to totally fresh-water. They include the guppy, *Lebistes reticulatus* (Arai, Cox and Fry, 1963), the plains killifish, *Fundulus kansae* (Hill and Carlson, 1970), and the cryprinodontid, *Idus melanotus* (Waede, in Schlieper, 1971).

Two euryhaline annelids, *Enchytraeus albidus* and *Nereis* sp. are protected from heat shock by increasing salinities (Ivleva, 1967; Kähler, 1970; Kinne, 1970) while the gastropods, *Eupleura caudata* and *Urosalpinx cinera*, were found to have highest mortality in a combination of high temperature and low salinity, 25 °C and 12.5‰ (Manzi, 1970). Schlieper, Flügel and Theede (1967) found that gill tissue of the mussel, *Mytilus edulis* was more heat resistant in 30‰ sea-water than in 15‰

In experiments whose purpose was to study salinity effects on thermal resistance, trends similar to those above were noted. Heat resistance is enhanced by the presence of salts in the water. A salinity increase from 0 to 5‰ protects the fresh-water *Planaria gonocephala* from heat shock (Schmitt, in Schlieper, 1971), while the brackish-water ciliate, *Zoothamnium hiketes*, is most resistant to high temperatures at its optimum salinity of 20‰ (Vogel, in Schlieper, 1971). Two marine forms, the hydroid, *Cordylophora caspia* (Kinne, 1958) and the turbellarian, *Convoluta roscoffensis* (Gompel and Legendre, in Schlieper, 1971) are more sensitive to heat in dilute sea-water.

Variability in the ionic composition of the aquatic medium seems to play a role in thermal resistance over a range of salinities. Calcium is one ion which is known to influence osmoregulation (Potts and Fleming, 1971; Lucu, 1973; Pic and Maetz, 1975) and it seems also to modify the thermal response although not in a predictable manner. *Mytilus edulis* in 15‰ supplemented with 200 mg l^{-1} calcium were as resistant to heat as animals in 30‰ (Schlieper, Flügel and Theede, 1967). Calcium had no effect on the cold tolerance of *M. edulis* but did increase freezing resistance of tissue from the bivalve, *Abra alba* (Theede, 1972). Kähler (1970) found that the addition of calcium, magnesium, sodium, or potassium decreased cold tolerance in cold-acclimated oligochaetes while calcium and potassium increased cold resistance in warm-acclimated worms. Magnesium and calcium decreased heat tolerance in warm-acclimated worms.

Various temperature and salinity combinations may have a differential effect on populations of the same species from different regions. Biggs and McDermott (1973) found one population of the hermit crab, *Pagurus longicarpus*, to show wider temperature tolerance than a second population, while the latter showed greater variation in salinity tolerance. Both groups were subjected to similar acclimation regimes.

Bradley (1975) has demonstrated that osmotically acclimated copepods (*Eurytemora affinis*) had different thermal limits at different salinities and at different seasons of the year. Resistance to high temperatures increased with salinity and animals collected in August showed a more marked response to salinity acclimation than March animals.

C. Larval Survival and Growth

In recent years a greater emphasis has been placed on analysing the survival and growth of larvae when subjected to various combinations of temperature and salinity. Data in this area have been graphically represented as response surface curves which can be used to predict the response of organisms in untested combinations of temperature and salinity. A few papers will serve as examples of this trend. Lough (1976), working on the larval dynamics of the Dungeness crab, *Cancer magister*, off the coast of Oregon, found a mass mortality of larval populations coinciding with unusually severe weather. This field observation agreed with the findings of earlier laboratory experiments of Reed as recalculated by Lough. Low salinity in conjunction with wider than normal temperatures appeared to influence larval survival, and low and high temperatures had a marked effect on larval mortality in marginally dilute salinities. Larvae of the mud-flat snail (*Nassarius obsoletus*) develop a greater salinity tolerance, and the influence of temperature–salinity interaction changes as they age which would probably enable adults to survive better in estuarine conditions (Vernberg and Vernberg, 1975). Christiansen and Costlow (1975) demonstrated that salinity and fluctuating temperatures also influenced survival of decapod crustacean larvae. In general each species shows a range of temperature–salinity combinations in which maximal survival occurs, and from this range increased or decreased temperatures and salinities increase mortality. Presumably one factor contributing to death would be osmoregulatory failure.

Temperature and salinity interact to influence growth and survival in most of the bivalve larvae and fish embryos tested with the exception of the mussels, *Adula califoriensis* and *Mytilus edulis* (Lough and Gonor, 1973a, 1973b; Lough, 1973). Survival of *Rangia cuneata*, *Crassostrea virginica* and *Mercenaria mercenaria* larvae is favoured in high salinity/high temperature or low salinity/ low temperature combinations with optimal growth in the upper portions of this range (Cain, 1973, 1974; Lough, 1975). Other larvae such as those of the geoduck clam, *Panope generosa* (Goodwin, 1973) display their broadest salinity tolerance near their optimum developmental temperature. Larval response to these environmental parameters is modified with age, either by increases in both temperature and salinity tolerance as in *Panope generosa*, *Rangia cuneata*, and *Mulinia lateralis* (Lough, 1975) or by increased low salinity tolerance accompanied by a decrease in temperature tolerance (*Crassostrea virginica*, *Mercenaria mercenaria*, and *Adula californiensis*).

Similar growth and survival studies have been performed on the larvae and embryos of a variety of fish including the English sole (Alderdice and Forrester, 1968), the petrale sole (Alderdice and Forrester, 1971), the Pacific herring (Alderdice and Velsen, 1971), the Pacific cod (Forrester and Alderdice, 1966), the killifish, *Fundulus heteroclitus* (Tay and Garside, 1975), the Atlantic menhaden (Lewis and Hettler, 1968), the Hawaiian fish, *Caranx mate* (Santerre, 1976), and the drum, *Bairdiella icistra* (May, 1974).

III. PHYSIOLOGICAL RESPONSES

A. Composition of Body Fluids

1. *Inorganic Ions*

Very little work has been done recently with temperature effects on the ionic composition of molluscs and annelids. Several papers on nereid polychaetes indicate that they are able to regulate their internal osmolarity well until they reach temperatures near freezing (Smith, 1955, 1957; Hohendorf, 1963). Among the molluscs, *Nassarius obsoletus* has increased haemolymph concentrations of potassium in the cold (Kasschau, 1975). Calcium is the ion which increased most with cold acclimation in the mussel, *Lamellidens marginalis*. Sodium and potassium rose somewhat while chloride declined (Rao, 1968). Lucu and Jelisavcic (1970) have studied temperature effects on the uptake of ^{131}Cs, an element which is reportedly handled like potassium, and they report no thermal influence on accumulation by *Mytilus galloprovincialis*.

Not only may temperature influence the osmoregulatory ability of an aquatic crustacean, but Lockwood (1960) also demonstrated a thermal effect on ionic regulation. The rate of sodium loss from the body of the isopod, *Asellus aquaticus*, is unaffected by temperatures ranging from 1 to 24 °C. However, the rate of active uptake of sodium increased as the temperature was increased. Although he was cautious about drawing a broad biogeographical generalization based on one species, he suggested that it is possible high temperatures might favour the invasion of marine animals into more brackish-water environments in the tropics, an idea suggested earlier by various workers (consult Lockwood, 1960, for a more detailed discussion). An inverse relationship between blood osmoconcentration and temperature and/or season was found in *Callinectes sapidus*, the blue crab (Ballard and Abbott, 1969), and for two related species of crabs of the genus *Hemigrapsus* (Dehnel, 1962). When acclimating the isopod, *Porcellio scaber*, to 3, 23, and 30 °C, Lindquist (1970) also reported an inverse relationship between blood osmotic pressure and acclimation temperature. Further, Engel and coworkers (1974) reported that sodium levels tended to be related directly to temperature at high salinity, but chloride concentration was unaffected by temperature when animals were exposed to high salinity. However, at low salinity both sodium and chloride levels were inversely and non-linearly related to temperature. In contrast, K^+ level was unaffected by temperature at either high or low salinity. Based on these data the authors suggested that Na^+ and Cl^- levels are regulated by different mechanisms at high and low salinities, and that the mechanisms of regulation are thermally influenced. As has been reported in adult blue crabs by Lynch, Webb and van Engel (1973), juvenile blue crabs have higher haemolymph salt concentrations in cold water than in warmer water of the same salinity (Leffler, 1975). Part of this increase is due to increased haemolymph Na^+ concentration. The permeability to salts of the crab, *Carcinus maenas*, changes with salinity. Although changes proceed more slowly at 5 °C than at 20 °C, temperature does not greatly influence the final permeability (Spaargaren, 1975). The ability of an isopod (*Sphaeroma serratum*) to regulate the sodium concentration in the haemolymph varied seasonally. Ionic regulation increases with low temperature and is important during the winter (Charmantier, 1975).

Wright (1975) reported that sodium regulation in two species of insects (*Chironomus dorsalis* and *Camptochironomus tentans*) is relatively insensitive to alterations in acclimatization temperature. The Q_{10} for sodium influx in *C. dorsalis* for the thermal range 0–10 °C = 1.31 and above 20 °C the Q_{10} value rises steeply to about 2.2. In *C. tentans* the whole body sodium concentration is not significantly different in animals acclimated to different temperatures (5, 12, 20, and 28 °C).

A large bibliography has been compiled regarding thermal effects on osmotic and ionic compositions of fish. We refer the reader to Houston's extensive discussion (Houston, 1973) of the literature before 1970. The generalization can be made that levels of the major extracellular ions are proportional to temperature in fresh-water fish and inversely proportional in salt-water animals. But at extremely high temperatures the trend reverses and ion levels shift back toward those of the environment. For example, goldfish increased plasma sodium and chloride levels as temperature increased from 5 to 25 °C, but at 35 °C the electrolyte levels dropped off (Murphy and Houston, 1974). At the other salinity extreme, sea-water alewives displayed elevated levels of plasma Na^+, Cl^-, and Ca^{2+} when heat stressed (Stanley and Colby, 1971).

The tendency for the major serum ions to shift toward environmental levels at temperature extremes can be explained by the fact that ionic gradients, particularly that of sodium, are maintained via transport enzymes which are subject to thermal modification. The analysis of temperature effects on enzyme activity in poikilothermic animals can thus predict shifts in blood ion levels. Fresh-water fish actively absorb ions to remain hyperosmotic to the environment. Extreme cold inhibits enzyme-mediated transport and sodium levels drop. High temperatures destabilize the system with the same result.

The opposite shift in blood electrolytes occurs in salt-water fish which must excrete ions in order to remain hyposmotic. Blood ion levels increase with both extreme cold and heat, indicating disturbance of the osmoregulatory system.

The shift in ion levels toward external values with cold has been variously explained as osmoregulatory breakdown or as an adaptive response which allows the animal to reduce metabolic demands. In view of the apparent mechanisms at work the adaptive response is probably not the actual change in ion levels, but the ability of the animal to tolerate the shift. The degree of tolerance varies, and some animals attempt to maintain constant blood osmotic pressure despite changes in ion concentration.

Umminger (1971a) has described four possible osmoregulatory responses of fresh-water fish to cold. Those fish which do not compensate simply tolerate the drop in serum electrolytes and osmolarity. The catfish, *Ictalurus nebulosus*, illustrates this case. Some fish partially compensate for decreases in sodium chloride by raising blood levels of organic constituents. For example, fresh-water-adapted *Fundulus heteroclitus* exhibited a 15% decrease in serum osmolarity at 0.1 °C, but an increase in glucose masked the true magnitude of the drop in electrolyte species (Umminger, 1971b). The third category is perfect compensation with no change in osmolarity, as illustrated by the carp (Houston and Madden, 1968), and the final category is overcompensation, as demonstrated in the goldfish by Umminger (1971b). In the latter case the increase in blood osmolarity was due to slightly elevated blood glucose and the build-up of an unknown component.

The response of brackish-water to cold appears to be more like that of salt-water animals than those from fresh-water. Two species of sculpin, *Myoxocephalus scorpius* and *M. quadricornus*, and the pike, *Esox lucius*, from 6‰ water in the Baltic all increased blood osmolarity, sodium, and chloride with acclimation to −0.1 °C in the winter (Oikari, 1975a, 1975b).

Salt-water fish require temperatures close to freezing before ion levels in the blood increase. Mullet showed no variation in serum composition over the range 7–15 °C, but did increase sodium and osmolarity after 8 days at 4 °C (Lasserre and Gallis, 1975). Umminger and Kenkel (1975) showed that increased sodium levels in *Fundulus heteroclitus* at −1 °C were accompanied by decreased activity of the chloride cells, supporting the concept of osmoregulatory failure as the underlying mechanism.

It has been suggested that cold death in fish results from the breakdown of osmoregulation. But in those cases where cold is lethal, the response is quite different from that of a cold-tolerant fish. Umminger (1971c) studied cold stress in sea-water *F. grandis*, a species limited in distribution to the southern United States. Serum osmolarity and glucose increased in the cold, but sodium and chloride decreased, possibly in an attempt to compensate for the increased osmolarity. The animals had sunken eyes and other signs of dehydration. The chill coma could be lessened by lowering the salinity. This case can be contrasted with the response of cod to freezing. In both 33‰ and 8‰ sea-water, serum salt concentrations shifted towards the environmental levels. This did not indicate osmoregulatory failure, however, since the animals survived freezing temperatures for 80 days (Jones and Scholes, 1974).

In marine fish the increase in freezing point depression of body fluids is due to higher ion levels, but organic antifreeze components have been reported for a variety of species including the marine sculpins, *Myoxocephalus scorpius* (Theede, 1969) and *M. quadricornus* (Oikari and Kristoffersson, 1973), Antarctic fishes (DeVries and Wohlschlag, 1969) and the winter flounder, *Pseudopleuronectes americanus* (Umminger, 1970). Serum osmolarity in this last species did not vary significantly in the laboratory for fish acclimated from 15 to −1 °C, yet the flounder was able to survive in a supercooled state. In winter flounder an antifreeze apparently replaces 26% of the sodium chloride present in serum and is responsible for the freezing point depression of the blood (Umminger, 1970).

2. *Organic Antifreezes*

The flounder antifreeze has been described by Duman and DeVries (1976) as three proteins ranging in weight from 6000 to 12 000 daltons and having a high alanine content. In this respect flounder antifreeze is similar to the protein and glycoprotein antifreezes described from Alaskan and Antarctic fish (Shier, Lin and DeVries, 1972; Raymond, Lin and DeVries, 1975). All of the antifreezes which have been described display thermal hysteresis; that is, the freezing point is several fractions of a degree lower than the melting point (Duman and DeVries, 1974a). DeVries (1971) has suggested that the glycoprotein antifreezes act by absorption onto the surface of ice crystals, thereby preventing their growth.

The production of antifreezes in the blood of Alaskan fishes can be induced by low temperatures over the period of a month, but the disappearance of the antifreeze requires three to five weeks of warm temperatures combined with long photoperiods (Duman and DeVries, 1974b). California populations of the same species were not able to produce antifreeze and had higher serum freezing points, which suggests that genetically distinct subgroups have evolved.

The presence of an 'antifreeze' in molluscs is suggested by the finding that the acclimation of intact *Mytilus edulis* to high salinity enhances freezing resistance, while acclimation of isolated gill preparations does not. The acquisition of this tolerance also required 24 hours longer exposure to high salinity than is needed for osmotic adjustment. This time course was found by Williams (1970) to coincide with the time course of increased cellular amino acid levels.

3. Amino Acids

The use of amino acids as intracellular osmotic effectors is a fairly new discovery (see Chapter 4 by Gilles). It is not surprising that amino acid concentrations are affected by temperature as well as by salinity, since most metabolic pathways which produce amino acids are temperature sensitive. The ciliate, *Tetrahymena pyriformis* has temperature-dependent uptake of L-phenylalanine with Q_{10}'s of 2.19 in the 12–20 °C range and 1.63 from 20–28 °C (Stephens and Kerr, 1962). Uptake of amino acids by the coelenterate, *Anemonia sulcata*, is also temperature dependent with a Q_{10} of 2. There is competition for transport among the various amino acids tested which suggests mediated uptake (Schlichter, 1974). The fishes, *Platypoecilus maculatus*, *Etroplus maculatus*, and *Xiphorus helleri*, and the mussel, *Lamellidens marginalis*, all showed increased amino acid levels with a temperature drop (Kinne, 1964; Roa, 1968). Cold-acclimated specimens of the euryhaline snail, *Nassarius obsoletus*, also have higher total free amino acids than warm-acclimated snails at any salinity tested which might explain the enhanced survival of these animals in low salinity and cold temperature (Kasschau, 1975).

The interaction of temperature and salinity in a complex manner to influence the intracellular free amino acid concentration of haemolymph osmolarities of *Crangon crangon* has been demonstrated by Weber and Van Marrewijk (1972). For example, in the regulation range of salinities the ninhydrin-positive substance (NPS) concentration at 5 °C is significantly higher than at other acclimation temperatures. Further, within the regulation range there is an inverse correlation between temperature and the concentration of NPS, but at high and at low salinities there is a direct correlation. Combinations of low salinity and low temperatures and high temperatures with high salinity are correlated respectively with strong decreases and increases of cellular NPS. Earlier, Duchateau and Florkin (1955) reported that for the muscle of the crab, *Eriocheir sinensis*, adapted to fresh-water at two temperature ranges (1–3 °C and 10–11 °C) the concentration of glycine, alanine, and arginine correlated inversely with temperature. The total free amino acid (FAA) concentration is higher at higher acclimation temperatures, particularly as a result of a marked increase in proline. In the muscle of a bivalve (*Mya arenaria*), an inverse temperature effect was noted in FAA concentration (DuPaul and Webb, 1970).

B. Other Responses

Temperature and salinity interact to affect not only growth, survival, and osmoregulation, but also a variety of other physiological functions such as respiration. Kinne (1958) found that these two factors affect the body configuration of the brackish-water hydroid, *Cordylophora caspia*, with the largest animals in low temperature and salinity.

It is not surprising that respiration is affected by changes in temperature and salinity since the amount of oxygen which will dissolve in water increases with decreasing temperature and salinity. The respiration rate of many marine invertebrates is dependent upon the amount of dissolved oxygen which is available (Lange, Staaland and Mostad, 1972). The oyster, *Crassostrea virginica* and the mussel, *Mytilus edulis*, have no significant temperature–salinity interaction on respiration over a wide range of temperature and salinity. But *Mercenaria* sp. and *Modiolus demissus* show significant interaction with oxygen consumption greatest in low temperature and salinity (Van Winkle, 1968). A study of the chiton, *Mopalia lignosa* found highest respiration in 100% sea-water over a range of temperatures, but a biphasic respiration curve in 90% sea-water with the shift at 13.5 °C (Lebsack, 1975).

Feng and Van Winkle (1975) have looked at heart beat in the oyster as an indicator of acclimation to salinity. At warmer temperatures they found that optimum salinity for heart beat was 14–20‰ which agrees with the finding of Vernberg, Schlieper and Schneider (1963) that oyster tissue survives heat stress best at lower salinities.

IV. EFFECTS OF POLLUTANTS

In recent years emphasis has been placed on the influence of various types of pollutants on the physiology of organisms including osmoregulation. The effects of pollutants on osmoregulation are discussed in Chapter 13. Only a few studies will thus be cited here to emphasize that pollutants are a part of the environment and must be taken into account when studying temperature and osmoregulation (see for instance Vernberg and Vernberg, 1974; Vernberg and coworkers (1977). Mercury decreases the survival time of fiddler crabs when exposed to thermal–osmoregulatory stress when compared to control animals (Vernberg, De Coursey and O'Hara, 1974). Mercury-depressed ion transport in several important osmoregulatory systems of teleosts (Renfro and coworkers, 1974). However, Caldwell (1974) found that although methoxychlor (an insecticide) decreased the resistance of crabs to reduced salinities, disruption of osmotic or ionic regulation is not a principal cause of this increased mortality.

Jones (1975) reported that certain heavy metals influenced the osmoregulatory ability of some estuarine and marine isopods. The marine species were more sensitive to heavy metals when the salinity was reduced than were the estuarine species. Increased temperature increased the toxicity of cadmium.

V. MECHANISMS OF ADAPTATION

One of the simplest means by which temperature affects osmoregulation is the alteration of membrane permeability leading to variations in water and ion fluxes. The increase in water flux with increasing temperature must then be balanced by concomitant increases in water excretion.

This system may be observed in its simplest form in the protozoans. The activity of the contractile vacuole increases with temperature until the animal dies (Kitching, 1948a). The question then arises, is the increased activity due to a direct thermal response on the contractile vacuole or is it compensation for the influx of water? Kitching (1967) ruled out a direct thermal effect on the contractile vacuole when he noted that a hypertonic solution stopped vacuolar activity as effectively at 30 °C as at 15 °C in *Carchesium aselli*. He also noted that the marine ciliate, *Vorticella marina*, swelled more rapidly with increased temperature when placed in 25% sea-water, suggesting an increase in the permeability of the cell membrane to water movement along the osmotic gradient (1948b). The idea of increased water influx is supported by detailed observations on the behaviour of the contractile vacuole with temperature changes in protozoans such as *Discophyra collini* (Kitching, 1967) and *Amoeba proteus* (Ahmad and Couillard, 1974).

Body permeability and urine output are usually proportional to temperature. Inulin clearance in the fresh-water clam, *Anodonta cygnea*, decreases with decreasing temperature with a Q_{10} of 2 (Potts, 1954), despite the protective influence of a shell.

Generally, fish respond to changes in temperature by a change in permeability with a Q_{10} of 2–4 (Houston, 1973).

The degree of permeability change with the change in temperature was found by Isaia (1972) to be higher in a fresh-water fish, *Carassius auratus* (Q_{10} = 2, 2.6) than in a salt-water fish, *Serranus scriba* and *S. cabrilla* which had a Q_{10} of 1.6.

The major site of permeability change is the gills (Motais and coworkers, 1969) since the integument is protected. There is some question as to whether temperature directly affects the physical characteristics of the membrane or whether alterations in blood flow vary the effective surface area of the membrane available for diffusion. The tremendous influx of water which occurs at higher temperatures is countered by increases in glomerular filtration rate (GFR) and urine output. Some fish are unable to modify their urine to any great extent so that high temperatures result in increased electrolyte loss for which the animal must compensate. This has been demonstrated by Malvin and coworkers (1970) for the fresh-water lamprey. The inability of the lamprey to modify its urine requires it to triple the recruitment of ions from the environment.

The effect of temperature on osmotic and ionic regulation in the goldfish has recently been examined by Mackay (1974). It appears from his data that urine flow is to some extent adaptive in nature and not merely a physiochemical process. Fish acclimated to 10 °C have a urine flow which is 3.2 times greater than that of 30 °C fish when both are measured at 20 °C. This change in urine flow can be extrapolated to mean a difference in gill permeability between warm- and cold-acclimated fish. Thus, the variation in urine flow of animals measured at an intermediate temperature may represent the effect of sudden temperature change on gill permeability as described by Motais and Isaia (1972) for the eel. They noted that sudden increases from the acclimation temperature to a higher temperature, T_1, caused an overshoot reaction so that the permeability at first was higher than that of an animal acclimated to T_1. An undershoot reaction was observed with sudden drops in temperature. The undershoot/overshoot phenomenon probably explains the 3.2 fold difference in urine flow of 10 and 30 °C fish measured at 20 °C. A similar response can also be seen in respiration studies of warm- and cold-acclimated animals tested at an intermediate temperature (Silverthorn, 1973).

Temperature effects in aquatic birds have not been studied to any great degree, probably because of their homeothermic nature. Staaland (1967) showed that local cooling to 25 °C of the salt gland in white Peking ducks decreased the volume of fluid without affecting its concentration, but the physiological significance of this work is questionable. Hughes (1968) showed that newly hatched terns showed the least cloacal fluid loss, and she postulated that this was a result of their higher water loss due to evaporative cooling. A study of penguins held at 18 and 28 °C revealed no significant difference in the osmotic and ionic composition of nasal and cloacal fluid (Oelofsen, 1973).

Thermal effects on water permeability of amphibians comprise many of the earliest experiments on osmoregulation, and interest in this area has continued (Schmidt-Nielsen and Forester, 1954; Dicker and Elliott, 1967; Miller, Standish and Thurman, 1968). Generally, water permeability is proportional to temperature, and the increased water uptake is compensated for by the kidneys. In extreme cold urine production is about 25% of the warm-acclimated rate (Schmidt-Nielsen and Forester, 1954; Miller, Standish and Thurman, 1968). The decrease in urine flow has two components: a GFR which is 14% of normal, and depressed tubular reabsorption of water. The decrease in GFR is due to intermittant glomerular filtration, i.e. a reduction in the number of operational glomeruli. This phenomenon has been described in a number of organisms besides the frog, including a fresh-water turtle (Dantzler and Schmidt-Nielsen, 1966) and several fish (Hickman, 1965; Mackay and Beatty, 1968).

Parsons and Lau (1976) have recently described an interesting difference between *in vivo* and *in vitro* changes of frog skin permeability with temperature. *In vivo* acute temperature changes in either direction proceed in two stages. When animals are dropped to 5 °C from 15 °C, the first response is a significant increase in water uptake: twice the 15 °C rate or six times the normal 5 °C rate. This increased water uptake results in plasma dilution within the first hour of acute cold exposure. By two or more hours the water uptake has dropped off to the 5 °C control rate. When frogs are moved from the cold to 15 °C, stage 1 response is no change for the first hour, followed by an increase to the control value. *In vitro* preparations show only the second responses, and water movement is constant with time. Parsons and Lau have suggested a temperature-sensitive osmoregulatory centre which uses a hormonal mediator to change the set point with changes in temperature.

As our understanding of physiological mechanisms grows the research emphasis has shifted from descriptive accounts of what happens to an analysis of why they happen. A major control point in sodium metabolism is the enzyme sodium–potassium adenosine triphosphatase (Na^+/K^+-ATPase) which has been described in some detail in frogs and fish.

In 1957 Snell and Leeman examined the temperature coefficients of sodium transport in the isolated frog skin. Transport increased linearly from 5–15 °C ($Q_{10} = 2$), but the slope dropped off slightly from 5–25 °C and became negative above 25 °C, probably due to instability of the system. They compared the energy dissipation of the enzyme to oxygen consumption at various temperatures and concluded that transport became more efficient as temperatures dropped. In 1963 Takenake found that a logarithmic plot of the free energy change of sodium transport in frog skin against temperature had two discrete slopes whose intersection lay between 10 and 15 °C.

The discontinuity in sodium transport described above by Takenaka was later found to result from the properties of the Na^+/K^+-ATPase. Enzymes from *Rana catesbiana* bladder had a Q_{10} of 3.7 below 13 °C and of 2.1 above 13 °C (Asano and coworkers, 1970). ATPase from warm-acclimated bullfrog kidney had a bimodal Arrhenius plot with the break at 10 °C (Tanaka and Teruya, 1973), while enzymes from frog epidermis had a discontinuity of 23 °C (Kawada, Taylor and Barker, 1975). Intestinal ATPase from the goldfish showed discontinuity of the Arrhenius plot of both the ouabain-sensitive and ouabain-insensitive fractions (Smith, 1967). It is interesting that the temperature at which the discontinuity occurs varies with acclimation temperature. Smith found that the ouabain-sensitive Na^+/K^+-ATPase of fish acclimated to 8 °C had its discontinuity at 12 °C. For 19 °C acclimated fish the discontinuity occurred at 16 °C, and for those acclimated to 30 °C it occurred at 21 °C.

In goldfish, intestinal enzyme activity decreased with increasing temperature compared to goldfish gill ATPase which increased activity with temperature (Murphy and Houston, 1974). Other studies on the intestinal enzyme showed that although sodium movement was inhibited 24 hours after a transfer from 16 to 30 °C, ATPase activity required 20 days to be inhibited. Smith and Ellory (1971) concluded from ouabain-binding experiments that the number of pump sites remains constant, and that the loss of activity is due to a change in the properties of the enzyme.

Tanaka and Teruya (1973) suggested that the alteration in the response of the enzyme to temperature resulted from a change in the degree of saturation of the lipid portion of the enzyme. Intestinal lipids from cold-acclimated goldfish were more unsaturated than those from warm-acclimated fish (Kemp and Smith, 1970). Similarly, more unsaturated fatty acids were found in the salt gland of spiny dogfish than in homeothermic marine birds (Bergh, Larson and Samuelson, 1975).

The theory that changes in saturation affect enzyme activity was confirmed by the very elegant experiments of Kimelberg and Papahadjopoulos (1974) in which they substituted various lipids onto the protein moiety of Na^+/K^+-ATPase. Their data indicated that the more unsaturated the fatty acid portion of the enzyme, the more activity will be retained at low temperature and the lower will be the discontinuity on the Arrhenius plot. If this theory can be expanded to all enzymes with a lipid moiety, then one reason for generalized increases in unsaturated lipids with cold acclimation is explained (cf. Hoar and Cottle, 1952; Lewis, 1962; Knipprath and Mead, 1965, 1966, 1968; Caldwell and Vernberg, 1970).

The ATPase system is a major mediator of sodium metabolism, particularly in sea-water animals which must excrete salt to remain hypo-osmotic to the environment. In fresh-water animals, however, the problem is one of acquisition of sodium. Maetz (1972, 1973) has presented evidence for a Na^+/NH_4^+ or Na^+/H^+ exchange across the gill of the goldfish and has examined the effects of temperature on this system. Intact fish subjected to a drop in temperature from 16 to 6 °C greatly reduced ammonia excretion (Q_{10} = 3.9). Sodium influx decreased more than efflux. The difference in the thermal effect on the two fluxes would explain the observed drop in serum sodium concentrations. Maetz felt that the temperature-sensitive step in this system was the metabolic pathways which produce ammonia. However, Payan and Matty (1975), working with a perfused isolated trout head, found that the magnitude of the reduction in permeability of the gills to ammonia with decreased temperature corresponded exactly to the decrease in perfusion which occurred with decreased temperature. These experiments are not comparable, however, and neither conclusion excludes the other.

The relationship between the induction of enzyme activity and the ecology of the animal has provided an interesting area of study. Sargent and coworkers (1975) looked at Na^+/K^+-ATPases in the gill of eels (*Anguilla anguilla*) acclimated to fresh- or salt-water and 5, 10 or 18 °C. Salt-water gills had highest ATPase levels at 5 and 10 °C due to increased amount of enzymes, while the activity in fresh-water fish was highest at 18 °C and was related to the quality rather than the quantity of enzymes. Arrhenius plots showed discontinuities at 20 °C for fresh-water preparations and at 12 °C for those from salt-water. All of these data fit nicely with the ecology of the eels which move from warm fresh-water regions into the cold ocean.

Another series of experiments on ATPase induction in steelhead trout and Coho salmon have been carried out by Zaugg and his associates (Zaugg, Adams and McLain, 1972; Adams, Zaugg and McLain, 1973; Zaugg and Wagner, 1973; Adams, Zaugg and McLain, 1975; Zaugg and McLain, 1976).

Increased levels of Na^+/K^+-ATPase are associated with the parr–smolt transformation which occurs just before seaward migration. Steelhead require increasing temperatures and advancing photoperiods for induction of ATPase activity, but temperatures over 11.3 °C and long photoperiods may subsequently decrease enzyme activity. Salmon are less sensitive to temperature, developing enzyme activity in temperatures up to 15 °C, and in these fish photoperioid is non-contributory. Cold temperatures or sea-water preserve activity while smolt held at high temperatures or in fresh-water lose enzyme activity.

Endocrine regulation of water and electrolyte balance is well documented in a variety of aquatic animals including crustaceans, fish, amphibians, and birds (see reviews by Kamemoto, 1976 and Bentley, 1971). Despite the wealth of information in this area very little research has focused on the thermal modification of endocrine function.

In fresh-water fish the adenohypophysial hormone prolactin retards sodium loss. The mechanism is uncertain, but probably is either an inhibition of gill Na^+/K^+-ATPase or stimulation of the active uptake mechanism. The bullhead, *Ictalurus melas*, was unable to survive hypophysectomy longer than 11 days at 20 °C, but at 11 °C 60% of the hypophysectomized fish were still alive after 30 days. The 11 °C fish displayed the hyponatremia associated with prolactin deficiency, but they were able to tolerate it (Chidambaram, Meyer and Hasler, 1972). The higher mortality at 20 °C may have resulted from increased water permeability for which the hypophysectomized fish were unable to compensate.

Lam and Hoar (1967) demonstrated the role of prolactin in the migration of stickleback from the ocean into fresh-water in the spring. They concluded that prolactin was necessary for the fish to osmoregulate under hypo-osmotic conditions. Umminger and Kenkel (1975) investigated the activity of prolactin-producing cells in *Fundulus heteroclitus* held at -1 °C in sea-water and 1 °C in fresh-water. They found that low temperatures seemed to inhibit prolactin release in fresh-water fish and stimulate it in salt-water animals. Their data coupled with those of Chidambaram, Meyer and Hasler (1972) suggest that temperature plays a crucial role in prolactin secretion.

Stress in fish such as that induced by handling and injection has been shown to affect serum electrolyte and glucose levels, probably via adrenocorticosteroids. Umminger and Gist (1973) have investigated the role of thermal acclimation in the response of goldfish to stress. They found that cold-acclimated fish had more pronounced hyperglycaemia and hyponatraemia than warm-acclimated fish. Simultaneous injections of cortisol further lowered serum sodium in the warm-acclimated fish but not in the cold, which apparently were displaying the maximum effect.

Neurohypophysial hormones with antidiuretic action, such as arginine vasotocin, have been widely studied in amphibians (see Chapter 6 and 10). The uptake of water under the influence of pituitary hormones is proportional to temperature, but the effect of hormones on body weight is inversely proportional (Boyd and Brown, 1938; Hong, 1957). At higher temperatures 'pituitrin' had no antidiuretic effect on the urine concentration of *Rana pipiens*, so that the water taken in through the skin was readily balanced by excretion of water from the kidney. In 16 °C frogs however 'pituitrin' caused a pronounced diminution of urine flow, resulting in net uptake of water. These data suggest that the overshoot response in frog skin permeability observed by Parsons and Lau (1976; in page 550 of this chapter) with a drop from 15 to 5 °C *in vivo* but not *in vitro* is due to vasotocin. Vasotocin effects would also explain the changes in urine flow which have been observed in cold frogs (Schmidt-Nielsen and Forester, 1954; Miller, Standish and Thurman, 1968).

REFERENCES

Adams, B. L., W. S. Zaugg, and L. R. McLain (1973). Temperature effect on parr–smolt transformation in steelhead trout (*Salmo gairdneri*) as measured by gill sodium–potassium stimulated adenosine triphosphatase. *Comp. Biochem. Physiol.*, **44A**, 1333–9.

Adams, B. L., W. S. Zaugg, and L. R. McLain (1975). Inhibition of salt water survival and Na–K-ATPase elevation in steelhead trout (*Salmo gairdneri*) by moderate water temperatures. *Trans. Am. Fish. Soc.*, **104**, 766–9.

Ahmad, M., and P. Couillard (1974). The contractile vacuole in *Amoeba proteus*: temperature effects. *J. Protozool.*, **21**, 330–6.

Alderdice, D. F., and C. R. Forrester (1968). Some effects of salinity and temperature on early development and survival of the English sole (*Parophrys vetulus*). *J. Fish. Res. Bd. Can.*, **25**, 495–521.

Alderdice, D. F., and C. R. Forrester (1971). Effects of salinity and temperature on embryonic development of the petrale sole (*Eopsetta jordani*). *J. Fish. Res. Bd. Can.*, **28**, 727–44.

Alderdice, D. F., and F. P. J. Velsen (1971). Some effects of salinity and temperature on early development of Pacific herring (*Clupea pallasi*). *J. Fish. Res. Bd. Can.*, **28**, 1545–62.

Arai, M. N., E. T. Cox, and F. E. J. Fry (1963). An effect of dilutions of seawater on the lethal temperature of the guppy. *Can. J. Zool.*, **41**, 1011–5.

Asano, Y., H. Matsui, K. Nagano, and M. Nakao (1970). ($Na^+ + K^+$)-ATPase from the frog bladder and its relationship to sodium transport. *Biochim. Biophys. Acta*, **219**, 169–78.

Ballard, B. S., and W. Abbott (1969). Osmotic accommodation in *Callinectes sapidus* Rathbun. *Comp. Biochem. Physiol.*, **29**, 671–87.

Bentley, P. J. (1971). *Endocrines and Osmoregulation*, Springer-Verlag, New York.

Bergh, C. H., G. Larson, and B. E. Samuelson (1975). Fatty acid and aldehyde composition of major phospholipids in salt gland of marine birds and spiny dogfish. *Lipids*, **10**, 299–302.

Biggs, D. C., and J. J. McDermott (1973). Variation in temperature–salinity tolerance between two estuarine populations of *Pagurus longicarpus* Say (Crustacea: Anomura). *Biol. Bull.*, **145**, 91–102.

Binyon, J. (1961). Salinity tolerance and permeability to water of the starfish *Asterias rubens* L. *J. Mar. Biol. Assoc. UK*, **41**, 161–4.

Boyd, E. M., and G. M. Brown (1938). Factors affecting the uptake of water by frogs when injected with extract of the posterior hypophysis. *Am. J. Physiol.*, **122**, 191–200.

Bradley, B. P. (1975). The anomalous influence of salinity on temperature tolerance of summer and winter populations of the copepod *Eurytemora affinis*. *Biol. Bull.*, **148**, 26–34.

Cain, T. D. (1973). The combined effects of temperature and salinity on embryos and larvae of the clam *Rangia cuneata*. *Mar. Biol.*, **21**, 1–6.

Cain, T. D. (1974). Combined effects of changes in temperature and salinity on early stages of *Rangia cuneata*. *Virginia J. Sci.*, **25**, 30–1.

Caldwell, R. S. (1974). Osmotic and ionic regulation in decapod Crustacea exposed to methoxychlor. In F. J. Vernberg and W. B. Vernberg (Eds), *Pollution and Physiology of Marine Organisms*, Academic Press, New York. pp. 197–224.

Caldwell, R. S., and F. J. Vernberg (1970). The influence of acclimation temperature on the lipid composition of fish gill mitochondria. *Comp. Biochem. Physiol.*, **34**, 179–91.

Charmantier, G. (1975). Variations saisonnieres des capacites ions regulatrices de *Sphaeroma serratum* (Febricius, 1787) (Crustacea, Isopoda, Flabellifera). *Comp. Biochem. Physiol.*, **50A**, 339–46.

Chidambaram, S., R. K. Meyer, and A. D. Hasler (1972). Effects of hypophysectomy, pituitary autografts, prolactin, temperature and salinity of the medium on survival and natremia in the bullhead, *Ictalurus melas*. *Comp. Biochem. Physiol.*, **43A**, 443–57.

Christiansen, M. E., and J. D. Costlow (1975). The effect of salinity and cyclic temperature on larval development of the mud-crab *Rhithropanopeus harrissi* reared in the laboratory. *Mar. Biol.*, **32**, 215–21.

Dantzler, W. H., and B. Schmidt-Nielson (1966). Excretion in freshwater turtle (*Pseudemys scripta*) and desert tortoise (*Gopherus agassizii*). *Am. J. Physiol.*, **210**, 198–210.

Dehnel, P. A. (1962). Aspects of osmoregulation in two species of intertidal crabs. *Biol. Bull.*, **122**, 208–27.

DeVries, A. L. (1971). Glycoproteins as biological antifreeze agents in Antarctic fishes. *Science*, **172**, 1152–5.

DeVries, A. L., and D. E. Wohlschlag (1969). Freezing resistance in some Antarctic fishes. *Science*, **163**, 1073–5.

Dicker, S. E., and A. B. Elliott (1967). Water uptake by *Bufo melanostictus* as affected by osmotic gradients, vasopressin, and temperature. *J. Physiol.*, **190**, 359–70.

Duchâteau, G., and M. Florkin (1955). Influence de la température sur l'état stationnaire du pool des acides aminés non protécques des muscles d' *Eriocheir sinensis* Milne Edwards. *Archs Int. Physiol. Biochim.*, **63**, 213–21.

Duman, J. G., and A. L. DeVries (1974a). Freezing resistance in the winter flounder *Pseudopleuronectes americanus*. *Nature*, **247**, 237–8.

Duman, J. G., and A. L. DeVries (1974b). The effects of temperature and photoperiod on antifreeze production in cold water fishes. *J. Exp. Zool.*, **190**, 89–97.

Duman, J. G., and A. L. DeVries (1976). Isolation, characterization, and physical properties of protein antifreezes from the winter flounder, *Pseudopleuronectes americanus*. *Comp. Biochem. Physiol.*, **54B**, 375–81.

DuPaul, W. D., and K. L. Webb (1970). The effect of temperature on salinity induced changes in the free amino acid pool of *Mya arenaria*. *Comp. Biochem. Physiol.*, **32**, 785–801.

Engel, D. W., E. M. Davis, D. E. Smith, and J. W. Angelovic (1974). The effect of salinity and temperature on the ion levels in the hemolymph of the blue crab, *Callinectes sapidus. Comp. Biochem. Physiol.*, **49A**, 259–66.

Eriksson, S., and B. Tallmark (1974). The influence of environmental factors on the diurnal rhythm of the prosobranch gastropod *Nassarius reticulatus* from a non-tidal area. *Zoon*, **2**, 135–42.

Feng, S. Y., and W. van Winkle (1975). The effect of temperature and salinity on the heart beat of *Crassostrea virginica. Comp. Biochem. Physiol.*, **50A**, 473–6.

Flügel, H. (1963). Elektrolytregulation und temperatur bei *Crangon crangon* L. und *Carcinus maenas* L. *Kieler Meeresforsch.*, **19**, 185–95.

Forrester, C. R., and D. F. Alderdice (1966). Effects of salinity and temperature on embryonic development of the Pacific cod (*Gadus macrocephalus*). *J. Fish Res. Bd. Can.*, **23**, 319–38.

Garside, E. T., and Z. K. Chin-Yuen-Kee (1972). Influence of osmotic stress on upper lethal temperatures in the cyprinodontid fish *Fundulus heteroclitus* (L.) *Can. J. Zool.*, **50**, 787–91.

Garside, E. T., and C. M. Jordan (1968). Upper lethal temperatures at various levels of salinity in the euryhaline cyprinodontids *Fundulus heteroclitus* and *F. diaphanus* after isosmotic acclimation. *J. Fish. Res. Bd. Can.*, **25**, 2717–20.

Gilles, R. (1975). Mechanisms of ion and osmoregulation. In O. Kinne (Ed.), *Marine Ecology*, Vol. II, part 2, Wiley Interscience, New York. pp. 259–347.

Goodwin, L. (1973). Effects of salinity and temperature on embryos of the geoduck clam (*Panope generosa* Gould). *Proc. Natl Shellfish Assoc.*, **63**, 93–5.

Grimm, A. S. (1969). Osmotic and ionic regulation in the shrimps *Crangon vulgaris* Fabr. and *Cangon allmani* Kinahan. PhD Thesis. University of Glasgow.

Haefner, P. A. (1969). Temperature and salinity tolerance of the sand shrimp *Crangon septemspinosa* Say. *Physiol. Zool.*, **42**, 388–97.

Hickman, C. P., Jr (1965). Studies on renal function in fresh-water teleost fish. *Trans. R. Soc. Can.*, **3**, 213–36.

Hill, L. G., and D. R. Carlson (1970). Resistance of the plains killifish *Fundulus kansae* (Cyprinodontidae) to combined stresses of temperature and salinity. *Proc. Oklahoma Acad. Sci.*, **50**, 75–8.

Hoar, W. S., and M. K. Cottle (1952). Some effects of temperature acclimatization on the chemical constitution of goldfish tissues. *Can. J. Zool.*, **30**, 49–54.

Hohendorf, K. (1963). Der Einfluß der temperature auf die Salzgehaltstoleranz und Osmoregulation von *Nereis diversicolor. Kieler Meeresforsch.*, **19**, 196–218.

Hong, S. K. (1957). Effects of pituitrin and cold on water exchanges of frogs. *Am. J. Physiol.*, **188**, 439–42.

House, C. R. (1974). *Water Transport in Cells and Tissues.* Edward Arnold, London.

Houston, A. H. (1973). Environmental temperature and the body fluid system of the teleost. In W. Chavin (Ed.), *Responses of Fish to Environmental Changes*, Charles C. Thomas, Springfield. pp. 87–162.

Houston, A. H., and J. A. Madden (1968). Environmental temperature and plasma electrolyte regulation in the carp, *Cyprinus carpis. Nature*, **217**, 969–70.

Hughes, M. R. (1968). Renal and extrarenal sodium excretion in the common tern *Sterna hirundo. Physiol. Zool.*, **41**, 210–19.

Hummon, W. D. (1975). Respiratory and osmoregulatory physiology of a meiobenthic marine gastrotrich *Turbanella ocellata* Hummon. *Cah. Biol. Mar.*, **16**, 255–68.

Isaia, J. (1972). Comparative effects of temperature on the sodium and water permeabilities of the gills of a stenohaline freshwater fish (*Carassius auratus*) and a stenohaline marine fish (*EERRANUS SCRIBA, Serranus cabrilla*).*J. Exp. Biol.*, **57**, 359–66.

Ivleva, I. V. (1967). The relation of tissue heat-resistance of Polychaetes to osmotic and temperature conditions of the enviromnent. In A. S. Troshin (Ed.), *The Cell and Environmental Temperature*, Pergamon Press, Oxford. pp. 232–7.
Jones, F. R. H., and P. Scholes (1974). The effect of low temperature on cod, *Gadus morhua*. *J. Cons Int. Explor. Mer.*, **35**, 258–71.
Jones, M. B. (1975). Synergistic effects of salinity, temperature, and heavy metals on mortality and osmoregulation in marine and estuarine isopods (*Crustacea*). *Mar. Biol.*, **30**, 13–20.
Jones, R. T., and K. S. Price Jr. (1974). Osmotic responses of spiny dogfish (*Squalus acanthias* L.) embryos to temperature and salinity stress. *Comp. Biochem. Physiol.*, **47A**, 971–9.
Jordan, C. M., and E. T. Garside (1972). Upper lethal temperatures of threespine stickleback, *Gasterosteus aculeatus* (L.) in relation to thermal and osmotic acclimation, ambient salinity, and size. *Can. J. Zool.*, **50**, 1405–11.
Kähler, H. H. (1970). Über den Einfluß er Adaptations temperature und des Salzgehaltes auf die Hitze- und Grefrierresistenz von *Enchytraeus albidus* (Oligochaeta). *Mar. Biol.*, **5**, 315–24.
Kamemoto, F. I. (1976). Neuroendocrine control of osmoregulation in decapod Crustacea. *Am. Zool.*, **16**, 141–50.
Kasschau, M. R. (1975). The relationship of free amino acids to salinity changes and salinity–temperature interactions in the mud-flat snail *Nassarius obsoletus*. *Comp. Biochem. Physiol.*, **51A**, 301–8.
Kawada, H., R. E. Taylor, Jr, and S. B. Barker (1975). Some biochemical properties of Na, K-ATPase in frog epidermis. *Comp. Biochem. Physiol.*, **50A**, 297–302.
Kemp, P., and M. W. Smith (1970). Effect of temperature acclimatization on the fatty acid composition of goldfish intestinal lipids. *Biochem. J.*, **117**, 9–15.
Kenny, R. (1969). The effects of temperature, salinity, and substrate on distribution of *Clymenella torquata* (Leidy) Polychaeta. *Ecology*, **50**, 624–31.
Kimelberg, H. K., and D. Papahadjopoulos (1974). Effects of phospholipid acyl chain fluidity, phase transitions and cholesterol on ($NA^+ + K^+$)-stimulated adenosine triphosphatase. *J. Biol. Chem.*, **249**, 1071–80.
Kinne, O. (1958). Adaptations to salinity variations—some facts and problems. In C. L. Prosser (Ed.), *Physiological Adaptation*, American Physiological Society, Washington. pp. 92–106.
Kinne, O. (1964). Physiology of estuarine organisms with special reference to salinity and temperature. In G. F. Lauff (Ed.), *Estuaries*, No. 83, American Association for the Advancement of Science, Washington. pp. 525–40.
Kinne, O. (1970). Temperature—Invertebrates. In O. Kinne (Ed.), *Marine Ecology*, Vol. 1, *Environmental Factors*, Wiley Interscience, London. pp. 407–514.
Kitching, J. A. (1948a). The physiology of contractile vacuoles. V. The effects of short-term variations of temperature on a feshwater peritrich ciliate. *J. Exp. Biol.*, **25**, 406–20.
Kitching, J. A. (1948b). The physiology of contractile vacuoles. VI. Temperature and osmotic stress. *J. Exp. Biol.*, **25**, 421–36.
Kitching, J. A. (1967). Contractile vacuoles, ionic regulation and excretion. In T. Chen (Ed.), *Research in Protozoology*, Vol. 1, Pergamon Press, New York. pp. 309–36.
Knipprath, W. G., and J. F. Mead (1965). Influence of temperature on the fatty acid pattern of muscle and organ lipids of the rainbow trout (*Salmo gairdneri*). *Fishery Indust. Res.*, **3**, 23–7.
Knipprath, W. G., and J. F. Mead (1966). Influence of temperature on the fatty acid

pattern of mosquitofish (*Gambusia affinis*) and guppies (*Lebistes reticulatus*). *Lipids*, **1**, 113–7.

Knipprath, W. G., and J. F. Mead (1968). The effect of the environmental temperature on the fatty acid composition and on the *in vivo* incorporation of 1-^{14}C-acetate in goldfish (*Carassius auratus* L.). *Lipids*, **3**, 121–8.

Lam, T. J., and W. S. Hoar (1967). Seasonal effect of prolactin on osmoregulation of the marine form (Trachurus) of the stickleback *Gasterosteus acculeatus*. *Can. J. Zool.*, **45**, 509–16.

Lange, R., H. Staaland, and A. Mostad (1972). The effect of salinity and temperature on solubility of oxygen and respiratory rate on oxygen-dependent marine invertebrates. *J. Exp. Mar. Biol. Ecol.*, **9**, 217–29.

Lasserre, P., and J.-L. Gallis (1975). Osmoregulation and differential penetration of two grey mullets, *Chelon labrosus* and *Liza ramada* in estuarine fish ponds. *Aquaculture*, **5**, 323–44.

Lawrence, J. M. (1975). Effect of temperature–salinity combinations on the functional well-being of adult *Lytechinus variegatus* (Echinodermata, Echinoidea). *J. Exp. Mar. Biol. Ecol.*, **18**, 271–6.

Lebsack, C. S. (1975). Effect of temperature and salinity on the oxygen consumption of the chiton *Mopolia lignosa*. *Veliger*, **18**, Suppl., 94–7.

Leffler, C. W. (1975). Ionic and osmotic regulation and metabolic response to salinity of juvenile *Callinectes sapidus* Rathbun. *Comp. Biochem. Physiol.*, **52A**, 545–9.

Lewis, R. M., and W. F. Hettler, Jr (1968). Effect of temperature and salinity on the survival of young Atlantic menhaden, *Brevoortia tyrannus*. *Trans. Am. Fish. Soc.*, **97**, 344–9.

Lewis, R. W. (1962). Temperature and pressure effects on the fatty acids of some marine ectotherms. *Comp. Biochem. Physiol.*, **6**, 75–89.

Lindquist, O. V. (1970). The blood osmotic pressure of the terrestrial isopods *Porcellio scaber* Latr. and *Oniscus asellus* L., with reference to the effect of temperature and body size. *Comp. Biochem. Physiol.*, **37**, 503–10.

Lockwood, A. P. M. (1960). Some effects of temperature and concentration of the medium on the ionic regulation of the isopod, *Asellus aquaticus* (L.). *J. Exp. Biol.*, **37**, 614–30.

Lockwood, A. P. M. (1962). The osmoregulation of Crustacea. *Biol. Rev.*, **37**, 257–305.

Lough, G. (1976). Larval dynamics of the Dungeness crab, *Cancer Magister*, off the central Oregon coast, 1970–71. *Fish. Bull.*, **74**, 1–23.

Lough, R. G. (1973). A re-evaluation of the combined effects of temperature and salinity on survival and growth of *Mytilus edulis* larvae using response surface techniques. *Proc. Natl Shellfish Assoc.*, **64**, 73–6.

Lough, R. G. (1975). A re-evaluation of the combined effects of temperature and salinity on survival and growth of bivalve larvae using response surface techniques. *US Natl Mar. Fish. Serv. Bull.*, **73**, 86–94.

Lough, R. G., and J. J. Gonor (1973a). A response-surface approach to the combined effects of temperature and salinity on the larval development of *Adula californiensis* (Pelecypoda: Mytilidae). I. Survival and growth of three and fifteen-day-old larvae. *Mar. Biol.*, **22**, 241–50.

Lough, R. G., and J. J. Gonor (1973b). A response-surface approach to the combined effects of temperature and salinity on the larval development of *Adula californiensis* (Pelecypoda: Mytilidae). II. Long-term larval survival and growth in relation to respiration. *Mar. Biol.*, **22**, 295–305.

Lowthion, D. (1974). The combined effects of high salinity and temperature on the survival of young *Limanda limanda*. *Mar. Biol.*, **25**, 169–75.

Lucu, C. (1973). Competitive role of calcium in sodium transport in *Carcinus mediterraneus* acclimated to low salinities. *Mar. Biol.*, **18**, 140–5.
Lucu, C., and O. Jelisavcic (1970). Uptake of ^{137}Cs in some marine animals in relation to temperature, salinity, weight and moulting. *Int. Rev. Sesamten Hydrobiol.*, **55**, 783–96.
Lynch, M. P., K. L. Webb, and W. A. van Engel (1973). Variations in serum constituents of the blue crab, *Callinectes sapidus*: chloride and osmotic concentrations. *Comp. Biochem. Physiol.*, **44**, 719–34.
Mackay, W. C. (1974). Effect of temperature on osmotic and ionic regulation in goldfish, *Carassius auratus*. *J. Comp. Physiol.*, **88**, 1–19.
Mackay, W. C., and D. D. Beatty (1968). The effect of temperature on renal function in the white sucker fish, *Catostomus* commersoni. *Comp. Biochem. Physiol.*, **26**, 235–45.
Maetz, J. (1972). Branchial Na$^+$ exchange and NH$_3$ excretion in the goldfish *Carassius auratus*. Effects of ammonia loading and temperature changes. *J. Exp. Biol.*, **56**, 601–20.
Maetz, J. (1973). Na$^+$/NH$_4^+$–Na$^+$/H$^+$ exchanges and NH$_3$ movement across the gill of *Carassius auratus*. *J. Exp. Biol.*, **58**, 255–75.
Malvin, R. L., E. Carlson, S. Legan, and P. Churchill (1970). Creatine reabsorption and renal function in the freshwater lamprey. *Am. J. Physiol.*, **218**, 1506–9.
Manzi, J. J. (1970). The combined effects of salinity and temperature on the feeding, reproductive and survival rates of *Eupleura caudata* (Say) and *Urosalpinx cinerea* (Say) (Prosobranchia: Muricidae). *Proc. Natl Shellfish Assoc.*, **60**, 7.
May, R. C. (1974). Effects of temperature and salinity on yolk utilization in *Bairdeilla icistra* (Jordan and Gilbert) (Pisces: Sciaenidae). *J. Exp. Mar. Biol. Ecol.*, **16**, 213–25.
Miller, D. A., M. L. Standish, and A. E. Thurman, Jr (1968). Effects of temperature on water and electrolyte balance in the frog. *Physiol., Zool.*, **41**, 500–6.
Mires, D., Y. Shak, and S. Shilo (1974). Further observations on the effect of salinity and temperature changes on *Mugil capito* and *M. cephalus* fry. *Bamidgen*, **26**, 104–9.
Motais, R., and J. Isaia (1972). Temperature-dependence of permeability to water and to sodium of the gill epithelium of the eel *Anguilla anguilla*. *J. Exp. Biol.*, **56**, 587–600.
Motais, R., J. Isaia, J. C. Rankin, and J. Maetz (1969). Adaptive changes of the water permeability of the teleostean gill epithelium in relation to external salinity. *J. Exp. Biol.*, **51**, 529–46.
Murphy, D. J., and S. K. Pierce (1975). The physiological basis for changes in the freezing tolerance of intertidal molluscs. I. Response to subfreezing temperatures and the influence of salinity and temperature acclimation. *J. Exp. Zool.*, **193**, 313–22.
Murphy, P. G., and A. H. Houston (1974). Environmental temperature and the body fluid system of the fresh-water teleost: V. Plasma electrolyte levels and branchial microsomal (Na$^+$–K$^+$) ATPase activity in thermally acclimated goldfish (*Carassius auratus*). *Comp. Biochem. Physiol.*, **47B**, 563–70.
Oelofsen, B. W. (1973). The influence of ambient temperature on the function of the nasal salt gland and the composition of cloacal fluid in the penguin *Spheniscus demersus*. *Zool. Afr.*, **8**, 63–74.
Oikari, A. (1975a). Hydromineral balance in some brackish-water teleosts after thermal acclimation, particularly at temperatures near zero. *Ann. Zool. Fenn.*, **12**, 215–29.
Oikari, A. (1975b). Seasonal changes in plasma and muscle hydromineral balance in

three Baltic teleosts, with special reference to the thermal response. *Ann. Zool. Fenn.*, **12**, 230–6.
Oikari, A., and R. Kristoffersson (1973). Plasma ionic and osmotic levels in *Myoxecephalus quadricornus* (L.) in brackish water during temperature acclimation, particularly to cold. *Ann. Zool. Fenn.*, **10**, 495–9.
Panikkar, N. K. (1940). Influence of temperature on osmotic behavior of some crustacea and its bearing on problems of animal distribution. *Nature*, **146**, 366–7.
Parsons, R. H., and Y.-T. Lau (1976). Frog skin osmotic permeability: Effect of acute temperature change *in vivo* and *in vitro*. *J. Comp. Physiol.*, **105B**, 207–17.
Payan, P., and A. J. Matty (1975). The characteristics of ammonia excretion by a perfused isolated head of trout, *Salmo gairdneri*: Effect of temperature and CO_2-free Ringer. *J. Comp. Physiol.*, **96B**, 167–84.
Pic, P., and J. Maetz (1975). Différences de potential transbranchial et flux ioniques chez *Mugil capito* adapté à l'eau de mer. Importance de l'ion Ca^{++}. *C.R. Acad. Sci.*, **280D**, 983–6.
Poljansky, G. I., and K. M. Sukhanova (1967). Some peculiarities in temperature adaptations of Protozoa as compared to multicellular poikilotherms. In A. S. Troshin (Ed.), *The Cell and Environmental Temperature*, Pergamon Press, Oxford. pp. 200–8.
Potts, W. T. W. (1954). Urine production in *Anodonta*. *J. Exp. Biol.*, **31**, 614–7.
Potts, W. T. W., and W. R. Fleming (1971). The effect of environmental calcium and ovine prolactin on sodium balance in *Fundulus kansae*. *J. Exp. Biol.*, **55**, 63–75.
Rao, K. P. (1968). Some biochemical mechanisms of low temperature acclimation in tropical poikilotherms. In A. S. Troshin (Ed.), *The Cell and Environmental Temperature*, Pergamon Press, Oxford. pp. 98–112.
Raymond, J. A., Y. Lin, and A. L. DeVries (1975). Glycoprotein and protein antifreezes in 2 Alaskan fishes. *J. Exp. Biol.*, **193**, 125–30.
Renfro, J. L., B. Schmidt-Nielsen, D. Miller, D. Benos, and J. Allen (1974). Methyl mercury and inorganic mercury: Uptake, distribution, and effect on osmoregulatory mechanisms in fishes. In F. J. Vernberg and W. B. Vernberg (Eds), *Pollution and Physiology of Marine Organisms*, Academic Press, New York. pp. 101–22.
Santerre, M. T. (1976). Effects of temperature and salinity on the eggs and early larvae of *Caranx mate* (Cuv. and Balenc.) (Pisces: Carangidae) in Hawaii. *J. Exp. Mar. Biol. Ecol.*, **21**, 51–68.
Sargent, J. R., A. J. Thomson, M. H. Dalgleish, and A. D. Dale (1975). Effects of temperature and salinity on the microsomal ($Na^+ + K^+$—dependent adenosine triphosphatase in the gills of the eel, *Anguilla anguilla* (L.). In H. Barnes (Ed.), *Proc. 9th Eur. Mar. Biol. Symp.*, Aberdeen University Press. pp. 463–474.
Schlichter, D. (1974). Der Einfluß physikalisher und chemischer Faktoren auf die Aufnahme in Meerwasser gelöster Aminosäuren durch Aktinien. *Mar. Biol.*, **25**, 279–90.
Schlieper, C. (1971). Physiology of brackish water. In A. Remane and C. Schlieper (Eds), *Biology of Brackish Water*, Wiley Interscience, New York. pp. 211–322.
Schlieper, C., H. Flügel, and H. Theede (1967). Experimental investigations of the cellular resistance ranges of marine temperate and tropical bivalves: Results of the Indian Ocean Expedition of the German Research Association. *Physiol. Zool.*, **40**, 345–61.
Schmidt-Nielsen, B., and R. P. Forester (1954). The effect of dehydration and low temperature on renal function in the bullfrog. *J. Cell. Comp. Physiol.*, **44**, 233–46.
Schoffeniels, E., and R. Gilles (1970). Osmoregulation in aquatic arthropods. In M. Florkin and B. T. Scheer (Eds), *Chemical Zoology*, Vol. V, Academic Press, New York. pp. 255–286.

Segal, E., and W. D. Burbanck (1963). Effects of salinity and temperature on osmoregulation in two latitudinally separated populations of an estuarine isopod, *Cyathura polita* (Stimpson). *Physiol. Zool.*, **36**, 250–63.
Shier, W. T., Y. Lin, and A. L. DeVries (1972). Structure and mode of action of glycoproteins from an Antarctic fish. *Biochim. Biophys. Acta*, **263**, 406–13.
Silverthorne, S. U. (1973). Respiration in eyestalkless *Uca* (Crustacea: Decapoda) acclimated to two temperatures. *Comp. Biochem. Physiol.*, **45A**, 417–20.
Smith, M. W. (1967). Influence of temperature acclimatization on the temperature-dependence and ouabain-sensitivity of goldfish intestinal adenosine triphosphatase. *Biochem. J.*, **105**, 65–71.
Smith, M. W., and J. C. Ellory (1971). Temperature-induced changes in sodium transport and Na^+/K^+—adenosine triphosphatase activity in the intestine of goldfish (*Carassius auratus* L.). *Comp. Biochem. Physiol.*, **39A**, 209–18.
Smith, R. I. (1955). Comparison of the level of chloride regulation by Nereis diversicolor in different parts of the geographic range. *Biol. Bull.*, **109**, 453–74.
Smith, R. I. (1957). A note on the tolerance of low salinities by nereid polychaetes and its relation to temperatures and reproductive habitat. *Ann. Biol.*, **33**, 93–107.
Snell, F. M., and C. P. Leeman (1957). Temperature coefficients of the sodium transport system of isolated frog skin. *Biochim. Biophys. Acta*, **25**, 311–20.
Spaargaren, D. H. (1971). Aspects of the osmotic regulation in the shrimps *Crangon crangon* and *Crangon allmanni*. *Neth. J. Sea Res.*, **5**, 275–335.
Spaargaren, D. H. (1972). Osmoregulation in the prawns *Palaemon serratus* and *Lysmata seticaudata* from the Bay of Naples. *Neth. J. Sea Res.*, **5**, 416–36.
Spaargaren, D. H. (1975). Changes in permeability in the shore crab, *Carcinus maenas*, as a response to salinity. *Comp. Biochem. Physiol.*, **51A**, 549–52.
Staaland, H. (1967). Temperature sensitivity of the avian salt gland. *Comp. Biochem. Physiol.*, **23**, 991–3.
Stanley, J. G., and P. J. Colby (1971). Effects of temperature on electrolytes balance and osmoregulation in the alewife (*Alosa pseudoharengus*) in fresh and sea water. *Trans. Am. Fish. Soc.*, **100**, 624–38.
Stephens, G. C., and N. S. Kerr (1962). Uptake of phenylalanine by Tetrahymena pyriforms. *Nature*, **194**, 1094.
Strawn, K., and J. E. Dunn (1967). Resistance of Texas salt- and fresh-water-marsh fishes to heat death at various salinities. *Texas J. Sci.*, **19**, 57–76.
Takenaka, T. (1963). Effects of temperature and metabolic inhibitors on the active sodium transport in frog skin. *Jap. J. Physiol.*, **13**, 208–18.
Tanaka, R., and A. Teruya (1973). Lipid dependence of activity-temperature relationship of (Na^+, K^+)—activated ATPase. *Biochim. Biophys. Acta*, **323**, 584–91.
Tay, K. L., and E. T. Garside (1975). Some embryogenic responses of mummichog, *Fundulus heteroclitus* (L.) (Cyprinodontidae), to continuous incubation in various combinations of temperature and salinity. *Can. J. Zool.*, **53**, 920–33.
Theede, H. (1965). Vergleichende experimentelle Untersuchungen über die zelluläre Gefrier-resistenz mariner Muscheln. *Kieler Meeresforsch.*, **31**, 153–66.
Theede, H. (1969). Experimentelle untersuchungen über physiologische Unterschiede bei Evertebraten und Fischen aus Meer- und Brachwasser. *Limnologica*, **7**, 119–28.
Theede, H. (1972). Vergleichende ökalogisch-physiologische Untersuchungen zur zellulären Kälteresistenz mariner Evertebraten. *Mar. Biol.*, **15**, 160–91.
Umminger, B. L. (1970). Effects of subzero temperatures and trawling stress on serum osmolarity in the winter flounder *Pseudopleuronectes americanus*. *Biol. Bull.*, **139**, 574–9.
Umminger, B. L. (1971a). Patterns of osmoregulation in freshwater fishes at temperatures near freezing. *Physiol. Zool.*, **44**, 20–7.

Umminger, B. L. (1971b). Chemical studies of cold death in the Gulf killifish, *Fundulus grandis*. *Comp. Biochem. Physiol.*, **39A**, 625-32.

Umminger, B. L. (1971c). Osmoregulatory role of serum glucose in freshwater-adapted killifish (*Fundulus heteroclitus*) at temperatures near freezing. *Comp. Biochem. Physiol.*, **38A**, 141-5.

Umminger, B. L. (1971d). Osmoregulatory overcompensation in the goldfish, *Carassius auratus*, at temperatures near freezing. *Copeia*, **1971**, 686-91.

Umminger, B. L., and D. H. Gist (1973). Effects of thermal acclimation on physiological responses to handling stress, cortisol and aldosterone injections in the goldfish, *Carassius auratus*. *Comp. Biochem. Physiol.*, **44A**, 967-77.

Umminger, B. L., and H. Kenkel (1975). Ionoregulatory role of prolactin in salt-water-adapted killifish at subzero temperatures. *Am. Zool.*, **15**, 796.

Van Winkle, W. (1968). The effects of season, temperature, and salinity on the oxygen consumption of bivalve gill tissue. *Comp. Biochem. Physiol.*, **26**, 69-80.

Vernberg, F. J., T. Calabrese, F. Thurberg, and W. B. Vernberg (1977). *Physiological Responses of Marine Biota to Pollutants*, Academic Press, New York. In press.

Vernberg, W. B., P. J. DeCoursey, and J. O'Hara (1974). Multiple environmental factor effects on physiology and behavior of the fiddler crab, *Uca pugilator*. In F. J. Vernberg and W. B. Vernberg (Eds), *Pollution and Physiology of Marine Organisms*, Academic Press, New York. pp. 381-425.

Vernberg, F. J., C. Schlieper, and D. E. Schneider (1963). The influence of temperature and salinity on ciliary activity of excised gill tissue of molluscs from North Carolina. *Comp. Biochem. Physiol.*, **8**, 271-85.

Vernberg, F. J., and W. B. Vernberg (1974). *Pollution and Physiology of Marine Organisms*, Academic Press, New York.

Vernberg, W. B., and F. J. Vernberg (1975). The physiological ecology of larval *Nassarius obsoletus* (Say). In H. Barnes (Ed.), *Proc. 9th Eur. Mar. Biol. Symp*, Aberdeen University Press, Aberdeen. pp. 179-90.

Weber, R. E., and D. H. Spaargaren (1970). On the influence of temperature on the osmoregulation of *Crangon crangon* and its significance under estuarine conditions. *Neth. J. Sea Res.*, **5**, 108-20.

Weber, R. E., and W. J. A. Van Marrewijk (1972). Free amino acids in the shrimp *Crangon crangon* and their osmoregulatory significance. *Neth. J. Sea Res.*, **5**, 391-415.

Williams, A. B. (1960). The influence of temperature on osmotic regulation in two species of estuarine shrimps (Penaeus). *Biol. Bull.*, **119**, 560-71.

Williams, R. J. (1970). Freezing tolerance in *Mytilus edulis*. *Comp. Biochem. Physiol.*, **35**, 145-61.

Wright, D. A. (1975). Sodium regulation in the larvae of *Chironomus dorsalis* and *Camptochironomus tentans*: the effect of salt depletion and some observations on temperature changes. *J. Exp. Biol.*, **62**, 121-39.

Zaugg, W. S., B. L. Adams, and L. R. McLain (1972). Steelhead migration: potential temperature effects as indicated by gill ATPase activities. *Science*, **176**, 415-6.

Zaugg, W. S., and L. R. McLain (1976). Influence of water temperature on gill sodium, potassium-stimulated ATPase activity in juvenile coho salmon (*Oncorhynchus kisutch*). *Comp. Biochem. Physiol.*, **54A**, 419-22.

Zaugg, W. S., and H. H. Wagner (1973). Gill ATPase activity related to parr-smolt transformation and migration in steelhead trout: influence of photoperiod and temperature. *Comp. Biochem. Physiol.*, **45B**, 955-65.

Chapter 12

Osmoregulation and Pollution of the Aquatic Medium

J. M. BOUQUEGNEAU AND R. GILLES

I. Introduction	563
II. Chlorinated Hydrocarbon Pesticides	564
III. Heavy Metals	566
A. Effects on osmoregulatory abilities	566
B. Mechanisms of action	571
C. Adaptive tolerance toward heavy metals intoxication	573
D. Conclusions	575
Acknowledgements	576
References	576

I. INTRODUCTION

The waters of several rivers, lakes and parts of the oceans (principally estuaries and coastal waters) have been heavily polluted for many years now. It is obvious that the study of the effects of toxic substances upon aquatic animals is essential to an understanding of the effects of pollution in aquatic ecosystems. In fact, while little is known about the effects of pollutants on these biotopes, a considerable amount of information has been published on their effects on individual organisms. These studies deal essentially with bioaccumulation, acute toxicity, enzyme activity, reproduction, teratology and some physiological functions such as respiration (see Waldichuk, 1974).

Considering their relative toxicity and biological effects, the most severe chemical pollutants can be divided into three groups: heavy metals, pesticides and polychlorinated biphenyls, oil and dispersants.

Works devoted to the effects of these three classes of compounds on osmoregulation are relatively recent and to our knowledge there are only very few data on the effects of oils and dispersants, except for the recent work by Anderson and Anderson (1975) who showed that the osmoregulatory abilities of the american oyster *Crassostrea virginica* is little affected by exposure to different oils.

This chapter will therefore essentially deal with the effects of polychlorinated hydrocarbon pesticides and heavy metals on osmoregulatory processes in aquatic animals.

II. CHLORINATED HYDROCARBON PESTICIDES

The current widespread presence of chlorinated insecticides, polychlorinated biphenyls and herbicides in world waterways has elicited much interest in the mechanism of their toxicity. The principal site of action of pesticides and more particularly of DDT in the vertebrate organism is the central nervous system (Aubin and Johansen, 1969) and there is a growing body of evidence that they act through interference with nerve transmission (Shanes, 1950; Narahashi and Haas, 1967). That interference occurs as a result of alterations in ion transport across the excitable membrane (Matsumura and O'Brien, 1966). Apart from this the nature of the pesticide-sensitive site on the nerve membrane remains unknown, but recently it has been reported by several authors that various chlorinated hydrocarbon insecticides inhibit nerve ATPases *in vitro*, and for example, Matsumura and Narahashi (1971) found a correlation between the degree of DDT inhibition of ATPases of the lobster nerve and the electrophysiological symptoms of DDT poisoning. The sensibility of animal tissue ATPase activities to chlorinated hydrocarbon pesticides is now well documented.

Koch (1969a) first reported observation of ATPase inhibition by chlorinated hydrocarbon insecticides in a subcellular particle fraction from rabbit brain homogenates. He and his coworkers extended this observation to inhibition of ATPases in insect tissue homogenate fractions and showed that DDT had little or no effect on the Na–K ATPase of cockroach nerve cord or honey bee brain but that chlordane did inhibit the activity of this enzyme (Koch, Cutkomp and Do, 1969). Koch (1969b) has also tested four chlorinated hydrocarbon pesticides (chlordane, lindane, DDT and dicofal (Kelthane)) on tissue homogenates from rabbits, chickens, lake trouts and cockroaches. The four pesticides inhibited both the $Na^+–K^+$ ATPase and Mg^{2+} ATPase enzyme systems, the Mg^{2+} ATPase being more sensitive than the $Na^+–K^+$ ATPase. DDT was the most effective inhibitor of the Mg^{2+} ATPase at low insecticide concentrations (1 ppm or less). At higher concentrations, Kelthane produced the greatest inhibition of both ATPase activities. Moreover, red blood cell ghosts differed from other tissues in that their $Na^+–K^+$ ATPase activity was insensitive to the insecticides.

Matsumura, Bratkowski and Patil (1969) and Matsumura and Patil (1969) found opposite results in numerous tissues from different sources and reported that DDT was a potent inhibitor of $Na^+–K^+$ ATPase in rat tissue preparations.

Recently, many other studies have shown that DDT and related pesticides as well as the polychlorinated biphenyls inhibit the activity of Na–K and Mg

ATPases in mammal, turtle and fish tissues (Cutkomp and coworkers, 1971; Yap and coworkers, 1971; Desaiah and coworkers, 1972; Desaiah and Koch, 1975; Witherspoon and Wells, 1975; Jackson and Gardner, 1973; Wells, Phillips and Murphy, 1974; Davis and Wedemeyer, 1971a, 1971b; etc.).

Payne, Herzberg and Howland (1973) inversely, present data indicating that low concentrations of aldrin, allethrin and DDT stimulate mouse mitochondrial ATPase in a manner similar to that of uncouplers of oxidative ATP phosphorylation.

Koch and coworkers (1972) have studied, *in vivo*, the effect of a long term exposure to Aroclors 1242 and 1254 on the ATPase activity in the fish, *Pimephales promelas*. Both inhibition and stimulation responses were observed depending on the tissue tested. Such effects were also observed during *in vitro* experiments (Yap and coworkers, 1971; Desaiah and coworkers, 1972). Although some differences exist between the *in vivo* and the *in vitro* results, the authors think there appears to be sufficient evidence in their studies to indicate that ATPase activity of rainbow trout.

Inversely, Davis, Friedhoff and Wedemeyer (1972) failed to demonstrate inhibition of ATPases following oral administration of DDT to rainbow trouts. However, Campbell, Leadem and Johnson (1974) recently demonstrated the inhibitory effects of p, p'DDT on branchial and renal Na^+-K^+ ATPase enzyme. There is little information on the mechanism of toxic action of the organochloric pesticides, the ATPase system has been proposed as a possible target site for these compounds.

Wells, Phillips and Murphy (1974) interpret the observed effect in relation to the ability of these compounds to alter the cellular membrane configuration by binding with the fat portion of the membrane. Since ATPase is a structural part of the membrane, its active site would be altered concomitantly.

Owing to the supposed importance of the glycoside-sensitive Na^+-K^+ ATPase in hydromineral regulation, it has been postulated that ATPase inhibition by pesticides may cause an alteration of ion transport processes in the cell membrane and then impair the osmoregulatory ability of various aquatic animals.

Friend, Haegele and Wilson (1973) studied the interference of DDE with extrarenal salt excretion in the mallard *Anas platyrhynchos*. DDE had no effect on the amount of sodium chloride excreted by birds previously maintained in salt-water, but markedly reduced excretion in fresh-water birds. Studies on DDE-fed sea birds (ducks, guillemots and puffins) however reveal that the DDE-induced osmoregulatory failure is apparently not the primary cause of the death of the intoxicated sea birds (Miller and coworkers, 1976). Nimmo and Blackman (1972) demonstrated that when living shrimps (*Penacus aztecus* and *P. duorarum*) were esposed to DDT, concentrations of some cations in the hepatopancreas became depressed.

DDT also induces a decrease in the absorption of water by the intestine in the eel *Anguilla rostrata* adapted to sea-water. This decrease appears to be correlated with the inhibition of the Na^+-K^+ ATPase that induces the pesticide in the intestine (Janicki and Kinter, 1971a, 1971b).

Nevertheless, these authors recognize the difficulty of demonstrating any inhibitory effects of DDT at the concentrations of this pollutant normally found in nature in polluted areas.

Kinter and coworkers (1972) have also shown the disruption of osmoregulation in the eel (*Anguilla rostrata*) and in the killifish (*Fundulus heteroclitus*) exposed to DDT and to polychlorinated biphenyls. These compounds increased plasma osmolarity while gut and gill ATPase activity is decreased (Eisler and Edmunds, 1966).

Endrin has also been observed to disrupt osmoregulation both in freshwater (Grant and Mehrle, 1970) and in a marine teleost. In the marine puffer (*Sphaeroides maculatus*) exposed to endrin, there is an increase in the blood concentrations of sodium, potassium and calcium. The content of these ions is decreased in the liver.

Thus organochlorine pollutants appear to disrupt osmoregulatory functions in various aquatic species. However, this effect does not seem to be the primary cause of the death which usually follows intoxication by chlorinated pesticides. As a matter of fact, studies on the short-term effects of DDT on osmoregulation in the marine teleost black surfperch (*Embiotoca jacksoni*) lead Waggoner and Zeeman (1975) to conclude that direct action of DDT on the nervous system seem a more likely cause of death than the disruption of osmoregulation which does not appear to the authors to be of sufficient magnitude.

However, Caldwell (1974) has shown that crabs (*Hemigrammus nudus*) exposed to metoxychlor at concentrations which are sublethal in acute tests exhibit a decreased resistance in tolerance to reduced salinities. Moreover he has shown that these exposures produce a partial inhibition of the gill Na^+-K^+ and Mg^{2+} ATPases in *Cancer magister* but has failed to demonstrate that osmoregulation was impaired by these treatments.

III. HEAVY METALS

A. Effects on Osmoregulatory Abilities

In fishes and crustaceans, the important surface of the gill tissue constitutes a main avenue for exchange with the environmental medium. It can thus be assumed that this tissue is one of the first which may have to suffer from toxicity of heavy metals (Backstrom, 1967; Bouquegneau, 1973a; Olson, Bergman and Fromm, 1973; etc.).

In the case of zinc, which is known to produce an inhibition of respiration involving suffocation and death of aquatic animals (Skidmore, 1964), Skidmore (1970) has studied both respiration and osmoregulation of rainbow trouts *Salmo gairdneri* with gills damaged by zinc sulphate. The oxygen utilization decreased seven-fold. However, the blood osmolarity as well as the blood concentrations of sodium, potassium, calcium and magnesium were largely unaffected by this pollutant. These results suggest that epithelial damage caused by zinc sulphate affect the gill respiratory processes but not their osmoregulation mechanisms. This is at variance with the results of Lewis and Lewis (1971) who showed that zinc (8–30 ppm as $ZnSO_4$) and copper (2.5–5 ppm as $CuSO_4$) decreased the osmolarity of blood serum of the channel catfish *Ictalurus punctatus* and of the golden shiner *Notemigonus crysoleucas*.

Recently, Strick and coworkers (1975) have observed a decrease of the Na^+ concentration but no significant variation of the K^+ level of the blood of rainbow trouts intoxicated with lethal doses of chromium. These authors also observed a higher incidence of swollen salt cells and marginal haemorrhages in gill lamellae as well as a severe necrosis of kidney tubules. Such injuries to the gills after exposure to heavy metals has also been described in the cases of copper, cadmium and mercury (Baker, 1969; Lindahl and Hell, 1971; Bouquegneau, 1975; Bubel, 1976).

Larsson (1975) has studied the effects of cadmium on plasma electrolytes in the flounder *Pleuronectes flesus* in short-term experiments. Cadmium had only slight effects on the major plasma electrolytes sodium and chloride. Similar results can be obtained with the european eel *Anguilla anguilla* (Bouquegneau and Noël-Lambot, unpublished results). Larsson (1975) however showed that the levels of other plasma ions are greatly modified after cadmium exposure: potassium and calcium decreased whereas the contents of Mg^{2+} and inorganic phosphate were significantly elevated. This indicates that exposure to high cadmium concentrations could bring about in teleosts severe injuries at the kidney level since this organ is known to be implicated in the regulation of the Ca^{2+} and Mg^{2+} blood levels.

Using sublethal doses of cadmium for 60 days in the winter flounder *Pseudopleuronectes americanus*, Calabresse and coworkers (1975) found a lowering of the rate of respiration but no significant haematological difference between controls and cadmium-exposed fish.

Meanwhile, McCarty and Houston (1976) showed that goldfishes (*Carassius auratus*) exposed to sublethal levels of cadmium (380 μg Cd^{2+} l^{-1}) for 50 days exhibit significant deviations in plasma sodium and chloride levels as well as in tissue sodium and water content.

Thurberg, Dawson and Collier (1973) have studied the effects of cadmium and copper on osmoregulation and oxygen consumption in two estuarine crabs: the green crab *Carcinus maenas* and the rock crab *Cancer irroratus*. After 48 hours exposure, the copper-exposed crabs exhibit loss of osmoregulatory function with increasing copper concentration (0.3 to 40 ppm) until normally hyperosmotic serum became isosmotic with the surrounding medium. Cadmium (from 0.12 to 8 ppm) elevated green crab serum above its normal hyperosmotic state. On the other hand, cadmium depresses gill-tissue oxygen consumption in both crab species, while copper had no effect.

The fresh-water snail *Lymnaea stagnalis* living in water containing 5×10^{-5} mol l^{-1} $CuSO_4$ also shows a breakdown of osmoregulatory function which is characterized by loss of NaCl and increase of water content in the kidney (Spronk and coworkers, 1971). The Na^+ losses are explained by these authors by an inhibition of active Na reabsorption.

McKim, Christensen and Hunt (1970) have observed changes in the blood of the brook trout *Salvelinus fontinalis* after exposure to copper. When fishes are exposed to copper concentrations of about 40 μg l^{-1} for 6 or 21 days, both the osmolarity and the plasma Cl^- content decrease. However after a longer exposure (337 days to copper concentrations of less than 32.5 μg l^{-1}), no more changes are observed in the blood ionic content. The disappearance of initial blood changes after extended exposure suggests the transient nature of these early responses.

Effects on osmoregulatory abilities have also been described following intoxication with mercury salts.

Exposure of rainbow trouts *Salmo gairdneri* to 10 ppb methylmercury for 4 to 12 weeks has no significant effects on plasma electrolytes concentration while it induces changes in gill metabolism (O'Connor and Fromm, 1975). In the same way, long-term intoxications with 50 ppb $HgCl_2$ do not alter the blood Cl^- content of the porcelain crab *Petrolisthes armatus* either acclimated or adjusting to different salinities (Roesijadi and coworkers, 1974). It would thus seem that mercury salts do not affect the osmoregulatory processes at work in these species. This is at variance with the results obtained by Renfro and coworkers (1974) who showed that the ability of sodium-depleted killifish *Fundulus heteroclitus* to take up Na^+ is inhibited following exposure to either methylmercury (5×10^{-6} mol l^{-1}) or $HgCl_2$ (4.6×10^{-6} mol l^{-1}). The effect of methylmercury appears to be easily reversible since it disappears within 3 hours after placing the intoxicated fishes in a medium which is free of the pollutant. On the contrary, Na^+ uptake would appear to be irreversibly blocked by $HgCl_2$. Such differences in effect are probably due to different properties of the inorganic and the organic mercury salts. Methylmercury, which is a lipophilic compound, could be rapidly washed out of the fishes while detoxication of fishes exposed to $HgCl_2$ requires a much longer period of time.

Bouquegneau (1973b, 1975) has compared the effects of mercury salts on osmoregulatory abilities of sea-water-adapted european eels *Anguilla anguilla*. The intoxication experiments have been carried out using various doses (10, 1, 0.1 and 0.01 ppm Hg in sea-water) of $HgCl_2$ or CH_3HgCl. 10 ppm Hg^{2+} proved to be a lethal concentration but 0.1 ppm had only a sublethal effect on eel.

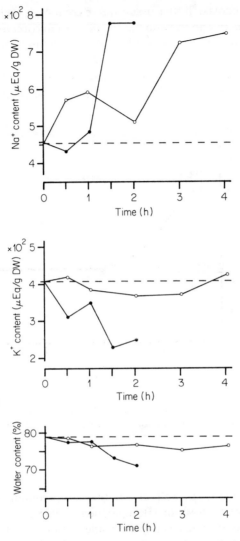

Figure 1 Changes in the Na$^+$, K$^+$ and water contents of the gills of sea-water acclimated eels intoxicated with 10 ppm Hg^{2+} (○, HgCl$_2$; ●, CH$_3$HgCl). The water content is given in % of the tissue wet weight

During short-term exposure of eels to HgCl$_2$, mercury is accumulating chiefly in the gills. Indeed, this tissue already contains more than 20 ppm Hg^{2+} after only a few hours of intoxication by 10 ppm mercury. Moreover, exposure to 10 ppm of Hg^{2+} salts causes important changes in the ionic content of the gill tissue (Fig. 1). The results are however rather different depending on whether inorganic or organic mercury salts are used. In the case of HgCl$_2$, there is a slight decrease in the tissue water content, a high and progressive increase in sodium, but no variation in the potassium content of the tissue. On the contrary, when CH_3HgCl is used, a larger decrease in water content can be observed. There is also a significant decrease in K^+, and a large increase in the Na^+ tissue content. The variations of chloride are quite similar to those of sodium. Identical observations have been reported on the plasma ionic levels

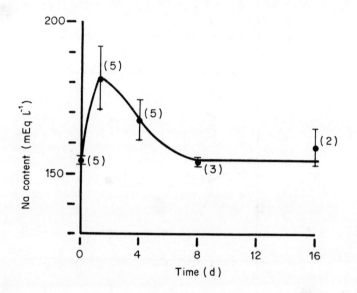

Figure 2 Na content of $HgCl_2$ intoxicated eels plasma. Eels are intoxicated with 0.1 ppm Hg. The error bars represent 2 × standard error

(Bouquegneau, 1973b) as well as on the ionic content of isolated gills incubated in polluted sea-water (Bouquegneau, 1975).

On the other hand, when using the sublethal dose of 0.1 ppm Hg, such effects on the gill ionic content become reversible within 8 days of intoxication. This is shown in the case of Na^+ in Fig. 2. Such results as well as those described by McKim, Christensen and Hunt (1970) clearly demonstrate the adaptive tolerance that can develop in fishes toward intoxication by salts of various heavy metals.

The results described up to now raise two important questions:

(1) What are the mechanisms responsible for the effects of heavy metals on the osmoregulatory abilities of various aquatic animals?
(2) What are the mechanisms implicated in the adaptive tolerance towards intoxication by heavy metals?

B. Mechanisms of Action

It seems that physiological effects of heavy metals can first be detected at the cell membrane level (Rothstein, 1960). Indeed they appear to alter its permeability characteristics. Heavy metals can also act upon enzymatic systems. Indeed, it is well known that metals can combine with enzymes in many ways, among which are sulphhydryl binding, chelation and salt formation. The binding of the metal on the enzyme molecule may influence activity in ways, ranging from activation to complete inhibition (Vallee and Ulmer, 1972; Jackim, 1974).

When dealing with effects of heavy metals on enzyme activity, care should be taken however before relating these effects to physiological modifications observed during intoxication. Heavy metals may indeed have a completely different effect on enzyme activity depending on whether they are assayed *in vitro* or *in vivo*. The work of Jackim (1974) gives a good illustration of this problem.

Jackim has studied the effects of a direct addition of toxic metal salt on the activity of enzyme preparations from unexposed fishes (usually *Fundulus heroclitus*). Then he exposed fishes to a toxic metal and measured the activity of the same enzymes extracted from organs of the intoxicated animals. His results show no consistent relationships between the direct *in vitro* effects of the metal on enzymes, which are usually inhibitory, and the effect of exposing the whole animal to the same pollutant. Moreover, the time of exposure can also influence the effect of the metal on the enzyme activity *in vivo*. Jackim thinks that such responses are the cause of many contradictory reports in the literature.

When considering more specifically osmoregulatory processes, the effects of heavy metals can *a priori* result from two types of action:

(1) An effect on the passive movements of ions. This should imply an effect of the heavy metals on the cell membrane permeability characteristics.
(2) An effect on the active transport processes. This implies either a direct effect of the pollutant on the carrier of the considered ion or a secondary effect due to the perturbation of another mechanism on which the transport process is dependent. Such secondary effects may for instance be related to hormonal actions or effects on the energy supply of the active transport system.

It does not seem that the modification in osmoregulatory abilities due to pollutants could be related to changes in the hormonal status of the animal. As a matter of fact, the effects of $HgCl_2$ and CH_3HgCl on the ionic levels in *Anguilla anguilla* can also be observed on gills isolated from non-intoxicated fishes after incubation in the presence of one of these two mercury salts (Bouquegneau, 1975).

On the other hand, incubation of isolated gills from *Anguilla anguilla* in

10 ppm methylmercury induces a necrosis of the tissue characterized by the formation of 'bag-like' structures which is concomitant to the apparition of a K^+ leakage (Bouquegneau, 1975). Lindahl and Hell (1970) also reported formation of 'bag-like' structures in the gills of *Leuciscus rutilus* upon short-term exposure to phenylmercuric hydroxide. It would thus seem that at least part of the modifications recorded in the blood concentration of various ions could be accounted for by changes in gill membrane structure, which would induce modification in ion permeability.

Renfro and coworkers (1974) have studied the effect of mercurials on Na^+ transport in the isolated bladder of *Pseudopleuronectes americanus*. Both $HgCl_2$ and CH_3HgCl appear to induce an inhibition of Na^+ active reabsorption. Such an effect can be correlated with a decrease in the Na^+–K^+ ATPase activity of the bladder. In the same way, the rupture of the ionic balance observed in *Anguilla anguilla* after exposure to $HgCl_2$ can be correlated with the inhibition of the activity of gill Na^+–K^+ ATPase (Bouquegneau, 1975).

Heavy metals are known to be potent inhibitors of ATPases. As shown by Skou (1963) this enzyme bears active –SH groups and therefore it can be expected to lose its activity in the presence of heavy metals. Many authors have now pointed out the sensitivity of these enzymes to heavy metals in various animal tissues (Jackim, 1974; Hasan, Vihko and Hernberg, 1967; Nechay, 1973; Renfro and coworkers, 1974).

Up to now studies on the physiological causes of the effects of heavy metals on the osmoregulatory abilities of various aquatic species have been only scanty. More information concerning the effects of the pollutants on ion fluxes in isolated tissues and in whole animals would obviously be needed before definitive conclusions could be drawn. It seems however from the few data available that heavy metals could act on both the passive and the active mechanisms of ion transport.

The studies undertaken to date have been essentially concerned with blood ionic concentration in some euryhaline fishes and crustaceans. Systematic studies on the effects of pollutants on the activity of the various structures implicated in osmoregulation, not only in fishes and crustaceans but also in the other zoological groups having aquatic representatives, would be of interest in order to draw a clearer picture of the ecological incidence of pollution of the aquatic environment by heavy metals.

Also of interest from the ecological standpoint would be the study of the synergistic action of heavy metals when combined with stressing conditions of another nature. It is indeed reasonable to assume that aquatic species in their natural environments will be more susceptible to heavy metals under stressful than under otherwise optimal conditions. For instance, the studies by Vernberg and Vernberg (1972) or Jones (1975) are illustrations of this problem with respect to temperature and salinity stresses.

Jones indeed showed that heavy metals appear more toxic to estuarine species of isopods which have to withstand osmoregulatory stresses than to marine species which have no osmoregulatory problems, so that marine isopods, living in their normal salinities are less susceptible to heavy metals than are estuarine species living in salinities near the lower limit of their normal salinity range. To our knowledge, nothing is known about the ecological implications of the relative toxicity of heavy metals on aquatic species withstanding other stressful conditions such as migration, development in media of different salinities and so on.

In this respect, it is worth noticing that pollution by heavy metals results in a delay in arrival and reduction in numbers of early-run salmon which constitute most if not all of the stock which reaches the headwaters of rivers (Saunders and Sprague, 1967). Whether this phenomenon is due to a synergistic effect of the pollutant and the osmotic stress that has to withstand the migrating fishes is not known however.

C. Adaptive Tolerance towards Heavy Metal Intoxication

As already stated in a previous part of this review (see p. 567) several studies point to the existence of mechanisms enabling aquatic species to deal with relatively high doses of heavy metals.

For instance, european eels (*Anguilla anguilla*) intoxicated with 0.1 ppm $HgCl_2$ show an increase in Na^+ blood content which is maximum after 2 days of exposure. However the Na^+ control concentration resumes control values after 8 days of intoxication (see Fig. 2).

Fishes gradually intoxicated with low sublethal doses of the pollutant can thus easily withstand doses which would prove lethal for fishes not previously intoxicated.

The mechanism implicated in such an adaptive tolerance to heavy metals could involve binding of the pollutant in a non-toxic form.

Metallothioneins are low molecular weight proteins mainly characterized by their high cysteine content and by the virtual absence of aromatic amino acids residues. It has been suggested that these polypeptides are implicated in various mammals in the binding of heavy metals in a non-toxic form (Weser, Donay and Rupp, 1973). This has been clearly demonstrated recently in the case of chicks intoxicated with Hg^{2+} by Suda and coworkers (1974). In the case of fishes, results of an intoxication experiment on possible binding of pollutant in a non-toxic form are given in Fig. 3.

Firstly, it can be seen that most of the mercury present in the acutely intoxicated animals is found in insoluble material. This is at variance with the situation encountered in chronically intoxicated animals where the major part of the pollutant is found in soluble fractions during the extraction procedure. Most of this mercury appears after chromatography on Sephadex G-75 in an elution fraction comprising between 630 and 810 ml. Hg is virtually absent from that fraction in acutely intoxicated fishes. Further studies on this fraction showed that the mercury is bound to a proteic compound which presents an

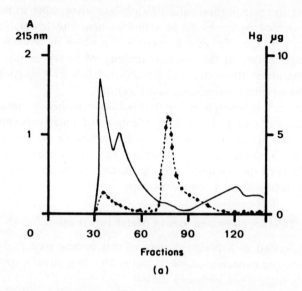

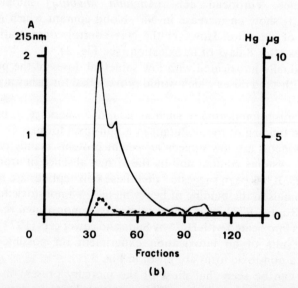

Figure 3 Elution profiles on Sephadex G-75 columns (5 × 50 cm) of the gill extracts prepared from 1 g gill tissue of (a) chronically and (b) acutely intoxicated eels. Hg^{2+} concentration is expressed in $\mu g/9$ ml fractions. From Bouquegneau, Gerday and Disteche (1975); reproduced by permission of Federation of European Biochemical Societies

amino acid pattern quite similar to that described for the so-called metallothioneins found in mammals. It thus appears that metallothioneins are implicated in fishes as in mammals in a process of detoxification of the mercury salts. Moreover they seem to be involved in the mechanism of adaptive tolerance we have described in these organisms.

Metallothioneins have been found recently in a variety of aquatic species (MacLean and coworkers, 1972; Coombs, 1975; Bouquegneau, Gerday and Disteche, 1975; Noël-Lambot, 1976). It may thus be that the role these proteins play in the adaptive tolerance shown to date only in a few species is of general occurrence in animals.

On the other hand, it is worth noticing that in the experiments reported in Fig. 3, no metallothioneins are found in the acutely intoxicated fishes. Although the amount of pollutant used is fairly high. This points to the adaptive nature of the activity of the mechanism implicated in the control of the level of these proteins in eel gills. Moreover, since no proteins are found in the animals after 5 hours of intoxication, it seems that the turnover rate of the metallothioneins is relatively slow. From preliminary experiments done in this laboratory, the half-life of the protein appears to be expressed in days instead of hours. Such a low turnover rate is however not incompatible with a regulation of the protein concentration exerted at the level of its synthesis and/or degradation. Further studies are needed to shed more light on this interesting problem.

Another possible protective agent against toxicity of mercury is the presence of selenium in the organisms. It has recently been shown that in livers of marine mammals mercury and selenium occurred together (Koeman and coworkers, 1973) and that their concentration ratio on a molar basis was found to be 1/1. It has been shown that tuna has a relatively high content of selenium and tends to accumulate additional selenium when mercury is present. Moreover, an amount of selenium comparable to that found in tuna, decreases methylmercury toxicity in rats when added to their diet (Ganther and coworkers, 1972).

D. Conclusions

Heavy metals can affect the osmoregulatory abilities of aquatic organisms. Such effects depend greatly upon several factors such as the nature of the heavy metal or the type of salt used, its concentration, the length of the intoxication period, the species studied or the synergistic effects of salinity and other stress factors.

In most studies, high concentrations of toxicants are used in order to produce symptoms of poisoning within relatively short periods of time. The data described above suggest the danger of such an approach since there is no evident correlation between the concentration of the pollutant in the organ and the effect produced. Whether the metal has been accumulated slowly or quickly makes an enormous difference and, at this point in time, it is unlikely that osmoregulatory malfunctioning may be significant at the environmental level except perhaps at the level of larvae or fragile species of the estuaries owing to the strong synergistic effect of salinity. Moreover, one should keep in mind that in laboratory experiments, the heavy metal is added in sea-water containing few or no suspended matter which, in nature, adsorbs a large part of the pollutant.

On the other hand, attention must be drawn to the fact that, depending on the animal studied, the effect of heavy metals can be totally different. For example, it seems that acute doses of mercury disturb respiration first in fresh-water fishes (Lindahl and Hell, 1970), but osmoregulation first in sea-water fishes, so that these fishes apparently die before any sign of anoxia can be detected (Bouquegneau, 1975).

The nature of the metal compound used is also very important: the effects on ionic regulation of sea-water-adapted eels are different whether organic bound or inorganic mercury is used.

Finally, the resistance to toxics shown by many aquatic organisms, as shown in the cases of copper and mercury, emphasizes the danger of the possible accumulation at sublethal doses, which may lead to toxic effects at higher trophic levels. However this could well not be the case since intoxication through alimentation is not always of great significance (cf. Bouquegneau, Noël-Lambot and Disteche, 1976).

Because adapted animals can contain high concentrations of pollutant, we can further conclude that the question of the existence of thresholds beneath which a pollutant can be considered non-toxic is very debatable indeed.

ACKNOWLEDGEMENTS

Part of the work presented in this paper has been aided by grant No. 2.4511.76 from the FRFC to Professors R. Gilles and A. Disteche.

Part of the work was carried out while participating in the concerted Belgian Universities action programme in Oceanology.

REFERENCES

Anderson, R. D., and J. W. Anderson (1975). Effects of salinity and selected petroleum hydrocarbons on the osmotic and chloride regulation of the american oyster, *Crassostrea virginica*. *Physiol. Zool.*, **48**, 420–30.
Aubin, A. E., and P. H. Johansen (1969). The effects of an acute DDT exposure on the spontaneous electrical activity of goldfish cerebellum. *Can. J. Zool.*, **47**, 163–6.
Backstrom, J. (1967). Distribution of mercury compounds in fish and birds. *Oïkos*, Suppl. No. 9, 30–1.
Baker, J. T. P. (1969). Histological and electron microscopical observations on copper poisoning in the winter flounder (*Pseudopleuronectes americanus*) *J. Fish. Res. Bd Can.*, **26**, 2785–93.

Bouquegneau, J. M. (1973a). Etude de l'intoxication par le mercure d'un poisson téléostéen *Anguilla anguilla*. I. Accumulation du mercure dans les organes. *Bull. Soc. R. Sc. Lg.*, **9–10**, 440–6.

Bouquegneau, J. M. (1973b). Etude de l'intoxication par le mercure d'un poisson téléostéen *Anguilla anguilla*. II. Effet sur l'osmorégulation. *Bull. Soc. R. Sc. Lg.*, **9–10**, 447–55.

Bouquegneau, J. M. (1975). L'accumulation du mercure et ses effets physiologiques chez *Anguilla anguilla* et *Myoxocephalus scorpius*. *Thèse de Doctorat en Sciences Zoologiques*, Université de Liège, Belgium.

Bouquegneau, J. M., Ch. Gerday, and A. Disteche (1975). Fish mercury-binding thionein related to adaptation mechanisms. *FEBS Lett.*, **55**, 173–7.

Bouquegneau, J. M., F. Noël-Lambot, and A. Disteche (1976). Le Problème de l'intoxication directe et indirecte par les métaux lourds. In J. C. J. Nihoul and A. Disteche (Eds), *Programme National de Recherche et de Développement—Environnement eau—Projet Mer—Rapport final*, Vol. 9, *Contamination de produits de la mer*, Services du Premier Ministre, Programmation de la Politique Scientifique, Bruxelles. pp. 266–92.

Bubel, A. (1976). Histological and Electron Microscopical Observations on the effects of different salinities and heavy metals ions on the gills of *Jaera nordmanni* (Rathke) (Crustacea, Isopoda). *Cell. Tiss. Res.*, **167**, 65–95.

Calabresse, A., F. P. Thurberg, M. A. Dawson, and D. R. Wenzloff (1975). Sublethal physiological stress induced by Cadmium and Mercury in the winter flounder, *Pseudopleuronectes americanus*. In J. H. Koeman and J. J. T. W. A. Strik (Eds), *Sublethal Effects of Toxic Chemicals in Aquatic Animals*, Elsevier, Amsterdam. pp. 15—21.

Caldwell, R. S. (1974). Osmotic and ionic regulation in decapod crustacea exposed to methoxychlor. In F. J. Vernberg and W. B. Vernberg (Eds), *Pollution and Physiology of Marine Organisms*, Academic Press, New York. pp. 197–223.

Campbell, R. D., T. P. Leadem, and D. W. Johnson (1974). The *in vivo* effect of p, p'DDT on Na^+-K^+-activated ATPase activity in rainbow trout (*Salmo gairdneri*). *Bull. Environ. Contamin. Toxicol.*, **11**, 425–8.

Clarckson, T. W. (1972). Recent advances in the toxicology of mercury with emphasis on the alkyl mercurials. *Critical Reviews of Toxicology*, Vol. 1, No. 2, Chemical Rubber, Cleveland, Ohio. pp. 203–34.

Coombs, T. L. (1975). The significance of multielement analyses in metal pollution studies. In A. D. McIntyre and C. F. Mills (Eds), *Ecological Toxicology Research. Effects of Heavy Metal and Organohalogen Compounds*, Plenum Press, New York, London. pp. 187–95.

Cutkomp, L. K., H. H. Yap, E. Y. Cheng, and R. B. Koch (1971). ATPase activity in fish tissue homogenates and inhibitory effects of DDT and related compounds. *Chem. Biol. Interactions*, **3**, 439–47.

Davis, P. W., and G. A. Wedemeyer (1971a). Na^+, K^+-activated-ATPase inhibition in rainbow trout: a site for organochlorine pesticide toxicity? *Comp. Biochem. Physiol.*, **40B**, 823–7.

Davis, P. W., and G. A. Wedemeyer (1971b). Inhibition by organochlorine pesticides of Na^+, K^+-activated adenosinetriphosphatase activity in the brain of rainbow trout. *Proc. West. Pharmacol. Soc.*, **14**, 47.

Davis, P. W., J. M. Friedhoff, and G. A. Wedemeyer (1972). Organochlorine insecticide, herbicide and polychlorinated biphenyl (PCB). Inhibition of NaK-ATPase in rainbow trout. *Bull. Environ. Contamin. Toxicol.*, **8**, 69–72.

Desaiah, D., and R. B. Koch (1975). Inhibition of ATPases activity in channel catfish brain by kepone and its reduction product. *Bull. Environ. Contamin. Toxicol.*, **13**, 153–8.

Desaiah, D., L. K. Cutkomp, H. H. Yap, and R. B. Koch (1972). Inhibition of oligomycin-sensitive and insensitive magnesium adenosine triphosphatase activity in fish by polychlorinated biphenyls. *Biochem. Pharmacol.*, **21**, 857–65.

Eisler, R., and P. Edmunds (1966). Effects of endrin on blood and tissue chemistry of a marine fish. *Trans. Am. Fish Soc.*, **95**, 153–9.

Friend, M., M. A. Haegele, and R. Wilson (1973). DDE: interference with extrarenal salt excretion in the mallard. *Bull. Environ. Contamin. Toxicol.*, **9**, 49–53.

Ganther, H. E., C. Goudie, M. L. Sunde, M. J. Kopecky, P. Wagner, O. H. Sang-Hwan, and W. G. Hoekstra (1972). Selenium: relation to decreased toxicity of methylmercury added to diets containing tuna. *Science*, **175**, 1122–4.

Grant, B. F., and P. M. Mehrle (1970). Chronic endrin poisoning in goldfish, *Carassius auratus*. *J. Fish. Res. Bd Can.*, **27**, 2225–32.

Hasan, J., V. Vihko, and S. Hernberg (1967). Deficient red cell membrane ($Na^+ + K^+$)-ATPase in lead poisoning. *Arch. Environ. Health*, **14**, 313–8.

Jackim, E. (1974). Enzyme responses to metals in fish. In F. J. Vernberg and W. B. Vernberg (Eds), *Pollution and Physiology of Marine Organisms*, Academic Press, New York. pp. 59–65.

Jackson, D. A., and D. R. Gardner (1973). The effects of some organochlorine pesticide analogs on salmonid brain ATPases. *Pestic. Biochem. Physiol.*, **2**, 377–82.

Janicki, R. M., and W. B. Kinter (1971a). DDT inhibits Na^+, K^+, Mg^{2+}-ATPase in the intestinal mucosa and gills of marine teleosts. *Nature New Biol.*, **233**, 148–9.

Janicki, R. H., and W. B. Kinter (1971b). DDT: disrupted osmoregulatory events in the intestine of the eel *Anguilla rostrata* adapted to seawater. *Science*, **173**, 1146–8.

Jones, M. B. (1975). Synergistic effects of salinity, temperature and heavy metals on mortality and osmoregulation in marine and estuarine isopods (Crustacea). *Mar. Biol.*, **30**, 13–20.

Kinter, W. B., L. S. Merkens, R. H. Janicki, and A. M. Guarino (1972). Studies on the mechanism of toxicity of DDT and polychlorinated biphenyls (PCBS): disruption of osmoregulation in marine fish. *Environ. Health Perspect.*, 169–73.

Hoch, E. B. (1969a). Chlorinated hydrocarbon insecticides: inhibition of rabbit brain ATPase activities. *J. Neurochem.*, **16**, 269.

Koch, R. B. (1969b). Inhibition of animal tissue ATPase activities by chlorinated hydrocarbon pesticides. *Chem. Biol. Interactions*, **1**, 199–209.

Koch, R. B., L. K. Cutkomp, and F. M. Do (1969). Chlorinated hydrocarbon insecticide inhibition of cockroach and honeybee brain ATPases. *Life Sci.*, **8**, 289–97.

Koch, R. B., D. Desaiah, H. H. Yap, and L. K. Cutkomp (1972). Polychlorinated biphenyls: effect of long-term exposure on ATPase activity in fish, *Pimephales promelas*. *Bull. Environ. Contamin. Toxicol.*, **7**, 87–92.

Koeman, J. M., W. H. M. Peeters, C. H. M. Koudstaal-Hol, P. S. Tjioe, and J. J. M. de Goeij (1973). Mercury-selenium correlations in marine mammals. *Nature*, **245**, 385–6.

Larsson, Å. (1975). Some biochemical effects of cadmium on fish. In J. H. Koeman and J. J. T. W. A. Strik (Eds), *Sublethal Effects of Toxic Chemicals on Aquatic Animals*, Elsevier, Amsterdam. pp. 3–13.

Lewis, S. D., and W. M. Lewis (1971). The effect of zinc and copper on the osmolality of blood serum of the channel catfish, *Ictalurus punctatus Refinesque*, and golden shiner, *Notemigonus crysoleucas Mitchill*. *Trans. Am. Fish. Soc.*, **4**, 639–43.

Lindahl, P. E., and C. E. B. Hell (1970). Effects of short-term exposure of *Leuciscus rutilus* L. (Pisces) to phenylmercuric hydroxyde. *Oikos*, **21**, 267–75.

MacLean, F. I., O. J. Lucis, Z. A. Shaikh, and E. R. Jansz (1972). The uptake and subcellular distribution of Cd and Zn in microorganisms. *Fedn Proc. Fedn Am. Socs Exp. Biol.*, **31**, 699.

Matsumura, F., and T. Narahashi (1971). ATPase inhibition and electrophysiological change caused by DDT and related neuroactive agents in lobster nerve. *Biochem. Pharmacol.*, **20**, 825–37.

Matsumura, F., and R. D. O'Brien (1966). Insecticide reaction with nerve-interactions of DDT with components of american cockroach nerve. *J. Am. Food Chem.*, **14**, 39–43.

Matsumura, F., and K. C. Patil (1969). Adenosine triphosphatase sensitive to DDT in synapses of rat brain. *Science*, **166**, 121–2.

Matsumura, F., T. A. Bratkowski, and K. C. Patil (1969). DDT: inhibition of an ATPase in the rat brain. *Bull. Environ. Contamin. Toxicol.*, **4**, 262.

McCarty, L. S., and A. H. Houston (1976). Effects of exposure to sublethal levels of cadmium upon water-electrolyte status in the goldfish (*Carassius auratus*). *J. Fish. Biol.*, **9**, 11–19.

McKim, J. M., G. M. Christensen, and E. P. Hunt (1970). Changes in the blood of brook trout (*Salvelinus fontinalis*) after short-term and long-term exposure to copper. *J. Fish. Res. Bd Can.*, **27**, 1883–9.

Miller, D. S., W. B. Kinter, D. B. Peakall, and R. W. Risebrough (1976). DDE feeding and plasma osmoregulation in ducks, guillemots and puffins. *Am. J. Physiol.*, **231** (2), 370–6.

Narahashi, T., and H. G. Hass (1967). DDT: interaction with nerve membrane conductance changes. *Science*, **157**, 1438–40.

Nechay, B. R. (1973). Action of mercury on renal sodium transport and adenosinetriphosphatase activity. In M. W. Miller and T. W. Clarkson (Eds), *Mercury, Mercurials and Mercaptans*, Charles C. Thomas, Springfield, Ill., pp. 111–23.

Nimmo, D. R., and R. R. Blackman (1972). Effects of DDT on cations in the hepatopancreas of penaeid shrimp. *Trans. Am. Fish. Soc.*, **3**, 547–9.

Noël-Lambot, F. (1976). Distribution of cadmium, zinc and copper in the mussel *Mytilus edulis*. Existence of cadmium-binding proteins similar to metallothioneins. *Experientia*, **32**, 324.

O'Connor, D. V., and P. O. Fromm (1975). The effect of methyl mercury on gill metabolism and blood parameters of rainbow trout. *Bull. Environ. Contamin. Toxicol.*, **13**, 406–11.

Olson, K. R., H. L. Bergman, and P. O. Fromm (1973). Uptake of methylmercuric chloride and mercuric chloride by trout: a study of uptake pathways into the whole animal and uptake by erythrocytes *in vitro*. *J. Fish. Res. Bd Can.*, **30**, 1293–9.

Payne, N. B., G. R. Herzberg, and J. L. Howland (1973). Influence of some insecticides on the ATPase of mouse liver mitochondria. *Bull. Environ. Contamin. Toxicol.*, **10**, 365–7.

Renfro, J. L., B. Schmidt-Nielsen, D. Miller, D. Benos, and J. Allen (1974). Methyl mercury and inorganic mercury: uptake, distribution, and effect on osmoregulatory mechanisms in fishes. In F. J. Vernberg and W. B. Vernberg (Eds), *Pollution and Physiology of Marine Organisms*, Academic Press, New York. pp. 101–22.

Roesijadi, G., S. R. Petrocelli, J. W. Anderson, B. J. Presley, and R. Sims (1974). Survival and chloride ion regulation of the porcelain crab *Petrolisthes armatus*. *Mar. Biol.*, **27**, 213–7.

Rothstein, A. (1960). Cell membrane as site of action of heavy metals. *Fed. Proc.*, **18**, 1026–38.

Saunders, R. L., and J. B. Sprague (1967). Effects of copper–zinc mining pollution on a spawning migration of atlantic salmon. *Wat. Res.*, **1**, 419–32.

Shanes, A. M. (1950). Electrical phenomena in nerve. II. Crab nerve. *J. Gen. Physiol.*, **33**, 75–102.

Skidmore, J. F. (1964). Toxicity of zinc compounds to aquatic animals, with special reference to fish. *Q. Rev. Biol.*, **39**, 227–48.

Skidmore, J. F. (1970). Respiration and osmoregulation in rainbow trout with gills damaged by zinc sulphate. *J. Exp. Biol.*, **52**, 481–94.

Skou, J. C. (1963). Studies on the Na^+–K^+ activated ATP hydrolysing enzyme system. The role of –SH groups. *Biochem. Biophys. Res. Commun.*, **10**, 79–84.

Spronk, N., F. G. Brinkman, R. J. van Hoek, and D. L. Knook (1971). Copper in *Lymnaea stagnalis* L. II. Effect on the kidney and body fluids. *Comp. Biochem. Physiol.*, **38A**, 309–16.

Strick, J. J. T. W. A., H. H. de Iongh, J. W. A. van Rijn van Alkemade, and T. P. Wuite (1975). Toxicity of chromium (VI) in fish, with special reference to organoweights, liver and plasma enzyme activities, blood parameters and histological alterations. In J. H. Koeman and J. J. T. W. A. Strick (Eds), *Sublethal Effects of Toxic Chemicals on Aquatic Animals*, Elsevier, Amsterdam. pp. 31–41.

Suda, T., N. Horiuchi, E. Ogata, I. Ezawa, N. Otaki, and M. Kimura (1974). Prevention by metallothionein of Cd-induced inhibition of vitamin D activation reaction in kidney. *FEBS Lett.*, **42**, 23–6.

Thurberg, F. P., M. A. Dawson, and R. S. Collier (1973). Effects of copper and cadmium on osmoregulation and oxygen consumption in two species of estuarine crabs. *Mar. Biol.*, **23**, 171–5.

Vallee, B. L., and D. D. Ulmer (1972). Biochemical effects of mercury, cadmium and lead. *A. Rev. Biochem.*, **41**, 91–128.

Vernberg, W. B., and J. Vernberg (1972). The synergistic effects of temperature, salinity and mercury on survival and metabolism of the adult fiddler crab, *Uca pugilator*. *Natl Oceanic Atmos. Admin. US Fish. Bull.*, **70**, 415–20.

Waggoner III, J. P., and M. G. Zeeman (1975). DDT: Short term effects on osmoregulation in black surfperch (*Embiotoca jacksoni*). *Bull. Environ. Contamin. Toxicol.*, **13**, 297–300.

Waldichuk, M. (1974). Some biological concerns in heavy metals pollution. In F. J. Vernberg and W. B. Vernberg (Eds), *Pollution and Physiology of Marine Organisms*, Academic Press, New York. pp. 1–57.

Wells, M. R., J. B. Phillips, and G. G. Murphy (1974). ATPase activity in tissues of the map turtle *Graptemys geographica* following *in vitro* treatment with aldrin and dieldrin. *Bull. Environ. Contamin. Toxicol.*, **11**, 572–6.

Weser, U., F. Donay, and H. Rupp (1973). Cd-induced synthesis of hepatic metallothionein in chicken and rats. *FEBS Lett.*, **32**, 171–4.

Witherspoon, E. G., Jr., and M. R. Wells (1975). Adenosine triphosphatase activity in brain, intestinal mucosa, kidney, and liver cellular fractions of the red-eared turtle following *in vitro* treatment with DDT, DDD and DDE. *Bull. Environ. Contamin. Toxicol.*, **14**, 537–44.

Yap, H. H., D. Desaiah, L. K. Cutkomp, and R. B. Koch (1971). Sensitivity of fish ATPases to polychlorinated biphenyls. *Nature*, **233**, 61–2.

Chapter 13

Osmoregulation and Ecology in Media of Fluctuating Salinity

R. GILLES AND CH. JEUNIAUX

I. INTRODUCTION	581
II. DIRECT EFFECTS OF SALINITY	584
A. On adult organisms	584
B. At different stages of the life cycle	593
III. INDIRECT EFFECTS OF SALINITY	597
A. Stress interactions	597
B. Species interactions	601
IV. CONCLUSIONS	603
REFERENCES	604

I. INTRODUCTION

Brackish and intertidal environments are probably among the most exacting and stressful aquatic biotopes; the establishment of animal communities in such habitats supposes various highly adapted physiological features (see Chapters 3, 4, 5 to 9).

Brackish waters enter the category of mixohaline waters which may range from oligohaline to polyhaline media. The salinity may therefore vary from about 0.5 to some 35‰, which is the average oceanic salinity (see Table 1).

Most of the brackish waters are found in restricted coastal regions such as estuaries or salt marshes and in partially land-locked seas such as the Caspian or the Baltic. In such large 'sea' areas, the salinity, although it may be far lower than in sea-water, is not subjected to as frequent and rapid changes as in estuaries. The physiological adaptations required for settlement in these two types of brackish biotopes will be basically the same however. The fauna found in both habitats may thus be relatively similar, being mainly dependent on the mean salinity of the medium provided the nature of the bottom is the same. Most of the time however, the ecological picture will be quite different in the various brackish-water ecotones, the more or less stressful salinity

Table 1. Salinity ranges and current denomination of species and waters.

Water denomination	Salinity range (‰)	Ecological typology of species		Number of species (species diversity of biocenoses)
Limmic	below 0.5	Stenohaline	Limmic euryhaline	+++++++
Oligohaline	0.5–5	Typical brackish water species		+++
	5–9			+
Mesohaline	9–18		Marine euryhaline	++++
Polyhaline	18–30			+++++
Oceanic	30–40	Stenohaline		+++++++

conditions that prevail nevertheless remaining the major ecological parameter.

The inhabitants of the intertidal zone may also be subjected to important osmotic stresses. Salinity in tide pools may indeed become very rapidly higher or lower than that of sea-water depending on the climatic conditions. Another important feature of this habitat is related to tidal movements. Many invaders of the intertidal zone which have only limited or no locomotive abilities and which are water dependent have to face the rapid and drastic changes in external medium conditions which are linked to the tidal rhythm.

Besides the problem of oxygen availability, the species of intertidal fauna such as mussels or barnacles will have to withstand temperature stress and dessication. With an exception for intertidal zones of cold regions where freezing may become an important physiological problem, temperature stress is essentially related to an osomotic problem: for high temperatures will increase the evaporative water loss and the risk of dessication.

The ability to tolerate osmotic stress is thus one of the important clues to settlement of a population in the intertidal zone and is probably the essential one required for succeeding in ecotones such as estuaries, coastal lagoons or mangroves. Since these last biotopes are considered more and more as possible zones for extensive aquaculture in temperate and tropical latitudes, special reference will be made in this review to the various effects of osmotic stress on populations inhabiting these media.

Salinity will affect aquatic organisms both directly and indirectly. Direct effects are related to the organismic and molecular abilities of animals, at any stage of their life cycle, to ensure proper water balance in their internal fluids, either intracellular or extracellular. Indirect effects are essentially of two types: osmotic shocks may produce modifications of the biocenotic composition of an ecosystem which will result in changes in the biotic background for the remaining forms; salinity stress may also become a limiting factor when acting together with one or several other stressing conditions. A species, which in the laboratory can easily withstand large fluctuations of the environmental salinity, may not survive even much smaller changes in the field if these changes are associated with other factors which may themselves constitute a stress. Examples of such factors are oxygen availability, temperature, presence of pollutants, changes in the physical or chemical nature of the substratum or in the nature of the diet and so on. Examples of interactions between the possibility to osmoregulate and the possibility to cope with temperature or pollution stresses are given in Chapters 12 and 13 of the present volume.

These indirect effects, when viewed in the framework of the survival of a population in a given medium, are obviously of the utmost importance. However they are difficult to apprehend and to quantify at the moment. Indeed an approach to the problem in field studies is almost impossible owing to the large number of parameters which may interact. To date, laboratory studies in ecophysiology have been largely directed towards the understanding of the different mechanisms of osmoregulation used by animals to ensure proper water balance. However it is our belief that this line of research should prove very rewarding in the future.

II. DIRECT EFFECTS OF SALINITY

A. On Adult Organisms

A major problem that all animals have to face is to maintain a cellular volume compatible with the different cell activities which are the support of life. As demonstrated at length in this volume, all animals have evolved molecular mechanisms which enable them to ensure a proper intracellular water balance in a given environment. Water homeostasis in media of fluctuating salinities such as brackish waters or the intertidal zone raises the essential problem of having mechanisms of osmoregulation powerful enough to respond to the sometimes extremely rough demands of the medium. The relatively small number of species which succeed in these ecotones when compared to the number of species found in fresh-water or marine habitats can be considered as a reflection of the difficulty of the problems encountered.

All animals living in such biotopes are necessarily euryhaline species to some extent; indeed they can survive in media of different salinities. They may be good osmoregulators or poor osmoregulators depending on the salinity range they can withstand. Good osmoregulators will rely on very efficient mechanisms of osmoregulation while poor osmoregulators will take advantage of several peculiarities which enable them, not to cope with, but to escape too large modifications in the salinity of the external medium for a more or less long period of time.

Osmoregulation in aquatic organisms can be effected in two different ways. The first consists of trying to keep the osmoconcentration of the extracellular fluids constant whatever the salinity of the external medium may be. The second is an attempt to maintain the intracellular fluid isosmotic to the extracellular fluid. Both methods tend towards the same goal which is to avoid water movements (loss or gain) at the cellular level.

It is not the purpose of this contribution to discuss the molecular mechanisms implicated in the control of the osmolarity of the extracellular and intracellular fluids. The reader interested in these processes is referred to Chapters 3, 4 and 5 of this treatise for detailed information. Rather, this chapter is devoted to examining the possible correlations between osmoregulation abilities and the distribution of species in different ecotones with important salinity fluctuations.

1. Osmoregulation of the Extracellular Fluid—Osmoregulators

In euryhaline species, the extracellular fluid osmolarity can evolve in different ways as a function of the osmolarity of the surrounding medium. This is illustrated in Fig. 1. Species able to maintain blood osmolarity at a more or less constant level independently of the osmolarity of the external medium are called homeosmotic. They enter the general category of the

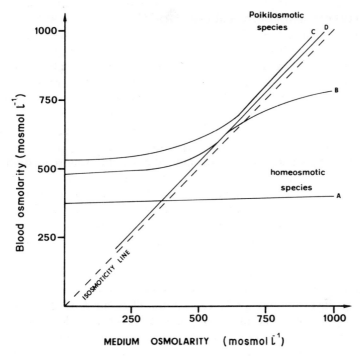

Figure 1 Schematic representation of different types of blood osmolarity evolution as a function of the osmolarity of the external medium. A, homeosmotic animals, hyper–hyporegulators; B,C,D, poikilosmotic animals: B, hyper–hyporegulators; C, hyper-regulators; D, osmoconformers

so-called hyper–hypo-osmotic regulators which are able to hyper-regulate the blood osmolarity in diluted media and to hypo-regulate it in concentrated ones. Hyper–hypo-regulation represents the most elaborate adaptation to salinity stress, it is the rule in teleost fishes and in higher aquatic euryhaline vertebrates; this mechanism also occurs in some crustaceans.

Only very few species have mechanisms of blood osmoregulation sufficiently effective to achieve homeosmoticity: examples are the prawn *Palaemonetes varians* and a few fishes such as *Anguilla anguilla*. These are extremely powerful osmoregulators and can cope with any fluctuation of the salinity of the external medium that normally occurs in nature. In the category of hyper–hypo-regulators, differences in the ecological niches of species can be pointed out depending on the efficiency of the osmoregulatory mechanisms. Both the thick- and the thin-lipped mullet can osmoregulate in both hyper- and hypo-osmotic media. However the thick-lipped mullet (*Chelon labrosus*) which is not as good an osmoregulator in hypo-osmotic media as the thin-lipped mullet (*Liza ramada*) will distribute mostly in polyhaline and mesohaline ponds while *L. ramada* is rather found is oligohaline and even fresh-water ponds (Lasserre and Gallis, 1975).

The second category illustrated in Fig. 1 comprises the hyperosmoregulators. They can maintain a more or less pronounced hyperosmotic state in dilute media. In more concentrated media, the blood osmolarity strictly follows the osmolarity of the external medium. Most euryhaline invertebrates are hyper-regulators: well-known examples are the crabs *Carcinus maenas* and *Eriocheir sinensis*. They are generally relatively weak osmoregulators and their natural distribution may be related to their respective capacities for regulation in hypo-osmotic media. The crab *Libinia emarginata* for instance which shows limited power for osmoregulation will be found in river mouths in salinities not lower than about 20‰; *Carcinus maenas* can be found in salinities down to 10‰ and even lower, and *Eriocheir sinensis* is successful in any medium from fresh-water to sea-water (Gilles, 1974, 1975a). Most more or less euryhaline fresh-water species, such as the crayfish *Astacus astacus*, belong to the same category (Duchateau and Florkin, 1961).

The last group of animals which can be found in media of fluctuating salinities is a group of osmoconformers. Osmoconformers are found among the marine invertebrates and the myxines. They have no power of blood osmotic regulation. Some of these species however have developed populations in estuaries and other habitats with extensive short-term salinity fluctuations; typical examples are various molluscs such as oysters and mussels and different sediment-dwelling invertebrates (These cases are discussed in more detail on p. 589.

It is apparent from the above discussion that media with fluctuating salinities have been invaded by organisms showing good, fair or even no power of osmoregulation of their extracellular fluids. This makes it obvious that blood osmoregulation is far from being the major element in limiting the distribution of animals in such media. Although correlations can be found between the distribution of some species and the efficiency of their blood osmoregulation process, this is far from being a general rule. For instance, the prawn *Palaemonetes varians* and the chinese crab *Eriocheir sinensis* can be found in any media ranging from fresh-water to sea-water. *P. varians* is among the most powerful osmoregulators since it can achieve almost complete homeosmoticity. On the contrary, *E. sinensis* can only maintain an hyperosmotic state in dilute media. A similar situation may be observed among fresh-water teleost fishes: *Culea inconstans* is a hyperosmoregulator showing a maximum salinity tolerance (TL_{50} 90 h) of 21‰, while *Fundulus diaphanus* is a powerful hyper–hypo-osmoregulator with a salinity tolerance of 34‰. These two species however occur together in the same fresh-water lakes and in the same alkaline ponds of North Dakota, the water osmolarity of which varied from 22 to 444 mosmol l^{-1} throughout a 2 year period (Ahokas and Duerr, 1975).

Osmoregulation of the intracellular fluid thus appears as an integral part of the osmoregulatory mechanisms which play a role in the distribution of species in media of fluctuating salinity.

2. *Osmoregulation of the Intracellular Fluid in Osmoregulators*

In a species such as *Eriocheir sinensis*, adaptation from sea-water to fresh-water results in a decrease in blood osmolarity of about 550 mosmol l^{-1}. If the osmolarity of the intracellular fluid were to remain at its level in the sea-water animal, i.e. 1100 mosmol l^{-1}, the cells would have to withstand a swelling pressure of about 12 atm (Gilles, 1974). This is of course a rough calculation made only on the vant'Hoff–Arrhenius equation, nevertheless it shows that the swelling pressure is enormous and that very effective regulation mechanisms must be at play to avoid the bursting out of the cells. As a matter of fact, when *E. sinensis* is passing from a sea-water to a fresh-water medium, there is first a swelling of the tissue; however this swelling is regulated very rapidly as shown by the evolution of the water content of the muscle tissue (Fig. 2). After the initial phase of swelling, the muscle tissue fluid comes into isosmotic equilibrium with the blood (Fig. 2). This demonstrates the existence of active processes implicated in the regulation of the intracellular fluid osmolarity (see Chapters 3 and 4).

It is worth noticing that the duration of the regulatory phase after hypo-osmotic shock (adaptation from a concentrated to a diluted medium) is quite different depending on the species studied. In *Eriocheir sinensis*, complete regulation is achieved in about 24 hours (Fig. 2). In *Carcinus maenas*, control volume is only resumed after some 70 hours (Harris, 1976). No regulation can be observed in isolated nerve cord of the lobster *Homarus vulgaris* after 5 hours of experiment while volume is completely regulated in nerves of *E. sinensis* after 4 hours of hypo-osmotic shock (Gilles, 1973). In the same way, isolated muscle fibres of the blue crab *Callinectes sapidus* resume their control volume in a few hours (Lang and Gainer, 1969) while in the same period of time, no regulation can be observed in isolated muscle fibres of the crayfish *A. fluviatilis* (Reuben, Girardier and Grundfest, 1964).

This indicates the possibility of correlations between the distribution of these species and the efficiency of their mechanisms of intracellular fluid isosmotic regulation. *H. vulgaris* is a stenohaline species, strictly restricted to sea-water. *A. fluviatilis*, although it can withstand, under laboratory conditions, brackish waters of about 15‰ for a couple of months (Duchateau-Bosson and Florkin, 1961), is limited in nature to fresh-water habitats. In these two species, no volume regulation can be observed at least on the time scale of the experiments performed. In *C. maenas* which distributes in waters of salinities down to 10‰ or lower, complete regulation is achieved after some 70 hours and in the excellent euryhaline species *E. sinensis* and *C. sapidus* which can live in any medium from fresh-water to sea-water, volume control is achieved very rapidly. It must however be said that the mechanisms

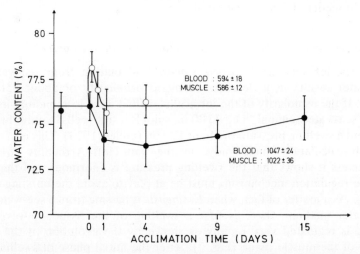

Figure 2 Changes in the water content of the muscle of *E. sinensis* during acclimation from fresh-water to sea-water (○) or from sea-water to fresh-water (●). Acclimation starts at time 0, indicated by an arrow. Control values for sea-water and fresh-water acclimated animals are not significantly different. The water content is expressed as % of the tissue wet weight. Blood and tissular fluid osmolarity measured at the end of the acclimation period are given in mosmoles per litre of kg tissue wet weight

of extracellular fluid osmotic regulation are a little more effective in *C. sapidus* than in *E. sinensis* or in *C. maenas*. *C. sapidus* can indeed maintain a blood hyperosmotic state which is higher in diluted media than can the two other species (Potts and Parry, 1964, Gérard and Gilles, 1972, Gilles, 1974). This might confer to that species some advantage over the two others for withstanding salinity fluctuations. However, the hyperosmotic state maintained by *C. maenas* and *E. sinensis* is of the same order of magnitude and *E. sinensis*, which has mechanisms of cell volume control more powerful than *C. maenas*, can invade more diluted media.

When considering Fig. 2, it is also worth noticing that there is an important difference in volume response efficiency following application of hypo- or hyperosmotic conditions. This difference may have ecological significance particularly for euryhaline osmoconformers and will be discussed later (p. 592).

3. Osmoconformers

Various species of osmoconformers have invaded media of fluctuating salinities. These animals have no power of osmotic regulation of extracellular fluid. Thus they rely entirely on their mechanisms of intracellular fluid isosmotic regulation to cope with changes in environmental salinity. These mechanisms have been studied mostly in molluscs which are typical examples of osmoconformers having colonized estuaries and littoral area (Staaland, 1970; Pierce, 1971; Bedford, 1971; Gilles, 1972a; see also Chapter 4 of this volume).

Osmoconformers do not establish communities in oligohaline or mesohaline waters, they are essentially marine polyhaline species and most of the time they are limited in their distribution to areas allowing temporary exposure to plain salinity conditions. The species which can colonize polyhaline waters permanently are probably restricted in their distribution by the efficiency of their mechanisms of intracellular fluid isosmotic regulation. Data concerning this problem are unfortunately only scanty. The platyasterid asteroid *Luidia clathrata* is found in Tampa Bay in Florida in salinities ranging from 18 to 27‰ as a result of seasonal changes in rainfall (Lawrence, 1973). Laboratory experiments show that specimens of *L. clathrata* can withstand salinities down to 14‰. Similarly, reproductively active populations of *Asterias rubens* are found in the Baltic Sea in salinities of 15–17‰ (Kowalski, 1955) and specimens can be acclimated in the laboratory to salinities of 15‰. In *A. rubens*, there remains only a slight increase in tissue water content after 5 days of acclimation to the diluted medium (Jeuniaux, Bricteux-Gregoire and Florkin, 1962). In the same way, tissue control volume is resumed in *L. clathrata* after 8 days following adaptation from 27‰ (Ellington and Lawrence, 1974). In these two species therefore, the mechanisms of intracellular fluid isosmotic regulation appear to be powerful enough to ensure proper volume control and thus settlement in polyhaline waters. However this is not always the case and many species, although currently found in estuaries or in the intertidal zone, show only poor cellular volume regulation. This is exemplified in Table 2 for several organisms which appear unable to resume their control volume even after long periods of acclimation to diluted media.

Many of these species, the distributions of which are limited to areas only temporarily diluted, will take advantage of several particularities which enable them to escape or to diminish the amplitude of an osmotic shock during a certain period of time.

Sediment-dwelling species such as the mollusc *Modiolus granosissimus* or the worms *Arenicola marina* and *Perinereis cultrifera* take advantage of the

Table 2. Water content (given in % wet weight) of the tissues of osmoconformers from estuaries or from the intertidal zone after acclimation to diluted media

Species	Tegula funebralis[a]		Modiolus granossissimus[b]		Modiolus squamosus[b]		Purpura lapillus[c]		Patella vulgata[c]		Scrobicularia plana[c]	
Salinity	100%SW	50%SW	100%SW	50%SW	36‰	22‰	100%SW	50%SW	100%SW	50%SW	100%SW	50%SW
Tissue water content	65.0	74.0	82.1	85.8	82.8	85.9	71.0	84.9	72.5	80.3	73.8	78.4
Acclimation time	10 days		3 weeks		3 weeks		3 weeks		3 weeks		3 weeks	

Species	Mytilus edulis[d]		Arenicola marina[e]		Ostrea edulis[f]		Gryphea angulatas[g]		Perinereis cultrifera[h]	
Salinity	100%SW	50%SW	100%SW	50%SW	100%SW	50%SW	100%SW	50%SW	100%SW	50%SW
Tissue water content	71.6	77.3	78.1	83.5	76.3	82.3	73.7	81.2	75.7	82.8
Acclimation time	3 weeks		3 days		3 days		1 day		1 day	

[a] Peterson and Duerr (1969).
[b] Pierce (1971).
[c] Hoyaux, Gilles and Jeuniaux (1976).
[d] Gilles (1972a).
[e] Duchateau-Bosson, Jeuniaux and Florkin (1961).
[f] Bricteux-Grégoire and coworkers (1964a).
[g] Bricteux-Grégoire and coworkers (1964b).
[h] Jeuniaux, Duchateau-Bosson and Florkin (1961).

fact that the salinity fluctuations are less important in the substrate than in the free water above. Dampening of salinity variations has been reported in estuarine waters a number of times. In the estuary of the River Pocasset (USA), Sanders, Mangelsdorf and Hampson (1965) reported that the salinity in the sediment below 5 cm is uniform at about 20.5‰ while it ranges from 2.3‰ to 29.3‰ during the tidal cycle (see also Melusky, 1968). Thus populations which inhabit sandy or muddy substrates are exposed to much smaller salinity fluctuations than can be expected from measurements in the free waters above. Moreover, these species generally remain inactive during part of the tidal cycle and resume activity at high tide, when the salinity of the estuaries reaches a maximum. Inactivity effectively contributes in maintaining stable salinity conditions in the sediment round about.

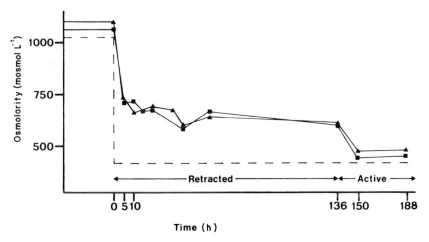

Figure 3 Changes in the osmolarity of the blood (■) and of the perivisceral fluid (▲) of *L. littorea* during the time course of direct acclimation from sea-water to 40% sea-water. The osmolarity is given in mosmol l^{-1}. Dashed line: external medium osmolarity. From Hoyaux and coworkers (1976); reproduced by permission of Pergamon Press

A period of inactivity is also the rule among the osmoconformers of the intertidal area during low tide. When the salinity is decreasing it is also found in several molluscs such as *Mytilus edulis, Ostrea edulis* or *Gryphaea angulata* which frequently colonize the mouths of estuaries. Such behaviour is illustrated in Fig. 3 for the case of the Gastropod *Littorina littorea*. It can be seen that as soon as the animal retracts into its shell, the pallial fluid can be maintained hyperosmotic to the external medium for a long time. When the animal resumes activity the blood and pallial fluid decrease in isosmotic equilibrium with the external medium. In many molluscan species, the periods of inactivity thus correspond to a phase of 'shell-closing' which helps the animals to withstand a sudden osmotic stress. This behaviour has been described many times for different molluscs (Avens and Sleigh, 1965; McAlister and Fisher, 1968; Gilles, 1972a; Hoyaux, Gilles and Jeuniaux, 1976). Bivalves such as *M. edulis* or *Scrobicularia plana* close their valves tightly, gastropods with an operculum such as *Littorina littorea* or *Purpura lapillus* retract strongly into their shells while species such as *Patella vulgata* or *Siphonaria pectinata* adhere firmly to the rock. Such a behaviour is certainly of ecological value since it allows the animals to invade different areas of fluctuating salinity by enabling them to escape from temporarily relatively important osmotic stresses.

A last point which may be worth mentioning here is that the relative speed of response of osmoregulatory mechanisms may become a limiting factor, particularly in osmoconformers. As shown in Fig. 2, volume regulation in tissues of the chinese crab *E. senensis* is very effective in responding to an hypo-osmotic shock. Volume control following an hyperosmotic stress is achieved far less efficiently. Indeed it takes about 15 days after transfer of this species from fresh-water to sea-water before volume regulation is completed.

Such differences in regulation efficiency have been observed in all the species studied up to now (the molecular bases of this phenomenon are discussed in Chapter 4). These differences are certainly of ecological importance in allowing the settlement of various species only in areas either where the salinity must change extremely slowly at least in the direction of an increase or where the salinity change in this direction must be far less important than that which may be expected on the basis of volume regulation performances following application of hypo-osmotic conditions.

For instance, *M. granosissimus* once completely acclimated from 36‰ to 3‰ will die within 2 weeks after being returned to 36‰. In the same way, the fresh-water shrimp *Palaemonetes paludosus* will hardly support rapid acclimation to 28‰. In this hyperosmotic medium, it will only show a weak recovery in abdominal muscle escape response after 100 hours. In these conditions, an important shrinkage of the muscle tissue is observed which persists even after 100 hours of experiment (Turner, Lowe and Lawrence, 1975).

For animals with very low efficiency of volume regulation in hyperosmotic conditions, which appears to be the case in many marine euryhaline osmoconformers, such a phenomenon makes the volume response and the possibility of complete acclimation almost unidirectional. The animals are indeed able to adapt quite easily to hypo-osmotic situations but, once fully adapted, can hardly support their previous, now hyperosmotic, environment.

Such a phenomenon may be part of a physiological mechanism which has some evolutionary significance. Possibly, the marine ancestors of the osmoconformers actually found in very diluted media were able to invade oligohaline or even fresh-waters but, once the move took place, it became almost irrevocable owing to the poor effectiveness of the volume control mechanisms at work following application of hyperosmotic conditions. For example, the european bivalve *Dreissena polymorpha*, originally confined to Black Sea brackish water, migrated during the last century into fresh-waters and into very diluted waters of the Baltic Sea. The Baltic Sea individuals exhibit a lower salinity tolerance when compared to those of Black Sea (Remane and Schlieper, 1971).

To summarize, distribution of populations of adult animals in media of fluctuating salinities cannot be related to the sole effectiveness of their mechanisms of extracellular fluid osmoregulation. Obviously, mechanisms of intracellular fluid osmoregulation as well as different behavioural and mechanical adaptations play an important part in the possibilities of colonization.

Moreover, it is worth noticing that some species, such as the prawn *P. varians*, which are extremely powerful osmoregulators (see p. 585) are characteristically found in brackish waters and are not common in the sea. On the other hand, cell volume regulation in salinities down to 20‰ is as effective in *Modiolus squamosus* (Pierce, 1971) than in other species such as *Mytilus edulis* or *M. granosissimus*. *M. squamosus* however distributes only in the subtidal environment of the oceanic sea-shores of Florida (USA) where the salinity remains stable at about 32‰ throughout the year, while the marsh dwelling species *M. granosissimus* and the intertidal species *M. edulis* have colonized polyhaline waters.

Other factors must therefore be at work in the limitation of species in fluctuating salinity ecotones. Notable among these are osmotic effects on different stages in the life cycle, species interactions and competition. These factors will now be briefly discussed.

B. At Different Stages of the Life Cycle

Successful establishment of a population in a given medium is only possible if the invaders can withstand the demands of the environment at any moment during their life cycles. This problem is particularly acute for animals experiencing ecotones with fluctuating salinities.

Salinity will indeed affect reproduction in areas where it underlies pronounced changes. As a matter of fact, changes in salinity are known to produce modifications not only in the reproduction rate of many species but also in the rates of growth and development of their embryos and larvae.

Reduced reproductive capacities and even complete sterility have been reported for several invertebrates inhabiting the Baltic Sea. Examples are the scyphozoan *Lucernaria quadricornis* the echinoderms *Asterias rubens* and *Ophiura albida* or the hybroid *Laomedia loveni* (Remane, 1940; Schütz, 1969). The euryhaline amphipod *Gammarus duebeni* loses its reproductive potential completely when in fresh-water (Hynes, 1954). In grey mullets and sea bass confined in low salinity enclosures, there is a lack of gamete production: the ovocyte development is blocked at stage III (Abraham, Blanc and Yashouv, 1966; Blanc-Livni and Abraham, 1970; Boisseau and coworkers, 1975). It seems that this effect on gamete production can be related to the latered hormonal status of the fish remaining permanently in waters of low salinity (Eckstein, 1975).

Spawning may also be affected by salinity conditions. Many invertebrates will deposit a reduced number of eggs or eventually none if the salinity goes below a given critical level. This limit is variable from species to species: for instance, it is about 27‰ for the oysters *Crassostrea gigas* and *C. virginica* and about 20‰ for the gastropods *Eupleura candata* and *Urosalpinx cinerea* (Manzi, 1970; Fujuya, 1970; Calabrese and Davis, 1970).

Salinity also seems to modify the reproduction effectiveness by inducing changes in sex ratio. This problem has not yet been studied in detail and there are only a few reported cases.

The archiannelid *Dinophilus gyrociliatus* deposits egg capsules which contain large and small eggs; the large eggs give rise to females while the small ones give rise to males. Salinity will influence the ratio of large eggs to small eggs and hence the sex ratio of the offspring of *D. gyrociliatus* (Traut, 1969). Salinity dependence of the sex ratio has also been observed in the amphipod *Gammarus duebeni*. In this case it appears that sex determination can be related to the salinity tolerance of a parasite (see p. 601).

Salinity also becomes a limiting ecological factor by affecting growth and development. As early as 1935, Clark noticed that in oysters, ontogenetic development proceeds normally in salinities ranging from 14.5 to 39‰ and that no swimming larvae can be obtained at lower salinities.

Thus, the limiting salinity varies in most cases depending on the stage in the life cycle and apparently there is no relation between the salinity required for growth and germ production. In the oyster *Crassostrea virginica* for instance, gonad maturation and spawning will be very reduced at salinities lower than 27‰. Very few eggs will yield the straight-hinge stage at 12.5‰ (Calabrese and Davis, 1970). In the same way, medusae of the coelenterate *Aurelia aurita* will produce viable germ cells in salinities below 6‰, however the scyphistoma stage will not develop in such a diluted medium (Segerstrale, 1969). In crustaceans, the length of zoe life as well as growth and development in the following larval stages is modified by a lowered salinity.

The reader is referred to the papers by Costlow and his colleagues who have studied this last problem in detail (see for instance Costlow, Bookhout and Monroe, 1966; Costlow, 1967).

More examples of salinity effects on larval growth and development are given in reviews by Kinne (1971) and Holliday (1971). These effects will become important limiting factors for many species. Typically, the osmoregulatory potential is minimal in eggs, develops in embryos and in larvae and come to full power in adults. Extensive salinity fluctuations or very low salinities will thus cause the number of benthic species with pelagic larvae to decrease to the benefit of species which release their offspring at advanced stages of development or provide them with some other kind of protection (eggs in burrows, gelatinous covering and so on). In most cases, however, benthic organisms can only maintain their populations in fluctuating salinity biotopes by larval reinforcements from more stable, saltier waters. It is worth noting in this context that in most invertebrates with no or only poor locomotion capabilities, it is often gonad maturation and spawning which show the narrowest limits of salinity tolerance. As already stated, spawning in mussels do not occur in salinities lower than 27‰ while the eggs can still reasonably develop at 20‰. This has ecological value since spawning at high salinities will give the eggs more chance to develop in better growth conditions.

Species with good locomotion capabilities have solved the problem of reproduction by effecting migrations towards media providing better conditions for reproduction and development. The transition from one medium to another appears to be triggered by physiological changes in osmoregulatory abilities mediated through hormonal action. Hormonal control of osmoregulation has been reviewed recently by Fontaine (1975) (see also Chapter 10 of this volume). On the other hand, Zaugg and McLain (1970) demonstrated seasonal changes including a spring rise in gill Na^+-K^+ ATPase of the coho salmon *Oncorhynchus kisutch*. The level of the enzyme decreases sharply during the summer if the fish remains blocked in fresh-water. Na^+-K^+ ATPase activity is implicated in the transport of Na^+ ions which constitute an essential part of the mechanisms of osmoregulation of the extracellular fluids (see Chapter 5). Moreover, ATPase levels in osmoregulatory organs are under the control of hormonal stimulation.

With this in mind and considering the results obtained with *O. kisutch*, it can be concluded that smolts experience seasonally physiological changes that are preparatory to seaward migration. However they should be able to readapt to their fresh-water environment if they do not reach the sea. Experiments by Baggerman (1960a, 1960b) clearly illustrate the ecological significance of such preparatory molecular modifications. She studied the salinity preferences of various salmonids and showed that, at variance with the situation encountered in other species, the fry of *O. kisutch* remains in fresh-water until the smolt stage which starts the spring migration to the sea. The fry of this species thus show a preference for waters of low salinities and smolts prefer salt-water in spring. However this preference is not found in late summer and autumn smolts, after the migration season has ended.

Preparatory physiological changes have also been reported in the eel *Anguilla rostrata*. This includes a rise in Na^+-K^+ ATPase in gills and intestine which might help the fish to cope successfully with the osmotic stress it will have to withstand during its catadromous migration. Corticosteroid secretion appears to be largely implicated in this process of physiological modifications

(Epstein, Cynamon and McKay, 1971). Catadromous migrations in eels thus also appear to be under the control of preparatory molecular adjustments. As far as anadromous migrations of elvers are concerned, this problem of salinity preference induced by preparative molecular adjustments has never been studied. However it seems that in the elvers of *A. vulgaris*, salinity preference may not be primarily involved in the inshore migration; instead, some substance present in fresh-waters would act as attractant (Creutzberg, 1961).

Preparatory molecular adjustments are apparently not restricted to the mechanisms of osmoregulation of the extracellular fluids. Salmonid smolts also show modifications in their muscle amino acids pool during catadromous migration. Amino acids are important osmotic effectors implicated in intracellular fluid osmoregulation (see Chapter 4) and the results obtained more particularly with parr and smolt stages of *Salmo salar* (Fontaine and Marchelidon, 1971) may indicate that preparatory physiological changes also affect this process of cell volume control. More experimental data are needed however to shed some light on this interesting problem.

To conclude, the direct effect of salinity on species distribution in media of fluctuating salinities is essentially related to the overall abilities of the organisms for osmoregulation. In such biotopes, three types of populations may be found which can be distinguished on the basis of their osmotic behaviour. In the first type, populations can reproduce actively in the ecotone; in species such as the brackish-water hydroid *Cordylophora caspia* or the amphipod *Gammarus duebeni*, reproduction growth and development will be the most effective in mesohaline waters. In *C. caspia* for instance maximum gonophore production is limited to salinities between 5‰ and 15‰ (Kinne, 1956) and in *G. duebeni*, the average number of eggs produced is higher in 10‰ than in 2‰ or 30‰ (Dennert and coworkers, 1969). In these organisms, distribution and population density will be dependent essentially on the limits of salinity tolerance allowed for reproduction. Adult forms will generally easily withstand salinities at which their reproduction is impaired. They will be quite successful in media in which salinity changes are effected slowly, on the basis of a large seasonal cycle.

The bulk of the species colonizing estuaries and other polyhaline waters is dependent on sea-water for reproduction. Spawning in benthic forms will only occur at salinities high enough to give the eggs a good chance of development. The other organisms will rely on migration to ensure reproduction.

The third type of population is constituted by temporary migrant, essentially catadromous and anadromous fishes which make their way into sea-water or fresh-water. These species normally show only limited aptitude to grow and eventually to survive for long periods in coastal lagoons, mangroves or estuaries. Even species like mullets or sea bass which are currently found in these biotopes show altered growth rates and sometimes infertility if they are herded in lagoonal ponds without the possibility of freely reaching the open sea (Lasserre, 1976). Distribution of these three types of population will also depend on the various indirect effects of salinity. Two major effects among these are discussed below.

III. INDIRECT EFFECTS OF SALINITY

A. Stress Interactions

Besides osmotic problems, the inhabitants of biotopes with fluctuating salinities will have to face other different stresses which are integral parts of the ecological framework of these media. Changes in temperature and oxygen availability may become important parameters in shallow estuarine waters, salt marshes, mangroves and semi-enclosed lagoonal ponds. These parameters even constitute essential conditions for life in the intertidal zone. The species living in this biotope will indeed be subjected to important modifications in temperature and oxygen availability following the tidal rhythm.

Moreover, these media, occupying finally very restricted areas, are the borderlines between land and ocean, fresh-water and sea-water and, as such, are subjected to an increasing pollution coming from both rivers and oceans.

It is obvious that, in any biotope, one stress is better withstood than two or more stresses added together. This becomes particularly important in ecotones with fluctuating salinities where the major stresses animals have to cope with are sometimes extremely rough.

The important temperature–osmoregulation and pollution–osmoregulation interactions are discussed at length in Chapters 12 and 13 of this volume, the present part of this review will thus be restricted to the presentation of a few examples of stress interactions and to a discussion of the incidence of such interactions on animal distributions.

1. *Temperature and Salinity*

Temperature will act on the mechanisms of ion transport implicated in osmoregulation as it acts on any chemical reaction. Carrier-mediated transport should thus be enhanced by an increase in temperature and decreased by a temperature drop. It might therefore be expected that osmoregulation of extracellular fluids is more effective at high temperature than at lower ones. The picture is far from being that simple. While various species can indeed withstand salinity fluctuations better at high temperatures, as for instance the shrimp *Crangon septempinosa* (Haefner, 1969), in many others high temperatures will narrow the salinity tolerance range. The gastropod *Nassarius reticulatus* will survive in 20–30‰ at 25 °C, but its salinity range will be widened to 10–40‰ at 5 °C (Eriksson and Tallmark, 1974). In the same way, the fish *Gadus morhua* can easily tolerate 8‰ at 2 °C but is unable to withstand this salinity if the temperature is raised to 10 °C. The arctic brackish-water amphipod *Onisimus affinis* tolerates high salinities better at low temperatures (Percy, 1975).

Conversely, salinity changes will affect the temperature tolerance of aquatic poikilotherms. For instance, the upper lethal temperature is higher in *Fundulus heteroclitus* at 32‰ than in fresh-water. The highest resistance to temperature stress is observed however at salinities giving isosmoticity between blood and the outside medium (Garside and Chin-Yuen-Kec, 1972). Maximum temperature tolerance offered at salinities at which animals are isosmotic with their environment appear to be a general rule not only in fishes but also in invertebrates. The euryhaline osmoconformer *Enchlytraeus albidus* is protected from heat shocks by increasing salinities (Kahler, 1970). More examples can be found in Chapter 12 of this volume. They serve as good illustrations of the fact that one stress is more easily withstood than two.

Animals thus tend to show the greatest temperature tolerance in media in which they have minimum osmotic problems. For euryhaline fishes, it will be in mesohaline waters of about 12–14‰. The other hyper–hypo-osmoregulators as well as hyper-regulators and osmoconformers will best support temperature stress in the medium which is isosmotic to their blood when in sea-water. However it must be noticed that in the numerous species which show no or only low power of extracellular osmoregulation, the mechanisms of isosmotic intracellular regulation brings about cell volume regulation and thus best adaptation in the medium they have previously experienced. As already stated, mussels of the species *M. granosissimus*, coming from sea-water, once adapted to 3‰ will be stressed to the point of death upon return to sea-water. In such species the medium offering the least stressful salinity conditions is therefore variable depending on the previous adaptation of the species. The same is true as far as temperature stress is concerned. Many poikilotherms possess mechanisms of metabolic thermocompensation (for reviews see Hochachka and Somero, 1973; Gilles, 1975) which will change the thermic preferendum to ranges related to the last acclimation conditions experienced.

2. *Dehydration and Salinity*

Two other problems related to temperature stress are freezing and dessication. Aquatic animals may have to face such problems in coastal waters where they are subjected to an exposure–immersion cycle related to the tidal rhythm.

Simple tolerance to freezing or formation of antifreeze molecules will help animals in withstanding the sometimes extremely low temperatures that they may have to face during the exposure time. The mechanisms of antifreeze formation and of tolerance to freezing are beyond the scope of this review. The reader interested may consult Hochachka and Somero (1973). The relative efficiency of such mechanisms will obviously by of importance in determining the distribution of animals in cold intertidal zones and estuaries.

Resistance to dessication implicates low water permeability of the integments and physiological adaptations allowing maximum water resorption in order to decrease water loss from the urine to a minimum. All terrestrial and semiterrestrial species possess such adaptations to a certain degree; with the exception of insects, aquatic invertebrates do not. Resistance to dessication in these last species living in the intertidal zone is therefore extremely limited. Some invaders of the intertidal zone have got round the problem of dessication by behavioural means; they are sand-dwellers, burrowers or algal-cryptic species. True resistance to dessication is only found in animals having shells, which therefore gives them the possibility of isolating themselves from the external medium, keeping as much water as they can inside their pallial cavity.

Typical species of this type are the molluscs (mainly bivalves) or the barnacles of the intertidal habitat. The period of time they can tolerate non-immersion is mainly dependent on their ability to tolerate anaerobic conditions and the hyperosmoconcentration of body fluids that occurs and progressively increases due to the inevitable, although small, water loss.

Anaerobic metabolism has been studied in mussels and oysters in which metabolic adaptations appear to allow relatively long periods of anaerobiosis. Briefly, anaerobic glucose catabolism does not lead to lactic acid but to succinic acid and derivatives, this gives a better ATP formation yield than normal anaerobic glycolysis (Gilles, 1972b; Hochachka and Somero, 1973). In some species of molluscs and barnacles, the anaerobic metabolism is helped by an air-gaping mechanism which allows the use of the mantle cavity as a lung. Air-gaping is one more physiological adaptation which seems to be implicated in the distribution of bivalves in the intertidal area. Most pelecypods usually close their valves when exposed to air; they are essentially limited to the low intertidal zone. The ribbed mussel *Modiolus demissus* with air-gaping for gaseous exchange (Lent, 1968) distributes up to the high intertidal habitat.

Tolerance to hyperconcentration of body fluids is directly related to the efficiency of the mechanisms of intracellular fluid osmoregulation (see Fig. 2 and p. 584). Apparently, cells support the shrinkage due to hyperosmotic conditions more easily than the swelling that occurs in hypo-osmotic situations. Unless under very severe conditions cellular osmo- and ionoregulation thus does not appear as an essential factor of distribution limitation in these intertidal species which can easily isolate themselves from the external medium.

3. *Oxygen Tension and Salinity*

Osmoregulation requires a great deal of energy and therefore tolerance to low oxygen tension may become a factor of ecological limitation for species inhabiting media with fluctuating salinities.

There is an enormous amount of literature on oxygen consumption in euryhaline species under osmotic stress or after a period of adaptation. These data have been discussed and reviewed many times (Kinne, 1964; 1971; Gilles, 1973, 1975; Schoffeniels and Gilles, 1970, 1972); however they are mostly of little ecological interest. Indeed, in many cases investigators have measured respiratory rates soon after the salinity change and without

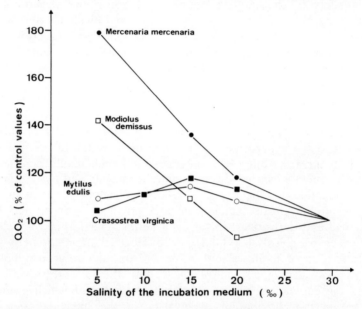

Figure 4 QO_2 of isolated gill pieces from four lamellibranch species as a function of salinity. QO_2 in sea-water (30‰s) is considered as 100%. After Van Winkle (1968); modified

sufficient knowledge about the time course of adaptation processes. Our present information may however allow two generalizations:

(1) Increased respiratory demands due to osmotic stresses can be reduced by the beneficial effects of other environmental factors; again this is an expression of the fact that one stress is better supported than the addition of two or more.
(2) In most invertebrates, the oxygen demand is the lowest in sea-water or in the medium to which they have been acclimated over extended periods; this can be related, at least partly, to the oxygen demand of the mechanisms at work in the cellular volume regulation process (Gilles, 1972c, 1973). From an ecological standpoint, it is worth noticing that osmoconformers which can distribute in rather large salinity gradients and which, therefore, probably have more effective mechanisms of intracellular osmoregulation do not show as important an increase in oxygen demand in low salinities as other species. This is illustrated in Fig. 4 for the case of four molluscs. The important O_2

consumption increase observed in *M. mercenaria* and *M. demissus* might be related to osmotic swelling due to poor cellular volume adjustment. More results are needed however before a more definitive statement can be made.

B. Species Interactions

Salinity probably affects species interactions in many ways. These problems however have not been investigated very well up to now. To our knowledge two types of interactions can be described with some certainty. The first type deals with sex determination in the euryhaline amphipod *Gammarus duebeni*, the second is concerned with trophic problems between benthophageous macrofauna and communities of euryhaline micro- and meiofauna.

As already stated, salinity may affect sex ratio directly in several cases (see p. 594). In the brackish-water amphipod *Gammarus duebeni*, this effect is not direct but apparently mediated through a species–species interaction as shown by Bulnheim (1969). In this species, a salinity of 30‰ tends to shift the sex ratio in favour of males. On the other hand, females infested by the microsporidian *Octosporea effeminans* give birth to daughters almost exclusively. Eggs free from microsporidans will differentiate into females or males according to genetic pattern and to photoperiodic conditions. In 30‰ salinity, females of *G. duebeni* will lay eggs free from microsporidians. The percentage of infested eggs will vary with the salinity of the environmental medium. Consequently the sex ratio is shifted in favour of females in low salinities and more males appear at high salinities. The sex-determination mechanism in *G. duebeni* is thus salinity dependent due to different salinity tolerances of host and parasite.

The main species–species interaction in media of fluctuating salinities is probably effected through trophic problems. Such trophic problems can occur as a result of competition between species for the same kind of food, as a result of modifications in the nature of the food supply or as a result of salinity-induced changes in feeding activity of organisms.

Activity in general and feeding activity in particular are influenced by salinity, at least during the acclimation time. In the lamellibranch filter-feeders, ciliary activity of the gills is related to salinity. In *Branchidontes recurvus* for instance, filtration rate decreased with decreasing salinity from 18‰ to 6‰ (Nagabhushanam and Sarojini, 1965). Moreover, the amplitude of the variations in filtration rate is different for different species depending on their osmoregulation abilities and on their pre-acclimation to a given environment. This is exemplified in Fig. 5 for the case of four bivalves. It can be seen first that the salinity range in which ciliary activity is maximum is shifted to lower salinities after acclimation to mesohaline waters. On the other hand, optimum ciliary activity is recorded in a much larger salinity range in the very tolerant *Mytilus edulis* than in the less tolerant *Cardium edule*.

Another type of salinity influence on feeding activity may be pointed out. It is known that free amino acids in sea-water may be an important source of nutrition for some invertebrates, due to active transepidermal absorption. The absorption of amino acids by the isolated gills of the bivalve *Mya arenaria* falls sharply when salinity of the external medium decreases (Stewart and Bamford, 1976).

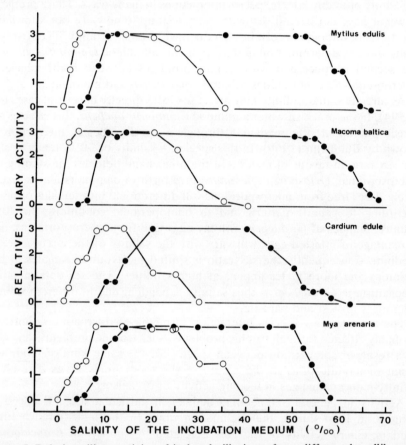

Figure 5 Relative ciliary activity of isolated gill pieces from different lamellibranch species as a function of salinity. Acclimation salinities are 6‰ (○) and 30‰ (●). Measurements are made 24 hours after transfer of the gill pieces into the different test salinities. After Theede and Lassig (1967); modified

Thus, some species may maintain better feeding activity than others under salinity stress. This may bring competition for food and eventually changes in species–species interactions in limiting conditions.

The boring snails, *Polinices heros* and *Polinices duplicata* are predators of the clam *Mya arenaria*. *P. heros* lives essentially on open coasts, while *P.*

duplicata can enter polyhaline waters. Both species have similar feeding rates in sea-water, consuming about three clams a day. When salinity decreases, *P. heros* reduces its feeding rate much more than *P. duplicata*. This can induce considerable modifications in the ecological equilibrium between the three species (Hanks, 1952). The same kind of problem exists when considering the relation between oyster (*Crassostrea virginica*), oyster drills gastropods (*Eupleura caudata* and *Urosalpinx cinerea*) and salinity. *C. virginica* is fairly euryhaline and can easily live in polyhaline waters of low salinities. Feeding rates of the drills are high in sea-water and decrease sharply in polyhaline waters, 12.5‰ is near the lower limit for feeding activity. The relation established between the oyster and its predators thus becomes salinity dependent.

Various microfaunal and meiofaunal communities are very successful in reproducing and growing in estuarine conditions. They have short generation times and their production/biomass ratio tends to be considerably higher than for macrofauna. Moreover they can be found in extremely dense populations in estuarine habitats (see Gerlack, 1971); for instance the euryhaline meiofauna can contribute up to 58% of total respiration recorded in some areas (Lassere, Renaud-Mornant and Castel, 1976). In this context, a significant part of the benthic primary production which is converted to meiofaunal mass might not be passed upwards to higher trophic levels. Microfaunal and meiofaunal populations are particularly active in shallow estuarine waters during the summer. Summer growth retardation observed in various species of the benthophageous macrofauna colonizing these media could therefore be related, at least partly, to a trophic competition between these two types of population.

IV. CONCLUSIONS

Osmotic stresses constitute an inescapable part of life in media such as estuaries, salt marshes, mangroves, land-locked seas or semi-enclosed lagoons. In these ecotones with fluctuating salinities, the distribution of organisms is primarily dependent on their ability for osmoregulation. Animals can cope with their osmotic problems either by behavioural means or by adaptive physiological processes.

Effective physiological mechanisms of osmoregulation either at the level of the extracellular fluids or at the level of the intracellular medium are of course necessary prerequisites to a successful invasion of mixohaline waters. However, in most cases, species distribution cannot be solely related to behavioural and physiological osmoregulatory adaptive processes. The ecological picture is thus, in many instances, completely different from what could be expected on the basis of results of laboratory studies on osmoregulation abilities. In the field, direct salinity effects on organisms are indeed concomitant to indirect effects. Among these indirect effects are interactions of salinity with other physical or chemical factors of the environment. Species-specific responses may also in turn induce changes in the equilibrium established between communities. These species–species interactions are also of importance in determining the distribution of organisms. The many aspects of salinity as an ecological entity frequently make it difficult to assess the specific causes of an organismic or a community response. These causes are sometimes difficult to size up at the moment. However they are of the utmost importance and their study is urgently needed in order to come to a clear understanding of the various parameters which govern animal distributions in media with fluctuating salinity.

While many marine ecologists have sought to understand animal distributions in terms of the environment as a whole, we have tried in this review to demonstrate that the individual and population responses to the environment are determined primarily by adaptive peculiarities of their physiology, even in such complicated processes as trophic competition. These adaptive features are almost impossible to isolate and to study in the framework of field researches and should be more easily fixed in laboratory studies.

It is our belief that the fundamental studies undertaken in this respect should prove very rewarding in the future and should also find a large field of application; indeed, the ecotones we have been dealing with and which show large variations in environmental factors are generally areas showing decreased number of species to the benefit of an increase in individual numbers. They can thus provide good farming grounds for many species if one knows how to manage them.

REFERENCES

Abraham, M., N. Blanc, and A. Yashouv (1966). Oogenesis in five species of grey mullets (Teleostei, Mugilidae) from natural and landlocked habitats. *Israël J. Zool.*, **15**, 155–72.

Ahokas, R. A., and F. G. Duerr (1975). Salinity tolerance and extracellular osmoregulation in two species of euryhaline Teleosts, *Culea inconstans* and *Fundulus diaphanus*. *Comp. Biochem. Physiol.*, **52A**, 445–8.

Avens, A. C., and M. A. Sleigh (1965). Osmotic balance in gastropod mollusc. I. Some marine and littoral gastropods. *Comp. Biochem. Physiol.*, **16**, 121–41.

Baggerman, B. (1960a). Factors in the diadromous migration of fish. *Symp. Zool. Soc. Lond.*, **1**, 33–60.

Baggerman, B. (1960b). Salinity preference, thyroid activity and the seaward migration of four species of Pacific salmon. *J. Fish. Res. Bd Can.*, **17**, 296–322.

Bedford, J. J. (1971). Osmoregulation in *Melanopsis trifasciata*. IV. The possible control of intracellular isosmotic regulation. *Comp. Biochem. Physiol.*, **40A**, 1015–27.

Blanc-Livni, N., and M. Abraham (1970). the influence of environmental salinity on the prolactin and gonadotropin-secreting regions in the pituitary of *Mugil* (Teleostei). *Gen. Comp. Endocrinol.*, **14**, 184–97.

Boisseau, J., P. Lasserre, J. L. Gallis, and P. Cassifour (1975). Aspects écophysiologiques de l'osmorégulation et de l'évolution génitale de poissons mugilidés en milieu lagunaire. *J. Physiol. (Paris)*, **70**, 669–70.

Bricteux-Grégoire, S., Gh. Duchateau-Bosson, Ch. Jeuniaux, and M. Florkin (1964a). Constituants osmotiquement actifs des muscles adducteurs d'*Ostrea edulis* adaptée à l'eau de mer ou à l'eau saumâtre. *Arch. Int. Physiol. Biochem.*, **72**, 267–75.

Bricteux-Grégoire, S., Gh. Duchateau-Bosson, Ch. Jeuniaux, and M. Florkin (1964b). Constituants osmotiquement actifs des muscles adducteurs de *Gryphea angulata* adaptée à l'eau de mer ou á l'eau saumâtre. *Arch. Int. Physiol. Biochem.*, **72**, 835–42.

Bulnheim, H. P. (1969). Zur analyse geschlechtsbestimmen der faktoren bei Gammarus duebeni (Crustacea: Amphipoda). *Zool. Anz.*, **32**, 244–60.

Calabrese, A., and H. C. Davis (1970). Tolerances and requirements of embryos and larvae of bivalve molluscs. *Helgoländer Wiss. Meeresunters*, **20**, 553–64.

Clark, A. E. (1935). The effects of temperature and salinity on the early development of the oyster. *Prog. Rep. Atl. Biol. Stn*, **16**, 10.

Costlow, J. D., Jr. (1967). The effect of salinity and temperature on survival and metamorphosis of megalops of the blue crab *Callinectes sapidus*. *Helgoländer Wiss. Meeresunters*, **15**, 84–97.

Costlow, J. D., Jr., C. G. Bookhout, and R. Monroe (1966). Studies on the larval development of the crab *Rhithropanopeus harrisii* gould. I: the effect of salinity and temperature on larval development. *Physiol. Zool.*, **39**, 81–100.

Creutzberg, F. (1961). On the orientation of migrating elvers (*Anguilla vulgaris* Turt) in a tidal area. *Neth. J. Sea Res.*, **1**, 257–338.

Dennert, H. G., A. L. Dennert, P. Kant, S. Pinkster, and J. H. Stock (1969). Upstream and downstream migration in relation to the reproductive cycle and to environmental factors in the amphipod *Gammarus zaddachi*. *Bijdr. Dierk.*, **39**, 11–43.

Duchateau-Bosson, Gh., and M. Florkin (1961). Change in intracellular concentration of free amino acids as a factor of euryhalinity in the crayfish *Astacus astacus*. *Comp. Biochem. Physiol.*, **3**, 245–9.

Duchateau-Bosson, Gh., Ch. Jeuniaux, and M. Florkin (1961). Rôle de la variation de la composante amino-acide intracellulaire dans l'euryhalinité d'*Arenicola marina* L. *Arch. Int. Physiol. Biochem.*, **69**, 30–5.

Eckstein, B. (1975). Possible reasons for the infertility of grey mullets confined to fresh water. *Aquaculture*, **5**, 9–17.

Ellington, W. R., and J. M. Lawrence (1974). Coelomic fluid volume regulation and isosmotic intracellular regulation by *Luidia clathrata*. (Echinodermata: asteroidea) in response to hyposmotic stress. *Biol. Bull.*, **146**, 20–31.

Epstein, F. H., M. Cynamon, and W. McKay (1071). Endocrine control of Na–K–ATPase and seawater adaptation in *Anguilla rostrata*. *Gen. Comp. Endocrinol.*, **16**, 323–8.

Eriksson, S., and W. Tallmark (1974). The influence of environmental factors on the diurnal rhythm of the prosobranch gastropod *Nassarius reticulatus* from a non tidal area. *Zoon*, **2**, 135–42.

Fontaine, M. (1975). Physiological mechanisms in the migration of marine and amphihaline fish. In F. S. Russel and M. Yonge (Eds), *Advances in Marine Biology*, Vol. 13, Academic Press, New York. pp. 241–355.

Fontaine, M., and J. Marchelidon (1971). Amino acid contents of the brain and the muscle of young salmon (*Salmo salar* L.) at parr and smolt stages. *Comp. Biochem. Physiol.*, **40A**, 127–34.

Fujiya, M. (1970). Oyster farming in Japan. *Helgoländer Wiss. Meeresunters*, **20**, 464–79.

Garside, E. T., and Z. K. Chin-Yuen-Kee (1972). Influence of osmotic stress on upper

lethal temperatures in the cyprinodontid fish *Fundulus heteroclitus* (L.). *Can. J. Zool.*, **50**, 787–91.
Gerard, J. F., and R. Gilles (1972). The free amino-acid pool in *Callinectes sapidus* (Rathbun) tissues and its role in the osmotic intracellular regulation. *J. Exp. Mar. Biol. Ecol.*, **10**, 125–36.
Gerlach, S. A. (1971). On the importance of marine meiofauna for benthos communities. *Oceanol.*, **6**, 176–90.
Gilles, R. (1972a). Osmoregulation in three molluscs: *Acanthochitona discrepans* (Brown), *Glycymeris glycymeris* (L.) and *Mytilus edulis* (L.). *Biol. Bull.*, **142**, 25–35.
Gilles, R. (1972b). Biochemical ecology of Mollusca. In M. Florkin and B. Scheer (Eds), *Chemical Zoology*, Vol. VII, Academic Press, New York, London. pp. 467–99.
Gilles, R. (1972c). Amino-acid metabolism and isosmotic intracellular regulation in isolated surviving axons of *Callinectes sapidus*. *Life Sci.*, **11**, 565–72.
Gilles, R. (1973). Oxygen consumption as related to the amino-acid metabolism during osmoregulation in the blue crab *Callinectes sapidus*. *Neth. J. Sea Res.*, **7**, 250–89.
Gilles, R. (1974). Metabolisme des acides aminés et contrôle du volume cellulaire. *Arch. Int. Physiol. Biochem.*, **82**, 423–589.
Gilles, R. (1975a). Mechanisms of iono and osmoregulation. In O. Kinne (Ed.), *Marine Ecology*, Vol. 2, Part 1, Wiley–Interscience, London, New York. pp. 259–347.
Gilles, R. (1975b). Mechanisms of thermoregulation. In O. Kinne (Ed.), *Marine Ecology*, Vol. 2, Part 1, Wiley–Interscience, London, New York. pp. 251–8.
Haefner, P. A. (1969). Temperature and salinity tolerance of the sand shrimp *Crangon septemspinosa* Say. *Physiol. Zool.*, **42**, 388–97.
Hanks, J. E. (1952). The effect of changes in water temperature and salinity on the feeding habits of the boring snails *Polinices heros* and *Polinices duplicata*. *Rep. Invest. Schellfish. Mass.*, **5**, 33–7.
Harris, R. R. (1976). Extracellular space changes in *Carcinus maenas* during adaptation to low environmental salinity. *J. Physiol.*, **258**, 31–2.
Hochachka, P. W., and G. N. Somero (1973). *Strategies of Biochemical Adaptation*. W. B. Saunders, Philadelphia, London.
Holliday, F. G. T. (1971). Salinity–Animals–Fishes. In O. Kinne (Ed.), Vol. 1, Part 2, Wiley–Interscience, London, New York. pp. 997–1033.
Hoyaux, J., R. Gilles, and Ch. Jeuniaux (1976). Osmoregulation in molluscs of the intertidal zone. *Comp. Biochem. Physiol.*, **53A**, 361–5.
Jeuniaux, Ch., S. Bricteux-Grégoire, and M. Florkin (1962). Regulation osmotique intracellulaire chez *Asterias rubens*. Rôle du glycocolle et de la taurine. *Cah. Biol. Mar.*, **3**, 107–13.
Jeuniaux, Ch., Gh. Duchateau-Bosson, and M. Florkin (1961). Variation de la composante amino-acide des tissus et euryhalinité chez *Perinereis cultrifera* Gr. et *Nereis diversicolor* (O. F. Müller). *J. Biochem.* (*Tokyo*), **49**, 527–31.
Kähler, H. H. (1970). Uber den Einflus der Adaptations temperature und des Salzgehaltes auf die Hitze-und Gefrierresistenz von *Enchytraeus albidus* (Oligochaeta). *Mar. Biol.*, **5**, 315–24.
Kenny, R. (1969). The effects of temperature, salinity and substrate on distribution of *Glymenella torquata* (Leidy) Polychaeta. *Ecology*, **50**, 624–31.
Kinne, O. (1956). Uber den Einfluss des Salzgehaltes und der Temperatur auf Wachstum, Form und Vermehrung bei dem Hydroidpolypen *Cordylophora caspia* (Pallas), Athecata, Clavidae. I. Mitteilung über den Einfluss des Salzgehaltes auf Wachstum und Entwicklung mariner, brackischer und limnischer Organismen. *Zool. J.* (*Physiol.*), **66**, 565–638.

Kinne, O. (1964). The effects of temperature and salinity on marine and brackish water animals. II. Salinity and temperature–salinity combinations. *Oceanogr. Mar. Biol. A. Rev.*, **2**, 281–339.
Kinne, O. (1971). Salinity–Animals–Invertebrates. In O. Kinne (Ed.), *Marine Ecology*, Vol. I, Part 2, Wiley–Interscience, London, New York. pp. 821–995.
Kowalski, R. (1955). Untersuchungen zuer Biologie des Seesternes *Asterias rubens* L. in Brackwasser. *Kiel Meeresforsch.*, **11**, 201–13.
Lang, M. A., and H. Gainer (1969). Volume control by muscle fibers of the blue crab. Volume readjustment in hypotonic salines. *J. Gen. Physiol.*, **53**, 323–41.
Lasserre, P. (1976). Metabolic activities of benthic microfauna and meiofauna: Recent advances and review of suitable methods of analysis. In I. N. McCave (Ed.), *The Benthic Boundary Layer: NATO Sci. Conf. Les Arcs, France*, Plenum Press, New York, pp. 95–142.
Lasserre, P., and J. L. Gallis (1975). Osmoregulation and differential penetration of two grey mullets *Chelon labrosus* (Risso) and *Liza ramada* (Risso) in estuarine fish ponds. *Aquaculture*, **5**, 323–44.
Lasserre, P., J. Renaud-Mornant, and J. Castel (1976). Metabolic activities of meiofaunal communities in a semi-enclosed lagoon. Possibilities of trophic competition between meiofauna and mugilid fish. In G. Persoone and H. Jaspers (Eds.), *Proc. 10th Eur. Symp. on Marine Biology*, Universa Press Wetteren, Belgium, Vol. 2, pp. 393–414.
Lawrence, J. M. (1973). Level, content and caloric equivalents of the lipid, carbohydrate and protein in the body components of *Luidia clathrata* (Echinodermata: Asteroidea: Platyasterida) in Tampa Bay. *J. Exp. Mar. Biol. Ecol.*, **11**, 263–74.
Lent, C. M. (1968). Air-gaping by the ribbed mussel *Modiolus demissus* (Dillwyn): Effects and adaptative significance. *Biol. Bull.*, **134**, 60–73.
Manzi, J. J. (1970). Combined effects of salinity and temperature on the feeding, reproductive, and survival rates of *Eupleura caudata* (Say) and *Urosalpinx cinerea* (Say) (Prosobranchia: Muricidae). *Biol. Bull.*, **138**, 35–46.
McAlister, R. O., and F. M. Fisher (1968). Response of the false limpet, *Siphonaria pectinata* Linnaeus (Gastropodo, pulmonata) to osmotic stress. *Biol. Bull.*, **134**, 96–117.
McLusky, D. S. (1968). Some effects of salinity on the distribution and abundance of *Corophium volutator* in the Ythan estuary. *J. Mar. Biol. Ass. UK*, **48**, 443–54.
Nagabhushanam, R., and R. Sarojini (1965). The rate of water propulsion by the mussel *Branchidontes recurvus* (Mollusca: Lamellibranchiata). *Ind. J. Physiol. All. Sci.*, **19**, 1–14.
Percy, J. A. (1975). Ecological physiology of arctic marine invertebrates. Temperature and salinity relationships of the amphipod *Onisimus affinis* H. J. Hansen. *J. Exp. Mar. Biol. Ecol.*, **20**, 99–117.
Peterson, M. B., and F. G. Duerr (1969). Studies on osmotic adjustment in *Tegula funebralis* (Adams, 1854). *Comp. Biochem. Physiol.*, **28**, 633–44.
Pierce, S. K. (1971). Volume regulation and valve movements by marine mussels. *Comp. Biochem. Physiol.*, **39A**, 103–17.
Potts, W. T. W., and G. Parry (1964). *Osmotic and Ionic Regulation in Animals*. Pergamon Press, Oxford.
Remane, A. (1940). Einführung in die zoologische Okologie der Nord—und Ostsee. *Tierwelt N.—u. Ost see*, 1a.
Remane, A., and C. Schlieper (1971). *Biology of Brackish Water*. Wiley, New York.
Reuben, J. P., L. Girardier, and H. Grundfest (1964). Water transfer and cell structure in isolated crayfish muscle fibers. *J. Gen. Physiol.*, **47**, 1141–74.
Sanders, H. L., P. C. Mangelsdorf, Jr., and G. R. Hampson (1965). Salinity and faunal

distribution in the Pocasset river, Massachusetts. *Limnol. Oceanogr.* **10**, *Suppl. R.* 216–29.

Staaland, H. (1970). Volume regulation in the common *Buccinum undatum* L. *Comp. Biochem. Physiol.*, **34**, 355–65.

Schoffeniels, E., and R. Gilles (1970). Osmoregulation in aquatic arthropods. In M. Florkin and B. Scheer (Eds), *Chemical Zoology*, Vol. V, Academic Press, New York, London. pp. 255–86.

Schoffeniels, E., and R. Gilles (1972). Ionoregulation and osmoregulation in Mollusca. In M. Florkin and B. Scheer (Eds), *Chemical Zoology*, Vol. VII, Academic Press, New York, London. pp. 393–420.

Schütz, L. (1969). Okologische Untersuchungen über die Benthosfauna in Nordostseekanal III Autokologie der vagilen und hemissessilen Arten im Bewuchs der Pfähle: Makrofauna. *Int. Rev. Ges. Hydrobiol.*, **54**, 553–92.

Segestrale, S. G. (1969). Biological fluctuations in the Baltic sea. *Prog. Oceanogr.*, **5**, 169–84.

Stewart, M. G., and D. R. Bamfort (1976). The effect of environmental factors on the absorption of amino-acids by isolated gill tissue of the bivalve *Mya arenaria* (L.). *J. Exp. Mar. Biol. Ecol.*, **24**, 205–12.

Theede, H., and J. Lassig (1967). Comparative studies on cellular resistance of bivalves from marine and brackish waters. *Helgoländer Wiss. Meeresunters*, **16**, 119–29.

Turner, R. L., E. F. Lowe, and J. M. Lawrence (1975). Isosmotic intracellular regulation in the freshwater palaemonid shrimp *Palaemonetes paludosus* (Crustacea: decapoda). *Physiol. Zool.*, **48**, 235–41.

Traut, W. (1969). Zur sexualität von *Dinophilus gyrociliatus* (Archiannelida). I. Der einfluss von Aussendingungen und genetischen faktoren auf das Geschlechtsverhältnis. *Biol. Zbl.*, **88**, 469–95.

van Winkle, W., Jr. (1968). The effects of season, temperature and salinity on the oxygen consumption of bivalve gill tissue. *Comp. Biochem. Physiol.*, **26**, 69–80.

Zaugg, W. S., and L. R. McLain (1970). Adenosine triphosphatase activity in gills of salmonids. Seasonal variations and salt water influence in coho salmon (*Oncorhynchus kisutch*). *Comp. Biochem. Physiol.*, **35**, 587–96.

Part E

Pathology of Extracellular Fluid Regulation in Man

Part I

Chapter 14

Pathology of Extracellular Fluid Regulation in Man: Hormonal Aspects

J. J. LEGROS

I. INTRODUCTION	611
II. CONTROL MECHANISMS OF WATER INGESTION AND EXCRETION	612
A. Regulation of water ingestion: thirst	612
B. Regulation of water excretion	613
III CLINICAL ASPECTS OF EXTRACELLULAR FLUID VOLUME	616
A. Symptoms and general physiopathology	616
B. Symptomatology and physiopathology of some disorders	618
IV. CONCLUSIONS	631
ACKNOWLEDGEMENTS	631
REFERENCES	631

I. INTRODUCTION

The clinical approach to the pathology of the extracellular fluid in the human is hampered by the complexity of interactions between the different systems, neurological, endocrine and renal, described in the preceding chapters: the sick organism will compensate for a deficit or excess of water and electrolytes by several mechanisms, some of which are not easily measured by current clinical techniques. Progress in the assay of certain hormones in recent years has allowed a better understanding of some of these phenomena.

In this chapter on the pathology of extracellular fluid we will first briefly discuss the mechanisms which govern the ingestion and excretion of water in humans. Subsequently we will analyse the clinical manifestations of changes in the distribution of water and electrolytes which will enable us to relate a series of signs and symptoms to the particular movement of water and electrolytes between the different compartments of the body.

Finally, we will give examples of some disorders of the metabolism of water and electrolytes whose underlying pathology, cellular or molecular, is known (or at least partially understood) in order to illustrate the link between clinical research and the understanding of fundamental regulatory mechanisms. We will limit our discussion to the posterior pituitary function with special references to neurophysin synthesis and secretion; moreover we shall not discuss here the differentiated diagnosis nor the treatment of those diseases which are beyond the scope of this review.

II. CONTROL MECHANISMS OF WATER INGESTION AND EXCRETION

The mechanisms controlling water movements in animals have been discussed at length in some of the preceding chapters. We shall limit ourselves to consider here only those mechanisms which are well established in man.

A. Regulation of Water Ingestion: Thirst

Since a recent review of the physiological mechanisms of thirst has been published by Fitzsimons (1976), we will only recall some of the major points. Thirst constitutes the only factor regulating water ingestion in the conscious man. Thirst may be defined either as 'the subjective sensation of longing for an unsalted drink' or as 'the state which causes a man or an animal to drink water or a hypotonic liquid' (Peters, 1975).

The first definition refers to the subjective aspect of thirst which can only be studied in man. The intensity of the sensation bears only a weak relationship to the quantities of water really needed by the body. Thus, some people in a state of hyponatraemia (intracellular dehydration, see below) present with intense thirst but only drink small quantities of water when given the opportunity to drink freely (Schluss and Zatuchin, 1949). Conversely, patients who have lost more water than electrolytes (extracellular dehydration, see below) have only a feeble sensation of thirst although they drink considerable quantities of water.

The subjective sensation of thirst appears to be related to intracellular dehydration of the nuclei of the diencephalic centres and may be noxious when it develops in non-dehydrated individuals who, for example, suffer from psychological disturbances (compulsive water drinking, psychogenic polydipsia–polyuria) or organic disease affecting the hypothalamus (meningioma, craniopharyngioma, pinealoma) (Morton and coworkers, 1974a), or in the little understood disturbances which may accompany uraemia. In the latter, we assume that it is the angiotensin and/or renin circulating in excess which are responsible for these phenomena. Several studies by Fitzsimons have demonstrated the direct stimulant action of these substances on the thirst centres in animals. Indirect clinical evidence to support this explanation is provided by the fact that bilateral nephrectomy (causing a profound reduction in renin and angiotensin levels) induced a rapid and dramatic reduction in the sensation of intense thirst.

The second definition of thirst is more *objective* as it permits quantification of the need for water by measurement of the amount of liquid ingested: it is this definition which is used for experimental purposes. The final result of the thirst mechanism will be to restore the extra- and intracellular osmolarities (and probably also the blood volume) to normal. The regulation of this behaviour depends on cholinergic and adrenergic mechanisms. Intracellular osmolarity seems to play only a minor role in this regulation.

B. Regulation of Water Excretion

1. *Antidiuretic Hormone (ADH)*

The natural antidiuretic hormone in man is arginine vasopressin (AVP) whose molecular weight is around 1100. It is mainly synthesized in the supraoptic nuclei but also partly in the paraventricular nuclei in the diencephalon. We shall not discuss its mechanism of action on cells, principally renal, which has been studied previously (see Chapters 9 and 10). Within the hypothalamo–pituitary axis vasopressin is associated by non-covalent linkage with neurophysin: a polypeptide carrier whose molecular weight is about 10 000 (see review in Walter, 1975; Archer, 1976). There are two principal neurophysins which have similar physicochemical properties but differ in their electrophoretic behaviour. They are labelled I and II according to their speed of migration. *In vitro* the two neurophysins can bind antidiuretic hormone as well as oxytocin.

In recent years, the radioimmunoassay of neurophysins has shown that they are liberated at the same time as the two neurohypophysial hormones, both *in vivo* and *in vitro* and that *in vivo*, there is a specific association between one neurophysin and one hormone. This assay may be used to study the pathophysiology of the neurohypophysis in man or animals (see reviews by Robinson, 1975a and Legros, 1975a).

The main factors responsible for activation of the neurohypophysial hormones are hyperosmolarity or hypovolaemia.

In *acute* conditions, when there is a competition between the two stimuli, the need to conserve blood volume appears to predominate over the need to maintain a constant osmolarity (see Menard and coworkers, 1973). In man, haemorrhage and hypotension tend to release much more antidiuretic hormone than does hyperosmolarity (Noble and Taylor, 1953).

Nevertheless, in *chronic* conditions it is the plasma osmolarity which appears to be mainly responsible for the control of the liberation of the hormones. There is a relationship between the plasma osmolarity and the level of circulating ADH in conditions of moderate activation of the hypophysis (Robertson and Athar, 1976,). The threshold for activation of the liberation of ADH is 281 mosmol kg^{-1} when the individual is normovolaemic and recumbant. It is significantly reduced to 278 mosmol kg^{-1} by the relatively mild hypovolaemia which occurs in the upper part of the body when changing to an upright position. Conversely, the threshold is increasing to 282 mosmol kg^{-1} if hypervolaemia occurs as a result of perfusion with hypertonic saline (Robertson and Athar, 1976). In normal man, variations in blood volume only result in minimal adjustment of the threshold for liberation of hormones in response to osmotic stimuli.

2. The Renin–Angiotensin–Aldosterone System

Aldosterone by its sodium-sparing action can cause a secondary retention of water. Further, angiotensin and renin, by direct action on the diencephalic nuclei, may induce the release of ADH (Bonjour and Malvin, 1970). The latter may lead in turn to direct inhibition of the secretion of renin by the kidney (Vander, 1968). Lastly, angiotensin itself, although only in very high concentration may have an antidiuretic effect similar to that of vasopressin. These actions on metabolism in man (Fressinaud, Corvol and Menard, 1975) nevertheless show that it is important to study variations in this hormone system in pathological conditions affecting the regulation of extracellular fluid.

3. Prolactin

Prolactin has been isolated recently from the anterior pituitary in man and the isolation was soon followed by the establishment of specific radioimmunoassay (Hwang, Goyda and Friesen, 1971; Bryant and coworkers, 1971; L'Hermitte and coworkers, 1972).

This hormone appears to exert a direct antidiuretic effect associated with the retention of sodium and potassium in man (Horrobin and coworkers, 1971). Buckman and coworkers (1973) have indirectly shown a relationship between circulating prolactin levels and the state of hydration in man by demonstrating a constant fall in the level of this hormone after the absorption of 20 ml of water per kg body weight in normal individuals and in patients suffering from functional hyperprolactinaemia. Nevertheless these results have not been confirmed by a recent study using a similar experimental protocol (Adler and coworkers, 1975). Furthermore, patients with anterior pituitary adenoma secreting excess prolactin do not show any unusual responses to the water load test (Legros, unpublished results).

The role of this hormone in osmoregulation in man would thus appear to be minimal.

4. Other Hormones

a. *Thyroid Hormones and Adrenal Steroids* These have an indirect action on water metabolism by modifying the tubular reabsorption of sodium (corticosteroids) and/or increasing the glomerular filtration rate. Further Aubry and coworkers (1965) established that cortisol, acting on the hypothalamus, causes an elevation of the threshold for release of ADH in response to osmotic stimuli in man.

b. *Oestrogens* Oestrogens also play a role in controlling water and electrolyte metabolism in women and perhaps in normal men. In the

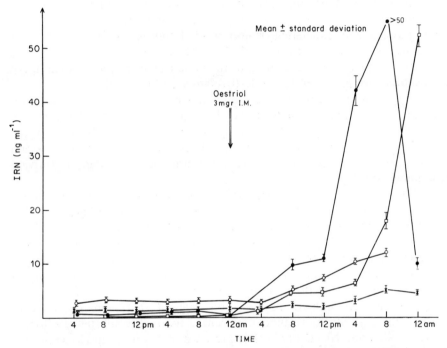

Figure 1 Effect of an intramuscular injection of 3 mg of oestriol on serum levels of immunoreactive neurophysins (IRN) in four normal men. Samples taken every 4 hours for 48 hours (24 hours control period and 24 hours after the injection). After Legros and Franchimont (1970); modified

renin–angiotensin–aldosterone system they augment the amount of substrate (angiotensinogen) which in turn leads to an increased activity of plasma renin, circulating angiotensin II and aldosterone excretion in the urine. Beside these effects, oestrogens also show direct antidiuretic properties and have a stimulant effect on the synthesis of mucopolysaccharides in the interstitial fluid, thus favouring the inhibition of water (see review in Christy and Shaver, 1974).

Finally, a stimulant effect of oestrogens on neurophysial function has also been described in normal man and woman (Legros and Franchimont, 1970; Robinson, Archer and Tolstoi, 1973; Cheng and Friesen, 1973). Such an effect is shown in Fig. 1; it has also been demonstrated in rat (Legros and

Grau, 1973). It appears to be due to direct action of these steroids on neurohypophysial tissue, a view supported by the partially inhibiting effect on oestrogen-induced neurophysin release of chlorpromazine, a central inhibitor of ADH release (Legros, Demoulin and Franchimont, 1975a). The presence of receptors which specifically bind oestrogens in neurohypophysial tissue may explain this action. An association between oestrogens and neurohypophysial function is also found in physiological conditions since in normal women there is a direct relationship between changing levels of neurophysins and circulating 17-β-oestradiol during the menstrual cycle (Legros, Frenchimont and Burger, 1975b). The levels of serum neurophysins are always lower in the first phase of the cycle (follicular phase) than in the two subsequent phases of the cycle (ovulatory and luteal phases).

c. *Progesterone* Progesterone in physiological doses, possesses a natriuretic effect on the kidney, rapidly masked by compensatory hyperaldosteronism (Landau and coworkers, 1955). Progesterone also partially inhibits the stimulating effect of oestrogens on neurohypophysial function (Legros and coworkers, 1973): this action was not confirmed by Robinson (private communication).

d. *Natriuretic Factor* The existence of a *natriuretic factor*, still called 'third factor' (De Wardener and coworkers, 1961; Cort and Lichardus, 1970) remains controversial. New techniques have enabled a factor to be isolated from bovine neurohypophysis which has been shown to be distinct from vasopressin, oxytocin, MSH and neurophysin. It has the property of binding *in vitro* to bovine neurophysins (Sedlakova and coworkers, 1974). It has not yet been isolated from the human hypophysis.

In contrast, preliminary investigations suggest that human renal cells possess the capacity to synthetize *in vitro* a natriuretic substance (Godon, 1975; Godon and Nizet, 1974). The presence of this polypeptide within the renal parenchyma might partly explain the important capacity for autoregulation of water and electrolyte demonstrated by Nizet (1969, 1973) in the isolated kidney of the dog. The role of this factor in human pathology has still to be determined.

e. *Non-hormonal Factors* We shall not discuss here the changes in the cardiovascular or nervous system which can lead to modification of the excretion of water essentially by causing variations in renal blood flow.

III CLINICAL ASPECTS OF EXTRACELLULAR FLUID VOLUME

A. Symptoms and General Physiopathology

The movements of water observed with changes in electrolytes in the extracellular compartment are summarised in Fig. 2.

Dehydration, as it bears primarily upon the extracellular compartment, is shown schematically in two categories according to whether the major loss is of water or electrolytes. If the major loss is from water compartment (diabetes insipidus), one observes a tendency to hypertonicity of the intravascular water and in compensation to the extracellular compartment: thus dehydration is essentially intracellular. In contrast, if the loss is mainly of electrolytes (renal tubular disorders, excess diuretics, excessive sweating,

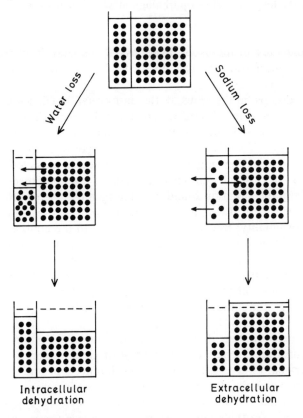

Figure 2 Scheme of the concentration of electrolytes (full circles) in the intracellular and extracellular compartments in state of water balance (above) and dehydration, intracellular (see left) or extracellular (see right). From Florkin and coworkers (1964); reproduced by permission of *Sciences et Lettres*

severe diarrhoea) this tends to cause extracellular dehydration hypotonicity with a consequent movement of water towards the intracellular compartment: there will be extracellular dehydration and a tendency to intracellular overhydration. The picture of extracellular dehydration is, in contrast, dominated by cardiovascular signs (tachycardia, hypotension) and biochemical changes (raised haematocrit, serum protein concentration and uraemia). Hyperosmolality and hypotension together cause activation of the renin–angiotensin–aldosterone system and of the neurohypophysial system causing marked oliguria (see a recent general review in Berl and coworkers, 1976).

B. Symptomatology and Physiopathology of some Disorders

1. *Central Diabetes Inspidus (DI)*

Central diabetes insipidus or hypophysial diabetes is characterized by polyuria–polydipsia (diuresis can be up to 15 l per day) due to deficient secretion of ADH.

The condition is not serious if the individual is conscious and if the hypothalamic thirst centre is intact. In this case, homeostasis of the extracellular milieu is maintained: there is no dehydration. In contrast if the individual is unconscious or if the thirst centre is unresponsive, rapid extracellular dehydration follows. This polyuria can be completely corrected by the action of exogenous ADH which demonstrates that the hormone receptor is normal as distinct from the situation found in nephrogenic diabetes insipidus (see p. 623). Urine density is always less than 1010 and urine osmolality always less than serum osmolality: free water is eliminated. In the conscious individual, blood osmolality is normal or slightly increased: it may rise rapidly in the conscious individual during the water deprivation test.

The level of circulating ADH is either undetectable (Husain and coworkers, 1973) or normal (Morton and coworkers, 1975a, 1975b) and do not increase during stimulation tests, indicating insufficiency of ADH reserve (Morton and coworkers, 1975a, 1975b).

ADH deficiency is due to an organic lesion in the hypothalamic–neurohypophysial region in 55% of cases. Lesions may be post-traumatic (cranial injury, neurosurgical interference), or due to compression of the hypothalamus (craniopharingioma, meningioma, internal hydrocephalus), or due to an invasive process (histyocytosis X, pinealoma with metastasis): the hypothalamic injury must be severe (more than 80%) and bilateral to cause the complete syndrome. During the water deprivation test the levels of neurophysins do not rise although there is marked dehydration as reflected by the hyperosmolality of the blood (see Fig. 3). In these cases there is a severe quantitative deficiency of the neurohypophysial material liberated. This is at variance with the situation observed in idiopathic diabetes insipidus (see p. 620).

When the injury is severe (post-trauma) the clinical features evolve classically in three phases: a transient hypothalamic shutdown leading first to a temporary diabetes insipidus, followed by a massive discharge of preformed hormone causing an oliguric state—the hypersecretion of ADH stage which some 8 days after the injury may be followed by true persistent diabetes insipidus.

Shridhar, Calvert and Ibbertson (1974) distinguished a syndrome of partial diabetes insipidus consisting of hypernatraemia with serum hyperosmolality and hypodypsia. The quantities of ADH liberated during a stimulation test should be normal as the abnormality lies at the level of hypothalamic osmoreceptors activated by a higher osmolality than in normal individuals. The syndrome is due to a disorder of osmoregulation whose pathology, organic or biochemical, is not known. The clinical picture bears a resemblance to that of neurogenic hypernatraemia in which there is always a detectable

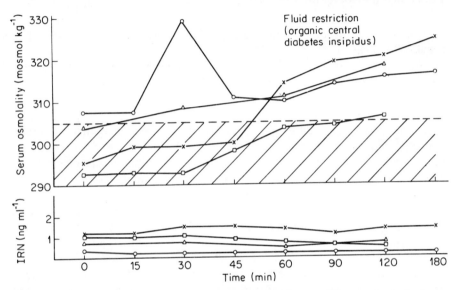

Figure 3 Serum osmolality and levels of neurophysins at different times during a water restriction test in four patients with organic diabetes insipidus. Modified from Legros (1976)

neurogical disorder (cerebral damage, tumour, hydrocephalus). The lesion always affects the hypothalamus which explains the hypodypsia and the relative lack of response of ADH.

In practice, it appears necessary to make a distinction between organic central diabetes insipidus *without* injury to the thirst centres with normal serum osmolality, and organic central diabetes insipidus *with* disturbance of the thirst centre (but sometimes without disturbance of consciousness) with an elevated serum osmolality. In those two categories, the ADH deficiency may be complete or partial. In the first case the 'free water clearance' is always positive which means that water is excreted in excess of electrolysis, even when an episode of dehydration is induced. In the second situation, it is possible to reduce the 'free water clearance' during dehydration tests but this reduction may be enhanced by the injection of exogenous ADH, thus suggesting an 'incomplete' storage of the hormone.

In 45% of cases, an organic cause for the central deficit cannot be found. This condition is thus called 'idiopathic'. It is rarely familial (4% of cases). Blotner (1958) suspected an autoimmune cause for the deficit by showing histological changes suggesting an antigen–antibody reaction in the hypothalamic nuclei. Antibodies to vasopressin (Roth and coworkers, 1966) and to neurophysins (Martin, 1971; Tissot-Berthet, Reinharz and Vallotton, 1975) may indeed be found in the blood of certain patients. However, their origin seems to be exogenous since in patients from a family with idiopathic central diabetes insipidus, antibodies are only present in those who had previously received therapy with posterior-hypophysial extract. They are never detectable in patients treated by synthetic hormone alone (Legros, 1976; Legros and Crabbé, unpublished results). One knows that synthetic hormone is less antigenic that neurophysins contained in glandular extracts used in pharmacological doses (Hurn and Landon, 1971). One could therefore postulate that treatment of transient diabetes insipidus by crude neurohypophysial extracts could induce the formation of an affection of immunological origin since bovine or porcine and human neurophysins have common antigenic properties (Legros, Hendrick and Franchimont, 1970; Watkins, 1971). This argument could also explain the phenomenon well known to clinicians of 'organized' transient diabetes insipidus: patients treated prematurely in practice rapidly develop the picture of true diabetes insipidus, which has been attributed until now to the hypothalamic–neurohypophysial axis 'going to sleep'.

In idiopathic disorders, the hormonal deficit seems to be more of genetic origin. The biochemical pattern resembles that seen in the rat (Brattleboro strain—Valtin, 1967) in which ADH and its associated neurophysins are deficient. In these animals, the deficiency may be situated either at the level of the stimuli responsible for the synthesis or liberation of ADH, or at the level of hormone–neurophysin complexes in the neurosecretory granules or finally at the level of the cleavage of the neurophysin–hormone complex.

We have recently shown that certain patients from a family with idiopathic diabetes insipidus can secrete the other neurohypophysial hormone (oxytocin), while Coculescu and Pavel (1973) demonstrated the presence of a closely related peptide (arginine vasotocin) in the cerebrospinal fluid of patients suffering from diabetes insipidus. Electrophoresis of the immunoreactive neurophysins circulating in these patients shows a lack of the major neurophysins migrating to the anode (hN_pI) normally associated with ADH within the normal neurohypophysis (Legros and Louis, 1973). On the other hand, the neurophysin associated with oxytocin (hN_pII) is detectable in patients who continue to secrete oxytocin. Further, when the neurohypophysis is stimulated by water deprivation test there is an increase in the level of the 'total immunoreactive neurophysins' and in three out of five

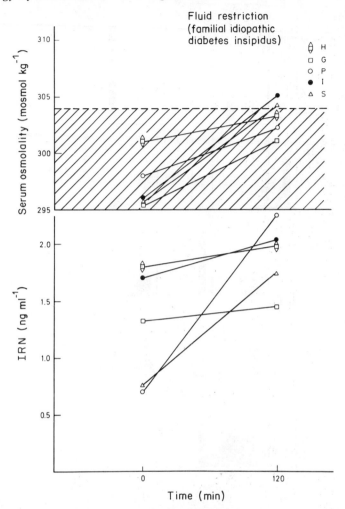

Figure 4 Serum osmolality and levels of total immunoreactive neurophysins before and after 2 hours of water restriction in five patients with familial idiopatic diabetes insipidus. After Legros and Crabbé, unpublished results

cases, a substance (or a group of substances) appears in the blood which paradoxically migrates to the cathode and reacts immunologically in the neurophysin radioimmunoassay (see Figs 4 and 5). This substance with paradoxical migration is absent from the blood of normal individuals but is, in contrast, present in the normal human neurohypophysis (Legros and Louis, 1973).

On the basis of these investigations, one may postulate that patients with idiopathic diabetes insipidus do not have a deficit in the mechanism of

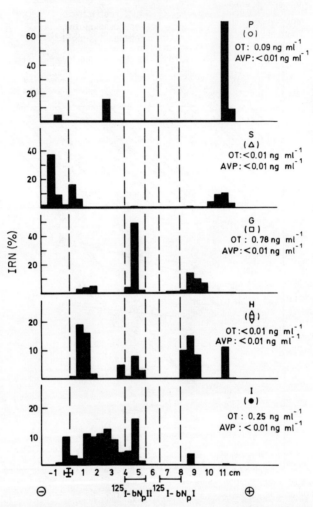

Figure 5 Electrophoretic migration pattern, on starch gel, of immunoreactive neurophysins present in the serum of five patients suffering from central idiopathic diabetes insipidus after 2 hours of dehydration. The gel was cut into $\frac{1}{2}$ cm strips. The migration of ^{125}I labelled bovine neurophysins (bN_p) I and II is shown for comparison. Results are expressed as a percentage of the total neurophysins present in the gel. There is a consistent lack of neurophysin I whereas neurophysin II, some neurophysin fractions and an immunoreactive substance showing paradixial migration to the cathode are present in variable quantities. The quantities of oxytocin and urinary ADH detected by radioimmunoassays is shown for each patient on the right. After Legros and Crabbé, unpublished results

hormone stimulation since dehydration results in liberation of neurohypophysial material. This fits with observations in Brattelboro rats where growth in the supraoptic and paraventricular nuclei has been observed during dehydration. In hereditary disorders the deficiency may involve either both hormones and both neurophysins, or one hormone (ADH) and one neurophysin (hN_pI). The biosynthesis of each hormonal system appears to be independent. The 'immunoreactive neurophysin' migrating to the cathode could be a common precursor of the neurophysins and of their hormones. The presence of such a precursor is postulated by Sachs and Takabatake (1974). This precursor might normally undergo enzymatic cleavage analogous to that seen in the conversion of pro-insulin to insulin leading to the liberation of active hormone on the one hand and neurophysins (waste products) on the other. In this hypothesis, one must admit the existence of a specific enzyme for the liberation of ADH and another for oxytocin, since one knows that both hormones and both neurophysins can be secreted separately. Such an enzymatic system has yet to be isolated from the neurohypophysis.

Although the pathogenesis of central organic diabetes insipidus is well defined the biochemical disorders responsible for idiopathic diabetes insipidus are not yet known with certainty. Nevertheless, there are several arguments for believing that the deficiency does not lie in the synthesis of a precursor common to the hormone and the neurophysins, nor in neurohypophysial stimuli, but is more likely to lie in the late stages preceding hormonal liberation. It may perhaps be related to a deficiency of a putative enzyme system responsible for cleavage of the hormone–neurophysin complex in the normal neurosecretory granules.

2. Nephrogenic Diabetes Insipidus

Primary nephrogenic diabetes insipidus is a congenital autosomal recessive condition affecting males (Cannon, 1955). It is characterized by polyuria–polydipsia (diabetes insipidus) due to a renal tubular disorder. The administration of exogenous ADH does not result in reduction of the 'free water clearance'. Moreover, the central osmoregulation system appears to be preserved in these patients; indeed, the levels of ADH are increased in states of dehydration and there is a relationship between plasma osmolality and the level of circulating ADH (see Robertson and coworkers, 1973) (Fig. 6). It may thus be considered that nephrogenic DI is due to a peripheral resistance to the hormone and its receptor or to a deficiency in the activation of cellular adenyl-cyclase.

Valtin and his coworkers have isolated a species of mice in which a genetic lesion analogous to nephrogenic DI can occur. Dousa and Valtin (1974) have shown that the activation of adenyl-cyclase by vasopressin in tubular cells from diseased mice was only 25% of that observed in controls. On the other hand, Fichman and Brooker (1972) showed in patients with nephrogenic DI, the absence of a rise in urinary cyclic AMP under the influence of exogenous ADH in contrast to that found in normal man. In fact, the deficiency is due to the lack of reaction between the hormone and its receptor rather than to a defect in a biochemical step implicated in the formation of cyclic-AMP. It has been showed by Dicker and Eggleton that the injection of exogenous vasopressin to patients suffering from nephrogenic DI results in an urinary excretion of 20 to 80% of the hormone in its biologically active form while the excretion is only 3 to 11% in normal individuals. It may thus be that the

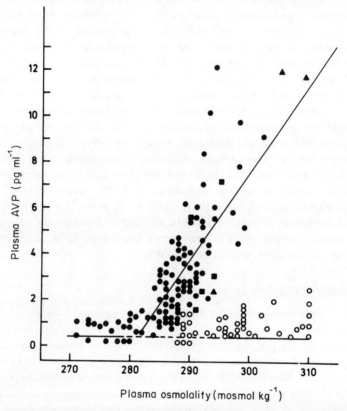

Figure 6 Levels of circulating ADH (ordinate) and plasma osmolality (abscissa) in: ●, 25 normal individuals; ▲, two patients with nephrogenic DI; ○, eight patients with central DI. From Robertson and coworkers (1973); reproduced by permission of the *Journal of Clinical Investigations*

hormone had not been fixed by its receptor (a stage prior to its utilization and metabolism) in the sick individual. However, it may also be that renal receptors could be saturated by the endogenous hormone present in large amounts thus preventing binding of exogenous hormone. In our present stage of knowledge we cannot be certain whether primary nephrogenic diabetes insipidus is due to a disorder of tubular receptors to ADH or to a disorder of the subsequent stage of renal enzymatic activity.

Primary nephrogenic diabetes insipidus must be distinguished from lesions secondary to either disturbances of water and electrolyte balance, or to other organic renal lesions, or the effect of certain drugs. Hypercalcaemia is the more important cause of disturbance of the renal concentrating mechanism: the mode of action is little understood. One knows that calcium inhibits binding of hormones to neurophysins *in vitro* and that there are 'immunoreactive neurophysins' in the kidney (Ginsburg and Jayanesa, 1968; Legros and coworkers, 1975c). However, these substances cannot be considered as hormone receptors because deaminated vasopressin which possesses normal or increased biological antidiuretic activity is not bound to neurophysin. One could nevertheless imagine that there is an analogy between the structure of neurophysins and of their receptors and that calcium, in supraphysiological amounts, inhibits hormone binding. This action might explain the inhibiting *in vitro* of the toad bladder (Petersen and Edelman, 1964). According to Jard and Bokaert (1975) the site of action of this ion is not at the level of the receptor: *in vitro* the total lack of calcium at a concentration of 10^{-6} mol l^{-1} restores the response whereas at higher concentrations (10^{-5} and 10^{-3} mol l^{-1}) it inhibits the response. In these three situations the affinity of the hormone for its receptor does not change significantly as evidenced by binding of tritiated hormone. According to these authors, the effects of calcium are therefore at the level of the linkage between the hormone and its receptor.

There are also neurohypophysial disturbances in chronic renal insufficiency. Neurophysin levels are elevated in patients suffering from renal impairment (Legros and Franchimont, 1972) and this is particularly marked in patients suffering from interstitial or tubular disorders (Legros, 1975b) (see Fig. 7). In these patients exogenous vasopressin perfusion (0.01 iu k^{-1}) does not lead to a reduction of the 'free water clearance' which occurs as a result of a failure of urine concentration. The chemical picture is analogous to that of primary nephrogenic diabetes insipidus. In a recent study we compared the levels of serum neurophysins and urinary excretion of neurophysins in two groups of patients suffering from mild renal insufficiency (creatinine clearance more than 30 ml min^{-1}). In those patients showing tubular disorders the levels of circulating neurophysins and excretion of urine neurophysins were significantly higher than in the group without tubular dysfunction (Legros, 1977). This would suggest that in the patients with tubular disorders there is hypersecretion of neurophysins due to the lack of renal response to antidiuretic hormone.

The mechanism responsible for this renal insensitivity is still not known. It is unlikely to be due to a hypothetical uraemic toxin because the creatine clearance is not significantly different in the two groups (63.19 ± 5.9 ml min^{-1} and 60.87 ± 9.68 ml min^{-1} respectively). For the same reason, it is difficult to accept the hypothesis that the polyuria might be due to a reduction in the number of functioning nephrons with consequent osmotic diuresis. This does not exclude the possibility that a reduction in the corticomedullary osmotic gradient secondary to a deficit in tubular cell function could be partly responsible for inactivation of the biological effect of ADH. The mechanism responsible for neurohypophysial activation in these patients also remains to be defined. It is unlikely to be due to an osmotic

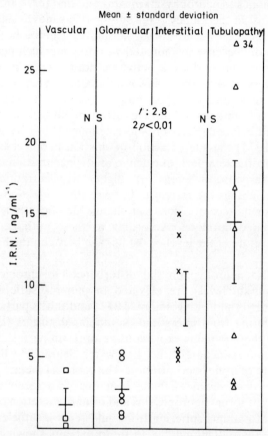

Figure 7 Levels of plasma neurophysins, in four groups of patients with moderated renal impairment classified according to their histological diagnosis. After Legros (1975b); modified

stimulus because the effective plasma osmolality is normal or low. It is possible, on the contrary that the tendency to hypovolaemia frequently found in these patients with tubular disease could be responsible for hormonal activity. Hypovolaemic stimuli swamping the osmotic regulatory mechanism might enhance the neurohypophysial response. Such a mechanism has also been postulated recently in patients with malignant hypertension where vasopressin levels have been found to be higher than in patients with benign hypertension whose vasopressin levels were low (Padfield and coworkers, 1976).

Finally, several drugs may also lead to the mixture of polyuria–polydipsia with resistance to exogenous ADH. The best known are lithium (used in psychiatry) and some antibiotics (certain tetracyclins). Direct stimulation of thirst centre, mainly by psychoactive lithium, has been suggested as the mechanism responsible for the polydipsia. However, it is clear that a nephrogenic action leading either to a decrease of the medullary gradient or to a lack of utilization of this gradient is also, in part, responsible for the polyuria (see a review in Singer and Forrest, 1976).

In summary, there are several conditions, congenital or acquired, which are accompanied by disordered urine concentration and neurohypophysial response. It is probable that there is more than one single pathological mechanism. An abnormality of the receptor–hormone link should be considered in primary DI and in some secondary disorders (organic disease, drugs). An enzyme deficiency for renal formation of adenyl-cyclase cannot be excluded in primary disorders and probably occurs in renal disorders, organic or secondary to ionic disturbances. Finally, one must not forget the possibility of modification of the corticomedullary renal gradient without which vasopressin is ineffective.

3. Ectopic ADH Syndrome (Schwartz and Bartter)

In patients suffering from neoplasm, usually oat cell tumour of the lung, Schwartz and coworkers (1957) recognized a clinical condition of overhydration, characterized by hypernatriuria and a concomitant hyponatraemia with low serum osmolality (may be as low as 220 mosmol kg^{-1}). The biochemical picture may be seen after excessive doses of exogenous vasopressin in normal individuals. These authors postulated that the major disturbances of osmoregulation was due to the autonomous secretion of vasopressin by the cancer.

The hypothesis proposed by Schwartz and coworkers (1957) has been confirmed by the demonstration of vasopressin in these tumours both by bioassay (Sawyer, 1967) and radioimmunoassay (Utiger, 1966).

The existence of neurophysins in cancer cells had been suspected on the basis of morphological criteria (George, Capen and Phillips, 1972) and has been confirmed by radioimmunoassay techniques (Hamilton, Upton and Amatruda, 1972; Legros and Louis, 1973).

The specificity of the association between vasopressin and neurophysin I and oxytocin with neurophysin II appears to be preserved in these tumours. In a tumour containing essentially oxytocin and little vasopressin it is neurophysin II which predominates whereas the reverse is found in tumours containing predominantly vasopressin (Legros, 1976) (see Fig. 8). This observation if confirmed in a larger number of cases suggests that in the dedifferentiation of neoplastic cells the association persists between the biosynthesis of one neurophysin and one hormone. This leads to the idea of a common genetic control of the neurophysin–hormone complex. Observations made in the course of idiopathic diabetes insipidus in the rat and in man (see p. 620) are also in agreement with this hypothesis.

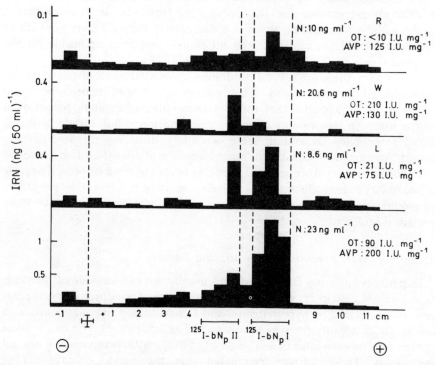

Figure 8 Electrophoretic migration pattern of immunoreactive neurophysins present in tumour extracts from patients with ectopic ADH secretion. The migration of ^{125}I labelled bovine neurophysins I and II is shown in order to locate the peak of immunoreactivity. The quantities of ADH and oxytocin in each tumour and the serum neurophysin concentration are noted on the right. After Legros (1976); modified

Levels of ADH in plasma (Beardwell, 1971; Morton and coworkers, 1975a) and urine (Merkelback and coworkers, 1975) are normal or elevated in these patients. The levels of neurophysins in blood are also normal or raised but can be low (Legros and Louis, 1973; Legros, 1976). In all cases, the inappropriate nature of hormone secretion is supported by the water load test according to Lee and coworkers (1961). Patients presenting the syndrome of inappropriate ADH secretion fail to excrete the excess of water during the five hours of the test (less than 80% of the excess) and the plasma osmolality falls consistently and sometimes may increase water intoxication.

Pathology of Extracellular Fluid Regulation in Man: Hormonal Aspects 629

This hypo-osmolality does not lead to reduction of the levels of vasopressin (Morton, Padfield and Forsling, 1975b) or neurophysins (Legros, 1975a) as it is the case in normal individuals.

Hypernatriuria persisting in spite of severe hyponatraemia is an important aspect of the clinical syndrome: the origin of increased elimination of sodium in the urine is still debated. A direct natriuretic effect of vasopressin *in vivo* seems to be excluded although the liberation of a natriuretic factor under the influence of hypervolaemia is still contested. A logical mechanism would be inhibition of the renin–aldosterone system due to hypervolaemia, however the resulting hyponatraemia constitutes a powerful stimulus to the secretion of renin which should counterbalance this effect. A recent study of mineralocorticoïd function in four patients suffering from a syndrome of ectopic ADH secretion demonstrated that the level of plasma aldosterone was normal and responded to the usual stimuli even though the plasma renin was abnormally low (Fischman, Michelekis and Morton, 1974). This dissociation, also observed in normal individuals treated with large doses of exogenous vasopressin, suggests the presence of a mechanism controlling aldosterone secretion independent of renin, the concentrations of sodium, potassium or ACTH. Since the result obtained by these authors failed to identify the factor responsible for hypernatriuria, perhaps it should be considered as a purely renal phenomenon.

Analogous to the conditions found in patients suffering from a syndrome of ectopic secretion of vasopressin a state of inappropriate ADH secretion called 'functional ADH hypersecretion' with overhydration and serum hypo-osmolality is suspected in certain other illnesses: hepatic cirrhosis, cardiac failure, neurological disorders, adrenal insufficiency, hypopituitarism or myxoedema. These disorders of osmoregulation have only rarely been studied in detail and hormonal estimations are usually lacking.

In certain cases, excess antidiuretic hormone might be excepted on the basis of our knowledge of physiology. This is the case of the 'post-commisurotomy' syndrome characterized by an abnormally high urine concentration in the days following surgery on the mitral valve. The hypersecretion of hormone might in these conditions, be explained by the brisk fall in pressure affecting the volume receptors of the left auricle. An effect on volume receptors could equally explain the increased vasopressin secretion suspected in certain patients suffering from cardiac failure, pneumothorax or other lung disorders. Finally, it is possible to find a disorder in the regulation of ADH secretion in certain neurological diseases (Hobson and English, 1963; Goodwin and Jenner, 1967; Linquette and coworkers, 1974) affecting the hypothalamic osmoregulation zones or nerve trunks supplying arterial baroreceptors or venous volume receptors.

In other conditions, essentially endocrine, the excess ADH is less clearly understood, e.g. hypopituitarism (Van't Hoff and Zilva, 1961; Davis and coworkers, 1969; Agus and Golberg, 1971), myxoedema (Chinitz and

Turner, 1965; Linquette, Lefebvre and Dessaint, 1973), adrenal failure (Ahmed and coworkers, 1967). Few posterior pituitary hormone estimations have been made in these cases. Only Ahmed and coworkers (1967) demonstrated the existence in large quantities of an antidiuretic substance in the serum of patients suffering from adrenal insufficiency. They also demonstrated a reduction in the quantity of this substance during glucocorticoïd replacement therapy. The bioassay used by these authors nevertheless does not exclude the possibility that the antidiuretic action could be due to circulating angiotensin which one knows is present in high concentration in these patients.

In a recent study we have examined the capacity to excrete a standard water load in 27 patients suffering from anterior pituitary failure. A disturbance of osmoregulation resulting in water intoxication with serum hypoosmolality by the end of the test was observed in 13 of these patients. The basal levels of blood neurophysin were normal in all cases and there was no significant fall in the serum level of neurophysin in the 13 patients in whom the test was abnormal in contrast with that observed in the other 14 patients. Further comparison of the remainder of the endocrine profile shows that 13 patients in whom the test was positive suffered from a greater impairment of adrenal function (17-OH steroids: 1.26 ± 0.31 mg/24 h) than the 14 others (2.03 ± 0.41 mg/24 h) ($2p < 0.05$).

It is also possible, as suggested by certain studies (see Dingman and Despointes, 1960; Aubry and coworkers, 1965; Travis and Share, 1971) that glucocortidoïd deficiency could be responsible for central disturbances of osmoregulation manifested by a lowered hypothalamic threshold to the liberation of neurohypophysial secretions. It is nevertheless possible that the direct action of glucocorticoïds at renal level might also account for the deficient elimination of water. Steroid deficiency results in a reduced GFR and this reduction could in its turn explain the persistence of unchanged circulating neurophysin levels: one knows that these substances, like ADH, are catabolized by the kidney (Johnson and coworkers, 1975) and we have recently shown a direct relationship between renal metabolic clearance of neurophysins and GFR (Legros and Nizet, unpublished results obtained on isolated kidney of dogs transplanted to the neck).

In summary, there are several conditions responsible for a syndrome of water intoxication manifested by serum hypo-osmolarity and urine hyperosmolality. This biochemical pattern may be achieved experimentally by injections of vasopressin: in certain cases an appropriate ADH secretion occurs either by biosynthesis in cancer tissue (Schwartz–Barrter syndrome) or secondary to a disturbance of one of the stages of hormone regulation (disorder of volume receptors or nerve trunks). In other cases, such as some endocrine deficiencies, the pathogenesis of disordered osmoregulation is still little understood. Thyroid or adrenal deficiency may by themselves lead to disturbances of renal haemodynamics resulting in reduced excretion of water.

It is equally probable that endocrine deficiency alters neurohypophysial regulation by direct action on hypothalamic osmoreceptors and/or by modifying the peripheral metabolism of ADH.

IV. CONCLUSIONS

There are numerous pathological conditions which are accompanied by disordered osmoregulation in man.

The study in depth of biological and hormonal changes seen in patients combined with fundamental experimental observations in whole animals or isolated cells, allows an understanding or at least, a rational approach to the mechanisms, cellular or molecular, underlying these disorders.

As we hope to have shown by a few examples, clinical observations and basic experimental studies are complementary. The study *in vitro* by the biochemist of isolated kidney cells should soon allow us to define the level of disturbance in patients suffering from nephrogenic diabetes insipidus: an abnormality of the hormone link at its receptor or a defect in cell enzyme systems? In the same way, the discovery of a state of excess ADH in certain patients must encourage continued research into the synthesis of analogues of vasopressin, capable of blocking hormone action at the level of the linkage between the vasopressin and its peripheral receptors in order to avoid the consequences, often fatal, of water intoxication.

On the other hand, certain observations made by clinicians in individual patients require consideration by the basic research worker and may orientate fundamental research. Is there an enzyme system responsible for cleavage of hormone–neurophysin in normal man as results obtained in idiopathic diabetes insipidus lead one to suppose? Does the same gene control the biosynthesis of the active hormone and its associated neurophysins as appears to be shown by results obtained in lung cancer? Are the neurophysins which are liberated at the same time as active hormones only waste products or do they play a role in circulating blood and affect the peripheral utilization of ADH?

ACKNOWLEDGEMENTS

We wish to thank Dr M. Tunbridge who agreed to translate the manuscript and Miss Hensens who typed it.

REFERENCES

Acher, R. (1976). Les neurophysines. Aspects moléculaires et cellulaires. *Biochim.*, **58**, 895–911.

Adler, R. A., G. L. Noel, L. Wartofsky, and A. G. Frantz (1975). Failure of oral water loading and intravenous hypotonic saline to suppress prolactin in man. *J. Clin. Endocrinol.*, **41**, 383–9.

Agus, Z. S., and M. Goldberg (1971). Role of antidiuretic hormone in the abnormal water diuresis of anterior hypopituitarism in man. *J. Clin. Invest.*, **50**, 1478–89.

Ahmed, A. B. J., B. C. George, C. Gonzales-Auvert, and J. F. Dingman (1967). Increased plasma arginine vasopressin in clinical adrenocortical insufficiency and its inhibition by glucosteroids. *J. Clin. Invest.*, **46**, 111–23.

Aubry, R. M., H. R. Nankin, A. M. Moses, and D. H. P. Streeten (1965). Measurement of the osmotic threshold for vasopressin release in human subjects and its modification by cortisol. *J. Clin. Endocrinol.*, **25**, 1481–92.

Beardwell, C. G. (1971). Radioimmunoassay of arginine vasopressin in human plasma. *J. Clin. Endocrinol.*, **33**, 154–260.

Berl, T., R. J. Anderson, K. M. McDonald, and R. W. Schrier (1976). Clinical disorders of water metabolism. *Kidney Int.*, **10**, 117–32.

Blotner, H. (1958). Primary or idiopathic diabetes insipidus: a system disease. *Metabolism*, **7**, 191–200.

Bonjour, J. P., and R. L. Malvin (1970). Stimulation of ADH release by the renin angiotensive system. *Am. J. Physiol.*, **218**, 1555–9.

Bryant, G. D., T. B. Silver, F. C. Greenwood, J. L. Pasteels, C. Robyn, and P. O. Hubinon (1971). Radioimmunoassay of a human pituitary prolactin in plasma. *Hormones*, **2**, 139–52.

Buckman, M. T., N. Kaminsky, M. Conway, and G. T. Peake (1973). Utility of L-dopa and water loading in evaluation of hyper-prolactinemia. *J. Clin. Endocrinol.*, **36**, 911–16.

Cannon, J. F. (1955). Diabetes insipidus: clinic and experimental studies with consideration of genetic relationship. *Arch. Int. Med.*, **96**, 215–72.

Cheng, K. W., and M. G. Friesen (1973). Studies of human neurophysin by radioimmunoassay. *J. Clin. Endocrinol.*, **36**, 553–60.

Chinitz, A., and F. L. Turner (1965). The association of primary hypothyroïdism and inappropriate secretion of the antidiuretic hormone. *Arch. Int. Med.*, **116**, 871–4.

Christy, N. P., and J. C. Shaver (1974). Estrogens and the kidney. *Kidney Int.*, **6**, 366–76.

Coculescu, M., and S. Pavel (1973). Arginine vasopressin-like activity of cerebrospinal fluid in diabetes insipidus. *J. Clin. Endocrinol.*, **36**, 1031–2.

Cort, J. H., and B. Lichardus (1970). *Regulation of Body Fluid Volumes by the Kidney*, Karger, Basel.

Davis, B. B., M. E. Bloom, J. B. Field, and D. H. Mintz (1969). Hyponatriema in pituitary insufficiency. *Metabolism*, **18**, 821–32.

De Wardener, H. E., I. H. Mills, W. F. Clapham, and C. J. Hayter (1961). Studies on the efferent mechanism of the sodium diuresis which follows the administration of intravenous saline in the dog. *Clin. Sci.*, **21**, 249–58.

Dingman, J. E., and R. Despoites (1960). Adrenal steroid inhibition of vasopressin from the neurohypophysial of normal subjects and patients with Addison's disease. *J. Clin. Invest.*, **39**, 1851–63.

Dousa, T. P., and H. Valtin (1974). Action of antidiuretic hormone in mice with inherited vasopressin-resistant urinary concentration defects. *J. Clin. Invest.*, **54**, 753–62.

Fischman, M. P., and G. Brooker (1972). Deficient renal cyclic adenose, 3'-,-5'-monophosphate production in nephrogenic diabetes insipidus. *J. Clin. Endocrinol.*, **35**, 35–47.

Fischman, M. P., A. M. Michelekis, and R. Horton (1974). Regulation of aldosterone in the syndrome of inappropriate antidiuretic hormone secretion (S.I.A.E.H.). *J. Clin. Endocrinol.*, **39**, 136–44.

Fitzsimons, J. T. (1976). The physiologic basis of thirst. *Kidney Int.*, **10**, 3–11.

Florkin, M., H. Van Cauwenberge, and P. Lefebvre (1964). *L'eau et les Electrolytes en Médecine Interne*, Sciences et Letres, Liège.
Fressinaud, P., P. Corvol, and P. Menard (1975). Dissociation de la rénine plasmitique et de l'hormone anti-diurétique urinaire dans divers états d'hydratation. In M. G. Chaumet and M. A. Gross (Eds), *Rein et Foie, Maladies de la Nutrition*, pp. 235–41.
George, J. M., C. C. Capen, and A. S. Phillips (1972). Biosynthesis of vasopressin *in vitro* and ultrastructure of a bronchogenic carcinoma. *J. Clin. Invest.*, **51**, 141–8.
Ginsburg, M., and K. Jayanesa (1968). The occurrence of antigen reacting with antibody to porcine neurophysin. *J. Physiol., Lond.*, **197**, 53–63.
Godon, J. P. (1975). Sodium and water retention in experimental glomerulonephritis: the urinary natriuretic material. *Nephron*, **14**, 382–9.
Godon, J. P., and A. Nizet (1974). Release by isolated dog kidney of a natriuretic material following saline loading. *Arch. Int. Physiol. Biochim.*, **84**, 309–11.
Goodwin, J. C., and F. A. Jenner (1967). The cyclical excretion of a sodium-thioglycollate-resistant antidiuretic factor by a periodic psychotic. *Endocrinol.*, **38**, xxiv.
Hamilton, B. P. M., G. V. Upton, and T. T. Amatruda (1972). Evidence for the presence of neurophysin in tumors producing the syndrome of inappropriate antidiuresis. *J. Clin. Endocrinol.*, **35**, 764–7.
Hobson, J. A., and J. T. English (1963). Self induced water intoxication, case study of chronical schizophrenic patient with physiological evidence of water retention due to inappropriate release of antidiuretic hormone. *Ann. Int. Med.*, **58**, 324–32.
Horrobin, D. F., I. A. Lloyd, A. Lipton, P. G. Burstyn, N. Durkin, and K. L. Muiruri (1971). Actions of prolactin on human renal function. *Lancet*, **ii**, 352–4.
Hurn, B. A. L., and J. Landon (1971). Antisera for radioimmunoassay. In K. E. Kirkham and W. M. Hunter (Eds), *Radioimmunoassay Methods*, Churchill-Livingstone, Edinburgh. pp. 121–42.
Husain, M. K., N. Fernando, M. Shapiro, A. Kagan, and S. M. Glick (1973). Radioimmunoassay of arginine vasopressin in human plasma. *J. Clin. Endocrinol.*, **37**, 616–25.
Hwang, P. H., H. Goyda, and H. Friesen (1971). A radioimmunoassay for human prolactin. *Proc. Natl Acad. Sci. USA*, **68**, 1902–6.
Jard, S., and J. Bockaert (1975). Stimulus-response coupling in neurohypophysial peptide target cells. *Physiol. Rev.*, **55**, 489–536.
Johnston, C. I., J. S. Hutchinson, B. J. Morris, and E. M. Dax (1975). Release and clearance of neurophysins and posterior pituitary hormones. *Ann. NY Acad. Sci.*, **248**, 272–80.
Landau, R. L., D. M. Bergenstal, K. Lugibihl, and M. E. Kascht (1955). Effects of progesterone in man. *J. Clin. Endocrinol.*, **15**, 1194–215.
Lee, W. Y., H. A. Grumer, D. Bronsky, and S. S. Waldstein (1961). Acute water loading as a diagnostic test for the inappropriate ADH syndrome. *J. Lab. Clin. Med.*, **58**, 937–1000.
Legros, J. J. (1975a). The radioimmunoassay of human neurophysins: contribution to the understanding of the physiopathology of neurohypophysial function. *Ann. NY Acad. Sci.*, **248**, 281–303.
Legros, J. J. (1975b). Blood levels of vasopressin neurophysin in patients with kidney diseases. In M. G. Chaumet and M. A. Gross (Eds), *Rein et Foie. Maladies de la Nutrition*, Vittel. pp. 635–43.
Legros, J. J. (1976). *Les Neurophysines. Recherches Méthodologiques Expérimentales et Chimiques*, Masson, Paris.
Legros, J. J. (1977). Urinary excretion of neurophysins in patient with kidney disease. Submitted for publication.

Legros, J. J., A. Demoulin, and P. Franchimont (1975a). Influence of chlorpromazine on positive and negative feedback mechanism of estrogens in man. *Psychoneuroendocrinology*, **1**, 158–98.

Legros, J. J., and P. Franchimont (1970). Influence de l'oestriol sur le taux de la neurophysine sérique chez l'Homme. Comparaison avec la capacité de fixation plasmatique des polypeptides post-hypophysaires marqués étudiés *in-vitro*. *C.R. Soc. Biol.*, Paris, **164**, 246–50.

Legros, J. J., and P. Franchimont (1972). Human neurophysin blood levels under normal, experimental and pathological conditions. *Clin. Endocrinol.*, **1**, 99–113.

Legros, J. J., and F. Louis (1973). Identification of a vasopressin–neurophysin and an oxytocin–neurophysin in man. *Neuroendocrinology*, **13**, 371–5.

Legros, J. J., P. Franchimont, and H. G. Burger (1975b). Variations of neurohypophysial function in normally cycling women. *J. Clin. Endocrinol.*, **91**, 54–9.

Legros, J. J., A. Govaerts, A. Demoulin, and P. Franchimont (1973). Interactions entre un dérivé progestatif (acétate de moréthistérone) et l'éthinyl ostradiol sur l'élimination urinaire de neurophysines, d'ocytocine et de vasopressine immunoréactives et sur le taux de neurophysine sérique I et II chez l'Homme normal. *C.R. Soc. Biol.*, Paris, **167**, 1668–72.

Legros, J. J., and J. D. Grau (1973). Effect of ethinyl-estradiol on neurohypophysial active compounds in rats. *Nature New Biol.*, **241**, 247–9.

Legros, J. J., J. C. Hendrick, and P. Franchimont (1970). Comparaison entre la composition en acides aminés et le comportement immunologique de la neurophysine bovine et d'une substance extraite parallèlement de la post-hypophyse huamine. *C.R. Soc. Biol.*, Paris, **164**, 2389–95.

Legros, J. J., F. Louis, U. Grötschel-Stewart, and P. Franchimont (1975c). Presence of immunoreactive neurophysin-like material in human target organs and pineal gland: physiological meaning. *Ann. NY Acad. Sci.*, **248**, 151–7.

L'Hermitte, M., P. Delvoye, J. Nonkin, M. Wekemans, and C. Robyns (1972). Human prolactin secretion as studied by radioimmunoassay: some aspects of its regulation. In A. R. Boyns and K. Griffiths (Eds), *Prolactin and Carcinogenesis*, Alpha Omega Alpha, Cardiff. pp. 81–100.

Linquette, M., P. Fossatt, J. Lefebvre, J. P. Cappoen, and C. Chopin (1974). Evolution d'un syndrome d'intoxication par l'eau au cours d'une psychose maniaco-dépressive. *Ann. Endocrinol.*, Paris, **35**, 127–38.

Linquette, M., J. Lefebvre, and J. P. Dessaint (1973). Les perturbations hydro-electrolytiques au cours de l'hypothyroïdie primaire. In *Hormones et Régulations Métaboliques*, Masson, Paris. pp. 105–16.

Martin, M. J. (1971). Demonstration of circulating antibodies to neurophysin in patients treated with pitressin. *J. Endocrinol.*, **49**, 553–54.

Menard, J., P. Fressinard, J. Breminer, A. Foliot, and P. Corvol (1973). Les méthodes de dosage de l'hormone antidiurétique: leur application à l'étude de la sécrétion de vasopressine. In *Hormones et Régulations Métaboliques*, Masson, Paris. pp. 61–80.

Merkelbach, U., P. Czernichow, R. C. Gaillard, and M. B. Vallotton (1975). Radioimmunoassay of (8-arginine)-vasopressin. II. Application to determination of antidiuretic hormone in urine. *Acta Endocrinol.*, Kbh., **80**, 453–64.

Morton, J. J., J. J. Brown, R. H. Chinn, R. F. Lever, and J. I. S. Robertson (1975a). A radioimmunoassay for plasma antidiuretic hormone and its application in case of hypopituitarism associated with a loss of thirst. In G. Peters, J. T. Fitzsimons and Peters-Haefli (Eds), *Control Mechanism of Drinking*, Springer, Heidelberg, New York. pp. 14–18.

Morton, J. J., P. L. Padfield, and M. L. Forsling (1975b). A radioimmunoassay for plasma arginine-vasopressin in men and dogs: application to physiological and pathological states. *J. Endocrinol.*, **65**, 411–24.

Nizet, A. (1969). The role of autonomous renal mechanisms in the control of sodium and water balance. *Urol. Nephrol.*, **1**, 361–71.
Nizet, A. (1973). The mechanisms of fast renal compensation. *Pflügers Arch.*, **341**, 209–17.
Noble, R. L., and N. B. G. Taylor (1953). Antidiuretic substances in human urine after haemorrhage, fainting, dehydration and acceleration. *J. Physiol., Lond.*, **122**, 220–37.
Padfield, P. L., A. F. Lever, J. J. Brown, J. J. Morton, and J. I. S. Robertson (1976). Changes of vasopressin in hypertension: cause or effect? *Lancet*, **1**, 1255–7.
Peters, G. (1975). Utilité et inutilité de la soif. In M. G. Chaumet and M. A. Gross (Eds), *Rein et Foie. Maladies de la Nutrition*, Vol. 16B, Vittel. pp. 143–8.
Peterson, M. J., and I. S. Edelman (1964). Calcium inhibition of the action of vasopressin on the urinary bladder of the toad. *J. Clin. Invest.*, **43**, 583–94.
Posner, J. B., H. E. Norman, R. J. Kossmann, and L. C. Scheinberg (1967). Hyponatraemia in acute polyneuropathy. *Arch. Neurol.*, **17**, 530–41.
Robertson, G. L., and S. Athar (1976). The interaction of blood osmolality and blood volume in regulating plasma vasopressin in man. *J. Clin. Endocrinol.*, **42**, 1298–305.
Robertson, G. L., E. A. Mahr, S. Athar, and T. Sinha (1973). Development and clinical application of a method for the radioimmunoassay of arginine vasopressin in human plasma. *J. Clin. Invest.*, **52**, 2340–52.
Robinson, A. G. (1975a). Radioimmunoassay of neurophysin proteins: utilization of specific neurophysin assays to demonstrate independent secretion of different neurophysins *in vivo*. *Ann. NY Acad. Sci.*, **248**, 246–56.
Robinson, A. G. (1975b). Isolation, assay and secretion of individual human neurophysins. *J. Clin. Invest.*, **55**, 360–7.
Robinson, A. G., D. F. Archer, L. F. Tolstoi (1973). Studies of neurophysin during oxytocin related events in women. *J. Clin. Endocrinol.*, **37**, 645–52.
Roth, J., S. M. Glick, L. A. Klein, and M. J. Peterson (1966). Antibody to vasopressin in man. *J. Clin. Endocrinol.*, **26**, 671–5.
Sachs, H., and Y. Takabatake (1964). Evidence for a precursor in vasopressin biosynthesis. *Endocrinology*, **75**, 943–8.
Sawyer, W. H. (1967). Pharmacological characteristics of the antidiuretic principle in a bronchogenic carcinoma from a patient with hyponatraemia. *J. Clin. Endocrinol.*, **27**, 1497–9.
Schluss, L. A., and J. Zatuchni (1949). Syndrome of salt depletion induced by a regimen of sodium restriction and sodium diuresis. *J. Am. Assoc.*, **139**, 1136–9.
Schwartz, W. B., W. Bennett, S. Curelops, and F. C. Bartter (1957). A syndrome of renal sodium loss and hyponatremia probably resulting from inappropriate secretion of antidiuretic hormone. *Am. J. Med.*, **23**, 529–42.
Sedlakova, E., Z. Pruzik, J. Skopkova, T. Barth, I. Kluh, and J. H. Cort (1974). Isolation of a tridecapeptide from natriuretic fractions of bovine posterior pituitary. *Eur. J. Clin. Invest.*, **4**, 285–92.
Shridhar, C. B., G. D. Calvert, H. K. Ibbertson (1974). Syndrome of hypernatraemia, hypodypsia and partial diabetes insipidus: a new interpretation. *J. Clin. Endocrinol.*, **38**, 891–901.
Singer, I., and J. V. Forrest (1976). Drug-induced states of nephrogenic diabetes insipidus. *Kidney Int.*, **10**, 82–95.
Tissot-Berthet, M. C., A. C. Reinharz, and M. B. Vallotton (1975). Radioimmunoassay of neurophysin I and II in human plasma. *Ann. NY Acad. Sci.*, **248**, 257–71.
Takabatake, Y., and H. Sachs (1964). Vasopressin biosynthesis: III *in vitro* studies. *Endocrinology*, **75**, 934–45.
Travis, R. H., and L. Share (1971). Vasopressin–Renin–Cortisol interrelations. *Endocrinology*, **89**, 246–53.

Utiger, R. D. (1966). Inappropriate antidiuresis and carcinoma of the lung: detection of arginine vasopressin in tumor extracts by immunoassay. *J. Clin. Endocrinol.*, **26**, 970–4.

Valtin, H. (1967). Hereditary hypothalamic diabetes insipidus in rats (Brattleborg strain). *Am. J. Med.*, **42**, 814–27.

Vander, A. J. (1968). Inhibition of renin release in the dog by vasopressin and vasotocin. *Circulat. Res.*, **23**, 605–9.

Van't Hoof, W., and J. F. Zilva (1961). Chromophobe adenoma and hyponatraemia. *Clin. Sci.*, **21**, 345–54.

Watkins, W. B. (1971). Neurophysins of the human pituitary gland. *J. Endocrinol.*, **51**, 595–6.

Walter, R. (Ed.) (1975). Neurophysins: carriers of peptide hormones. *Ann. NY Acad. Sci.*, **248**, 512.

Addendum Section

Bringing to completion a multi-author treatise usually takes a long time. There is often a delay of at least two years between the first contact with the contributors and the issue of the printed book. This means that the newest information contained in such a volume is about three years old at the time of issue.

The addendum section presented in this book has been conceived to provide the authors with a means of updating their contribution if necessary. Rather than simply offering new data, its first aim is to give a brief and critical presentation of selected information which might bring the reader to new ideas or concepts. Being edited and printed at the end of the overall publication procedure, this section is thus essentially devoted to what has been considered by the authors as new and important advances in the field they covered.

R. GILLES

Addendum to Chapter 1

Structure and Properties of Water in the Cells

D. A. T. DICK

Since this chapter was written a useful symposium on water structure and transport in biology has appeared (Richards and Franks, 1977). This is contributed mainly by physical chemists and deals with protein solvation, water structure, physical chemistry of water transport, and cryobiology.

Two recent papers have clarified the effect of haemoglobin on the osmotic behaviour of erythrocytes. Hladky and Rink (1978) have measured the changes in external pH, erythrocyte volume, and membrane potential when the concentration of the external unbuffered NaCl medium is doubled. In all cases the changes were inconsistent with significant variations in Cl content of the cells; they concluded that the data of Gary-Bobo and Solomon (1968) showing such variations must be incorrect. Hladky and Rink consequently rejected Gary-Bobo and Solomon's hypothesis of concentration-dependent variations in the charge on the haemoglobin molecule; non-ideality must result simply from variation in the osmotic coefficient of haemoglobin (Dick, 1966). In explaining this variation, Hladky and Rink have suggested that interaction between haemoglobin molecules in concentrated solutions creates a negative fluid pressure within the cell. This is an attractively concrete way of looking at the phenomenon; it will be useful if it aids a general understanding of what is taking place. Two points must be borne in mind, however: (1) The concept of negative fluid pressure cannot be used to derive quantitative estimates of the entropy of mixing as can the concept of excluded volume, which is therefore more satisfactory for thermodynamic analysis; (2) in any case the anomalous entropy of mixing, whether visualized in terms of excluded volume or of negative fluid pressure, is only one, though probably the main one, of a number of effects producing the anomalous osmotic coefficient of haemoglobin; other effects such as water binding (or strictly preferential hydration) and ion binding must still be remembered even though they are probably only of secondary importance.

Freedman and Hoffman (1977) have recently shown that when an erythrocyte is rendered permeable to cations by means of nystatin its internal buffering capacity does not change with cell volume; this result also contradicts Gary-Bobo and Solomon's hypothesis of concentration-dependent variations of haemoglobin molecular charge.

REFERENCES

Dick, D. A. T. (1966). *Cell Water*, Butterworths, London.

Freedman, J. C., and J. F. Hoffman (1977). Hemoglobin charge and membrane potentials in human red cells at varying volumes. *Proc. XXVII Int. Cong. Physiol. Sciences*; Published by the Congress Secretary's Office-Pitié-Salpetrière-Paris. Abstract 694, p. 238.

Gary-Bobo, C. M., and A. K. Solomon (1968). Properties of hemoglobin solutions in red cells. *J. gen. Physiol.*, **52**, 825–853.

Hladky, S. B., and T. J. Rink (1978). Osmotic behaviour of human red cells: an interpretation in terms of negative intracellular fluid pressure. *J. Physiol.*, **274**, 437–446.

Richards, R. E. and F. Franks (1977). A discussion on water structure and transport in biology. *Phil. Trans. R. Soc. Lond.* B278, 1–205.

Addenda to Chapter 3

Intracellular Inorganic Osmotic Effectors

G. RORIVE and R. GILLES

(A) General Comments

G. Rorive and R. Gilles

Since the preparation of this chapter, an important review on NMR studies of water and ions within cells has been published (Shporer and Civan, 1977). The NMR data discussed in this review strongly suggest that fractional immobilization of intracellular Na^+ is very slight (1%). On the other hand, the enhancement of the rates of nuclear magnetic relaxation observed when studying intracellular Na^+ appears to be non-specific; it could be best explained by the existence of a condensed phase of cations associated with the surface of charged polyelectrolyte macromolecules. This condensation would not immobilize Na^+, whose freedom of motion appears to remain very large.

From studies on the effect of hypo-osmotic shocks on renal cortex slices from rats, rabbits, and guinea pigs (Hughes and MacKnight, 1976), it appears that intracellular K^+ does not participate in the volume regulation process occurring in this tissue under such conditions. On the contrary, an important decrease in tissue Na^+ content takes place within the first two minutes following application of the hypo-osmotic conditions. There is however no further modification in the Na^+ level. These results can be interpreted in the light of the hypothesis presented in Chapter 3 considering two distinct phases in the overall volume control process. Indeed, it can be considered that the fast decrease in Na^+ is implicated in an early phase of volume swelling limitation but not in the subsequent slower phase of volume regulation. If it was the case, a slower, further decrease should have been observed following the initial fast drop in Na^+ level. Further experiments on rat kidney cortex slices recently run in our laboratory confirm this interpretation. As shown in Fig. 1, application of hypo-osmotic conditions brings about an important decrease in the intracellular level of Na^+, K^+, and ninhydrin positive substances (NPS). These changes are concomitant with the rapid dilution of the intracellular fluid which can account for most of the early decrease in K^+ but not in Na^+ and NPS. These results moreover show that the NPS content is further decreasing during the volume regulation phase. Volume regulation would rather be achieved through control of osmotic effectors other than ions among which amino acids would play an important part. In such a view, and when considering physiological conditions, could the so-called 'inert' components postulated by Kleinzeller in the following comment be organic osmotic effectors of low molecular weight?

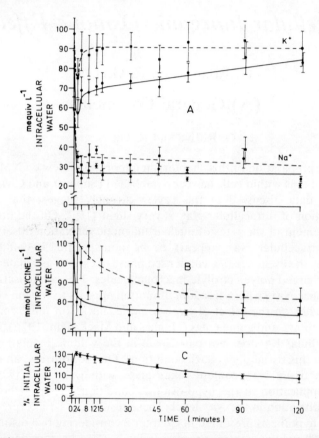

Figure 1 Effect of a hypo-osmotic shock on the intracellular level of Na^+ and K^+ (A), of ninhydrin positive substances (B) and on the cellular hydration (C) of rat kidney cortex slices. The hypo-osmotic stress is achieved by modification of the NaCl content of the incubation saline (control: 128 mmol l^{-1}-hypo-osmotic shock: 67 mmol l^{-1})—Broken line in A and B: values corrected for changes in cellular hydration. After M. Henrard-Seel (unpublished).

REFERENCES

Hughes, P. M., and A. D. C. MacKnight (1976). The regulation of cellular volume in renal cortical slices incubated in hypo-osmotic medium. *J. Physiol.*, **257**, 137–154.

Shporer, M., and M. M. Civan (1977). The state of water and alkali cations within the intracellular fluids: the contribution of NMR spectroscopy. In Bronner, F. and Kleinzeller, A. (Eds) *Current Topics in Membranes and Transport*, Vol. 9, Academic Press, New York, pp. 1–69.

(B) Comments on the ouabain-insensitive component of cell volume control

A. Kleinzeller
Department of Physiology, School of Medicine, University of Pennsylvania, Philadelphia, 19104, USA

Efforts to elucidate the mechanism of the ouabain-insensitive component of cell volume control have yielded some discrepant results. Based on studies in which the phenomenon was investigated by first loading cells with NaCl at 0 °C, subsequent switching-on of metabolism by aerobic incubation at more physiological temperatures indicated that the ouabain-insensitve transport system lacks cationic specificity (Kleinzeller, 1972). On the other hand, metabolizing cells exposed to variations of external osmolarity appeared to regulate cell volume by a controlled flux of K^+; some other studies, however, suggested that Na^+, rather than K^+, was involved (compare, for example, Kregenow (1971) and Schmidt and McManus (1977) for duck erythrocytes). To some extent such discrepancies may be accounted for by the actual design of experiments. This point will be illustrated by some hitherto unpublished observations of Kleinzeller and Knotkova.

Hughes and MacKnight (1976) found that upon placing renal cortical slices at a steady state of tissue components (O_2, 25 °C, isotonic balanced medium) into media made hypotonic by reduction of $[Na^+]_0$ the ouabain-insensitive cell volume control system was primarily operative by variations in cell Na^+. We have carried out similar experiments but maintained external Na^+ and K^+ rigorously constant, varying external osmolarity by using solutes assumed not to be actively transported. Figure 1 shows that with Li^+ as the osmotic agent, the changes in cell volume were primarily produced by a net K^+ (and Cl^-) flux, tissue Na^+ remaining constant. Not shown are data demonstrating that the adjustment of tonicity between the cells and the medium is produced also by net fluxes of Li^+. Figure 2 demonstrates the same point by a different experimental approach. Here, the tissue was first loaded with solutes and water at 0 °C and then incubated (15 °C, O_2) in media of different osmolarity at constant $[Na^+]_0$ and $[K^+]_0$ until a new steady state was reached. It will be seen that the net extrusion of Na^+ was independent of variations of external osmolarity. The osmotically dependent changes in fluxes of K^+ were greatest with LiCl, whereas sucrose had no effect. The insert, in which the ratio Na^+

extruded/K$^+$ accumulated was plotted against external osmolarity shows that the greater the membrane permeability for the 'inert' osmotic component (Li$^+$, choline, Tris, mannitol) the smaller the net flux of K$^+$ and hence the greater the flux ratio.

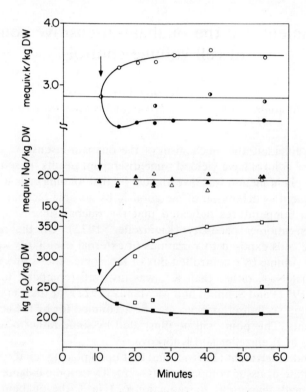

Figure 1 The effect of external osmolarity (LiCl) on the steady-state levels of tissue water and cations in renal cortical slices. Slices were first brought to a steady state of tissue components by incubation (25 °C, O$_2$) in isotonic saline (308 mosmol l^{-1}) of the Krebs–Ringer phosphate type containing (mmol l$^-$): 104 Na, 6.4 K; 45 Li. Subsequently, the tissue was transferred into salines in which the osmolarity was varied by the concn. of Li, i.e. 221 mosmol l^{-1} (0 Li, open symbols), 308 mosmol l^{-1} (45 mmol l^{-1} Li, symbols half shaded ◐, ▲, ◨): 428 mosmol l^{-1} (105 mmol l^{-1} Li, solid symbols). Values in mequiv. cations and kg H$_2$O per kg tissue dry wt.

It is concluded that in such studies the role of the 'inert' component, be it cationic, anionic or uncharged, cannot be neglected. The data are consistent with the view that the cellular volume response to variations in external osmolarity lacks specificity and is determined (in part) by the respective permeabilities of the bulk ionic species across a membrane which is very leaky for water.

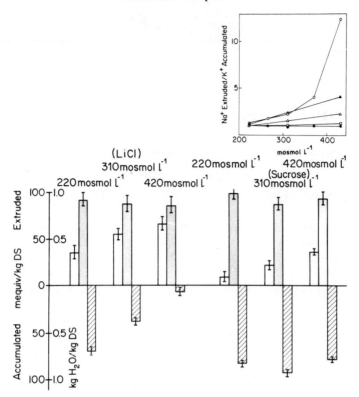

Figure 2 The effect of external osmolarity on the net fluxes of tissue water and cations in slices of renal cortex. Slices were loaded at 0 °C with electrolytes and water in a Ca-free isotonic saline containing (mmol l⁻¹): 104 Na, 6.4 K; 45 Li. The tissue was then incubated 45 min. (25 °C, O_2) into salines of varying osmolarity containing 104 mol l⁻¹ Na and 6.4 mmol l⁻¹ K, the osmotic agents being LiCl and sucrose. The respective osmolarities of the incubation media are indicated above the bars. The bars indicate net fluxes (in mequiv. or kg/kg D.W.). Open: H_2O; full: Na; cross-hatced: K. Insert. The net flux ratio (Na^+ extruded/K^+ accumulated) as a function of external osmolarity. Experimental conditions as above; osmotic agents: o, LiCl; ▲, choline chloride; △ Tris chloride; □, urea; ● mannitol.

REFERENCES

Kleinzeller, A. (1972). Cellular transport of water. In Hokin, L. E. (Ed.) *Metabolic Pathways*, Academic Press, New York, pp. 91–113.

Kregenow, F. M. (1971). The response of duck erythrocytes on hypertonic media: further evidence of a volume-controlling mechanism. *J. Gen. Physiol.*, **58**, 396–412.

Schmidt III, W. F., and T. J. McManus (1977). Ouabain-insensitive salt and water movements in duck red cells: I. Kinetics of cation transport under hypertonic conditions. *J. Gen. Physiol.*, **70**, 59–79.

Hughes, P. M. and A. D. C. MacKnight (1976). The regulation of cellular volume in renal cortical slices incubated in hyposmotic medium. *J. Physiol.*, **257**, 137–154.

Addendum to Chapter 6

Control Mechanisms in Amphibians

V. Koefoed-Johnsen

Since this chapter was written a number of papers have been published which significantly further our knowledge of osmoregulation in amphibia.

Hence the question of localizing the Na transport compartment of the frog skin seems fairly well settled. By means of electron microprobe analysis Rick *et al.* (1978a) have measured the intracellular Na and K concentrations in individual cells and cell types of each of the cell layers in the frog skin epithelium. Under all experimental conditions the various cells constituting the different layers show identical changes with two exceptions: (1) the mitochondria-rich cells which are transepithelial Na-transporting cells (can be inhibited by ouabain) but are not sensitive to amiloride; and (2) the gland cells which are very slightly sensitive to ouabain. Consequently the frog skin epithelium can be regarded as a functional syncytium as suggested by Ussing and Windhager (1964), and Farquhar and Palade (1964). The authors further state that all their results are compatible with the transcellular two-barrier concept (Koefoed-Johnsen and Ussing, 1958) as their measured low cellular Na concentrations, together with recent measurements of intracellular potentials (Nagel, 1976; Helman and Fisher, 1977), can provide electrochemical gradients that allow net uptake of sodium by passive forces even from solutions containing less than 0.1 mmol l^{-1} Na.

Lithium accumulation in the intracellular space of frog skin seems to be possible also on the basis of a passive entrance step at the outer border as estimated from measurements of intracellular potentials in skins where the Li concentration in the outside bathing solution has been varied from 1 to 25 mmol l^{-1} (Nagel, 1977).

Electron microprobe analysis similar to that carried out in frog skin has been made in toad bladder, too, and the conclusion is the same: toad bladder epithelial cells form a functional syncytium, the only exception may be the mitochondria-rich cells (Rick *et al.*, 1978b). Furthermore, in contrast to previous results obtained from scraped off epithelial cells (MacKnight *et al.*, 1975a) the microprobe analysis provides evidence for the hypothesis that the cellular Na-transport pool derives exclusively from the mucosal medium.

Further support for the idea of identifying the sodium transport compartment with all cell layers in the frog skin epithelium comes from radioautographic studies of ^{3}H-ouabain binding sites (Mills and DiBona, 1977). The results show that the Na-K-ATPase which generally is equated with the sodium pump is distributed on the inward facing membranes of all living epithelial cells. The density seems to be highest in *Str. spinosum*, and this is the case whether ^{3}H-ouabain is added from the inside or from the outside (the latter being poosible when hyperosmolarity on the outside has opened the tight seals).

Observations of a linear correlation in split frog skin epithelia between ^{3}H-ouabain binding and short-circuit current inhibition with a zero intercept in skins with high s.c.c. indicate further that all Na-K-ATPase molecules are involved in active Na-transport, and thus support the idea of a functional syncytium (Cala et al., 1978). These authors found, furthermore, that the number of ouabain molecules bound to the frog skins is not correlated to their initial s.c.c. values, suggesting that the spontaneous skin-to-skin variations in short-circuit current are not related to the number of functional pump sites but rather to their turnover rate.

The mitochondria-rich cells seem to have special properties not only in frog skin, but also in toad bladder. When ionic lanthanum (10^{-3} mol l^{-1}) is applied to the serosal surface of the isolated toad bladder the lateral and basal membranes of the mitochondria-rich cells are preferentially 'stained', possibly by displacing calcium in the membranes (Strum, 1977). Like calcium, lanthanum inhibits ADH-induced waterflow, but not the increasing effect of ADH on Na transport. Prolonged treatment with lanthanum causes a swelling of the other epithelial cells while the mitochondria-rich cells are unaltered. The specific binding of lanthanum by the mitochondria-rich cells indicates that the plasma membranes of these cells are distinctly different from those of the other toad-bladder epithelial cells. This difference might—the author suggests—be critically related to the transporting and permeability properties of the membranes, and/or to the response of the mitochondria-rich cells to hormones such as ADH and aldosterone.

REFERENCES

Cala, P. M., N. Cogswell, and L. J. Mandel (1978). Binding of ^{3}H-ouabain to split frog skin. The role of Na-K-ATPase in the generation of short-circuit current. *J. Gen. Physiol.*, **71**, 347–367.

Helman, S. I., and R. S. Fisher (1977). Microelectrode studies of the active Na-transport pathway of frog skin. *J. Gen. Physiol.*, **69**, 571–604.

Mills, J. W., and D. R. DiBona (1977). On the distribution of Na-pump sites in the frog skin. *J. Cell. Biol.*, **75**, 968–973.

Nagel, W. (1976). The intracellular electrical potential profile of the frog skin epithelium. *Pflügers Arch.*, **365**, 135–143.

Nagel, W. (1977). Influence of lithium upon the intracellular potential of frog skin epithelium. *J. Membr. Biol.*, **37**, 347–359.

Rick, R., A. Dörge, E. von Arnim, and K. Thurau (1978a). Electron microprobe analysis of frog skin epithelium: Evidence for a syncytial sodium transport compartment. *J. Membr. Biol.*, **39**, 313–331.

Rick, R. A., A. Dörge, A. D. C. MacKnight, A. Leaf, and K. Thurau (1978b). Electron microprobe analysis of the different epithelial cells of toad urinary bladder. Electrolyte concentrations at different functional states of transepithelial sodium transport. *J. Membr. Biol.*, **39**, 257–271.

Strum, J. M. (1977). Lanthanum 'staining' of the lateral and basal membranes of the mitochondria-rich cell in toad bladder epithelium. *J. Ultrastructure Res.*, **59**, 126–139.

Addendum to Chapter 7

Control Mechanisms in Reptiles

W. A. Dunson

Shoemaker and Nagy (1977) have briefly reviewed osmoregulation in reptiles.

The first scanning electron microscopical study of the surfaces of sea turtle salt-gland cells has demonstrated numerous plications along the intercellular channels (Ellis and Goertemiller, 1976). These authors also localized p-nitrophenyl phosphatase along the lateral cell membranes. An ultrastructural study of the nasal gland of an Australian skink (*Tiliqua rugosa*) has cast doubt on its presumed function as a salt gland (Saint Girons et al., 1977). However it seems likely that various populations of this species differ in their ability to excrete salt extracloacally. Peaker (1978) has suggested that a South American tortoise (*Testudo carbonaria*) has an orbital salt gland. Yet the maximum concentration of the most abundant ion (K) was only 260 mmol l^{-1}, and it is very likely that unspecialized exocrine glands are responsible for this secretion. Much stronger evidence has been presented for the discovery of a new type of salt gland in the estuarine homalopsid snake *Cerberus rhynchops* (Dunson and Dunson, submitted). In this case the salt gland is believed to be the rostral premaxillary gland, which is not homologous with the posterior sublingual salt glands of marine hydrophiid and acrochordid snakes.

Some estuarine natricine snakes apparently lack salt glands, yet are capable of prolonged survival in sea water. Differences in Na influx and skin water permeability seem to be the most distinctive physiological features separating estuarine and fresh-water races of *Nerodia fasciata* in Florida (Dunson, submitted a). Both races prefer to be in fresh water when given a choice (Zug and Dunson, in press). A Florida Keys mud turtle (*Kinosternon b. bauri*) apparently can live in coastal environments only by seeking out fresh water microhabitats (Dunson, submitted b).

A desert snake (*Malpolon monspessulanus*) has been found to have an extremely low rate of total evaporative water loss (Dunson et al., 1976). The dermal component of water loss is especially low; this may be related to the spreading on the skin of a fluid secreted by the enlarged nasal glands.

Maderson et al. (1978) have suggested that α-keratin is the major barrier to water loss in lizard skin. However this theory does not account for the presumed low permeability of pure β-keratin in turtle scutes.

REFERENCES

Dunson, W. A. The relation of sodium and water balance to survival in sea water of estuarine and fresh-water races of the snake *Nerodia fasciata* (Submitted for publication, a).

Dunson, W. A. Salinity tolerance and osmoregulation of the Key mud turtle, *Kinosternon b. bauri* (Submitted for publication, b).

Dunson, W. A., and M. K. Dunson. A possible new salt gland in a marine homalopsid snake (Submitted for publication).

Dunson, W. A., M. K. Dunson and A. Keith (1976). The nasal gland of the Montpelier snake: fine structure, secretion composition, and a possible role in reduction of dermal water loss. *J. Exp. Zool.*, **203**, 461–473.

Ellis, R. A., and C. C. Goertemiller, Jr. (1976). Scanning electron microscopy of intercellular channels and the localization of ouabain sensitive *p*-nitrophenyl phosphatase activity in the salt-secreting lacrymal glands of the marine turtle *Chelonia mydas*. *Cytobiol.*, **13**, 1–12.

Maderson, P. F. A., A. H. Zucker, and S. I. Roth (1978). Epidermal regeneration and percutaneous water loss following cellophane stripping of reptile epidermis. *J. Exp. Zool.*, **204**, 11–32.

Peaker, M. (1978). Excretion of potassium from the orbital region in *Testudo carbonaria:* a salt gland in terrestrial tortoises? *J. Zool., Lond.*, **184**, 421–422.

Saint Girons, H., M. Lemire and S. D. Bradshaw (1977). Structure de la glande nasale externe de *Tiliqua rugosa* (Reptilia, Scincidae) et rapports avec sa fonction. *Zoomorphologie*, **88**, 277–288.

Shoemaker, V. H., and K. A. Nagy (1977). Osmoregulation in amphibians and reptiles. *Ann. Rev. Physiol.*, **39**, 449–471.

Zug, D. A. and W. A. Dunson. Salinity preference in fresh water and estuarine snakes (*Nerodia sipedon* and *N. fasciata*). *Florida Sci.* (in press).

Addendum to Chapter 8

Control Mechanisms in Birds

M. Peaker

Two new views on the mechanism of salt gland secretion have recently been published, both based on the supposed leakiness of the tight junctions of the tubule and the distribution of Na^+/K^+-ATPase on the baso-lateral membranes. Although the published electron micrographs from freeze-fractured material do not appear convincing to the author, Ellis, Goertemiller, and Stetson (1977) propose that the peripheral cells at the blind end of the tubule produce an isotonic secretion, that the principal cells pump sodium into the intercellular channels, and that water then passes osmotically from the lumen through the junctions leaving a hypertonic solution to be secreted. There are a number of problems with this hypothesis including the fact that the relatively small number of undifferentiated peripheral cells would have to produce a large volume of isotonic fluid very rapidly. Moreover, if the junctions are truly leaky, sodium would pass into the lumen and osmotic water flow would not occur. Ernst and Mills (1977) have suggested that the principal cells pump sodium into the intercellular space, and that sodium enters the lumen by crossing the junction; they propose that chloride enters the lumen by permeating the apical membrane. Again there are difficulties in accepting this view, not least of which is the failure of sucrose, when present in the blood, to appear in the secretion (Peaker and Hanwell, 1974, see main text; M. C. Neville and M. Peaker, unpublished). Moreover, neither of these suggested mechanisms explains the presence of the enormous basal infoldings which are such a feature of avian salt glands.

REFERENCES

Ellis, R. A., C. C. Goertemiller, and D. L. Stetson (1977). Significance of extensive 'leaky' cell junctions in the avian salt gland. *Nature, Lond.*, **268,** 555–556.

Ernst, S. A., and J. W. Mills (1977). Basolateral plasma membrane localization of ouabain-sensitive sodium transport sites in the secretory epithelium of the avian salt gland. *J. Cell. Biol.,* **75,** 74–94.

Addendum to Chapter 9

Control Mechanisms in Mammals

M. ABRAMOW

Further studies of the factors governing the transglomerular passage of macromolecules confirmed the important role of the electric charge of the molecules (and hence of glomerular wall) in the sieving properties of the glomerulus. The fractional clearances of polycationic forms of dextrans were shown to be significantly higher than those of the polyanionic dextran sulphate or neutral dextrans in Murich-rats (Bohrer et al., 1978).

Continuing attention has been given to the function of the countercurrent multiplier in the inner medulla. The validity of the passive models has been strengthened by the finding that the inner medullary descending limb of the loop in the rabbit is even less permeable to Na and urea than was previously believed (Abramow and Cogan, 1978). At the same time a reinvestigation of the urinary concentrating mechanism in *Psammomys obesus* confirmed that significant NaCl addition occurs in the descending limb. However a larger proportion of water extraction than previously believed may contribute to the process of osmotic equilibration (Jamison, Roinel, and de Rouffignac, 1978).

The detailed analysis of the complex functions of the distal nephron according to its anatomical components was brought one step further by Imaï (1978) who succeeded in perfusing the connecting tubule of the rabbit *in vitro*. The potential difference (P.D.) in this segment (-28 mV, lumen negative) was insensitive to DOCA pretreatment in contrast to the more distal collecting tubule. It decreased when incubated with ADH or isoproterenol. In this regard the connecting segment is more sensitive (by two orders of magnitude) to isoproterenol than the collecting tubule. On the other hand, the latter is tenfold more sensitive to ADH than the connecting tubule. These functional segmentations are in agreement with the biochemical findings of hormone dependent adenylate cyclase at the same nephron levels.

Progress in the understanding of the physiology of the mammalian nephron continues to benefit from the now well established technique of perfusing isolated tubules *in vitro*, as attested by a recent review (Grantham, Irish, and Hall, 1978).

Addendum: Chapter 9

REFERENCES

Abramow, M., and E. Cogan, (1978). Transport characteristics of the descending limb of the loop of Henle in the inner medulla. In R. Robinson (Ed.) *Proc. VII Int. Congress Nephrol-Montréal. G-12.* Kidney Int.

Bohrer, M. P., C. Baylis, H. D. Humes, R. Glassock, C. R. Robertson, and B. M. Brenner (1978). Permselectivity of the glomerular capillary wall. Facilitated filtration of circulating polycations. *J. Clin. Invest.*, **61,** 72–78.

Grantham, J. J., J. M. Irish III, and D. A. Hall (1978). Studies of isolated renal tubules *in vitro. Ann. Rev. Physiol.,* **40,** 249–277.

Imaï, M. (1978). The connecting tubule: a functional subdivision of the rabbit distal nephron perfused *in vitro.* In R. Robinson (Ed.) *Proc. VII Int. Congress Nephrol.-Montréal. R-3.* Kidney Int.

Jamison, R. L., Roinel, N. and de Rouffignac, C. (1978). Urinary concentrating mechanism in *Psammomys obesus.* In R. Robinson (Ed.) *Proc. VII Int. Congress Nephrol.-Montréal. Q-5.* Kidney Int.

Addendum to Chapter 11

Temperature and Osmoregulation in Aquatic Species

F. J. VERNBERG and S. U. SILVERTHORN

General Comments

An extensive review of the effects of temperature on salt tolerance in crustacea has recently been published by Dorgelo (1976).

Heat death

A detailed histological study of heat death in the banded killifish, *Fundulus diaphanus*, suggests that hypoxia was the primary cause of death, although osmoregulatory failure was a contributing factor (Rombough and Garside, 1977). Extensive damage, including subepithelial oedema, was seen in the gills. Degenerative tubular changes occurred in the kidneys, which would also adversely affect ionic regulation.

Antifreezes

A molluscan antifreeze has been isolated from the mussel, *Mytilus edulis* (Theede *et al.*, 1976). Initial studies indicate it is a glycoprotein with a molecular weight greater than 10,000 daltons.

Numerous additional studies have been made on the protein and glycoprotein antifreezes of fish (Ahmed *et al.*, 1976; Lin *et al.*, 1976; Tomimatsu *et al.*, 1976; DeVries and Lin, 1977; Raymond and DeVries, 1977). Despite chemical differences the two types of antifreeze are functionally similar. They apparently inhibit ice formation by hydrogen-binding to the ice; this disrupts crystal formation without affecting the bulk properties of the ice.

ATPases

A further study on ATPase induction during salt-water acclimation in eels has been published by Thomson *et al.* (1977). When salt-water eels caught at 5 °C were acclimated to 12 °C, their gill lipids became more saturated and the discontinuity of the Arrhenius plot for gill ATPase shifted from 12 to 20 °C. A discontinuity at 20 °C is normal for fresh water eels, which inhabit waters averaging 12 °C. These data suggest that temperature rather than salinity change may be the important factor in altering ATPase activity during seaward migration in the wild.

REFERENCES

Ahmed, A. I., Y. Yeh, D. T. Osuga, and R. E. Feeney (1976). Antifreeze glycoproteins from Antarctic fish: inactivation by borate. *J. Biol. Chem.*, **251**, 3033–3036.

DeVries, A. L., and Y. Lin (1977). Structure of a peptide antifreeze and mechanism of adsorption to ice. *Biochim. Biophys. Acta*, **495**, 388–392.

Dorgelo, J. (1976). Salt tolerance in crustacea and the influence of temperature on it. *Biol. Rev.*, **51**, 255–291.

Lin, Y., J. A. Raymond, J. G. Duman and A. L. DeVries (1976). Compartmentalization of NaCl in frozen solutions of antifreeze glycoproteins. *Cryobiology*, **13**, 334–340.

Raymond, J. A., and A. L. DeVries (1977). Adsorption inhibition as a mechanism of freezing resistance in polar fishes. *Proc. Natl. Acad. Sci. USA*, **74**, 2589–2593.

Rombough, P. J., and E. T. Garside (1977). Hypoxial death inferred from thermally induced injuries at upper lethal temperatures in the banded killifish, *Fundulus diaphanus*. *Can. J. Zool.*, **55**, 1705–1719.

Thomson, A. J., J. R. Sargent, and J. M. Owen (1977). Influence of acclimatization temperature and salinity on ($Na^+ + K^+$)-dependent adenosine triphosphatase and fatty acid composition in the gills of the eel (*Anguilla anguilla*). *Comp. Biochem. Physiol.*, **56B**, 223–228.

Tomimatsu, Y., J. R. Scherer, Y. Yeh, and R. E. Feeney (1976). Raman spectra of a solid antifreeze glycoprotein and its liquid and frozen aqueous solutions. *J. Biol. Chem.*, **251**, 2290–2298.

Taxonomic Index

Abra alba, 541
Acrochordus granulatus, 305
Adula californiensis, 542
Aequipecten irradians, 539
Agama agama, 477
Aipysurus eydouxii, 305
Aipysurus fuscus, 305
Aipysurus laevis, 305
Alligator sp, 279
Alligator mississipiensis, 276
Amblyrhynchus cristatus, 305
Ambystoma gracile, 225
Ambystoma mexicanum, 233
Ambystoma tigrinum, 234
Amia calva, 420
Amoeba proteus, 549
Amphibolurus maculosus, 277, 316
Amphibolurus ornatus, 277, 478
Amphiuma tigrinum, 228
Anas platyrhynchos, 565
Anemonia sulcata, 547
Anguilla anguilla, 163, 183, 188, 189, 421, 473, 552, 567, 571, 585
Anguilla japonica, 475
Anguilla rostrata, 565, 595
Anodonta cygnea, 115, 549
Anolis carolinensis, 276
Aphanius dispar, 161, 204
Apis cerastes, 289
Archirus lineatus, 193
Arenicola marina, 118, 589, 590
Artemia salina, 140, 161, 163, 185, 188, 189, 192, 193, 203
Asellus aquaticus, 543
Astacus astacus, 586
Astacus fluviatilis, 163, 169, 183, 587
Astacus pallipes, 173
Asterias rubens, 118, 538, 589, 593
Aurelia aurita, 594

Bairdiella icistra, 543
Barytelphusa guerini, 139

Blennius pholis, 193
Branchidontes recurvus, 601
Bufo arenarum, 476
Bufo bufo, 225, 234, 255
Bufo cognatus, 232
Bufo marinus, 232, 253, 424, 476
Bufo melanostictus, 230
Bufo punctatus, 230, 423, 424, 476
Bufo viridis, 227

Cacatua roseicapilla, 333, 335
Caiman sclerops, 286
Callinectes sapidus, 117, 120, 123, 124, 125, 126, 127, 128, 140, 166, 180, 543, 587
Callyptocephallea sp, 226
Callyptocephallea gayi, 226
Camptochironomus tentans, 544
Cancer irroratus, 567
Cancer magister, 166, 183, 542, 566
Caranx mate, 543
Carassius auratus, 163, 173, 183, 420, 476, 549, 567
Carchesium aselli, 549
Carcinus maenas, 129, 131, 133, 139, 163, 166, 173, 175, 183, 200, 201, 210, 544, 567, 586
Cardisoma guanhumi, 180
Cardium edule, 601
Caretta sp, 279
Caretta caretta, 276, 294, 299
Carettochelys insculpta, 311
Cerberus rhynchops, 294, 300, 301
Chelon labrosus, 586
Chelonia mydas, 276, 277, 299, 305, 479
Chelydra sp, 279
Chelydra serpentina, 276
Chironomus dorsalis, 544
Chrysemys picta, 477
Chrysemys scripta, 286, 292
Clupea harengus, 474
Coluber constrictor, 276

Taxonomic Index

Coluber ravergieri, 289
Conolophus subcristatus, 305
Convoluta roscoffensis, 541
Cordylophora caspia, 541, 548, 596
Cottus scorpius, 205
Crangon allmanni, 538
Crangon crangon, 185, 537, 547
Crangon septemspinosa, 539, 597
Crassostrea gigas, 594
Crassostrea virginica, 134, 539, 542, 548, 563, 594, 603
Crocodylus acutus, 294, 298
Crocodylus porosus, 298, 304
Crotalus atrox, 289
Crotalus scutellatus, 289
Culea inconstans, 587
Cyathura polita, 539

Dermochelys coriacea, 311
Dinophilus gyrociliatus, 594
Dipsosaurus dorsalis, 286, 287, 305, 306, 309, 477
Discophyra collini, 549
Dreissena polymorpha, 593

Elaphe obsoleta, 276
Embiotoca jacksoni, 566
Enchytraeus albidus, 541, 598
Epatretus stoutii, 163, 201
Eriocheir sinensis, 116, 120, 122, 123, 125, 132, 133, 138, 140, 142, 144, 163, 166, 173, 175, 183, 184, 547, 586, 592
Esox lucius, 545
Etroplus maculatus, 547
Eupagurus bernhardus, 163
Eupleura caudata, 541, 594, 603
Eurytemora affinis, 542

Fundulus diaphanus, 587
Fundulus grandis, 545
Fundulus heteroclitus, 173, 194, 540, 543, 553, 566, 571
Fundulus kansae, 541

Gadus morhua, 540, 597
Gammarus duebeni, 167, 173, 210, 593, 596, 601
Gammarus pulex, 168, 173, 183
Gillichthys mirabilis, 189, 193
Gopherus sp, 279
Gopherus agassizii, 289
Gryphea angulata, 590, 591

Hemigrapsus sp, 543
Hemigrapsus nudus, 566
Hemiscyllium plagiosum, 209
Hippocampus erectus, 193
Homarus sp, 183, 201
Homarus gammarus vulgaris, 120, 587
Hydrophis elegans, 305
Hydrophis semperi, 277

Ictalurus melas, 553
Ictalurus nebulosus, 545
Ictalurus punctatus, 567
Idus melanotus, 541
Iguana iguana, 286

Kinosternon subrubrum, 276

Lamellidens marginalis, 543, 547
Lampetra fluviatilis, 419
Laomedia loveni, 593
Lapemis hardwickii, 305, 311
Larus glaucescens, 338
Laticauda semifasciata, 276, 479
Latimeria chalumnae, 206
Lebistes reticulatus, 541
Lepidochelys olivacea, 276
Leptodactylus sp, 226
Leptodactylus ocellatus, 238
Leuciscus rutilus, 572
Libinia emarginata, 118, 163, 586
Limanda limanda, 541
Littorina littorea, 591
Liza ramada, 586
Lophortyx gambelii, 326, 328
Lucernaria quadricornis, 593
Luidia clathrata, 589
Lymnea stagnalis, 568
Lysmata seticaudata, 538
Lytechinus variegatus, 539

Macrophtalmus japonicus, 140
Maja verrucosa, 163, 164
Malaclemys sp, 279
Malaclemys terrapin, 277, 286, 294, 300, 305, 309
Marinogammarus finmarchicus, 173
Melanopsis trifasciata, 125, 132
Melopsittacus undulatus, 326
Mercenaria mercenaria, 542
Mercenaria sp, 548
Metapenaeus bennetlae, 188
Metopaulias depressus, 168
Miamensis avidus, 118

Taxonomic Index

Modiolus demissus, 539, 548, 599
Modiolus granasissimus, 589, 590, 592, 598
Modiolus modiolus, 123, 540
Modiolus squamosus, 590, 593
Mogalia lignosa, 548
Mugil capito, 189, 193, 540
Mulinia lateralis, 542
Mya arenaria, 547, 601
Myoxocephalus quadricornus, 545
Myoxocephalus scorpius, 545
Mytilus edulis, 118, 541, 542, 546, 548, 590, 591, 601
Mytilus galloprovincialis, 543
Myxine glutinosa, 118, 183

Nassarius obsoletus, 542, 543, 547
Nassarius reticulatus, 539, 597
Natrix cyclopion, 289, 477
Natrix fasciata, 289, 294, 300
Natrix natrix, 477
Natrix rhombifera, 276
Natrix sipedon, 294, 427
Necturus maculosus, 229, 231, 425
Notemigonus crysoleucas, 567

Octosporea effeminans, 601
Onchorhynchus kisutch, 595
Onchorhynchus nerka, 474
Onisimus affinis, 597
Ophiura albida, 593
Opsanus beta, 193
Opsanus tau, 182
Orconectes limosus, 125, 140
Ostrea edulis, 590, 591

Pachygrapsus crassipes, 185, 189, 193, 201
Pagurus longicarpus, 541
Palaemon serratus, 538
Palaemonetes paludosus, 592
Palaemonetes varians, 189, 193, 201, 585, 593
Palaemonetes vulgaris, 140
Panope generosa, 542
Paralichthys lethostigma, 201, 202
Parartemia zeitziana, 185, 188, 203
Paratelplusa hydrodromus, 140
Passerculus sandwichensis, 326
Patella vulgata, 590, 592
Pelamis platurus, 276, 277, 286, 299, 305, 309, 313, 479
Penaeus aztecus, 565

Penaeus duorarum, 188, 539, 565
Perinereis cultrifera, 589, 590
Petrolisthes armatus, 568
Pholis gunnelus, 188, 189, 193
Pimephales promelas, 565
Pituophis catenifer, 289
Platychthys flesus, 163, 183, ·188, 189, 190, 193, 201
Platypoecilus maculatus, 547
Pleuronectes flesus, 120, 123, 567
Poecilia latipinna, 173
Polinices duplicata, 603
Polinices heros, 502
Porcellana platycheles, 163, 211
Porcellio scaber, 543
Potamon niloticus, 163, 168, 183, 184
Protopterus aethiopicus, 420
Psammomys obesus, 352, 397, 401, 435
Pseudemys scripta, 427, 479
Pseudopleuronectes americanus, 120, 202, 546, 567, 572
Pseudothelphusa jouyi, 163, 168, 173, 182, 183
Pugettia producta, 163, 183, 201
Purpura lapillus, 590, 592

Rana catesbiana, 229, 231, 256, 425, 551
Rana clamitans, 229
Rana crancivora, 206, 227, 231, 426
Rana esculenta, 224, 226, 231, 234, 255, 425, 476
Rana pipiens, 225, 226, 227, 249, 255, 554
Rana ridibunda, 228
Rana temporaria, 120, 225, 229, 234, 249, 255, 476
Rangia cuneata, 542

Salamandra maculosa, 233
Salmo gairdneri, 163, 173, 183, 188, 189, 193, 201, 472, 567, 568
Salmo salar, 596
Salmo trutta, 472
Salvelinus fontinalis, 567
Sauromalus hispidus, 283
Sauromalus obesus, 276, 286, 289, 305
Scaphiopus couchi, 231, 445
Scrobicularia plana, 590, 592
Scyliorhinus canicula, 209
Sepia officinalis, 115
Serranus sp, 188, 193
Serranus cabrilla, 549
Serranus scriba, 189, 549

Sesarma plicatum, 139
Siphonaria pectinata, 592
Spalerosophis cliffordi, 289
Sphaeroides maculatus, 566
Sphaeroma serratum, 544
Squalus acanthias, 201, 207, 209, 541
Struthio camelus, 344

Taeniopygia castanotis, 344
Tegula funebralis, 590
Terrapene carolina, 276, 286
Testudo graeca, 479
Testudo hermanni, 276
Tetrahymena pyriformis, 118, 547
Thamnophis sauritus, 289
Thamnophis sirtalis, 276
Tilapia mossambica, 189
Trachysaurus rugosus, 277, 427, 477
Trionyx spiniferus, 275, 276, 292, 311
Trionyx triunguis, 297

Triturus alpestris, 231
Turbanella ocellata, 539

Uca sp, 188, 189, 193
Uca pugnax, 201
Uma notata, 287
Uma scoparia, 286
Uromastix acanthinurus, 314
Uromastix aegyptia, 278
Urosalpinx cinera, 541, 594, 603

Varanus gouldii, 429
Varanus semiremax, 305
Vipera palaestinae, 289
Vorticella marina, 549

Xenopus laevis, 288, 230, 232, 233, 425
Xiphorus helleri, 547

Zoothamnium hiketes, 541

Subject Index

Acetamide, 251
Acetazolamide, 177, 227
Acetylcholine, 341
Actinomycin D, 377, 469, 485, 488, 502
Active transport
 adrenocortical hormones and, 473, 488
 amino acids (in cell volume control), 124
 cell volume control and, 89, 124
 chloride
  Cl^-/HCO_3^- exchange pump, 171, 227, 340, 365
  in amphibians, 227
  in birds, 340
  in fishes, 171
  in frog skin, 238
  in mammalian Henle's loop, 371
 Curie–Prigogine principle, 63
 ions amino acids oxidation and, 144
 neurohypophysial hormones and, 436, 442
 sodium
  amino acids oxidation and, 144
  cell volume control and, 90, 93, 96, 98, 101
  cortical collecting tubules (mammals) and, 386
  electrogenic, 64, 254, 339
  fluxes, 315, 341, 366
  Na^+/H^+ exchange pump, 144, 171, 227, 340, 552
  Na^+/K^+ exchange pump, 90, 98, 240, 247, 334, 339
  Na^+/NH_4^+ exchange pump, 144, 170, 226, 552
  ouabaïn insensitive, 96, 97, 101, 102
Adaptative tolerance (to heavy metals), 573
Adenohypophysis, 234
Adenyl cyclase, 373, 454, 471, 623
ADH, 249, 253, 386, 613, 618
Adrenal steroids, 614

see also Adrenocortical hormones
Adrenalectomy, 377, 463, 473, 477, 479, 480, 483, 497
Adrenaline, 422
Adrenocortical hormones, 376, 463
 chemical composition of, 464
 effects in amphibians, 476
 effects in birds, 480
 effects in fishes, 471
 effects in mammals, 483
 effects in reptiles, 477
 effects on cells of responsive tissues, 487
 molecular basis of action, 487
 site of production and release, 467
Adrenocortical insufficiency, 463, 476, 477, 481, 483
Adrenocorticosteroids, 332, 337, 553
Adrenocorticotrophin (ACTH), 453, 467, 473
Alanine aminotransferase, 134
Aldosterone, 234, 253, 258, 332, 377, 388, 465, 467, 472, 477, 478, 480, 484, 488, 490, 492, 495, 501, 505, 614, 629
Amiloride, 177, 226, 241, 245, 254, 386
Amino acids
 as intracellular osmotic effectors, 113
 changes in blood level, 138
 $^{14}CO_2$ production from, 128
 content in tissues, 115, 117, 122
 metabolism (in cell volume control), 125
 osmotic role in different species, 118
 oxidation and Na^+ transport, 144
 proteins and (equilibrium), 125
 salinity–temperature effect, 547
 transport, see Active transport; Passive transport
Ammonia
 and Na^+ transport, 143, 170, 226, 552
 excretion, 139, 171, 226, 229, 552

661

Ammonia–(contd)
  level in blood and tissues, 140
3′,5′ (cyclic) AMP, 141, 241, 257, 341, 388, 445, 452, 460, 467
Amphenone B, 479
Amphotericin B, 241, 442, 445, 451
Angiotensin II, 467, 477
Antennal gland, 167, 181
Antidiuresis–antidiuretic effect, 233, 234, 356, 375, 415, 419, 424, 428, 430, 431, 445, 481, 614
Antifreezes, 546
Antinatriuresis, 257, 423, 436, 472, 477, 480, 483
Arginine vasopressin, 415, 419, 430, 436, 442, 454, 613
  see also Vasopressin
Arginine vasotocin, 233, 328, 332, 337, 415, 419, 424, 427, 430, 454, 554
Aspartate aminotransferase, 133, 134
ATPases
  Ca^{2+} activated ATPase, 100
  cell volume control and, 100
  effect of heavy metals on, 572
  effect of pesticides on, 564
  level during fish migrations, 595
  Na^+/K^+ activated, 101, 195, 242, 254, 256, 309, 506, 551
  ouabaïn insensitive Na^+/K^+, 101
Aspartocin, 415
Atropine, 343
AVT, 233

Betaine, 116
Betamethasone, 473
Bound water, 10, 87
Bowman's space, 350
Brunn effect, 423

Capacitance (frog skin), 242
Carbonate hydrolyase (carbonic anhydrase), 227
Carriers
  see Passive transport; Active transport
  and mediated diffusion, 51
Cation binding, 88
Cation compartmentation, 88
Cation transport, see Passive transport; Active transport
Chloride cell, 192, 422
Chloride movements and fluxes, see Active transport; Passive transport

Chloride permeability, see Passive transport
Choline chloride, 227
Cloaca, 290, 315, 324, 334, 424, 427, 479
CO_2 production
  C1/C6 ratio from glucose, 130
  from glucose pyruvate, amino acids, 126, 128
Coefficients, see also specific headings
  diffusion, 31, 48, 56
  mutual diffusion, 31
  osmotic, 19
  osmotic: molal, 20
  osmotic: permeability, 29, 250
  osmotic: values, 33
  partition, 49
  permeability, 49, 58
  reflection, 24, 59, 61, 74
  self-diffusion, 32
Colchicine, 257, 445, 459
Cold tolerance, 540, 545
Compensation, partial-perfect, 545
Contractile vacuoles, 25, 549
Corticopapillary gradient, 394
Corticosterone, 234, 331, 464, 472, 477, 478, 480, 490, 496
Corticotrophin, 234, 477
Cortisol, 234, 464, 472, 479, 480
Cortisone, 465, 472
Countercurrent multiplication system, 369, 393
Cryptic Na^+/K^+ pump, 98
  see also Active transport
Cyclohexamide, 469
Cystectomy, 427
Cytochalasin B, 257, 391
Cytosolic receptor, 492, 495

11-Dehydrocorticosterone, 464, 492
Deoxycorticosterone, 464, 472, 484, 496
11-Deoxycortisol, 464, 473
Diabetes insipidus, 397, 431, 618
Diffusion, see also Water movements; Passive transport; Pore theory
  coefficient, 31, 48, 56
  effect of neurohypophysial hormones on, 448
  exchange diffusion, 64, 188, 386
  Fick's law, 35, 48, 56, 383
  Goldman equation, 58
  Hodgkin–Horowicz equation, 58
  mediated diffusion, 51

Subject Index

Nernst–Planck equation, 56
Schlögl's equation, 56
simple diffusion, 47
single-file diffusion, 64
Distribution of aquatic species, 587
Diuresis (diuretic effect), 374, 421, 426, 432, 480, 618
2,4-DNP, 90
Doca-escaped state, 388
Donnan equilibrium (Gibbs–Donnan, Nernst–Donnan), 55, 84
Drinking and drinking rate, 188, 203, 284, 344, 431, 475, 612

Ecology of aquatic blood osmoregulators, 587
Ecology of osmoconformers, 589
Ecophysiology of estuarine species, 581
Ectopic ADH syndrome, 627
Electroosmosis, 26
Endoplasmic reticulum, 246
Enzymes, see also specific headings
    activity, 134
    effects of heavy metals, 571
    level of, 133
Epinephrine, see Adrenaline
Ethacrynic acid, 97, 372
Evaporative water loss, 285, 286, 344
Excluded volume, 22
Exocytosis, 419, 460

Feeding activity, 601
Fick's law, 35, 48, 56, 383
    see also Diffusion
9α-Fluorocortisol, 487, 499
Freezing, resistance to, 540, 545
Furosemide, 102, 372

Gamete production (in aquatic species), 594
Gills, 143, 171, 176, 192, 422, 475, 529, 595
Glomerular filtration and glomerular filtration rate, 200, 231, 233, 327, 350, 352, 353, 420, 424, 427, 430, 472, 477, 549
Glomerular recruitment, 354
Glomerulo-tubular balance, 362
Glumitocin, 415
Glutamate, metabolic utilization of, 129
Glutamate dehydrogenase, 133, 134, 135
3-Glycerophosphate dehydrogenase, 134
Glyoxylate reductase, 134

Gonads maturation (in aquatic species), 594
Granular cells, 445, 507
Growth (in aquatic species), 594
Guanidinium compounds, 245

Heat resistance, 541
Heavy metals, 548, 566
    adaptative tolerance to, 573
    mechanisms of action, 571
Hexose monophosphate shunt activity, 130
Hormones,
    effect of various hormones, see specific headings
    in invertebrates cell volume regulation, 141
Hydraulic conductivity, 60, 360, 363, 370, 379
    see also Water movements
Hydroosmosis (hydroosmotic effect), see Water movements
Hydroosmotic receptor, 235
Hydrophobic bonds, 9
Hydrostatic pressure, 60, 85, 243, 350
    see also Water movements
1-Hydroxycorticosterone, 465
18-Hydroxycorticosterone, 465, 482
Hypercalcaemia, 625
Hypernatraemia, 619
Hyperprolactinaemia, 614
Hypervolaemia, 613
Hypodypsia, 619
Hyponatraemia, 553, 612, 627, 629
Hypophysectomy, 234, 553
Hypopituitarism, 629
Hypothalamus, 612
Hypovolaemia, 613

Ice,
    density, 7
    hexagonal ice lattice, 6
    latent heat of vaporization, 7
    latent heat of fusion, 7
    structure, 7
Idiopathic diabetes insipidus, 620
Inactivity in osmoconformers, 591
Intercellular system, 243, 245, 248, 358
Intestine, 344, 475, 487, 492
Ions,
    control of enzyme activity by, 133
    pumps, see Active transport

Ions–(contd)
  transport, see Passive transport; Active transport
Ionization (concentration dependency), 22
17α-Isoaldosterone, 489, 499
Isocitrate dehydrogenase, 134
Isotocin, 418

Juxta glomerular apparatus, 353

Kidney
  collecting system, 378, 394, 433
    cortical collecting tubule, 378
    papillary collecting duct, 386
  distal convoluted tubule, 374
  effects of hormones on, 421, 425, 427, 431, 453, 473, 476, 478, 480, 496
  inner medulla, 368, 394
  kidney slices, 85, 96, 98
  loop of Henle, 325, 355, 393
    thick ascending limb, 371
    thin limb, 367
  nephron, see specific heading
  outer medulla, 368, 394
  proximal convoluted tubule, 355
  proximal straight tubule, 366
  role in amphibians, 231
    in birds, 323, 331
    in fishes, 181, 183, 199
    in mammals, 350
    in reptiles, 315

Lactic dehydrogenase, 134
Larval stages, 594
Lithium, 242, 246
Lymphocytes, 120
Lysine vasopressin, 415, 423, 436, 453
  see also Vasopressin

Macula densa, 353, 376
Malate dehydrogenases, 134
Malate hydrolyase, 134
Mannitol, 249, 357
Mechano-chemical hypothesis, 99
Medullary circulation, 401
Mesotocin, 415, 420, 425, 427, 429
Metabolic water, 287, 344
Metacholine, 341
Metallothioneins, 573
Metamorphosis (in amphibians), 229, 233
Methoxychlor, 548

6α-Methylpregnenolone, 499
Metyrapone, 473, 482
Microfilaments, 391, 459
Microtubules, 390, 459
Migrations, 595
Myxoedema, 629

Natriferic effect, see Antinatriuresis
Natriferic receptor, 235
Natriuretic factor, 616
Nephrogenic diabetes insipidus, 623
Nephrons, 325, 352, 424, 429, 486, 495
Neurohypophysectomy, 430
Neurohypophysial hormones, 354, 415
  chemical composition, 416
  effects in amphibians, 423
  effects in fishes, 469
  effects in mammals, 431
  effects in reptiles, 427
  effects on cells of responsive tissues, 435
  molecular basis of action, 452
  phylogenetic distribution, 417
Neurohypophysis, 234, 418, 432
Neurophysins, 613, 620, 625, 627
Non-solvent volume, 12, 19, 84
Nuclear magnetic resonance studies
  of ions binding, 89
  of water, 12
Nuclear receptors, 492, 495

Oestrogens, 615
Oil, 564
Oncotic pressure, 350
Ornithin–urea cycle, 229
Osmoregulatory abilities,
  and ecology of aquatic species, 584
  and oxygen tension, 599
  and pollutants, 563
  and species interactions, 601
  and stresses interactions, 597
  and temperature, 537, 597
Osmotic coefficient, 19
  molal, 20
  values, 33
Osmotic equilibrium
  in cells, 21
  thermodynamic treatment, 16
Osmotic permeability, 250, 340
Osmotic pressure, 17, 59
  and cell swelling, 86, 111
  and chemical potential, 17
  and non-solvent volume, 18

Subject Index

Osmotic swelling, 86, 111
Ouabaïn, 91, 93, 101, 180, 241, 249, 254, 256, 357, 386, 444, 506
Overhydration, 627
Oxaloacetate decarboxylase, 134
Oxidative metabolism,
 and active transport, 255
Oxygen consumption, 127, 140, 141, 255, 436, 548, 568
Oxygen tension and salinity interactions, 599
Oxytocin, 415, 421, 425, 427, 429, 454, 620

Paraventricular nuclei, 418
Passive transport, see also Water movements; Diffusion
 amino acids, 122
 cell volume control and, 89, 91, 92
 ions, 190, 292, 294, 300, 371, 379, 386, 442, 444
 membrane potentials and, 54
 Nernst–Donnan equilibrium and, 55
Pesticides, 564
Pinocytosis, 25
Pitressin, see Vasopressin
Podophyllotoxin, 257, 459
Pollutants
 adaptative tolerance to, 573
 heavy metals, 548, 566, 571, 573
 methoxychlor, 548
 oils, 564
 pesticides, 564
 polychlorinated biphenyls, 564
 salinity interactions, 548
Polydipsia-polyuria, 612, 618
Ponder's R, 12, 19, 20, 21, 84
Pore theory, 34
 see also Diffusion; Water movements; Passive transport of ions
 and neurohypophysial hormones, 448
 and water movements in cortical (collecting tubules), 383
 in glomerular filtration, 351
 pore radius (kidney tubules), 383
Potential difference
 in fish gills, 176, 193
 in frog skin, 227, 236, 237, 239, 242
 in mammalian kidney, 360, 370
 in marine osmoconformers, 162
Pregnenolone, 469
Pre-optic nucleus, 418
Progesterone, 616

Prolactin, 331, 332, 337, 472, 553, 614
Propionyl CoA pathway, 131
Prostaglandin E1, 390, 445, 451
Proteins,
 amino acids (equilibrium), 125
 changes in blood content, 142
 contractile and mechano-chemical hypothesis, 99
 contractile proteins in cell volume regulation, 87, 100
 hydration of, 10, 11
 in peritubular medium, 361
 synthesis and adrenocortical hormones effects, 501
Proteins kinase, 388, 455, 460, 469
Puromycin, 488
Pyridine nucleotides
 control of glutamate dehydrogenase activity, 135
 in progesterone synthesis, 471
 oxido-reduction level, 129

Rectal gland, 206, 209
Rectum, 324
Reflection coefficient, (see also Diffusion) 24, 59, 61, 74, 365, 370
Renal insufficiency (chronic), 625
Renal portal blood supply, 330
Renin, 353, 629
Renin-angiotensin, 234, 614
Reproductive capacities (in aquatic species), 593
Resistance to desiccation, 598

Salinity and oxygen tension interactions, 599
Salinity–temperature interactions, 537, 597
Salinity–temperature–pollutants interactions, 548
Salt gland,
 adaptative hypertrophy, 342
 and hormones, 479, 481, 490
 birds, 331, 335
 hyperplasia, 343
 reptiles, 301
 secretory mechanism, 338
 temperature effect, 550
Schwartz and Bartter syndrome, 627
Seasonal distribution, 539
Secretory reflex, 337
Sediment-dwelling species, 589
Serine hydrolyase, 134

Sex ratio (in aquatic species), 594, 601
'Shell-closing' mechanism, 592
Short circuit current, 236, 252, 436, 486, 506
Shunt conductance, 249, 359
Smolt, 595
Sodium channels, 245
Sodium movements and fluxes (see Active transport; Passive transport)
Solvent drag, 251, 340, 449
 see also Water movements
Spawning, 594, 596
Spironolactone, 478, 482, 487, 489, 500
Standing gradient model, 359
Sterility in aquatic species, 593
Succinate dehydrogenase, 134
Supraoptic nucleus, 418

Taurine, 116
Temperature–salinity interactions, 537, 597
Temperature–salinity–pollutants interactions, 548
Theophylline, 454
Thermal hysteresis, 546
Thiourea, 251
Thirst (subjective and objective), 612
Thyroid hormones, 614
Thyroxin, 472
Transepithelial potentials, see Potential difference
Trimethylamine oxide, 116, 206
Tubular receptors, 624
Tubulin, 459

Unstirred layers, 32, 50, 251, 383, 449
Urate salts, 290, 316, 324, 329
Urea, 206, 227, 229, 249, 281, 393
Uremia, 426
Uric acid, 329, 334
Urinary bladder, 184, 232, 233, 252, 256, 315, 423, 436, 472, 477, 480, 483
 see also Diuresis; Antidiuresis
Urine, 182, 231, 316, 324, 332, 393
Urine flow, 162, 201, 415, 420, 425, 427, 429, 431, 477, 480, 549, 554

Valitocin, 415
Vasopressin, 257, 372, 388, 429, 432, 454, 459, 488
Vinblastine, 257, 459

Volume regulation in cells
 and amino acids, 113
 and amino acid catabolism, 127
 and cryptic Na^+/K^+ pump, 98
 and inorganic ions, 83
 and ions passive movements, 91, 92
 and mechano-chemical hypothesis, 99
 and ouabaïn insensitive pumps, 96, 97, 101, 102
 and water structure, 86

Water
 content of cells (measurement), 13, 14, 15
  of nucleus, 14
  of mitochondria, 14
  of microsomes, 15
 continuum theories of structure, 8
 density, 7
 electrolyte hydration, 9
 expansion on freezing, 7
 flickering cluster model, 8
 hydrodynamic hydration, 10
 in association in living cells, 12
 in association with macromolecules, 10
 in solutions of non-polar solutes, 9
 isopiestic hydration, 11
 mixture models, 8
 NMR studies of, 12
 non-solvent volume, 12, 18, 87
 of crystallization, 22
 perfect solution, 18
 preferential hydration, 10, 22
Water molecule
 dielectric constant, 6
 dipole moment, 5
 hydrogen bond, 5
 structure
  in cell volume control, 86
  of aqueous solutions, 9
  of ice, 7
  of liquid water, 7
Water movements
 activation energy, 37, 38
 bulk flow, 34, 250
 cell volume regulation and, 84
 equations, 26
 Fick's law, 35, 48, 56
 hormones and, see specific hormonal headings
 hydraulic conductivity, 60, 360, 370, 379
 hydroosmosis, 390, 419, 431, 445

hydrostatic pressure, 60, 85
in amphibians, 230
in artificial membranes, 36
in birds, 332
in crustaceans and fishes, 157, 162, 210
in frog skin, 250
in mammals, 351
in reptiles, 286
measurements, osmotic, 30
traur, 32
microtubules and, 390, 459
non-linear osmosis, 32
ouabaïn insensitive, 101
Poiseuille's law, 34, 383
pore radius in kidney tubules, 383
pore theory, 34
solvent drag, 251, 340, 449
temperature salinity effect, 549
unstirred layers and, 32, 50
Water permeability, 27, 424, 427
acclimation temperature effect, 550
coefficients in animal cells, 33
coefficients in artificial membranes, 17
in amphibians, 230, 444
in fishes and crustaceans, 163
in mammalian kidney, 360, 370, 379, 382
in reptiles, 314

Zonula occludentes, 242, 259